# A Research Guide to Cartographic Resources

# A Research Guide to Cartographic Resources

## Print and Electronic Sources

Eva H. Dodsworth

ROWMAN & LITTLEFIELD
Lanham • Boulder • New York • London

Published by Rowman & Littlefield
A wholly owned subsidiary of The Rowman & Littlefield Publishing Group, Inc.
4501 Forbes Boulevard, Suite 200, Lanham, Maryland 20706
www.rowman.com

Unit A, Whitacre Mews, 26-34 Stannary Street, London SE11 4AB

Copyright © 2018 by Rowman & Littlefield

*All rights reserved.* No part of this book may be reproduced in any form or by any electronic or mechanical means, including information storage and retrieval systems, without written permission from the publisher, except by a reviewer who may quote passages in a review.

British Library Cataloguing in Publication Information Available

**Library of Congress Cataloging-in-Publication Data Available**

ISBN 9781538100837 (cloth: alk. paper) | ISBN 9781538100844 (electronic)

∞™ The paper used in this publication meets the minimum requirements of American National Standard for Information Sciences—Permanence of Paper for Printed Library Materials, ANSI/NISO Z39.48-1992.

Printed in the United States of America

# Contents

| | | |
|---|---|---|
| Preface | | vii |
| 1 | What Are Cartographic Resources? | 1 |
| 2 | Directory of Cartographic Collections | 7 |
| 3 | Atlases: Historic and Thematic | 53 |
| 4 | Topographic, City, and Town Maps | 93 |
| 5 | Railroad Maps | 115 |
| 6 | Geological Maps | 123 |
| 7 | Nautical Charts | 131 |
| 8 | Fire Insurance Plans | 203 |
| 9 | Air Photos | 205 |
| 10 | Bird's-Eye View Maps | 223 |
| 11 | GIS Data Sources | 241 |
| 12 | GIS Subject Guides | 267 |
| 13 | Handbooks and Manuals | 289 |
| 14 | Cartographic and GIS Dictionaries | 309 |
| 15 | Bibliographies | 311 |
| 16 | Gazetteers | 323 |
| 17 | Citing Cartographic Material | 337 |
| 18 | Professional Map and GIS Associations | 343 |
| Appendix A. Example of a Map Library Collection: University of Waterloo | | 347 |
| Appendix B. Depository Library Programs | | 415 |
| Index | | 473 |
| About the Author | | 479 |

# Preface

The interdisciplinary uses of traditional cartographic resources and modern geographic information system (GIS) tools allow for the analysis and discovery of information across a wide spectrum of fields. Whether maps, surveys, or plans, cartographic resources have been used by geographers, historians, planners, engineers, archaeologists, and the general public for years. They tell a story of people, places, and things in both the past and the present. Whether the visual resource be a study of the impact of railroad development, a model for flood prevention, or the history of a building, it can offer rich information to both the novice and experienced researcher.

There are as many different types of cartographic resources as there are uses for them. They differ in geographic extent, theme, size, scale, date, and accessibility. *A Research Guide to Cartographic Resources* navigates the numerous American and Canadian cartographic resources available in print and online, offering information for locating and accessing them. Dozens of cartographic material types are highlighted and summarized, along with lists of map libraries, geospatial centers, and related professional associations. This volume combines the traditional and historical collections of cartography with modern GIS-based maps and geospatial datasets.

Written for field experts, scholars, instructors, faculty members, graduate students, new students, librarians, independent researchers, mapmakers, policy makers, surveyors, and government representatives, *A Research Guide to Cartographic Resources* consists of eighteen chapters, two appendices, and a detailed index that includes place names. Structured in a manner consistent with most reference guides, this publication includes cartographic categories such as atlases, dictionaries, gazetteers, handbooks, maps, plans, GIS data, and other related material.

The resource formats discussed in this guide include monographs, atlases, maps, plans, surveys, aerial imagery, and GIS data. The resources included in this guide have been selected based on the year of publication, their presence in catalog records such as the Library of Congress and WorldCat, and their ability to be discovered on library and government websites. The specific criteria used in each resource type are outlined in the corresponding chapters. With many annotated listings, this volume provides a level of detail not found in any other resource. Resources are arranged by title and author, place of publication, publisher, date of publication, and URL, with annotations following the citation where applicable.

Chapter 1 introduces readers to the variety of cartographic materials available in libraries and online, offering background information and the uses of each map type. Chapter 2 consists of an updated list of local geographic and cartographic collections in Canada and in the United States, including university and public libraries. The list offers library names, web addresses, and the geographic coverage of each cartographic collection.

Chapters 3 through 10 focus on traditional cartographic resources. Chapter 3 introduces readers to atlases, such as road atlases, historical atlases, and thematic atlases, listing them according to their geographic coverage. It includes atlases published between the 1995 and 2016. Chapter 4 examines maps that showcase both urban and rural areas, such as topographic maps and city and town maps (including cadastral and land ownership maps). Each map type is discussed in detail with individual published maps listed according to their geographic area. Due to the thematic nature of these maps, only historical ones, published from the late 1600s to the 1970s, are listed. Chapter 5 lists railroad maps published in the nineteenth century and made available from the Library of Congress and WorldCat catalogs. This rich collection of maps is organized by province or state, highlighting the development of railroad mapping, capturing and illustrating travel and settlement growth, and illuminating industrial and agricultural development. Chapter 6 focuses on geological maps published by the Geological Survey of Canada, provincial geological survey agencies, and the United States Geological Survey (USGS). Links to print and digital maps are included, offering readers easy access to information about rock formations, soil types, and other geological features. Chapter 7 moves away from land resources and introduces readers to nautical charts, maps of the shoreline, and seafloor maps. The two agencies in North America that publish, update, and distribute nautical charts are the Canadian Hydrographic Service (CHC), part of the Department of Fisheries and Oceans, and the Office of Coast Surveys, part of the National Oceanic and Atmosphere Administration (NOAA). The chapter lists of all the nautical charts currently available for download from either the CHC or NOAA, categorizing them by body of water. Chapter 8 focuses on fire insurance plans, large-scale maps of communities that offer very detailed information about individual buildings. Fire insurance plans are available for almost all populated cities in Canada and the United States, and therefore publishing the complete list is not possible. Union lists that have been previously published for individual provinces and states are available, so these lists have been referenced. Users can visit the Library of Congress website to determine if their plans of interest have been digitized; otherwise, the closest academic or public library can be consulted for access to the plans in print.

Moving away from maps and plans, chapter 9 offers an introduction to aerial photography—that is, photographs taken from an airplane. More recent photography is easily accessible through online products such as Google Earth and Google Maps, and many historical, printed photos are available from map libraries and via online resources such as the National Air Photos Library and the USGS EarthExplorer. This chapter includes a list of U.S. air photos available from EarthExplorer, organized by state. Chapter 10 introduces readers to bird's-eye view maps, an elevated oblique-angled view of a geographic area from above, often showcasing street patterns, individual buildings, and major landscapes. The Library of Congress has the most impressive collection of over fifteen hundred bird's-eye view maps, providing easily accessible free images of the maps. This chapter lists these maps, as well as Canadian maps available from Library and Archives Canada and university collections across the country.

Chapter 11 introduces users to geospatial resources. Geographic information systems (GIS) have been used by many organizations to develop digital and online cartographic products. Almost every province and state has a data clearinghouse, offering the public base or topographic datasets. Many counties and cities also offer their data online; these data tend to be more detailed, focusing on city or county services. This chapter provides a list of recommended GIS data sources available for Canada, the United States, and internationally.

Academic libraries organize all of the geospatial resources available for their users by creating a GIS subject guide or a GIS research guide. Chapter 12 provides a directory of over 130 GIS subject guides, organized by province and state, listing important resources that are included in the guides. This list can be used to determine which guides to visit for access to information about the institution's GIS training program, lists of GIS software, data, and other related online resources. The subject guides were explored to see whether they provided a map or data search engine, offering the capability to search the database for resources directly on the website or through a link on the website. The directory of GIS subject guides also include, wherever applicable, a link to the library's cartographic guide. Cartographic resources often provide important historical information that geospatial data do not offer, so when conducting visual research, both spatial and printed resources should be consulted.

Chapter 13 provides readers with a list of handbooks, manuals, guides, and how-to resources for the fields of cartography and GIS that specifically cover instructions, principles, and techniques. Many of these resources include a combination of older and more recent publications addressing topics such as map interpretation, map design, principles of cartography (that is, projections), cartographic tool techniques, air photo tools, cartographic citations, and many GIS-related topics, such as how to use GIS to solve problems and to make decisions. Readers may find the section on GIS software to be useful because it documents early GIS software programs and lists a variety of products, including open-source options and industry standard software like Esri's ArcGIS, for the purposes of instruction and disseminating technical procedures.

Chapter 14 lists dictionaries and glossaries that define terms specific to cartography, GIS, GIS software, and related topics such as geoinformatics, geostatistics, computer cartography, geospatial databases, spatial and network analysis, geospatial data, geodesy, surveying, photogrammetry, aerial photography, and remote sensing. The listed dictionaries are in English and have been compiled by researching library catalogs, online bookstores, and cartographic and GIS-related websites, including book publishers.

Chapter 15 consists of cartographic bibliographies, or carto-bibliographies, books that include bibliographical lists of cartographic publications in formats like monographs, maps, and air photos. Chapter 16 offers readers a list of national and regional gazetteers. Also known as a geographical index or dictionary, a gazetteer is an alphabetical listing of place names, or toponyms, and geographical features, quite often used by researchers not only as a reference but as rich historical evidence of the people from the past.

Chapter 17 provides guidelines for citing a variety of cartographic resources. Examples of both print and electronic resources are included. Chapter 18 consists of map and GIS associations in Canada and the United States, compiled using a large number of resources related to cartography, geomatics, surveying, data, and GIS.

There are millions of maps published and made available through North American libraries. Many maps are not catalogued in library systems because of the sheer quantity of map sheets available. For any given map series, there may be tens of thousands of maps that compose a collection covering an entire country. Producing a list that captures the entire collection is often not possible. Researchers may wonder, however, what a typical map collection looks like. How many maps cover a city or a lake? How many air photos does it take to capture coverage for a specific town? Who publishes the maps? How often are they revised? Appendix A offers a snapshot of a southwest Ontario academic map library print collection. Although not a very large collection, it offers a representation of maps of the local area, North America, and the world, varying in themes and topics. Map themes include air photos, fire insurance plans, geology, road maps, history, human and physical geography, land use, meteorology, and more.

Many of the maps that this local library, and others, collect and offer to their users are provided to them from government organizations as part of a depository map program. In the United States, the Federal Depository Library Program (FDLP) offers federal government publications to over one thousand libraries. In Canada, the Depository Services Program (DSP) offers Canadian government publications to over one hundred libraries. Appendix B lists all Canadian and U.S. libraries that are a part of these programs.

The purpose of this guide is to make efficient work of the material required to meet the information needs of those interested in researching cartographic-related resources, as well as to help those interested in developing a comprehensive collection in these subject areas and gaining an understanding of what materials are being collected and housed in specific map libraries and geospatial centers. This guide is a one-stop shop for cartographic resources—past and present—that can be used across many fields. Many resources are available directly from libraries, so library names and addresses are listed for those who need to visit in person. Website addresses are listed for those who prefer to access digital materials. Almost all of the resources listed in this guide are categorized by geography down to the county level, offering researchers an efficient way to locate resources for specific areas of interest.

# 1
# What Are Cartographic Resources?

A unique combination of science, art, and technology, cartography presents a well-researched subject in a visual format. Practically every day, groups and individuals employ cartographic resources for personal and professional uses, such as analyzing maps in the news or entering destinations in GPS units or mobile map apps. Finding the quickest routes, discovering alternative directions, and locating specific points of interests on a map are among the most common uses of cartography today. Online and digital interactive maps are not the traditional static resources that may come to mind when considering the use of cartographic materials, but nevertheless they provide users with answers to "where?" Because of the interest in using maps and related technology, more and more resources are being made available for personal, academic, and professional research.

Although some of the earliest maps date back to over fourteen thousand years, scholars did not start to value maps until just after World War I. Libraries began purchasing and acquiring maps, eventually creating their own "map library" spaces within institutions. Many professional associations, journals, conferences, and institutions embrace cartography as well. The digital revolution popularized this field even more as maps started becoming available digitally on CDs, as scanned images, and then eventually as products of geographic information systems (GIS). Cartography has evolved to be more dynamic, interactive, up-to-date, and artistic, inviting noncartographers to dabble with mapmaking technology.

Cartography, regardless of medium (print maps, atlases, globes, aerial imagery, or GIS), involves data that have been collected, evaluated, processed, drawn, reproduced, and presented graphically. Many researchers, including geographers, planners, historians, genealogists, archaeologists, and engineers, spend hours analyzing specific maps to gather and understand information about land, water, people, the past, and the future. There are as many uses for cartographic resources as there are types of them. Fortunately, these resources are quite accessible; for years, libraries have collected, managed, and preserved them. In the last several years, libraries and archives have been scanning their collections, digitizing maps and sharing them online. Maps can be viewed online or downloaded to be used offline; they can be custom built or drawn using online or desktop GIS mapping applications. Today there are many options for accessing and working with cartographic maps and digital files. There are also many government websites that offer interactive maps and GIS datasets related to neighborhood mapping, land use, property boundaries, recreation, and more. Files are commonly downloadable in a database format, a Google Earth format (KML), or a GIS format (SHP).

This chapter introduces readers to cartographic materials, whether available in libraries or online and whether scanned from print collections or created using a GIS program. Understanding the different maps available will help researchers determine which map will best suit their needs and hence help answer their questions related to urban growth, landscape change, navigation on land or water, land use, and much more.

## TOPOGRAPHIC MAPS

One of the most popular cartographic resources are topographic maps, an all-purpose, general representation of the earth's surface, including features like elevation, transportation (streets, railroads, bridges), buildings, forests, and water (lakes, ponds, rivers). Topographic maps also have cartographic elements, including scale, projection, and directional arrows that point toward true north, magnetic north, and grid north. The maps identify the publisher or corporate author of the map, such as the United States Geological Survey (USGS) in the United States or the Ministry of Natural Resources in Canada. Topographic maps are typically published as part of a series and at a variety of specific scales so they can be used for many different applications. Topographic maps are often used for recreational purposes, such as hiking, canoeing, fishing, and orienteering, but they are also popular with geographers and planners to study city and street name changes, the city development and related transportation networks, and so forth. Topographic maps are also used by industry and government for urban planning, emergency management, and establishing legal boundaries and land ownership.

In the United States, the most widely known topographic map series is the USGS National Mapping Program (NMP), created from the mid-1940s through the early 1990s, which produced 1:24,000 scale, 7.5-minute topographic maps. This map series, today known as the Historical Topographic Map Collection (HTMC), includes fifty-three thousand maps sheets for the conterminous United States. Since 2006, all new maps have been produced in a native digital form, the US Topo Quadrangle Series, and have been made available in both digital and print formats.

Topographic map coverage of Canada is based on the National Topographic System (NTS), established in 1927. Maps were printed in a variety of scales, including 1:25,000, 1:50,000, 1:125,000, 1:250,000, 1:500,000, and 1:1,000,000 (which covers the entire country). Today, many updated digital versions are available online.

## CITY AND TOWN MAPS

City and town maps depict urban areas, recording the evolution and growth of cities over time. These maps capture the evolution of the city or town at a specific point in time, often offering historical information on the area's economic activities, educational and religious facilities, transportation systems, and parks, to name a few. City and town maps were often drawn to propose new subdivisions or parks, as well as to display the general "plan" of the city, such as the locations of water mains, stop valves, hydrology, street pavement, street carline, cemeteries, trees, and shrubs. Popular with historians, planners, and genealogists, these maps allow users to analyze geographic boundary changes, geographic township shifts, and city and street name changes, as well as the location of an ancestor's property. Larger scale maps like township and city maps offer quite a bit of detail as they often focus on communities and neighborhoods and are typically produced at a 1:5,000 scale.

Included within this map category are cadastral maps, land ownership plans originally compiled for taxation purposes that show boundaries and ownership of land including owner names, parcel numbers, certificate of title numbers, positions of existing structures, concession and lot numbers, and more. *Cadaster* is a technical term for records showing the extent, value, and ownership of land; however, cadastral maps can reveal much more about a plot of land, at times even including the distribution of vegetation, like marshes and woodlands. Many of these maps were drawn by hand, an art that no longer exists. Today, the information on these maps and on other city and town maps has been replaced by GIS datasets and, to some extent, topographic maps.

## ATLASES

An atlas is a collection of information in a graphic format, consisting of text, tables, graphs, and accompanying maps used to describe an area. The purpose of an atlas is to provide a comprehensive look at a geographical region, whether it is the world, a country, a state or province, a county or city, or specific areas of interest such as a watershed or park. Almost every school, college, university, and household at some point has owned an atlas, and, to this day, many individuals do not travel without one. Many children and adults are familiar with the desk or reference atlas, but there are several types of atlases, including those that cater to specific research interests. Most atlases are still printed on paper in a book format, but electronic atlases are becoming popular in academia and research institutions. Printed road maps are being replaced by GPS systems in vehicles and handheld devices. Although atlases span a wide variety of themes and topics, they are categorized into three main types: general, thematic, and road.

General atlases contain maps that show physical and political features of individual countries or groups of countries, highlighting topics such as physical geography, economic activities, demographics, and climate. In addition, these atlases often include other related information—for example, air photos, satellite imagery, indexes of place names, and cartographic elements such as geographic coordinates and internal locational grids. General atlases typically portray a specific geographic area and are composed of small- to medium-scale maps.

Thematic atlases cover topics of a specific subject—for instance, geology, hydrology, military history, ornithology, climatology, railroads, land ownership, land use, forestry, and much more. All subject atlases, also known as regional atlases, concentrate on a specific area. A region or area can depict the size of a small community, or it can describe a group of states or a continent. These atlases display thematic information using a series of medium- to large-scale maps.

Road atlases, which are the most common atlases, offer maps that show roads and highways for defined geographic areas. Details in a road map often include the type of road depicted, whether a major highway, secondary road, or gravel road. Although quite popular for travelers and vacationers, road atlases are also quite useful for geographers, who use them to study settlement patterns and population densities.

## GEOLOGICAL MAPS

Geological maps represent the distribution of rock formations beneath the soil and other materials on the earth's surface; such maps range from black-and-white line drawings to full-color maps. Unlike ordinary maps, however, geological maps include information that allows users to assess not only the location of particular rocks and the areas they cover, but also their geological history. Many kinds of geological maps exist, including those emphasizing surficial features, bedrock and sediment, subsurface rocks, and geophysical features such as heat flow and gravity. Many libraries house geological maps in a variety of formats, including stand-alone folded maps, folded maps within books, and, most popular, separate flat maps stored in map cabinets.

The Geological Survey of Canada (GSC), the country's oldest scientific organization, is a Canadian federal government agency that is part of the Earth Sciences section of Natural Resources Canada, and it is responsible for conducting and publishing geological surveys for Canada. Geological maps have been published since 1842, and in the past several years, many have been scanned or updated digitally and shared freely online for easier access. Most provinces and territories in Canada have provincial governments that are responsible for the documentation and public dissemination of geoscience information. All of them provide information about their geological map publications and offer access to online digital versions. Most maps are available in image format or PDF, as well as in GIS mapping files (GeoTiffs or Shapefiles).

Established in 1879, the United States Geological Survey (USGS) is a scientific agency of the U.S. government studying the U.S. landscape, its natural resources, and the natural hazards that threaten it. USGS geologic maps are readily available to be searched, viewed, and downloaded from a number of sources. Many states have their own state geological surveys providing citizens with documents and maps related to geology.

## PUBLIC LAND SURVEYS

Public land surveys represent the legal lay of the land, including details such as parcel ownership and building footprints. The surveys provide information about how the land was used, by whom, and during what time period. Researchers can delve further into this resource to learn about early vegetation and patterns of growth to better understand changes in the landscape. Offered at large scales, this survey of public domain lands is the basis for most land transfers and land ownership details today. With almost one and a half billion acres surveyed, many areas of the United States and Canada have been mapped with land boundaries at the township level; however, the surveys often come with plat maps and field survey notes, providing additional information about the property and surrounding areas such as vegetation, rocks, forests, rivers, build-up areas, and the like. These notes are in the custody of the Bureau of Land Management Office, though many individual states offer scanned versions.

## NAUTICAL CHARTS

A nautical chart, or hydrological map, is a map of the shoreline and seafloor, providing water depths, latitude and longitude, topographical features, locations of hazards to navigation, aids to navigation, and other related features. Similar to the public land surveys, the water is surveyed for what is seen on and in the water. In the early 1800s, these reports were taken by people who went into the ocean to collect the information, gaining much knowledge about animal life, fishing practices, and shipwrecks, eventually leading to the discoveries of waterways.

By law, mariners must use nautical charts to plot courses, to determine their positions, and to avoid dangers. Traditional nautical charts have been printed on paper, and, more recently, they have been digitized and made freely available as PDF files. Today, many vessels include navigation technology consisting of a digital system with navigational data that not only replaces the traditional paper version but offers additional details and features.

The two agencies in North America that publish, update, and distribute nautical charts are the Canadian Hydrographic Service (CHC), part of the Department of Fisheries and Oceans, and the Office of Coast Surveys, part of the National Oceanic Atmosphere Administration (NOAA). CHC covers all three of Canada's coastlines plus major lakes and inland waterways. NOAA maintains nautical charts and publications for U.S. coasts and the Great Lakes.

## FIRE INSURANCE PLANS

A fire insurance plan is a large-scale map of a community that offers very detailed information about individual buildings that existed at the time of the map drawing. The original purpose of the plan was to record potential fire risks of individual commercial, industrial, and residential buildings, including building material and the location of water and fire alarm boxes. Using a variety of symbols and colors, the plan reveals the construction type of each building exterior, as well as the building height, the use of the building (residential, commercial, and so forth), and, for many nonresidential buildings, the organization and commercial names. Related structures such as fire hydrants and water or oil pipes are also available. Street information, such as street name, number, and street width, is printed on the map.

Due to its large-scale details, researchers find fire insurance plans incredibly valuable, offering information that cannot be obtained elsewhere. Reviewing and studying a fire insurance plan for a city or town will often help the researcher narrow the construction date for a building; will provide the historical street name and house number (often street names and house numbers change with future development); may highlight any building zoning changes, additions, or constructions to a house (such as a side garage);

may highlight if a building has been demolished; and may indicate potential buried oil tanks or other possible contaminated resources.

In Canada, the most extensive coverage was produced between 1848 and 1910 by Charles E. Goad, who published around thirteen hundred plans of Canadian cities. In 1918, the Underwriters' Survey Bureau was established and acquired exclusive rights to update, revise, and reprint the Goad plans. Over the years, with amalgamation with other Canadian associations, the Canadian Underwriter's Association centralized the production of all plans, creating the Insurers Advisory Organization of Canada (IAO), and eventually ceased all plan production in 1975. Today, many Canadian libraries house Goad and Underwriters' Survey Bureau plans, published during the early to mid-twentieth century.

In the United States, the Sanborn Map Company has been creating and publishing fire insurance plans since 1866, offering users coverage of over twelve thousand U.S. cities and towns. The Library of Congress houses the largest collection of Sanborn maps, offering over twenty-five thousand sheets from over three thousand cities online.

## AIR PHOTOS

Air photos, or aerial photography, are photographs of the ground taken from an airplane. Some of the first air photos of towns in North America were taken in the 1920s and were used to assist land managers in the evaluation of agricultural resources. Traditionally, black-and-white photos were taken, and cartographers deciphered the gray tones and patterns of the land surface. Today, photos continue to be captured, but usually with digital cameras rather than on film. Photos taken with drones are also becoming more and more common. Digital air photos are easily viewed using online products like Google Earth and Google Maps; however, many municipalities are also offering digital air photos or orthoimagery from their websites. Older air photos are available from many academic map libraries as well as from the National Air Photo Library in Canada, and the U.S. Geological Survey in the United States. Because of the large interest in photos, many have been scanned and made available digitally as well. There are several federal agencies that produce aerial photography, such as the U.S. Army Corps of Engineers, the Bureau of Indian Affairs, Earth Resources Observation Satellites (EROS), the Bureau of Reclamation, the Bureau of Land Management, and the U.S. Fish and Wildlife Service.

Air photos offer a source of information about the earth's landform, vegetation, and built-up areas, making them a rich resource for researchers interested in exploring land changes, monitoring agriculture and forestry, studying the history of properties, and examining natural disasters.

## BIRD'S-EYE VIEW MAPS

A bird's-eye view map is an elevated oblique-angled view of a geographic area from above, often drawn with the use of an aerial photograph. These maps, also known as panoramic maps, were a very popular cartographic form used to depict U.S. and Canadian cities and towns during the late nineteenth and early twentieth centuries. They are an artist's representation of what a specific geographic area looked like at a particular point in time and are generally not drawn to scale. Most such maps showcase street patterns, individual buildings, and major landscapes. They may be one of the only visual resources that provides large-scale detailed drawings of residential and industrial buildings, trees, gardens, horse-drawn carriages, even ships and trains. The details seen in these drawings enable researchers to see whether homes had garages, additions, or trees, as well as to determine the street names at the time of production. Because of its easy readability, this map style was also sometimes used to depict areas for planned development. Land developers and real estate agents used these maps to demonstrate future growth possibilities to potential buyers. Advances in lithography and photolithography made the distribution of these prints possible, so they ended up becoming an extremely popular way to capture neighborhoods in the 1800s. Researchers today are very fortunate to have online access to the history of the towns so depicted.

The Library of Congress has the most impressive collection of over fifteen hundred panoramic, or bird's-eye view maps, providing easily accessible free images of the maps. Most of the maps have been drawn by artists Albert Ruger, Thaddeus Mortimer Fowler, Lucien R. Burleigh, Henry Wellge, and Oakley H. Bailey. These five artists prepared more than 55 percent of the panoramic maps in the Library of Congress.

## GAZETTEERS

The gazetteer, also known as a geographical index or dictionary, is an alphabetical listing of place names, or toponyms, and geographical features. Gazetteers are commonly used to locate the areas that the names are associated with, and when place names have changed several times, an up-to-date gazetteer is the ideal resource to identify locations. Because gazetteers represent the places that existed during a specific time, describing counties and their people, they are often used for historical studies. Throughout the years as populations increased, gazetteers started leaving the sketches of towns out and only listed point-blank facts, such as place name and geographic coordinate information. The gazetteers that are published in print are not equivalent to those made available online today. Whether at the national or regional levels, most printed gazetteers paint a picture of the geographical area at that time—the landowners, the businesses, and the transportation routes that play a role in defining what that place is. In many cases, rich historical information can be derived just from the printed advertisements alone.

There are three styles of gazetteers: alphabetical list, dictionary, and encyclopedic. The alphabetical list consists of place names and locations, often including geographic coordinates. The dictionary-style gazetteer includes geographic coordinates and descriptions of places as they relate to other places spatially. Sometimes demographic information is included, as well as a pronunciation guide, similar to a word dictionary. Encyclopedic gazetteers include similar information but also offer considerably more geographic detail, often in the form of articles. The latter two styles may concentrate on specific themes, like county or state history, businesses, hydrology (such as a gazetteer of rivers and lakes), parks, land ownership, and the like. All gazetteers cover a specific geographic area—some are specific to counties, whereas others may cover the entire country.

# 2
# Directory of Cartographic Collections

Finding and locating specific cartographic resources can be a difficult task, as many print cartographic collections are neither searchable online nor cataloged within their library collections. Researchers, students, and the general public may be in search of information related to their city, province, or state, such as points of interests, names of water bodies, soil and geological compositions, elevation numbers, aerial coverage, and so on. For regional, state, or provincial information, it is recommended to start research at a local institution or organization—for example, a university library, a public library, an archive, or a government research center. Since libraries continue to offer material in their original print format, visits to the library location may be useful. An email or phone call to library staff will often help identify which resources are required and what options are available for access.

In the past several decades, several cartographic directories have been created to help users find local resources. Many have been produced for specific countries, using surveys to determine the extent and availability of collections. The most recent publication for Canadian resources, *Directory of Canadian Map Collections*, 7th edition (Leitch, 1999), was published in 1999 by the Association of Canadian Map Libraries and Archives. The most recent publication for American resources, the *Guide to U.S. Map Resources* (Thiry, 2005), was published in 2005 by the Map and Geography Round Table of the American Library Association, though an updated version is currently in the works. World directories of map collections have also been published, such as the *World Directory of Map Collections* (Dubreuil, 1993), published by the International Federation of Library Associations and Institutions (IFLA). Such directories offer lists of cartographic collections included in university, college, public, and corporate libraries, archives, museums, and historical societies, covering geographic areas that extend from the city level to cosmic space. Most of these collections are available through university libraries, often in government documents sections, special collections departments, or map libraries and GIS centers within institutional library systems.

In the last decade, however, many libraries have either continued to acquire paper maps for their collections or, alternatively, released them out of their collections in large quantities. In addition, tracking addresses on the Web have proved problematic, with URLs changing for over 90 percent of the libraries listed in these older directories. This chapter provides an updated list of geographic and cartographic collections that includes local geographic areas, such as city and state or provincial coverage. Each individual listing was checked for the individual organization, current library name, web address (URL), location of cartographic collection (map library, GIS center, geology department, and the like), and geographic coverage. In addition, in order to create a more comprehensive list, a web search was conducted of many university and public libraries that were not included in the original directories. A list of libraries was obtained from the World Libraries Online website (http://www.lib-web.org), as well as the Western Association of Map Libraries Principal Region Map Collections (http://www.waml.org/maplibs.html). Because many map collections are not actively promoted through institutional websites, information was also sought through individual emails and phone calls to librarians to inquire specifically about their map collections and geographic coverage.

While searching and comparing past directories to information present on the Web, it became apparent that many institutions no longer shelve their map collections within their general collections in their library or in specifically designated government information sections. Instead, the institutions have created independent map departments or map libraries. In recent years these map departments or map libraries have morphed into geospatial departments or centers because of the more demanding need for geospatial information, commonly known as GIS, or geographic information systems. Geospatial collections, as with many other library collections, are composed of material actively sought out and purchased, donated, or received through the federal depository systems.

## FEDERAL DEPOSITORY LIBRARY PROGRAM

In 2012 the Canadian government discontinued the distribution of paper maps, primarily because of budget constraints. It took two years to fully transition from paper to electronic maps, and therefore it was not until 2014 that the depository program officially stopped the production of printing and the distribution of hard copies to libraries. The United States Federal Depository Library Program (FDLP) has not converted all of their paper into electronic formats, however, and a large selection of cartographic printed material is automatically shipped to participating libraries. Union lists and lists of classes are available from the FDLP website (https://www.fdlp.gov/file-repository/). The U.S. Federal Depository Library Program includes the following government agencies:

- Department of Agriculture
  http://www.usda.gov/wps/portal/usda/usdahome
- National Agriculture Library
  https://www.fdlp.gov/file-repository/
- Forest Service
  http://www.fs.fed.us
- Natural Resources Conservation Service
  http://www.nrcs.usda.gov/wps/portal/nrcs/site/national/home/
- Economic Research Service
  http://www.ers.usda.gov
- Department of Commerce
  https://www.commerce.gov
- National Institute of Standards and Technology (NIST)
  https://www.nist.gov
- Census Bureau
  https://www.census.gov
- National Oceanic and Atmospheric Administration
  http://www.noaa.gov
- National Weather Service
  http://www.weather.gov
- National Centers for Environmental Prediction
  http://www.ncep.noaa.gov
- National Ocean Service
  http://oceanservice.noaa.gov
- Great Lakes Environmental Research Laboratories
  https://www.glerl.noaa.gov
- Bureau of Economic Analysis
  http://www.bea.gov
- U.S. Department of Defense
  http://www.defense.gov
- National Geospatial Intelligence Agency
  https://www.nga.mil
- U.S. Army Corps of Engineers
  http://www.usace.army.mil
- United States Military Academy West Point
  http://www.westpoint.edu
- U.S. Air Force
  http://www.af.mil
- U.S. Department of Energy
  http://energy.gov
- Federal Energy Regulatory Commission
  https://www.ferc.gov
- U.S. Energy Information Administration
  https://www.eia.gov
- Western Area Power Administration
  https://www.wapa.gov
- National Energy Technology Laboratory (NETL)
  https://www.netl.doe.gov
- U.S. Department of Health and Human Services
  http://www.hhs.gov
- Cartographic Guidelines for Public Health
  http://www.cdc.gov/dhdsp/maps/gisx/resources/cartographic_guidelines.pdf
- Indian Health Service
  https://www.ihs.gov
- U.S. Department of Homeland Security
  https://www.dhs.gov
- Federal Emergency Management Administration (FEMA)
  https://www.fema.gov
- U.S. Department of the Interior
  https://www.doi.gov
- U.S. Geological Survey
  https://www.usgs.gov
- Bureau of Indian Affairs
  http://www.bia.gov
- National Park Service
  https://www.nps.gov/index.htm
- U.S. Fish and Wildlife Service
  https://www.fws.gov
- Bureau of Land Management
  http://www.blm.gov/wo/st/en.html
- Bureau of Ocean Energy Management
  http://www.boem.gov
- U.S. Department of Labor
  https://www.dol.gov
- Occupational Safety and Health Administration (OSHA)
  https://www.osha.gov
- Executive Office of the President
  https://www.whitehouse.gov
- Central Intelligence Agency
  https://www.cia.gov/index.html
- U.S. Department of Transportation
  https://www.transportation.gov
- Bureau of Transportation Statistics
  http://www.rita.dot.gov/bts/
- Federal Highway Administration
  https://www.fhwa.dot.gov
- Federal Railroad Administration
  http://www.fra.dot.gov

A list of libraries that are part of the FDLP can be found in appendix B. The directory in this chapter presents a comprehensive catalog of U.S. and Canadian organizations that offer regional and state or provincial cartographic resources, as opposed to only depository resources. The website addresses are current as of October 2017.

## CARTOGRAPHIC RESOURCES

This section introduces the various cartographic subject resources available in libraries as well as the top collections in libraries that house them. Some organizations such as public libraries have a modest collection of local cartographic material, whereas others are known around the world for their large quantity of map holdings. Large university library map collections will typically collect maps that cover a wide subject matter, including the following:

- aerial photos
- topographic maps
- aeronautical charts
- historical cartography
- roads
- railroads
- land ownership
- land use
- agriculture
- soil
- geology
- fire insurance plans
- celestial
- city
- forestry
- hydrology
- climatology
- demographics
- hydrology
- campus
- projections

These maps are often available as flat maps, but sometimes they are folded or in a book. Some, like wall maps, are oversized; others are raised relief maps; and still others, like aerial photographs, are made of different materials, such as photographic paper. Although many of these maps can be found in large map collections, there are also unique collections located in specific organizations or institutions. For example, the Institute of Oceanography carries hydrological-themed maps such as nautical charts, bathymetric charts, geodetic surveys, and mineral resources.

A web search examined the largest print map collections in both the United States and Canada. The U.S. National Archives (https://www.archives.gov/publications/general-info-leaflets/26-cartographic.html) has the largest collection, totaling over 15 million maps, charts, aerial photos, architectural drawings, patents, and shipping plans. The next largest collection is held by the Library of Congress (https://www.loc.gov/maps/collections/), with over 5.5 million maps and 80,000 atlases. The list below further breaks down the collections by the numbers.

- National Air Photo Library: 6,000,000 air photos
  http://www.nrcan.gc.ca/earth-sciences/geomatics/satellite-imagery-air-photos/9265
- Library and Archives Canada (Ottawa): 3,000,000 architectural drawings, plans, and maps
  http://www.bac-lac.gc.ca/eng/about-us/about-collection/Pages/about.aspx
- University of British Columbia: 800,000 maps
  http://collections.library.ubc.ca
- University of California, Los Angeles (Henry J. Bruman Collection): 750,000 maps
  http://guides.library.ucla.edu/maps
- United States Geological Survey: 700,000 maps
  http://library.usgs.gov/libhistory.html
- University of Wisconsin, Milwaukee (AGS Libraries): 520,000 maps, 11,000 atlases
  http://uwm.edu/libraries/agsl/cartographic/
- University of Alberta: 500,000 maps
  http://guides.library.ualberta.ca/c.php?g=301218&p=2009722
- University of California, Santa Barbara: 500,000 maps, 2.3 million aerial photos, 5,000 atlases
  http://www.library.ucsb.edu/mil/collections
- University of Florida (George A. Smathers Libraries): 500,000 maps, 300,000 aerial photos and satellite images
  http://cms.uflib.ufl.edu/maps/Index.aspx
- State Archives of Michigan: 500,000 maps
  http://www.michigan.gov/mhc/0,4726,7-282-61083-332218--,00.html
- California State University, Northridge: 500,000 maps
  http://www.csun.edu/geography-map-library/collections
- New York Public Library (Lionel Pincus and Princess Firyal Map Division): 433,000 maps
  https://www.nypl.org/about/divisions/map-division

What follows is a directory of cartographic collections in Canada and the United States.

## CANADA

### Alberta

*Calgary*

Glenbow Museum Library
Collection: Maps
http://ww2.glenbow.org/search/libraryMapSearch.aspx

University of Calgary
Taylor Family Digital Library
Spatial and Numeric Data Services
Collection: Maps and GIS
http://libguides.ucalgary.ca/map_internet_resources

*Edmonton*

University of Alberta
Cameron Library
The William C. Wonders Map Collection

Collection: Maps and GIS
http://maps.library.ualberta.ca/index.cfm

*Lethbridge*

Lethbridge Public Library
Collection: Maps
http://www.lethlib.ca

*Red Deer*

Red Deer Public Library
Downtown Branch
Collection: Maps
https://www.rdpl.org

*Sherwood Park*

Strathcona County Library
Collection: Maps
http://www.sclibrary.ab.ca

## British Columbia

*Burnaby*

British Columbia Institute of Technology Library
Collection: Maps
http://libguides.bcit.ca/c.php?g=48697

Burnaby Public Library
Collection: Maps
Burnaby, British Columbia

Simon Fraser University
W. A. C. Bennett Library
Collection: Maps and GIS
Burnaby, British Columbia
http://www.lib.sfu.ca/find/other-materials/data-gis/gis/map-collection

*Chilliwack*

Fraser Valley Regional Library
Maple Ridge Library
Collection: Maps
https://fvrl.bibliocommons.com/locations/MR

*Coquitlam*

Coquitlam Public Library
Poirier Branch
Collection: Maps
http://www.library.coquitlam.bc.ca

*Gibsons*

Gibsons Public Library
Collection: Maps
http://gibsons.bc.libraries.coop

*Kamloops*

Thompson-Nicola Regional District Library System
Kamloops Library
Collection: Maps
https://www.tnrd.ca/content/library-system

*Nanaimo*

Vancouver Island Regional Library
Harbourfront Branch
Collection: Maps
http://virl.bc.ca/branches/nanaimo-harbourfront

*New Westminster*

New Westminster Public Library
Collection: Maps
http://www.nwpl.ca

*Prince George*

Prince George Public Library
Collection: Maps
http://www.pgpl.ca

*Squamish*

Squamish Public Library
Collection: Maps
http://squamish.bc.libraries.coop

*Vancouver and Region*

City of Vancouver Archives and Records
Collection: Maps
Vancouver, British Columbia
http://searcharchives.vancouver.ca

North Vancouver District Public Library
Lynn Valley Branch
Collection: Maps
North Vancouver, British Columbia
http://www.nvdpl.ca

University of British Columbia
Geographic Information Centre
Collection: Maps and GIS
Vancouver, British Columbia
http://gic.geog.ubc.ca/resources/map-collection/

University of British Columbia
Irving K. Barber Learning Centre
Collection: Maps
Vancouver, British Columbia
http://guides.library.ubc.ca/maps-atlases

Vancouver Public Library
Central Library
Collection: Maps
Vancouver, British Columbia
http://guides.vpl.ca/mapsatvpl

*Victoria*

Greater Victoria Public Library
Central Branch
Collection: Maps
http://gvpl.ca

Royal BC Museum
Collection: Maps
http://royalbcmuseum.bc.ca/archives-collections/maps/

University of Victoria
Mearns Centre—McPherson Library
Maps and GIS Department
Collection: Maps and GIS
http://www.uvic.ca/library/locations/home/map/

*Whistler*

Whistler Public Library
Collection: Maps
Whistler, British Columbia
http://www.whistlerlibrary.ca

## Manitoba

*Brandon*

Brandon University
John Langton Tyman Map Library
John Langton Tyman Geography Reading Room (located in the John E. Robbins Library)
Collection: Maps
http://www3.brandonu.ca/library/default.htm

*Winnipeg*

Archives of Manitoba
Collection: Maps
http://www.gov.mb.ca/chc/archives/index.html

Manitoba Education Library
Collection: Maps
http://www.edu.gov.mb.ca/k12/mel/

Royal Aviation Museum of Western Canada
Collection: Maps
http://www.wcam.mb.ca

University of Manitoba
Elizabeth Dafoe Library
Collection: Maps
http://libguides.lib.umanitoba.ca/dafoe

University of Winnipeg
Map Library
Centennial Hall Mezzanine
Collection: Maps
http://maplibrary.uwinnipeg.ca

Winnipeg Public Library
Millennium Library
Collection: Maps
Winnipeg, Manitoba
http://wpl.winnipeg.ca/library/branchpages/branch.aspx?mill

## New Brunswick

*Fredericton*

Provincial Archives of New Brunswick
Collection: Maps
http://archives.gnb.ca/archives/?culture=en-CA

University of New Brunswick
Harriet Irving Library
Map Room
Collection: Maps and GIS
http://www.lib.unb.ca/gddm/maps/

*Moncton*

Université de Moncton
Campus de Moncton
Bibliothèque Champlain (Champlain Library)
Cartothèque (Map Library)
Collection: Maps and GIS
http://www.umoncton.ca/umcm-bibliotheque-champlain/cartotheque

Université de Moncton
Campus de Moncton
Bibliothèque Champlain (Champlain Library)
Service de Référence (Reference Service)
Collection: GIS
http://www.umoncton.ca/umcm-bibliotheque-champlain/services-donnees-geospatiales

## Newfoundland

*Carmanville*

Carmanville Public Library
Collection: Maps
http://townofcarmanville.ca/index.php?option=com_content&view=article&id=26&Itemid=53

*Gander*

Gander Public and Resource Library
Collection: Maps
http://www.nlpl.ca/component/jumi/library.html?site_id=ggr

*Grand Falls–Windsor*

Harmsworth Public Resource Library
Collection: Maps
http://www.nlpl.ca/component/jumi/library.html?site_id=cgf

*Harbour Breton*

Harbour Breton Public Library
Collection: Maps
http://www.nlpl.ca/component/jumi/library.html?site_id=chb

*Saint Alban's*

St. Alban's Public Library
Collection: Maps
http://www.nlpl.ca/component/jumi/library.html?site_id=csa

*Saint John's*

Memorial University
Queen Elizabeth II Library
Map Room
Collection: Maps and GIS
http://www.library.mun.ca/researchtools/mapsdatagovdocs/maps/

Provincial Archives of Newfoundland and Labrador
Colonial Building
Collection: Maps
http://www.newfoundlandlabrador.com/PlanYourTrip/Detail/11784376

Newfoundland and Labrador Public Libraries
Arts and Culture Centre
Collection: Maps
http://www.nlpl.ca

## Nova Scotia

*Amherst*

Nova Scotia Geomatics Centre
Collection: GIS
http://www.nsgc.gov.ns.ca

*Halifax*

Dalhousie University
Killam Memorial Library
GIS Centre
Collection: Maps and GIS
http://libraries.dal.ca/find/map-collection.html

Halifax Public Libraries
Central Library
Collection: Maps
http://www.halifaxpubliclibraries.ca/branches/locations/halifax-central-library.html

Public Archives of Nova Scotia
Collection: Maps
https://archives.novascotia.ca

Saint Mary's University
Geography Department (Burke Building)
Map Library
Collection: Maps
https://libanswers.smu.ca/friendly.php?slug=faq/130413

*Lawrencetown*

Nova Scotia Community College
Annapolis Valley Campus
J. B. Hall Library
Collection: Maps
http://www.cogs.ns.ca

*New Glasgow*

Pictou-Antigonish Regional Library
New Glasgow Library
Collection: Maps
https://parl.catalogue.library.ns.ca

*Sydney*

Cape Breton University
Beaton Institute
Collection: Maps
https://www.cbu.ca/library/search-find/data-and-statistics/
https://www.cbu.ca/library/research/mapping-resources/

*Wolfville*

Acadia University
Earth and Environmental Science
Geology Department
Collection: Maps
http://libguides.acadiau.ca/geol/geology/maps

## Ontario

*Bowmanville*

Clarington Public Library
Collection: Maps
http://www.clarington-library.on.ca

*Cambridge*

Cambridge Public Library
Collection: Maps
http://www.cambridgema.gov/cpl/aboutus

*Cornwall*

Cornwall Public Library
Collection: Maps
Cornwall, Ontario
http://library.cornwall.on.ca

*Georgetown*

Halton Hills Public Library
Collection: Maps
Georgetown, Ontario
http://www.hhpl.on.ca

*Haileybury*

Haileybury
Northern College Library
Collection: Maps
Haileybury, Ontario
http://libguides.northernc.on.ca/home

*Hamilton*

McMaster University
Mills Memorial Library
Lloyd Reeds Map Collection
Collection: Maps and GIS
http://library.mcmaster.ca/maps/Map_Collection

*Kingston*

Kingston Frontenac Public Library
Central Branch
Collection: Map
https://www.kfpl.ca

Queen's University
Douglas Library
The Map and Geospatial Data Collection
Collection: Maps and GIS
http://library.queensu.ca/geo/collections_resources

*Kitchener*

Kitchener Public Library
Collection: Maps
http://www.kpl.org

*Lindsay*

Kawartha Lakes Public Library
Lindsay Library
Collection: Maps
https://olco.ent.sirsidynix.net/client/en_US/kawarthalibrary

*London*

London Public Library
Collection: Maps
http://www.londonpubliclibrary.ca

Western University
D. B. Weldon Library
Map and Data Centre
Collection: Maps and GIS
https://www.lib.uwo.ca/madgic/

*Mississauga*

Mississauga Library System
Collection: Maps
https://miss.ent.sirsidynix.net/client/en_US/mlsathome/

*Niagara Falls*

Niagara Falls Public Library
Collection: Maps
https://my.nflibrary.ca/iguana/www.main.cls?sUrl=home

*Oshawa*

Oshawa Public Library
McLaughlin Branch
Collection: Maps
http://www.oshawalibrary.on.ca

*Ottawa*

Carleton University
Carleton University Library
Maps, Data and Government Information Centre (MADGIC)
Collection: Maps and GIS
https://library.carleton.ca/find/maps

Earth Sciences Information Centre
Collection: Maps
http://www.nrcan.gc.ca/earth-sciences/resources/maps

Library and Archives Canada
Collection: Maps
http://www.archives.ca

Ottawa Public Library
Collection: Maps
https://biblioottawalibrary.ca/en

University of Ottawa
Morisset Hall
Geographic, Statistical, and Government Information Centre
Collection: Maps and GIS
http://biblio.uottawa.ca/en/gsg-centre

*Peterborough*

Trent University
Bata Library
Maps, Data, and Government Information Centre (MADGIC)
Collection: Maps and GIS
https://www.trentu.ca/library/madgic

*Rainy River*

Rainy River Public Library
Collection: Maps
http://www.rainyriverlibrary.com

*St. Catharines*

Brock University
James A. Gibson Library
Map, Data, and GIS Library (MDG)
Mackenzie Chown Complex Room
Collection: Maps and GIS
https://brocku.ca/library/collections/MDG

St. Catharines Public Library
Collection: Maps
http://www.stcatharines.library.on.ca

*Sudbury*

Laurentian University
Map Library
Collection: Maps
https://biblio.laurentian.ca/research/guides/geospatial-data-lu

*Thunder Bay*

Lakehead University
Department of Geography
Lakehead University Map Library
Collection: Maps and GIS
https://www.lakeheadu.ca/academics/departments/anthropology/resources-tools-links/lu-map-library

Thunder Bay Public Library
Collection: Maps
http://www.tbpl.ca

*Timmins*

Timmins Public Library
Collection: Maps
http://tpl.timmins.ca

*Toronto and Region*

Archives of Ontario
Cartographic Records Collection
Collection: Maps
Toronto, Ontario
http://www.archives.gov.on.ca/en/maps/index.aspx

Ryerson University
Library
The Geospatial Map and Data Centre (GMDC)
Collection: Maps and GIS
Toronto, Ontario
http://www.ryerson.ca/library/info/collect.htm
https://library.ryerson.ca/gmdc/

Toronto Reference Library
Special Collections, Genealogy and Maps Centre
Collection: Maps
Toronto, Ontario
http://www.torontopubliclibrary.ca/history-genealogy/

Toronto Reference Library
Urban Affairs Library
Collection: Maps
Toronto, Ontario
http://www.mtrl.toronto.on.ca/centres/mual/index.html

University of Toronto at Scarborough
V. W. Bladen Library
Collection: Maps and GIS
Scarborough, Ontario

https://utsc.library.utoronto.ca/geospatial-data-gis-and-maps-utsc

University of Toronto
Erindale Library
Collection: Maps and GIS
Mississauga, Ontario
http://library.utm.utoronto.ca/datagis

University of Toronto
Robarts Library
Map and Data Library
Collection: Maps and GIS
Toronto, Ontario
http://mdl.library.utoronto.ca
http://library.utm.utoronto.ca/datagis

University of Toronto
Thomas Fisher Rare Book Library
Collection: Maps
Toronto, Ontario
http://fisher.library.utoronto.ca

York University
Scott Library
Map Library
Collection: Maps and GIS
North York, Ontario
http://www.library.yorku.ca/web/map/

*Waterloo*

University of Waterloo
Dana Porter Library
Geospatial Centre
Collection: Maps and GIS
https://uwaterloo.ca/library/geospatial/

Waterloo Public Library
Collection: Maps
http://www.wpl.ca

Wilfrid Laurier University
Arts Building
The Geography Resource Centre
Collection: Maps and GIS
https://library.wlu.ca/locations/geography-resource-centre

*Windsor*

University of Windsor
Leddy Library
Government Documents
Collection: Maps and GIS
Windsor, Ontario
http://leddy.uwindsor.ca/maps

## Prince Edward Island

*Charlottetown*

Public Archives and Records Office of Prince Edward Island
Collection: Maps
https://www.princeedwardisland.ca/en/home

## Quebec

*Chicoutimi*

Université du Québec à Chicoutimi
Bibliothéque Paul-Èmile-Boulet (The Paul-Èmile-Boulet Library)
La Cartothèque Se (The Map Library)
Collection: Maps and GIS
http://libguides.uqac.ca/cartotheque
http://uqactualite.uqac.ca/acces-aux-donnees-geospatiales-gouvernementales-a-la-bibliotheque/

*Gatineau*

Bibliothéque et Archives nationale du Québec
BAnQ Gatineau
Collection: Maps
Gatineau, Quebec
http://www.banq.qc.ca/archives/entrez_archives/centres_archives/ca_outaouais.html

*Montreal*

Bibliothéque nationale du Québec
Service des Collections Spéciales
Collection: Maps
http://iris.banq.qc.ca/iris.aspx
http://www.banq.qc.ca/collections/cartes_plans/description/

Institut National de la Recherche Scienfigique
Centre Urbanisation Culture Société
Collection: Maps
http://www.ucs.inrs.ca/ucs

McGill University
Department of Geography
Collection: Maps
http://www.mcgill.ca/geography/

McGill University
McLennan Library
Rare Books and Special Collections
Collection: Maps and GIS

http://www.mcgill.ca/library/branches/rarebooks/special-collections

Stewart Museum
Collection: Maps
http://www.stewart-museum.org/en/

Université du Québec à Montréal
Map Library (La Cartothèque)
Collection: Maps and GIS
http://www.bibliotheques.uqam.ca/cartotheque

*Rimouski*

Université du Québec à Rimouski
Rimouski Campus Library
Map Library
Collection: Maps
http://biblio.uqar.ca/renseignements/categorie/riki-cartotheque/

*Sainte Foy*

Université Laval
Pavillon Jean-Charles-Bonenfant
GéoStat Center (Geographical and Statistical Information)
Collection: Maps and GIS
Sainte Foy, Quebec
http://www.bibl.ulaval.ca/services/centregeostat

*Sherbrooke*

Bishop's University
John Bassett Memorial Library
Map Room
Collection: Maps
Sherbrooke, Quebec
http://www.ubishops.ca/library/library-information-services/collections/

Université de Sherbrooke
Brother-Théode Library
Geospatial Data, Maps, and Aerial Photographs
Collection: Maps and GIS
https://www.usherbrooke.ca/biblio/trouver-des/cartes-et-photographies-aeriennes/cartes/
https://www.usherbrooke.ca/biblio/trouver-des/donnees-geospatiales/

*Trois-Rivières*

Université du Québec à Trois-Rivières
Roy-Denommé Library
Collection de cartes et de données géospatiales (Collection of Maps and Geospatial Data)
Collection: Maps and GIS
http://www.uqtr.ca/biblio/services/carto_collections.shtml

## Saskatchewan

*Regina*

Regina Public Library
Collection: Maps
http://www.reginalibrary.ca/prairiehistory/

University of Regina
Archives and Special Collections
Collection: Maps
http://www.uregina.ca/library/services/archives/

University of Regina
Department of Geography and Environmental Studies
Map Library
Collection: Maps and GIS
http://www.uregina.ca/arts/geography/facilities/map-library.html

*Saskatoon*

Provincial Archives of Saskatchewan
Collection: Maps
http://www.saskarchives.com

Saskatoon Public Library
Frances Morrison Library
Collection: Maps
http://www.saskatoonlibrary.ca/branch/frances-morrison-central-library

## Yukon

*Whitehorse*

Yukon Archives
Collection: Maps
http://yukondigitallibrary.ca

# UNITED STATES

## Alabama

*Auburn*

Auburn University
Ralph Brown Draughon (RBD) Library
Map Collection
Collection: Maps and GIS
http://libguides.auburn.edu/subject/govdocs

*Birmingham*

Birmingham Public Library
Linn-Henley Research Library

Collection: Maps
http://www.bplonline.org

Samford University
Samford University Library
Maps Collection
Collection: Maps and GIS
http://library.samford.edu
https://library.samford.edu/special/maps-collection.html

*Huntsville*

Huntsville-Madison County Public Library
Collection: Maps
https://hmcpl.org/departments/hhr

*Mobile*

University of South Alabama
Marx Library
Maps/GIS Collection (Government Documents)
http://www.southalabama.edu/departments/library/
http://libguides.southalabama.edu/govmaps

*Montgomery*

Alabama Department of Archives and History
Collection: Maps
http://www.archives.state.al.us

*Tuscaloosa*

Geological Survey of Alabama
Library
Collection: Maps
http://www.gsa.state.al.us

University of Alabama
Department of Geography (Farrah Hall)
The University Map Library
Collection: Maps and GIS
https://geography.ua.edu/about/facilities/

**Alaska**

*Anchorage*

Alaska Resources Library and Information Services
Collection: Maps
http://www.arlis.org

Bureau of Land Management
Alaska State Office
Collection: Maps
Anchorage, Alaska
http://www.blm.gov/ak/st/en.html

University of Alaska, Anchorage
UAA/APU Consortium Library
Archives and Special Collections
Collection: Maps
http://www.lib.uaa.alaska.edu

*Fairbanks*

University of Alaska, Fairbanks
Elmer E. Rasmuson Library
Government Documents and Maps
Collection: Maps
http://uaf.edu
http://library.uaf.edu/govdocs

*Juneau*

Alaska State Library
Alaska Historical Collections
The Map Collection
Collection: Maps
http://library.alaska.gov/hist/hist.html
http://library.alaska.gov/hist/collections.html

*Ketchikan*

Ketchikan Public Library
Collection: Maps
http://www.firstcitylibraries.org

**Arizona**

*Cottonwood*

Cottonwood Public Library
Collection: Maps
http://www.ctwpl.info/services.asp

*Flagstaff*

Coconino County Public Library
Main Library
Collection: Maps
http://flagstaffpubliclibrary.org

Northern Arizona University
Cline Library
Special Collections and Archives
Government Documents
Collection: Maps and GIS
http://library.nau.edu/speccoll/exhibits/sca/collect/maps/
https://library.nau.edu/general_information/government.html

*Kingman*

Mohave Museum of History and Arts
Collection: Maps
http://www.mohavemuseum.org

*Mesa*

Mesa Public Library
Main Library
Collection: Maps
http://www.mesalibrary.org

*Phoenix*

Arizona Public Library
Burton Barr Branch
Collection: Maps
http://www.phoenixpubliclibrary.org

Arizona State Library, Archives and Public Records
Collection: Maps
http://www.azlibrary.gov/sla

Arizona State Library, Archives and Public Records
Archives and Records Management
Collection: Maps
http://www.azlibrary.gov/arm/research-archives/maps-arizona-state-archives

Maricopa County Assessor's Office
Collection: Maps
http://mcassessor.maricopa.gov/gis

Phoenix Public Library
Burton Barr Central Library
Collection: Maps
http://www.phxlib.org

*Prescott*

Prescott Public Library
Collection: Maps
http://www.prescottlibrary.info

Sharlot Hall Museum
Archives
Collection: Maps
http://www.sharlot.org/archives/index.html

Yavapai College
Prescott Campus Library
Collection: Maps
https://www.yc.edu/v5content/library/prescott-spaces.htm

*Tempe*

Arizona State University
Hayden Library
Collection: Maps and GIS
https://lib.asu.edu/hayden

Arizona State University
Noble Science and Engineering Library
The Map and Geospatial Hub
Collection: Maps and GIS
https://lib.asu.edu/noble/collections
https://lib.asu.edu/noble

*Tucson*

Arizona Geological Survey
Headquarters
Collection: Maps
http://www.azgs.az.gov

Arizona State Museum
University of Arizona
Collection: Maps
http://www.statemuseum.arizona.edu/library

Pima Public Library
Joel D. Valdez Main Library
Collection: Maps
http://www.library.pima.gov

University of Arizona
Library
Collection: Maps and GIS
http://www.library.arizona.edu/about/libraries/mapcoll.html

*Yuma*

Arizona Western College
Academic Library
Government Information Resources
Collection: Maps and GIS
http://www.azwestern.edu/library/

**Arkansas**

*Conway*

University of Central Arkansas
Torreyson Library
Archives Map Collection
Collection: Maps
http://www.uca.edu
http://libguides.azwestern.edu/gov_info/maps

*Fayetteville*

University of Arkansas
University Libraries
Mullins Library
Geographic Information Systems and Maps
http://libinfo.uark.edu/gis/

*Little Rock*

Arkansas Archives
Collection: Maps
http://www.ark-ives.com

Arkansas Geological Survey
Vardelle Parham Geology Center
Collection: Maps
http://www.geology.ar.gov/home/index.htm

Arkansas State Library
Collection: Maps
http://www.library.arkansas.gov/Pages/default.aspx

Central Arkansas Library System
Main Library
Collection: Maps
http://www.cals.lib.ar.us

University of Arkansas at Little Rock
Donald Reynolds Center for Business and Economic Development
UALR GIS Applications Laboratory
Collection: Maps and GIS
https://argis.ualr.edu/OnlineArkansas/Home/Location

*Russellville*

Arkansas Tech University
Pendergraft Library and Technology Center
Special Collections
Collection: Maps
http://library.atu.edu

## California

*Arcata*

Humboldt State University
University Library
Atlas and Map Collection
Collection: Maps and GIS
http://library.humboldt.edu/about/collections/atmapcoll.html

*Bakersfield*

Kern County Library
Beale Memorial Library
Geology/Mining/Petroleum Room
Collection: Maps
http://kerncountylibrary.org/beale-memorial-library/

*Berkeley*

Berkeley Public Library
Collection: Maps
http://www.berkeleypubliclibrary.org

University of California, Berkeley
The Bancroft Library
Collection: Maps and GIS
http://www.lib.berkeley.edu/libraries/bancroft-library

University of California, Berkeley
Department of Geography
Map Collection
Collection: Maps
http://geography.berkeley.edu/resources/collections/

University of California, Berkeley
Earth Sciences and Map Library
Collection: Maps
http://www.lib.berkeley.edu/libraries/earth-sciences-library

*Beverly Hills*

Beverly Hills Public Library
Collection: Maps
http://www.beverlyhills.org/exploring/beverlyhillspubliclibrary/historicalcollection/

*Chico*

California State University, Chico
Meriam Library
Map Collection
Collection: Maps
http://www.csuchico.edu/lbib/maps/index.html

*Claremont*

Claremont University Center, Claremont Colleges
Harvey/Mudd Library
Geographic Information System Services
Collection: Maps and GIS
http://libraries.claremont.edu/#gsc.tab=0
https://www.hmc.edu

*Davis*

University of California, Davis
Shields Library
The UCD Map Collection
Collection: Maps and GIS
http://guides.lib.ucdavis.edu/Maps

*Fresno*

California State University, Fresno
Henry Madden Library
Map and Aerial Photographs
Collection: Maps and GIS
https://library.fresnostate.edu/find/maps-and-aerial-photographs

Fresno County Public Library
Heritage Center
Collection: Maps
http://www.fresnolibrary.org

*Fullerton*

California State University, Fullerton
Humanities Building
GIS Research Center
Collection: GIS
http://hss-geogwebsrvr.fullerton.edu/gisrc/index.html

California State University, Fullerton
Pollak Library
Government Documents
The Roy V. Boswell Collection (part of Special Collections)
Collection: Maps
http://libraryguides.fullerton.edu/govdocs
http://www.library.fullerton.edu/boswellmaps/dev/BoswellArticle.php

Fullerton Public Library
Collection: Maps
http://fullertonlibrary.org

*Irvine*

University of California, Irvine
Langson Library
Collection: Maps and GIS
https://www.lib.uci.edu/collections

*La Jolla*

Scripps Institution of Oceanography
University of California, San Diego
Map and Chart Collection
Collection: Maps
https://scripps.ucsd.edu

University of California, San Diego
Geisel Library
The Data and GIS Lab
Collection: Maps and GIS
https://ucsd.libguides.com/data-gis-lab

*Los Angeles*

Los Angeles Public Library
Central Library
Collection: Maps
http://www.lapl.org/collections-resources/visual-collections/map-collection

Loyola Marymount University
William H. Hannon Library
Department of Archives and Special Collections
Collection: Maps and GIS
http://library.lmu.edu/archivesandspecialcollections/

University of California, Los Angeles
Charles E. Young Research Library
Henry J. Bruman Map Collection
Collection: Maps and GIS
http://digital2.library.ucla.edu/viewItem.do?ark=21198/zz0028tr8x
http://guides.library.ucla.edu/maps

*Menlo Park*

United States Geological Survey
Menlo Park Library
Collection: Maps
http://library.usgs.gov/menlib.html

*Mill Valley*

Mill Valley Public Library
Lucretia Little History Room
Collection: Maps
http://www.millvalleylibrary.org

*Monterey*

Monterey Public Library
California History Room
Collection: Maps
http://www.monterey.org/library/History-Room

*Napa*

Napa City County Library
Collection: Maps
http://countyofnapa.org/library/

*Nevada City*

Nevada County Library System
Doris Foley Library for Historic Research
Collection: Maps
https://www.mynevadacounty.com/nc/library/Pages/Doris-Foley-Library-for-Historical-Research.aspx

*Oakland*

Oakland Public Library
Collection: Maps
http://oaklandlibrary.org
http://www.oaklandlibrary.org/online-resources/government-resources

*Palmdale*

Palmdale City Library
Collection: Maps
http://www.cityofpalmdale.org/Library

*Pasadena*

California Institute of Technology
Geological and Planetary Sciences Library
Collection: Maps
http://library.caltech.edu

*Pleasant Hill*

Contra Costa County Library
Pleasant Hill Library
Collection: Maps
http://ccclib.org

*Redwood City*

Redwood City Public Library
Collection: Maps
http://www.redwoodcity.org/departments/library

*Riverside*

University of California, Riverside
Orbach Science Library
Earthquake Processes, Geophysics, and Tectonics
Collection: Maps and GIS
https://library.ucr.edu/libraries/orbach-science-library/collections

University of California, Riverside
Orbach Science Library
Map Collection, Geospatial Services, Map Reading Room, and Digital Map Lab
Collection: Maps and GIS
http://guides.lib.ucr.edu/c.php?g=171041&p=3515014

University of California, Riverside
Orbach Science Library
Water Resources Collections and Archives
Collection: Maps
http://library.ucr.edu/collections/water-resources-collections-archives

*Sacramento*

California State Library
California History Section
Collection: Maps
http://www.library.ca.gov/calhist/index1.html

California State Library
Stanley Mosk Library and Courts Building
Government Publications Section
Collection: Maps
http://www.library.ca.gov/gps/index.html

*San Bruno*

The National Archives at San Francisco
Collection: Maps
https://www.archives.gov/san-francisco/

*San Diego*

San Diego State University
SDSU Library and Information Access
Collection: Maps and GIS
http://library.sdsu.edu

*San Francisco*

San Francisco State University
College of Health and Social Sciences
Geographic Information Systems (GIS) Specialty Center
Collection: GIS
https://csugis.sfsu.edu/

San Francisco State University
J. Paul Leonard Library
Maps and Atlas Collection
Collection: Maps
http://geog.sfsu.edu
http://library.sfsu.edu/map-atlas-collection
http://csugis.sfsu.edu

California Historical Society
Collection: Maps
http://www.californiahistoricalsociety.org

David Rumsey Map Collection
Collection: Maps
http://www.davidrumsey.com

Presidio Trust
Collection: Maps
http://www.presidio.gov

San Francisco Maritime National Historical Park
Collection: Maps
https://www.nps.gov/safr/learn/historyculture/library-collections.htm

San Francisco Municipal Railway
San Francisco War Memorial and Performing Arts Center
Collection: Maps
http://sfwmpac.org/herbst-theatre

San Francisco Public Library
Collection: Maps
http://www.sfpl.org

*San Jose*

San Jose State University
Dr. Martin Luther King Jr. Library
Collection: Maps
http://www.sjlibrary.org

San Jose State University
Research Library and Archives
Map Collection
Collection: Maps
http://www.historysanjose.org/research/library

*San Leandro*

City of San Leandro Public Library
Main Library
Collection: Maps
http://www.sanleandro.org/depts/library/

*San Luis Obispo*

California Polytechnic State University
Robert E. Kennedy Library
Data and GIS Services
Collection: Maps and GIS
http://www.lib.calpoly.edu

*San Marino*

Huntington Library
Rare Books and Manuscripts Departments
Collection: Maps
http://hdl.huntington.org/cdm/landingpage/collection/p15150coll4

*Santa Barbara*

Santa Barbara Museum of Natural History
Museum Library
Collection: Maps
https://www.sbnature.org/crc/50.html

University of California, Santa Barbara
UCSB Library
Interdisciplinary Research Collaboratory
Collection: Maps and GIS
http://www.library.ucsb.edu/mil/collections
https://www.library.ucsb.edu/interdisciplinary-research-collaboratory

*Santa Clara*

Santa Clara City Library
Collection: Maps
http://santaclaraca.gov/government/departments/library

*Santa Cruz*

University of California, Santa Cruz
McHenry Library
Map Collection
Collection: Maps and GIS
http://guides.library.ucsc.edu/maps

*Santa Rosa*

Sonoma County Library
Central Santa Rosa Library
Collection: Maps
http://www.sonomalibrary.org

*Stanford*

Stanford University Libraries
Branner Earth Sciences Library and Map Collections
Stanford Geospatial Center
Collection: Maps and GIS
http://library.stanford.edu/branner/map-collections-guides
https://library.stanford.edu/collections/stanford-geological-survey-collection

*Tiburon*

Belvedere Tiburon Library
Collection: Maps
https://www.beltiblibrary.org

*Ventura*

Museum of Ventura County
Research Library
Collection: Maps
http://venturamuseum.org/research-library/

**Colorado**

*Boulder*

Boulder Public Library
Carnegie Branch Library for Local History
Collection: Maps
https://boulderlibrary.org/locations/carnegie/

University of Colorado, Boulder
Jerry Crail Johnson Earth Sciences and Map Library
Collection: Maps and GIS
http://www.libraries.colorado.edu
http://www.colorado.edu/libraries/libraries/earth-sciences-map-library

*Colorado Springs*

Colorado College
Tutt Library
Government Documents
Collection: Maps
https://www.coloradocollege.edu/library/gov-docs/
http://coloradocollege.libguides.com/c.php?g=286907&p=1909679

Pikes Peak Library District
Penrose Library
Special Collections
Collections: Maps
http://ppld.org/regional-history-and-genealogy

University of Colorado, Colorado Springs
Kraemer Family Library
Government Information
Collection: Maps and GIS
http://www.uccs.edu/~library/

*Denver*

Colorado Historical Society
Stephen H. Hart Library and Research Center
Collection: Maps
http://www.historycolorado.org/researchers/stephen-h-hart-library-and-research-center

Denver Public Library
Western History Collection
Collection: Maps
https://history.denverlibrary.org

National Archives at Denver
Denver Federal Center
Collection: Maps
http://www.archives.gov/denver/

United States Geological Survey
Denver Library
Collection: Maps
www.library.usgs.gov
http://library.usgs.gov/denlib.html

*Durango*

Fort Lewis College
John F. Reed Library
Government Documents
Collection: Maps and GIS
https://library.fortlewis.edu
http://subjectguides.fortlewis.edu/govdocs

*Fort Collins*

Colorado State University
Morgan Library
Colorado State University Geospatial Centroid
Collection: Maps and GIS
http://lib.colostate.edu
https://gis.colostate.edu

*Golden*

Colorado School of Mines
Arthur Lakes Library
Collection: Maps
http://library.mines.edu

*Greeley*

University of Northern Colorado
James A. Michener Library
Government Publications
Collection: Maps and GIS
http://www.unco.edu/library

*Lakewood*

Jefferson County Public Library
Lakewood Library
Collection: Maps
http://jefferson.lib.co.us

*Louisville*

Louisville Public Library
Collection: Maps

http://www.louisvilleco.gov/government/departments/louisville-library

*Loveland*

Loveland Public Library
Collection: Maps
http://www.ci.loveland.co.us/library

*Pueblo*

Pueblo City-County Library
Western Research Map Collection
Collection: Maps
http://www.pueblolibrary.org

## Connecticut

*Fairfield*

Fairfield Historical Society and Museum
Collection: Maps
http://www.fairfieldhistory.org/library-collections/collections/maps/

*Hartford*

Connecticut Historical Society
Museum and Library
Collection: Maps
http://www.chs.org

Connecticut State Library
History and Genealogy Unit
Collection: Maps
http://ctstatelibrary.org
http://libguides.ctstatelibrary.org/hg/home

Trinity College
Watkinson Library and College Archives
http://www.trincoll.edu/LITC/Watkinson/Pages/default.aspx

*New Haven*

Yale University
Yale University Library
Beinecke Rare Book and Manuscript Library
Collection: Maps and GIS
http://web.library.yale.edu/maps

*Storrs*

University of Connecticut
Homer Babbidge Library
Map and Geographic Information Center (MAGIC)
Collection: Maps and GIS
http://magic.lib.uconn.edu

*Stratford*

Stratford Library
Collection: Maps
http://stratfordlibrary.org

## Delaware

*Dover*

Delaware Public Archives
Collection: Maps
http://archives.delaware.gov/index.shtml

*Newark*

University of Delaware
Hugh M. Morris Library
Student Multimedia Design Center
Collection: Maps and GIS
http://guides.lib.udel.edu/GIS

## District of Columbia

District of Columbia Public Library
Martin Luther King Jr. Memorial Library
Washingtoniana and Black Studies
Collection: Maps
http://www.dclibrary.org

Folger Shakespeare Library
Collection: Maps
http://www.folger.edu

Library of Congress
Collection: Maps
http://www.loc.gov/rr/geogmap/

National Archives and Records Administration
Collection: Maps
http://www.archives.gov/publications/general-info-leaflets/26-cartographic.html

National Geographic Library
Collection: Maps
http://www.ngslis.org

Smithsonian
National Air and Space Museum
Planetary Maps
Collection: Maps
https://airandspace.si.edu

World Bank
Library and Archives Development
Collection: Maps
http://www.worldbank.org/en/about/archives

**Florida**

*Boca Raton*

Florida Atlantic University
S. E. Wimberly Library
Government Documents
Collection: Maps
http://www.fau.edu/library
http://libguides.fau.edu/gov_documents-boca

*Coral Gables*

University of Miami
Otto G. Richter Library
Special Collections
Collection: Maps
Coral Gables, Florida
http://scholar.library.miami.edu/maps/

University of Miami
Otto G. Richter Library
GIS Lab
Collection: GIS
https://sp.library.miami.edu/subjects/gis

*DeLand*

Stetson University
DuPont-Ball Library
Government Documents Department
Collection: Maps
https://www2.stetson.edu/library/

*Gainesville*

University of Florida
George A. Smathers Library
Map and Imagery Library
Collection: Maps and GIS
http://cms.uflib.ufl.edu/maps/Index.aspx

*Miami*

Historical Museum of Southern Florida
Charlton W. Tebeau Research Library
Collection: Maps
http://www.historical-museum.org

Miami-Dade Public Library System
Main Library
Collection: Maps
http://www.mdpls.org

University of Miami
Rosenstiel School of Marine and Atmospheric Science Library
Collection: Maps and GIS
http://library.miami.edu/rsmaslib/

*Sarasota*

Sarasota County Libraries
Historical Resources
Collection: Maps
https://www.scgov.net/Archives/Pages/CartographicDrawing.aspx

*Tallahassee*

Florida State University
R. M. Strozier Library
Map Collection
Collection: Maps
http://guides.lib.fsu.edu/maps

*Tampa*

University of South Florida
Tampa Library
Florida Studies Center (Special Collections)
Collection: Maps
http://www.lib.usf.edu/special-collections/
http://www.lib.usf.edu/florida-studies/collections/

**Georgia**

*Americus*

Georgia Southwestern State University
James Earl Carter Library
Government Documents and Publications
Collection: Maps
http://www.gsw.edu/~library/

*Athens*

University of Georgia Libraries
Map and Government Information Library
Collection: Maps and GIS
http://www.libs.uga.edu/magil/

*Atlanta*

Atlanta-Fulton Public Library
Central Library

Special Collections
Collection: Maps
http://www.afpls.org/central-library/66-special-collections

Emory University
Robert W. Woodruff Library
Emory Center for Digital Scholarship
Collection: Maps and GIS
http://web.library.emory.edu/library-materials/collections/index.html
http://digitalscholarship.emory.edu/

Georgia Institute of Technology
Library
Collection: Maps
http://www.library.gatech.edu

Georgia State University
University Library
Collection: Maps and GIS
http://research.library.gsu.edu/geography

*Augusta*

Augusta State University
Reese Library
Special Collections
Collection: Maps
http://www.augusta.edu/library/reese/

*Carrollton*

University of West Georgia
Ingram Library
Maps Collection (part of Government Documents)
Collection: Maps
http://www.westga.edu/library/

*Macon*

Middle Georgia Archives
Washington Memorial Library
Map Collection
Collection: Map Collection
http://www.bibblib.org

*Milledgeville*

Georgia College
Ina Dillard Russell Library
Georgia College Special Collections
Collection: Map Collection
http://www.gcsu.edu/library

*Rome*

Sara Hightower Regional Library System
Rome/Floyd County Public Library
Collection: Maps
http://rome.shrls.org

## Hawaii

*Honolulu*

Hawaii State Archives
Map Collection
Collection: Maps
http://ags.hawaii.gov/archives/

Hawaii State Public Library System
Collection: Maps
http://www.librarieshawaii.org/serials/databases.html

University of Hawaii at Manoa
University of Hawai'i at Mānoa Library Map Collection
MAGIS (Maps, Aerial Photographs, and GIS)
Collection: Maps and GIS
http://guides.library.manoa.hawaii.edu/magis

*Laie*

Brigham Young University
Joseph F. Smith Library
Special Collections
Collection: Map and GIS
http://library.byuh.edu/

## Idaho

*Boise*

Boise State University
Albertsons Library
Hollenbaugh Map Collection
Collection: Maps and GIS
http://guides.boisestate.edu/c.php?g=74218&p=476350

Idaho State Historical Society
Idaho State Archives
Collection: Maps
https://history.idaho.gov/idaho-state-archives

*Moscow*

University of Idaho
University of Idaho Library
University of Idaho Library Cartographic Collections
Collection: Maps and GIS
http://www.lib.uidaho.edu

## Illinois

*Carbondale*

Southern Illinois University, Carbondale
Morris Library
Geospatial Resources (includes Map Library and GIS services)
Collection: Maps and GIS
http://www.lib.siu.edu/maps-gis

*Champaign*

Illinois State Geological Survey
Geological Samples Library
Collection: Maps
http://www.isgs.illinois.edu/geological-samples-library

*Charleston*

Eastern Illinois University
Booth Library
Government Documents
Collection: Maps and GIS
http://www.library.eiu.edu

*Chicago*

Alder Planetarium
Astronomy Department
Collection: Maps
http://www.adlerplanetarium.org

Chicago Historical Society
Research Center
Collection: Maps
http://www.chicagohistory.org

Chicago Public Library
Harold Washington Library Center
Collection: Maps
http://www.chipublib.org/special-collections/

Illinois Institute of Technology
Paul V. Galvin Library
Government Documents
Collection: Maps
http://guides.library.iit.edu/govdocs

Newberry Library
Map Collection
Collection: Maps
http://www.newberry.org/maps-travel-and-exploration

University of Chicago
The University of Chicago Library
Map Collection
Collection: Maps and GIS
https://www.lib.uchicago.edu/collex/?view=collections&format=Maps
https://www.lib.uchicago.edu/e/collections/maps/

University of Illinois, Chicago
Richard J. Daley Library
Special Collections and University Archives
Collection: Maps and GIS
http://library.uic.edu/help/article/2990/visit-special-collections-and-university-archives

*De Kalb*

Northern Illinois University
Founders Memorial Library
Map Collection
Collection: Maps and GIS
http://libguides.niu.edu/Maps

*Edwardsville*

Southern Illinois University Edwardsville
Lovejoy Library
Map Collection
Collection: Maps and GIS
http://www.siue.edu/lovejoylibrary/govdocs/maps/index.shtml

*Evanston*

Northwestern University
Northwestern University Library
Government and Geographic Information Collection
Collection: Maps and GIS
http://www.library.northwestern.edu/libraries-collections/government-collection/maps-gis/index.html

*Macomb*

Western Illinois University
Malpass Library
Map Collection
Collection: Maps and GIS
http://www.wiu.edu/libraries/govpubs/map_collection/

*Normal*

Illinois State University
Milner Library
Map Collection
Collection: Maps and GIS
http://library.illinoisstate.edu

*Oglesby*

Illinois Valley Community College
Jacobs Library
Special Collections
Collection: Maps
https://www.ivcc.edu/libraryhome.aspx

*Peoria*

Bradley University
Cullom-Davis Library
Government Documents
Collection: Maps
http://www.bradley.edu/library/

*Quincy*

Quincy Public Library
Collection: Maps
http://qupl.ent.sirsi.net/client/default/

*Rock Island*

Augustana College
Thomas Tredway Library
Loring Map Collection
Collection: Maps
http://www.augustana.edu/library

*Springfield*

Abraham Lincoln Presidential Library and Museum
Collection: Maps
https://www.illinois.gov/alplm/library/Pages/default.aspx

Illinois State Archives
Margaret Cross Norton Building
Collection: Maps
https://www.cyberdriveillinois.com/departments/archives/home.html

*Urbana*

University of Illinois, Urbana-Champaign
Illinois State Library
Scholarly Commons
Collection: Maps and GIS
https://www.library.illinois.edu/sc/gis/

University of Illinois, Urbana-Champaign
Rare Book and Manuscript Library
Collection: Maps
http://www.library.illinois.edu/rbx/

The Urbana Free Library
Champaign County Historical Archives
Collection: Maps
http://urbanafreelibrary.org/tags/champaign-county-historical-archives

**Indiana**

*Anderson*

Anderson Public Library
Collection: Maps
http://www.and.lib.in.us

*Bloomington*

Indiana University, Bloomington
Department of Geological Sciences
Collection: Maps
http://www.geology.indiana.edu/department/index.html

Indiana University, Bloomington
Herman B Wells Library
Government Information, Maps, and Microform Services
Collection: Maps and GIS
https://libraries.indiana.edu/gimms

Indiana University, Bloomington
Herman B Wells Library
The Wells Library Map Collection
Collection: Maps and GIS
https://libraries.indiana.edu/maps-and-gis

*Evansville*

University of Evansville
Libraries
Collection: Maps
https://www.evansville.edu/libraries/

*Fort Wayne*

Indiana University–Purdue University, Fort Wayne
Walter E. Helmke Library
Collection: Maps
http://www.lib.ipfw.edu

*Greencastle*

DePauw University
Prevo Science Library
Collection: Maps and GIS
http://www.depauw.edu/libraries/

*Greensburg*

Greensburg Public Library
Collection: Maps
http://www.greensburglibrary.org/

*Hanover*

Hanover College
Duggan Library
Special Collections
Collection: Maps
http://library.hanover.edu

*Indianapolis*

Indiana Historical Society
William Henry Smith Memorial Library
Collection: Maps
http://www.indianahistory.org

Indiana State Library
Collection: Maps
http://www.in.gov/library/3551.htm

Indiana University–Purdue University, Indianapolis
Herman B Wells Library
Government Information, Maps, and Microform Services
Collection: Maps and GIS
http://www.ulib.iupui.edu
https://libraries.indiana.edu/gimms

Indianapolis Public Library
Central Library
Collection: Maps
http://www.indypl.org/

*Muncie*

Ball State University
Bracken Library
GIS Research and Map Collection (GRMC)
Collection: Maps and GIS
http://cms.bsu.edu/academics/libraries/collectionsanddept/archives
http://cms.bsu.edu/academics/libraries/collectionsanddept/gisandmaps

*New Albany*

New Albany–Floyd County Public Library
Collection: Maps
http://nafclibrary.org

*Richmond*

Earlham College
Science and Technology Commons (STC)
Collection: Maps
http://library.earlham.edu/STC

*Terre Haute*

Indiana State University
Indiana State Library
Map Collection
Collection: Maps and GIS
http://library.indstate.edu/about/units/archives/research_collections.htm
https://www.in.gov/library/3551.htm

*Valparaiso*

Valparaiso University
Christopher Center for Library and Information Resources
Map Collection
Collection: Maps
http://library.valpo.edu/archives/index.html

*West Lafayette*

Purdue University
The Purdue Earth and Atmospheric Sciences Map Room
Earth and Atmospheric Sciences (EAS) Library
Collection: Maps
http://guides.lib.purdue.edu/mapresources

## Iowa

*Ames*

Iowa State University
Parks Library
Media Center
Map Collection
Collection: Maps and GIS
http://www.lib.iastate.edu/research-tools/collections-areas/media-center

*Cedar Falls*

University of Northern Iowa
Rod Library
Government Documents and Maps Collection
Collection: Maps and GIS
https://www.library.uni.edu/collections/gov-docs

*Grinnell*

Grinnell College
Burling Library
Special Collections and Archives
Collection: Maps
https://www.grinnell.edu/libraries/about/collections

*Iowa City*

State Historical Society of Iowa, Department of Cultural Affairs
Manuscripts, Maps, and Ephemera
Collection: Maps
https://iowaculture.gov/history/research/collections/manuscripts-maps-ephemera

University of Iowa
Sciences Library
Collection: Maps
http://guides.lib.uiowa.edu/geo

University of Iowa
University of Iowa Main Library
Map Collection
Collection: Maps and GIS
http://www.lib.uiowa.edu/maps

*Sioux City*

Sioux City Public Library
Wilbur Aalfs (Main) Library
Collection: Maps
http://www.siouxcitylibrary.org/wilbur-aalfs-library/

**Kansas**

*Emporia*

Emporia State University
William Allen White Library
Information and Instructional Services
Collection: Maps and GIS
http://www.emporia.edu/libsv
http://academic.emporia.edu/abersusa/maplibrary.htm

Emporia State University
William Allen White Library
The Emporia State University Federal Depository Map Library
Collection: Maps
http://academic.emporia.edu/abersusa/maplibrary.htm

*Hays*

Fort Hays State University
Forsyth Library
Collection: Maps
https://www.fhsu.edu/library/

*Lawrence*

University of Kansas
Anschutz Library
Thomas R. Smith Map Collections
Collection: Maps and GIS
https://lib.ku.edu/map-collection
https://spencer.lib.ku.edu/collections/special-collections
https://lib.ku.edu/map-collection

University of Kansas
Kenneth Spencer Research Library
Special Collections
Collection: Maps
https://spencer.lib.ku.edu/collections/special-collections

*Manhattan*

Kansas State University
Hale Library
Map Collection
Government Publications Division
Collection: Maps and GIS
http://www.lib.k-state.edu/government-collections

*Overland Park*

Johnson County Library
Central Resource Library
Collection: Maps
http://www.jocolibrary.org

*Pittsburg*

Pittsburg State University
Leonard H. Axe Library
Government Documents
Collection: Maps and GIS
http://axe.pittstate.edu/docs/

*Topeka*

Kansas State Library
Collection: Maps
http://kslib.info

Kansas State Historical Society
Library and Archives Division
Collection: Maps
http://www.kshs.org

*Wichita*

Wichita Public Library
Reference Services Division
Collection: Maps
http://www.wichitalibrary.org

Wichita State University
Ablah Library
Special Collections and University Archives
Collection: Maps and GIS
https://libraries.wichita.edu/ablah/index.php/8-special-collections

**Kentucky**

*Berea*

Berea College
Hutchins Library
Collection: Maps
http://libraryguides.berea.edu

*Bowling Green*

Western Kentucky University
Helm-Cravens Library
Government Documents and Law Information
Collection: Maps
http://www.wku.edu/library

*Cold Spring*

Campbell County Public Library
Collection: Maps
http://www.cc-pl.org

*Frankfort*

Commonwealth of Kentucky
Collection: Maps
Kentucky Mine Mapping Information System
http://www.minemaps.ky.gov

Kentucky Historical Society
Thomas D. Clark Library and Special Collections
Collection: Maps
http://www.history.ky.gov

*Highland Heights*

Northern Kentucky University
Steely Library
Map Collection
Collection: Maps and GIS
http://library.nku.edu

*Lexington*

University of Kentucky
Margaret I. King Library
Science and Engineering Library
Map Collection
Collection: Maps and GIS
http://libguides.uky.edu/c.php?g=222914&p=1476173

*Louisville*

University of Louisville
William F. Ekstrom Library
Maps
Collection: Maps and GIS
http://louisville.edu/library/ekstrom/tour/maps.html

*Morehead*

Morehead State University
Camden-Carroll Library
Map Collection
Collection: Maps
http://www.moreheadstate.edu/library

*Newport*

Newport Branch of Campbell County Public Library
Collection: Maps
http://www.cc-pl.org/locations/newport-branch

*Richmond*

Eastern Kentucky University
John Grant Crabbe Library
Government Documents
Collection: Maps
http://library.eku.edu/john-grant-crabbe-library

**Louisiana**

*Baton Rouge*

Louisiana State University
Department of Geography and Anthropology
Cartographic Information Center (CIC)
Collection: Maps
http://www.cic.lsu.edu/collections.htm

Louisiana State University
Middleton Library
Government Documents/Microforms Department
Collection: Maps
http://www.lib.lsu.edu/collections/govdocs

State Library of Louisiana
Collection: Maps
http://www.state.lib.la.us/library-collections/louisiana-collection

*Lake Charles*

McNeese State University
Frazar Memorial Library
Government Information Department
Collection: Maps
http://library.mcneese.edu/Govt_Info.html

*New Orleans*

Louisiana State Museum
Collection: Maps
http://www.crt.state.la.us/louisiana-state-museum/collections/historical-center/maps/

New Orleans Public Library
Collection: Maps
http://nutrias.org/spec/speclist.htm

United States Department of the Interior, Minerals and Management Service
Library

Collection: Maps
http://www.boem.gov/Contact-Us/

*Ruston*

Prescott Memorial Library
Collection: Maps
http://www.latech.edu/library/govdocs/maps.php

## Maine

*Augusta*

Maine State Archives
Collection: Maps
http://www.maine.gov/sos/arc/research/special.html

Maine State Library
Collection: Maps
http://www.maine.gov/msl/

*Orono*

University of Maine
Raymond H. Fogler Library
Government Documents Department
Collection: Maps and GIS
http://libguides.library.umaine.edu/GovernmentDocuments

*Portland*

Maine Historical Society
Collection: Maps
Brown Library
http://www.mainehistory.org

## Maryland

*Baltimore*

Enoch Pratt Free Library
Central Library
Collection: Maps
Maryland Department/Map Collection
http://www.prattlibrary.org/locations/

Johns Hopkins University
Milton S. Eisenhower Library
Collection: Maps and GIS
Baltimore, Maryland
http://guides.library.jhu.edu/gis

*College Park*

Nixon Presidential Library and Museum
National Archives at College Park Cartographic Section
Collection: Maps
https://www.nixonlibrary.gov/index.php

University of Maryland, College Park
McKeldin Library
Government Information, Maps and GIS Services
Collection: Maps and GIS
http://www.lib.umd.edu/special
https://www.lib.umd.edu/gov-info-gis

*Frostburg*

Frostburg State University
Lewis J. Ort Library
Government Documents
Collection: Maps
http://www.frostburg.edu/lewis-ort-library/
http://libguides.frostburg.edu/govdocs-main

*Rockville*

Montgomery College
Special Collections
Collection: Maps
http://cms.montgomerycollege.edu/libraries/

## Massachusetts

*Amherst*

University of Massachusetts, Amherst
W. E. B. Du Bois Library
Map Collection
Collection: Maps and GIS
http://www.library.umass.edu/locations/du-bois-library/maps/

*Boston*

Boston Public Library
Central Library
Norman Leventhal Map Center
Collection: Maps
http://www.bpl.org/research/nblmapcenter.htm

Bostonian Society
Library and Special Collections
Collection: Maps
http://www.bostonhistory.org

State Library of Massachusetts
Collection: Maps
http://www.mass.gov/anf/research-and-tech/oversight-agencies/lib/

*Cambridge*

Harvard University
Harvard College Library
Pusey Library
Collection: Maps and GIS
http://hcl.harvard.edu/libraries/maps/

Massachusetts Institute of Technology
Hayden Library
Map Room
Collection: Maps and GIS
https://libraries.mit.edu/hayden/
https://libraries.mit.edu/hayden/hayden-collections/
https://libguides.mit.edu/gis

*Chestnut Hill*

Boston College
Thomas P. O'Neill Jr. Library
Collection: Maps and GIS
http://libguides.bc.edu/oneill/home

*Dartmouth*

University of Massachusetts, Dartmouth
Claire T. Carney Library
Map Collection
Collection: Maps and GIS
http://www.lib.umassd.edu

*Springfield*

Springfield City Library
Central Library
Collection: Maps
http://www.springfieldlibrary.org

## Michigan

*Allendale*

Grand Valley State University
James H. Zumberge Library
Collection: Maps
http://libguides.gvsu.edu/govdoc

*Ann Arbor*

University of Michigan
Stephen S. Clark Library
Harlan Hatcher Graduate Library
Collection: Maps
http://www.lib.umich.edu/clark-library

University of Michigan
William L. Clements Library
Map Collection
Collection: Maps and GIS
http://clements.umich.edu/maps.php

*Detroit*

Detroit Public Library
Main Library
Special Collections
Collection: Maps
http://www.detroit.lib.mi.us/special-collections

*East Lansing*

Michigan State University
Main Library
Map Library
Collection: Maps and GIS
http://www.lib.msu.edu/map/

*Flint*

Flint Public Library
Collection: Maps
Flint, Michigan
http://fpl.info

*Grand Rapids*

Grand Rapids Public Library
Grand Rapids History and Special Collections Center
Collection: Maps
http://www.grpl.org/research/

*Houghton*

Michigan Technological University
J. R. Van Pelt and John and Ruanne Opie Library
Collection: Maps
http://www.mtu.edu/library/

*Kalamazoo*

Kalamazoo Public Library
Central Branch
Collection: Maps
http://www.kpl.gov

Western Michigan University
Dwight B. Waldo Library
Maps and Atlases
Collection: Maps and GIS
http://www.wmich.edu/library
http://www.wmich.edu/library/collections/maps

*Lansing*

Library of Michigan
Collection: Maps
http://www.michigan.gov/libraryofmichigan

State Archives of Michigan
Collection: Maps
http://www.michigan.gov/mhc/0,4726,7-282-61083---,00.html

*Mount Pleasant*

Central Michigan University
Department of Geography and Environmental Studies, Engineering and Technology Building Center for Geographic Information Science (CGIS)
Collection: Maps and GIS
https://www.cmich.edu/library/Pages/default.aspx

*Ypsilanti*

Eastern Michigan University
Eastern Michigan University Library
Map Library
Collection: Maps and GIS
http://guides.emich.edu/maps

## Minnesota

*Bemidji*

Bemidji State University
A. C. Clark Library
Government Publications
Collection: Maps and GIS
http://www.bemidjistate.edu/library
http://www.bemidjistate.edu/library/government/

*Collegeville*

Saint John's University
Saint John's University Library
Government Documents
Collection: Maps
http://www.csbsju.edu/libraries

*Duluth*

University of Minnesota, Duluth
Kathryn A. Martin Library
Government Documents
Collection: Maps
http://www.d.umn.edu/lib/

*Mankato*

Minnesota State University, Mankato
Memorial Library
The Mary T. Dooley Map Library
Collection: Maps and GIS
http://www.lib.mnsu.edu
http://lib.mnsu.edu/collections/maps/index.html
https://libguides.mnsu.edu/mapsarea

*Minneapolis*

East View Cartographic
Company Headquarters
East View Geospatial
Collection: Maps
http://geospatial.com

Hennepin County Library
Minneapolis Public Library
Collection: Maps
http://www.hclib.org/about/locations/minneapolis-central

University of Minnesota
O. M. Wilson Library
John R. Borchert Map Library
Collection: Maps and GIS
https://www.lib.umn.edu/borchert

University of Minnesota
Science and Engineering Library
Geologic Map Collection
Collection: Maps
https://www.lib.umn.edu/walter/collections

*Northfield*

Carleton College
Laurence McKinley Gould Library
Map Collection
Collection: Maps and GIS
http://apps.carleton.edu/campus/library/about/collections/maps/

*Saint Cloud*

Saint Cloud State University
James W. Miller Learning Resources Center
Government Documents/Maps
Collection: Maps and GIS
http://www.stcloudstate.edu/library/
https://stcloud.lib.mnscu.edu/subjects/guide.php?subject=governmentpublications

*Saint Paul*

Minnesota Geological Survey
Collection: Maps
http://www.mngs.umn.edu

Saint Paul Public Library
George Latimer Central Library
Collection: Maps
http://www.sppl.org

*Saint Peter*

Gustavus Adolphus College
Folke Bernadotte Memorial Library
Special Collections
Collection: Maps
http://www.gustavus.edu/library

## Mississippi

*Jackson*

Mississippi Department of Archives and History
Collection: Maps
http://www.mdah.ms.gov/new/

## Missouri

*Joplin*

Missouri Southern State University, Joplin
George A. Spiva Library
Archives and Special Collections
Collection: Maps
http://www.mssu.edu/academics/library/archives-historical-manuscripts-guide.php
https://www.mssu.edu/academics/library/find-government-documents.php

*Rolla*

Missouri University of Science and Technology
Curtis Laws Wilson Library
A. C. Spreng Map Room
Collection: Maps
http://library.mst.edu

*Saint Louis*

Saint Louis County Library
Collection: Maps
http://www.slcl.org

Saint Louis Public Library
Collection: Maps
http://www.slpl.org/index.asp

Saint Louis University
Pius XII Memorial Library
Government Documents
Collection: Maps and GIS
http://lib.slu.edu
http://lib.slu.edu/research/government-documents

Washington University
Rudolph Hall
GIS Research Studio
Collection: Maps and GIS
http://www.library.wustl.edu
http://lib.slu.edu/research/government-documents

*Springfield*

Southwest Missouri State University
Duane G. Meyer Library
Collection: Maps and GIS
http://libraries.missouristate.edu/meyer.htm
https://guides.library.missouristate.edu/c.php?g=701786&p=4979943

*Webster Groves*

Webster Groves Public Library
Collection: Maps
http://www.wgpl.org

## Montana

*Bozeman*

Montana State University, Bozeman
Roland R. Renne Library
http://guides.lib.montana.edu/maps

*Butte*

Montana Tech
Library
Collection: Maps
http://www.mtech.edu/library

*Great Falls*

Great Falls Public Library
Collection: Maps
http://www.greatfallslibrary.org

*Helena*

Montana Historical Society
Research Center
Collection: Maps
http://mhs.mt.gov

Helena State Library
Collection: Maps
http://home.msl.mt.gov

*Libby*

Libby County Public Libraries
Collection: Maps
http://www.lincolncountylibraries.com/libby-library/

*Missoula*

Missoula Public Library
Collection: Maps
http://www.missoulapubliclibrary.org/collection/montana-collection

University of Montana
Maureen and Mike Mansfield Library
Map Collection
Collection: Maps and GIS
http://www.lib.umt.edu
http://libguides.lib.umt.edu/maps

## Nebraska

*Crete*

Doanne College
Perkins Library
Collection: Maps
http://www.doane.edu/library

*Fremont*

Midland University
Luther Library
Collection: Maps
https://www.midlandu.edu/landing-page/luther-library

*Kearney*

University of Nebraska, Kearney
Calvin T. Ryan Library
Government Documents Department
Collection: Maps
http://guides.library.unk.edu/govdocs

University of Nebraska, Kearney
Department of Geography
Geography Map Library
Collection: Maps and GIS
http://www.unk.edu/academics/geography/index.php

*Lincoln*

Lincoln City Libraries
Bennett Martin Public Library
Collection: Maps
http://lincolnlibraries.org

Nebraska Department of Natural Resources
Data Bank
Collection: Maps
http://www.dnr.ne.gov/databank

Nebraska Department of Roads
Map Library
Collection: Maps
http://www.roads.nebraska.gov/travel/map-library/

Nebraska Library Commission
Nebraska Publications Clearinghouse
Collection: Maps
http://nlc.nebraska.gov/govdocs/index.aspx

Nebraska State Historical Society
Library/Archives
Collection: Maps
http://www.nebraskahistory.org/lib-arch/services/refrence/la_pubs/index.shtml

University of Nebraska, Lincoln
Conservation and Survey Division/SNR
School of Natural Resources
Collection: Maps
http://snr.unl.edu/csd/

University of Nebraska, Lincoln
C. Y. Thompson Library
Collection: Maps
http://libraries.unl.edu/cy-thompson-library

University of Nebraska, Lincoln
Geology Library
Collection: Maps
http://libraries.unl.edu/geology-library

University of Nebraska, Lincoln
Love Library
Reference Collection
Collection: Maps
http://libraries.unl.edu/architecture-library

*Norfolk*

Norfolk Public Library
Collection: Maps
http://www.ci.norfolk.ne.us/library

*Omaha*

Douglas County Historical Society Library
Collection: Maps
http://www.omahahistory.org

Omaha Public Library
W. Dale Clark Library
Collection: Maps
http://www.omaha.lib.ne.us

University of Nebraska, Omaha
Department of Geography and Geology
Collection: Maps
http://www.unomaha.edu/college-of-arts-and-sciences/geography/

University of Nebraska, Omaha
Dr. C. C. and Mabel L. Criss Library
Government Documents
Collection: Maps
http://libguides.unomaha.edu/c.php?g=100463&p=650231

## Nevada

*Carson City*

Carson City Library
Collection: Maps
http://www.carsoncitylibrary.org

Nevada State Library and Archives
Collection: Maps
http://nsla.nv.gov/Archives/Nevada_State_Archives/

*Reno*

Nevada Historical Society
Collection: Maps
http://www.nevadaculture.org

University of Nevada, Reno
DeLaMare Library
Mary B. Ansari Map Library
Collection: Maps and GIS
http://www.delamare.unr.edu/Maps

University of Nevada, Reno
Special Collections and Archives Department
Collection: Maps
https://library.unr.edu/SpeColl/Index

## New Hampshire

*Concord*

New Hampshire Historical Society
Library
Special Collections
Collection: Maps
https://www.nhhistory.org

*Durham*

University of New Hampshire
Dimond Library
The Class of 1943 Map Room
Collection: Maps and GIS
https://www.library.unh.edu/find/maps-geospatial-data

University of New Hampshire
Dimond Library
Government Information Collections
Collection: Maps and GIS
https://www.library.unh.edu/find/government-information

*Hanover*

Dartmouth College
Berry Library
Evans Map Room
Collections: Maps and GIS
http://www.dartmouth.edu/~library/maproom/?mswitch-redir=classic

Dartmouth College
Kresge Physical Sciences Library
Collections: Maps
http://www.dartmouth.edu/~library/kresge/

*Keene*

Historical Society of Cheshire County
Collections: Maps
http://www.hsccnh.org

## New Jersey

*Camden*

Rutgers University, Camden Campus
Alexander Library
Government Documents
Collections: Maps and GIS
http://www.libraries.rutgers.edu/robeson

*Hillside*

Hillside Public Library
Collections: Maps
http://www.hillsidepl.org

*Newark*

Newark Public Library
Collections: Maps
http://www.npl.org/pages/collections/njic.html#archives

*Piscataway*

Rutgers University
Library of Science and Medicine
Collections: Maps
https://www.libraries.rutgers.edu/lsm

*Princeton*

Princeton University
Lewis Library
Map and Geospatial Information Center
Collection: Maps and GIS
http://library.princeton.edu/collections/pumagic

*Trenton*

New Jersey State Library
Collection: Maps
http://www.njstatelib.org

**New Mexico**

*Albuquerque*

University of New Mexico
Centennial Science and Engineering Library
Map and Geographic Information Center (MAGIC)
Collection: Maps and GIS
http://library.unm.edu/about/magic.php

*Hobbs*

Hobbs Public Library
Collection: Maps
http://hobbspubliclibrary.org

*Santa Fe*

New Mexico State Library
Collection: Maps
http://www.stlib.state.nm.us

*Socorro*

New Mexico Tech, New Mexico Bureau of Geology and Mineral Resources
Geological Information Center Library
Collection: Maps
http://geoinfo.nmt.edu/libraries/gic/home.html

**New York**

*Albany*

New York State Library
Collection: Maps
http://www.nysl.nysed.gov/mssc/maps.htm

*Brooklyn*

Pratt Institute
Library
Collection: Maps
https://www.pratt.edu/academics/academic-resources/academic-library/

*Buffalo*

Buffalo and Erie County Public Library
Central Library
Collection: Maps
http://www.buffalolib.org

University at Buffalo, State University of New York
Buffalo Lockwood Memorial Library
Map Collection
Collection: Maps
http://library.buffalo.edu/maps/

*Huntington Station*

South Huntington Public Library
Collection: Maps
https://www.shpl.info

*Ithaca*

Cornell University
John M. Olin Library
Map and Geospatial Information Collection
Collection: Maps
https://olinuris.library.cornell.edu/collections/maps

*Jamaica*

Queens Library
Central Library
Collection: Maps
http://www.queenslibrary.org

*Liverpool*

Liverpool Public Library
Collection: Maps
http://lpl.org

*New York*

Columbia University
Lehman Library Map Collection
Collection: Maps and GIS
http://library.columbia.edu/locations/maps.html

New York Historical Society
New York Historical Society Library
Collection: Maps
http://www.nyhistory.org/library

New York Public Library
Stephen A. Schwarzman Building
Map Division
Collection: Maps
https://www.nypl.org/about/locations/schwarzman

*Stony Brook*

Stony Brook University
Frank Melville Jr. Memorial Library
Special Collections and University Archives
Collection: Maps and GIS
http://www.stonybrook.edu/commcms/libspecial/collections/maps.html

*Syracuse*

Syracuse University
Bird Library
The Map Room
Collection: Maps and GIS
https://library.syr.edu/about/departments/Access_Services/Service_Areas/mgi.php

**North Carolina**

*Boone*

Appalachian State University
Belk Library
Special Collections
Collection: Maps and GIS
http://collections.library.appstate.edu/subjects/maps

*Buies Creek*

Campbell University
Wiggins Memorial Library
Government Documents
Collection: Maps
http://www.lib.campbell.edu
http://www.lib.campbell.edu/government-documents

*Chapel Hill*

University of North Carolina at Chapel Hill
Davis Library
Davis Library Research Hub
Collection: Maps and GIS
http://library.unc.edu/science/
http://library.unc.edu/services/data/

*Charlotte*

Public Library of Charlotte, Mecklenburg County
Main Library
North Carolina Room Map Collection
Collection: Maps
https://www.cmlibrary.org

University of North Carolina at Charlotte
J. Murrey Atkins Library
Government Maps
Collection: Maps
https://library.uncc.edu/atkins/governmentdocuments/maps

University of North Carolina at Charlotte
J. Murrey Atkins Library
Research Data Services
Collection: GIS
http://guides.library.uncc.edu/c.php?g=173282&p=1142057

*Cullowhee*

Western Carolina University
Hunter Library
Maps
Collection: Maps and GIS
http://www.wcu.edu/hunter-library/index.asp
http://researchguides.wcu.edu/maps

*Durham*

Duke University
Bostock Library
Data and Visualization Services
Collection: GIS
http://guides.library.duke.edu/c.php?g=289607&p=1930490
https://library.duke.edu/data/gis
https://guides.library.duke.edu/maps

Duke University
Perkins Library
Map Collection
Collection: Maps
https://guides.library.duke.edu/maps

*Greensboro*

Greensboro Public Library
North Carolina Collection
Collection: Maps
http://www.greensboro-nc.gov/index.aspx?page=959

University of North Carolina at Greensboro
Walter Clinton Jackson Library
Reference Collection (part of the Research, Outreach, and Instruction Department)
Collection: Maps
https://library.uncg.edu/info/depts/reference/
https://library.uncg.edu/collection/collection_description.aspx

*Greenville*

East Carolina University
Joyner Library
Government Documents and Microforms Department (houses Joyner Library Map Collection)
Collection: Maps and GIS
http://www.ecu.edu/cs-lib/govdoc/maps.cfm

*Raleigh*

North Carolina Department Office of Archives and History
State Archives of North Carolina
Collection: Maps
http://www.history.ncdcr.gov

North Carolina Geological Survey
Collection: Maps
https://deq.nc.gov/about/divisions/energy-mineral-land-resources/north-carolina-geological-survey/

North Carolina State University
Natural Resources Library
Collection: Maps and GIS
http://www.lib.ncsu.edu/nrl

State Library of North Carolina
Collection: Maps
http://www.mass.gov/anf/research-and-tech/research-state-and-local-history/maps-atlases-and-plans.html

*Salisbury*

Catawba College
Corriher-Linn-Black Library
Collection: Maps
http://libweb.catawba.edu

*Wilmington*

University of North Carolina, Wilmington
William Madison Randall Library
Special Collections and University Archives
Collection: Maps
http://library.uncw.edu

*Winston-Salem*

Moravian Archives
Collection: Maps
http:// moravianarchives.org

**North Dakota**

*Bismarck*

North Dakota State Library
Collection: Maps
http://www.library.nd.gov

State Historical Society of North Dakota
Collection: Maps
http://www.history.nd.gov/archives/index.html

*Fargo*

North Dakota State University
Main Library
Government Documents
Collection: Maps and GIS
https://library.ndsu.edu/locations/main-library

*Minot*

Minot State University
Gordon B. Olson Library
Government Documents and Maps
Collection: Maps
http://www.minotstateu.edu/library/collection_dev.shtml

**Ohio**

*Akron*

University of Akron
Department of Geography and Planning
Collection: Maps and GIS
http://www.uakron.edu/geography/indexOld.dot

University of Ohio
Bierce Library
Government Documents
Map Room
Collection: Maps and GIS
http://www.uakron.edu/libraries/about/gov-docs.dot

*Athens*

Ohio University
Alden Library
Government Documents and Maps Collection
Collection: Maps and GIS
https://www.library.ohiou.edu/about/collections/government-documents/

*Bowling Green*

Bowling Green State University
Jerome Library
Government Documents
Collection: Maps
http://www.bgsu.edu/library.html

*Cincinnati*

Cincinnati Museum Center
Cincinnati Historical Library and Archives
Collection: Maps
http://www.cincymuseum.org/library

Public Library of Cincinnati and Hamilton County
Government Resources
Collection: Maps
http://www.cincinnatilibrary.org/resources/government.asp?section=1&category=23

University of Cincinnati
Geology-Mathematics-Physics Library
Willis G. Meyer Map Collection
Collection: Maps and GIS
http://www.libraries.uc.edu/gmp.html

*Cleveland*

Case Western Reserve University
Kelvin Smith Library
Center for Statistics and Geospatial Data
Collection: Maps and GIS
http://library.case.edu/ksl/

Cleveland Public Library
Main Library
Collection: Maps
http://cpl.org/thelibrary/subjectscollections/map-collection/

Cleveland State University
Michael Schwartz Library
Collection: Maps
http://www.ulib.csohio.edu

*Columbus*

Columbus Metropolitan Library
Collection: Maps
http://www.columbuslibrary.org

Ohio Department of Natural Resources
Geographic Information Systems
Collection: Maps and GIS
http://geospatial.ohiodnr.gov

Ohio Historical Connection
Ohio History Center
Collection: Maps
http://www.ohiohistory.org

Ohio State University
Orton Memorial Library of Geology
Collection: Maps
https://library.osu.edu/about/locations/geology-library/

Ohio State University
Thompson Library
Gardner Family Map Room
Collection: Maps and GIS
https://library.osu.edu/about/locations/thompson-library/

State Library of Ohio
Government Information Services
Collection: Maps
http://www.winslo.state.oh.us/LPD/fed_state_docs

*Dayton*

Dayton Metro Library
Collection: Maps
http://www.daytonmetrolibrary.org

Wright State University
Dunbar Library
Government Documents
Collection: Maps
http://www.libraries.wright.edu

*Euclid*

Euclid Public Library
Collection: Maps
http://www.euclidlibrary.org

*Geneva*

Western Reserve Historical Society
Collection: Maps
http://www.wrhs.org

*Granville*

Denison University
William Howard Doane Library
Richard H. Mahard Map Library
Collection: Maps and GIS
http://denison.edu/campus/library
http://libguides.denison.edu/maps#s-lg-box-6354762

*Kent*

Kent State University
University Libraries
The Map Library
Collection: Maps and GIS
http://www.library.kent.edu/map-library

*Newark*

Ohio State University, Newark and Central Ohio Technical College
Thompson Library
Gardner Family Map Room
Collection: Maps
https://library.osu.edu/about/locations/newark-campus-warner-library/

*Oxford*

Miami University
Brill Science Library
Map Collections
Collection: Maps and GIS
http://www.users.miamioh.edu/messnekr/brill/

*Tiffin*

Heidelberg College
Beeghly Library
Government Documents
Collection: Maps
https://www.heidelberg.edu/academics/resources-and-support/beeghly-library

*Toledo*

Toledo—Lucas County Public Library
Main Library
Collection: Maps
http://www.toledolibrary.org

University of Toledo
William S. Carlson Library
Map Collection
Collection: Maps and GIS
http://www.utoledo.edu/library/carlson/

*Westerville*

Otterbein College
Courtright Memorial Library
Government Documents
Collection: Maps
http://www.otterbein.edu/public/Library

## Oklahoma

*Durant*

Southeastern Oklahoma State University
Henry G. Bennett Memorial Library
Government Documents
Collection: Maps
http://www.se.edu/library/government-information/

*Oklahoma City*

Oklahoma Department of Libraries
Collection: Maps
http://libraries.ok.gov/welcome/

Oklahoma History Center
Collection: Maps
http://www.okhistory.org/research/maps

University of Central Oklahoma
Chambers Library
Map Collection
Collection: Maps
Edmond, Oklahoma
http://library.uco.edu
https://libraries.ou.edu/resources-subject/geography-gis

*Tulsa*

University of Tulsa
McFarlin Library
University Archives
Collection: Maps
https://utulsa.edu/mcfarlin-library/
https://utulsa.edu/mcfarlin-library/special-collections/the-collections/

## Oregon

*Corvallis*

Oregon State University
The Valley Library
Map Collection
Collection: Maps and GIS
http://guides.library.oregonstate.edu/subject-guide/286-Government-Information

*Eugene*

University of Oregon
Knight Library
Collection: Maps and GIS
http://library.uoregon.edu

University of Oregon
Map and Aerial Photography Library
Collection: Maps
http://library.uoregon.edu/map-library

*La Grande*

Eastern Oregon University
Pierce Library
Collection: Maps
http://library.eou.edu

*Pendleton*

Blue Mountain Community College Library
Map Collection
Collection: Maps
http://www.bluecc.edu/academics/library-1048

*Portland*

Lewis and Clark College
Aubrey R. Watzek Library
Special Collections
Collection: Maps
http://library.lclark.edu/

Multnomah County Library
Central Library
Collection: Maps
http://www.multcolib.org

Natural Resources Information for the Pacific Northwest
StreamNet Regional Library
Collection: Maps
http://www.streamnetlibrary.org

Oregon Department of Geology and Mineral Industries
Collection: Maps
http://www.oregongeology.com

Oregon Historical Society
Research Library
Collection: Maps
http://www.ohs.org/research-and-library/

Portland State University
Brandon Price Millar Library
Government Information and Maps
Collection: Maps and GIS
http://library.pdx.edu/research/
http://guides.library.pdx.edu/maps

*Salem*

Oregon State Library
Government Information and Library Services
Collection: Maps
http://www.oregon.gov/osl

**Pennsylvania**

*Bethlehem*

Lehigh University
E. W. Fairchild Martindale (EWFM) Library
Library and Technology Services
Collection: GIS
http://library.lehigh.edu
https://lts.lehigh.edu/services/geographic-information-systems-gis

Moravian College
Earth Science Collection
Collection: Maps
https://www.moravian.edu/physics

*Bryn Mawr*

Bryn Mawr College
Canaday Library
Collection: Maps
http://www.brynmawr.edu/geology

*Harrisburg*

Pennsylvania Historical and Museum Commission
Pennsylvania State Archives
Collection: Maps
http://www.phmc.state.pa.us/bah/dam/mg/mg11.htm

State Library of Pennsylvania
Maps Collection
Collection: Maps
http://www.statelibrary.pa.gov/Pages/search.aspx?searchBox=maps%20collection

*Kutztown*

Kutztown University
Rohrback Library
Maps Department
Collection: Maps and GIS
http://library.kutztown.edu

*New Castle*

New Castle Public Library
Collection: Maps
http://www.ncdlc.org

*Philadelphia*

Drexel University
Ewell Sale Stewart Library
James Bond Map Room
Collection: Maps and GIS
http://www.ansp.org/research/library/

Free Library of Philadelphia
Map Collection
Collection: Maps
https://libwww.freelibrary.org/digital/collection/maps

Historical Society of Pennsylvania
Library Division
Archives
Collection: Maps
https://hsp.org

Library Company of Philadelphia
Print and Photograph Department
Collection: Maps
http://www.librarycompany.org/collections/prints/index.htm

Senator John Heinz History Center
Collection: Maps
http://www.heinzhistorycenter.org

*Pittsburgh*

University of Pittsburgh
Hillman Library
Digital Scholarship Commons
Collection: GIS
http://www.library.pitt.edu/geographic-spatial-data-services

University of Pittsburgh
Hillman Library
Microforms, Government Documents, and Maps
Collection: Maps
http://www.library.pitt.edu/maps

*University Park*

Pennsylvania State University
Central Pattee Library
Donald W. Hamer Center for Maps and Geospatial Information
Collection: Maps and GIS
http://www.libraries.psu.edu/maps

*West Chester*

West Chester University
Francis Harvey Green Library
Government Documents and Maps
Collection: Maps
http://www.wcupa.edu/viceProvost/library.fhg/mather.asp

**Rhode Island**

*Kingston*

University of Rhode Island
The University Library
Government Publications
Collections: Maps and GIS
http://web.uri.edu/library/government-publications/

*Providence*

Providence Public Library
Collections: Maps
http://www.provlib.org

Rhode Island College
James P. Adams Library
U.S. Government Information Resources
Collections: Maps
http://www.ric.edu/adamslibrary

Rhode Island Historical Society
Graphics Collections
Collections: Maps
http://www.rihs.org/library/graphics-collections/

**South Carolina**

*Anderson*

Anderson County Public Library
Collections: Maps
http://www.andersonlibrary.org

*Charleston*

Charleston County Public Library
Main Library
Collections: Maps
http://www.ccpl.org

Charleston Library Society
Collections: Maps
http://www.charlestonlibrarysociety.org/

The Citadel, Military College of South Carolina
Daniel Library
Collections: Maps
http://library.citadel.edu/home

South Carolina Historical Society
Collections: Maps
http://www.schistory.org

*Clemson*

Clemson University
R. M. Cooper Library
Center for Geospatial Technologies (GIS Lab)
Collections: GIS
http://libraries.clemson.edu

Clemson University
R. M. Cooper Library
Government Documents
Collections: Maps
http://libraries.clemson.edu
https://clemson.libguides.com/c.php?g=230493&p=1529195

*Columbia*

South Carolina Department of Archives and History
Reference Services
Collection: Maps
http://scdah.sc.gov/Pages/default.aspx

University of South Carolina
Thomas Cooper Library
The Map Library (Government Information and Maps)
Collection: Maps and GIS
http://library.sc.edu/maps/mapSC.html

*Conway*

Coastal Carolina University
Kimbel Library
Collection: Maps
http://www.coastal.edu/intranet/library/

*Georgetown*

Georgetown County Library
Collection: Maps
http://georgetowncountylibrary.sc.gov/Pages/default.aspx

*Greenville*

Furman University
James B. Duke Library
Collection: Maps
http://libguides.furman.edu/library/home

Greenville County Library System
Collection: Maps
http://www.greenvillelibrary.org

*Lexington*

Lexington County Public Library System
Collection: Maps
http://www.lex.lib.sc.us

*Rock Hill*

Winthrop University
Ida Jane Dacus Library
Louise Pettus Archives and Special Collections
Collection: Maps
http://www2.winthrop.edu/dacus/

*Spartanburg*

Spartanburg County Public Library
Government Department
Collection: Maps
http://www.infodepot.org/newpages/GovDoc.asp

**South Dakota**

*Brookings*

South Dakota State University
Hilton M. Briggs Library
Government Documents Department
Collection: Maps
http://www.sdstate.edu/hilton-m-briggs-library

South Dakota State University
SDSU Census Data Center
Hansen Hall
Collection: GIS
https://www.sdstate.edu/sociology-rural-studies/census-data-center

*Pierre*

South Dakota State Historical Society
Collection: Maps
http://history.sd.gov/archives/

*Rapid City*

South Dakota School of Mines and Technology
Devereaux Library
Collection: Maps
http://www.sdsmt.edu/library/

*Vermillion*

University of South Dakota
I. D. Weeks Library
Government Publications

Collection: Maps
http://www.usd.edu/library

## Tennessee

*Chattanooga*

Chattanooga Public Library
Local History Map Collection
Collection: Maps
http://chattlibrary.org

Tennessee Valley Authority
Map and Photo Records
Collection: Maps
https://tva.com

*Clarksville*

Austin Peay State University
GIS Center Information Services
Collection: GIS
http://library.apsu.edu
http://apsugis.org

*Cookeville*

Tennessee Technological University
Angelo and Jennette Volpe Library and Media Center
Collection: Maps and GIS
https://www.tntech.edu/library

*Johnson City*

East Tennessee State University
Charles C. Sherrod Library
Document/Law/Maps
Collection: Maps
http://sherrod.etsu.edu/research/govt

*Knoxville*

Knox County Public Library
East Tennessee History Center
Calvin M. McClung Historical Collection
Collection: Maps
http://www.knoxlib.org/local-family-history/calvin-m-mcclung-historical-collection

Knoxville/Knox County Metropolitan Planning Commission
Library
Collection: Maps
http://www.knoxmpc.org

United States Department of Interior
Office of Surface Mining, Reclamation and Enforcement
Knoxville Field Office Geographic Information System
Collection: Maps
http://www.osmre.gov

University of Tennessee, Knoxville
James D. Hoskins Library
Map Collection
Collection: Maps and GIS
https://www.lib.utk.edu/special/
https://libguides.utk.edu/map

*Memphis*

Memphis County Public Library and Information Center
Memphis and Shelby County Room
Collection: Maps
http://www.memphislibrary.org

University of Memphis
Ned R. McWherter Library
Special Collections
Collection: Maps
http://www.memphis.edu/libraries/
http://www.memphis.edu/libraries/special-collections/resources.php

*Nashville*

Tennessee Secretary of State
Archives Development Program
Collection: Maps
http://sos.tn.gov/products/tsla/archives-development-program

Vanderbilt University
Science and Engineering Library
Map Room
Collection: Maps and GIS
http://www.library.vanderbilt.edu/science/

## Texas

*Abilene*

Abilene Christian University
Brown Library
Collection: Maps
http://www.acu.edu/academics/library

Hardin-Simmons University
Rupert and Pauline Richardson Library
J. W. Williams Map Collection
Collection: Maps
http://www.hsutx.edu/library/fg-ul.html

## Angleton

Brazoria County Historical Museum
Adriance Library and Research Center
Collection: Maps
http://brazoriacountytx.gov/departments/museum/collections

## Arlington

University of Texas at Arlington
The University of Texas Arlington Library (Central Library)
The Virginia Garrett Cartographic History Library (Special Collections)
Collection: Maps and GIS
http://library.uta.edu
http://library.uta.edu/special-collections/about
http://library.uta.edu/collections/cartographic-history-collections

## Austin

Texas State Library and Archives
Historic Map Collection
Collection: Maps
https://www.tsl.texas.gov/apps/arc/maps/

University of Texas at Austin
Benson Latin American Collection
Rare Books and Manuscript Division
Collection: Maps
http://www.lib.utexas.edu/benson/collections/rare-books-and-manuscripts

University of Texas at Austin
Perry-Castaneda Library
Map Room
Collection: Maps and GIS
http://www.lib.utexas.edu/pcl/collections/map_room

University of Texas at Austin
Walter Geology Library
The Tobin Map International Geological Collection
Collection: Maps
http://www.lib.utexas.edu/geology/collections/maps

## Canyon

West Texas A&M University
Cornette Library
Government Documents
Collection: Maps
http://www.wtamu.edu/library/documents/docsmaps.shtml

## College Station

Texas A&M University
Sterling C. Evans Library
Maps/GIS Department
Collection: Maps and GIS
http://evans.library.tamu.edu/index.html
http://library.tamu.edu/services/map_gis/about.html

## Dallas

Dallas Public Library
Dallas History and Archives
Collection: Maps
https://dallaslibrary2.org/dallashistory/

Dallas Public Library
J. Eric Jonsson Central Library
Map Collection
Collection: Maps
https://dallaslibrary2.org/central/onlineExhibits.php

Southern Methodist University
DeGolyer Library
Map Collection
Collection: Maps
http://www.smu.edu/CUL/DeGolyer/Collections/Digital-Collections

Southern Methodist University
Fondren Library
Edwin J. Foscue Map Library
Collection: Maps
https://www.smu.edu/CUL/Libraries/Foscue

## Denton

University of North Texas
Eagle Commons Library
Map Collection (part of the Government Documents department)
Collection: Maps and GIS
http://www.library.unt.edu/government-information-connection

## El Paso

University of Texas at El Paso
Department of Civil Engineering
Regional Geospatial Service Center
Collection: Maps and GIS
http://gis.utep.edu

University of Texas at El Paso
The University of Texas at El Paso Library
Government Documents and Maps
Collection: Maps
http://libraryweb.utep.edu/about/departments/documents/index.php

*Houston*

Lunar and Planetary Institute
Library
Collection: Maps
http://www.lpi.usra.edu/library

Rice University
Fondren Library
GIS/Data Center (GDC)
Collection: Maps and GIS
http://library.rice.edu

Rice University
Fondren Library
Kelley Center for Government Information, Data and Geospatial Services
Collection: Maps and GIS
https://library.rice.edu/gov

*Lubbock*

Texas Tech University
University Library
Map Collection (component of Government Documents)
Collection: Maps and GIS
https://www.depts.ttu.edu/library/collections/maps/

*Nacogdoches*

Stephen F. Austin State University
Ralph W. Steen Library
East Texas Digital Archives
Collection: Maps
http://digital.sfasu.edu

*Richardson*

University of Texas at Dallas
Eugene McDermott Library
Maps
Collection: Maps and GIS
http://www.utdallas.edu/library

*San Antonio*

Daughters of the Republic of Texas Library
http://www.drtl.org

Saint Mary's University
Louis J. Blume Library
Government Information
Collection: Maps
http://lib.stmarytx.edu/maps

*Waco*

Baylor University
Jesse H. Jones Library
Government Documents
Collection: Maps
http://www.baylor.edu/lib/centrallib/

**Utah**

*Logan*

Utah State University
Merrill-Cazier Library
Government Information Department
Collection: Maps and GIS
http://libguides.usu.edu/govinfo/

Utah State University
University Libraries
Special Collections and Archives
Collection: Maps
https://archives.usu.edu

*Ogden*

Weber State University
Steward Library
Government Documents Collection
Collection: Maps and GIS
https://library.weber.edu/collections/government_documents

*Provo*

Brigham Young University
Harold B. Lee Library
Harold B. Lee Library Geospatial Services and Training
Collection: GIS
http://guides.lib.byu.edu/c.php?g=246326&p=3227075

Brigham Young University
Harold B. Lee Library
Science and Maps
Collection: Maps
https://lib.byu.edu/collections/

*Salt Lake City*

Family History Library
Collection: Maps
http://www.familysearch.org

University of Utah
J. Willard Marriott Library
GIS Services
Collection: GIS
https://www.lib.utah.edu/services/geospatial/

University of Utah
J. Willard Marriott Library
Collection: Maps
http://www.lib.utah.edu/collections/special-collections/

Utah Division of State History
Utah Department of Heritage and Arts
Collection: Maps
https://heritage.utah.gov/history

Utah Geological Survey
Department of Natural Resources Library
Collection: Maps
http://utsl.sirsi.net/uhtbin/cgisirsi/?ps=kfzy5ai6LB/DNR/64730068/2/6

## Vermont

*Burlington*

University of Vermont
Bailey/Howe Library
Government Information and Maps
Collection: Maps and GIS
https://library.uvm.edu/research/maps

*Woodstock*

Woodstock Historical Society
Collection: Maps
http://woodstockhistorical.org/library-resources-woodstock-history-center/

## Virginia

*Blacksburg*

Virginia Tech
Carol M. Newman Library
Geospatial Data Services
Collection: GIS
https://lib.vt.edu/collections/geospatialdata.html

*Bridgewater*

Bridgewater College
Alexander Mack Library
Collection: Maps
http://libguides.bridgewater.edu/c.php?g=22598&p=133624

*Charlottesville*

University of Virginia
University of Virginia Library
Map Collection
Collection: Maps and GIS
http://guides.lib.virginia.edu/maps

*Fairfax*

George Mason University
Fenwick Library
International Government Documents & Treaties
Digital Scholarship Center (DiSC)
http://infoguides.gmu.edu/igo
http://sca.gmu.edu/collections-subject.php
https://library.gmu.edu/search/data

George Mason University
Fenwick Library
Special Collections Research Center
Collection: Maps and GIS
http://sca.gmu.edu/collections-subject.php

*Fredericksburg*

Mary Washington College
Simpson Library
Government Documents
Collection: Maps
http://libraries.umw.edu/?simpson

*Reston*

United States Geological Survey, Reston
USGS Library
Cartographic Information Services
Collection: Maps
http://library.usgs.gov

*Richmond*

Library of Virginia
Archival and Information Services Division/Map Collection
Collection: Maps
http://www.lva.virginia.gov

University of Richmond
Boatwright Memorial Library
Galvin Rare Book Room
Carole Weinstein International Center
Collection: Maps
http://library.richmond.edu

University of Richmond
Carole Weinstein International Center
Spatial Analysis Lab
Collection: GIS
https://blog.richmond.edu/sal/about/

*Williamsburg*

College of William and Mary
Earl Gregg Swem Library
Geology Department Library
https://libraries.wm.edu/

Colonial Williamsburg Foundation
Collection: Maps
http://library.richmond.edu

## Washington

*Anacortes*

Anacortes History Museum
Collection: Maps
www.anacorteshistorymuseum.org

*Bellingham*

Western Washington University
The Spatial Analysis Lab (SAL)
Environmental Studies Building
Collection: GIS
https://huxley.wwu.edu/sal/spatial-analysis-lab-sal

Western Washington University
Wilson Library
Collection: Maps
http://libguides.wwu.edu/mapcollection

*Bremerton*

Kitsap History Museum
JFK Library
Collection: Maps
http://research.ewu.edu/maps_atlases

*Cheney*

Eastern Washington University
John F. Kennedy Library
Archives and Special Collections
Collection: Maps
http://research.ewu.edu/maps_atlases

*Colfax*

Whitman County Library
Colfax Branch
Collection: Maps
http://www.whitco.lib.wa.us

*Ellensburg*

Central Washington University
James E. Brooks Library
Government Publications, Maps, and Microforms
Collection: Maps
http://www.lib.cwu.edu/Archives-Special-Collections

*Fox Island*

Fox Island Historical Society
Map Room
Collection: Maps
https://foxislandmuseum.org

*Olympia*

Evergreen State College
Daniel J. Evans Library
Collection: Maps
http://evergreen.edu/library/collection.htm

Washington State Department of Ecology
Department of Ecology
Collection: Maps
http://www.ecy.wa.gov/services/gis/maps/maps.htm

*Port Townsend*

Jefferson County Historical Society
Research Center
Collection: Maps
http://www.jchsmuseum.org

*Pullman*

Washington State University
Frances Penrose Owen Science and Engineering Library
Map Room
Collection: Maps and GIS
http://libraries.wsu.edu/about/owen

*Seattle*

National Oceanic and Atmospheric Administration
NOAA Seattle Library
Collection: Maps
http://www.wrclib.noaa.gov

Seattle Public Library
Collection: Maps
http://www.spl.org

University of Washington
Suzzallo and Allen Library
Map Collection and Cartographic Information Services
Collection: Maps and GIS
http://www.lib.washington.edu/maps

University of Washington
Suzzallo and Allen Library
Special Collections
Collection: Maps
Seattle, Washington
http://www.lib.washington.edu/specialcoll

*Tacoma*

University of Puget Sound
Collins Memorial Library
Government Documents
Collection: Maps
http://research.pugetsound.edu/c.php?g=304317&p=2031955

*Walla Walla*

Whitman College
Penrose Library
Collection: Maps
https://library.whitman.edu

## West Virginia

*Charleston*

West Virginia State Archives
Collection: Maps
http://www.wvculture.org/history/archives/wvsamenu.html

*Morgantown*

West Virginia University
The West Virginia GIS Technical Center
Department of Geology and Geography (Brooks Hall)
Collection: Maps and GIS
http://wvgis.wvu.edu/about/about.php

West Virginia University
West Virginia University Downtown Campus Library
Map Collection (part of Government Information Services)
Collection: Maps
https://lib.wvu.edu/downtown/
http://libguides.wvu.edu/govdocs/maps

## Wisconsin

*Beloit*

Beloit College
Col. Robert Morse Library
Collection: Maps
https://www.beloit.edu/library/about/
https://www.beloit.edu/library/find/Referenceresource/mapresources/

*Eau Claire*

University of Wisconsin, Eau Claire
William D. McIntyre Library
Collection: Maps
http://libguides.uwec.edu/gov

*Green Bay*

University of Wisconsin, Green Bay
Cofrin Library
Government Publications
Collection: Maps
http:://www.uwgb.edu/library

*Kenosha*

Kenosha Public Library
Southwest Branch
Collection: Maps
https://www.mykpl.info

*La Crosse*

La Crosse Public Library
Collection: Maps
http://www.lacrosselibrary.org

University of Wisconsin, La Crosse
Murphy Library
Map Room
Collection: Maps
http://www.uwlax.edu/murphylibrary/

*Madison*

University of Wisconsin, Madison
American Geological Society Library, third floor
Collection: Maps and GIS
http://uwm.edu/libraries/agsl/

University of Wisconsin, Madison
Geography Library
Collection: Maps and GIS
https://www.library.wisc.edu/geology/
https://www.library.wisc.edu/geography/

Wisconsin Historical Society
Library
Collection: Maps
http://www.wisconsinhistory.org/Content.aspx?dsNav=N:1166

*Milwaukee*

Milwaukee Public Library
Collection: Maps
http://www.mpl.org

University of Wisconsin, Milwaukee
American Geographical Society Library
Collection: Maps
http://uwm.edu/libraries/agsl/

*Oshkosh*

University of Wisconsin, Oshkosh
Department of Geology
Collection: Maps
http://www.uwosh.edu/geology/

University of Wisconsin, Oshkosh
Polk Library
Government Information
Collection: Maps
Oshkosh, Wisconsin
http://www.uwosh.edu/library/collections/government

*Sheboygan*

Mead Public Library
Collection: Maps
http://www.meadpl.org

*Stevens Point*

University of Wisconsin, Stevens Point
Learning Resources Center
GIS Center
Collection: GIS
http://www.uwsp.edu/cols-ap/GIS/Pages/default.aspx

*Superior*

University of Wisconsin, Superior
Jim Dan Hill Library
Lake Superior Maritime Collections
Collection: Maps and GIS
http://library.uwsuper.edu/maritime

# Wyoming

*Cheyenne*

Wyoming State Library
Statewide Information Services
Collection: Maps
http://library.wyo.gov/collections

*Laramie*

University of Wyoming
Brinkerhoff Geology Library
Map Room
Collection: Maps and GIS
http://libguides.uwyo.edu/maps

University of Wyoming
Libraries
Grace Raymond Hebard Collection
Collection: Maps
http://libguides.uwyo.edu/maps

*Riverton*

Central Wyoming College
Brinkerhoff Geology Library
Collection: Maps
http://www.cwc.edu/library/

*Rock Springs*

Western Wyoming Community College
Hay Library
Collection: Maps
https://www.westernwyoming.edu/academics/library/

## REFERENCES

Dubreuil, L. 1993. *World Directory of Map Collections.* International Federation of Library Associations and Institutions/IFLA Publications. Munich: K. G. Saur.

FDLP. 2016. "FDLP File Repository." October 1, 2016. https://www.fdlp.gov/file-repository/.

Leitch, M. 1999. *Directory of Canadian Map Collections.* Seventh edition. Ottawa, ON: Association of Canadian Map Libraries and Archives.

Lib-web.org. 2016. "World Libraries Online." October 1, 2016. http://www.lib-web.org/.

Thiry, C. 2005. *Guide to U.S. Map Resources.* Third edition. Lanham, MD: Scarecrow / Map & Geography Round Table.

Western Association of Map Libraries. 2015. "WAML Principal Region Map Collections." October 1, 2016. http://www.waml.org/maplibs.html.

# 3
# Atlases: Historic and Thematic

An atlas is a collection of information in a graphic format, consisting of text, tables, graphs, and accompanying maps used to describe an area. It provides a comprehensive look at a geographical region, whether the world, a country, a state or province, a county or city, or specific areas of interest such as a watershed or park. Almost every school, college, university, and household at some point has owned an atlas, and, to this day, many individuals do not travel without one. Many children and adults are familiar with the desk or reference atlas, but there are several types of atlases, such as those that cater to specific research interests. Most atlases are still printed on paper in book format, but electronic atlases are becoming popular in academia and research institutions. Printed road maps are being replaced by GPS systems in vehicles and handheld devices. Although atlases span a wide variety of topics, they are categorized into three main types: general, thematic, and road.

General atlases contain maps that show physical and political features of individual countries or groups of countries, highlighting topics such as physical geography, economic activities, demographics, and climate. In addition, these atlases often include other related information, including air photos, satellite imagery, indexes of place names, and cartographic elements such as geographic coordinates and internal locational grids. General atlases typically portray a specific geographic area and are composed of small- to medium-scale maps.

Thematic atlases cover a specific subject, such as geology, hydrology, military history, ornithology, climatology, railroads, land ownership, land use, forestry, and much more. All subject atlases, also known as regional atlases, concentrate on a specific area, which can range from a small community, to a continent. These atlases display thematic information using a series of medium- to large-scale maps.

Road atlases, which are the most common, offer maps that show roads and highways for defined geographic areas. Details in a road map often include the type of road depicted, whether a major highway, secondary road, or gravel road. Although quite popular for travelers and vacationers, road atlases are also quite useful for geographers, who use them to study settlement patterns and population densities.

This chapter will briefly discuss road atlases but will focus primarily on historical atlases and other thematic atlases. Historical atlases have been specifically highlighted because of their popularity with researchers and local communities. In keeping with the traditional organization, the atlases are categorized by scale: continent, nation, state or province, and county. The atlases listed in this chapter represent those published between 1995 and 2016 that have been catalogued in Worldcat (https://www.worldcat.org), one of the most comprehensive library catalogs.

## ROAD ATLASES

The top publishers of American road atlases include the following:

- America Automobile Association (AAA)
- American Map Corporation
- DeLorme
- Hammond
- Mapquest
- Rand McNally
- Michelin
- National Geographic
- State Farm

Many of these organizations, such as the American Automobile Association (AAA), American Map Corporation, and Michelin, publish atlases each year that cover North America and specifically the United States. Rand McNally publishes the largest number of atlases, offering roads of North America as "road maps," "pocket atlases," and "travel guides." The following is a title list of atlases published less regularly.

*1997 Traveler's Guide and Road Atlas: U.S., Canada, Mexico, Caribbean, Central America.* Phoenix, AZ: Best Western International, 1997.

*America's Scenic Drives: Travel Guide and Atlas,* by W. C. Herow. Lenexa, KS: Roundabout, 1997, 2000.

*Canada and U.S.A. = Canada & É.U.: Road Atlas with Special Events Directory.* Oshawa, ON: MapArt, 1995.

*Complete Atlas of Route 66,* by B. Moore and R. Cunningham. Laughlin, NV: Route 66 Magazine, 2003.

*Firefly Atlas of North America: United States, Canada and Mexico.* Richmond Hill, ON: Firefly Books, 2006.

*Hammond Pocket Road Atlas: United States, Canada, Mexico.* Union, NJ: Hammond World Atlas, 2002.

*Hammond Road Atlas and Vacation Guide: United States, Southern Canada.* Maplewood, NJ: Hammond, 1996.

*National Geographic Road Atlas: United States, Canada, Mexico.* Mississauga, ON: MapQuest, 1999.

*Pan-Americana North Travel Atlas: Alaska to Panama: Whitehorse, Vancouver, Seattle, San Francisco, Los Angeles, San Diego, Tijuana, Mexicali, La Paz,* by Stefanie Siegmund and Lan Joyce. Richmond, BC: ITMB, 2013.

*Road Atlas: Canada, USA., Mexico = Atlas routier: Canada, Etats Unis, Mexique.* Oshawa, ON: Peter Heiler, 2007.

*Road Atlas: Canada, USA, Mexico.* Oshawa, ON: MapArt, 2003.

*Road Atlas: United States–Canada–Mexico.* ExxonMobil Travel, 2004.

*Road Atlas: United States, Canada, Mexico.* Litiz, PA: GeoNova, 2008.

*RoadMaster Standard Road Atlas: United States, Canada, Mexico: 2003.* New York: Barnes & Noble Books, 2002.

*Traveler's Atlas, North America,* by D. Dailey. Hauppauge, NY: Barron's Educational Series, 2009.

*United States, Canada, Mexico Road Atlas.* New York: Gousha, 1995.

*UniversalMAP Large Print Road Atlas: United States, Canada, Mexico.* Williamston, MI: UniversalMAP, 2000.

*World Atlas: USA, Southern Canada and Alaska.* Basingstoke, UK: GeoCenter International UK, 1997.

## NORTH AMERICAN ATLASES

### North American Historical Atlases

*America Discovered: A Historical Atlas of North American Exploration,* by Derek Hayes. Vancouver: Douglas & McIntyre, 2004.

*Atlas of Indians of North America,* by Gilbert Legay. Hauppauge, NY: Barron's, 1995.

*Atlas of North American Exploration: From the Norse Voyages to the Race to the Pole,* by W. H. Goetzmann and G. Williams. Norman: University of Oklahoma Press, 1998.

*Atlas of the North American Indian,* by C. Waldman and M. Braun. New York: Checkmark Books, 2009.

*Historical Atlas of Native Americans,* by I. Barnes. New York: Chartwell, 2010.

*Historical Atlas of North America before Columbus,* by F. Ramen. New York: Rosen, 2005.

*Historical Atlas of the North American Railroad,* by Derek Hayes. Vancouver, BC: Douglas & McIntyre, 2010.

*Historical Atlas of the Pacific Northwest: Maps of Exploration and Discovery: British Columbia, Washington, Oregon, Alaska, Yukon,* by Derek Hayes. Seattle: Sasquatch Books, 1999.

*Illustrated Atlas of Native American History: Traces the Movement of North America's Native Peoples from Prehistoric Times to the Present Day,* by Samuel Crompton. Edison, NJ: Chartwell Books, 1999.

*Mapping a Continent: Historical Atlas of North America, 1492–1814,* by R. Litalien et al. Georgetown, ON: McGill-Queen's University Press, 2007.

*Penguin Historical Atlas of North America,* by E. Homberger. New York: Penguin, 1995.

*Settling of North America: The Atlas of the Great Migrations into North America from the Ice Age to the Present,* by H. H. Tanner. New York: Macmillan, 1995.

### North American Thematic Atlases

*Atlas of North America,* edited by H. J. de Blij. New York: Oxford University Press, 2005.

*Atlas of North American Railroads,* by B. Yenne. St. Paul, MN: MBI, 2005.

*Atlas of Short-Duration Precipitation Extremes for the Northeastern United States and Southeastern Canada,* by M. McKay, D. S. Wilks, and Northeast Regional Climate Center. Ithaca, NY: Northeast Regional Climate Center, Cornell University, 1995.

*Les Grands lacs: atlas écologique et manuel des ressources,* by K. Fuller et al. Toronto, ON: Government of Canada, 1995.

*Pictorial Atlas of the United States and Canada,* by K. B. Smith. Baltimore, MD: Ottenheimer, 1996.

*Professional Railroad Atlas of North America: United States, Canada, Mexico.* Austin, TX: DeskMap Systems, 1999, 2004, 2012.

*Running Press Atlas of North America: A Practical Guide to the United States, Canada, and North America.* Philadelphia, PA: Running Press, 1995.

*Steam Powered Video's Comprehensive Railroad Atlas of North America,* by M. Walker. Faversham, UK: Steam Powered Publishing, 1995.

*Temperature and Salinity Atlas for the Scotian Shelf and the Gulf of Maine,* by B. Petrie. Dartmouth, NS: Bedford Institute of Oceanography, 1996.

*Western Gem Hunters Atlas: Rock Locations from California to the Dakotas and British Columbia to Texas,* by H. Cyl Johnson. Susanville, CA: Cy Johnson, 1996.

# CANADIAN NATIONAL ATLASES

## Canadian Historical Atlases

*Atlas of Canada: The Story of a Country through Maps, Photographs, History and Culture*, by J. Geiger. Glasgow, UK: HarperCollins, 2014.

*Atlas of the Acadian Settlement of the Beaubassin, 1660 to 1755: The Great Marsh, Tintamarre and Le Lac*, by P. Surette. Sackville, NB: Tantramar Heritage Trust, 2005.

*Canadian Military Atlas: The Nation's Battlefields from the French and Indian Wars to Kosovo*, by M. Zuehlke and C. S. Daniel. Vancouver, BC: Douglas & McIntyre, 2006.

*Concise Historical Atlas of Canada*, by B. Moldofsky et al. Toronto, ON: University of Toronto Press, 1998.

*Electoral Atlas of the Dominion of Canada (1895)*. Ottawa, ON: Library and Archives Canada, 2004.

*Historical Atlas of Canada: A Thousand Years of Canada's History in Maps*, by Derek Hayes. Vancouver, BC: Douglas & McIntyre, 2002.

*Historical Atlas of Canada: Canada's History Illustrated with Original Maps*, by Derek Hayes. Vancouver, BC: Douglas & McIntyre, 2006, 2015.

*Historical Atlas of Canada*, vol. 1, by C. Harris and G. J. Matthews. Toronto, ON: University of Toronto Press, 2004.

*Historical Atlas of Canada*, col. 3: *Addressing the Twentieth Century*, by D. W. Holdsworth and D. Kerr. Toronto, ON: University of Toronto Press, 2016.

*Historical Atlas of the Maritime Provinces 1878*, by J. Dawson. Halifax, NS: Nimbus, 2005.

*Lines of Country: An Atlas of the Railway and Waterway History in Canada*, by C. Andrae and G. J. Matthews. Erin, ON: Boston Mills, 1997.

## Canadian Thematic Atlases

*Aquaculture Atlas of Canada: Atlas d'aquaculture du Canada*. Ottawa, ON: Aquaculture Management Directorate, Fisheries and Oceans Canada, 2009.

*Atlas of Canada*. Montreal, QC: Reader's Digest Association, 1995.

*Atlas of Canada: Final Report*, by M. Rodney. Ottawa, ON: Natural Resources Canada, 2003.

*Atlas of Canada: The Story of a Country through Maps, Photographs, History and Culture*, 2014.

*Atlas of Canada and the World*, by N. Hillmer and J. F. Ryan. Vancouver, BC: Whitecap Books, 1999.

*Atlas of Common Benthic Foraminiferal Species for Quaternary Shelf Environments of Western Canada*, by R. T. Patterson, S. M. Burbidge, and J. L. Luternauer. Ottawa, ON: Geological Survey of Canada, 1998.

*Atlas of Ocean Currents in Eastern Canadian Waters*, by Y. S. Wu and C. L. Tang. Dartmouth, NS: Fisheries and Oceans Canada, 2011.

*Atlas of Migmatites*, by E. W. Sawyer. Ottawa, ON: NRC Research Press, 2008.

*Atlas of the Canadian Columbia*, by T. A. Weaver and W. D. Layman. Wenatchee, WA: Columbia River Visions, 2010.

*Atlas of Tidal Currents: St. Lawrence Estuary, from Cap de Bon-Désir to Trois-Rivières = Atlas des courants de marée: estuaire du Saint-Laurent, du cap de Bon-Désir à Trois-Rivières*. Ottawa, ON: Fisheries and Oceans Canada, 1997.

*Canada and the World: An Atlas Resource*, by R. Morrow and G. J. Matthews. Scarborough, ON: Prentice-Hall Ginn Canada, 1996.

*Canada Road Atlas = Canada atlas routier*. Oshawa, ON: Canadian Cartographic Corporation, 2015.

*Canadian Atlas: Our Nation, Environment and People*. Montreal, QC: Reader's Digest, 2004.

*Canadian Atlas of F.S.A. Postal Areas*. Scarborough, ON: Compusearch, 2001.

*Canadian National Parks Atlas*. Ottawa, ON: Parks Canada, 1996.

*Canadian Oxford Junior Atlas*, by P. Wiegand. Toronto, ON: Oxford University Press, 2000.

*Canadian Oxford School Atlas*, by P. Healy and S. Pisani. 8th edition. Don Mills, ON: Oxford University Press, 2004.

*Canadian Oxford School Atlas*, by B. J. Smith and D. Francis. Toronto, ON: Oxford University Press, 1998, 2009.

*Canadian Oxford World Atlas*. Toronto, ON: Oxford University Press, 2008.

*Classroom Atlas of Canada and the World*. Markham, ON: Rand McNally, 2011.

*Classroom Atlas of Canada and the World: Teacher's Guide*. Markham, ON: Rand McNally, 2006.

*Complete Road Atlas of Canada*. Montreal, QC: Reader's Digest, 2002.

*Discover Canada's Watersheds*. Ottawa, ON: Natural Resources Canada, 2006.

*Geist Atlas of Canada: Meat Maps and Other Strange Cartographies*, by M. Edwards. Vancouver, BC: Arsenal Pulp, 2006.

*Geological Atlas of the Beaufort-Mackenzie Area = Atlas géologique de la région de Beaufort-Mackenzie*, by J. Dixon. Ottawa, ON: Geological Survey of Canada, 1996.

*Illustrated Atlas of the Dominion of Canada: Containing Authentic and Complete Maps of All the Provinces, the North-West Territories and the Island of Newfoundland*. Toronto, ON: H. R. Page, 2007.

*Landforms and Surface Materials of Canada: A Stereoscopic Airphoto Atlas and Glossary*, by J. D. Mollard. Regina, SK: Mollard, 1996.

*Le grand atlas du Canada et du monde: [document cartographique] / avec la collaboration de Rodolphe De Koninck*, by K. R. De. Saint-Laurent, QC: ERPI De Boeck, 2002.

*Monthly Climatological Atlas of Surface Atmospheric Conditions of the Northwest Atlantic*, by B. M. DeTracey and C. L. Tang. Dartmouth, NS: Fisheries & Oceans Canada, 1997.

*Nelson Canadian Atlas*, by L. G. Greenham and B. Moldofsky. 3rd edition. Scarborough, ON: Nelson Thompson Learning, 2000.

*Nitrate, Silicate and Phosphate Atlas for the Gulf of St. Lawrence*, by D. Brickman and B. Petrie. Dartmouth, NS: Department of Fisheries and Oceans, Bedford Institute of Oceanography, 2003.

*Nystrom Atlas of Canada and the World*, by J. R. Chalk. Indianapolis, IN: Nystrom, 2003, 2009.

*Nystrom Canadian Desk Atlas*. Chicago: Nystrom, 2000.

*Philip's Canadian World Atlas*. London, UK: Philip's, 2008.

*Photographic Atlas: Images of Canada; Canadian Landscapes*, by M. Gagné. Montreal, QC: Photographic Atlas, 1995.

*Rand McNally Canada and World Atlas*. Markham, ON: Rand McNally, 2002.

*Rand McNally Manitoba, Saskatchewan, Northwest Ontario: Provincial Road Atlas*. Markham, ON: Rand McNally Canada, 2003.

*Sea Ice Climatic Atlas, Northern Canadian Waters, 1971–2000: Atlas climatique des glaces de mer, eaux du nord canadien, 1971–2000*. Ottawa, ON: Canadian Ice Service, 2002.

*Thematic Tactile Atlas of Canada*. Ottawa, ON: Mapping Services Branch, 2002.

## CANADIAN PROVINCIAL ATLASES

### Alberta

*Alberta Back Road Atlas*. Oshawa, ON: MapArt, 2010.

*Alberta: Cities and Towns Street Atlas*, by C. White. Calgary, AB: MapArt, 1996.

*Alberta: Surface and Subsurface*, by F. J. Hein. Edmonton, AB: Alberta Geological Survey, 2000.

*Atlas of Alberta Lakes*, by P. Mitchell and E. Prepas. Edmonton: University of Alberta Press, 2000, 2009.

*Atlas of Lithofacies of the McMurray Formation Athabasca Oil Sands Deposit, Northeastern Alberta: Surface and Subsurface*, by F. J. Hein, D. K. Colterill, and H. Berhane. Calgary, AB: Alberta Energy and Utilities Board, 2000.

*Atlas of the North Saskatchewan River Watershed in Alberta*. Edmonton, AB: North Saskatchewan Watershed Alliance Society, 2012.

*British Columbia, Alberta: Deluxe Road Atlas*. Markham, ON: Canadian Cartographics, 2010.

*Calgary and Southern Alberta Atlas*. Oshawa, ON: Peter Heiler, 2006.

*Edmonton-Calgary Corridor Groundwater Atlas*. Edmonton, AB: Alberta Geological Survey, 2011.

*Grid Road Atlas of Alberta*, vol. 2. Camrose, AB: Wildrose Mapping & Atlas, 1997.

*Last Great Intact Forests of Canada: Atlas of Alberta*, by P. Lee. Edmonton, AB: Global Forest Watch Canada, 2009.

*Last Great Intact Forests of Canada: Atlas of Alberta. Part 2, What Are the Threats to Alberta's Forest Landscapes?*, by P. Lee. Edmonton, AB: Global Forest Watch Canada, 2009.

*Map File: Resource Service Atlas of Alberta*. Edmonton, AB: CJI Technology, 2001.

*Northern Alberta Conservation Atlas*. Ottawa, ON: Nature Conservancy of Canada and Global Forest Watch Canada, 2014.

*Northern Alberta Conservation Atlas*. Toronto, ON: Nature Conservancy of Canada, 2000.

### British Columbia

*Atlas of British Columbia: People, Environment, and Resource Use*, by A. L. Farley. Vancouver, BC: University of British Columbia Press, 2014.

*B.C. Coastal Recreation Kayaking and Small Boat Atlas*, by J. Kimantas. North Vancouver, BC: Whitecap Books, 2011.

*British Columbia: A New Historical Atlas*, by Derek Hayes. Vancouver, BC: Douglas & McIntyre, 2010.

*British Columbia, Alberta: Deluxe Road Atlas*. Markham, ON: Canadian Cartographics, 2010.

*British Columbia Road and Recreational Atlas*. Victoria, BC: Informap, 2001.

*British Columbia Road Atlas*. Oshawa, ON: MapArt, 2009.

*Geography of Wellness and Well-Being across British Columbia: A Supplement to the British Columbia Atlas of Wellness*, by B. McKee et al. Victoria, BC: Department of Geography, University of Victoria, 2009.

*Islands in the Salish Sea: A Community Atlas*, by J. Stevenson and S. Harrington. Surrey, BC: TouchWood Editions, 2009.

*Kennedy Watershed Atlas Series*, vol. 1: *Watershed Overview, a Working Atlas*, by K. D. Hyatt, M. R. S. Johannes, and C. L. K. Robinson. Lantzville, BC: Northwest Ecosystem Institute, 2000.

*Lower Fraser River Stream Inventory Atlas*. Vancouver, BC: Fisheries and Oceans Canada, Habitat and Enhancement Branch, 1996.

*Marine Atlas of Pacific Canada*. Vancouver, BC: British Columbia Marine Conservation Analysis, 2011.

*Southwestern British Columbia Road and Recreational Atlas*. Victoria, BC: Informap, 2004.

*Toward a Web-Based Multimedia Atlas of British Columbia*, by J. J. Fowler and C. P. Keller. Victoria, BC: University of Victoria.

### Manitoba

*Historical Atlas of the East Reserve: Illustrated*, by E. N. Braun and G. R. Klassen. Winnipeg, MB: Mennonite Historical Society, 2015.

### New Brunswick

*New Brunswick Atlas: Atlas Nouveau-Brunswick.* Halifax, NS: Nimbus, 2002.

### Newfoundland and Labrador

*Atlas of Newfoundland and Labrador*, by G. E. McManus et al. St. John's, NL: Breakwater Books, 2008.

*Temperature Climate Atlas for the Inshore Regions of Newfoundland and Labrador*, by E. B. Colbourne, P. Stead, and S. Narayanan. St. John's, NL: Department of Fisheries and Oceans, 1996.

### Northwest Territories

*Sahtu Atlas: Maps and Stories from the Sahtu Settlement Area in Canada's Northwest Territories*, by J. Auld. Norman Wells, NWT: Sahtu GIS Project, 2005.

### Nova Scotia

*Nova Scotia Atlas.* Halifax, NS: Formac, 2001, 2006.

*Nova Scotia Street and Road Atlas.* Halifax, NS: Nimbus, 2006.

*Resource Guide and Ecological Atlas for Conducting Research in Nova Scotia's Wilderness Areas and Nature Reserves*, by R. P. Cameron and L. Helmer. Halifax: Nova Scotia Environment and Labour, Environmental and Natural Areas Management Division, 2004.

### Ontario

*Atlas of Early Pioneers of Niagara Peninsula*, by C. Taylor. St. Catharines, ON: House of Dwyer, 2002.

*Atlas of the Breeding Birds of Ontario, 2001–2005*, by M. D. Cadman. Port Rowan, ON: Bird Studies Canada, 2007.

*Concise Historical Atlas of Canada*, by W. G. Dean et al. Toronto, ON: University of Toronto Press, 2016.

*Eastern and Northern Ontario.* Oshawa, ON: Peter Heiler, 2004.

*Frontenac, Lennox and Addington Atlas and History*, by W. S. Herrington. Manotick, ON: Archive CD Books Canada, 2007.

*Golden Horseshoe: Barrie, Brampton . . . , St. Catharines, Toronto.* ON: MapArt, 2000.

*Illustrated Historical Atlas of Frontenac, Lennox and Addington Counties, Ontario*, by C. R. Allen and J. H. Meacham. Manotick, ON: Archive CD Books Canada, 2007.

*MapArt Ontario Town and Country Atlas.* Oshawa, ON: Peter Heiler, 1999.

*Northern Bruce Peninsula Ecosystem Community Atlas*, by B. Cundiff and R. Czok. Toronto, ON: CPAWS Wildlands League, 2005.

*Ontario and the Great Lakes: Road Atlas and Travel Guide.* Greenville, SC: Michelin Travel, 2004.

*Ontario Pocket Street Atlas: Includes 27 Downtown and City Plans, Distance Chart, Points of Interest.* Whitby, ON: MapArt, 2006.

*Ontario: Provincial Pocket Road Atlas.* Markham, ON: Rand McNally, 2006.

*Ontario Provincial Road Atlas.* Markham, ON: Rand McNally Canada, 1999, 2004.

*Ontario Road and Recreational Atlas.* Victoria, BC: Phototype Compsoing, 2000.

*Rand McNally Ontario Provincial Road Atlas.* Markham, ON: Rand McNally Canada, 2000.

*Southern Ontario Recreational Atlas*, by W. Winter and L. J. Wells. Victoria, BC: Informap, 1998.

*Ontario Town and Country Atlas.* Oshawa, ON: Map MEDIA, 2008.

*Western Ontario: Windsor, London, Kitchener, Brantford, Guelph, Barrie: Includes over 120 Cities and Towns.* Oshawa, ON: MapArt, 1996.

*Wind Resource Information [Ontario].* Toronto, ON: Ministry of Natural Resources, 2005.

### Prince Edward Island

*Historical Atlas of Prince Edward Island: The Way We Saw Ourselves*, by J. W. MacNutt. Halifax, NS: Formac, 2009.

*Illustrated Historical Atlas of Prince Edward Island.* Charlottetown, PEI: Prince Edward Island Museum and Heritage Foundation, 1995.

### Quebec

*Atlas historique du Québec: population et territoire*, by S. Courville. Sainte-Foy, QB: Presses de l'Université Laval, 1996.

*Historical and Genealogical Atlas of Quebec from Antiquity to the 20th Century*, by R. J. Quintin and B. L. Pren-Mendiola. Pawtucket, RI: Quintin, 1995.

*Quebec Road Atlas and Travel Guide.* Greenville, SC: Michelin, 2004.

### Saskatchewan

*Chortitza Colony Atlas: Altklonie = Mennonite Old Colony*, by H. Bergen. Saskatoon, SK: Mennonite Historical Society of Saskatchewan, 2004.

*Geological Atlas of Saskatchewan.* Regina, SK: Saskatchewan Energy and Mines, 2000.

*Saskatchewan Back Road Atlas.* San Francisco, CA: CCC, 2014.

*Saskatchewan Street Atlas.* Oshawa, ON: MapArt, 2009.

## CANADIAN COUNTY ATLASES

### Alberta

*Beaver County*

*County of Beaver No. 9: Rural Atlas and Directory.* Camrose, AB: Wildrose Mapping & Atlas, 1995.

*Lethbridge County*

*County of Lethbridge No. 26: Rural Atlas and Directory.* Calgary, AB: Wildrose Mapping & Atlas, 1999.

*Rocky View County*

*Calgary Pocket Street Atlas: Includes Detailed City Maps.* Oshawa, ON: MapArt, 2007.

*St. Paul County*

*County of St. Paul, No. 19: Rural Atlas and Directory.* Calgary, AB: Rural Mapping & Atlas Inc., 1996.

*Strathcona County*

*Edmonton Pocket Street Atlas.* Oshawa, ON: Canadian Cartographics, 2014.

*Westlock County*

*Municipal District of Westlock, No. 92: Rural Atlas and Directory.* Calgary, ON: Rural Mapping & Atlas, 1996.

*Wetaskiwin County*

*Rural Atlas and Directory for County of Wetaskiwin No. 10.* Camrose, AB: Wildrose Mapping & Atlas, 1997.

**British Columbia**

*Vancouver and Area*

*Central Vancouver Island Watercourse Atlas*, by M. R. S. Johannes and J. Cleland. Lantzville, BC: Northwest Ecosystem Institute, 2002.

*Greater Vancouver and Fraser Valley Street Atlas.* Oshawa, ON: MapArt, 2004.

*Greater Vancouver Pocket Street Atlas.* Oshawa, ON: Canadian Cartographics Corporation, 2014.

*Historical atlas of Vancouver and the Lower Fraser Valley*, by Derek Hayes. Vancouver, BC: Douglas & McIntyre, 2005.

*Journeys through the Neighbourhood: Our Community Atlas; Grandview Woodland, Vancouver.* Vancouver, BC: Our Own Backyard, 1998.

*West Coast of Vancouver Island Cruising Atlas*, by G. McNutt, B. Nelson, and C. Nelson. Shoreline, WA: Evergreen Pacific, 2002.

**Manitoba**

*Winnipeg Pocket Street Atlas.* Oshawa, ON: Canadian Cartographics, 2012.

**Nova Scotia**

*Halifax County*

*Halifax Regional Municipality Atlas.* Halifax, NS: Nimbus, 2008.

**Ontario**

*Carleton County*

*Atlas Outaouais Ottawa—Hull and vicinity. Outaouais Ottawa-Hull et environs atlas*, by L. Fournier, L. Bernard, and G. Cartotek. St.-Laurent, QC: Cartotek Geo, 1999.

*Illustrated Historical Atlas of the County of Carleton, including City of Ottawa, Ont*, by H. Belden. Campbellford, ON: Wilson's, 1997.

*Frontenac County*

*Kingston Street Atlas Map Book.* Kingston, ON: Corp. of the City of Kingston, 2007.

*Haldimand County*

*Illustrated Historical Atlas of the Counties of Haldimand and Norfolk.* Campbellford, ON: Wilson's, 1997.

*Hastings County*

*Heritage Atlas of Hastings County*, by O. French. Belleville, ON: County of Hastings, 2006.

*Illustrated Historical Atlas of Hastings and Prince Edward Counties, Ontario.* Campbellford, ON: Wilson's, 1997, 2006.

*Huron County*

*Illustrated Historical Atlas of Huron County, Ontario.* Goderich, ON: Huron County Branch, OGS, 1996.

*Lanark County*

*Illustrated Historical Atlas of the County of Lanark, Ontario, 1880.* Stirling, ON: Wilson's, 2003.

*Index, Illustrated Atlas, Lanark County, 1880, Renfrew County, 1881: H. Beldon [i.e. Belden] & Co., Toronto, Reprinted by Ross Cumming, Port Elgin, Ontario and H.F. Walling Map by D.P. Putnam, Prescott, C.W., 1863*, by M. R. Livingston, R. Cumming, and H. F. Walling. Prescott, ON: Livingston, 1995.

*Lincoln County*

*Illustrated Historical Atlas of the Counties of Lincoln and Welland, Ont.* Toronto, ON: H. R. Page, 2008.

*Middlesex County*

*Illustrated Historical Atlas of the County of Middlesex, Ont.* Toronto, ON: Hammerburg Productions, 2008.

*Muskoka County*

*Guide Book and Atlas of Muskoka and Parry Sound Districts*, by W. E. Hamilton, S. Penson, and J. Rogers. Erin, ON: Boston Mills, 2000.

*Norfolk County*

*Illustrated Historical Atlas of the Counties of Haldimand and Norfolk.* Campbellford, ON: Wilson's, 1997.

*Peel County*

*Illustrated Historical Atlas of the County of Peel*, by J. H. Pope, H. C. O. Peel, and M. F. Walker. Campbellford, ON: Wilson's, 2000.
*Regional Planning Atlas.* Brampton, ON: Region of Peel Planning Department, 2000.

*Prince Edward County*

*Illustrated Historical Atlas of Hastings and Prince Edward Counties, Ontario.* Campbellford, ON: Wilson's, 1997, 2006.
*Illustrated Historical Atlas of the Counties of Hastings and Prince Edward, Ont: Compiled, Drawn and Published from Personal Examinations and Surveys.* Campbellford, ON: Fifth Line, 2006.

*Renfrew County*

*Index, Illustrated Atlas, Lanark County, 1880, Renfrew County, 1881: H. Beldon [i.e. Belden] & Co., Toronto, Reprinted by Ross Cumming, Port Elgin, Ontario and H. F. Walling Map by D. P. Putnam, Prescott, C.W., 1863*, by M. R. Livingston, R. Cumming, and H. F. Walling. Prescott, ON: Livingston, 1995.

*Victoria County*

*Belden's Illustrated Historical Atlas of the County of Victoria, Ontario, 1881*, by E. Phelps. Ancaster, ON: Alexander, 2000.

*Welland County*

*Illustrated Historical Atlas of the Counties of Lincoln and Welland, Ont.* Toronto, ON: H. R. Page, 2008.

*Wentworth County*

*Imperial Atlas of 1903 for the County of Wentworth*, by B. Kingdon. Hamilton, ON: Hamilton Branch, Ontario Genealogical Society, 1998.
*Imperial Atlas of 1903 for the County of Wentworth: Barton Township*, by B. Kingdon. Hamilton, ON: Hamilton Branch, Ontario Genealogical Society, 1998.
*Imperial Atlas of 1903 for the County of Wentworth: Beverly Township*, by B. Kingdon. Hamilton, ON: Hamilton Branch, Ontario Genealogical Society, 1998.
*Imperial Atlas of 1903 for the County of Wentworth: Binbrook Township*, by B. Kingdon. Hamilton, ON: Hamilton Branch, Ontario Genealogical Society, 1998.
*Imperial Atlas of 1903 for the County of Wentworth: Flamboro East Township*, by B. Kingdon. Hamilton, ON: Hamilton Branch, Ontario Genealogical Society, 1998.
*Imperial Atlas of 1903 for the County of Wentworth: Glanford Township*, by B. Kingdon. Hamilton, ON: Hamilton Branch, Ontario Genealogical Society, 1998.
*Imperial Atlas of 1903 for the County of Wentworth: Saltfleet Township*, by B. Kingdon. Hamilton, ON: Hamilton Branch, Ontario Genealogical Society, 1998.
*Wentworth County: The Imperial Atlas of 1903 for the City of Hamilton*, by B. Kingdon. Hamilton, ON: Hamilton Branch, Ontario Genealogical Society, 1996.
*Wentworth County Mini-Atlas, 1791–1875.* Jordan Station, ON: M. Parnall, 1998.

*York County*

*Historical Atlas of Toronto*, by Derek Hayes. Vancouver. BC: Douglas & McIntyre, 2008.
*Illustrated Historical Atlas of York County, Ontario.* Campbellford, ON: Wilson's, 1996.
*Perly's Bluemap Atlas, Metropolitan Toronto and Vicinity.* Toronto, ON: Perly's Maps, 1995.

## AMERICAN NATIONAL ATLASES

### American Historical Atlases

*American Military Pocket Atlas (1776)*, by Robert Sayer and John Bennett. Hanover, NH: Dartmouth College Library, 2010.
*Atlas Accompanying the Original Journals of the Lewis and Clark Expedition, 1804–1806, Being Facsimile Reproductions of Maps Chiefly by William Clark*, by M. Lewis, W. Clark, and R. G. Thwaites. Scituate, MA: Digital Scanning, 2001.
*Atlas of American History*, by R. H. Ferrell and R. Natkiel. New York: Facts on File, 1995.
*Atlas of American History*, by G. B. Nash. New York: Infobase, 2006.

*Atlas of American History.* Skokie. IL: Rand McNally, 1999, 2005.
*Atlas of American Migration,* by S. A. Flanders. New York: Facts on File, 1998.
*Atlas of American Military History,* by J. C. Bradford. New York: Oxford University Press, 2003.
*Atlas of American Military History,* by S. Murray. New York: Checkmark Books, 2004.
*Atlas of Early Maps of the American Midwest: Part 1,* by W. R. Wood. Springfield: Illinois State Museum, 1983.
*Atlas of Early Maps of the American Midwest: Part 2,* by W. R. Wood. Springfield: Illinois State Museum, 2001.
*Atlas of Historical County Boundaries,* by J. H. Long et al. New York: Charles Scribner's Sons, 1996.
*Atlas of Lewis and Clark in Missouri,* by J. Harlan and J. Denny. Columbia: University of Missouri Press, 2003.
*Atlas of the American Civil War: Army Organization,* by J. C. Nelson. New York: Createspace, 2011.
*Atlas of the Historical Geography of the United States,* by C. O. Paullin and J. K. Wright. Ann Arbor, MI: Carnegie Institution of Washington, 2006.
*Dent Atlas of American History,* by M. Gilbert. New York: Routledge, 1995.
*Family Tree Historical Maps Book: A State-by-State Atlas of U.S. History, 1790–1900,* by A. Dolan. Tecumseh, MI: Family Tree, 2014.
*Historical Atlas of Colonial America,* by J. Axelrod-Contrada. New York: Rosen, 2005.
*Historical Atlas of Native Americans,* by Ian Barnes. New York: Chartwell Books, 2010.
*Historical Atlas of the American Revolution,* by I. Barnes and C. Royster. New York: Routledge, 2000.
*Historical Atlas of the American West, with Original Maps,* by Derek Hayes. Berkeley: University of California Press, 2009.
*Historical Atlas of the United States,* by M. C. Carnes. New York: Routledge, 2003.
*Historical Atlas of the United States* by M. C. Carnes. Hoboken, NJ: Taylor and Francis, 2013.
*Historical Atlas of the United States, with Original Maps,* by Derek Hayes. Vancouver, BC: Douglas & McIntyre, 2006.
*Historical Atlas of U.S. Presidential Elections 1788–2004.* Washington, DC: CQ Press, 2006.
*Mapping America's Past: A Historical Atlas,* by M. C. Carnes, J. A. Garraty, and P. Williams. New York: Henry Holt, 1996.
*National Geographic Historical Atlas of the United States,* by R. M. Fisher. Washington, DC: National Geographic Society, 2004.
*Naval Institute Historical Atlas of the U.S. Navy,* by C. L. Symonds and W. J. Clipson. Annapolis, MD: Naval Institute Press, 1995.
*New Historical Atlas of Religion in America,* by E. S. Gaustad et al. Oxford: Oxford University Press, 2001.
*Nystrom Atlas of Our Country's History: Student Activities.* Chicago: Nystrom, 2002.
*Nystrom Atlas of United States History.* Chicago: Nystrom, 2002.
*Railroad Atlas of the United States in 1946.* Vol. 1, *The Mid-Atlantic States,* by R. C. Carpenter. Baltimore, MD: Johns Hopkins University Press, 2003.
*Railroad Atlas of the United States in 1946.* Vol. 2, *New York and New England,* by R. C. Carpenter. Baltimore, MD: Johns Hopkins University Press, 2005.
*Railroad Atlas of the United States in 1946.* Vol. 3, *Indiana, Lower Michigan and Ohio,* by R. C. Carpenter. Baltimore, MD: Johns Hopkins University Press, 2008.
*Railroad Atlas of the United States in 1946.* Vol. 4, *Illinois, Wisconsin and Upper Michigan,* by R. C. Carpenter. Baltimore, MD: Johns Hopkins University Press, 2011.
*Railroad Atlas of the United States in 1946.* Vol. 5, *Iowa and Minnesota,* by R. C. Carpenter. Baltimore, MD: Johns Hopkins University Press, 2013.
*Routledge Atlas of American History,* by M. Gilbert. London: Routledge, 1995, 2002, 2006, 2009.
*Routledge Historical Atlas of the American Railroads,* by J. F. Stover. New York: Routledge, 1999.
*Routledge Historical Atlas of Presidential Elections,* by Y. Mieczkowski. New York: Routledge, 2001.
*Routledge Historical Atlas of Religion in America,* by B. E. Carroll. New York: Routledge, 2000.
*Routledge Historical Atlas of Religion in America,* by B. E. Carroll. Hoboken, NJ: Taylor and Francis, 2013.
*Routledge Historical Atlas of Women in America,* by S. Opdycke. New York: Routledge, 2000.
*Thoroughfare for Freedom: The Second Atlas of the Cumberland Settlements, 1779–1804,* by J. Masters and B. Puryear. Gallatin, TN: Warioto, 2011.
*United States History Atlas.* Santa Barbara, CA: Maps.com, 2008.
*United States Military Road Atlas,* by R. J. Crawford, W. R. Crawford, and A. C. Crawford. Falls Church, VA: Military Living Publications, 1995.
*West Point Atlas of American Wars.* Vol. 2, *1900–1918,* by V. J. Esposito. New York: Henry Holt, 1997.

**American Thematic Atlases**

*Allyn & Bacon Social Atlas of the United States,* by W. H. Frey, A. B. Anspach, and A. B. DeWitt. Boston, MA: Pearson / Allyn & Bacon, 2008.
*Atlas of American Agriculture: The American Cornucopia,* by R. Pillsbury and J. W. Florin. New York: Simon & Schuster Macmillan, 1996.
*Atlas of American Diversity,* by L. H. Shinagawa and M. Jang. Walnut Creek, CA: AltaMira, 1998.
*Atlas of America's Polluted Waters.* Washington, DC: Environmental Protection Agency, 2000.
*Atlas of Natural America.* Washington, DC: National Geographic Society, 2000.
*Atlas of Our Country,* by C. Novosad. Chicago: Nystrom, 1996.

*Atlas of Poverty in America: One Nation, Pulling Apart 1960–2003*, by A. Glasmeier. Hoboken, NJ: Taylor and Francis, 2014.

*Atlas of Southern Trails to the Mississippi*, by C. Eldridge. Chesapeake, OH: Carrie Eldridge, 1999.

*Atlas of the 2008 Elections*, by S. D. Brunn et al. Lanham, MD: Rowman & Littlefield, 2011.

*Atlas of the 2012 Elections*, by J. C. Archer et al. Lanham, MD: Rowman & Littlefield, 2014.

*Atlas of the United States*, by H. J. de Blij. New York: Oxford University Press, 2008.

*Atlas of Trails West of the Mississippi River*, by C. Eldridge. Chesapeake, OH: C. Eldridge, 2001.

*Atlas of Urban Expansion*, by S. Angel. Cambridge, MA: Lincoln Institute of Land Policy, 2012.

*Census Atlas of the United States: Census 2000 Special Reports*, by T. A. Suchan. Washington, DC: U.S. Census Bureau, 2006.

*Color Landform Atlas of the United States*, by R. Sterner. Baltimore, MD: Johns Hopkins University, 1995.

*Genesee-Finger Lakes Regional Atlas*. Rochester, NY: Genesee Transportation Council, 2005.

*Geologic Atlas of the United States*. College Station: Texas A&M University, 2000.

*Illustrated Atlas of the United States*, by K. Lye. Broxbourne, UK: Regency House, 1997.

*Interactive Web-Based Atlas for the Spokane Valley-Rathdrum Prairie Aquifer*, by J. M. Saltemnberger. San Diego, CA: San Diego State University, 2011.

*Lake Champlain Basin Atlas*, by Lake Champlain Basin Program. South Burlington, VT: Northern Cartographic, 1999.

*Macmillan Color Atlas of the States*, by M. T. Mattson. New York, Macmillan, 1996.

*Midwestern Climate Center Soils Atlas and Database*, by S. E. Hollinger. Champaign, IL: Illinois State Water Survey, 1995.

*National Geographic Picture Atlas of Our Fifty States*. Louisville, KY: American Printing House for the Blind, 1995.

*Snowstorms across the Nation: An Atlas about Storms and Their Damages*, by S. A. Changnon and D. Changnon. Asheville, NC: National Climatic Data Center, 2008.

*State Atlas of Political and Cultural Diversity*, by W. Lilley and L. J. DeFranco. Washington, DC: Congressional Quarterly, 1996.

*State-by-State Atlas*, by J. Ciovacco, K. Feeley, and K. Behrens. New York: DK, 2009.

*Ultimate Yellowstone Park and Surrounding Area Atlas and Travel Encyclopedia*, by M. Dougherty. Bozeman, MT: Ultimate, 2006.

*United States Reference Atlas*. Santa Barbara, CA: Maps.com, 2001.

*U.S. Railroad Traffic Atlas*, by H. Ladd. Orange, CA: Ladd, 1997.

*Where the Great River Rises: An Atlas of the Upper Connecticut River Watershed in Vermont and New Hampshire*, by R. A. Brown. Hanover, NH: Dartmouth College Press, 2009.

*Women's Atlas of the United States*, by T. Fast and C. C. Fast. New York: Facts on File, 1995.

## AMERICAN STATE ATLASES

### Alabama

*Alabama Atlas and Gazetteer*. Skokie, IL: Rand McNally, 2000.

*Alabama: State Recreation Atlas*. Washington, D.C: National Geographic, 2012.

*Ecologic Atlas of Upper Cretaceous Ostracods of Alabama*, by T. M. Puckett. Tuscaloosa: Geological Survey of Alabama. 1996.

*Historical Atlas of Alabama*, by W. C. Remington and T. J. Kallsen. Tuscaloosa: University of Alabama, 1997.

*Historical Atlas of Alabama*. Vol. 1, *Historical Locations by County*, by W. C. Remington. Tuscaloosa: University of Alabama-Tuscaloosa, 2010.

*Historical Atlas of Alabama*. Vol. 2, *Cemetery Locations by County*, by W. C. Remington. Tuscaloosa: University of Alabama-Tuscaloosa, 1999, 2014.

*Historical Atlas of Alabama: Excerpts from Volume 1, Historical Locations by County, and Volume 2, Cemeteries by County: Chambers, Clay, Coosa, Randolph, Talladega, and Tallapoosa Counties*. Tuscaloosa: University of Alabama-Tuscaloosa, 2000.

*Historical Atlas of Alabama: Excerpts from Volume 1, Historical Locations by County, and Volume 2, Cemeteries by County: Conecuh County*. Tuscaloosa: University of Alabama-Tuscaloosa, 1999.

### Alaska

*Alaska Atlas and Gazetteer*. Yarmouth, ME: DeLorme Mapping, 2000.

*Alaska in Maps: A Thematic Atlas*, by P. H. Partnow. Fairbanks: University of Alaska Fairbanks, 1998.

*Glaciers of North America: Glaciers of Alaska*, by B. F. Molina, R. R. Krimmel, and W. F. Manley. Washington, DC: U.S. Government Printing Office, 2008.

*Teacher's Guide for Alaska in Maps: A Thematic Atlas*, by P. H. Partnow. Fairbanks: University of Alaska Fairbanks, 1998.

### Arkansas

*Arkansas: An Illustrated Atlas*, by T. Paradise. Fayetteville: University of Arkansas Press, 2012.

*Arkansas Atlas and Gazetteer*. Yarmouth, ME: DeLorme Mapping, 2004.

*Arkansas Outdoor Atlas: A Sportsman's Guide to Public Access Facilities in Arkansas*. Little Rock: Arkansas Game and Fish Commission, 1996.

*Arkansas Township Atlas: 1819–1930*, by R. P. Baker. Hot Springs: Arkansas Genealogical Society, 2006.

*Civil War Arkansas: A Military Atlas*, by R. Puckett and R. Kelley. Helena, AR: Arkansas Toothpick, 2016.

## California

*Atlas of California: Mapping the Challenges of a New Era*, by R. Walker and S. K. Lodha. Berkeley: University of California Press, 2013.

*Atlas of California Presidential Elections by County, 1932–1992*, by A. W. Miller. Richmond, VA: Klipsan, 1995.

*Atlas of Coastal Ecosystems in the Western Gulf of California: Tracking Limestone Deposits on the Margin of a Young Sea*, by M. E. Johnson and J. Ledesma-Vázquez. Tucson: University of Arizona Press, 2009.

*Atlas of Human Adaptation to Environmental Change, Challenge, and Opportunity: Northern California, Western Oregon, and Western Washington*, by H. H. Christensen. Portland, OR: U.S. Department of Agriculture, Forest Service, Pacific Northwest Research Station, 2000.

*Atlas of Social and Economic Conditions and Change in Southern California*, by T. L. Taettig, D. M. Elmer, and H. H. Christensen. Portland, OR: U.S. Department of Agriculture, 2001.

*Atlas of Southern California*, by M. J. Dear and H. Sommer. Los Angeles: University of Southern California, 1996.

*Atlas of Yellowstone*, by W. A. Marcus et al. Berkeley: University of California Press, 2012.

*California Atlas: A Geography of the Golden State*. Portola: California Geography Associates, 2006.

*Distributional Atlas of Fish Larvae and Eggs in the Southern California Bight Region, 1951–1998*, by H. G. Moser. La Jolla: California Department of Fish and Game, 2001.

*Draft Atlas of the North Coast Study Region (Alder Creek/Point Arena to the California-Oregon Border)*. Sacramento: California Marine Life Protection Act Initiative, 2009.

*Historical Atlas of California: With Original Maps*, by Derek Hayes, CA: University of California Press, 2007.

*Northern California Atlas and Gazetteer*. Yarmouth, ME: DeLorme Mapping, 1997, 2000.

*Southern and Central California Atlas and Gazetteer: Detailed Topographic Maps; Outdoor Recreation; Back Roads, Outdoor Recreation*. Yarmouth, ME: DeLorme Mapping, 2005.

*Southern California and Central Atlas and Gazetteer*. Yarmouth, ME: DeLorme Mapping, 2000.

## Colorado

*Colorado: A Historical Atlas*, by T. J. Noel and C. Zuber-Mallison. Norman: University of Oklahoma Press, 2015.

*Colorado Atlas and Gazetteer: Topo Maps of the Entire State, Public Lands, Back Roads*. Freeport, ME: DeLorme Mapping, 1999.

*Colorado Ground-Water Atlas*, by A. Aikin. Lakewood: Colorado Ground-Water Association, 2000.

*Colorado Road and Recreation Atlas*. Medford, OR: Benchmark Maps, 2007.

*Lakes of the Sangres in 3D: Indexed Atlas with Over 200 Color Photographs*, by W. Rychlik. Colorado Springs, CO: Mother's House, 2012.

## Connecticut

*Fairfield, Litchfield, New Haven Counties Connecticut: Street Atlas*. Maspeth, NY: Hagstrom Map, 2007.

## Florida

*Atlas of Artificial Reefs in Florida*, by D. W. Pybas. Miami: Florida Sea Grant College and Florida Sea Grant Extension Program, 1997.

*Atlas of Maritime Florida*, by R. C. Smith. Gainesville: University Press of Florida, 1997.

*Atlas of Race, Ancestry, and Religion in 21st-Century Florida*, by M. D. Winsberg and J. Ueland. Gainesvillle: University Press of Florida, 2002, 2006.

*Central Florida Street Atlas: Counties of Seminole, Orange, and Osceola*. Orlando, FL: Map & Globe Store, 2003.

*Distribution of Submerged Aquatic Vegetation in the Lower St. Johns River: 1998 Atlas*, by D. Dobberfuhl. Palatka, FL: St. Johns River Water Management District, 2002.

*Distribution of Submerged Aquatic Vegetation in the Lower St. Johns River: 2004 Atlas*, by D. Dobberfuhl and C. Hart. Palatka, FL: St. Johns River Water Management District, 2006.

*Florida, Atlas of Historical County Boundaries*, by P. T. Sinko, K. F. Thorne, and J. H. Long. New York: Charles Scribner, 1997.

*Florida Atlas and Gazetteer, 2010*. Freeport, ME: DeLorme Mapping, 2010.

*Florida Atlas and Gazetteer, 2012*. Freeport, ME: DeLorme Mapping, 2012.

*Florida: State Recreation Atlas*. Washington, DC: National Geographic, 2012.

*Florida State Road Atlas*. Long Island, NY: American Map, 2006.

*Florida State Road Atlas 1999*. Miami, FL: Trakker Maps, 1999.

*Mapmobility Florida Road Atlas.* Toronto, ON: Dun-Map, 2003.

*Water Resources Atlas of Florida,* by J. R. Anderson, P. A. Kraft, and E. A. Fernald. Tallahassee: Institute of Science and Public Affairs, Florida State University, 1998.

## Georgia

*Georgia: State Recreation Atlas.* Washington, DC: National Geographic, 2012.

*Georgia Atlas and Gazetteer.* Yarmouth, ME: DeLorme Mapping, 2000.

*Maps of Chickamauga: An Atlas of the Chickamauga Campaign, including the Tullahoma Operations, June 22–September 23, 1863,* by D. A. Powell and D. A. Friedrichs. New York: Savas Beatie, 2009.

## Hawaii

*Atlas of Hawai'i,* by S. P. Juvik, J. O. Juvik, and T. R. Paradise. Honolulu: University of Hawaii Press, 1998.

*Hawaii Atlas and Gazetteer.* Yarmouth, ME: DeLorme Mapping, 1999.

## Idaho

*Idaho Atlas and Gazetteer.* Yarmouth, ME: DeLorme Mapping, 1998.

*Ultimate Idaho Atlas and Travel Encyclopedia,* by K. E. Hill and E. Dougherty. Bozeman, MT: Ultimate, 2005.

## Illinois

*Atlas of Historical County Boundaries, Illinois,* by G. DenBoer and J. H. Long. New York: Simon & Schuster, 1997.

*Illinois Atlas and Gazetteer.* Yarmouth, ME: DeLorme Mapping, 1996, 2000, 2006.

*Illinois Land Cover: An Atlas.* Springfield, IL: Department of Natural Resources, 1996.

*Illinois Travel Atlas: Complete State Road Map Coverage; Easy to Read Metropolitan Street Maps; State and City Map Indexing; Points of Interest; Mileage Charts.* Wood Dale, IL: American Map, 1996.

## Indiana

*Indiana: Atlas of Historical County Boundaries,* by P. T. Sinko and H. H. Long. New York: Charles Scribner's Sons, 1996.

*Indiana Atlas and Gazetteer.* Yarmouth, ME: DeLorme Mapping, 1998.

*Indiana Atlas and Gazetteer: Detailed Topographic Maps; Back Roads, Recreation Sites, GPS Grids, Boat Ramps, Campgrounds, Places to Explore.* Yarmouth, ME: DeLorme Mapping, 2009.

## Iowa

*Atlas of Iowa.* Roseville, MN: Park Genealogical Books, 2003.

*Iowa, Atlas of Historical County Boundaries,* by G. DenBoer and J. H. Long. New York: Charles Scribner's Sons, 1998.

*Iowa Sportsman's Atlas: Back Roads and Outdoor Recreations, Containing Transportation Maps of Iowa's 99 Counties, the Public Lands of Those Counties, Their Outdoor Opportunities and Amenities, Hotels, Motels and Attractions.* Lytton, IA: Sportsman's Atlas, 2010.

## Kansas

*Geophysical Atlas of Selected Oil and Gas Fields in Kansas,* by N. L. Anderson and D. E. Hedke. Lawrence: Kansas Geological Survey and Kansas Geological Society, 1995.

*Kansas Atlas and Gazetteer.* Freeport, ME: DeLorme Mapping, 2003.

## Kentucky

*Atlas of Kentucky,* by R. Ulack, K. Raitz, and G. Pauer. Lexington: University of Kentucky Press, 1999.

*Genealogical Atlas of Kentucky,* by C. M. Franklin. Indianapolis, IN: Ye Olde Genealogie Shoppe, 1997.

*Kentucky Atlas and Gazetteer.* Yarmouth, ME: DeLorme Mapping, 2005.

## Louisiana

*Historical Atlas of Louisiana,* by C. R. Goins and J. M. Caldwell. Norman: University of Oklahoma Press, 1995.

## Maine

*Historical Atlas of Maine,* by S. J. Hornsby, R. W. Judd, and M. J. Hermann. Orono: University of Maine Press, 2015.

*Maine Atlas and Gazetteer.* Freeport, ME: DeLorme Mapping, 1997.

*Maine Atlas and Gazetteer: Topo Maps plus GPS Grids.* Freeport, ME: DeLorme Mapping, 1999.

## Maryland

*Maryland State Archives Atlas of Historical Maps of Maryland, 1608–1908,* by E. C. Papenfuse, J. M. Coale, and E. C. Papenfuse. Baltimore, MD: Johns Hopkins University Press, 2003.

## Massachusetts

*Atlas of the Quabbin Valley, Past and Ware River Diversion*, by J. R. Greene. Athol, MA: J & P Printing, 1996.

*Massachusetts Atlas and Gazetteer*. Yarmouth, ME: DeLorme Mapping, 2009.

*Southeastern Massachusetts Natural Resources Atlas: A Regional Guide to Biodiversity*, by P. M. Cavanagh and K. H. Grey. Manomet, MA: Manomet Center for Conservation Sciences, 2004.

*Universal Atlas of Metropolitan Boston and Eastern Massachusetts*. Stoughton, MA: Universal Publishing, 2009.

## Michigan

*Coastal Atlas: Western Upper Peninsula of Michigan*, by K. J. Stoker et al. Houghton, MI: Army Corps of Engineers, 1998.

*Dartmouth Atlas of Health Care in Michigan*, by J. E. Wennberg. Hanover, NH: Trustees of Dartmouth College, 2000.

*Michigan Atlas and Gazetteer*. Freeport, ME: DeLorme Mapping, 1995, 2001.

*Michigan Off Road Guide and Travel Atlas*. Dearborn, MI: Off Road Venture, 2008.

## Minnesota

*King's King Size Twin Cities Metro Street Atlas*. Brooklyn Park, MN: King Maps & Design, 2016.

*King's Twin Cities Metro Street Atlas, 2013*. Minneapolis, MN: Lawrence Group, 2013.

*Landscapes of Minnesota: A Geography*, by J. F. Hart and S. S. Ziegler. St. Paul: Minnesota Historical Society Press, 2008.

*Minnesota Atlas of Historical County Boundaries*, by J. H. Long and G. DenBoer. New York: Charles Scribner's Sons, 2000.

*Minnesota on the Map: A Historical Atlas*, by D. A. Lanegran and C. L. Urness. St. Paul: Minnesota Historical Society Press, 2008.

## Mississippi

*Mississippi Coast Street Atlas*. Baton Rouge, LA: Stinson Map, 1999.

## Missouri

*Gallup Map: Kansas City Multi-County Metro Coverage Atlas*. Kansas City, MO: Gallup Map, 1998.

## Montana

*Montana Atlas and Gazetteer*. Yarmouth, ME: DeLorme Mapping, 1997.

*Montana Ground-Water Atlas*. Helena: Montana State Library, 1997.

*Ultimate Montana Atlas and Travel Encyclopedia: Yellowstone Gateway Edition*, by M. Dougherty, H. Pfeil-Dougherty, and K. Hill. Arlington, TX: Ultimate Press, 2002, 2009.

## Nebraska

*Atlas of Seward County, Nebraska*. Minneapolis, MN: Title Atlas, 1996.

*Atlas of the Sand Hills*, by A. S. Bleed and C. Flowerday. Lincoln: University of Nebraska-Lincoln, 1998.

## Nevada

*Nevada Atlas and Gazetteer*. Freeport, ME: DeLorme Mapping, 1996.

*Nevada Ghost Towns and Mining Camps: Illustrated Atlas*, by S. W. Paher and N. Murbarger. Las Vegas: Nevada Publishing, 1999.

## New Hampshire

*New Hampshire Atlas*. Keene, NH: Keene State College, 2004.

*New Hampshire Atlas and Gazetteer*. Freeport, ME: DeLorme Mapping, 1996.

*New Hampshire Atlas and Gazetteer: Detailed Topographic Maps; Back Roads, Recreation Sites, City Street Maps, GPS Grids*. Yarmouth, ME: DeLorme Mapping, 2005.

*Where the Great River Rises: An Atlas of the Upper Connecticut River Watershed in Vermont and New Hampshire*, by R. A. Brown. Hanover, NH: Dartmouth College Press, 2009.

## New Jersey

*New Jersey Atlas and Gazetteer*. Yarmouth, ME: DeLorme Mapping, 1999.

## New York

*Atlas of New York: Legacies of the Eric Canal*, by P. Anthamatten. New York: New York Geographic Alliance, 2014.

*Central New York Street Atlas*. Round Lake, NY: Jimapco, 2000.

*New York: The Photographic Atlas; an Aerial Tour of All Five Boroughs and More*. New York: Harper Resource, 2004.

*New York Atlas and Gazetteer*. Yarmouth, ME: DeLorme Mapping, 2001.

*New York State Urban Areas Atlas*. Albany: New York State Departent of Transportation, 1998, 2000.

## North Carolina

*Historical Atlas of the Haw River*, by M. Chilton. Carrboro, NC: M. Chilton, 2008.

*North Carolina Atlas: Portrait for a New Century*, by D. M. Orr and A. W. Stuart. Chapel Hill: University of North Carolina Press, 2000.

*North Carolina, Atlas of Historical County Boundaries: A Project of the Dr. William M. Scholl Center for Family and Community History, the Newberry Library*, by G. DenBoer and J. H. Long. New York: Charles Scribner's Sons, 1998.

*North Carolina Atlas and Gazetteer: Topo Maps of the Entire State*. Yarmouth, ME: DeLorme Mapping, 1997.

*Shifting Shorelines: A Pictorial Atlas of North Carolina Inlets*, by W. J. Cleary and T. P. Marden. Raleigh: North Carolina Sea Grant, 1999.

## Ohio

*Atlas of Historical County Boundaries: Ohio*, by J. H. Long and P. T. Sinko. New York: Charles Scribner's Sons, 1998.

*Atlas of the State of Ohio from Surveys under the Direction of H. F. Walling*, by H. F. Walling and H. S. Stebbins. Kokomo, IN: Selby, 1999.

*Ohio Atlas and Gazetteer*. Freeport, ME: DeLorme Mapping, 1995.

*Ohio, Atlas of Historical County Boundaries: A Project of the Dr. William M. Scholl Center of Family and Community History, the Newberry Library*, by P. T. Sinko and J. H. Long. New York: Charles Scribner's Sons, 1998.

*Ohio Coastal Atlas*. Columbus: Ohio Coastal Management Program, 1998.

## Oklahoma

*Atlas of Oklahoma Climate*, by H. L. Johnson and C. E. Duchon. Norman: University of Oklahoma Press, 1995.

*Historical Atlas of Oklahoma*, by C. R. Goins and D. Goble. Norman: University of Oklahoma Press, 2006.

*Oklahoma Atlas and Gazetteer*. Yarmouth, ME: DeLorme Mapping, 1998.

## Oregon

*Archaeoclimatology Atlas of Oregon: The Modeled Distribution in Space and Time of the Past Climates of Oregon*, by R. A. Bryson, K. M. E. DeWall, and A. Stenger. Salt Lake City: University of Utah Press, 2009.

*Atlas of Oregon Gubernatorial Elections by County, 1934–1994*, by A. W. Miller. Richmond, VA: Klipsan, 1995.

*Historical Atlas of Washington and Oregon*, by D. Hayes and I. Cheung. Berkeley: University of California Press, 2011.

*Oregon: An Atlas of Oregon's Greatest Hiking Adventures*, by L. Dunegan. Guilford, CT: Globe Pequot, 2001.

## Pennsylvania

*Early Landowners of Pennsylvania*, by S. C. MacInnes and A. MacInnes. Apollo, PA: Closson, 2008.

*Pennsylvania Atlas and Gazetteer*. Yarmouth, ME: DeLorme Mapping, 2000, 2007.

## South Carolina

*Dutch Fork: An Atlas of the Dutch Fork of Newberry District, S.C*, by C. W. Nichols. Fernandina Beach, FL: Wolfe, 2001.

*Historical Atlas of the Rice Plantations of Georgetown County and the Santee River*, by S. C. L. Hurley, M. L. Thacker, and A. L. Baldwin. Columbia: South Carolina Department of Archives and History for the Historic Ricefields Association, 2001.

*South Carolina Atlas*. Columbia: South Carolina Department of Natural Resources, 2006.

*South Carolina Atlas and Gazetteer*. Yarmouth, ME: DeLorme Mapping, 2003.

*South Carolina Atlas of Historical County Boundaries*, by G. DenBoer, J. H. Long, and K. F. Thorne. New York: Charles Scribner's Sons, 1997.

*South Carolina: State Road Atlas*. Rixeyville, VA: RLM Enterprises, 2003.

*South Carolina Waterways: As They Appear in Mill's Atlas with Bridges, Ferries, and Fords*, by M. D. Cropper and R. Mills. West Jordan, UT: Genealogical Services, 1999.

## Tennessee

*First Southwest: The Third Atlas, the Cumberland and Duck River Settlements, Tennesseans Expand Our Nation South and West*, by J. Masters and B. Puryear. Gallatin, TN: Warioto Press, 2012.

*Founding of the Cumberland Settlements: The First Atlas, 1779–1804*, by D. Drake, J. Masters, and B. Puryear. Gallatin, TN: Warioto, 2009.

*Landscape Atlas of the Chesapeake Bay Watershed*, by K. H. Ritters, J. D. Wickham, and K. B. Jones. Norris: Tennessee Valley Authority, 1996.

*Tennessee Atlas of Historical County Boundaries*, by J. H. Long and P. T. Sinko. New York: Charles Scribner's Sons, 2000.

## Texas

*Historical Atlas of Texas Methodism*, by W. C. Hardt and J. W. Hardt. Garland, TX: CrossHouse, 2008.

*Michelin Texas Road Atlas*. Greenville, SC: Michelin Travel, 2004.

*School Atlas of Texas*. Austin: Southwest Texas State University, 2001.

*Texas: A Historical Atlas*, by A. R. Stephens and C. Zuber-Mallison. Norman: University of Oklahoma Press, 2010.

*Texas Atlas and Gazetteer: Detailed Maps of the Entire State.* Freeport, ME: DeLorme Mapping, 1995, 1997, 2005.

*Texas Health Atlas*, by L. E. Estaville, K. Egan, and A. A. Galaviz. College Station: Texas A&M University Press, 2012.

*Texas Water Atlas*, by L. E. Estaville and R. A. Earl. College Station: Texas A&M University Press, 2008.

## Vermont

*Vermont Atlas*. Keene, NH: Keene State College, 2005.

*Vermont Atlas and Gazetteer*. Freeport, ME: DeLorme Mapping, 1996.

*Where the Great River Rises: An Atlas of the Upper Connecticut River Watershed in Vermont and New Hampshire*, by R. A. Brown. Hanover, NH: Dartmouth College Press, 2009.

## Virginia

*Blackwater, Nottoway and Meherrin Rivers Atlas: Rediscovering River History on the Chowan River and Its Branches*, by W. E. Trout and J. Turner. Lexington: Virginia Canals and Navigations Society, 2006.

*Falls of the James Atlas: Historic Canal and River Sites on the Falls of the James with a Special Supplement on the Tuckahoe Creek Navigation*, by W. E. Trout, J. Moore, and G. D. Rawls. Richmond: Virginia Canals and Navigations Society, 1995.

*Hike America, Virginia: An Atlas of Virginia's Greatest Hiking Adventures*, by B. Burnham and M. Burnham. Guilford, CT: Globe Pequot, 2001.

*Historical Atlas of West Virginia*, by F. S. Riddel. Morgantown: West Virginia University Press, 2008.

*Holston, Clinch and Powell's Rivers Atlas: Rediscovering River History in Far Southwest Virginia*, by W. E. Trout and N. R. Trout. Lexington: Virginia Canals and Navigations Society, 2013.

*James River Batteau Festival Trail: A Guide to the James River and Its Canal, from Lynchburg to Richmond*, by W. E. Trout. Lexington: Virginia Canals and Navigations Society, 1997.

*Rappahannock Scenic River Atlas: A Virginia Canals and Navigations Society River Atlas Project*, by W. E. Trout. Lexington: Virginia Canals and Navigations Society, 2004.

*Rivanna Scenic River Atlas: A Virginia Canals and Navigations Society River Atlas Project*, by W. E. Trout. Lexington: Virginia Canals and Navigations Society, 2002.

*Roanoke/Staunton River Atlas: Rediscovering River History along the Roanoke/Staunton and Its Branches*, by W. E. Trout. Lexington: Virginia Canals and Navigations Society, 2006.

*Upper James Atlas: Rediscovering River History in the Blue Ridge and Beyond*, by W. E. Trout. Lexington: Virginia Canals and Navigations Society, 2001.

*Virginia Atlas and Gazetteer*. Freeport, ME: DeLorme Mapping, 1995, 1999, 2000.

*Virginia State Road Atlas*. Alexandria, VA: ADC, 1996.

## Washington

*Historical Atlas of Washington and Oregon*, by D. Hayes and I. Cheung. Berkeley: University of California Press, 2011.

## West Virginia

*Hardesty's Biographical Atlas of Upshur County, WV*, by W. Cochran and H. H. Hardesty. Parkersburg, WV: Wes Cochran, 2010.

## Wisconsin

*9 County Street Atlas: East Central Wisconsin; Complete Street Level Detail for the Counties of Brown, Calumet, Door, Fond du Lac, Kewaunee, Manitowoc, Outagamie, Sheboygan, and Winnebago*. Milwaukee, WI: Milwaukee Map Service, 2004.

*Historical Atlas of Wisconsin: Embracing Complete State and County Maps, City and Village Plats*. Janesville, WI: S. Van Vechten, 1999.

*Seeger's 10 County Atlas: Milwaukee and Southeastern Wisconsin Street Locator*. Reedsburg, WI: Midwest Map, 2004.

*Southeastern Wisconsin 6 County Street Atlas*. Milwaukee, WI: Milwaukee Map Service, 2011, 2015.

*Wind Atlas of Wisconsin*, by P. N. Knox. Madison: University of Wisconsin, 1996.

*Wisconsin Atlas and Gazetteer*. Freeport, ME: DeLorme Mapping, 1995.

*Wisconsin's Past and Present: A Historical Atlas*. Madison: University of Wisconsin Press, 1998, 2002.

## Wyoming

*Ultimate Wyoming Atlas and Travel Encyclopedia*, by M. Dougherty and H. P. Dougherty. Bozeman, MT: Ultimate Press, 2003.

*Wyoming Atlas and Gazetteer*. Yarmouth, ME: DeLorme Mapping, 1998.

*Wyoming Atlas and Gazetteer: Topo Maps of the Entire State, Public Lands, Back Roads*. Yarmouth, ME: DeLorme Mapping, 2000.

*Wyoming Climate Atlas*, by J. Curtis and K. Grimes. Laramie: Office of the Wyoming State Climatologist, 2004.

## AMERICAN COUNTY ATLASES

### Alabama

*Baldwin County*

*Street and Road Atlas Baldwin County, Alabama.* Mobile, AL: Keith Map Service, 1999.
*Land Atlas and Plat Book, Baldwin County, Alabama, 1988.* Rockford, IL: Rockford Map, 2000.

*Butler County*

*Land Atlas and Plat Book, Butler County, Alabama, 1985.* Rockford, IL: Rockford Map, 2000.

*Chilton County*

*Land Atlas and Plat Book, Chilton County, Alabama, 1989.* Rockford, IL: Rockford Map, 2000.

*Clarke County*

*Land Atlas and Plat Book, Clarke County, Alabama.* Rockford, IL: Rockford Map, 1998.

*Coffee County*

*Land Atlas and Plat Book, Coffee County, Alabama, 1987.* Rockford, IL: Rockford Map, 2000.

*Conecuh County*

*Land Atlas and Plat Book, Conecuh County, Alabama, 1994.* Rockford, IL: Rockford Map, 2000.

*Crenshaw County*

*Land Atlas and Plat Book, Crenshaw County, Alabama, 1986.* Rockford, IL: Rockford Map, 2000.

*Cullman County*

*Land Atlas and Plat Book, Cullman County, Alabama, 1982–1983.* Rockford, IL: Rockford Map, 2000.

*Dale County*

*Land Atlas and Plat Book, Dale County, Alabama, 1992.* Rockford, IL: Rockford Map, 2000.

*Lauderdale County*

*Land Atlas and Plat Book, Lauderdale County, Alabama.* Rockford, IL: Rockford Map, 2004.

*Lee County*

*Land Atlas and Plat Book, Lee County, Alabama.* Rockford, IL: Rockford Map, 2001.

*Marengo County*

*Land Atlas and Plat Book, Marengo County, Alabama.* Rockford, IL: Rockford Map, 2000.

*Mobile County*

*Land Atlas and Plat Book, Mobile County, Alabama, 1984.* Rockford, IL: Rockford Map, 2000.
*Metro-Mobile Street and Road Atlas Mobile and Mobile County, Alabama.* Mobile, AL: Keith Map Service, 1999.
*Street and Road Atlas Mobile County, Alabama.* Mobile, AL: Keith Map Service, 2002.

*Monroe County*

*Land Atlas and Plat Book, Monroe County, Alabama: 2005.* Rockford, IL: Rockford Map, 2005.

*Walker County*

*Land Atlas and Plat Book, Walker County, Alabama, 1985.* Rockford, IL: Rockford Map, 2000.

*Washington County*

*Land Atlas and Plat Book, Washington County, Alabama, 1989.* Rockford, IL: Rockford Map, 2000.

*Wilcox County*

*Land Atlas and Plat Book, Wilcox County, Alabama.* Rockford, IL: Rockford Map, 2000.

### Arizona

*Cochise County*

*Atlas of Historic Maps of Cochise County, Arizona*, by B. Smith. Tucson, AZ: B. Smith, 1997.

*Phoenix Metropolitan Area*

*Metropolitan Phoenix Street Atlas.* Phoenix, AZ: Phoenix Mapping Service / Wide World of Maps, 2014.
*Metropolitan Phoenix Street Atlas: 2010–2011.* Phoenix, AZ: Phoenix Mapping Service / Wide World of Maps, 2010.

### Arkansas

*Jackson County*

*Atlas and Plat Book of Jackson County, Arkansas.* Conway, AR: Arkansas Research, 2003.

## California

*Napa County*

*Atlas Peak: A History of a Napa County Settler, 1870–1902*, by C. E. Setty. Napa, CA: Cecelia Elkington Setty, 2004.

*Napa Valley Historical Ecology Atlas: Exploring a Hidden Landscape of Transformation and Resilience*, by R. Grossinger. Berkeley: University of California Press, 2012.

*Riverside County*

*Corona City Atlas*. Corona, CA: City of Corona Information Services, 2000.

*San Diego County*

*Demographic Atlas San Diego/Tijuana: Atlas demográfico [San Diego/Tijuana]*, by R. Arnaiz et al. La Jolla: University of California, San Diego, 1995.

*San Francisco County*

*Infinite City: A San Francisco Atlas*, by R. Solnit, B. Pease, and S. Siegel. Berkeley: University of California Press, 2010.

*Sonoma County*

*Illustrated Atlas of Sonoma County, California*. Santa Rosa, CA: Reynolds and Proctor, 1998.

*Yolo County*

*Illustrated Atlas and History of Yolo County, Cal: Containing a History of California from 1513 to 1850, a History of Yolo County from 1825 to 1880, with Statistics of Agriculture, Education, Churches, Elections, Lithographic Views of Farms, Residences, Mills, Portraits of Well-Known Citizens, and the Official County Map*. McMinnville, TN: Isha Books, 2013.

## Colorado

*Eagle County*

*Eagle County Tax Map Atlas*. Eagle, CO: Eagle County Assessor's Office, 2007.

## Connecticut

*Middlesex County*

*County Atlas of Middlesex, Connecticut*. Salt Lake City: Genealogical Society of Utah, 2005.

## Florida

*Bay County*

*Bay County Florida GIS Atlas, 2011*. Panama City, FL: Bay County, 2011.

*Bay County Florida Atlas, 2012*. Panama City, FL: Bay County, 2012.

*Bradford County*

*Land Atlas and Plat Book, Bradford County, Florida, 1977*. Rockford, IL: Rockford Map, 2000.

*Brevard County*

*Land Atlas and Plat Book, Brevard County, Florida, 1978*. Rockford, IL: Rockford Map, 2000.

*Charlotte County*

*Land Atlas and Plat Book, Charlotte County, Florida, 1977*. Rockford, IL: Rockford Map, 2000.

*Citrus County*

*Experian Realty Atlas, Citrus County, Florida Tax Map*. Anaheim, CA: Experian National Headquarters, 1998.

*Land Atlas and Plat Book, Citrus County, Florida, 1980–1981*. Rockford, IL: Rockford Map, 2000.

*Collier County*

*Land Atlas and Plat Book, Collier County, Florida, 1983*. Rockford, IL: Rockford Map, 2000.

*De Soto County*

*Land Atlas and Plat Book, De Soto County, Florida, 1976*. Rockford, IL: Rockford Map, 2000.

*Escambia County*

*Land Atlas and Plat Book, Escambia County, Florida, 1985*. Rockford, IL: Rockford Map, 2000.

*Gadsden County*

*Atlas and Plat Book, Gadsden County, Florida, 1978*. Rockford, IL: Rockford Map, 2000.

*Hardee County*

*Land Atlas and Plat Book, Hardee County, Florida, 1979*. Rockford, IL: Rockford Map, 2000.

*Hernando County*

*Land Atlas and Plat Book, Hernando County, Florida, 1980–1981*. Rockford, IL: Rockford Map, 2000.

*Highlands County*

*Land Atlas and Plat Book, Highlands County, Florida, 1980–1981.* Rockford, IL: Rockford Map, 2000.

*Hillsborough County*

*Land Atlas and Plat Book, Hillsborough County, Florida.* Rockford, IL: Rockford Map, 2000.

*Indian River County*

*Atlas and Plat Book, Indian River County, Florida, 1978.* Rockford, IL: Rockford Map, 2000.

*Jefferson County*

*Land Atlas and Plat Book, Jefferson County, Florida, 1976.* Rockford, IL: Rockford Map, 2000.

*Lake County*

*Land Atlas and Plat Book, Lake County, Florida, 1989.* Rockford, IL: Rockford Map, 2000.

*Lee County*

*Land Atlas and Plat Book, Lee County, Florida, 1976.* Rockford, IL: Rockford Map, 2000.

*Madison County*

*Land Atlas and Plat Book, Madison County, Florida, 1977.* Rockford, IL: Rockford Map, 2000.

*Marion County*

*Land Atlas and Plat Book, Marion County, Florida, 1983.* Rockford, IL: Rockford Map, 2000.

*Orange County*

*Land Atlas and Plat Book, Orange County, Florida, 1981.* Rockford, IL: Rockford Map, 2000.

*Osceola County*

*Land Atlas and Plat Book, Osceola County, Florida, 1976.* Rockford, IL: Rockford Map, 2000.

*Palm Beach County*

*Land Atlas and Plat Book, Palm Beach County, Florida, 1976.* Rockford, IL: Rockford Map, 2000.

*Pasco County*

*Land Atlas and Plat Book, Pasco County, Florida, 1981.* Rockford, IL: Rockford Map, 2000.

*Putnam County*

*Land Atlas and Plat Book, Putnam County, Florida, 1979.* Rockford, IL: Rockford Map, 2000.

*Street Atlas of Putnam County Florida.* St. Petersburg, FL: MAPSource, 1999.

*Seminole County*

*Land Atlas and Plat Book, Seminole County, Florida, 1980–1981.* Rockford, IL: Rockford Map, 2000.

*St. Lucie County*

*Land Atlas and Plat Book, St. Lucie County, Florida, 1977.* Rockford, IL: Rockford Map, 2000.

*Sumter County*

*Atlas and Plat Book, Sumter County, Florida, 1976.* Rockford, IL: Rockford Map, 2000.

*Suwannee County*

*Atlas and Plat Book, Suwannee County, Florida, 1975.* Rockford, IL: Rockford Map, 2000.

## Georgia

*Rockdale County*

*TRW REDI Property Data Realty Atlas, Rockdale County, Georgia.* Anaheim, CA: TRW REDI, 1996.

## Idaho

*Bonneville County*

*Land Atlas and Plat Book, Bonneville County, Idaho, 1982.* Rockford, IL: Rockford Map, 2000.

*Madison County*

*Land Atlas and Plat Book, Madison County, Idaho, 1982.* Rockford, IL: Rockford Map, 2000.

## Illinois

*Bond County*

*Combined Atlases of Bond County, Illinois, 1875 and 1900.* Mt. Vernon, IN: Windmill, 1998.

*Boone County*

Boone County, Illinois: Land Atlas and Plat Book. Rockford, IL: Rockford Map, 200.

*Bureau County*

Bureau County, Illinois: Land Atlas and Plat Book, 1996. Rockford, IL: Rockford Map, 1996.

*Cass County*

Land Atlas and Plat Book, Cass County, Illinois, 1987. Rockford, IL: Rockford Map, 2000.

*Champaign County*

Land Atlas and Plat Book, Champaign County, Illinois: 2005. Rockford, IL: Rockford Map, 2005.

*Christian County*

Land Atlas and Plat Book, Christian County, Illinois: 2005. Rockford, IL: Rockford Map, 2005.

*Cook County*

Geologic Atlas of Cook County for Planning Purposes, by H. E. Leetaru, M. L. Sargent, and D. R. Kolata. Champaign, IL: Department of Natural Resources, 2004.

*Crawford County*

Land Atlas and Plat Book, Crawford County, Illinois, 1980–1981. Rockford, IL: Rockford Map, 2000.

*Cumberland County*

Land Atlas and Plat Book, Cumberland County, Illinois, 1982. Rockford, IL: Rockford Map, 2000.

*DeKalb County*

Land Atlas and Plat Book, DeKalb County, Illinois. Rockford, IL: Rockford Map, 1999.

*DeWitt County*

Land Atlas and Plat Book: DeWitt County, Illinois. Rockford, IL: Rockford Map, 1996.

*DuPage County*

DuPage and Kane Counties Atlas. Wood Dale, IL: American Map, 1995.

*Edwards-Wabash County*

Atlas and Plat Book, Edwards-Wabash County, Illinois, 1978. Rockford, IL: Rockford Map, 2000.

*Ford County*

Land Atlas and Plat Book, Ford County, Illinois, 1996. Rockford, IL: Rockford Map, 2000.

*Greene County*

Land Atlas and Plat Book, Greene County, Illinois. Rockford, IL: Rockford Map, 2004.

*Grundy County*

Grundy County, Illinois: Land Atlas and Plat Book, 1998. Rockford, IL: Rockford Map, 1998.

*Jefferson County*

Land Atlas and Plat Book, Jefferson County, Illinois, 1995. Rockford, IL: Rockford Map, 2000.

*Jersey County*

Atlas Map of Jersey County, Illinois: Compiled, Drawn and Published from Personal Examinations and Surveys, by K. Ballard. Carrollton, IL: Jersey County Historical Society, 1995.

*Johnson County*

Land Atlas and Plat Book, Johnson County, Illinois, 1978. Rockford, IL: Rockford Map, 2000.

*Kane County*

DuPage and Kane Counties Atlas. Wood Dale, IL: American Map, 1995.

*Knox County*

Knox County, Illinois, Land Atlas and Plat Book: 2012. Rockford, IL: Rockford Map, 2012.

*La Salle County*

1876 La Salle County Atlas and 1870 La Salle County Map: Together with a Combined Index. Ottawa, IL: LaSalle County Genealogy Guild, 2000.

LaSalle County, Illinois: Land Atlas and Plat Book, 2000. Rockford, IL: Rockford Map, 2000.

*Lake County*

Lake and McHenry Counties Atlas. Wood Dale, IL: American Map, 1995.

*Lake County, Illinois: Land Atlas and Plat Book.* Rockford, IL: Rockford Map, 2000.

*Lake County, Illinois Illustrated Atlas of 1885 Index*, by A. P. Hapke. Libertyville, IL: Lake County Genealogical Society, 1999.

Lee County

*Combination Atlas Map of Lee County, Illinois.* Dixon, IL: Lee County Genealogical Society, 2003.

*Lee County, Illinois: Land Atlas and Plat Book, 2000.* Rockford, IL: Rockford Map, 2000.

Logan County

*Logan County, Illinois, Land Atlas and Plat Book: 2013.* Rockford, IL: Rockford Map, 2013.

Macoupin County

*Land Atlas and Plat Book, Macoupin County, Illinois, 1989.* Rockford, IL: Rockford Map, 2000.

Madison County

*Standard Atlas of Madison County, Illinois: Including a Plat Book of the Villages, Cities and Townships of the County.* Chicago: Geo. A. Ogle, 2003.

Marshall-Putnam Counties

*Land Atlas and Plat Book Marshall-Putnam Counties, Illinois: 2007.* Rockford, IL: Rockford Map, 2007.

Mason County

*Land Atlas and Plat Book, Mason County, Illinois.* Rockford, IL: Rockford Map, 1997.

McHenry County

*Lake and McHenry Counties Atlas.* Wood Dale, IL: American Map, 1995.

*Land Atlas and Plat Book, McHenry County, Illinois.* Rockford, IL: Rockford Map, 2004.

McLean County

*Land Atlas and Plat Book, McLean County, Illinois.* Rockford, IL: Rockford Map, 2000.

Mercer County

*Land Atlas and Plat Book, Mercer County, Illinois, 1979.* Rockford, IL: Rockford Map, 2000.

*McLean County Combined Indexed Atlases, 1856–1914*, by W. P. LaBounty. Bloomington, IL: McLean County Historical Society, 2006.

*Mercer Co. Atlas and Plat Book, 1980–1981.* Rockford, IL: Rockford Map, 2000.

Monroe County

*Land Atlas and Plat Book, Monroe County, Illinois, 1989.* Rockford, IL: Rockford Map, 2000.

Ogle County

*Ogle County, Illinois: Land Atlas and Plat Book, 1998.* Rockford, IL: Rockford Map, 1998.

Piatt County

*Land Atlas and Plat Book, Piatt County, Illinois: 2002.* Rockford, IL: Rockford Map, 2002.

Pope and Hardin Counties

*Land Atlas and Plat Book, Pope-Hardin Counties, Illinois.* Rockford, IL: Rockford Map, 2000.

Richland County

*Land Atlas and Plat Book, Richland County, Illinois, 1988.* Rockford, IL: Rockford Map, 2000.

Rock Island County

*Land Atlas and Plat Book, Rock Island County, Illinois.* Rockford, IL: Rockford Map, 2000.

Saline County

*Land Atlas and Plat Book, Saline County, Illinois.* Rockford, IL: Rockford Map, 2003.

Schuyler County

*Atlas Map of Schuyler County, Illinois: Compiled, Drawn, and Published from Personal Examinations and Surveys.* Rushville, IL: Schuyler County Historical Museum, 2000.

Scott County

*Land Atlas and Plat Book, Scott County, Illinois, 1988.* Rockford, IL: Rockford Map, 2000.

Shelby County

*Land Atlas and Plat Book, Shelby County, Illinois.* Rockford, IL: Rockford Map, 2003.

*St. Clair County*

*Land Atlas and Plat Book, St. Clair County, Illinois, 1992.* Rockford, IL: Rockford Map, 2000.

*Stephenson County*

*Land Atlas and Plat Book Stephenson County, Illinois.* Rockford, IL: Rockford Map, 1997.

*Union County*

*Land Atlas and Plat Book, Union County, Illinois, 1978.* Rockford, IL: Rockford Map, 2000.

*Warren and Henderson Counties*

*Land Atlas and Plat Book, Warren-Henderson Counties, Illinois.* Rockford, IL: Rockford Map, 2004.

*Whiteside County*

*Whiteside County, Illinois: 1872 Atlas and Index with Individual Illustrated Township Plats Including Complete Name and Landmark Indexes.* Sterling, IL: Sterling-Rock Falls Historical Society, 2009.

*Whiteside County, Illinois: Land Atlas and Plat Book, 1996.* Rockford, IL: Rockford Map, 1996.

*Will County*

*Land Atlas and Plat Book, Will County, Illinois: 2005.* Rockford, IL: Rockford Map, 2005.

*Will County, Illinois: Land Atlas and Plat Book, 2000.* Rockford, IL: Rockford Map, 2000.

*Will County, Illinois: Land Atlas and Plat Book, 2010.* Rockford, IL: Rockford Map, 2010.

*Winnebago County*

*Welcome to Winnebago County, Illinois.* Elgin, IL: Village Profile, 2008.

*Woodford County*

*Land Atlas and Plat Book, Woodford County, Illinois.* Rockford, IL: Rockford Map, 2000.

**Indiana**

*Boone County*

*Combination Atlas Map of Boone County, Indiana: Compiled, Drawn and Published from Personal Examinations and Surveys by Kingman Brothers.* Lebanon, IN: Boone County Historical Society, 2000.

*Brown County*

*Land Atlas and Plat Book, Brown County, Indiana, 1976.* Rockford, IL: Rockford Map, 2000.

*Clinton County*

*Clinton County, Indiana Historical Atlas*, by S. H. Irick. Kokomo, IN: Shelby, 1998.

*Daviess County*

*Atlas and Plat Book, Daviess County, Indiana, 1977.* Rockford, IL: Rockford Map, 2000.

*Dearborn County*

*Dearborn County Indiana Atlas*, by C. A. Ridlen. Indianapolis, IN: Ye Olde Genealogie Shoppe, 2003.

*Decatur County*

*Triennial Atlas and Plat Book, Decatur County, Indiana, 1971.* Rockford, IL: Rockford Map, 2000.

*Fountain County*

*Land Atlas and Plat Book, Fountain County, Indiana, 1976.* Rockford, IL: Rockford Map, 2000.

*Franklin County*

*Atlas of Franklin Co., Indiana: Atlas and World War Historical Data of Franklin County, Indiana 1925*, by J. L. Stewart and F. E. Bartlett. Brookville, IN: Franklin County Historical Society, 2003.

*Gibson County*

*Atlas and Plat Book, Gibson County, Indiana, 1975.* Rockford, IL: Rockford Map, 2000.

*Illustrated—Standard Atlas of Gibson County, Indiana.* Princeton, IN: Hammond & Tillman, 2001.

*Hamilton County*

*Land Atlas and Plat Book, Hamilton County, Indiana, 1988.* Rockford, IL: Rockford Map, 2000.

*Hendricks County*

*Land Atlas and Plat Book, Hendricks County, Indiana, 1986.* Rockford, IL: Rockford Map, 2000.

*Howard County*

*Combination Atlas Map of Howard County, Indiana.* Kokomo, IN: Selby, 1996.

*Jay and Blackford Counties*

*Historical Hand-Atlas, Illustrated: Containing Twelve Farm Maps, and History of Jay County, Indiana.* Chicago: H. H. Hardesty, 2015.

*Triennial Farm Atlas and Resident's Directory, Jay-Blackford County, Indiana, 1967.* Rockford, IL: Rockford Map, 2000.

*Jefferson County*

*Index of Land Owners on the Jefferson County, Indiana, Plat Maps for the year 1900*, by V. L. Shatley et al. Madison, IN: Jefferson County Historical Society, 2003.

*Land Atlas and Plat Book, Jefferson County, Indiana, 1989.* Rockford, IL: Rockford Map, 2000.

*Jennings County*

*Land Atlas and Plat Book, Jennings County, Indiana, 1989.* Rockford, IL: Rockford Map, 2000.

*Knox County*

*Illustrated Historical Atlas of Knox County, Indiana.* Vincennes, IN: Ewing Printing, 2003.

*La Grange County*

*Tri-Annual Atlas and Plat Book, La Grange County, Indiana.* Rockford, IL: Rockford Map, 2000.

*Marion County*

*Atlas of Indianapolis and Marion County, Indiana*, by H. B. Fatout and G. Bohn. Mt. Vernon, IN: Windmill, 2000.

*Marshall County*

*1881 History Combined with 1908, 1922 Atlases, Marshall County, Indiana*, by D. McDonald. Plymouth, IN: The Marshall County Historical Society, 2001.

*Monroe County*

*Monroe County, Indiana.* Evansville, IN: Ryle, 2008.

*Pike County*

*Land Atlas and Plat Book, Pike County, Indiana, 1983.* Rockford, IL: Rockford Map, 2000.

*Ripley County*

*Land Atlas and Plat Book, Ripley County, Indiana, 1987.* Rockford, IL: Rockford Map, 2000.

*Spencer County*

*Illustrated Historical Atlas of Spencer County, Indiana: From Actual Surveys*, by B. N. Griffing. Evansville, IN: D. J. Lake, 2015.

*St. Joseph County*

*Illustrated Historical Atlas of St. Joseph Co., Indiana.* Chicago: Higgins, Belden, 2014.

*Wabash County*

*Atlas of Wabash County Indiana: To Which Is Added a Township Map of the State of Indiana, Also an Outline and Rail Road Map of the United States*, by H. Paul. Mt. Vernon, IN: Windmill, 1998.

*Warrick County*

*Land Atlas and Plat Book, Warrick County, Indiana, 1977.* Rockford, IL: Rockford Map, 2000.

## Iowa

*Alachua County*

*Land Atlas and Plat Book, Alachua County, Florida, 1981.* Rockford, IL: Rockford Map, 2000.

*Allamakee County*

*Land Atlas and Plat Book, Allamakee County, Iowa, 1983.* Rockford, IL: Rockford Map, 2000.

*Audubon County*

*Atlas of Audubon County, Iowa: Containing Maps, Plats of the Townships, Rural Directory, Pictures of Farms and Families, Articles about History, Etc.* Battle Lake, MN: Title Atlas, 2001.

*Benton County*

*Land Atlas and Plat Book, Benton County, Iowa, 1982.* Rockford, IL: Rockford Map, 2000.

*Boone County*

*Land Atlas and Plat Book, Boone County, Iowa, 1979.* Rockford, IL: Rockford Map, 2000.

*Bremer County*

*Atlas of Bremer County, Iowa: Containing Maps, Plats of the Townships, Rural Directory, Pictures of Farms and Families, Articles about History, Etc.* Battle Lake, MN: Title Atlas, 1997.

### Buchanan County

*Land Atlas and Plat Book, Buchanan County, Iowa, 1982.* Rockford, IL: Rockford Map, 2000.

### Buena Vista County

*Atlas of Buena Vista County, Iowa: Containing Maps, Plats of the Townships, Rural Directory, Pictures of Farms and Families, Articles about History, Etc.* Battle Lake, MN: Title Atlas, 2004.

*Land Atlas and Plat Book, Buena Vista County, Iowa, 1982.* Rockford, IL: Rockford Map, 2000.

### Butler County

*Land Atlas and Plat Book, Butler County, Iowa, 1977.* Rockford, IL: Rockford Map, 2000.

*Millennium Atlas of Butler County, Iowa: Containing Maps, Plats of the Townships, Rural Directory, Pictures of Farms and Families, Articles about History, Etc.* Battle Lake, MN: Title Atlas, 2000.

### Calhoun County

*Land Atlas and Plat Book, Calhoun County, Iowa, 1982–1983.* Rockford, IL: Rockford Map, 2000.

### Cass County

*Atlas of Cass County, Iowa: Containing Maps, Plats of the Townships, Rural Directory, Pictures of Farms and Families, Articles about History, Etc.* Battle Lake, MN: Title Atlas, 2003.

### Cedar County

*Land Atlas and Plat Book, Cedar County, Iowa, 1983.* Rockford, IL: Rockford Map, 2000.

### Cherokee County

*Atlas of Cherokee County, Iowa: Containing Maps, Plats of the Townships, Rural Directory, Pictures of Farms and Families, Articles about History, Etc.* Battle Lake, MN: Title Atlas, 2004.

### Chickasaw County

*Land Atlas and Plat Book, Chickasaw County, Iowa, 1977.* Rockford, IL: Rockford Map, 2000.

### Clay County

*Atlas of Clay County, Iowa: Containing Maps, Plats of the Townships, Rural Directory, Pictures of Farms and Families, Articles about History, Etc.* Battle Lake, MN: Title Atlas, 2003.

### Clayton County

*Land Atlas and Plat Book, Clayton County, Iowa, 1983.* Rockford, IL: Rockford Map, 2000.

### Clinton County

*Clinton County, Iowa Plat Books.* Cedar Rapids, IA: Heritage Microfilm, 2002.

*Land Atlas and Plat Book, Clinton County, Iowa, 1990.* Rockford, IL: Rockford Map, 2000.

### Crawford County

*Atlases of Clinton County, Iowa: 1874 and 1905.* Cedar Rapids, IA: Heritage Microfilm, 2002.

*Combined Atlases and Plat Book, Clinton County, Iowa: 1874 Atlas, 1894 Plat Book, 1905 Atlas, 1925 Atlas.* Mt. Vernon, IN: Windmill, 2000.

*Land Atlas and Plat Book, Crawford County, Iowa, 1981.* Rockford, IL: Rockford Map, 2000.

### Decatur County

*Land Atlas and Plat Book, Decatur County, Iowa, 1978.* Rockford, IL: Rockford Map, 2000.

### Delaware County

*Land Atlas and Plat Book, Delaware County, Iowa, 1982.* Rockford, IL: Rockford Map, 2000.

### Dickinson County

*Land Atlas and Plat Book, Dickinson County, Iowa, 1981.* Rockford, IL: Rockford Map, 2000.

### Fayette County

*Land Atlas and Plat Book, Fayette County, Iowa, 1980–1981.* Rockford, IL: Rockford Map, 2000.

### Floyd County

*Land Atlas and Plat Book, Floyd County, Iowa, 1978.* Rockford, IL: Rockford Map, 2000.

### Fremont County

*Atlas and Directory, Fremont County, Iowa, 1970.* Rockford, IL: Rockford Map, 2000.

### Hamilton County

*Land Atlas and Plat Book, Hamilton County, Iowa, 1982.* Rockford, IL: Rockford Map, 2000.

### Harrison County

*Illustrated Atlas of Harrison County, Iowa*, by C. R. Allen and F. S. Allen. Mount Vernon, IN: Windmill, 1996.

### Henry County

*Land Atlas and Plat Book, Henry County, Iowa, 1982*. Rockford, IL: Rockford Map, 2000.

### Howard County

*Land Atlas and Plat Book, Howard County, Iowa, 1978*. Rockford, IL: Rockford Map, 2000.

### Humboldt County

*Land Atlas and Plat Book, Humboldt County, Iowa, 1982*. Rockford, IL: Rockford Map, 2000.

### Ida County

*Atlas of Ida County, Iowa: Containing Maps, Plats of the Townships, Rural Directory, Pictures of Farms and Families, Articles about History, Etc.* Battle Lake, MN: Title Atlas, 2005.

### Jasper County

*Land Atlas and Plat Book, Jasper County, Iowa, 1976*. Rockford, IL: Rockford Map 2000.

### Jefferson County

*Land Atlas and Plat Book, Jefferson County, Iowa, 1982*. Rockford, IL: Rockford Map, 2000.

### Johnson County

*Land Atlas and Plat Book, Johnson County, Iowa, 1981–1982*. Rockford, IL: Rockford Map, 2000.

### Jones County

*Land Atlas and Plat Book, Jones County, Iowa, 1983*. Rockford, IL: Rockford Map, 2000.

### Keokuk County

*Land Atlas and Plat Book, Keokuk County, Iowa, 1981*. Rockford, IL: Rockford Map, 2000.

### Lee County

*Land Atlas and Plat Book, Lee County, Iowa, 1982*. Rockford, IL: Rockford Map, 2000.

### Louisa County

*Land Atlas and Plat Book, Louisa County, Iowa, 1978*. Rockford, IL: Rockford Map, 2000.

### Lyon County

*Atlas of Lyon County, Iowa: Containing Maps, Plats of the Townships, Rural Directory, Pictures of Farms and Families, Articles about History, Etc.* Battle Lake, MN: Title Atlas, 1998.

### Mahaska County

*Land Atlas and Plat Book, Mahaska County, Iowa, 1981*. Rockford, IL: Rockford Map, 2000.

### Marion County

*Land Atlas and Plat Book, Marion County, Iowa, 1977*. Rockford, IL: Rockford Map, 2000.

### Mitchell County

*Land Atlas and Plat Book, Mitchell County, Iowa, 1977*. Rockford, IL: Rockford Map, 2000.

### Monona County

*Land Atlas and Plat Book, Monona County, Iowa, 1981–1982*. Rockford, IL: Rockford Map, 2000.

### Monroe County

*Land Atlas and Plat Book, Monroe County, Iowa, 1977*. Rockford, IL: Rockford Map, 2000.

### Muscatine County

*Land Atlas and Plat Book, Muscatine County, Iowa, 1977*. Rockford, IL: Rockford Map, 2000.

### O'Brien County

*Atlas of O'Brien County, Iowa: Containing Maps, Plats of the Townships, Rural Directory, Pictures of Farms and Families, Articles about History, Etc.* Battle Lake, MN: Title Atlas, 1998.

*Land Atlas and Plat Book, O'Brien County, Iowa, 1981*. Rockford, IL: Rockford Map, 2000.

### Osceola County

*Atlas of Osceola County, Iowa: Containing Maps, Plats of the Townships, Rural Directory, Pictures of Farms and Families, Articles about History, Etc.* Battle Lake, MN: Title Atlas, 1999.

### Palo Alto County

*Atlas of Palo Alto County, Iowa: Containing Maps, Plats of the Townships, Rural Directory, Pictures of Farms and Families, Articles about History, Etc.* Battle Lake, MN: Title Atlas, 2000.

### Plymouth County

*1996 Plymouth County Atlas and Farm Directory.* Harlan, IA: R. C. Booth Enterprises, 1996.

*Atlas and Directory, Plymouth County, Iowa, 1972.* Rockford, IL: Rockford Map, 2000.

*Atlas of Plymouth County, Iowa: Containing Maps, Plats of the Townships, Rural Directory, Pictures of Farms and Families, Articles about History, Etc.* Battle Lake, MN: Title Atlas, 1998.

### Pocahontas County

*Atlas of Pocahontas County, Iowa: Containing Maps, Plats of the Townships/Precincts, Rural Directory, Pictures of Farms and Families, Articles about History, Etc.* Battle Lake, MN: Title Atlas, 2006.

*Land Atlas and Plat Book, Pocahontas County, Iowa, 1977.* Rockford, IL: Rockford Map, 2000.

### Polk County

*Land Atlas and Plat Book, Polk County, Iowa, 1983.* Rockford, IL: Rockford Map, 2000.

### Pottawattamie County

*Illustrated Atlas of Pottawattamie County, Iowa*, by C. R. Allen. Council Bluffs, IA: Pottawattamie County Genealogical Society, 1995.

*Standard Atlas of Pottawattamie County, Iowa, 1902: Including a Plat Book of the Villages, Cities and Townships of the County, Map of the State, United States and World.* Mt. Vernon, IN: Windmill, 1997.

### Poweshiek County

*Index 1927 Poweshiek County, Iowa, Land Atlas.* Montezuma, IA: Poweshiek County Historical and Genealogical Society, 2011.

*Land Atlas and Plat Book, Poweshiek County, Iowa, 1982.* Rockford, IL: Rockford Map, 2000.

### Sac County

*Atlas and Plat Book, Sac County, Iowa, 1976.* Rockford, IL: Rockford Map, 2000.

*Atlas of Sac County, Iowa: Containing Maps, Plats of the Townships/Precincts, Rural Directory, Pictures of Farms and Families, Articles about History, Etc.* Battle Lake, MN: Title Atlas, 2005.

### Sioux County

*Atlas of Sioux County, Iowa: Containing Maps, Plats of the Townships, Rural Directory, Pictures of Farms and Families, Articles about History, Etc.* Battle Lake, MN: Title Atlas, 1997.

### Tama County

*Land Atlas and Plat Book, Tama County, Iowa, 1982.* Rockford, IL: Rockford Map, 2000.

### Union County

*1995 Plat Maps of Union County, Iowa: With Township Plats Corrected to March 1995.* Afton, IA: Afton Star Enterprise, 1995.

### Wapello County

*Land Atlas and Plat Book, Wapello County, Iowa, 1982–1983.* Rockford, IL: Rockford Map, 2000.

### Warren County

*Land Atlas and Plat Book, Warren County, Iowa, 1977.* Rockford, IL: Rockford Map, 2000.

### Washington County

*Land Atlas and Plat Book, Washington County, Iowa, 1978.* Rockford, IL: Rockford Map, 2000.

### Webster County

*Land Atlas and Plat Book, Webster County, Iowa, 1981.* Rockford, IL: Rockford Map, 2000.

### Winneshiek County

*Land Atlas and Plat Book, Winneshiek County, Iowa, 1980–1981.* Rockford, IL: Rockford Map, 2000.

*Winneshiek County, Iowa: Reprints of the 1874 Plat Book and Excerpts from the 1875 Iowa Illustrated Historical Atlas.* Decorah, IA: Decorah Genealogy Society and Winneshiek County Historical Society, 2000.

### Woodbury County

*Land Atlas and Plat Book, Woodbury County, Iowa, 1981.* Rockford, IL: Rockford Map, 2000.

*Worth County*

Land Atlas and Plat Book, Worth County, Iowa, 1981. Rockford, IL: Rockford Map, 2000.

## Kansas

*Atchison County*

Combined Standard Atlas of Atchison County, Kansas: 1887, 1903, 1925. Mt. Vernon, IN: Windmill, 2001.

*Bourbon County*

Illustrated Historical Atlas of Bourbon County, Kansas. Fort Scott: Old Fort Genealogical Society of Southeastern Kansas, 2001.

*Brown County*

Brown County, Kansas Atlas of 1887, by J. A. Ostertag and E. Ostertag. St. Joseph, MO: J. & E. Ostertag, 1995.

*Jefferson County*

Combined Standard Atlas of Jefferson County, Kansas, 1887, 1899, 1916. Mt. Vernon, IN: Windmill, 2001.

*Miami County*

Illustrated Historical Atlas of Miami County, Kansas. Paola, KS: Miami County Genealogical Society, 1996.

## Kentucky

*Bath County*

Atlas of Bath and Fleming Counties, Kentucky: From Actual Surveys, by J. M. Lathrop and J. H. Summers. Philadelphia, PA: D. J. Lake, 1995.
Index, 1884 Atlas, Bath County, Kentucky, by D. L. Heflin. Georgetown, KY: D. L. Heflin, 1995.

*Bracken County*

Atlas of Bracken and Pendleton Counties, Kentucky, by J. M. Lathrop and J. H. Summers. Philadelphia, PA: D. J. Lake, 1998.

*Fleming County*

Atlas of Bath and Fleming Counties, Kentucky: From Actual Surveys, by J. M. Lathrop and J. H. Summers. Philadelphia, PA: D. J. Lake, 1995.
Index, 1884 Atlas, Fleming County, Kentucky, by D. L. Heflin. Georgetown, KY: D. L. Heflin, 1995.

*Grant County*

Atlas of Grant County, Kentucky, 1858: Property Lines, Roads and Geographic Features Plotted from Deeds, Surveys, Wills and Court Orders of Grant County and Its Predecessor Counties, by T. H. Hutzelman. Erie, PA: T. H. Hutzelman, 1999.

*Lexington County*

Historic Lexington: An Atlas of Selected Landscapes. Lexington: University of Kentucky, 1998.

*Pendleton County*

Atlas of Bracken and Pendleton Counties, Kentucky, by J. M. Lathrop and J. H. Summers. Philadelphia, PA: D. J. Lake, 1998.

## Louisiana

*New Orleans*

Historic New Orleans Cemetery Lafayette: #1 atlas, by F. Hatfield and S. M. Perry. New Orleans, LA: Digital Cottage, 1997.
Older Americans Demographic Atlas. Part 2, Jefferson Parish, Louisiana. Metairie, LA: Jefferson Council on Aging, 1998.
Unfathomable City: A New Orleans Atlas, by R. Solnit. Berkeley: University of California Press, 2013.

## Maryland

*Baltimore County*

Baltimore County Atlas, 1877. Baltimore, MD: Baltimore County Historical Society, 2002.

*Carroll County*

1916 Atlas of Carroll County, Maryland: District Maps and Alphabetical Listing of Rural Landowners and Permanent Tenants. Westminster, MD: Carroll County Genealogical Society, 2002.

*Frederick County*

Atlas of Frederick County, Maryland, by D. J. Lake. Frederick, MD: Frederick County Landmarks Foundation, 1999.

*Prince George's County*

Atlas of Prince George's County, Maryland 1861: Adapted from Martenet's Map of Prince George's County, Maryland, by S. J. Martenet. Riverdale, MD: Prince George's County Historical Society, 1996.

## Massachusetts

*Barnstable County*

*Historical and Genealogical Atlas and Guide to Barnstable County, Mass. (Cape Cod)*, by M. H. Gibson. Falmouth, MA: Falmouth Genealogical Society, 1995.

## Michigan

*Alger County*

*Land Atlas and Plat Book, Alger County, Michigan.* Rockford, IL: Rockford Map, 2000.

*Allegan County*

*Allegan County Historical Atlas and Gazetteer*, by K. Lane. Douglas, MI: Pavilion, 1998.

*Barry County*

*Barry County Michigan: Land Atlas and Plat Book 2007.* Hastings, MI: J-AD Graphics, 2007.

*Bay County*

*Land Atlas and Plat Book, Bay County, Michigan.* Rockford, IL: Rockford Map, 2000.

*Benzie County*

*Land Atlas and Plat Book, Benzie County, Michigan.* Rockford, IL: Rockford Map, 2000.

*Branch County*

*Combined 1829–1929 Land Ownership Atlas of Branch County, Michigan: Showing Original Purchasers Starting in 1829, and Land Ownership in the Years of 1858, 1872, 1894, 1909, 1915 and 1929.* Coldwater, MI: Branch County Genealogical Society, 1996.

*Cass County*

*Atlas of Cass County, Michigan, 1872*, by D. J. Lake et al. Salt Lake City: Genealogical Society of Utah, 2005.

*Cheboygan County*

*Land Atlas and Plat Book, Cheboygan County, Michigan, 1989–1992.* Rockford, IL: Rockford Map, 2000.

*Chippewa County*

*Land Atlas and Plat Book, Chippewa County, Michigan.* Rockford, IL: Rockford Map, 2000.

*Clinton County*

*Land Atlas and Plat Book, Clinton County, Michigan.* Rockford, IL: Rockford Map, 2000.

*Delta County*

*Land Atlas and Plat Book, Delta County, Michigan.* Rockford, IL: Rockford Map, 2004.

*Dickinson County*

*Land Atlas and Plat Book, Dickinson County, Michigan.* Rockford, IL: Rockford Map, 2000.

*Genesee County*

*Land Atlas and Plat Book, Genesee County, Michigan.* Rockford, IL: Rockford Map, 2000.

*Gratiot County*

*1914 Atlas for Gratiot County, Michigan*, by H. Guernsey and C. Douglas. Ithaca, MI: Gratiot County Historical and Genealogical Society, 2007.

*Ingham County*

*Land Atlas and Plat Book, Ingham County, Michigan.* Rockford, IL: Rockford Map, 2000.

*Ionia County*

*Atlas of Ionia County, Michigan: From Recent and Actual Surveys and Records*, by L. P. Fox. Lake Odessa, MI: Ionia County Genealogical Society, 2002.

*Jackson County*

*Land Atlas and Plat Book, Jackson County, Michigan 1996.* Rockford, IL: Rockford Map, 1996.

*Kalamazoo County*

*Kalamazoo County, Michigan, Land Atlas and Plat Book, 1996.* Rockford, IL: Rockford Map, 1996.

*Kent County*

*Land Atlas and Plat Book, Kent County, Michigan.* Rockford, IL: Rockford Map, 2000.

*Keweenaw County*

*Land Atlas and Plat Book, Keweenaw County, Michigan.* Rockford, IL: Rockford Map, 2000.

*Lake County*

Land Atlas and Plat Book, Lake County, Michigan. Rockford, IL: Rockford Map, 2000.

*Marquette County*

Land Atlas and Plat Book, Marquette County, Michigan. Rockford, IL: Rockford Map, 2000.

*Monroe County*

1876 Township Plat Map Includes Index and 1880 Federal Census Index Includes Census of Berlin Township Monroe County Michigan, by R. Grassley. Monroe County, MI: Genealogical Society of Monroe County, 1995.

1876 Township Plat Map, Includes Index, and 1880 Federal Census Index, Includes Census of Whiteford Township, Monroe County, Michigan, by S. M. Bartlett. Monroe, MI: Genealogical Society of Monroe County, 1995.

County Atlas of Monroe, Michigan: From Recent and Actual Surveys and Records under the Superintendence of S. M. Bartlett. Monroe, MI: Genealogical Society of Monroe County, 1996.

*Oakland County*

Oakland County Street Atlas. Pontiac, MI: Oakland County Development and Planning Division, 1997.

*Ottawa County*

Land Atlas and Plat Book, Ottawa County, Michigan. Rockford, IL: Rockford Map, 2000.

*Sanilac County*

Land Atlas and Plat Book, Sanilac County, Michigan. Rockford, IL: Rockford Map, 2004.

*Schoolcraft County*

Land Atlas and Plat Book, Schoolcraft County, Michigan. Rockford, IL: Rockford Map, 2000.

*Shiawassee County*

Combined 1859 Wall Map, 1875 Atlas, 1895 Plat Book Atlas and 1915 Atlas of Shiawassee County Michigan, Indexed. Owosso, MI: Shiawassee County Historical Society, 2000.

*Washtenaw County*

Land Atlas and Plat Book, Washtenaw County, Michigan. Rockford, IL: Rockford Map, 2005.

**Minnesota**

*Benton County*

Land Atlas and Plat Book, Benton County, Minnesota, 1982. Rockford, IL: Rockford Map, 2000.

*Big Stone County*

Atlas and Plat Book, Big Stone County, Minnesota, 1975. Rockford, IL: Rockford Map, 2000.

*Blue Earth County*

Blue Earth County, Minnesota: Emergency Road Atlas. St. Cloud, MN: GeoComm, 2002.

Land Atlas and Plat Book, Blue Earth County, Minnesota, 1983. Rockford, IL: Rockford Map, 2000.

*Brown County*

Atlas of Brown County Minnesota: Containing Maps, Plats of the Townships, Rural Directory, Pictures of Farms and Families, Articles about History, Etc. Battle Lake, MN: Title Atlas, 1999.

Land Atlas and Plat Book, Brown County, Minnesota, 1976. Rockford, IL: Rockford Map, 2000.

*Carlton County*

Land Atlas and Plat Book, Carlton County, Minnesota: 2005. Rockford, IL: Rockford Map, 2005.

*Carver County*

Land Atlas and Plat Book, Carver County, Minnesota, 1987. Rockford, IL: Rockford Map, 2000.

*Cass County*

Land Atlas and Plat Book, Cass County, Minnesota, 1985. Rockford, IL: Rockford Map, 2000.

*Chisago County*

Chisago County Land Atlas and Plat Book. St. Cloud, MN: Cloud Cartographics, 1999.

Land Atlas and Plat Book, Chisago County, Minnesota, 1985. Rockford, IL: Rockford Map, 2000.

*Clay County*

Land Atlas and Plat Book, Clay County, Minnesota, 1980–1981. Rockford, IL: Rockford Map, 2000.

Faribault County

*Land Atlas and Plat Book, Faribault County, Minnesota, 1979.* Rockford, IL: Rockford Map, 2000.

Fillmore County

*Land Atlas and Plat Book, Fillmore County, Minnesota, 1983–85.* Rockford, IL: Rockford Map, 2000.

Freeborn County

*Land Atlas and Plat Book, Freeborn County, Minnesota, 1987.* Rockford, IL: Rockford Map, 2000.

Goodhue County

*Land Atlas and Plat Book, Goodhue County, Minnesota.* Rockford, IL: Rockford Map, 2000.

Grant County

*Land Atlas and Plat Book, Grant County, Minnesota, 1978.* Rockford, IL: Rockford Map, 2000.

Houston County

*Land Atlas and Plat Book, Houston County, Minnesota, 1984.* Rockford, IL: Rockford Map, 2000.

Isanti County

*Land Atlas and Plat Book, Isanti County, Minnesota, 1982.* Rockford, IL: Rockford Map, 2000.

Jackson County

*Atlas and Plat Book, Jackson County, Minnesota, 1975.* Rockford, IL: Rockford Map, 2000.

Kanabec County

*Land Atlas and Plat Book, Kanabec County, Minnesota, 1985.* Rockford, IL: Rockford Map, 2000.

LeSueur County

*Land Atlas and Plat Book, LeSueur County, Minnesota, 1980–1981.* Rockford, IL: Rockford Map, 2000.

Lyon County

*Land Atlas and Plat Book, Lyon County, Minnesota, 1979.* Rockford, IL: Rockford Map, 2000.

Mahnomen County

*Land Atlas and Plat Book, Mahnomen County, Minnesota, 1976.* Rockford, IL: Rockford Map, 2000.

McLeod County

*Atlas and Plat Book, McLeod County, Minnesota, 1975.* Rockford, IL: Rockford Map, 2000.

Meeker County

*Atlas and Plat Book, Meeker County, Minnesota, 1975.* Rockford, IL: Rockford Map, 2000.

Mille Lacs County

*Land Atlas and Plat Book, Mille Lacs County, Minnesota, 1978.* Rockford, IL: Rockford Map, 2000.

Nicollet County

*Land Atlas and Plat Book, Nicollet County, Minnesota, 1978.* Rockford, IL: Rockford Map, 2000.

Oakdale Township

*History of Oakdale Township: Vol. 1*, by T. G. Armstrong. Oakdale, MN: Oakdale Lake Elmo Historical Society, 1996.

Olmsted County

*Land Atlas and Plat Book, Olmsted County, Minnesota, 1983–84.* Rockford, IL: Rockford Map, 2000.

Otter Tail County

*Otter Tail County Emergency Services Atlas.* St. Cloud, MN: Cloud Cartographics, 1999.

Pennington County

*Land Atlas and Plat Book, Pennington County, Minnesota, 1976.* Rockford, IL: Rockford Map, 2000.

Pipestone County

*Land Atlas and Plat Book, Pipestone County, Minnesota, 1977.* Rockford, IL: Rockford Map, 2000.

Pope County

*Land Atlas and Plat Book, Pope County, Minnesota, 1977.* Rockford, IL: Rockford Map, 2000.

*Red Lake County*

Land Atlas and Plat Book, Red Lake County, Minnesota, 1976. Rockford, IL: Rockford Map, 2000.

*Redwood County*

Land Atlas and Plat Book, Redwood County, Minnesota, 1977. Rockford, IL: Rockford Map, 2000.

*Rock County*

Land Atlas and Plat Book, Rock County, Minnesota, 1977. Rockford, IL: Rockford Map, 2000.

*Rose Township*

Historical Atlas of Early Rose Township: Includes Falcon Heights, Lauderdale, Roseville and Parts of St. Paul, by S. Fudenberg. Roseville, MN: S. Fudenberg, 1996.

*Roseau County*

Atlas of Roseau County, Minnesota: Containing Maps, Plats of the Townships, Rural Directory, Pictures of Farms and Families, Articles about History, Etc., by L. Wood. Battle Lake, MN: Title Atlas, 1998.

*Sherburne County*

Land Atlas and Plat Book, Sherburne County, Minnesota, 1980. Rockford, IL: Rockford Map, 2000.

*Sibley County*

Land Atlas and Plat Book, Sibley County, Minnesota, 1978. Rockford, IL: Rockford Map, 2000.

*St. Louis County*

Land Atlas and Plat Book, St. Louis County, Minnesota: 2012. Duluth, MN: St. Louis County Planning and Community Development Department, 2012.

*Stearns County*

Stearns County, Minnesota: 2005 Plat Book and Map Atlas: 150 Year Anniversary Commemorative Issue Celebrating the Stearns County Sesquicentennial. St. Cloud, MN: Stearns County Public Works Department, 2005.

*Steele County*

Land Atlas and Plat Book, Steele County, Minnesota, 1983. Rockford, IL: Rockford Map, 2000.

*Stevens County*

Land Atlas and Plat Book, Stevens County, Minnesota. Rockford, IL: Rockford Map, 2000.

*Todd County*

Land Atlas and Plat Book, Todd County, Minnesota, 1981. Rockford, IL: Rockford Map, 2000.

*Wadena County*

Land Atlas and Plat Book, Wadena County, Minnesota, 1977. Rockford, IL: Rockford Map, 2000.

*Waseca County*

Land Atlas and Plat Book, Waseca County, Minnesota, 1976. Rockford, IL: Rockford Map, 2000.

*Washington County*

Land Atlas and Plat Book, Washington County, Minnesota, 1986. Rockford, IL: Rockford Map, 2000.

*Yellow Medicine County*

Atlas of Yellow Medicine County, Minnesota: Containing Maps, Plats of the Townships, Rural Directory, Pictures of Farms and Families, Articles about History, Etc. Battle Lake, MN: Title Atlas, 2001.

## Mississippi

*Hancock County*

Hancock County, Mississippi Hurricane Surge Atlas. Washington, DC: Federal Emergency Management Agency, 2001.

*Harrison County*

Harrison County, Mississippi Hurricane Surge Atlas. Washington, DC: Federal Emergency Management Agency, 2001.

*Jackson County*

Jackson County, Mississippi Hurricane Surge Atlas. Washington, DC: Federal Emergency Management Agency, 2001.

*Lauderdale County*

Land Atlas and Plat Book, Lauderdale County, Mississippi. Rockford, IL: Rockford Map, 2000.

*Wayne County*

*Land Atlas and Plat Book, Wayne County, Mississippi.* Rockford, IL: Rockford Map, 2000.

## Missouri

*Adair County*

*Land Atlas and Plat Book, Adair County, Missouri, 1981.* Rockford, IL: Rockford Map, 2000.

*Andrew County*

*Land Atlas and Plat Book, Andrew County, Missouri, 1987.* Rockford, IL: Rockford Map, 2000.

*Atchison County*

*Land Atlas and Plat Book, Atchison County, Missouri, 1977.* Rockford, IL: Rockford Map, 2000.

*Audrain County*

*Land Atlas and Plat Book, Audrain County, Missouri, 1986.* Rockford, IL: Rockford Map, 2000.

*Barry County*

*Land Atlas and Plat Book, Barry County, Missouri, 1984.* Rockford, IL: Rockford Map, 2000.

*Barton County*

*Land Atlas and Plat Book, Barton County, Missouri, 1978.* Rockford, IL: Rockford Map, 2000.

*Bates County*

*Land Atlas and Plat Book, Bates County, Missouri, 1982.* Rockford, IL: Rockford Map Publihers, 2000.

*Benton County*

*Land Atlas and Plat Book, Benton County, Missouri, 1985.* Rockford, IL: Rockford Map, 2000.

*Boone County*

*Standard Atlas of Boone County, Missouri.* Columbia: Genealogical Society of Central Missouri, 2006.

*Caldwell County*

*Atlas of Caldwell County, Missouri 1917.* Kingston, MO: Caldwell County Historical Society, 2010.

*Cape Girardeau County*

*Atlas and Plat Book, Cape Girardeau County, Missouri, 1976.* Rockford, IL: Rockford Map, 2000.

*Carroll County*

*Land Atlas and Plat Book, Carroll County, Missouri, 1985.* Rockford, IL: Rockford Map, 2000.

*Standard Atlas of Carroll County, Missouri: Including a Plat Book of the Villages, Cities and Townships of the County.* Carroll County, MO: Carroll County Historical Society, 1999.

*Cedar County*

*1908 Atlas and Plat Book of Cedar County, Missouri,* by C. Snider et al. Cedar County, MO: Geo A. Ogle, 1998.

*Land Atlas and Plat Book, Cedar County, Missouri, 1980.* Rockford, IL: Rockford Map, 2000.

*Standard Atlas of Cedar County, Missouri: Including a Plat Book of the Villages, Cities and Townships of the County, Map of the State, United States and World. Analysis of the System of U.S. Land Surveys, Digest of the System of the Civil Government, Etc.* Springfield, MO: Greene County Archives and Records Center, 1998.

*Chariton County*

*Land Atlas and Plat Book, Chariton County, Missouri, 1985.* Rockford, IL: Rockford Map, 2000.

*Christian County*

*1912 Plat Book of Christian County, Missouri (Part of Greene County up to 1859),* by C. Snider et al. Springfield, MO: Greene County Archives and Records Center, 1998.

*Clark County*

*Land Atlas and Plat Book, Clark County, Missouri, 1985.* Rockford, IL: Rockford Map, 2000.

*Clay County*

*Clay County Historical Atlas.* Liberty, MO: Clay County Archives and Historical Library, 2011.

*Clinton County*

*Land Atlas and Plat Book, Clinton County, Missouri, 1977.* Rockford, IL: Rockford Map, 2000.

*Cole County*

*Land Atlas and Plat Book, Cole County, Missouri, 1985.* Rockford, IL: Rockford Map, 2000.

*Cooper County*

Land Atlas and Plat Book, Cooper County, Missouri, 1978. Rockford, IL: Rockford Map, 2000.

*Crawford County*

Land Atlas and Plat Book, Crawford County, Missouri, 1982. Rockford, IL: Rockford Map, 2000.

*Dade County*

Standard Atlas of Dade County, Missouri: Including a Plat Book of the Villages, Cities and Townships of the County, Map of the State, United States and World. Analysis of the System of U.S. Land Surveys, Digest of the System of the Civil Government, Etc. Springfield, MO: Greene County Archives and Records Center, 1998.

*Daviess County*

Land Atlas and Plat Book, Daviess County, Missouri, 1985. Rockford, IL: Rockford Map, 2000.

*DeKalb County*

Land Atlas and Plat Book, DeKalb County, Missouri, 1980. Rockford, IL: Rockford Map, 2000.

*Dunklin County*

Land Atlas and Plat Book, Dunklin County, Missouri, 1988. Rockford, IL: Rockford Map, 2000.

*Gasconade County*

Land Atlas and Plat Book, Gasconade County, Missouri, 1983. Rockford, IL: Rockford Map, 2000.

*Grundy County*

Land Atlas and Plat Book, Grundy County, Missouri, 1978. Rockford, IL: Rockford Map, 2000.

*Holt County*

Illustrated Historical Atlas Map, Holt County MO: Combined Atlases of Holt Co., Missouri 1877, 1898, 1918, and 1952. Maysville, MO: Holt County Historical Society, 2000.

*Howard County*

Land Atlas and Plat Book, Howard County, Missouri, 1978. Rockford, IL: Rockford Map, 2000.

*Howell County*

Atlas and Plat Book, Howell County, Missouri, 1979. Rockford, IL: Rockford Map, 2000.

*Iron County*

Atlas and Plat Book, Iron County, Missouri, 1978. Rockford, IL: Rockford Map, 2000.

*Jefferson County*

Hallemann's Interpretation of the 1898 Standard Atlas of Jefferson County, Missouri, by D. Hallemann. Jefferson County, MO: Jefferson County Historical Society, 1998.

Standard Atlas of Jefferson County Missouri: Including a Plat Book of the Villages, Cities and Townships of the County 1898, by B. Gagnon. De Soto, MO: Jefferson County Historical Society, 2004.

*Johnson County*

Land Atlas and Plat Book, Johnson County, Missouri, 1986. Rockford, IL: Rockford Map, 2000.

*Lawrence County*

Illustrated Historical Atlas of Lawrence County, Missouri. Springfield, MO: Greene County Archives and Records Center, 1999.

*Lewis County*

1996 Twin Cities Metropolitan Area Street Atlas. Williamston, MI: Universal Maps, 1996.

Land Atlas and Plat Book, Lewis County, Missouri, 1985. Rockford, IL: Rockford Map, 2000.

*Lincoln County*

Atlas and Plat Book of Lincoln County, Missouri: Containing Outline Map of the County, Graveled Road Map of County, Plats of All the Townships with Owners' Names, Missouri State Map, Map of the United States, Map of the World, by C. H. Vice. Des Moines, IA: Kenyon, 2015.

Land Atlas and Plat Book, Lincoln County, Missouri, 1983. Rockford, IL: Rockford Map, 2000.

New Atlas of Lincoln County, Missouri, 1938, by A. J. Brown. Troy, MO: Review, 2014.

Standard Atlas of Lincoln County, Missouri: Including a Plat Book of the Villages, Cities and Townships of the County, Map of the State, United States and World. Chicago: Geo. A. Ogle, 2014.

*Linn County*

Land Atlas and Plat Book, Linn County, Missouri, 1981. Rockford, IL: Rockford Map, 2000.

*Livingston County*

Land Atlas and Plat Book, Livingston County, Missouri, 1985. Rockford, IL: Rockford Map, 2000.

*Macon County*

Land Atlas and Plat Book, Macon County, Missouri, 1986. Rockford, IL: Rockford Map, 2000.

*Madison County*

Atlas and Plat Book, Madison County, Missouri, 1976. Rockford, IL: Rockford Map, 2000.

*Maries County*

Atlas and Plat Book, Maries County, Missouri, 1976. Rockford, IL: Rockford Map, 2000.

*Mercer County*

Illustrated Historical Atlas of Mercer County, Missouri: 1877. Princeton, MO: Mercer County Genealogical and Historical Society, 2010.

*Miller County*

Land Atlas and Plat Book, Miller County, Missouri, 1979. Rockford, IL: Rockford Map, 2000.

*Mississippi County*

Land Atlas and Plat Book, Mississippi County, Missouri, 1981–1982. Rockford, IL: Rockford Map, 2000.

*Moniteau County*

Descriptive Atlas of Moniteau County, Missouri 1900. Chicago: Acme, 2000.
Land Atlas and Plat Book, Moniteau County, Missouri, 1982. Rockford, IL: Rockford Map, 2000.
Standard Atlas of Moniteau County, Missouri: Including a Plat Book of the Villages, Cities and Townships of the County. California, MO: Moniteau County Historical Society, 1997.

*Monroe County*

Land Atlas and Plat Book, Monroe County, Missouri, 1980–1981. Rockford, IL: Rockford Map, 2001.

*Montgomery County*

Land Atlas and Plat Book, Montgomery County, Missouri, 1984. Rockford, IL: Rockford Map, 2001.

*Morgan County*

Land Atlas and Plat Book, Morgan County, Missouri, 1983. Rockford, IL: Rockford Map, 2000.

*Mormon County*

Northeast of Eden: A Historical Atlas of Missouri's Mormon County, by J. Hamer. Mirabile, MO: Far West Cultural Center, 2004.

*New Madrid County*

Land Atlas and Plat Book, New Madrid County, Missouri, 1984. Rockford, IL: Rockford Map, 2000.

*Pemiscot County*

Land Atlas and Plat Book, Pemiscot County, Missouri, 1978. Rockford, IL: Rockford Map, 2000.

*Perry County*

Atlas and Plat Book, Perry County, Missouri, 1975. Rockford, IL: Rockford Map, 2000.

*Phelps County*

Land Atlas and Plat Book, Phelps County, Missouri, 1985. Rockford, IL: Rockford Map, 2000.

*Pike County*

Land Atlas and Plat Book, Pike County, Missouri, 1982. Rockford, IL: Rockford Map Publishers, 2000.

*Platte County*

Standard Atlas of Platte County, Missouri [1907]: Including a Plat Book of the Villages, Cities and Townships . . . Patron's Directory, Reference Business Directory. Salt Lake City: Genealogical Society of Utah, 1995.

*Polk County*

Plat Book of Polk County Missouri. Springfield, MO: Greene County Archives and Records Center, 1997.

*Ralls County*

Land Atlas and Plat Book, Ralls County, Missouri, 1981. Rockford, IL: Rockford Map, 2000.

*Randolph County*

Land Atlas and Plat Book, Randolph County, Missouri, 1986. Rockford, IL: Rockford Map, 2000.

*Ray County*

*Land Atlas and Plat Book, Ray County, Missouri, 1982.* Rockford, IL: Rockford Map, 2000.

*Reynolds County*

*Atlas and Plat Book, Reynolds County, Missouri, 1979.* Rockford, IL: Rockford Map, 2000.

*Scott County*

*Land Atlas and Plat Book, Scott County, Missouri, 1982.* Rockford, IL: Rockford Map, 2000.

*Shelby County*

*Land Atlas and Plat Book, Shelby County, Missouri, 1981.* Rockford, IL: Rockford Map, 2000.

*St. Charles County*

*Land Atlas and Plat Book, St. Charles County, Missouri, 1986.* Rockford, IL: Rockford Map, 2000.

*St. Clair County*

*Land Atlas and Plat Book, St. Clair County, Missouri, 1986.* Rockford, IL: Rockford Map, 2000.

*St. Francois County*

*Land Atlas and Plat Book, St. Francois County, Missouri, 1980.* Rockford, IL: Rockford Map, 2000.

*Ste. Genevieve County*

*Land Atlas and Plat Book, Ste. Genevieve County, Missouri, 1985.* Rockford, IL: Rockford Map, 2000.

*Stoddard County*

*Land Atlas and Plat Book, Stoddard County, Missouri, 1979.* Rockford, IL: Rockford Map, 2000.

*Stone County*

*Land Atlas and Plat Book, Stone County, Missouri, 1985.* Rockford, IL: Rockford Map, 2000.

*Texas County*

*Atlas and Plat Book, Texas County, Missouri, 1978.* Rockford, IL: Rockford Map, 2000.

*Warren County*

*Land Atlas and Plat Book, Warren County, Missouri, 1984.* Rockford, IL: Rockford Map, 2000.

*Webster County*

*Land Atlas and Plat Book, Webster County, Missouri, 1986.* Rockford, IL: Rockford Map, 2000.

**Nebraska**

*Douglas and Sarpy Counties*

*Atlas of Douglas and Sarpy Counties, Nebraska, 1920: Containing Maps of the Townships and Farmer's Directory.* Puyallup, WA: Anderson, 2001.

*Richardson County*

*Atlas of Richardson County, Nebraska, 1896.* Lincoln: Nebraska State Genealogical Society, 2001.

**New Hampshire**

*Belknap County*

*Old Maps of Belknap County, New Hampshire, in 1892.* Fryeburg, ME: Write Stuff, 2006.

*Carroll County*

*Old Maps of Carroll County, New Hampshire in 1892.* Fryeburg, ME: Write Stuff, 2006.

*Cheshire County*

*Old Maps of Cheshire County, New Hampshire in 1892.* Fryeburg, ME: Write Stuff, 2006.

*Grafton County*

*Old Maps of Grafton County, New Hampshire in 1892.* Fryeburg, ME: Write Stuff, 2006.

*Merrimack County*

*Old Maps of Merrimack County, New Hampshire, in 1892.* Fryeburg, ME: Write Stuff, 2006.

*Rockingham County*

*Old Maps of Rockingham County, New Hampshire in 1892.* Fryeburg, ME: Write Stuff, 2006.

*Sullivan County*

*Old Maps of Sullivan County, New Hampshire in 1892.* Fryeburg, ME: Write Stuff, 2006.

## New Jersey

*Bergen County*

*Every-Name Index, Atlas of Bergen County, New Jersey*, by A. H. Walker. Englewood, NJ: Bergen Historic Books, 1996.

*Burlington County*

*Combination Atlas Map of Burlington County, New Jersey: Compiled, Drawn and Published from Personal Examinations and Surveys*, by J. D. Scott and M. M. Pernot. Pemberton, NJ: Burlington County Historic Trust, 2002.

*Gloucester County*

*Combination Atlas Map of Salem and Gloucester Counties, New Jersey: Compiled, Drawn and Published from Personal Examinations and Surveys*, by Everts & Stewart. Woodbury, NJ: Gloucester County Historical Society, 2001.

*Salem County*

*Combination Atlas Map of Salem and Gloucester Counties, New Jersey: Compiled, Drawn and Published from Personal Examinations and Surveys*, by Everts & Stewart. Woodbury, NJ: Gloucester County Historical Society, 2001.

*Somerset County*

*Atlas of Somerset County, New Jersey: From Recent and Actual Surveys and Records*. Somerville, NJ: Somerset County Historical Society, 1995.

## New Mexico

*Albuquerque*

*Address Atlas*. Albuquerque, NM: AGIS, 1999.
*Environmental Justice Atlas and Data Book for the Albuquerque Metropolitan Planning Area*. Albuquerque: Middle Rio Grande Council of Governments of New Mexico, 2001.

*Socorro County*

*A Regional Geochemical Atlas for Part of Socorro County, New Mexico*, by J. M. Watrus. Socorro: New Mexico Institute of Mining and Technology, 1998.

## New York

*Chautauqua County*

*1903 Chautauqua County Atlas*. Chautauqua County, NY: Chautauqua County Historical and Genealogical Society, 2002.

*Jefferson County*

*Land Atlas and Plat Book, Jefferson County, New York, 1987*. Rockford, IL: Rockford Map, 2000.

*New York City*

*Historical Atlas of New York City: A Visual Celebration of 400 Years of New York City's History*, by E. Homberger. New York: Henry Holt, 1998, 2005, 2016.
*Nonstop Metropolis: A New York City Atlas*, by R. Solnit and J. Jelly-Schapiro. Oakland: University of California Press, 2016.

## North Carolina

*Mecklenburg County*

*Charlotte-Mecklenburg Street Atlas*. Blue Bell, PA: Kappa Map Group, 1996.

*Orange County*

*The Land Grant Atlas of Old Orange County*, by M. Chilton. Carrboro, NC: Hand Made Books, 2012.

*Union County*

*Union County Street Atlas: Complete Street Guide for the Entire County*. Charlotte, NC: Map Shop, 1998.

*Wilmington–New Hanover County*

*Wilmington–New Hanover County Atlas*. Columbia, SC: Accurate Maps, 1997.

## North Dakota

*Morton County*

*Standard Atlas of Morton County, North Dakota*. Bismarck, ND: Bismarck Mandan Historical and Genealogical Society, 1999.

## Ohio

*Adams County*

*Caldwell's Illustrated Historical Atlas of Adams County, Ohio: 1797–1880*, by J. A. Caldwell. Mt. Vernon, IN: Windmill, 1997.

*Athens County*

*Centennial Atlas of Athens County, Ohio: Containing Complete Maps of the County and Each of Its Townships and Villages*, by F. W. Bush. Athens: Ohio University Press, 1996.

### Clark County

*County of Clark, Ohio, an Imperial Atlas and Art Folio, including Chronological Chart, Statistical Tables, and Description of Surveys, 1894.* Milford, OH: Little Miami, 2002.

### Clermont County

*Clermont County, Ohio, 1891 Atlas and History: With Comprehensive Index,* by A. M. Whitt and D. J. Lake. Mt. Vernon, IN: Windmill, 2000.

### Clinton County

*Combined 1903 Atlas and 1859 and 1953 Wall Maps of Clinton County, Ohio.* Mt. Vernon, IN: Whippoorwill, 1999.

*Illustrated Historical Atlas of Clinton County, Ohio,* by Lake, Griffing, and Stevenson. Mt. Vernon, IN: Windmill, 1996.

### Coshocton County

*Atlas of Coshocton County, Ohio: From Actual Surveys by and under the Directions of D. J. Lake,* by C. O. Titus and D. J. Lake. Coshocton: Ohio Genealogical Society, 1997.

### Cuyahoga County

*Atlas of Cuyahoga County, Ohio from Actual Surveys by and under the Directions of D. J. Lake, C.E.: From the Cleveland Memory Project,* by D. J. Lake. Philadelphia: Titus, Simmons & Titus, 2004.

*Atlas of Townships of Coshocton County, Ohio: Circa 1938,* by M. Kinkade. Coshocton, OH: Coshocton County Genealogical Society, 1999.

### Defiance County

*Historical Atlas Illustrated 1876,* by C. H. Jones and T. F. Hamilton. Defiance: Ohio Genealogical Society, 2007.

### Fulton County

*Pioneers around Delta, Ohio,* by V. K. Seaman. Denver, CO: V. G. Miller, 1997.

### Hocking County

*Atlas and Plat Book, Hocking County, Ohio, 1975.* Rockford, IL: Rockford Map, 2000.

### Jackson County

*1875 Atlas of Jackson County, Ohio,* by D. J. Lake. Jackson: Jackson County Chapter of the Ohio Genealogical Society, 2011.

### Lucas County

*Illustrated Historical Atlas of Lucas and Part of Wood Counties, Ohio.* Bowie, MD: Ohio Genealogical Society, 1999.

### Madison County

*Caldwell's Atlas of Madison Co., Ohio: From Actual Surveys,* by H. Cring and J. A. Caldwell. Condit, OH: J. A. Caldwell, 2014.

### Miami County

*Miami County, Ohio 1931 Atlas.* Troy, OH: Miami County Historical and Genealogical Society, 2003.

### Montgomery County

*Montgomery County, Ohio, Cemetery Atlas: By Townships,* by D. R. Bowman, A. Johnson, and R. E. Johnson. Dayton: Montgomery County Chapter of the Ohio Genealogical Society, 2002.

### Murrow County

*Morrow County, Ohio Historical Atlas: A Combined Reprint of the 1857 Map of Harwood and Watson; 1871 Atlas by C. O. Titus and the 1901 Atlas by Thad E. Buck.* Mount Gilead: Ohio Genealogical Society, 2000.

### Orange County

*Land Grant Atlas of Old Orange County,* by M. Chilton. Carrboro, NC: Hand Made Books, 2012.

### Putnam County

*History of Putnam County, Ohio, 1880: Illustrated, Containing Outline Map, Fifteen Farm Maps and a History of the County; Lithographic Views of Buildings—Public and Private; Portraits of Prominent Men; General Statistics; Miscellaneous Matters.* Mt. Vernon, IN: Windmill, 1995.

### Ross County

*Plat Book and Land Atlas Ross County, Ohio: Including Township Maps with Landowner's Name County and City Maps.* Kettering, OH: Mercury, 1995.

### Sandusky County

*Sandusky County, Ohio Landowner Atlas, 1903.* Fremont, OH: Sandusky County Kin Hunters, 2004.

*Union County*

*Atlas of Union County, Ohio: From Records and Original Surveys*, by A. S. Mowry. Washington, DC: Harrison, Sutton & Hare, 1998.

*Vinton County*

*Land Atlas and Plat Book, Vinton County, Ohio, 1987.* Rockford, IL: Rockford Map, 2000.

*Plat Book and Land Atlas, 2001 Vinton County, Ohio: Including Township Maps with Landowner's Name, County and City Maps.* Clayton, OH: Mercury, 2001.

*Plat Book and Land Atlas 2007 Vinton County, Ohio.* Clayton, OH: Mercury, 2007

*Wood County*

*Wood County, Ohio Atlas: Landowner Maps 1871 and 1858, Tract Book 1830–1850, Road and River Tracts, Canal Land Purchases.* Bowling Green, OH: Ohio Genealogical Society, 1999.

## Oklahoma

*Muskogee County*

*Land Atlas and Plat Book, Muskogee County, Oklahoma, 1981.* Rockford, IL: Rockford Map, 2000.

## Oregon

*Multnomah County*

*Portland Railroad Atlas: 1963 and 2010.* Portland, OR: M. C. Byrnes, 2011.

## Pennsylvania

*Adams County*

*Atlas and Plat Book, Adams County, Pennsylvania, 1977.* Rockford, IL: Rockford Map, 2000.

*Allegheny County*

*Atlas of Pittsburgh.* Pittsburgh: Department of City Planning, 1995.

*Armstrong County*

*Land Atlas and Plat Book, Armstrong County, Pennsylvania, 1985.* Rockford, IL: Rockford Map, 2000.

*Beaver County*

*Land Atlas and Plat Book, Beaver County, Pennsylvania, 1988.* Rockford, IL: Rockford Map, 2000.

*Berks County*

*Early Landowners of Pennsylvania: Atlas of Township Warrantee Maps of Berks County, PA*, by S. C. MacInnes. Apollo, PA: Closson, 2006.

*Township Map of Berks County, Pennsylvania, from Actual Surveys*, by L. Fagan and H. F. Bridgens. Reading, PA: Berks County Genealogical Society, 2006.

*Blair County*

*Land Atlas and Plat Book, Blair County, Pennsylvania, 1986.* Rockford, IL: Rockford Map, 2000.

*Bradford County*

*Land Atlas and Plat Book, Bradford County, Pennsylvania, 1989.* Rockford, IL: Rockford Map, 2000.

*Cambria County*

*Land Atlas and Plat Book, Cambria County, Pennsylvania.* Rockford, IL: Rockford Map, 1997.

*Cameron County*

*Land Atlas and Plat Book, Cameron County, Pennsylvania, 1989.* Rockford, IL: Rockford Map, 2000.

*Centre County*

*Land Atlas and Plat Book, Centre County, Pennsylvania, 1988.* Rockford, IL: Rockford Map, 2000.

*Clarion County*

*Caldwell's Illustrated, Historical, Combination Atlas of Clarion County, Pennsylvania: From Actual Surveys*, by J. A. Caldwell and H. Cring. Butler, PA: Mechling Associates, 1998.

*Land Atlas and Plat Book, Clarion County, Pennsylvania, 1995.* Rockford, IL: Rockford Map, 2000.

*Clearfield County*

*Land Atlas and Plat Book, Clearfield County, Pennsylvania, 1995.* Rockford, IL: Rockford Map, 1995.

*Clinton County*

*1862 Historical Atlas of Clinton County, PA: Originally Published as a Wall Map by Way, Palmer & Company*, by H. F. Walling. Lock Haven, PA: Clinton County Genealogical Society, 2005.

*Elk County*

*Land Atlas and Plat Book, Elk County, Pennsylvania, 1990.* Rockford, IL: Rockford Map, 2000.

*Erie County*

*Land Atlas and Plat Book, Erie County, Pennsylvania, 1987.* Rockford, IL: Rockford Map, 2000.

*Fayette County*

*Atlas and Plat Book, Fayette County, Pennsylvania, 1978.* Rockford, IL: Rockford Map, 2000.

*Forest County*

*Land Atlas and Plat Book, Forest County, Pennsylvania.* Rockford, IL: Rockford Map, 1995.

*Greene County*

*Early Landowners of Pennsylvania: Atlas of Township Warrantee Maps of Greene County, PA*, by S. C. MacInnes and A. MacInnes. Apollo, PA: Closson, 2005.
*Land Atlas and Plat Book, Greene County, Pennsylvania, 1985.* Rockford, IL: Rockford Map, 2000.

*Indiana County*

*Atlas of Indiana Co., Pennsylvania.* Indiana, PA: Historical and Genealogical Society of Indiana County, 1999.

*Jefferson County*

*Land Atlas and Plat Book, Jefferson County, Pennsylvania, 1989.* Rockford, IL: Rockford Map, 2000.

*Lancaster County*

*Illustrated Historical Atlas of Lancaster County.* East Petersburg, PA: Historic Arts, 2006.
*Lancaster Co. Pennsylvania: Street Atlas.* Blue Bell, PA: Kappa Map Group, 2012.

*Lawrence County*

*Land Atlas and Plat Book, Lawrence County, Pennsylvania, 1987.* Rockford, IL: Rockford Map, 2000.

*Lycoming County*

*Land Atlas and Plat Book, Lycoming County, Pennsylvania, 1986.* Rockford, IL: Rockford Map, 2000.

*McKean County*

*Land Atlas and Plat Book, McKean County, Pennsylvania, 1986.* Rockford, IL: Rockford Map, 2000.

*Northampton County*

*Upper Mount Bethel Twp., Northampton County, Pennsylvania Beer's Atlas, 1874.* Philadelphia: D. G. Beers, 2002.

*Potter County*

*Land Atlas and Plat Book, Potter County, Pennsylvania, 1986.* Rockford, IL: Rockford Map, 2000.

*Susquehanna County*

*Land Atlas and Plat Book, Susquehanna County, Pennsylvania, 1989.* Rockford, IL: Rockford Map, 2000.

*Warren County*

*Land Atlas and Plat Book, Warren County, Pennsylvania, 1989.* Rockford, IL: Rockford Map, 2000.

*Washington County*

*Early Landowners of Pennsylvania: Atlas of Township Warrantee Maps of Washington County, PA*, by S. C. MacInnes. Apollo, PA: Closson, 2004.
*Land Atlas and Plat Book, Washington County, Pennsylvania, 1989.* Rockford, IL: Rockford Map, 2000.
*Pennsylvania Historic Towns of Washington County*, by P. Connors. Monongahela, PA: Historic Towns, 1996.

*Wayne County*

*Land Atlas and Plat Book, Wayne County, Pennsylvania, 1989.* Rockford, IL: Rockford Map, 2000.

*Westmoreland County*

*Early Landowners of Pennsylvania: Atlas of Township Patent Maps of Westmoreland County, PA*, by S. C. MacInnes and A. MacInnes. Apollo, PA: Closson, 2007.

*Wyoming County*

*Land Atlas and Plat Book, Wyoming County, Pennsylvania, 1988.* Rockford, IL: Rockford Map, 2000.

*York County*

*Atlas and Plat Book, York County, Pennsylvania, 1976.* Rockford, IL: Rockford Map, 2000.

## South Carolina

*Newberry County*

*The Dutch Fork: An Atlas of the Dutch Fork of Newberry District, S.C*, by C. W. Nichols. Fernandina Beach, FL: Wolfe, 2001.

## South Dakota

*Campbell County*

*Standard Atlas of Campbell County, South Dakota: Including a Plat Book of the Villages, Cities and Townships of the County, Map of the State, United States and World*, by J. S. Huber. Bowdle, SD: Prairie Family, 2002.

*Day County*

*Standard Atlas of Day County, South Dakota: Including a Plat Book of the Villages, Cities and Townships of the County*, by J. S. Huber. Bowdle, SD: Prairie Family, 2002.

*Edmunds County*

*Atlas of Edmunds County, South Dakota: Compiled and Drawn from Official Records and Special Survey*, by J. S. Huber. Bowdle, SD: Prairie Family, 2001.

*Faulk County*

*Standard Atlas of Faulk County, South Dakota, Including a Plat Book*, by J. S. Huber. Bowdle, SD: Prairie Family, 2000.

*Lincoln County*

*Standard Atlas of Lincoln County, South Dakota: Including a Plat Book of the Villages, Cities and Townships of the County . . . Patrons Directory, Reference Business Directory*, by J. S. Huber. Bowdle, SD: Prairie Family, 2004.

*McPherson County*

*Standard Atlas of McPherson County, South Dakota: Including a Plat Book of the Villages, Cities and Townships of the County . . . Patrons Directory, Reference Business Directory*, by J. S. Huber. Bowdle, SD: Prairie Family, 2002.

*Minnehaha County*

*Land Atlas and Plat Book, Minnehaha County, South Dakota, 1981*. Rockford, IL: Rockford Map, 2000.

*Pennington County*

*Eastern Pennington County Atlas 2000: An Atlas of Property Ownership in Eastern Pennington County, South Dakota*, by B. Newland and B. P. R. Roose. Hermosa, SD: Spizzirri, 2001.

*Roberts County*

*Standard Atlas of Roberts County, South Dakota: Including a Plat Book of the Villages, Cities and Townships of the County, Map of the State, United States and World*, by J. S. Huber. Bowdle, SD: Prairie Family, 2004.

## Tennessee

*Madison County*

*Madison County and Jackson, Tennessee Street Atlas*. Racine, WI: Seeger Map, 2004.

*Obion County*

*Land Atlas and Plat Book, Obion County, Tennessee, 1987*. Rockford, IL: Rockford Map, 2000.

*Roane County*

*Cemetery Atlas for the Thirty-Four Atomic Energy Commission, Depart of Energy Conservation Cemeteries in Roane County, Tennessee*. Roane County, TN: Avery Trace Chapter, NSDAR, 2007.

## Texas

*Harris County*

*Houston Harris County Atlas, 2004*. Houston, TX: Key Maps, 2003.
*Houston Street Atlas*. Chicago: Rand McNally, 2006.
*Key Map: Houston, Harris County Atlas*. Houston, TX: Key Maps, 1997.

## Utah

*Salt Lake City*

*Salt Lake City Statistical Atlas 1998*. Salt Lake City: Salt Lake City Corporation, 1998.

## Washington

*Lewis County*

*Lewis County Road Atlas*. Chehalis, WA: Lewis County Public Works Department, 2002.

*Mason County*

*Mason County 2010 Road Atlas*. Shelton, WA: Mason County, 2010.

## Wisconsin

*Barron County*

*Historical Atlas of Barron County, Wisconsin: Including a Plat Book of the Villages, Cities and Townships of the County with Landowner's Index*. Cameron, WI: Blue Hills Genealogical Society, 2005.

*Brown County*

Land Atlas and Plat Book, Brown County, Wisconsin, 2012. Rockford, IL: Rockford Map, 2012.

*Buffalo County*

Land Atlas and Plat Book, Buffalo County, Wisconsin. Rockford, IL: Rockford Map, 2004, 2012, 2015.

*Chippewa County*

Land Atlas and Plat Book, Chippewa County, Wisconsin, 1989. Rockford, IL: Rockford Map, 2000.

*Clark County*

Land Atlas and Plat Book, Clark County, Wisconsin, 1994. Rockford, IL: Rockford Map, 2000.

*Douglas County*

Land Atlas and Plat Book, Douglas County, Wisconsin 2011. Superior, WI: Douglas County Land Records, 2011.
Land Atlas and Plat Book, Douglas County, Wisconsin 2013. Superior, WI: Douglas County Land Records, 2013.

*Dunn County*

Land Atlas and Plat Book, Dunn County, Wisconsin, 1993. Rockford, IL: Rockford Map, 2000.
Land Atlas and Plat Book, Dunn County, Wisconsin, 2015. Rockford, IL: Rockford Map, 2015.

*Fond Du Lac County*

Land Atlas and Plat Book, Fond Du Lac County, Wisconsin. Rockford, IL: Rockford Map, 2005.
Land Atlas and Plat Book, Fond du Lac County, Wisconsin 2011. Rockford, IL: Rockford Map, 2011.
Land Atlas and Plat Book, Fond du Lac County, Wisconsin 2013. Rockford, IL: Rockford Map, 2013.

*Forest County*

Land Atlas and Plat Book, Forest County, Wisconsin, 1983. Rockford, IL: Rockford Map, 2001.

*Grant County*

Land Atlas and Plat Book, Grant County, Wisconsin, 1985. Rockford, IL: Rockford Map, 2000.

*Kenosha-Racine Counties*

Land Atlas and Plat Book, Kenosha-Racine Counties, Wisconsin, 1986. Rockford, IL: Rockford Map, 2000.

*Kewaunee County*

Land Atlas and Plat Book, Kewaunee County, Wisconsin, 1989. Rockford, IL: Rockford Map, 2000.

*Lafayette County*

Combined Standard Atlas of Lafayette County, Wisconsin: 1874, 1895, 1916. Evansville, IN: Lafayette County Genealogical Society, 2006.
Land Atlas and Plat Book, Marquette County, WI. Milwaukee: Milwaukee Map Service, 2012.

*Oconto County*

Land Atlas and Plat Book, Oconto County, Wisconsin, 2003. Rockford, IL: Rockford Map, 2003.

*Oneida County*

Land Atlas and Plat Book, Oneida County, Wisconsin, 2000. Rockford, IL: Rockford Map, 2000.

*Outagamie County*

Land Atlas and Plat Book, Outagamie County, Wisconsin, 1989. Rockford, IL: Rockford Map, 2000.

*Pepin County*

Land Atlas and Plat Book, Pepin County, Wisconsin, 1981–1982. Rockford, IL: Rockford Map, 2000.

*Portage County*

Land Atlas and Plat Book, Portage County, Wisconsin. Rockford, IL: Rockford Map, 2014.

*Sauk County*

Land Atlas and Plat Book, Sauk County, Wisconsin. Rockford, IL: Rockford Map, 2000.

*Sawyer County*

Land Atlas and Plat Book, Sawyer County, Wisconsin. Rockford, IL: Rockford Map, 2005.

*Sheboygan County*

Land Atlas and Plat Book, Sheboygan County, Wisconsin, 2015. Rockford, IL: Rockford Map, 2015.
Sheboygan County 2010 Illustrated Atlas. Gwinn, MI: RedBarn Plat Books and Cartographic Services, 2010.

*Taylor County*

Land Atlas and Plat Book, Taylor County, Wisconsin, 1981. Rockford, IL: Rockford Map, 2000.

*Vernon County*

*Land Atlas and Plat Book, Vernon County, Wisconsin.* Rockford, IL: Rockford Map, 2012.

*Vilas County*

*Land Atlas and Plat Book, Vilas County, Wisconsin.* Rockford, IL: Rockford Map, 2005.

*Walworth County*

*Combination Atlas Map of Walworth County, Wisconsin.* Mt. Vernon, IN: Windmill, 1995.

*Washburn County*

*Land Atlas and Plat Book, Washburn County, Wisconsin, 1988.* Rockford, IL: Rockford Map, 2000.

*Washington-Ozaukee County*

*Land Atlas and Plat Book, Washington-Ozaukee County, Wisconsin 1995.* Rockford, IL: Rockford Map, 2000.

*Winnebago County*

*Land Atlas and Plat Book, Winnebago County, Wisconsin, 2003.* Rockford, IL: Rockford Map, 2003.

# 4
# Topographic, City, and Town Maps

Prior to the late eighteenth century, maps were drawn by hand by private mapmakers. Today, the responsibility rests on national survey bodies, where printed maps follow certain standards related to features, scales, and overall look and feel. Historical maps that were once hand-drawn by cartographers may vary in accuracy and detail, as the "art" of the map is simply one person's interpretation of a geographical location. These maps, however, provide insight into the lives of people at a certain time. Details such as churches, post offices, trees and shrubs, and agricultural crops may be the only evidence of a past society. Such maps are gems for researchers and are easily available through physical and online libraries.

Maps are available in a variety of formats and materials spanning hundreds or, more rarely, thousands of years. This chapter exams individual paper sheet maps. However, maps can be found in bound atlases, in books and journals, as plates, as digital images on CD-ROMS, on websites, or as customized map products generated by using GIS software. This chapter looks at a number of maps that showcase both urban and rural areas: topographic maps and city and town maps, including cadastral and land ownership maps. Each map type will be discussed in detail below.

## TOPOGRAPHIC MAPS

Topographic maps are general-purpose maps that are detailed and accurate graphic representations of both human-made and natural features on the earth's surface, including toponyms, administrative boundaries, and cultural, hydrographical, relief, and vegetation-based information. They are printed in a variety of scales and can be used for many applications. Topographic maps are often used for recreational purposes, such as hiking, canoeing, fishing, and orienteering, but they are also used by industry and government for urban planning, managing emergencies, and establishing legal boundaries and land ownership.

Topographic maps are available for many countries, offered by mapping programs that create topographic map series. In the United States, the most widely known topographic map series is the United States Geological Survey (USGS) National Mapping Program (NMP), published from the mid-1940s through the early 1990s, which produced 1:24,000 scale, 7.5-minute topographic maps. Depending on latitude, the area portrayed by the 7.5-minute series ranges from forty-nine to sixty-four square miles. This map series includes fifty-three thousand map sheets for the conterminous United States and today is referred to as the Historic Topographic Map Collection (HTMC) (USGS, 2014). Since 2006, all new maps have been produced in a native digital form, the U.S. Topo Quadrangle Series, and have been made available in both digital and print format. Digitized versions of the HTMC and the U.S. Topo Quadrangles are available for download through the National Map Viewer (https://nationalmap.gov).

Topographic map coverage of Canada is based on the National Topographic System (NTS), established in 1927. Maps were printed in a variety of scales, including 1:50,000, 1:125,000, 1:250,000, 1:500,000, and 1:1,000,000 (covers the entire country). In 1952 a larger scale, 1:25,000, was added into the system for military and urban use (Marsh, 1999). Many depository libraries in Canada (see appendix B) still have this historical collection of topographic maps; however, today only two of the NTS scales are left in the system, 1:50,000 and 1:250,000. The 1:250,000 series consists of 914 sheets and was completed in 1971. By 1995, the 1:50,000 series covered all Canadian provinces, the Yukon, and mainland Northwest Territories. Today, updated digital versions are available through GeoGratis (http://geogratis.cgdi.gc.ca).

A list of 1:50,000 topographic maps published by Natural Resources Canada, organized by province, follows:

- Alberta, 1948–2002
- British Columbia, 1947–2012

- New Brunswick, 1948–2012
- Newfoundland and Labrador, 1953–2013
- Northwest Territories, 1932–2013
- Nunavut, 1932–2013
- Ontario, 1942–2011
- Prince Edward Island, 1978–2011
- Quebec, 1948–2013
- Saskatchewan, 1948–2012
- Yukon, 1949–2010

Topographic maps in print can be found at many depository libraries, the list of which is available in appendix B. The more recent digital versions can be accessed through a number of websites, listed below. There are numerous websites that offer a gateway to the maps. The end results are the same (that is, published topo maps), but navigation differs from site to site.

## Canadian National Topographic System (NTS) Maps

The Atlas of Canada—Toporama
http://atlas.gc.ca/toporama/en/index.html
Produced by Natural Resources Canada (NRCan), this free product covers the entire area of Canada's landmass and provides georeferenced topographic maps in GeoTIFF format.

CanMatrix—Print Ready
http://www.nrcan.gc.c2/earth-sciences/geography/topographic-information/maps/10995
These digital files were produced by scanning the 1:25,000, 1:50,000, 1:125,000, 1:250,000, 1:500,000, and 1:1,000,000 scale NTS map sheets.

CanTopo
http://www.nrcan.gc.c2/earth-sciences/geography/topographic-information/maps/10995
CanTopo is a GIS product from Natural Resources Canada and has offered updated topographic information on a quarterly basis since June 2009.

CanVec
http://www.nrcan.gc.c2/earth-sciences/geography/topographic-information/maps/10995
CanVec is a GIS product from the National Topographic Data Base (NTDB) and the GeoBase initiative, offering users maps, data, and an opportunity for web mapping. Data include features such as watercourses, urban areas, railways, roads, vegetation, and relief.

Geogratis
http://geogratis.cgdi.gc.ca
GeoGratis is an Internet portal website provided by Natural Resources Canada, offering free topographic data, including over 350 updated or new topographic map sheets in 2016 alone.

## U.S. National Topographic Maps

DataGov
https://catalog.data.gov/dataset/usgs-us-topo-map-collection
USGS 7.5 Minute Quadrangle Maps include additional layers such as orthoimagery, roads, geographic names, contours, hydrography, and more. The maps are available as layered GeoPDF documents.

Historical Topographic Map Collection (HTMC)
https://nationalmap.gov/historical/
The goal of the National Map's Historical Topographic Map Collection is to provide a digital repository of all editions and all scales of USGS standard topographic maps published between 1884 and 2006, including the USGS topographic map series. Paper copies of the maps have been scanned and made available as GeoPDF documents, GeoTiff and JPEG images, as well as KMZ files (for Google Earth use). These files can be accessed from TopoView (https://ngmdb.usgs.gov/maps/TopoView/).

National Geographic PDF Quads
http://www.natgeomaps.com/trail-maps/pdf-quads
National Geographic offers a web interface allowing the public to find, download, and print 1:24,000 7.5-minute USGS topographic maps.

TopoView
https://ngmdb.usgs.gov/maps/Topoview/viewer/#4/40.00/107.51
This interface offers a quick and easy way to search for and download USGS maps published between 1882 and 2006, in a variety of formats such as GeoPDF, GeoTiff, JPEG, and KMZ. This is a beta application developed by the USGS National Geologic Map Database project (NGMDB) in collaboration with the National Geospatial Program (NGP).

University of Texas at Austin
University of Texas Libraries
http://www.lib.utexas.edu/maps/topo/topo_us.html, http://www.lib.utexas.edu/maps/topo/250k/
This site offers a large number of scanned USGS maps for download, sorted by state. Map publication year and scale are also included.

U.S. Topo Maps
https://nationalmap.gov/ustopo/
Published since 2006, U.S. Topo maps are the current generation of USGS topographic maps, published as PDF (as well as GeoPDF) documents rather than in the traditional paper map format. U.S. Topo maps include base data from the National Map such as roads, hydrography, contours and shaded relief, boundaries, woodlands, structures, air photos, and more. These maps are modeled on the USGS topo-

graphic maps but are mass-produced using GIS databases on a repeating cycle.

USGS Historical Topographic Map Explorer
http://historicalmaps.arcgis.com/usgs/
This mapping interface is a collaboration between USGS and Esri, offering users a map to navigate and search in order to locate USGS maps published between 1882 and 2006.

USGS Store Map Locator and Downloader
https://store.usgs.gov/
This site offers access to USGS maps from 1882 to the present, including the maps described above.

## References

Marsh, H. James. 1999. *The Canadian Encyclopedia: Year 2000 Edition.* Toronto, ON: McClelland & Stewart.

USGS. 2014. "US Topo Map and Historical Topographic Map Users Guide." http://www.lib.vt.edu/help/handouts/databases/national-maps-quickstart.pdf.

## CITY AND TOWN MAPS

City and town maps depict urban areas, recording the evolution and growth of cities over time. These maps capture the development of the city or town at a specific point in time, often offering historical information on the area's economic activities, educational and religious facilities, transportation systems, and parks, to name a few. City and town maps were often drawn to propose new subdivisions or parks, as well as to display the general "plan" of the city, such as locations of water mains, stop valves, hydrology, street pavement, street carline, cemeteries, trees, and shrubs.

This map category includes cadastral maps, land ownership plans that show boundaries and ownership of land including owner names, parcel numbers, certificate of title numbers, positions of existing structures, lot numbers, and more. Many of these maps were drawn by hand, an art that no longer exists. Today, the information on these maps, and on other city and town maps, has been replaced by GIS datasets and, to some extent, topographic maps. The resources listed in this chapter are all historical, published from the late 1600s to the 1970s. These maps are a fabulous resource for capturing moments in time before towns expanded into large metropolitan cities.

This chapter offers city and town maps sorted by province and state. The Canadian collection was discovered using the library catalog of Library and Archives Canada. There were certainly fewer results available from this site than from the enormous database offered by the U.S. Library of Congress's map collection. All American map titles listed are available to be viewed and downloaded in image format from the Library of Congress. The titles of the maps explain the information portrayed on the map, so annotating the titles was not necessary.

## Canada

### Alberta

*City of Calgary Showing the Parts of the City Included in the Federal Electoral Districts of Calgary North and Calgary South According to the Representation Act, 1952.* Ottawa, ON: Office of the Surveyor General, 1952.

*Edmonton Area (including the City of St. Albert).* Edmonton, AB: Surveys and Mapping Branch, 1979.

*Manitoba, Map Showing Disposition of Lands.* Ottawa, ON: The Branch, 1924.

*Map of Manitoba, Saskatchewan & Alberta: Showing the Number of Quarter Sections Available for Homestead Entry, Also Number of Quarter Sections Privately Owned and Unoccupied,* by F. C. C. Lynch. Ottawa, ON: Department of the Interior, 1922.

*Platbook of the City of Calgary, Alberta, Canada.* Harlingen, TX: Carson Map, 1976.

*Southern Alberta: Map Showing Disposition of Lands.* Toronto, ON: Department of the Interior, Canada, 1921.

*Southwest Alberta, Southeast B.C.: Backgrounds, Parks, Lodges, Points of Interest.* Cochrane, AB: Gem Trek, 1999.

### British Columbia

*British Columbia, Plan of Township 18, Range 21 West of Sixth Meridian.* Ottawa, ON: Department of the Interior, 1911.

*British Columbia Railway Belt: Map Showing Disposition of Lands.* Ottawa, ON: The Department, 1913.

*City of Penticton: Penticton City Map and Restaurant Guide.* Penticton, BC: Penticton Chamber of Commerce, 1976.

*City of Vancouver: Federal Electoral Districts, the Representation Act, 1952.* Ottawa, ON: Office of the Surveyor General, 1953.

*Map of the Northern Interior of British Columbia: Showing Undeveloped Areas.* Victoria, BC: Provincial Bureau of Information, 1910.

*Victoria: Showing Administrative Boundaries.* Ottawa, ON: Department of National Defence, 1935.

*Western Guide's Plan of the City of Vancouver.* Vancouver, BC: Western Guide, 1915.

### Manitoba

*City of Brandon Showing Boundary and Area of Residential Growth.* Brandon, MB: City of Brandon Engineering Department, 1973.

*City of Winnipeg: Proposed Electoral District Boundaries.* Ottawa, ON: Surveys and Mapping Branch, 1973.

*City of Winnipeg: Shewing Plans Registered on Parts of D.G.S. Lots 1 & 2, Parish of Kildonan, and 43 to 45, Parish of St. John*, by R. C. McPhillips. Winnipeg, MB: McPhillips, 1893.

*City of Winnipeg and Environs: Federal Electoral Districts, the Representation Act, 1952*. Ottawa, ON: Office of the Surveyor General, 1953.

*Historical Map Commemorating the 100th Annual Meeting of the Association of Manitoba Land Surveyors: Thursday May 15th, 1980*. Winnipeg, MB: Surveys and Mapping Branch, 1980.

*Map of Manitoba, Saskatchewan & Alberta: Showing the Number of Quarter Sections Available for Homestead Entry, Also Number of Quarter Sections Privately Owned and Unoccupied*, by F. C. C. Lynch. Ottawa, ON: Department of the Interior, 1922.

*Map of Part of the Province of Manitoba Shewing Dominion Lands Surveyed, and Lands Disposed Of*, by J. Johnston. Ottawa, ON: Dominion lands Office, 1881.

*Plan of Township No. 20, Range 2 East of Principal Meridian*. Ottawa, ON: Department of the Interior, 1903.

New Brunswick

*City of Fredericton*. Ottawa, ON: Public Archives of Canada, 1978.

*Plan of the Country around the City of Fredericton*, by T. G. Loggie. Halifax, NS: McAlpine, 1902.

Newfoundland

*Newfoundland and Labrador, Canada: Touring and Exploring Map*. St. John's, NL: Department of Tourism, Culture and Recreation, 2006.

Nova Scotia

*City of Dartmouth*. Dartmouth, NS: Maritime Resource Management Service, 1979.

*City of Halifax, Nova Scotia 1890*, by D. D. Currie. Halifax, NS: Scotia Sales and Marketing, 1973.

*Halifax, Halifax County, Nova Scotia, Scale 1:1,000*. Summerside, PE: Council of Maritime Premiers, 1982.

*South East Coast of Nova Scotia*, by J. F. W. Des Barres. London: J. F. W. Des Barres, 1781.

Ontario

*Blyth's Guide Map of Ottawa and Locations of Suburbs*. Toronto, ON: Bylth's Guide, 1900.

*City of Hamilton and Vicinity: Federal Electoral Districts, the Representation Act, 1952*. Ottawa, ON: Surveys and Mapping Branch, 1952.

*City of London, Canada*. Toronto, ON: Toronto Lithographing, 1893.

*City of Toronto*. Toronto, ON: Graphic Publishers, 1953.

*Early Settlement Patterns in Selected Townships in the North Bay Area*, by K. H. Topps, R. S. Brozowski, and D. L. Rees. North Bay, ON: Department of Geography, Nipissing University College, 1981.

*Map of Part of Keewatin Shewing Dominion Land Surveys to 31st December, 1876*. Ottawa, ON: Dominion Lands Office, 1877.

*Map of Part of the Huron and Ottawa Territory, Ontario, Showing the Districts of Parry Sound, Muskoka and Nipissing, County of Haliburton, Etc*. Ottawa, ON: Ontario Department of Lands, Forests and Mines, 1909.

*Map of the City of Hamilton*, by T. J. Kirk. Hamilton, ON: J. W. Tyrell, 1922.

*Map of the City of London: And Suburbs of London East, London West and London South*, by A. O. Graydon. Toronto, ON: Joseph Doust, 1886.

*Map of the City of Toronto*, by T. Deavin. Toronto, ON: Miller Litho, 1929.

*Map of the Indian Reserve Consisting of the Township of Tuscarora and a Part of the Township of Onondaga in the County of Brant and Part of the Township of Oneida in the County of Haldimand in the Province of Ontario*, by O. Robinson. Ottawa, ON: Indian Affairs, 1875.

*Map of the Vicinity of Niagara Falls: From Actual Surveys*. Niagara, ON: S. Geil and J. L. Delp, 1853.

*Map of Windsor, Ontario, Showing Lot Numbers and Names of the Settlers along the South Shore of the Detroit River*, by A. Iredell. Windsor: Ontario Genealogical Society, 1990.

*Muskoka and Nipping County of Haliburton*. Toronto, ON: Department of Crown Lands, 1899.

*New City Map, Smith Falls, Ontario*. Ottawa, ON: Pathfinder Air Surveys, 1962.

*New Map of Toronto Metropolitan Area: Including the 12 Adjoining Municipalities*, by V. Guyenoswki. Toronto, ON: S. A. Renouf Advertising, 1955.

*North Bay, Ontario, Properties Federal = North Bay, Ontario, propriétés fédérales*. Ottawa, ON: Public Works Canada, 1978.

*Ontario South: Showing County and Township Boundaries*. Ottawa, ON: Hydrographic and Map Service, 1944.

*Ottawa-Hull and Vicinity*, by J. Gréber. Ottawa, ON: National Capital Planning Service, 1946.

*Plan of Part of the Township of Nipissing in the Parry Sound District*, by H. Lillie. Toronto, ON: Department of Crown Lands, 1878.

*Plan of the City and Liberties of Kingston: Delineating Severally the Wards and Lots with the Streets, Wharves, and Principal Buildings, 1850*, by T. F. Gibbs and H. Scobie. Ottawa, ON: Association of Canadian Map Libraries and Archives, 1992.

*Plan of the City of Toronto*. Toronto, ON: W. S. Johnston, 1905.

*Plan of the Residue Township of Nipissing, Parry Sound District*, by L. Tallon. Toronto, ON: Department of Crown Lands, 1881.

*Plan of the Town of Renfrew: The Property of William Ross, Esq., on Lots 13 in the 1st and 2nd Con. of the Township of Horton*, by D. Kennedy. Toronto, ON: Maclear, 1854.

*Plan of the Township of Grand Calumet Island*. Ottawa, ON: Department of Crown Lands, 1987.

*Plan of the Township of Gurd, in the Parry Sound District*, by J. W. Fitzgerald. Toronto, ON: Department of Crown Lands, 1878.

*Plan of the Township of Mills*, by T. Byrne. Toronto, ON: Department of Crown Lands, 1877.

*Plan of the Township of Patterson*, by J. W. Fitzgerald. Toronto, ON: Department of Crown Lands, 1878.

*Plan of the Township of Waters, Algoma District*, by W. R. Burke. Toronto, ON: Copp Clark Co., 1883.

*Plan Showing Cadastral Lot No. 34 on Chateauguay River*, by J. H. Sullivan. Ottawa, ON: Archives nationales du Canada, 1977.

*Proposed Development of Central Area Ottawa*. Ottawa, ON: Town Planning Commission, 1926.

*Township of Gloucester: Schedule A, Sheet One to Zoning By-law No. 26 of 1960*. Ottawa, ON: McRostie, 1963.

*Tremaine's Map County of York, Canada West*. Toronto, ON: G. R. Tremaine, 1972.

*Tremaine's Map of Oxford County, Canada West*, by W. G. Wonham. Kingston, ON: Tremaine, 1857.

*Tremaine's Map of the Counties of Lincoln and Welland, Canada West*, by G. R. Tremaine. Toronto, ON: G. R. Tremaine, 1862.

*Tremaine's Map of the County of Durham, Upper Canada*, by J. Shier and J. F. Ward. Toronto, ON: Geo. C. Tremaine, 1861.

*Tremaine's Map of the County of Elgin, Canada West*, by G. R. Tremaine. Toronto, ON: Tremaine, 1864.

*Tremaine's Map of the County of Halton, Canada West*, by G. R. Tremaine. Oakville, ON: Tremaine, 1858.

*Tremaine's Map of the County of Ontario, Upper Canada*, by J. Shier. Toronto, ON: Tremaine, 1860.

*Tremaine's Map of the County of Peel: Canada West*, by G. R Tremaine. Toronto, ON: Tremaine, 1859.

*Tremaine's Map of the County of Prince Edward, Upper Canada*, by J. F. Ward. Toronto, ON: Tremaine, 1863.

*Tremaine's Map of the County of Waterloo, Canada West*, by G. M. Tremaine. Toronto, ON: Geo. R. & G. M. Tremaine, 1861.

*Tremaine's Map, Upper Canada*, by G. R. Tremaine. Toronto, ON: Geo. C., Geo. R. & G. M. Tremaine, 1862.

*Zoning for Township of Hanmer*. Ottawa, ON: Planning and Development Department, 1978.

## Quebec

*Cadastral Plan of the City of Quebec and St. Sauveur, Village Compiled from the Original Plans Deposited at the Crownlands Department*, by P. Cousin. New York: H. H. Lloyd, 1875.

*Cadastral Plans, City of Montreal*, by F. W. Blaiklock et al. Montreal, QC: Department of Public Works, 1880.

*City of Montreal, Map of Central Portion*, by A. Oligny. Montreal, QC: Oligny, 1946.

*Correct Plan of the Environs of Quebec, and of the Battle Fought on the 13th September, 1759: Together with a Particular Detail of the French Lines and Batteries, and Also of the Encampments, Batteries and Attacks of the British Army, and the Investiture of That City under the Command of Vice Admiral Saunders, Major General Wolfe, Brigadier General Monckton, and Brigadier General Townsend*, by T. Jefferys. London: Engineers of the Army, 1760.

*Map of the City of Montreal*. Montreal, QC: Royal Bank of Canada Collection, 1845.

*Map of the Island of Montreal 1890*, by H. Malingre. Montreal, QC: Canada Bank Note, 1890.

*Plan of the Township of Mansfield*, by J. Robertson. Montreal, QC: Crown Lands Office, 1847.

*Plans in Detail of the Survey Made in 1835 of the Lands in the District of Three Rivers and County of Sherbrooke, Lower Canada Sold by His Majesty's Government to the British American Land Company*, by J. Hughes. Sherbrooke, QC: British American Land, 1835.

## Saskatchewan

*Map of Manitoba, Saskatchewan & Alberta: Showing the Number of Quarter Sections Available for Homestead Entry, Also Number of Quarter Sections Privately Owned and Unoccupied*, by F. C. C. Lynch. Ottawa, ON: Department of the Interior, 1922.

*Plan of the Town of Naicam, Sask*. Saskatoon, SK: Department of Public Works, 1958.

*Saskatchewan, Map Showing Disposition of Lands*. Ottawa, ON: Department of the Interior, 1924.

*Saskatchewan, Plan of Township No. 16, Range 26 West of Second Meridian*. Ottawa, ON: Department of the Interior, 1922.

*Saskatoon, Plan of the Subdivision of Legal Subdivision 10, Sec. 22, Tp. 36, R. 5, W. 3rd Mer*, by H. B. Proudfoot. Halifax, NS: Northwestern Land and Investment, 1908.

*Sectional Map of Saskatchewan, Canada, Showing the Lands of the Saskatoon and Western Land Company, Limited, at Regina, Saskatchewan, October, 12th to 17th*. Ottawa, ON: Saskatoon & Western Land, 1908.

## Yukon

*Official Community Plan*. Whitehorse, YT: City of Whitehorse, 1971.

*Plan of Dawson City: Retracing of "Dawson and Klondike [townsites],"* by J. Gibbon. Dawson City, YT: James Gibbon, 1899.

## United States of America

### Alabama

*Plan of Part of the Rivers Tombecbe, Alabama, Tensa, Perdido, & Scambia in the Province of West Florida; with a Sketch of the Boundary between the Nation of Upper Creek Indians and That Part of the Province Which Is Contiguous Thereto, as Settled at the Congresses at Pensacola in the Years 1765 & 1771*, by D. Taitt and J. Stuart. St. Louis, MO: D. Taitt, 1771.

*Township and Sectional Map of Mobile County, State of Alabama: Compiled and Drawn from the Most Reliable and Recent Data: Showing the Railroads, Public Roads, School Houses, Old Spanish & Other Grants*, by H. Fondé. St. Louis, MO: Photo. Eng., 1895.

### Arkansas

*Map of the City of Little Rock and Argenta, Arkansas: Compiled from Official Sources and Actual Surveys*. Little Rock, AR: Rickon, Gibb & Duff, 1888.

### California

*Map of Part of San Francisco, California, April 18, 1908: Showing Buildings Constructed and Buildings under Construction during Two Years after Fire of April 18, 1906*. San Francisco: Punnett Brothers, 1908.

*Map of the City of San Jose*. San Jose, CA: James A. Clayton, 1886.

*Map of the County of El Dorado, California: Compiled from the Official Records and Surveys*. San Francisco: Punnett Brothers, 1895.

*Map of the County of Los Angeles, California: Compiled from U.S. Land Surveys, Records of Private Surveys, and from Other Reliable Sources*, by J. H. Wildy. New York: Julius Bien, 1877.

*Map of the County of San Joaquin: Compiled from the United States Surveys, the Maps and Records of the County Surveyor and State Engineer, 1895*, by H. T. Compton. Los Angeles: Britton & Rey, 1895.

*Map of the Old Portion of the City Surrounding the Plaza, Showing the Old Plaza Church, Public Square, the First Gas Plant and Adobe Buildings, Los Angeles City, March 12th, 1873*, by A. G. Ruxton. Los Angeles: A. G. Ruxton, 1873.

*Map Showing Portions of Alameda and Contra Costa Counties, City and County of San Francisco, California, Carefully Compiled from Official and Private Maps, Surveys and Data*, by T. Wagner. San Francisco: Britton & Rey, 1894.

*Map Showing Territory Annexed to the City of Los Angeles, California*, by H. Hamlin. Los Angeles: City Engineer, 1916.

*Official Map of Colusa County, California: Compiled and Drawn from Official Surveys and Records*. San Francisco: Britton & Rey, 1885.

*Official Map of "Chinatown" in San Francisco*, by W. B. Farwell, J. E. Kunkler, and E. B. Pond. San Francisco: Bosqui Eng. & Print, 1998.

*Official Map of San Diego County, California: Compiled from Latest Official Maps of U.S. Surveys, Railroad and Irrigation Surveys, County Records, and Other Reliable Sources*, by T. D. Beasley. San Diego, CA, 1889.

*Official Map of San Francisco*, by M. Eddy, A. Zakreski, and F. Michelin. San Francisco: W. M. Eddy, A. Zakreski, and F. Michelin, 1849.

*Official Map of Sonoma County, California: Compiled from the Official Maps in the County Assessor's Office, with Additions and Corrections to June 1st, 1900*, by L. E. Ricksecker. San Francisco: W. B. Walkup, 1900.

*Official Map of the City of Brawley, Imperial County, California*. Brawley, CA: City of Brawley, 1927.

*Official Map of the County of Butte, California: Carefully Compiled from Actual Surveys*, by J. McGann. San Francisco: Britton, Rey, 1877.

*Official Map of the County of Napa, California: Compiled from the Official Records and Latest Surveys*, by O. H. Buckman. San Francisco: Punnett Bros, 1895.

*Official Map of the County of Santa Clara, California: Compiled from U.S. Surveys, County Records, and Private Surveys and the Tax-List of 1889, by Order of the Hon. Board of Supervisors*. San Jose, CA: Herrmann Bros., 1890.

*Official Map of the County of Solano, California: Showing Mexican Grants, United States Government and Swamp Land Surveys, Present Private Land Ownerships, Roads and Railroads*, by E. N. Eager. Solano County, CA: E. N. Eager, 1890.

*Official Map of Trinity County, California: Compiled from Government and Local Surveys*, by H. L. Lowden. San Francisco: H. S. Crocker, 1894.

*Official Map of Yuba County, State of California: Compiled and Drawn from Official Recs. and Actual Surveys*, by J. M. Doyle. San Francisco: Britton & Rey, 1887.

*Plat Book of San Diego County, California*, by W. E. Alexander. Los Angeles: Pacific Plat Book, 1912.

*Street and Section Map of the Los Angeles Oil Fields, California*. Baltimore, MD: A. Hoen, 1906.

### Colorado

*Map of Denver, Colorado: From Authentic Sources*, by E. H. Kellogg and J. H. Bonsall. Denver, CO: Worley & Bracher, 1871.

*Map of Larimer County, Colorado: A.D. 1883: Showing Public Roads, Irrigating Canals, Rail Roads, Road Districts, School Districts, Voting Precincts, and Complete

*Topography*, by H. P. Handy. Colorado: H. P. Handy, 1884.

### Connecticut

*Geer's New Map of the City of Hartford: From the Latest Surveys*, by E. Geer and J. R. Hawley. Hartford, CT: Geer, 1859.

*Map of New Milford, Litchfield Co., Conn.*, by L. Fagan. Litchfield, CT: L. Fagan, 1852.

*Map of the City of Bridgeport, Conn.* Philadelphia: Collins & Clark, 1850.

*Map of the Town of Canaan, Connecticut.* Canaan, CT: L. Fagan, 1853.

*Map of the Town of Goshen, Litchfield County, Connecticut*, by E. M. Woodford. Litchfield, CT: E. M. Woodford, 1852.

*Map of the Town of New Hartford, Litchfield Co., Conn*, by L. Fagan. Litchfield, CT: L. Fagan, 1852.

*Map of the Town of Norfolk, Litchfield County, Ct.*, by L. Fagan. Litchfield, CT: L. Fagan, 1853.

*Map of the Town of Salisbury, Litchfield Co., Connecticut*, by L. Fagan. Litchfield, CT: L. Fagan, 1853.

*Map of the Town of Washington, Litchfield Co., Conn.*, by L. Fagan. Litchfield, CT: L. Fagan, 1879.

*Plan of New Haven*, by A. Doolittle. New Haven, CT: A. Doolittle, 1817.

*Plan of the City of Hartford: From a Survey Made in 1824*, by D. St. John and N. Goodwin. Hartford, CT: D. St. John and N. Goodwin, 1824.

*Plan of the Town of New Haven: With All the Buildings in 1748 Taken by the Hon. Gen. Wadsworth of Durham to Which Are Added the Names and Professions of the Inhabitants at That Period: Also the Location of Lots to Many of the First Grantees*, by W. Lyon and J. Wadsworth. New Haven, CT: T. Kensett, 1806.

*View of Seymour, Conn. 1879.* Boston, MA: O. H. Bailey, 1879.

### Florida

*Map of Marion County, Florida: From U.S. Surveys and Other Official and Original Sources Showing All Lands Belonging to the U.S. State Railroad and Disston to Oct. 15th*, by J. W. Bushnell, A. T. Williams, and F. Bourquin. Miami, FL: J. W. Bushnell and A. T. Williams, 1885.

*Map of the City of Miami and Environs: Showing the Greater Miami Development and the Estimated Expansion for 1935*, by D. Sauer and L. Seghy. Miami, FL: A. Douglas Printing, 1925.

*Plan du port de St. Augustin dans la Floride*, by J. N. Bellin. Paris: J. N. Bellin, 1764.

*Plan of St. Augustine*, by Lieut. Birch. St. Augustine, FL: Lieut. Birch, 1819.

*Plan of the Inlet, Strait, & Town of St. Augustine*, by S. Roworth. St. Augustine, FL: S. Roworth, 1760.

### Georgia

*Plan of the City of Savannah: With a Drawing of the Part of the City Burnt in the Dreadful Fires of the 26 November & 6 December, 1796*, by M. S. Carson. Savannah, GA: Marian S. Carson Collection, 1796.

*Vincent's Subdivision Map of the City of Atlanta, Dekalb County, State of Georgia: Showing All the Lots, Blocks, Sections*, by E. A. Vincent. Savannah, GA: Edward A. Vincent, 1852.

### Illinois

*City of Chicago*, by G. Watson. New York: Gaylord Watson, 1871.

*Map Showing the Burnt District in Chicago: Published for the Benefit of the Relief Fund.* Saint Louis, MO: R. P. Studley, 1872.

*Souvenir Map of the World's Columbian Exposition at Jackson Park and Midway Plaisance, Chicago, Ill, U.S.A. 1893.* Chicago: A. Zeese, 1892.

### Indiana

*Map of Floyd County, Indiana: Showing the Townships Sections Divisions and Farm Lands with the Owners' Names and Number of Acres, Together with Roads, Rivers, Creeks, Railroads*, by P. O'Beirne. New York: P. O'Beinre, 1859.

*Map of Floyd County, Indiana: Showing Townships, Ranges, Sections & Farm Lines with the Owners Names and Number of Acres*, by G. M. Smith. Philadelphia: E. P. Noll, 1882.

### Kansas

*Edward's Map of Douglas Co., Kansas*, by J. P. Edwards. Quincy, IL: John P. Edwards, 1887.

*Edward's Map of Johnson Co., Kansas*, by J. P. Edwards. Quincy, IL: John P. Edwards, 1886.

*Edward's Map of Wyandott County, Kansas*, by J. P. Edwards and W. Bracher. Philadelphia: Bourquin, 1885.

*Map of Leavenwoth County, Kansas*, by F. C. Waite. Leavenworth, KS: F. C. Waite, 1894.

*Map of McPherson County, Kansas*, by H. A. Rowland. Chicago: Rand McNally, 1898.

*Map of the County of Greenwood, Kansas*, by J. Hoenscheidt. Milwaukee: J. Knauer & Co. Lith, 1877.

*Map of Wyandotte Co., Kansas*, by G. M. Hopkins. Philadelphia: G. M. Hopkins, C.E., 1887.

### Kentucky

*Map of Jefferson County, Kentucky: Showing the Names of Property Holders, Division Lines of Farms, Position of Houses, Churches, School-Houses, Roads, Water-Courses, Distances, and the Topographical Features of the County: Distinctly Exhibiting the Country around the Falls of the Ohio, Including New Albany and Jeffersonville, Indiana*, by G. T. Bermann. Louisville, KY: Bergmann, 1858.

*Map of Marion and Washington Counties, Ky: From Actual Surveys and Official Records*, by D. G. Beers and J. Lanagan. Philadelphia: D. G. Beers, 1877.

*Map Showing Property of Beaver Creek Consolidated Coal Co. in Floyd, Knott and Magoffin Counties, Kentucky*, by F. W. Gesling. Huntington, WV: Beaver Creek Consolidated Coal, 1910.

### Louisiana

*Livingston Parish, Louisiana*, by J. Lynch. New Orleans: Surveyor General's Office, 1870.

*Map of the Parish of Avoyelles and Part of Rapides, Louisiana: From the United States Survey*. New Orleans: McCerren, Landry & Powell, 1860.

*Map of the Parish of Caldwell, Louisiana: From United States Surveys*. New Orleans: McCerren, Landry & Powell, 1860.

*Map of the Parish of Carroll, Louisiana: From the United States Surveys*. New Orleans: McCerren, Landry & Powell, 1860.

*Map of the Parish of Catahoula, Louisiana: From United States Surveys*. New Orleans: McCerren, Landry & Powell, 1860.

*Map of the Parish of Concordia, Louisiana: From United States Surveys*. New Orleans: McCerren, Landry & Powell, 1860.

*Map of the Parish of Franklin, Louisiana: From the United States Surveys*. New Orleans: McCerren, Landry & Powell, 1860.

*Map of the Parish of Morehouse, Louisiana: From United States Surveys*. New Orleans: McCerren, Landry & Powell, 1860.

*Map of the Parish of Ouachita, Louisiana*. New York: Pudney & Russell, 1858.

*Map of the Parish of Washita, Louisiana: From United States Surveys*. New Orleans: McCerren, Landry & Powell, 1860.

*Map of the Town of Alexandria, Louisiana*, by R. W. Bringhurst. Indianapolis, IN: Braden & Burford, 1872.

*Map Showing the Location, Acreage and Name of Owner Together with the General Topographical Features of All the Land Situated in Madison Parish, La.* St. Louis, MO: Aug. Gast Bank Note & Litho., 1891.

*Norman's Plan of New Orleans & Environs, 1845*, by H. Möllhausen and B. M. Norman. New Orleans: Shields & Hammond, 1845.

*Plan de la Nouvelle Orleans*, by J. N. Bellin. Paris: J. N. Bellin, 1764.

*Plan of Baton Rouge and Adjoining Properties on the Mississippi River*, by V. Pintado. New Orleans: V. Pintado, 1805.

*Plan of New Orleans*, by P. Pittman. London, UK: P. Pittman, 1770.

*Plan of the City of New Orleans and Adjacent Plantations*, by C. L. Trudeau. New Orleans: H. Wehrmann, 1875.

### Maine

*Map of Lakenwild, Washington County, Maine: Owned by Nathan S. Read: 1886*, by B. Gardner and N. S. Read. Philadelphia: J. L. Smith, 1886.

*Map of the City of Belfast, Waldo Co., Maine*, by D. Osborn et al. Philadelphia: E. M. Woodford, 1855.

*Map of the Timber Lands in Oxford and Franklin Counties, Maine, Coos County, New Hampshire: Showing the Different Townships with Their Allotments, 1899*, by A. M. Carter and E. S. Bryant. Alfred, ME: Berlin Mills, 1899.

*Plan of Alfred, Maine*, by J. H. Bussell. Alfred, ME: Joshua H. Bussell, 1845.

*Plan of the Town of Turner, Formerly Silvester Plantation: Out Side Lines and Part of the Lotts Ran Out*, by S. Getchel. Turner, ME: S. Getchel, 1770.

*Plan Showing Some of the Original Lots in the Town of Union, Knox County, Maine, Formerly Sterlingtown*, by R. R. Holmes and E. Jennison. Union, ME: Raynold R. Holmes, 1774.

### Maryland

*Lloyd's Baltimore Elevated Building Map: Of the Five Hundred Million Business District of Baltimore*. New York: Friedenwald, 1891.

*Map of Frederick County, Md.: Accurately Drawn from Correct Instrumental Surveys of All County Roads*, by I. Bond. Baltimore, MD: I. Bond, 1860.

*Map of the State of Maryland Laid Down from an Actual Survey of All the Principal Waters, Public Roads, and Divisions of the Counties Therein; Describing the Situation of the Cities, Towns, Villages, Houses of Worship and Other Public Buildings, Furnaces, Forges, Mills, and Other Remarkable Places; and of the Federal Territory; as Also a Sketch of the State of Delaware Shewing the Probable Connexion of the Chesapeake and Delaware Bays*, by D. Griffith and J. Thackara. Philadelphia: J. Vallance, 1795.

*Martenet's Map of Prince George's County, Maryland*, by S. J. Martenet. Baltimore: S. J. Martenet, 1861.

*Plan of the City of Baltimore*, by L. Fielding. Baltimore: L. Fielding, 1822.

*Plan of the City of Baltimore, Maryland*. Baltimore: Lloyd van Derveer, 1851.

*Plan of the Town of Baltimore and Its Environs*, by A. P. Folie and J. Poupard. Baltimore: A. P. Folie and J. Poupard, 1792.

*Plan Showing That Part of "Onion's Addition to Sweaty Banks": Which Was Conveyed by Thomas B. Onion to Charles Ridgley, September 12, 1794, and Now Found to Contain about 323 Acres of Land: Clear of the Claims of Adjoining Owners*. Harford County, MD: W. E. Somerville and Thomas B. Onion, 1794.

*Rough Plan of the Defences of the Harbour of Annapolis in Maryland*, by W. Tatham and W. H. Winder. Annapolis, MD: W. Tatham, 1814.

*Section No. 2, Chevy Chase, Maryland*, by T. J. Fisher and N. Peters. Washington, DC: Thos. J. Fisher, 1909.

*Survey of Annapolis*, by H. Ridgely. Annapolis, MD: H. Ridgely, 1690.

Massachusetts

*City of Fall River*. Boston: Sampson, Davenport, 1874.

*Map of Boston and Its Vicinity*, by J. G. Hales and E. Gillingham. Boston: J. G. Hales, 1820.

*Map of Boston for 1881*, by D. Sampson. Boston: Davenport, 1881.

*Map of Boston in the State of Massachusetts: 1814*, by J. G. Hales and T. Wightman. Boston: J. G. Hales, 1814.

*Map of Cambridge, Mass*, by J. Hayward. Boston: Eddy's Lithography, 1838.

*Map of Franklin County, Massachusetts: Based on the Trigonometrical Survey of the State; Details from Original Surveys*, by H. F. Walling. Boston: Smith & Ingraham, 1858.

*Map of Great Barrington, Mass*, by J. W. Curtiss. Great Barrington, MA: Crowley & Lunt, 1919.

*Map of Mendon, Worcester County, Mass.*, by N. Nelson and W. H. Taft. Boston, MA: N. Nelson, 1831.

*Map of the City of Boston and Vicinity*, by Boston, MA: Sampson, Murdock, 1896.

*Map of the City of Cambridge, Mass.*, by W. Mason, J. Hayward, and O. Pelton. Cambridge, MA: Mason William, 1857.

*Map of the City of Lowell: Surveyed in 1841 by Order of the Municipal Authorities*, by I. Beard, A. J. Hoar, and G. W. Boynton. Boston: I. A. Beard, 1841.

*Map of the City of Springfield, Massachusetts*. Springfield, MA: L. J. Richards, 1905.

*Map of the City of Worcester, Worcester Co., Mass.* Worcester, MA: Drew Allis, 1878.

*Map of the Town of Attleborough, Bristol County, Massachusetts*, by H. F. Walling. Attleborough, MA: H. F. Walling, 1850.

*Map of the Town of Concord, Middlesex County Mass.* Boston: H. F. Walling, 1852.

*Map of the Town of Holden*. Boston: Pendleton's Lithogy, 1832.

*Map of the Town of Medfield*, by H. F. Walling and A. Kollner. Philadelphia: A. Kollner, 1852.

*Map of Wareham*, by S. Bourne. Boston: Pendleton's, 1832.

*Map of Watertown, Mass.: Surveyed by Order of the Town, 1850*, by S. D. Eaton et al. Boston: Tappan & Bradford's Lith, 1850.

*New Plan of Ye Great Town of Boston in New England in America with the Many Additional Buildings & New Streets to the Year 1743*, by W. Price. Boston: W. Price, 1743.

*Plan of Boston: From Actual Survey*, by O. Carleton. Boston: O. Carleton, 1805.

*Plan of Falmouth Heights, Falmouth Mass., April 1st 1873*. Boston: J. H. Bufford's Lith, 1873.

*Plan of Lagoon Heights, Martha's Vineyard, Mass*, by R. L. Pease. Boston: J. H. Bufford's Lith, 1873.

*Plan of Sections 1, 2 & 3, East Boston*. Boston: Pendleton's Lithogy, 1836.

*Plan of the Town of Dorchester*, by T. W. Davis and A. Meisel. Boston: A. Meisel, 1870.

*Plat & Environs of Lagoon Heights, Cottage City, Mass.: Showing Property Owned by the Lagoon Heights Land Co., 53 State St., 610 Exchange B'd'g., Boston, Mass.* Boston: O. H. Bailey, 1887.

*Town of Boston in New England*, by J. Bonner et al. Boston: George G. Smith, 1835.

Michigan

*Map of the City of Detroit in the State of Michigan*, by J. Farmer. Detroit: Farmer, 1835.

*Plat Book of Kalamazoo County, Michigan: Compiled from Surveys and the Public Records of Kalamazoo County, Michigan*. Kalamazoo: Glen C. Wheaton, 1928.

Minnesota

*Davison's Map 25 Miles around Minneapolis 1881*, by D. C. Wright. Minneapolis: Davison C. Wright, 1881.

*Map of Blue Earth County, Minnesota: Drawn from Actual Surveys and the County Records*. Minneapolis: Warner & Foote, 1879.

*Map of Brown County, Minn.: From Personal Examination and Public Record*, by L. G. Vogel. New Ulm, MN: L. G. Vogel, 1900.

*Map of Carver County, Minnesota: Drawn from Actual Surveys and the County Records*. Minneapolis: Warner & Foote, 1880.

*Map of Freeborn County, Minnesota: Drawn from Actual Surveys and the County Records*. Red Wing, MN: Warner & Foote, 1878.

*Map of Martin County, Minnesota: From Personal Examinations and Public Records*. Mankato, MN: M. B. Haynes, 1887.

*Map of McLeod County, Minnesota: Drawn from Actual Surveys and the County Records.* Minneapolis, MN: Warner & Foote, 1880.

*Map of Murray County, Minnesota: Compiled and Drawn from a Special Survey and Official Records*, by E. F. Peterson. Vermillion, SD: E. Frank Peterson, 1898.

*Map of Nicollet County, Minn.: From Personal Examination and Public Records.* Mankato, MN: Haynes & Woodard, 1885.

*Map of Steele County, Minnesota: Drawn from Actual Surveys and the County Records.* Minneapolis: Warner & Foote, 1879.

*Map of the City of Saint Paul, Capital of Minnesota.* St. Paul: Miller & Boyle's Lith, 1852.

*Map of Watonwan County, Minnesota: Compiled and Drawn from a Special Survey and Official Records.* Vermillion, SD: E. Frank Peterson, 1898.

*Outline Map of Minneapolis and St. Paul.* Saint Paul: Blodgett & Moore, 1897.

*Plat of Cottonwood County, Minnesota*, by W. A. Peterson. Saint Paul: Pioneer Press, 1898.

*Sentinel Map of Martin County, Minnesota: From Personal Examinations and Public Records.* Rockford, IL: Hunter & VanValkenburgh, 1901.

*Mississippi*

*Map of Yazoo County, Miss.: Containing the Public Roads, Schools, Churches, Precincts, Farms, Residences*, by J. W. Mercer. Yazoo City, MS: Mercer & Fontaine, 1874.

*Missouri*

*Bellefontaine Cemetery, St. Louis, Mo.* St. Louis: Jno. McKittrick, 1875.

*City of St. Louis*, by J. Hutawa. Saint Louis: J. Hutawa, 1870.

*King's Land Owners Map of Gasconade County, Missouri: Carefully Compiled from Personal Examinations and Surveys*, by G. H. King. St. Louis: George H. King, 1875.

*Map of Madison County, Missouri, Showing the Lands of B. B. Cahoon, of Fredericktown, Madison County, Missouri*, by B. B. Cahoon. Fredericktown, MO: B. B. Cahoon, 1882.

*Map of Pawnee Reservation Showing Allotments.* Kansas City, MO: Hudson-Kimberly, 1893.

*Map of Wayne County, Missouri, Showing Lands Therein Owned by B. B. Cahoon, of Fredericktown, Madison County, Missouri*, by B. B. Cahoon. Fredericktown, MO: B. B. Cahoon, 1882.

*Pharus-Map World's Fair St. Louis, 1904.* St. Louis: Pharus, 1904.

*New Hampshire*

*Map of Manchester, N.H.*, by J. B. Sawyer. Manchester, NH: John B. Clarke, 1879.

*Map of the City of Manchester, N.H.: Compiled from Recent Surveys in the Engineers Office, Manchester Water Works*, by J. T. Fanning. Boston: Sampson, Davenport, 1873.

*Plan of Exeter Village, New Hampshire*, by J. Dow. Boston: J. H. Bufford & Co.'s Lithography, 1845.

*Plan of the Town of Stratham*, by P. Merrill. Stratham, NH: P. Merrill, 1793.

*New Jersey*

*Atlantic Highlands, New Jersey 1894.* Boston: O. H. Bailey, 1894.

*Ed. H. Radcliffe's Frankford, Pa. Business Map of Bristol, Bordentown, Burlington, and Mount Holly*, by E. H. Radcliffe. Frankford, PA: Ed. H. Radcliffe, 1870.

*Farm Map of Alexandria Township, Hunterdon Co., N.J.*, by M. Hughes. Philadelphia: Lith. of Friend & Aub, 1860.

*Holbrook's Map of the City of Newark, New Jersey.* Newark, NJ: A. S. Holbrook, 1875.

*Map of Cumberland Co., New Jersey: From Actual Surveys*, by S. N. Beers et al. Philadelphia: A. Pomeroy, 1862.

*Map of Hillside Park near Atlantic Highlands, N.J.*, by G. Cooper. Atlantic Highlands, NJ: J.C. Nobles, 1882.

*Map of Newark and East Newark, N.J. from the Most Authentic Surveys.* New York: C. B. Graham, 1836.

*Map of Roselle, N.J.: from Actual Surveys*, by Wm. J. Lansley and R. A. Welcke. Elizabeth, NJ: Wm. J. Lansley, 1887.

*Map of the Borough of Woodbury, N.J.*, by J. Pierson, Z. Woodbury, and P. S. Duval. Woodbury, NJ: P. S. Duval, 1854.

*Map of the City of Newark, State of New Jersey*, by S. Dod. Newark, NJ: B. T. Pierson's Directory, 1859.

*Map of the City of New Brunswick, New Jersey*, by D. Ewen. New Brunswick, NJ: City of New Brunswick, 1836.

*Map of the City of Paterson, N.J.*, by J. H. Goetschius. New York: Joseph E. Crowell, 1871.

*Map of the City of Perth Amboy, N.J.: February 1836*, by F. W. Brinley. New York: Baker, 1836.

*Map of the City of Trenton, New Jersey*, by J. C. Sidney and A. Kollner. Philadelphia: M. Dripps, 1849.

*Map of the Residence & Park Grounds, near Bordentown, New Jersey: Of the Late Joseph Napoleon Bonaparte, Ex-King of Spain.* New York: Miller's Lith, 1847.

*Map of the Settled Parts of the Vineland Tract*, by A. F. Wrotnowski and C. K. Landis. Vineland, NJ: A. F. Wrotnowski, 1867.

*Plan of the Island of Burlington: And a View of the City from the River Delaware*, by W. R. Birch. Philadelphia: M. Carey's, 1797.

*New York*

*Brooklyn Hills Improvement Co.'s Plat No. 1: Property at Woodhaven, L.I.; Brooklyn Hills Improvement Co.'s*

*Map Showing Relative Position of Property to Prominent Points in Cities of New York & Brooklyn*, by J. Wells. New York: Brooklyn Hills Improvement Company, 1888.

*Buffalo New York 1851*. Ithaca, NY: Historic Urban Plans, 1975.

*Chester Hill, Mount Vernon, Westchester Co., New York; 13 Miles from the Grand Central Depot, 42nd St., New York, on the Line of the New Haven Rail Road*, by F. Mayer. New York: Ferdinand Mayer, 1890.

*City of Albany, New York*, by R. H. Bingham. Albany, NY: Sampson, Davenport, 1874.

*City of New-York*, by D. H. Burr. New York: J. H. Colton, 1834.

*City of Syracuse*, by H. D. Borden and R. Griffin. Syracuse, NY: Sage, Sons, 1868.

*Colton's New York City: Brooklyn, Jersey City, Hoboken, Etc.* New York: G. W. and C. B. Colton, 1879.

*Elementary Schools of Manhattan*. New York: Board of Education, 1960.

*Elmira, N.Y.*. Newark, NJ: Landis & Alsop, 1901.

*Enlarged and Complete Map of the Village of Lockport: Drawn from the Latest Maps on File and Other Authentic Surveys*, by J. P. Haines. Buffalo, NY: Hall & Mooney, Lith, 1845.

*Hinrichs' Guide Map of the Central Park*, by O. Hinrichs. New York: Mayer, Merkel & Ottmann, Lithographers, 1875.

*Hitchcock's Second Plan—Homes for the People at West Flushing: Corona, Queens County, Long Island, N.Y.*, by B. W. Hitchcock. New York: B. W. Hitchcock Music, 1980.

*Map B of Part of the Laurel Hill Tract: Situated on Newtown Creek opposite Greenpoint & Brooklyn, Belonging to Augustus Rapelye*, by J. Twiss et al. New York: G. Hayward Lith., 1856.

*Map of Canajoharie, Montgomery Co., New York*. New York: T. & J. Slator, C.E., 1857.

*Map of Cazenovia, Madison Co., N.Y.* New York: Henry Hart, 1852.

*Map of Cottage Sites at Dearman: Never before Offered, Being Those Reserved by the Proprietors but Now Sold in Order to Close the Interest of the Different Parties Engaged in Establishing This Beautiful Village*. New York: Miller's Lith, 1850.

*Map of Dexter in Jefferson County N.Y.*, by J. Denison. New York: A. E. Baker's Lith, 1836.

*Map of Dutchess Co., New York: From Actual Surveys*, by C. G. Bachman et al. Philadelphia: John E. Gillette, 1858.

*Map of Hamilton, Madison County, New York*. Philadelphia: Benj. A. Clark, 1858.

*Map of Kings County: With Parts of Westchester, Queens, New York & Richmond: Showing Farm Lines, Sounding*, by M. Dripps. New York: M. Dripps, 1872.

*Map of Long Island City, Queens Co., N.Y.*, by E. Whitney. Astoria, NY: E. Whitney, 1876.

*Map of Marsh Lots 10, 11, 12, 13, 14, and 15 at Rockaway Beach, Borough of Queens, City of New York, N.Y.* New York: City Surveyor, 1907.

*Map of New-York City*. Buffalo, NY: Matthews, Northrup, 1889.

*Map of New York Island and Vicinity*, by H. Stevens and W. Heath. New York: H. Stevens and W. Heath, 1840.

*Map of Pleasantville in Queens County, N.Y.: 25 Miles from the City of New York*, by G. C. Burst. New York: Narine, 1887.

*Map of Property Situated in the Village of Flushing, Queens County, L.I., Belonging to Gabriel Winter*, by E. Smith and G. Winter. New York: Perkins Sun Litho., 1850.

*Map of Rochester from a Correct Survey*, by V. Gill, J. Child, and J. F. Morin. New York: City of New York, 1832.

*Map of Saratoga Springs, Saratoga Ct., New York*, by J. Bevan. New York: John Began, 1859.

*Map of the Borough of Queens*. New York: Williams Map & Guide, 1923.

*Map of the City of Brooklyn, L.I.: Showing the Streets as at Present Existing with the Buildings and the Intended Canal and Other Works: Also the Village of Williamsburgh*, by J. F. Harrison et al. New York: Lith'y of A. Kellner, H. Camp's Lith'c Steam Press, 1850.

*Map of the City of New-York: Burr to Fiftieth St*, by J. F. Harrison. New York: Engraved and printed at Kollner's Lithographic Establishment, 1852.

*Map of the City of New-York: Extending Northward to Fiftieth St*, by J. F. Harrison, M. Dripps, and A. Kollner. New York: Kollner's Lithographic, 1852.

*Map of the City of New York and Island of Manhattan, as Laid Out by the Commissioners Appointed by the Legislature, April 3d, 1807 Is Respectfully Dedicated to the Mayor, Aldermen and Commonalty Thereof*, by W. Bridges and P. Maverick. New York: W. Bridges and P. Maverick, 1811.

*Map of the City of Syracuse*, by C. H. Wadsworth. Syracuse, NY: Andrew Boyd, 1871.

*Map of the City of Utica*, by J. H. Francis. Boston: J. H. Francis, 1873.

*Map of the Incorporated Village of Poughkeepsie, Dutchess County, State of New York*. Poughkeepsie, NY: Henry Whinfield, 1834.

*Map of the New Subway of Greater New York: Interborrow System*. New York: Manhattan Publishing, 1918.

*Map of the New York City Subway System*, by S. J. Voorhies. New York: Union Dime Savings Bank, 1954.

*Map of the Town of Amboy*, by A. H. Foster and S. Foster. New York: Lith. of F. Heppenheimer, 1855.

*Map of the Village of East New York Elmhurst Showing Places of Historical Interest*, by J. H. Innes. New York: City History Club, 1908.

*Map of the Village of East New York, Kings County, and Part of the Town of Jamaica, Queens County, Long Island, New*

*York*, by J. C. E. Hinrichs. New York: J. C. E. Hinrichs, Civil Engineer & Surveyor, 1871.

*Map of the Village of Flushing, Queens County, New York: 1894*, by G. A. Roullier and R. A. Welcke. Flushing, NY: G. A. Roullier, 1894.

*Map of the Village of Jamaica, Queens County, N.Y.: Showing Every Lot and Building*, by M. Dripps. New York: M. Dripps, 1876.

*Map of the Village of West Flushing in the Township of Newtown, Queens Co. Long Island, New York: Comprising the Parts Laid Out and Described on Sectional Maps No. 1, 2, 3, and 4*. New York: Barker & Elliott, 1854.

*Map of Troy, Also West Troy, and Green Island*, by H. F. Walling. Troy, NY: Sampson, Davenport, 1873.

*Map of Village Lots and Cottage Sites at Dearman, Westchester Co.: Adjacent to the Hudson River Station & Piermont Ferry Depot*. New York: Miller's Lith, 1850.

*Map of Woodsburgh, Long Island*, by S. Mosher. New York: Henry Seibert, 1871.

*Map Showing Proposed Waterfront Improvements, Terminals & Railroad Installations: At South Brooklyn, Gowanus Bay, and Erie Basin, Borough of Brooklyn*. New York: City of New York, Department of Docks and Ferries, 1911.

*Matthews-Northrup New Map of the City of Buffalo*. Buffalo, NY: J. N. Matthews, 1916.

*Merchants' Association Hotel and Theater Map*. New York: Merchants Association of New York, 1906.

*New York City and Environs*, by A. P. Lindenkohl and P. Witzel. New York: Westermann, 1860.

*New Yorke: 1695*, by J. Miller. New York: John Miller, 1695.

*Official Map of Long Island City, Queens Co. N.Y.* Long Island City, NY: Julius von Hunerbein, 1893.

*Old Jamaica Village*, by J. H. Innes. New York: J. H. Innes, 1908.

*Plan of Forts Green, Laurence & Swift and Lines of Intrenchments Constructed in the Vicinity of Brooklyn for the Defence of the City of New York*. New York: Office of the Chief of Engineers and the U.S. War Department, 1814.

*Plan of Old Flushing Showing the Historic Sites*, by J. H. Innes. New York: J. H. Innes, 1908.

*Routes of the Interborough Rapid Transit Company*. New York: Interborough Rapid Transit Company, 1924.

*Salem, N.Y.* Troy, NY: L. R. Burleigh, 1889.

*Sandy Hill, N.Y.* Troy, NY: L. R. Burleigh, 1884.

*Street Map of Fulton, N.Y.: With Street Numbers*, by A E. Worden. Fulton, NY: A. E. Worden, 1903.

*View of Oneida, N.Y.: 1874*. New York: O. H. Bailey, 1874.

## Ohio

*Bryan's New Map of the City of Portsmouth and Part of Clay Township Scioto Co., Ohio*, by R. A. Bryan. Philadelphia: J. L. Smith, 1889.

*Map of Erie & Part of Ottowa Counties, Ohio: Showing the Sections, Farms, Lots and Villages*, by P. Nunan et al. Philadelphia: Philip Nunan, 1863.

*Map of Hamilton County, Ohio: Compiled and Executed from Official Records Surveys and Conveyances up to the Date of Completion*, by G. Moessinger and F. Bertsch. Cincinnati, OH: Geo. Moessinger & Fred Bertsch, 1884.

*Map of Hamilton County, Ohio: Exhibiting the Various Divisions and Sub Divisions of Land with the Name of the Owners & Number of Acres in Each Tract together with the Roads, Canals, Streams, Towns throughout the County*, by A. W. Gilbert and E. O. Weed. Cincinnati, OH: A. W. Gilbert, 1856.

*Map of Huron County, Ohio: Showing the Farms & Original Lots in Each Township with Names of Proprietors, Also Plans of Villages & Business Directory*, by P. Nuan. Pittsburgh, PA: Krebs, 1859.

*Map of the City of Cincinnati*, by J. Gest and W. Haviland. Cincinnati, OH: Joseph Gest, 1838.

*Map of the City of Portsmouth and Wayne Township, Scioto County, Ohio*, by R. A. Bryan. Portsmouth, OH: N. W. Evans, 1868.

*Map of Spring Grove Cemetery, Adjoining Cincinnati: Showing All Burial Lots with Their Numbers and Sections, as Laid Out to Date, together with about 2000 Names of Owners of the Largest Sized Lots, and Location of a Number of the Principal Monuments, Etc.*, by M. and R. Burgheim. Cincinnati, OH: Robert Clarke, 1883.

## Oregon

*Map of Multnomah County, Oregon: Compiled from County Records, Railroad Surveys, and Other Official Data*, by R. A. Habersham. New York: Julius Bien & Co. Photo. Lith, 1889.

## Pennsylvania

*Beers, Ellis & Soule's Map of Venango Co., Penn.: From Actual Surveys*, by E. Beers. New York: Beers, Ellis & Soule, 1865.

*Draught of a Part of the York Farm Tract Situate in Norwegian Township, Schuylkill County: As Divided into Lots in Aug. 1847*, by S. Lewis. Philadelphia: S. Lewis, 1847.

*Lloyd's Map of the Great Oil Region of Allegheny River, Cherry & Cherry-Tree Runs, and Pithole Creek: In Cornplanter, Cherry-Tree & Allegheny Townships, Venango County, Pennsylvania*, by J. T. Lloyed. New York: J. T. Lloyd, 1864.

*Map of Allegheny County, Pennsylvania: From Actual Surveys*, by A. Lee and O. Krebs. Pittsburgh, PA: Otto Krebs, 1883.

*Map of Berks County, Pennsylvania: From Surveys*, by M. S. Henry et al. Reading, PA: M. S. Henry, 1854.

*Map of Cambria Co., Pennsylvania: From Actual Surveys & Official Records*, by D. G. Beers et al. Philadelphia: A. Pomeroy, 1867.

*Map of Clearfield Co., Pennsylvania: From Actual Surveys & Official Records*, by D. G. Beers et al. Philadelphia: A. Pomeroy, 1866.

*Map of Clinton County, Pennsylvania: From Actual Surveys*, by H. F. Walling, K. Volkmar, and F. W. Beers. New York: Palmer, 1862.

*Map of Crawford Co., Pennsylvania: From Actual Surveys*, by F. W. Beers. Philadelphia: A. Pomeroy & S. W. Treat, 1865.

*Map of Cumberland County, Pennsylvania: From Actual Surveys*, by H. F. Bridgens. Philadelphia: H. F. Bridgens, 1858.

*Map of Delaware County, Pennsylvania: From Original Surveys, with the Farm Limits*, by J. W. Ash, K. Gustavus, and R. P. Smith. Philadelphia: Robert P. Smith, 1848.

*Map of Erie County, Pennsylvania: From Actual Surveys*, by J. Chace, O. McLeran, and I. W. Moore. Philadelphia: Isaac W. Moore, 1855.

*Map of Lehigh County, Pennsylvania: From Original Surveys*, by G. A. Aschbach and M. H. Traubel. Allentown, PA: G. A. Aschbach, C.E., 1862.

*Map of McKean Co., Pennsylvania: From Actual Surveys and Official Records*, by D. G. Beers et al. Philadelphia: D. G. Beers, 1871.

*Map of Mercer Co., Pennsylvania: From Actual Surveys*, by G. M. Hopkins et al. Philadelphia: A. Pomeroy and S. W. Treat, 1860.

*Map of Northampton Co., Pennsylvania: From Actual Surveys*, by G. M. Hopkins. Philadelphia: Smith, Gallup, 1860.

*Map of Northumberland County, Pennsylvania: From Actual Surveys*, by G. M. Hopkins et al. Montandon, PA: J. A. J. Cummings, 1874.

*Map of Potter Co., Penna.: From Recent and Actual Surveys and Official Records*, by F. W. Beers and W. A. Crosby. New York: J. W. Vose, 1893.

*Map of Schuylkill County, Pennsylvania: From Actual Surveys*, by W. Scott et al. Philadelphia: James D. Scott, 1864.

*Map of the County of Pike, Pennsylvania: Shewing the Location and Form of the Original Surveys with the Numbers by Which They Are Designated on the Commissioner's Books of Said County: Also the Townships, Streams, Roads, Plank Roads, Railroads, Canals, and Principal Places*, by J. T. Cross. New York: Charles Vinten Lithog, 1856.

*Map of the Original Warrants of Portions of Warren & Forest Counties: Carefully Compiled from the Original Drafts and Warrants of the Department of Internal Affairs of Pennsylvania*, by J. C. Gardiner. Harrisburg, PA: Department of Internal Affairs, 1881.

*Plan of the Lots Laid Out at Pittsburg and the Coal Hill*, by H. Hill. Pittsburgh, PA: J. Hills, 1787.

*Property and Insurance Atlas of the City of Reading, Berks County, Penna.: From Official Records and Actual Surveys.* Reading, PA: Forsey Breou, 1884.

*Survey of the President, Porcupine, and Redfield Petroleum Co.'s Lands: Containing in All 8,400 Acres Situate in President and Pinegrove Townships, Venango County, Pennsylvania*, by C. E. Slator. New York: Waters-Son, 1865.

South Carolina

*Complete Map and Sketch of Laurens County, S.C.* South Carolina: Kyzer and Hellams, 1883.

*Map of Marion County, South Carolina: A Complete Map Showing the Townships, Public Roads & Principle Residences, besides Other Things Not Found on Any Other Map of the County*, by P. Y. Bethea and H. Thomas. South Carolina: P. Y. Bethea, 1882.

South Dakota

*Map of Aurora County, South Dakota: Compiled and Drawn from a Special Survey and Official Records*, by E. F. Peterson. Vermillion, SD: E. Frank Peterson, 1900.

*Map of Bon Homme County, South Dakota: Compiled and Drawn from a Special Survey and from Official Records.* Vermillion, SD: Rowley & Peterson, 1893.

*Map of Brookings County, South Dakota: Compiled and Drawn from a Special Survey and from Official Records.* Vermillion, SD: E. P. Noll, 1897.

*Map of Clark County, South Dakota: Compiled and Drawn from a Special Survey and Official Records.* Vermillion, SD: E. Frank Peterson, 1900.

*Map of Codington County, South Dakota: Compiled and Drawn from a Special Survey and Official Records*, by E. F. Peterson. Vermillion, SD: E. Frank Peterson, 1898.

*Map of Deuel County, South Dakota: Compiled and Drawn from a Special Survey and Official Records*, by E. F. Peterson. Vermillion, SD: E. Frank Peterson, 1899.

*Map of Grant County, South Dakota: Compiled and Drawn from a Special Survey and Official Records.* Vermillion, SD: E. Frank Peterson, 1899.

*Map of Hamlin County, South Dakota: Compiled and Drawn from a Special Survey and Official Records.* Vermillion, SD: E. Frank Peterson, 1897.

*Map of Hanson County, South Dakota: Compiled and Drawn from a Special Survey and Official Records.* Vermillion, SD: Rowley & Peterson, 1893.

*Map of Kingsbury County, South Dakota: Compiled and Drawn from a Special Survey and Official Records.* Vermillion, SD: E. Frank Peterson, 1899.

*Map of Lake County, South Dakota: Compiled and Drawn from a Special Survey and Official Records.* Vermillion, SD: E. Frank Peterson, 1899.

*Map of Lincoln County, South Dakota: Compiled and Drawn from Official Records and a Special Survey.* Vermillion, SD: E. Frank Peterson, Rowley & Peterson, 1893.

*Map of McCook County, South Dakota: Compiled and Drawn from a Special Survey and Official Records.* Vermillion, SD: E. Frank Peterson, 1900.

*Map of Miner County, South Dakota: Compiled and Drawn from a Special Survey and Official Records.* Vermillion, SD: E. Frank Peterson, 1898.

*Map of Moody County, South Dakota: Compiled and Drawn from a Special Survey and Official Records*, by E. F. Peterson. Vermillion, SD: E. Frank Peterson, 1896.

*Map of Spink County, South Dakota: Compiled and Drawn from a Special Survey and Official Records*, by E. F. Peterson. Vermillion, SD: E. Frank Peterson, 1899.

*Map of Turner County, South Dakota: Compiled and Drawn from a Special Survey and from Official Records.* Vermillion, SC: E. P. Noll, 1893.

*Map of Union County, South Dakota: Compiled and Drawn from Official Records and a Special Survey.* Vermillion, SD: E. Frank Peterson, 1892.

*Sanborn County, South Dakota: Compiled and Drawn from a Special Survey and Official Records.* Vermillion, SD: E. Frank Peterson, 1900.

Tennessee

*Map of Davidson County, Tennessee*, by W. B. Southgate. Nashville, TN: W. W. Southgate, 1900.

*Map of Gibson County, Tenn.: From Actual Surveys and Official Records*, by D. G. Beers. Philadelphia: D. G. Beers, 1877.

*Map of Madison County, Tenn.: From Actual Surveys and Official Records*, by D. G. Beers. Philadelphia: D. G. Beers, 1877.

*Map of Marshall County, Tenn*, by W. M. Carter. Lewisburg, TN: Rand McNally, 1899.

*Map of Maury Co., Tennessee: From New and Actual Surveys*, by D. G. Beers. Philadelphia: D. G. Beers, 1878.

*Map of Montgomery County, Tennessee: From Actual Surveys and Official Records*, by D. G. Beers. Philadelphia: D. G. Beers, 1877.

*Map of Rutherford County, Tenn.: From Actual Surveys*, by D. G. Beers. Philadelphia: D. G. Beers, 1878.

*Map of Shelby County, Tennessee*, by M. T. Williamson. Memphis, TN: Strobridge, 1888.

*Map of Sumner Co., Tennessee: From New and Actual Surveys*, by D. G. Beers. Philadelphia: D. G. Beers, 1878.

*Map of the City of Memphis: Including Fort Pickering and Hopefield, Ark.; Together with the Original Grants and Their Subdivisions*, by W. E. Rucker. Cincinnati, OH: Klauprech & Menzel's Lith., 1858.

*Map of the City of Nashville and Vicinity*, by W. F. Foster. Nashville, TN: W. F. Foster, 1877.

Texas

*Map of the City of Fort Worth and Vicinity*, by W. B. King. Fort Worth, TX: W. B. King, 1887.

*Map of the City of San Antonio, Bexar County: Including Suburbs Both North and South*, by N. Tengg. San Antonio, TX: Nic Tengg, 1924.

*Map of the Post of Fort Brown, Brownsville, Texas*, by L. M. Haupt and E. R. S. Canby. Brownsville, TX: L. M. Haupt and E. R. S. Canby, 1877.

*Nelson & White's Official Map of Beaumont: Compiled from Official Data and Reliable Surveys.* Beaumont, TX: Nelson & White, 1902.

*Pocket Map Showing the Railroads, Street Railways, Manufactories, Deep Water Connections, Blocks and Subdivisions of the City of Houston.* Houston, TX: Wm. M. Thomas, 1890.

*Rullman's Map of the City of San Antonio*, by J. D. Rullmann. San Antonio, TX: Rullman & Stappenbeck, 1890.

Vermont

*Map of Addison County, Vermont*, by H. F. Walling. Boston: Baker & Tilden, 1857.

*Map of Orange County, Vermont*, by H. F. Walling. New York: Baker & Tilden, 1858.

*Map of the City of Rutland, Vt.* New York: Tuttle Publishing, 1910.

*Map of the Counties of Franklin and Grand Isle, Vermont: From Actual Surveys*, by H. F. Walling. Boston: Baker, Tilden, 1857.

*Map of the Counties of Orleans, Lamoille, and Essex, Vermont*, by H. F. Walling. New York: Loommis & Way, 1859.

*Map of the County of Essex, Vermont.* New York: F. W. Beers, 1878.

*Map of the Town of Plymouth, Windsor County, Vermont*, by C. A. Scott et al. Boston: Albert D. Hager, 1859.

*Map of the Village of Highgate, Franklin County, State of Vermont*, by A. Martin. New York: Baker's Lithogy, 1836.

*Map of Washington County, Vermont*, by H. F. Walling. New York: Baker & Tilden, 1858.

*Map of Windsor County, Vermont*, by H. Doton. Pomfret, VT: Hosea Doton, 1856.

*McClellan's Map of Windham County, Vermont*, by J. Chace. Philadelphia: C. McClellan, 1856.

*Plan of the City of Burlington Shewing Water & Sewer Service.* Burlington, VT: City of Burlington, 1855.

*Plan of the Fairhaven Slate Quarry Estate, the Property of the Allen Slate Company, Vermont*, by C. S. Richardson. New York: Wm. W. Rose, Lith, 1856.

*Scott's Map of Rutland County, Vermont*, by I. W. Moore et al. Philadelphia: Owen McLeran & James D. Scott, 1854.

Virginia

*Colonial Land Patents and Grantees: Calfpasture Rivers, Augusta County, Virginia*, by M. Leitch. Staunton, VA: Meredith Leitch, 1947.

*First Addition to Carlin Springs, Alexandria Co., Va.* Washington, DC: Norris Peters, 1890.

*Leesburg's Old and Historic District*, by D. Parry. Leesburg, VA: Debi Parry, 2013.

*Lower Parish of Nansemond County, Va. with Adjoining Portions of Norfolk County: Elizabeth City Shire 1634, New Norfolk County 1636, Upper Norfolk County 1637, Nansemond County 1642*, by J. Granbery. Virginia: J. H. Granbery, 1948.

*Map of Elizabeth City*, by E. A. Semple, W. Ivy, and C. Hubbard. Philadelphia: E. W. Smith, 1892.

*Map of Henrico County, Virginia: Showing Portions of Chesterfield County Also City of Richmond*. Richmond, VA: T. Crawford Redd, 1911.

*Map of Holmes Island and the Adjoining Upland: The Property of R. B. Mason Esq.*, by F. Macveagh and R. B. Mason. Washington, DC: F. Macveagh and R. B. Mason, 1800.

*Map of the Upper District of Henrico County; Map of the Lower District of Henrico County*, by J. T. Redd. Richmond, VA: A. Hoen, 1887.

*Map of Washington Co., Virginia and Contiguous Territory: Showing Mineral Resources, Agricultural Areas, Water-Powers, Railway Facilities, Roads, Streams, Educational Institutions, Names and Residences of Many Land-Owners, Post-offices*. Philadelphia: J. L. Smith Map, 1890.

*Plat of Curtis & Burdett's Subdivision of Carlin Springs, Alexandria County, Virginia: 260 Ft. above Sea Level*, by H. W. Newby. Washington, DC: Baxter & MacGowan, 1887.

*Survey of a Plot in Lancaster and Northumberland Counties Bounded by the Chesapeake Bay, Rappahnnock River, and the Potomac River*, by W. Ball. Washington, DC: W. Ball, 1736.

Washington

*Anderson's New Guide Map of the City of Seattle and Environs, Washington*. Seattle, WA: O. P. Anderson, 1890.

*Foot Traffic in Seattle, Washington, Wednesday, April 26th, 1922*. Seattle, WA: Kroll Map, 1922.

*Map of Port Angeles, Washington Territory*. Port Angeles, WA: W. J. Ware, 1864.

*Map Showing the Townsite of Port-Angeles and Vicinity*. Port Angeles, WA: W. J. Ware, 1891.

Washington, D.C.

*5th Plan, from 7th East to 13th Street and G Street South to East Capitol Street: S.E. Washington D.C.*, by H. C. Gauss. Washington, DC: Association of the Oldest Inhabitants of the District of Columbia, 1794.

*16th Plan from E Street South to W and from 3d West to 4th Street East: Washington D.C.*, by H. C. Gauss. Washington, DC: Association of the Oldest Inhabitants of the District of Columbia, 1798.

*A. P. Fardon & E. B. Townsend's Subdivision to Be Known as Woodridge: Being Tracts Called Barbadoes & Scotland Enlarged in the District of Columbia*, by W. H. Grant and A. P. Fardon. Washington, DC: A. P. Fardon, 1887.

*Baist's Map of the Vicinity of Washington D.C.*, by G. Wm. Baist. Philadelphia: G. Wm. Baist, 1904.

*Blanchard's Guide Map of Washington*, by I. H. Blanchard. New York: Isaac H. Blanchard, 1901.

*Blue Plains and Part of Belleview Surveyed and Subdivided: Washington D.C.*, by T. Berry and R. Coyle. Washington, DC: T. Berry, 1868.

*Boyd's Map of the City of Washington and Suburbs, District of Columbia*, by Wm. H. Boyd. Washington, DC: Wm. H. Boyd, 1884.

*Burrville, Being Part of the Sheriff Estate, at the North East Part of the District of Columbia*, by A. C. Tabor, M. D. Dean, and C. A. McEuen. Washington, DC: Charles A. McEuen, 1880.

*Cadastral Map of Beatty and Hawkins's Addition to Georgetown, Washington D.C.*, by W. Smith. Washington, DC: W. Smith, 1755.

*Cadastral Map of Eastern Part of LeDroit Park and Blocks Northward to the Soldiers' Home, Washington D.C.*, by J. F. Thos. Washington, DC: Thomas J. Fisher & Co., Real Estate Brokers, 1899.

*Cadastral Map of Massachusetts Avenue N.W. and Adjacent Parts of American University Park and Spring Valley, Washington D.C.* Washington, DC: Office of the Surveyor, 1910.

*Cadastral Map of Part of Brightwood, Washington D.C.*, by A. Klakring. Washington, DC: A. Klakring, 1930.

*Cadastral Map of Part of S.W. Washington D.C. Showing Land Tracts and Buildings Belonging to Notley Young*, by N. King. Washington, DC: N. King, 1796.

*Cadastral Map of "Peter's Square," Georgetown, Washington D.C.*, by W. Peter. Washington, DC: W. Peter, 1814.

*Cadastral Survey Map of a Land Tract Fronting Tennallytown Road North of Pierce's Mill Road, Tenleytown, Washington D.C.*, by Wm. H. Benton. Washington, DC: Wm. Benton, 1891.

*Cadastral Survey Map of Charles Carroll Jr.'s Land in Central Washington D.C. in 1793*, by J. F. Priggs. Washington, DC: Joseph M. Toner, 1793.

*Cadastral Survey Map of Land Tracts in Central Washington D.C. ca. 1793*, by J. F. Priggs. Washington, DC: J. F. Priggs, 1793.

*Cadastral Survey Map of Land Tracts in Washington D.C. near the Anacostia River*, by W. Hickey. Washington, DC: W. Hickey, 1840.

*Cadastral Survey Map of Pleasant Plains, Washington D.C.*, by G. Fenwick. Washington, DC: Office of the Surveyor, 1804.

*Cadastral Survey Map of Squares A and B Adjacent to Pennsylvania Avenue, N.W., Washington D.C.: New Part of the Mall*, by F. C. De Krafft. Washington, DC: Office of the Surveyor, 1822.

*Cadastral Survey Map of the Hayman Property, Georgetown, Washington D.C.* Washington, DC: Chesapeake and Ohio Canal, 1835.

*Cadastral Topographical Map of Part of N.W. Washington D.C.*, by A. Bastert, J. Enthoffer, and A. Peterson. Washington, DC: A. Bastert, 1872.

*Cliffbourne: Washington D.C.*, by T. J. Fisher. Washington, DC: Thomas J. Fisher, 1899.

*Comprehensive Plan for the National Capital, Mass Transportation Plan*. Washington, DC: United States National Capital Planning Commission, 1974.

*Comprehensive Plan for the National Capital, Parks and Recreation Facilities Diagram No. 1, National Open Space System 1970/1985*. Washington, DC: United States National Capital Planning Commission, 1974.

*Copy from Actual Survey of the 3 Sisters: Potomac River, Washington D.C.*, by H. W. Brewer. Washington, DC: H. W. Brewer, 1893.

*Correct Map of the City of Washington: Capital of the United States of America: Lat. 38.53 n., Long. 0.0*, by W. J. Stone and P. Force. Washington, DC: Davis & Force, 1820.

*Dermott or Tin Case Map of Washington 1797–8*, by J. R. Dermott. Washington, DC: United States Coast and Geodetic Survey Office, 1797.

*Development of the Central Area West and East of the Capitol—Washington D.C. 1941*. Washington, DC: United States National Capital Park and Planning Commission, 1941.

*Division of Part of a Tract of Land in Washington County D.C. Called "Mount Pleasant": As the Same Was Made by the Devisees of Robert Peter Deceased, and the Allotment Thereof Made March 8th 1809*, by B. D. Carpenter. Washington, DC: B. D. Carpenter, 1809.

*Estate of Samuel Blodget, Jr.: One of the Founders of the City of Washington, D.C.: Jamaica, Washington D.C.*, by L. Blodget and S. Blodget. Washington, DC: B & S Blodget, 1870.

*Fairview Heights Subdivision of Parts of Scott's Ordinary Terra Firma and Alliance Tracts*, by J. E. Beall et al. Washington, DC: Breckinridge Family, and John E. Beall, 1890.

*Fernwood Heights: Washington D.C.* Washington, DC: T. J. Fisher, 1900.

*Fort Stevens Park, General Plan: Washington D.C.* Washington, DC: U.S. Department of the Interior, National Park Service, National Capital Parks, 1935.

*Geographical, Statistical, and Historical Map of the District of Columbia*, by H. C. Carey and I. Lea. Philadelphia: H. C. Carey and I. Lea, 1822.

*Hall and Elvans' Subdivision of Meridian Hill, Washington County, D.C.: Sept. 1867*, by C. H. Bliss. Washington, DC: Hall's Real Estate Exchange, 1867.

*Historical Map of the City of Washington, District of Columbia: View of the City & Location of the Houses in the Year 1801–02; The Beginning of Washington*, by A. C. Harmon. Washington, DC: A. C. Harmon, 1801.

*Le'Droit Park as Recorded in the Surveyor's Office*. Washington DC: A. L. Barber, 1880.

*L'Efant and F. D. Owen*. Washington, DC: Corps of Engineers, U.S. Army, Office of Public Buildings and Grounds, 1791.

*The Mall and Vicinity, Washington: 1927*. Washington, DC: Public Buildings Commission, 1927.

*The Mall as Proposed by Pierre L'Enfant 1790: From the Original*. Washington DC: Corps of Engineers, U.S. Army, Office of Public Buildings and Grounds, 1790.

*The Mall, Washington, D.C.: Plan Showing Development to 1914 in Accordance with the Recommendations of the Park Commission of 1901*. Washington, DC: United States Senate Park Commission, 1914.

*Map Exhibiting the Property of the U.S. in the Vicinity of the Capitol: Colored Red, with the Manner in Which It Is Proposed to Lay Off the Same in Building Lots, as Described in the Report to the Sup't of the City to Which This Is Annexed*. Washington, DC: Office of the Surveyor, 1815.

*Map of Brightwood Park in the District of Columbia: 1890*, by H. B. Looker and Wm. H. Benton. Washington, DC: H. B. Looker and Wm. H. Benton, 1890.

*Map of Carroll's Land: Central Washington, D.C.*, by J. F. Priggs and C. Carroll. Washington, DC: J. F. Priggs and C. Carroll, 1793.

*Map of Carrollsburg Showing Its Street and Lot System with the Later Streets and Lots of Washington D.C. Superimposed*, by W. Forsyth. Washington, DC: Office of the Surveyor, 1858.

*Map of District North of Washington D.C.*, by H. L. Abbot and F. W. Vaughan. Washington, DC: H. L. Abbot and F. W. Vaughan, 1859.

*Map of Georgetown, D.C., April 1934: Present Names of Streets Black, Former Names of Streets Red*, by T. R. Fullalove. Washington, DC: T. R. Fullalove, 1934.

*Map of Georgetown in the District of Columbia*, by W. Bussard and W. Harrison. Washington, DC: W. Bussard and W. Harrison, 1830.

*Map of Georgetown in the District of Columbia: Prepared from Surveys and Other Data under an Act of the Legislature Approved Dec'r 28th 1871*, by W. Forsyth and H. T. Taggart. Washington, DC: W. Forsyth, 1871.

*Map of George Town, 1751*, by A. C. Harmon and J. F. A. Priggs. Washington, DC: A. C. Harmon and J. F. A. Prigggs, 1758.

*Map of George Truesdell's Addition to the City of Washington: Being a Subdivision of a Tract of Land Known as Eckington, Which Tract Is a Part of the Original Tract Called Youngsborough: May 1887*, by W. Forsyth, W. T. O'Bruff, and G. Truesdell. Washington, DC: Office of the Surveyor, 1887.

*Map of Greater Washington*, by Wm. J. Rogers. Washington, DC: Engineering Department, 1922.

*Map of Kalorama Heights, Washington, D.C.*, by H. K. Vielé et al. Washington, DC: Thomas J. Fisher, 1890.

*Map of Meridian Hill Subdivision, Washington, D.C., Showing Selected Residential Lots*. Washington, DC: Breckinridge Family, 1880.

*Map of National Zoological Park, D.C.* Washington, DC: U.S. Coast and Geodetic Survey, and U.S Geological Survey, 1889.

*Map of Part of the Potomac River, from the Head of Tide Water to Alexandria: With Its Lateral Canals, Soundings, and Topography*, by W. M. Fairax. Washington, DC: Alexandria Canal Company, 1835.

*Map of Part of Washington D.C. Central Business District in the Vicinity of the Center Market and Tiber Creek*, by H. C. Gauss. Washington, DC: Association of the Oldest Inhabitants of the District of Columbia, 1800.

*Map of Property Purchased from the United States by Wm. T. Dove, Esq., South Half of Square No. 3 in the City of Washington: Colored Red, Also the Origl. Boundary of the Potomac River*, by W. Forsyth and W. T. Dove. Washington, DC: W. Forsyth, 1854.

*Map of Rock Creek Cemetery near Washington, D.C.* Washington, DC, 1884.

*Map of Streetcar Lines in Anacostia and Surrounding Districts, S.E. Washington D.C.* Washington, DC: Board of Commissioners and United States Congress, 1904.

*Map of Survey of Part of Pleasant Plains knows as "Bellevue,"* by S. C. Hill. Washington, DC: Surveyor's Office, 1870.

*Map of Survey of Whitney Close, Divided into Squares 1 to 11 Inclusive, in the County of Washington, District of Columbia: Recorded in Book "County No. 6," Pages 62, 63, Surveyor's Office D.C., Dec. 9, 1886*. Washington, DC: B. H. Warner, 1886.

*Map of That Part of the City of Washington on Which Is Situated the House and Grave-Yard Belonging to the Heirs of John Davidson*, by J. M. Toner and N. King. Washington, DC: J. J. Toner and N. King, 1794.

*Map of the Cities of Washington and Georgetown D.C.*, by W. H. Morrison and O. H. Morrison. Washington, DC: W. H. and O. H. Morrison, 1876.

*Map of the Cities of Washington and Georgetown, D.C.: Exhibiting the Public Buildings, Principal Hotels, Churches, Etc.* Washington, DC, 1860.

*Map of the City of Washington: Showing the Sub-Divisions, Grades, and the General Configuration of the Ground in Equidistances from 5 to 5 Feet Altitude*, by A. Bastert, J. Enthoffer, and P. H Donegan. Washington, DC: U.S. Coast Survey, 1872.

*Map of the City of Washington, D.C.: Showing Public Buildings and Places of Interest*. New York: Hart & Von Arx, 1885.

*Map of the City of Washington and Environs*, by A. G. Gedney and W. H. Morrison. Washington, DC: W. H. Morrison, 1884.

*Map of the City of Washington and Surroundings*, by F. R. Fava Jr. and B. L. Walker. Washington, DC: Bartow L. Walker, 1890.

*Map of the City of Washington and Surroundings Showing Recent Subdivisions*, by A. G. Gedney. Washington, DC: W. H. Lowdermilk, 1887.

*Map of the City of Washington Showing Project for Re-Arrangement of Water Distribution System*, by C. D. Cole. Washington, DC: District of Columbia, Engineer Department, 1929.

*Map of the City of Washington Showing the Public Reservations under Control of Office of Public Buildings and Grounds*, by J. Stewart and J. M. Wilson. Washington, DC: U. S. Office of the Public Buildings and Grounds, 1894.

*Map of the City of Washington Showing the System of Water Supply & Distribution: To Accompany the Annual Report of Cap't J. L. Lusk, Corps of Eng'rs U.S.A.*, by J. L. Lusk. Washington, DC: U.S. Army Corps of Engineers, 1888.

*Map of the City of Washington Showing United States Reservations*, by T. Bingham et al. Washington, DC: U.S. Senate, 60th Congress, 2nd Session, 1908.

*Map of the District of Columbia*. Washington, DC: United States Federal Housing Administration and United States Works Progress Administration, 1937.

*Map of the District of Columbia: 1901*, by J. G. Langdon. Washington, DC: Commission on the Improvement of the Park System, 1901.

*Map of the District of Columbia and Adjacent Portions of Maryland and Virginia*, by W. Schoepf et al. Washington, DC: Thos. J. Fisher, 1891.

*Map of the District of Columbia, October 1907: Showing Present and Proposed Streets: Compiled from Latest Data at the Office of the Engineer Commissioner, District of Columbia*. Washington, DC: Chesapeake & Potomac Telephone; Norris Peters, 1907.

*Map of the District of Columbia Showing Public Reservations and Possessions and Areas Recommended to Be Taken as Necessary for New Parks and Park Connections*, by R. A. Outhet. Washington, DC: United States Senate Committee on the District of Columbia, 1901.

*Map of the District of Columbia Showing Real Estate Valuation Areas*. Washington, DC: United States Congress, 1912.

*Map of the Eastern Section of the City of Washington in the District of Columbia*, by F. R. Fava Jr. and J. H. Walter. Washington, DC: John H. Walter, 1890.

*Map of the Permanent System of Highways, District of Columbia*. Washington, DC: Office of the Surveyor and Office of the Engineer Commissioner, 1921.

*Map of the White House Grounds, Washington D.C.* Washington, DC, 1800.

*Map of Tract of Land Called Portroyal, Formerly the Property of Joseph Comb in the City of Washington*, by G. Fenwick and W. Forsyth. Washington, DC: Office of the Surveyor, 1790.

*Map of Washington, D.C.* Boston: Walker Lith., 1910.

*Map of Washington, D.C., and Suburbs: Showing the Latest Streets and All the New Railway and Street-Car Routes*, by F. L. Averill. Washington, DC: The Platoon, 1892.

*Map of Washington D.C. Metropolitan Area Showing Roads and Ferries as of 1792*. Washington, DC: U.S. Geological Survey, 1792.

*Map of Washington D.C. Showing Wood, Concrete, and Stone Street Pavements*, by W. J. Stone. Washington, DC: W. J. Stone, 1872.

*Map Showing Location of Public Buildings, Washington, D.C.* Washington, DC: National Family Register, 1908.

*Map Showing Properties under the Jurisdiction of the Architect of the Capitol, Washington, D.C.*, by J. G. Stewart and J. F. Robinson. Washington, DC: United States Architect of the Capitol, 1967.

*Map Showing Route of District of Columbia Suburban Railway: Sept. 1892*, by D. J. Howell. Washington, DC: D. J. Howell, 1892.

*Map Showing Section of Washington, D.C.*, by G. Robertson. Washington, DC: National Grand Lodge, 1913.

*Map Showing Stellwagen & Wolf Trustees' Subdivision of the Schuetzen Park near Washington, Dist. of Col.*, by W. Schoepf et al. Washington, DC: Latimer & Sloan, 1892.

*Map Showing Suburban Subdivisions of the District of Columbia*. Washington, DC: District of Columbia, Engineer Department, 1895.

*Map Showing Water Mains, Stop Valves, Fire Plugs, and Elevations in the City of Washington, D.C.* Washington, DC: Water Registrars Office, 1876.

*Marietta Park in the District of Columbia*, by J. P. Day. Washington, DC: Joseph P. Day, 1921.

*Massachusetts Avenue Park and Massachusetts Avenue Heights, Washington, D.C.*, by D. J. Howell and J. W. Thompson. Washington, DC: John W. Thompson, 1921.

*Memorials and Museums Master Plan: Washington, D.C.* Washington, DC: National Capital Planning Commission, 2010.

*Morrison's Map of the Country about Washington*, by J. R. D. Morrison and A. G. Gedney. Washington, DC: J. R. D. Morrison, 1888.

*Mount Pleasant S.P. Browns Suburban Subdivision: Part of N.W. Washington D.C.*, by S. P. Brown. Washington, DC: S. P. Brown, 1885.

*National Capital, Washington City, D.C.* Baltimore: Sachse, 1883.

*New Map, Washington, D.C.: Compiled from Official Surveys and Best Authorities*, by Wm. A. Flamm. Baltimore: Union Engineering & Surveying, 1900.

*No. 1. Survey for John Davidson; No. 2. Lots Surveyed on 1st Street between Fred. & Fayette Streets: Georgetown, Washington D.C.*, by H. T. Taggart. Washington, DC: H. T. Taggart, 1843.

*Oak View D.C., Former Suburban Residence of Ex. Pres*, by E. A. Greenough. Washington, DC: Thomas J. Fisher, 1890.

*One Hundred Years: Washington D.C. and Metropolitan Area*. Fort Belvoir: United States War Department, Engineer Bureau, and Intelligence United States Army, 1865.

*Palisades of the Potomac, Scenery Unsurpassed: Embracing the Additions to Washington of Drovers Rest, White Haven, Toronto Heights, and River View, Adjoining Washington and Extending along the Potomac River on Both Sides of the Conduit Road for Five Miles to the Little Falls*, by H. Brewer et al. Washington, DC: Palisades of the Potomac Land Improvement, 1890.

*Partial Cadastral Map of the Area Immediately South of the Mall, S.W. Washington D.C.*, by H. C. Gauss. Washington, DC: Association of the Oldest Inhabitants of the District of Columbia, 1794.

*Plan of a Tract of Land Called Port Royal, Covered by Part of the City of Washington*, by W. Forsyth. Washington, DC: Office of the Surveyor, 1790.

*Plan of LeDroit Park, Washington, D.C.*, by D. McClelland, J. J. Albright, and A. L. Barber. Washington, DC: D. McClelland, J. J. Albright, and A.L. Barber, 1880.

*Plan of Lots for Sale at Montello, D.C.*, by L. C. Loomis. Washington, DC, 1890.

*Plan of New City in the District of Columbia: Half a Mile from Washington*. Washington, DC: Office of the Surveyor, 1873.

*Plan of Part of the City of Washington on Which Is Exhibited the Division of Lots: Made under a Commission from the High Court of Chancery of the State of Maryland*, by N. King. Washington, DC: Nicholas King, 1805.

*Plan of Part of the City of Washington: On Which Is Shewn the Squares, Lots, &c., Divided between William Prout Esq'r and the Commissioners of the Federal Buildings, Agreeably to the Deed of Trust*, by N. King. Washington, DC: N. King, 1800.

*Plan of Proposed Subdivision of Part of Mount Pleasant and Pleasant Plains in the District of Columbia: For Mrs. E.J. Stone*, by W. Forsyth. Washington, DC: Thos. E. Waggaman, 1871.

*Plan of Smithsonian Park: Part of the Mall, Washington D.C.* Washington, DC: United States Office of Public Buildings and Grounds, 1882.

*Plan of Square No. 378: Showing the True Course of the Creek Running through It at the Time It Was Changed in Eighteen Hundred & Twenty Six, & the Present Direction, since the Change*, by F. C. De Krafft et al. Washington, DC: Office of the Surveyor, 1828.

*Plan of Squares Marked A. & B. in Mr. Bulfinch's Plan, Divided in Lots: Now Part of the Mall, N.W. Washington D.C.* Washington, DC: Office of the Surveyor, 1822.

*Plan of the City Intended for the Permanent Seat of the Government of the United States: Projected Agreeable to the Direction of the President of the United States, in Pursuance of an Act of Congress, Passed on the Sixteenth Day of July, MDCCXC, "Establishing the Permanent Seat on the Bank of the Potowmac,"* by P. C. L'Enfant. Washington, DC: United States Commissioner of Public Buildings, 1791.

*Plan of the City of Washington and Territory of Columbia*, by W. H. Lizars and D. Lizars. Washington DC: Lizars, 1819.

*Plan of the City of Washington in the Territory of Columbia: Ceded by the States of Virginia and Maryland to the United States of America and by Them Established as the Seat of Their Government, after the Year MDCCC*, by A. Ellicott and S. Hill. Washington, DC: A. Ellicott and S. Hill, 1792.

*Plan of Wharfing: Anacostia River Waterfront, Washington D.C.*, by B. Latrobe. Washington, DC: B. H. Latrobe, 1817.

*Plat Compiled from Official Records for Thomas J. Fisher & Co., Real Estate Brokers: 1324–F St. N.W., Washington D.C.: Parts of Mount Pleasant, Meridian Hill, and Columbia Heights, Washington D.C.* Washington, DC: Thos. J. Fisher, 1899.

*Plat of Part of Pleasant Plains and Lemar's Outlet: Being an Addition to Chapins Brown's Sub-Division of S.P. Brown's Homestead*, by J. C. Lang and A. P. Brown. Washington, DC: J. C. Lang, 1885.

*Plat of Part of the Estate of the Late William East, Deceased: In Squares Nos. 10, 11, 12, S. of 12, 62 & 63, with the Wharfs Fronting along Said Property: Washington D.C.*, by W. Forsyth. Washington, DC: W. Forsyth, 1856.

*Plat of Proposed Reservation at Fort Stevens, near Brightwood, D.C.* Washington, DC: Office of the Surveyor, 1900.

*Plat of Survey and Subdivision of "Washington Heights": By the Commissioners Appointed in Equity Cause No. 9912.* Washington, DC: Office of the Surveyor, 1888.

*Plat of Survey of Land Conveyed by Wm. D.C. Murdock to His Daughters Being Part of "St. Philip & Whitehaven" in Washington County, District of Columbia 1873*, by B. D. Carpenter and W. D. C. Murdock. Washington, DC: Wm. D. C. Murdock, 1873.

*Plat of Survey of Wm. M. Rapley's Property Fronting Square West of Sq. 471: Showing the Docks and Wharves Described in Lease and Also the Part Occupied by Stephenson & Bro.: Washington D.C.*, by W. Forsyth, and M. Rapley. Washington, DC: Surveyor's Office D.C., 1887.

*Plat of the Sub-Division of Part of a Tract of Land Called "Fletchalls Goodwill": Situated near Brightwood, Washington County D.C.*, by B. C. Carpenter. Washington, DC: B. D. Carpenter, 1881.

*Plat of the Survey and Subdivision of the Estate of the Late Clark Mills Esq. in Washington County D.C.*, by B. D. Carpenter. Washington, DC: Duncanson Bros, 1884.

*Plat of the Survey of the Paper Mill Tract: As Conveyed by Edgar Patterson to Eli Williams, Chas. & Daniel Carroll, by Deed, July 26th 1811: Recorded Liber A.C. Folio 93. 1876*, by B. D. Carpenter. Washington, DC: B. D. Carpenter.

*Plat of the Tracts of Land called "Mill Seat," "Philadelphia," and "Frogland," from the Original Patents as Located in Georgetown D.C. 1885*, by B. D. Carpenter. Washington, DC: B. D. Carpenter, 1885.

*Plat of the Tracts of Land Comprising the Tract Known as "Kalorama": Part of Original Tracts of Land Called "Widow's Mite" and "Pretty Prospect" Situated in the District of Columbia: Showing What Part of "Kalorama" the Tunnel for the Extension of Washington Aqueduct Passes through 1886*, by B. D. Carpenter. Washington, DC: B. D. Carpenter, 1886.

*Plat Showing Proposed U.S. Military Reservation at Fort Stevens D.C.*, by H. B. Looker and M. J. Hale. Washington, DC: H. B. Looker, 1902.

*Property Survey of That Part of Prince Georges County, Maryland, Later to Become Central Washington D.C.*, by J. F. Priggs. Washington, DC: J. F. Priggs, 1760.

*Proposed Development of Propagating Gardens, Green-Houses & Nursery: Washington D.C*, by T. Bingham et al. Washington, DC: U.S. Army Corps of Engineers, Office of Public Buildings and Grounds, 1901.

*Proposed Urban Design Concepts.* Washington, DC: United States National Capital Planning Commission, 1972.

*Public Buildings and Public Parks in the District of Columbia: Under the Jurisdiction of the Director.* Washington, DC: United States Office of Public Buildings and Public Parks of the National Capital, 1932.

*Public Buildings in the District of Columbia.* Washington, DC: United States National Park Service, Branch of Buildings, 1937.

*Receivers' Subdivision of Part of a Tract of Land in County of Washington, D.C., Called Friendship, St. Philip, and*

*Jacob, and Re-Survey of Jacob*, by W. D. C. Murdock. Washington, DC: Green & Williams, 1872.

*Richmond Park: Washington D.C.*, by W. F. Matteson. Washington, DC: William F. Matteson, 1908.

*Roose's New Map of Washington and Its Suburbs*, by W. S. Roose and N. Peters. Washington, DC: W. S. Roose, 1895.

*Sherman's Subdivision: Of Part of Columbia Heights, Washington D.C.* Washington, DC: Board of Commissioners D.C., 1885.

*Sketch of a Survey Made for Col. W'm Hickey, under a Warrant from the United States, to Take up Vacant Land: Made to Lay before His Lawyers to Determine Some Legal Points: Part of Washington D.C. Adjacent to the Anacostia River*, by L. Carbery and W. Hickey. Washington, DC: L. Carbery, 1840.

*Sketch of Washington in Embryo: viz., Previous to Its Survey by Major L'Enfant, 1792*, by E. F. Faehtz et al. Washington, DC: U.S. Capitol Centennial Committee, 1792.

*Sketch Plan for Landscaping the Grounds of the President's House, ca. 1802–05*, by B. H. Latrobe, R. Mills, and T. Jefferson. Washington, DC: B. H. Latrobe, R. Mills, and T. Jefferson, 1802.

*Square No. 118: Washington D.C.*, by H. C. Gauss. Washington, DC: Association of the Oldest Inhabitants of the District of Columbia, 1800.

*Square No. 602, City of Washington. 1799.* Washington, DC: Office of the Surveyor, 1799.

*Square No. 653, City of Washington*, by J. R. Dermott. Washington, DC: J. R. Dermott, 1794.

*Square No. 667, City of Washington. 1799.* Washington, DC: Office of the Surveyor, 1799.

*Statistical Map No. 10 Showing the Location of Street Railways, City of Washington*, by W. T. Rossell. Washington, DC: Norris Peters, 1892.

*Subdivision of Fort Lincoln Heights, District of Columbia*, by J. C. Lang and T. A. Mitchell. Washington, DC: J. C. Lang, 1892.

*Subdivision of Property in the County of Washington, D.C.*, by W. Forsyth. Washington, DC: Kilbourn & Latta, 1869.

*Subdivision of the Howard University Lands.* Washington, DC: Board of Commissioners, District of Columbia, 1885.

*Suggestion for a Car Line Route along the Site of the Old James Creek Canal.* Washington, DC: City of Washington, 1918.

*Survey Map of Friendship Tract, Washington D.C.*, by H. C. Gauss. Washington, DC: Oldest Inhabitants of the District of Columbia, 1800.

*Survey of Parts of Lots Nos. 251 & 252 in Beattie & Hawkins Addition to Georgetown D.C.: Owned by Peter Dill Esq. 1882*, by H. W. Brewer. Washington, DC: H. W. Brewer, 1882.

*Survey of the Original Landholdings in Washington D.C. in the Area Bounded by The Mall, North Capitol St., Tiber Creek, N St. N.W., and Seventh St. N.W.*, by L. Carbery. Washington, DC: Office of the Surveyor, 1867.

*Territory of Columbia*, by A. Ellicott. Washington, DC: A. Ellicott, 1793.

*Thos. E. Waggaman and Jno. Ridout, Trustees,' Addition to the City of Washington, Formerly Called "Woodley Park": Lots Extended to Center of Avenues*, by B. D. Carpenter, J. F. Waggaman, and W. Forsyth. Washington, DC: John F. Waggaman, 1888.

*Todd & Brown's Subdivision of Part of "Pleasant Plains & Mt. Pleasant," Suburbs of Washington D.C.*, by W. Forsyth. Washington, DC: W. Forsyth, 1868.

*Topography of the Federal City, 1791*, by D. Hawkins. Washington DC: D. Hawkins, 1791.

*Townsend's Subdivision to Be Known as Woodridge: Being Tracts Called Barbadoes & Scotland Enlarged in the District of Columbia.* Washington, DC: E. B. Townsend, Real Estate Broker, 1887.

*Trees and Shrubs of Lafayette Square: Washington D.C.* Washington, DC: United States Department of the Interior, National Park Service, National Capital Parks, 1942.

*Trimmer's Subdivision of North Columbia Heights, Washington, D.C.: Dec. 1901*, by Wm. H. Benton. Washington, DC: Wm. H. Benton, 1901.

*True Copy of Plat as Recorded in Gov. Shepherd, Page 141: Part of Pleasant Plains, Washington D.C.*, by W. Forsyth. Washington, DC: W. Forsyth, 1860.

*United States City Plans 1:36,000. Washington and Vicinity.* Washington, DC: United States Army Map Service, 1955.

*United States National Arboretum.* Washington, DC: United States Department of Agriculture, 1978.

*Washington and Suburbs, District of Columbia, Showing Permanent System of Highways*, by F. Weller and A. Perley. Baltimore: Wm. A. Flamm, 1902.

*Washington in 1800*, by W. B. Bryan. Washington, DC: W. B. Bryan, 1800.

*Washington, the Nation's Capital.* Washington, DC: National Capital Parks, National Park Service, 1998.

*William Corcoran's Subdivision of Lots: Between High & Congress & Bridge & Gay Streets, Georgetown D.C. 1839*, by W. W. Corcoran. Washington, DC: W. W. Corcoran, 1839.

### West Virginia

*Farm Line Map of Cass, Grant & Clay Dist's. Monongalia Co. and Part of Paw Paw Dist. Marion Co., W. Va.* Morgantown, WV: M.V.E., 1915.

*Farm Line Map of Marshall Co., West Virginia*, by J. B. Hogg. Uniontown, PA: J. B. Hogg, 1909.

*Insurance Map of Charleston, W. Va.* Charleston, WV: Freedley & Morris, 1874.

*Map of Raleigh County and Western Part of Summers County Showing Topography*, by I. C. White. Wheeling: West Virginia Geological and Economic Survey, 1916.

*Map of the Property of the Western Mining & Manufacturing Co. Situate on Drody's Creek, Coal River Peytona, Boone County, Western Virginia.* Philadelphia: Sinclair's Lith., 1852.

*Palmer's Farm Map of Brooke County, W. Va*, by J. C. Palmer. Wellsburg, WV: John C. Palmer Jr., 1905.

# 5
# Railroad Maps

Rail lines have been mapped since the first tracks were laid in North America, and over time these maps progressed to illustrating the entire railroad track system in the United States and Canada, linking the Atlantic and Pacific Oceans as well as parts of the two countries together. The development of steam-powered railways in the nineteenth century revolutionized transportation and played a key role in the process of industrialization. For the first time, large quantities of goods such as coal, grain, and even motor vehicles were being transported from coast to coast, opening manufacturing sites, creating employment, and building cities. Knowing the locations of railways and rail stops was immensely important, so maps that allowed the measurement of distance between stops and important points of interests were of great value.

Railroad maps vary in area coverage, content, scale, and purpose; some maps focus on the major railway systems, displaying several on one map, whereas others highlight particular segments of a line as it relates to a town or road. Railway maps have been used to conduct surveys, to present future planned routes, and to find the best route of travel for shipping and other business purposes.

The following list of railroad maps, separated by provinces and states, has been compiled using the Library of Congress's online catalog of nineteenth-century digitized U.S. historical railroad maps and using WorldCat for Canadian content. The Library of Congress has a fabulous selection of railroad maps that highlight the development of railroad mapping, capturing and illustrating travel and settlement growth, as well as industrial and agricultural development.

## CANADA

### Canada-Wide

*Canada Railroad Systems.* Austin, TX: DeskMap Systems, 2007.
*Map (on Mercator's Projection) Showing the Several Proposed Lines of Railroad in the North American Provinces of Nova Scotia, New Brunswick and Part of Canada.* Ireland: Irish University Press, 1976.
*Sketch of Part of Canada Shewing the Railroads, Completed and Projected.* Montreal, QC: Montreal Herald, 1850.

### Alberta

*Alberta Railroads.* Chicago: Rand McNally, 1911.

### British Columbia

*British Columbia Railroads.* Chicago: Rand McNally, 1911.
*Railroad Map of Greater Vancouver, B.C.* Vancouver, BC: Canadian Historical Association, Pacific Coast Branch, 1971.
*Railroad Proposals for the Roberts Bank Port Area, British Columbia.* Vancouver, BC: Lower Mainland Regional Planning Board, 1968.
*Railroad Routes in the Southern Coastal Region of British Columbia in 1892.* Victoria, BC: J. Cowie, 1892.
*Railroads of the Lower Mainland of British Columbia, 1882–1992.* Vancouver, BC: Canadian Railroad Historical Association, Pacific Coast Division, 1992.
*Victoria Lumber & Manufacturing Company Logging Railroad on Robertson River.* Victoria, BC: Victoria Lumber & Manufacturing, 1928.

### Manitoba

*Manitoba Railroads.* Chicago: Rand McNally, 1909.
*Railroad and County Map of Dakota & Manitoba*, by G. F. Cram. Chicago: Geo. F. Cram, 1888.

### New Brunswick

*Colton's Railroad & Township Map of the State of Maine, with Portions of New Hampshire, New Brunswick & Canada*, by C. C. Hall. New York: J. H. Colton, 1855.

## Nova Scotia

*Province of Nova Scotia, Showing Railroads*, by M. Murphy. Halifax, NS: McAlpine Publishing, 1903.

## Ontario

*Map Shewing the Water and Railroad Communication between Lake Ontario & Lake Champlain*, by R. H. Pease. Albany, NY: Lith. of Richd. H. Pease., 1850.
*New Railroad, Post-Office, Township and County Map, with Distances.* Toronto: Province of Ontario, 1899.
*Rand McNally & Co.'s New Railroad and County Map of the Province of Ontario.* Toronto, ON: Rand McNally, 1877.
*Regional Niagara: Railroads.* St. Catharines, ON: Planning and Development Dept., 1975.

## Quebec

*Concord & Montreal Railroad and Its Principal Connections.* Boston: Rand Avery Supply, 1900.
*Gosford Wooden Railroad, Québec*, by G. R. Baldwin. Montreal, QC: G. R. Baldwin, 1871.
*Nouvelle carte de la province de Québec avec distances indiquant chemins de fer, bureaux de poste et comtés = New Map of the Province of Quebec Showing Railroads with Distances between Stations, Post-Offices and Counties*, by G. F. Cram. Montreal, QC: F. Cram, 1903.
*Plan of Proposed Line of Railroad on the North Shore of the River St. Lawrence: Between the Cities of Quebec and Montreal, April 1852*, by E. Staveley. Montreal, QC: Leggo, 1852.
*Plan of the Proposed Quebec & L. St. John Railroad*, by J. J. Rickon. Montreal: Quebec and Lake St. John Railway Company, 1875.
*Quebec, railroads.* Chicago: Rand McNally, 1910.
*Watson's New County, Railroad and Distance Map, of the Quebec*, by G. Watson. New York: Gaylor Watson, 1870.

## Saskatchewan

*Railroads of Saskatchewan.* Ottawa, ON: Natural Resources Intelligence Service, 1920.

## UNITED STATES OF AMERICA

## Alabama

*Map of Georgia & Alabama Exhibiting the Post Offices, Post Roads, Canals, Rail Roads*, by D. H. Burr. London: D. H. Burr, 1839.
*Map Showing the Line of the Alabama & Tennessee River Rail Road and Its Proposed Extensions; Exhibiting Also the Contiguous Mineral Deposits and Zone of Production.* New York: Alabama and Tennessee River Railroad, 1867.
*Map Showing the Line of the New Orleans, Mobile & Chattanooga Railroad, and Also the Chief Agricultural and Mineral Districts of the State of Alabama.* New York: Mobile New Orleans, 1867.
*Map Showing the N.E. & S.W. Alabama R.R. with Its Connections Also the Principal Routes between New York and New Orleans.* Richmond, VA: Hoyer & Ludwig, 1857.

## Alaska

*Map of the All American Route Showing Proposed Railroad and U.S. Government Mail Road to the Yukon*, by B. F. Millard. Seattle, WA: Central Alaska Transportation & Trading, 1899.

## Arizona

*Railroad and County Map of Arizona*, by G. F. Cram. New York: G. F. Cram, 1887.

## Arkansas

*Colton's Railroad & Township Map of Arkansas Compiled from the U.S. Surveys and Other Authentic Sources*, by D. F. Shall. New York: D. F. Shall, 1854.
*Cram's Township and Rail Road Map of Arkansas*, by G. F. Cram. Chicago: G. F. Cram, 1895.
*Map of Franklin County, Arkansas; Showing the Land Grant of the Little Rock & Fort Smith Railway.* Washington, DC: Little Rock & Fort Smith R.R., 1893.
*Maps Showing the Connections of the Little Rock and Fort Smith Railroad and Its Land Grant.* New York: Little Rock & Fort Smith R.R., 1873.

## California

*Map Showing the Location of Sacramento Valley Railroad, Cal*, by T. D. Judah. San Francisco: T. D. Judah, 1854.
*Map Showing the Projected Route of the Potomac and Ohio Railway.* New York: Potomac and Ohio Railway, 1874.
*New Commercial and Topographical Rail Road Map & Guide of California and Nevada.* New York: Asher & Adams, 1874.
*New Enlarged Scale Railroad and County Map of California Showing Every Railroad Station and Post Office in the State.* Chicago: Rand McNally, 1883.
*Railroad Map of the Central Part of California, and Part of Nevada*, by C. Bielawski, J. D. Hoffman, and A. Poett. San Francisco: C. Bielawski, 1865.
*Route of the Pacific and Atlantic Rail Road between San Francisco, & San Jose*, by Wm. J. Lewis. San Francisco: Pacific and Atlantic Railroad, 1851.

## Colorado

*Indexed Map of Colorado Showing the Railroads in the State, and the Express Company Doing Business over Each, Also Counties and Rivers.* Chicago: Rand McNally, 1879.

*Map of the Denver & Rio Grande Railway, Showing Its Connections and Extensions Also the Relative Position of Denver and Pueblo to All the Principal Towns and Mining Regions of Colorado and New Mexico,* by S. W. Eccles. Chicago: Denver and Rio Grande Railway, 1881.

## Connecticut

*Map of the Railroads of Connecticut,* by S. D. Tilden. Hartford, CT: Railroad Commissioners, 1893.

## Florida

*Drew's New Map of the State of Florida, Showing the Townships by the U.S. Surveys, the Completed & Projected Railroads, the Different Railroad Stations and Growing Railroad Towns. The New Towns on the Rivers and Interior, and the New Counties, up to the Year 1874,* by C. Drew. Jacksonville, FL: C. Drew, 1874.

*Map of Part of Alabama & Florida, Showing the Route of the Proposed Columbus & Pensacola Rail Road, Accompanying the Report of Major J. D. Graham, U.S. Topographical Engr. Feb. 6th, 1836; Drawn Chiefly from the Original Surveys in the Gen. Land Office at Washington,* by W. R. Palmer. Washington, DC: W. R. Palmer, 1836.

*Sketch Map of Northeastern Florida Showing the Florida Railroad and Proposed Connections Jan. 1860,* by P. Koerner and W. Oscar. New York: Florida Railroad, 1860.

## Georgia

*Indexed Railroad and County Map of Georgia,* by G. F. Cram. Chicago: G. F. Cram, 1883.

*Map of the Country Embracing the Various Routes Surveyed for the Western & Atlantic Rail Road of Georgia,* by J. F. Cooper. Chicago: Western and Atlantic Railroad, 1837.

## Idaho

*Cram's Township and Railroad Map of Idaho,* by G. F. Cram. Chicago: G. F. Cram, 1896.

## Illinois

*Cooke & Co's Railway Guide for Illinois Shewing All the Stations with Their Respective Distances Connecting with Chicago.* Chicago: D. B. Cooke, 1855.

*G. Woolworth Colton's Railroad Map of Illinois.* New York: G. Woolworth Colton, 1861.

*Galbraith's Railway Mail Service Maps, Illinois,* by F. H. Galbraith. Chicago: United States Railway Mail Service, 1897.

## Indiana

*A. J. Johnson's Map of Indiana Showing the Rail Roads and Townships Compiled from the Latest & Best Authorities,* by A. J. Johnson. Chicago: A. J. Johnson, 1858.

*Cram's Township and Rail Road Map of Indiana,* by G. F. Cram. Chicago: G. F. Cram, 1888.

*Map Showing the Line of the Louisville, New Albany, and St. Louis Air Line Railroad and Its Connections.* New York: G.W. & C.B. Colton, 1872.

*Railroad Map of Indiana,* by T. A. Morris. New York: T. A. Morris, 1850.

*Section of Colton's Large Map of Indiana with the Fort Wayne and Southern Rail Road Marked upon It, as Located Also a Map of the United States Showing Road and Its Connections Together with a Profile of the Ohio River and Lands Adjoining and a Section of The Double Track Rail Road Tunnel under the Ohio River at Louisville, Kentucky & Jeffersonville, Indiana for the Year 1855,* by W. J. Holman. New York: Fort Wayne and Southern Railroad, 1855.

## Iowa

*Railroad Map of Iowa.* Chicago: Railroad Commissioners, 1881.

## Kansas

*Galbraith's Railway Mail Service Maps, Kansas,* by F. H. Galbraith. Chicago: United States Railway Mail Service, 1897.

*Map Showing the Atchison, Topeka & Santa Fé Rail Road and Its Auxiliary Roads in the State of Kansas.* New York: G. W. & C. B. Colton, 1886.

*New Sectional Map of the State of Kansas Showing the Route of the Union Pacific Railway—E. D. to Denver City. Col. and Complete System of Projected Rail Roads. Information Compiled & Collected from Departments of the Government at Washington, D.C. and Other Authentic Sources,* by C. Du Bois. Washington, DC: C. Du Bois, 1867.

## Kentucky

*Map of the Virginia, Kentucky, and Ohio Railroad Connecting the Railroads of Virginia with the Railroads of Kentucky on the Shortest Route East and West from the Mississippi Valley to the Atlantic Ocean.* New York: G. W. & C. B. Colton, 1881.

## Maine

*Map of the Railroads of the State of Maine Accompanying the Report of the Railroad Commissioners*, by W. A. Allen. Augusta, ME: W. A. Allen, 1899.

## Maryland

*Experimental Survey for the Eastern Shore Rail Road, Maryland*, by W. H. Emory, J. McClelland, and J. Kearney. Baltimore: Eastern Shore Railroad, 1853.

## Massachusetts

*Map of Part of New Hampshire and Massachusetts, Showing the Location of the Wilton and Other Railroads*. New York: Wilton Railroad, 1847.

*Map of the Boston & Woonsocket Rail Road Routes Compiled from the State Map, and the Plans of the Different Surveys Returned to the Joint Standing Committee on Railroads & Canals*, by E. W. Bouvé. Boston: E. W. Bouvé, 1847.

*Map of the Electric Railways of the State of Massachusetts Accompanying the Report of the Railroad Commissioners, 1899*, by G. H. Walker. Boston: Geo. H. Walker, 1899.

*Plan of a Survey for the Proposed Boston and Providence Rail-Way*, by H. James. Boston: Annin & Smith, 1828.

*Plan of Railroads North and East of Boston, with the Projected Railroads from Danvers, Georgetown & Gloucester; Showing the Situation of the Towns & Villages, Their Distance from Boston & Number of Inhabitants*, by A. Lewis. Boston: A. Lewis, 1850.

*Rail Road & Township Map of Massachusetts*, by A. Williams. Boston: A. Williams, 1879.

## Michigan

*Galbraith's Railway Mail Service Maps, Michigan*, by F. H. Galbraith. Chicago: United States Railway Mail Service, 1897.

*Map of Michigan Showing the Toledo, Ann Arbor, & North Michigan Railway and Connecting Lines*. New York: G. W. & C. B. Colton, 1886.

*Official Map of Michigan, Railroad, Township and Sectional, Prepared under the Direction of the Commissioner of Railroads*. Chicago: Cram & Stebbins, 1885.

*Railroad Map of Michigan Prepared for the Commissioner of Railroads*. Philadelphia: G. W. & C. B. Colton, 1876.

*Railroads in Michigan, with Steamboat Routes on the Great Lakes. Drawn and Engraved for Doggett's Railroad Guide & Gazetteer*, by J. Doggett. Chicago: J. Doggett, 1848.

## Minnesota

*Railroad and Post Office Map of Minnesota and Wisconsin*. New York: H. H. Lloyd, 1871.

*Township and Railroad Map of Minnesota Published for the Legislative Manual, 1874*, by A. J. Reed. New York: A. J. Reed, 1874.

## Missouri

*Commissioners Official Railway Map of Missouri*. St. Louis: Higgins, 1888.

*Galbraith's Railway Mail Service Maps, Missouri*. Chicago: F. H. Galbraith, 1897.

*New Commercial and Topographical Rail Road Map & Guide of Missouri*. New York: Asher & Adams, 1872.

## Montana

*Map of Central Montana, the Montana Railroad, September 1, 1899*, by J. F. Polley. Chicago: Montana Railroad, 1899.

## Nebraska

*Cram's Rail Road and Township Map of Nebraska*, by G. F. Cram. Chicago: G. F. Cram, 1879.

*Map and Profile of First 40 Miles of Union Pacific Rail Road Eastern Division Extending West from Boundary between States of Missouri and Kansas*, by J. R. Gillis. Washington, DC: Union Pacific Railroad, 1865.

*Map Showing the Different Routes Surveyed for the Union Pacific Rail Road between the Missouri River and the Platte Valley*, by J. R. Gillis. Washington, DC: Union Pacific Railroad, 1865.

*New Commercial and Topographical Rail Road Map & Guide of Nebraska*. New York: Asher & Adams, 1874.

*Railway Map of Nebraska Issued by State Board of Transportation 1889*, by W. W. Alt. Wahoo, NE: W. W. Alt, 1889.

*Union Pacific Rail Road, Map of a Portion of Nebraska Territory, Showing Surveys and Location of Lines*, by A. Dey. New York: Union Pacific Railroad, 1865.

## New Hampshire

*Map of Part of New Hampshire and Massachusetts, Showing the Location of the Wilton and Other Railroads*. Chicago: Wilton Railroad, 1847.

*Railroad Map of New Hampshire Accompanying Report of the Railroad Commissioners, 1894*. Boston: Railroad Commissioners, 1894.

## New Jersey

*Map of the Northern Rail Road of New Jersey*, by W. S. Sneden et al. New York: New York Lith. of Robertson, Seibert & Shearman, 1859.

*Map of the Rail Roads of New Jersey, and Parts of Adjoining States*, by J. A. Anderson. New York: J. A. Anderson, 1869.

*Map of the Rail Roads of New Jersey 1887*, by J. T. Van Cleef and J. B. Betts. New York: J. B. Betts, 1887.

## New York

*Colton's New Township Railroad Map of New York with Parts of Adjoining States & Canada.* New York: Colton, 1883.

*Cram's Township and Rail Road Map of New York.* Chicago: G. F. Cram, 1888.

*Map and Guide of the Elevated Railroads of New York City*, by H. I. Latimer. New York: Manhattan Railway, 1881.

*Map Exhibiting the Experimental and Located Lines for the New-York and New-Haven Rail-Road*, by P. Anderson. New York: Snyder & Black Lithogrs, 1845.

*Map of Long Island Showing the Long Island Railroad and Its Leased Lines.* New York: G. W. & C. B. Colton, 1882.

*Map of the Canals and Railroads for Transporting Anthracite Coal from the Several Coal Fields to the City of New York*, by W. Lorenz. Baltimore: W. Lorenz, 1856.

*Map of the Hudson River Rail Road from New York to Albany*, by G. Snyder, W. C. Moore, and R. Haering. New York: Hudson River Railroad, 1848.

*Map of the Rail Roads of the State of New York Showing the Stations, Distances & Connections with Other Roads*, by T. Pentingale. Buffalo, NY: T. Petingale, 1858.

*Map of the Route of the Proposed New York & Erie Railroad, as Surveyed in 1834*, by B. Wright. New York: B. Wright and New York and Erie Railroad, 1834.

*Map of the State of New-York Showing Its Water and Rail Road Lines*, by D. Vaughan. Albany, NY: D. Vaughan, 1855.

*Map of the Williamsport and Elmira Railroad with Its Connections.* Elmira, NY: Elmira Railroad, 1859.

*Map Showing the Geneva & Hornellsville Railroad and Its Connections.* New York: G. W. & C. B. Colton, 1875.

*Map Showing the Location of the N.Y. & Oswego Midland R.R. with Existing and Proposed Connection, January 1st 1869*, by V. R. Richmond. New York: Oswego Midland Railroad, 1869.

*Map Showing the Route & Connections of the Central Rail Road Extension Company of Long Island.* New York: Central Railroad Extension Company of Long Island, 1873.

*Map Showing the Sodus Point & Southern Railroad and Its Connections.* New York: G. W. & C. B. Colton, 1872.

*Railroad Map of New England & Eastern New York Compiled from the Most Authentic Sources*, by J. H. Goldthwait. Boston: Redding, 1849.

*Tunison's Railroad, Distance, and Township Map of New York from Latest Surveys*, by E. L. Tunison. Brooklyn, NY: E. L. Tunison, 1898.

## North Carolina

*Maps Showing the Norfolk, Albermarle & Atlantic Railroad and Its Connections.* New York: G. W. & C. B. Colton, 1891.

*Railroad Map of North Carolina, 1900, Examined and Authorized by the North Carolina Corporation Commission*, by H. C. Brown. Chicago: H. C. Brown, 1900.

*Railroads in Virginia and Part of North Carolina, Drawn and Engraved for Doggett's Railroad Guide & Gazetteer*, by J. Doggett. Chicago: J. Doggett, 1848.

## North Dakota

*Correct Map of Dakota Compiled from United States and Territorial Surveys Nov. 1, 1882.* Chicago: Rand McNally, 1882.

*Official Railroad Map of Dakota Issued by the Railroad Commissioners, November 1st, 1886.* Chicago: Rand McNally, 1886.

## Ohio

*Cleveland and Toledo Rail-Road 1856.* Cincinnati, OH: Cleveland & Toledo Railroad, 1855.

*Colton's Railroad & Township Map of the State of Ohio*, by G. W. Colton. New York: J. M. Atwood, 1854.

*E. C. Bridgman's New Reversible Railroad Distance and Township Map of Ohio and United States Compiled from the Most Authentic Sources.* New York: E. C. Bridgman, 1873.

*Hillsborough & Cincinnati Rail-Road Map Extending from Hillsborough, Highland Co. to the Coal Field at Jackson, Jackson Co*, by L. Jacobi. Cincinnati, OH: Cincinnati Railroad, 1852.

*Map of Rail Road Line between Loveland and Cincinnati.* Cincinnati, OH: Marietta and Cincinnati Railroad, 1860.

*Map of the Marietta and Pittsburgh Railroad and Its Connections.* New York: Marietta and Pittsburgh Railroad, 1871.

*Map of the Rail Road Surveys between Hillsborough & Chillicothe*, by E. Morris. Cincinnati, OH: Cincinnati Onken's Lith., 1851.

*Map of the Surveys of the Cincinnati Railway, W. A. Gunn, Ch. Eng.*, by R. Hinman. Cincinnati, OH: H. Russell, 1873.

*Map Showing the Route and Connections of the Bellaire, Zanesville and Cincinnati Railway.* New York: Zanesville and Cincinnati Railway Bellaire, 1883.

*Map Showing the Route and Connections of the Wheeling and Cincinnati Mineral Railway.* New York: Wheeling and Cincinnati Mineral Railway, 1882.

*Railroad & Township Map of Ohio.* New York: Ensign, Bridgman & Fanning, 1854.

*Rail Road Map of Ohio 1873*, by O. W. Gray. Philadelphia: O. W. Gray, 1873.

*Railroad Map Showing the Lands of the Standard Coal and Iron Co. Situated in the Hocking Valley, Ohio, and Their Relation to the Markets of the North and West.* New York: G. W. & C. B. Colton, 1881.

## Oregon

*Colton's Township Map of Oregon & Washington Territory.* New York: G. W. & C. B. Colton, 1880.

*Indexed Map of Oregon Showing the Railroads in the State and the Express Company Doing Business over Each, Also, Counties, Lakes & Rivers.* Chicago: Rand McNally, 1876.

## Pennsylvania

*Barringtons New and Reliable Railroad Map and Shippers & Travellers Guide of Pennsylvania, Showing the Name of Every City, Town and Village in the State, with Nearest Rail Road Station,* by J. Duncan. Philadelphia: J. Duncan, 1860.

*Map and Profile of the Gettysburg Rail Road as Surveyed by Order of the Legislature of Pennsylvania, 1839,* by H. R. Campbell. Philadelphia: Gettysburg Railroad, 1839.

*Map Exhibiting That Portion of the State of Pennsylvania Traversed by the Surveys for a Continuous Rail Road from Harrisburg to Pittsburg Made under the Direction of Charles L. Schlatter, C.E. in the Year 1839 and 1840,* by C. Cramer and J. T. Bowen. Philadelphia: J. T. Bowen, 1840.

*Map of the Catawissa, Williamsport & Erie Rail Road; Showing Its Connection with the North and West by the Williamsport and Elmira Rail Road, & Its Connection with Philada. & New York by the Reading & Lehigh Valley R. Rd,* by T. Kimber. Philadelphia: Williamsport and Erie Railroad, 1856.

*Map of the Pennsylvania Railroad and Its Connections,* by S. C. Patterson. Philadelphia: Pennsylvania Railroad, 1889.

*Map of the Pennsylvania Rail Road, from Harrisburg to Pittsburg [sic] and of the Columbia & Lancaster & Harrisburg R.R.s from Philadelphia to Harrisburg,* by H. Haupt. Philadelphia: Pennsylvania Railroad, 1855.

*Map of the Philadelphia & Erie Rail Road the City and Harbor of Erie Its Western Terminus, and the State of Pennsylvania Showing the Different Rail-Road Connections, from the Latest Surveys Constructed and Drawn by John F. Burgin,* by J. F. Burgin. Buffalo, NY: Philadelphia and Erie Railroad, 1862.

*Map of the Philadelphia and Erie Railway, Branches and Connecting Lines,* by H. W. Gwinner. Philadelphia: Philadelphia and Erie Railroad, 1871.

*Map of the Proposed Lock Haven & Tyrone Rail Road,* by J. M. McMinn. Philadelphia: Lock Haven Railroad, 1858.

*Map of the Rail Roads of Pennsylvania and Parts of Adjoining States,* by J. A. Anderson. Philadelphia: J. A. Anderson, 1871.

*Map Showing the Rail Road Connection between Pottsville & Sunbury through the Schuylkill Mahanoy and Shamokin Coal Fields, July 9th 1852,* by P. W. Sheafer. Philadelphia: P. W. Sheafer, 1852.

*Map Showing the Seaboard, Pennsylvania and Western Railroad and Its Connections.* New York: G. W. & C. B. Colton, 1884.

*Plan and Profile of the Danville and Pottsville Rail Road,* by D. K. Kennedy and W. B. Lucas. Philadelphia: Pottsville Rail Road, 1831.

*Plan of the West-Philadelphia Rail-Road,* by H. R. Campbell. Philadelphia: Lehman & Duval Lithrs, 1835.

*Topographical Plan & Profile of the Philadelphia and Reading Rail Road,* by R. B. Osborne. Philadelphia: Philadelphia & Reading Railroad, 1838.

## Rhode Island

*Indexed Map of Rhode Island Showing the Railroads in the State, and the Express Company Doing Business over Each, Also Counties, Townships, Lakes, Rivers, Islands, Etc.* Chicago: Rand McNally, 1875.

## South Carolina

*Map Showing the Location of the Charleston & Savannah R.R. May, 1856,* by E. Walker. New York: Charleston and Savannah Railroad, 1856.

*Railroad Map of South Carolina,* by J. Hotchkiss. Stanton, VA: J. Hotchkiss, 1880.

*South Carolina Railroads.* Chicago: Rand McNally, 1900.

## Tennessee

*Boones Map of the Black Diamond System of Railways,* by A. E. Boone. Knoxville, TN: Black Diamond System, 1896.

*Map of the Nashville, Chattanooga and St. Louis Ry.; and Connections,* by W. L. Danley. Buffalo, NY: W. L. Danley, 1889.

*New Enlarged Scale Railroad and County Map of Tennessee Showing Every Railroad Station and Post Office in the State.* Chicago: Rand McNally, 1888.

## Texas

*Map of the State of Texas Showing the Line and Lands of the Texas and Pacific Railway Reserved and Donated by the State of Texas, 1873.* New York: G. W. & C. B. Colton, 1873.

*Map Showing the Proposed Railroad between Laredo and Corpus Christi and Its Connections with Mexico,* by J. Bien. New York: Corpus Christi and Rio Grande Railway, 1873.

*Texas Railroads.* Chicago: Rand McNally, 1900.

## Vermont

*Coffin's New Rail-Road Map of Vermont Accompanying Report of the Board of Railroad Commissioners, 1896.* Boston: Coffin, 1896.

*Map of the Western Vermont Rail Road and Connecting Lines*, by Wm. B. Gilbert. New York: Western Vermont Railroad, 1851.

## Virginia

*Indexed County and Railroad Pocket Map and Shippers Guide of West Virginia, Accompanied by a New and Original Compilation and Ready Reference Index, Showing in Detail the Entire Railroad System.* Chicago: Rand McNally, 1898.

*Map of the Manassas Gap Railroad and Its Extensions; September, 1855*, by T. Dwyer. Baltimore: Manassas Gap Railroad, 1855.

*Map of the Proposed Line of Rail Road Connection between Tide Water Virginia and the Ohio River at Guyandotte, Parkersburg and Wheeling*, by W. Vaisz. Philadelphia: Virginia Board of Public Works, 1852.

*Map of the Routes Examined and Surveyed for the Winchester and Potomac Rail Road, State of Virginia, under the Direction of Capt. J. D. Graham, U.S. Top. Eng., 1831 and 1832*, by A. A. Humphreys. New York: Potomac Railroad, 1832.

*Map of the Virginia Central Rail Road Showing the Connection between Tide Water Virginia, and the Ohio River at Big Sandy, Guyandotte and Point Pleasant*, by W. Vaisz. Philadelphia: Virginia Central Railroad Company, 1852.

*Map of the Virginia Central Railroad, West of the Blue Ridge, and the Preliminary Surveys, with a Profile of the Grades*, by L. F. Citti. Richmond, VA: Lith. of L. F. Citti 1860.

*Map of the Virginia, Kentucky, and Ohio Railroad Connecting the Railroads of Virginia with the Railroads of Kentucky on the Shortest Route East and West from the Mississippi Valley to the Atlantic Ocean.* New York: G. W. & C. B. Colton, 1881.

*Map Showing the Albemarle & Pantego Railroad and Its Connections.* New York: Albemarle and Pantego Railroad, 1887.

*Maps Showing the Connections of the Northern and Southern West Virginia Railroad, with the Three Grand Trunk Railways Which Unite the Atlantic Seaboard with the Ohio River.* New York: Northern and Southern West Virginia Railroad, 1873.

*Map Showing the Fredericksburg & Gordonsville Rail Road of Virginia, Leading from Fredericksburg, via Orange C.H., to Charlottesville, Where It Connects with the Chesapeake & Ohio R.R. and the Extension of the Orange & Alexandra R.R. to Lynchburg.* New York: G. W. & C. B. Colton, 1869.

*Map Showing the West Virginia Midland Railway and Its Connections.* New York: West Virginia Midland Railroad, 1883.

*Maps Showing the Norfolk, Albermarle & Atlantic Railroad and Its Connections.* New York: G. W. & C. B. Colton, 1891.

*Railroads in Virginia and Part of North Carolina, Drawn and Engraved for Doggett's Railroad Guide & Gazetteer*, by J. Doggett. Chicago: J. Doggett, 1848.

## Washington

*Colton's Township Map of Oregon & Washington Territory.* New York: G. W. & C. B. Colton, 1880.

## Washington, D.C.

*Cram's Township and Railroad Map of Washington*, by G. F. Cram. Chicago: G. F. Cram, 1896. *Map of the Located Route of the Metropolitan Rail Road and the Adjacent Country Comprising the District of Columbia and the Counties of Montgomery, Frederick, and Washington in the State of Maryland*, by Wm. R. Hutton. Washington, DC: Metropolitan Railroad Company, 1855.

## Wisconsin

*Fox River Valley R.R. in Wisconsin with Its Connections.* Milwaukee, WI: Fox River Valley Railroad, 1857.

*Map of the La Crosse and Milwaukee Rail Road and Connections*, by J. H. Colton. New York: Milwaukee Railroad, 1855.

*Map of the Milwaukee & Superior Rail Road and Its Connections.* New York: Milwaukee and Superior Railroad, 1857.

*Map of the Wisconsin Central Line and Connections.* Buffalo, NY: Wisconsin Central Railroad Company, 1888.

*Map Showing the Route of the Proposed Rail Road from the Copper and Iron Mining District of Lake Superior to Connect with Rail Roads Built or Being Constructed in the State of Wisconsin as Adopted by the Citizens of Ontonagon and Marquette Counties Mich. at Public Meetings Held in November and December 1855.* New York: W. Endicott, 1855.

*Railroad and Post Office Map of Minnesota and Wisconsin.* New York: H. H. Lloyd, 1871.

*Township Map of Wisconsin Showing the Milwaukee & Horicon Rail Road and Its Connections*, by J. Vilet. New York: Milwaukee and Horicon Railroad, 1857.

## Wyoming

*Cram's Township and Railroad Map of Wyoming*, by G. F. Cram. Chicago: G. F. Cram, 1895.

# 6
# Geological Maps

Geological maps represent the distribution of rock formations beneath the soil and other materials on the earth's surface; such maps range from black-and-white line drawings to full-color maps. Standardized colors and symbols are usually used, especially in publications from the United States Geological Survey (USGS). Many kinds of geological maps exist, including those emphasizing surficial features, bedrock and sediment, subsurface rocks, and geophysical features such as heat flow and gravity. Many libraries house geological maps in a variety of formats, including stand-alone folded maps, folded maps within books, and most popular, separate flat maps stored in map cabinets.

Geological mapping is conducted mostly by government agencies at the national, provincial, and state levels. This chapter focuses on maps published by the Geological Survey of Canada, provincial geological survey agencies, and the USGS. Because of the sheer number of maps available, this section provides links for finding paper or online maps. The information was compiled by visiting and reviewing the national, provincial, and state geological survey online websites.

## GEOLOGICAL SURVEY OF CANADA

The Geological Survey of Canada (GSC), the country's oldest scientific organization, is a Canadian federal government agency that is part of the Earth Sciences section of Natural Resources Canada; it is responsible for conducting and publishing geological surveys for Canada. Geological maps have been published since 1842, and in the past several years, many have been scanned or updated digitally and shared freely online for easier access.

As with any map series, indexes are the starting point for finding map sheets of interest. For older print maps, most library catalogs can be searched for print indexes such as the *Index of Publications of the Geological Survey of Canada (1845–1958)*, published in 1972. Older material can also be accessed via GeoRef, an online bibliographic database that provides access to geological materials. Thousands of geological maps can be accessed through the GSC's GEOSCAN (http://geoscan.nrcan.gc.ca/geoscan-index.html). Users should use the advanced search feature to find the "A" Series maps, which focus on geology, or the Canadian Geoscience maps. The GSC has published the following maps:

- A-Series maps (geology), 1909–2016
- G-Series maps (geophysics), 1947–1999
- Canadian Geoscience maps, 2010–2017
- Preliminary maps, 1935–1992
- Multicolored maps, 1849–1977

## Ontario Geological Survey

The Ontario Geological Survey (OGS) has been collecting and publishing geological features for over 120 years. Common map scales available are 1:250,000, 1:1,000,000, and 1:2,000,000. All OGS publications can be ordered from the Publication Sales Office (http://www.infogo.gov.on.ca/infogo/home.html) or downloaded from the OGS Publication database (listed below). OGS maps are part of the following series:

### A Series

The "A" series maps, recording magnetic surveys by the Ontario Department of Mines, were issued from 1950 to 1959.

### AR Series

The "AR" maps, accompanying the annual reports of the Ontario Bureau of Mines and Ontario Department of Mines, were issued from 1891 to 1944.

### G Series

This series of aeromagnetic maps of Canada provides coverage of standard quadrangles of the National Topographic System (NTS).

*OFM Series*

Open File Maps were maps intended for quick release and issued without editing.

*P Series*

The current series of preliminary maps, usually uncolored, has been produced since 1956.

*S Series*

This series of maps was produced by the Ontario Department of Lands and Forest in 1965. There are only four maps in this series.

*1900 Series*

Initially issued separately from annual reports, these maps later replaced the AR series. They were issued from 1931 to 1960.

*2000 Series*

The current series of final, colored maps has been produced since 1961. These maps commonly accompany reports.

*5000 Series*

This is a series of engineering geology maps.

*80000 Series*

The current series of geophysical and geochemical maps has been produced since 1979. Earlier geophysical and geochemical maps were issued as A, 2000, or P series maps.

**Geology Ontario**

Geology Ontario is one of the best sources for geological maps issued by the Ontario Geological Survey. The GeologyOntario search tool allows users to search five databases: the Assessment File Research Image (AFRI) database, the Abandoned Mines Information System (AMIS) database, the Ontario Drill Hole database (ODHD), the Mineral Deposit Inventory (MDI) database, and the Ontario Geological Survey Publications (PUB) database. The files come in database and GIS file format.

**Ontario Geological Survey OGSEarth**

http://www.ontario.ca/ogsearth
OGSEarth offers geological data collected by the Mines and Minerals Division, viewable by using Google Earth. Datasets are available on the following topics:

- Abandoned Mines
- Activity Report–Mineral Exploration (RGP)
- Aggregate Test Results
- Aggregate Resources
- Assessment Files
- Bedrock Geology
- Drill Holes
- Elevation
- Geoscience Theses
- Geotechnical Boreholes
- Gravity and Aeromagnetic Data
- Mineral Deposits (MDI)
- Mining Claims
- Precambrian Bedrock Magnetic Susceptibility Database
- Quaternary Geology

**Other Canadian Provincial Geological Survey Maps**

Most provinces and territories in Canada have provincial governments that are responsible for the documentation and public dissemination of geoscience information. All of them provide information about their geological map publications and offer access to online digital versions. Most maps are available in image formats or as PDFs, as well as in GIS mapping files (GeoTiffs or Shapefiles). Some maps can be downloaded, and others can only be viewed on the screen. This section lists all provincial resources freely available to the public.

**Alberta**

Alberta Geological Survey Open Data Portal
http://geology-ags-aer.opendata.arcgis.com
This portal provides access to all Alberta Geological Survey reports, maps, and digital datasets. Users can visualize, query, and download bedrock geology, topography, surficial geology, sediment thickness, and glacial landforms.

**British Columbia**

B.C. Digital Geology Maps
http://www.empr.gov.bc.ca/Mining/Geoscience/PublicationsCatalogue/DigitalGeologyMaps/Pages/default.aspx
This site offers 1:250,000 scale bedrock geological maps covering the province of British Columbia, available in GIS Shapefile format.

MapPlace
http://www.MapPlace.ca
MapPlace is a web service that allows browsing and visualizing geological data, including bedrock and surficial geology, publications, geochemistry, geophysical surveys, and more.

**Manitoba**

Manitoba Resource Development Division
http://www.gov.mb.ca/iem/geo/index.html

This online resource offers an interactive map portal to view, query, and download geological data such as bedrock and surficial geology, including a 3-D geological model of the Phanerozoic section in southwest Manitoba.

## New Brunswick

Provincial Bedrock Geology Maps
http://www2.gnb.ca/content/gnb/en/departments/erd/energy/content/minerals/content/BedrockGeologyMaps_1–50–000.html
Access to provincial bedrock geology maps at scales of 1:20,000, 1:50,000, and 1:100,000 is available in GIS Shapefiles and PDFs.

## Newfoundland and Labrador

GeoFiles
http://gis.geosurv.gov.nl.ca/minesen/geofiles/
This repository of documents about the province's geosciences includes maps.

GeoScience Online
http://gis.geosurv.gov.nl.ca
This web resource offers interactive maps of geology, geophysics, geochemistry, mineral occurrences, and mineral land tenure.

## Northwest Territories

Gateway
http://gateway.nwtgeoscience.ca
Gateway provides a user-friendly way to browse and download geological publications.

GoMap
http://ntgomap.nwtgeoscience.ca
This interactive mapping platform allows users to create custom geological maps.

Northwest Territories Geological Survey Web Application
http://www.nwtgeoscience.ca/search-and-explore-our-data
This web application provides search, display, and download opportunities for Northwest Territories geological publications.

## Nova Scotia

Interactive Maps
https://novascotia.ca/natr/meb/geoscience-online/maps-interactive.asp
This site provides access to geological interactive maps with the ability to view, query, and print maps, images, and databases.

NovaScan
https://novascotia.ca
NovaScan is a searchable database of geoscience maps, publications, and downloadable GIS datasets.

Nova Scotia Geoscience Atlas
https://fletcher.novascotia.ca/DNRViewer/?viewer=Geoscience
This interactive map offers seamless geologic coverage of Nova Scotia.

## Nunavut

Nunavut Geoscience
http://nunavutgeoscience.ca
Nunavut Geoscience is a joint initiative of the Canada-Nunavut Geoscience Office (CNGO), Aboriginal Affairs and Northern Development Canada (AANDC), the Government of Nunavut (GN), Natural Resources Canada (NRCan), and Nunavut Tunngavik Incorporated (NTI) to provide industry, government, and the public with geoscience information of Nunavut. The website offers a portal for searching and downloading geological publications.

## Prince Edward Island

Prince Edward Island GIS Data Layers
http://www.gov.pe.ca/gis/index.php3?number=1021412&lang=E
This website offers surficial geology map layers for download along with comprehensive metadata.

## Quebec

SIGÉOM
http://sigeom.mines.gouv.qc.ca/signet/classes/I1102_index Accueil?l=f
This spatial information system offers geological data for the province of Quebec collected in the last 150 years. SIGÉOM offers an interactive web mapping service (WMS) that allows the data to be loaded directly into GIS software, as well as access to over eighty thousand records, reports, and maps.

## Saskatchewan

GeoAtlas
http://gisappl.saskatchewan.ca/HTML5Ext/index.html?viewer=GeoAtlas
The Saskatchewan Mining and Petroleum GeoAtlas is the survey's web service that offers browsing, visualizing, and creating custom geological maps. Over one hundred databases are linked through the atlas, including bedrock and surficial geology maps.

## Yukon

Yukon Geological Survey Integrated Data System
http://www.geology.gov.yk.ca/

The site provides access to Yukon Geological Survey publications, including bedrock and surficial geology maps.

## UNITED STATES GEOLOGICAL SURVEY

Established in 1879, the United States Geological Survey (USGS) is a scientific agency of the U.S. government studying the U.S. landscape, its natural resources, and the natural hazards that threaten it. USGS maps are readily available to be searched, viewed, and downloaded from a number of sources. The National Geological Map Database (https://ngmdb.usgs.gov/ngmdb/ngmdb_home.html) is perhaps one of the most comprehensive online sources, offering an online catalog of over ninety-five thousand geological maps and reports, dating back to the 1800s. Users can view and download in a variety of formats, such as PDF, GeoTiff, KMZ, and GIS files. Documents and maps can be searched by keyword, theme, location, scale, date, format, and publisher. The USGS published maps for state and territories during the years indicated below.

- Alabama: 1932–2017
- Alaska: 1957–2017
- American Samoa: 1924–2016
- Arizona: 1842–2017
- Arkansas: 1857–2017
- California: 1842–2017
- Colorado: 1842–2017
- Connecticut: 1842–2017
- Delaware: 1857–2017
- District of Columbia: 1857–2017
- Federated States of Micronesia: 1954–2005
- Florida: 1857–2017
- Georgia: 1857–2017
- Guam: 1951–2016
- Hawaii: 1857–2017
- Idaho: 1842–2017
- Illinois: 1851–2017
- Indiana: 1839–2017
- Iowa: 1847–2017
- Kansas: 1857–2017
- Kentucky: 1856–2017
- Louisiana: 1857–2017
- Maine: 1857–2017
- Maryland: 1856–2017
- Marshall Islands: 1948–2009
- Massachusetts: 1857–2017
- Michigan: 1857–2017
- Minnesota: 1852–2017
- Mississippi: 1847–2017
- Missouri: 1853–2017
- Montana: 1842–2017
- Nebraska: 1852–2017
- Nevada: 1842–2017
- New Hampshire: 1823–2017
- New Jersey: 1836–2017
- New Mexico: 1842–2017
- New York: 1838–2017
- North Carolina: 1825–2017
- North Dakota: 1857–2017
- Northern Mariana Islands: 1955–2016
- Ohio: 1856–2017
- Oklahoma: 1857–2017
- Oregon: 1842–2017
- Pennsylvania: 1876–2017
- Puerto Rico: 1899–2017
- Republic of Palau: 1948–2005
- Rhode Island: 1857–2017
- South Carolina: 1843–2017
- South Dakota: 1857–2017
- Tennessee: 1951–2017
- Texas: 1857–2017
- U.S. Minor Outlying Islands: 1956–2013
- Utah: 1842–2017
- Vermont: 1845–2017
- Virginia: 1856–2017
- Virgin Islands: 1922–2015
- Washington: 1842–2017
- West Virginia: 1856–2017
- Wisconsin: 1847–2017
- Wyoming: 1842–2017

## STATE GEOLOGICAL SURVEYS

Many states have their own state geological surveys providing citizens with documents and maps related to geology. Below is a list of these surveys with links to their maps and data. If a state does not offer maps or only has USGS-published maps, it was not included in the list.

### Alabama

Alabama Geologic Mapping
http://www2.gsa.state.al.us/gsa/geo_mapping.html
Thirty-five geologic maps of 7.5-Minute (1:24,000) quadrangles are available for download as PDF or GIS digital files.

### Arizona

Arizona Geological Survey Online Map and Database Services
http://www.azgs.az.gov/map_services.shtml
This excellent resource provides a wide number of online and interactive maps and databases relating to the state geology of Arizona. Maps from 1915 are available as PDF files,

and multiple ArcGIS Online Story Maps feature Arizona's geology.

## Arkansas

Arkansas Geological Survey Maps and Data
http://www.geology.ar.gov/catalog/mapsdata.htm
This resource offers a large number of geologic maps of Arkansas, including 1:24,000, 1:62,500, and 1:100,000 scales, as well as county and state park geologic maps in PDF and KMZ formats. An interactive web map featuring surface geology is also available.

## California

Geologic Mapping Geologic Maps
http://www.conservation.ca.gov/cgs/information/geologic_mapping
A number of maps are available for download, including the 2010 State Geologic Map at 1:750,000 scale, as well as 1:250,000 scale regional maps. Interactive maps are also available.

## Connecticut

Connecticut Geological Survey Geologic Maps
http://www.ct.gov/deep/cwp/view.asp?a=2701&q=519804&deepNav_GID=1641
Over forty regional 1:24,000 and 1:31,680 scale bedrock and surficial geology maps are available as PDFs, as are georeferenced maps that can be accessed using the AVENZA PDF Maps navigation application.

## Florida

Florida Geological Survey—Data and Maps
http://www.dep.state.fl.us/geology/gisdatamaps/
A significant collection of historical and current maps are available to view, zoom, and share. A separate interactive mapping application is available, as are GIS files for download.

## Georgia

Georgia Geological Survey Maps
http://epd.georgia.gov/georgia-geologic-survey-maps
Fewer than twenty geologic maps are available for download as PDF files.

## Idaho

Idaho Geology
http://www.idahogeology.org/Products/MapCatalog/
A searchable index is available for over five hundred available geological maps that can be ordered in print or downloaded as PDF files. A large number of GIS datasets are also available.

## Illinois

Illinois State Geological Survey
http://isgs.illinois.edu/maps
A large number of quadrangle, county, and regional maps are available for download. GIS datasets offer county coverage of surficial and bedrock geology.

## Indiana

Indiana Geological Survey
https://igs.indiana.edu
This site provides primarily digital products in the form of interactive maps and GIS datasets. Map layers include bedrock and surficial geology, hydrology, and imagery.

## Iowa

Iowa Geological Survey
http://www.iihr.uiowa.edu/igs/
This is a searchable database for Iowa Geological Survey maps, available to be downloaded as PDF files.

## Kansas

Kansas Geological Survey
http://www.kgs.ku.edu/General/geologyMaps.html
A few maps are available for download; however, most need to be ordered at a cost.

## Kentucky

Kentucky Geological Survey
http://kgs.uky.edu/kgsweb/main.asp
Although geological sheets are not readily available for download, the Kentucky Geological Survey does offer GIS vector data for individual 7.5-minute quadrangles. The datasets include faults, geologic units, contours, fossil locations, and economic points. The site also offers a number of interactive mapping applications, using map services such as ArcGIS.com.

## Louisiana

Louisiana Geological Survey
http://www.lsu.edu/lgs/
This site offers a number of geological maps for purchase. GIS data are not currently being offered.

### Maine

Maine Geological Survey
http://www.maine.gov/dacf/geology/index.html
A large number of printed maps can be purchased, although many of them (over four thousand) have been digitized and made available freely online. Both bedrock and surficial maps are available in a variety of scales, ranging from 1:24,000 to 1:250,000. These maps are available in GIS data format and are viewable in web map applications.

### Maryland

Maryland Geological Survey
http://www.mgs.md.gov
This excellent resource provides access to over seven hundred maps as PDF and GIS files. Map publication dates range from 1900 to 2016.

### Massachusetts

Massachusetts Geological Survey
http://mgs.geo.umass.edu
Although there are a few state maps available for viewing, most are published by the USGS.

### Michigan

Michigan Geological Survey
http://wmich.edu/geologysurvey
Maps are available for order and purchase. A comprehensive GIS website displays geological data, available for purchase as well.

### Minnesota

Minnesota Geological Survey Online Map Service
http://www.mngs.umn.edu/service.htm
A few maps are available for download as GIS files and can be viewed via a web map application. Printed maps can be ordered and purchased.

### Mississippi

Ministry Department of Environmental Equality
http://www.deq.state.ms.us/MDEQ.nsf/page/Geology_home
A number of maps are available for order and purchase.

### Missouri

Missouri Geological Survey
https://dnr.mo.gov/geology/index.html
A large number of maps may be ordered and purchased online. Each map is viewable online in small image format.

### Montana

Montana Bureau of Mines and Geology
http://www.mbmg.mtech.edu
A large number of maps are available for viewing at 1:100,000 and 1:250,000 scales. Print maps can be ordered and purchased as well. The 1:500,000 scale maps are viewable via a web mapping service.

### Nebraska

Nebraska Geological Survey
http://snr.unl.edu/csd/
A large number of 1:250,000 series bedrock maps are available for download as image files and GIS files. A handy index makes the search and download process quick and simple.

### Nevada

Nevada Bureau of Mines and Geology
http://www.nbmg.unr.edu
A large number of maps are available for download, with scales ranging from 1:24,000 to 1:500,000; however, most must be purchased.

### New Hampshire

New Hampshire Geological Survey
http://www.des.nh.gov/organization/commissioner/gsu/index.htm
A large number of 7.5-minute quad and 15-minute quad maps are available for purchase.

### New Jersey

New Jersey Geological and Water Survey
http://www.state.nj.us/dep/njgs/index.html
This older website offers a small selection of geological maps for download. Surficial and bedrock geology in GIS Shapefile format is also available at 1:100,000 scale.

### New Mexico

New Mexico Bureau of Geology and Mineral Resources
http://geoinfo.nmt.edu
Maps are available for purchase or viewing via an interactive mapping program.

### New York

New York State Museum Geology Resources
http://www.nysm.nysed.gov/research-collections/geology/resources

A large number of geological maps and data are available for download. Thousands of unpublished maps are available by visiting the New York State Museum.

## North Carolina

North Carolina Geological Survey
http://deq.nc.gov/about/divisions/energy-mineral-land-resources/north-carolina-geological-survey
Many maps, including regional and open-file maps, are available for download as PDF, GeoPDF, MrSID, or GIS Shapefile. Some require payment.

## North Dakota

North Dakota Geologic Survey
https://www.dmr.nd.gov/ndgs/
Maps can be ordered and purchased. Maps are not available for viewing or downloading online.

## Ohio

Ohio Geological Survey
http://geosurvey.ohiodnr.gov
Many geological maps are available freely for download.

## Oklahoma

Oklahoma Geological Survey
http://www.ou.edu/ogs/
Many maps are available for download as PDF files, and GIS files are available for the Oklahoma Geologic Quadrangles map series.

## Oregon

Oregon Department of Geology and Mineral Industries
http://www.oregongeology.org/sub/default.htm
Several map series are available for download as PDF files, including the Geologic Map Series and the Geologic Quadrangle Maps. Interactive maps and GIS data are also available.

## Pennsylvania

Pennsylvania Geological Survey
http://www.dcnr.state.pa.us/topogeo/index.aspx
A significant number of maps are freely available for download, including open files and county maps. Interactive maps and GIS datasets are also available.

## Rhode Island

Department of Geosciences
http://web.uri.edu/geo/
Maps are available for purchase only.

## South Carolina

South Carolina Geological Survey
http://www.dnr.sc.gov/geology/
A large number of maps are available for download as PDF files. GIS datasets are available in a variety of scales, including the 1:24,000 quadrangles.

## South Dakota

South Dakota Geological Survey
http://www.sdgs.usd.edu
Maps are available for download in PDF format, as well as in GIS format in a variety of scales, including 1:24,000, 1:62,500, and 1:250,000. Interactive maps are not yet available but will be coming soon.

## Tennessee

Tennessee Geological Survey
http://www.tn.gov/environment/section/geo-geology
Maps are available for purchase only.

## Texas

State Geological Survey
http://www.beg.utexas.edu/outreach/state-geological-survey
Maps are available for purchase. A small number of GIS datasets are available for download.

## Utah

Utah Geological Survey
https://geology.utah.gov
An interactive geologic map portal provides access to several map layers, including 1:24,000, 1:250,000 and 1:500,000 scale geologic maps. Quadrangle maps are also available for download as a PDF file and as GIS files. Many of these, however, are only available for purchase.

## Vermont

Division of Geology and Mineral Resources
http://dec.vermont.gov/geological-survey
Maps are available for purchase only.

## Virginia

Division of Geology and Mineral Resources
https://www.dmme.virginia.gov/dgmr/mapspubs.shtml
Paper maps can be ordered and purchased; however, there is access to free data via the Virginia Geologic Information Catalog (VGIC), which offers geologic maps, rock and core samples, and coal and mines data.

## Washington

Washington Geological Survey
http://www.dnr.wa.gov/geology
Geological maps are available for download as PDF files at the scales of 1:24,000, 1:100,000, 1:250,000, and 1:500,000. Surface geology can be downloaded in GIS format as well.

## West Viginia

West Virginia Geological and Economic Survey
http://www.wvgs.wvnet.edu
Most maps are available for purchase.

## Wisconsin

Wisconsin Geological and Natural History Survey
http://wgnhs.uwex.edu
A very large collection of bedrock and surficial geological maps are available for free download. Georeferenced images of the 1:100,000 map series are also accessible.

## Wyoming

Wyoming State Geological Survey
http://www.wsgs.wyo.gov
A large number of maps are available for free download, or print copies can be purchased. GIS datasets include 1:500,000 scale maps.

# 7
# Nautical Charts

A nautical chart is a map of the shoreline and seafloor, providing water depths, latitude and longitude, topographical features, locations of hazards to navigation, aids to navigation, and other related features. By law, mariners must use nautical charts to plot courses, determine their positions, and avoid dangers to arrive safely at their destination. Traditional nautical charts have been printed on paper, and more recently the charts have been digitized and made freely available as PDF files. Today, many vessels include navigation technology consisting of a digital system with navigational data that not only replaces the traditional paper version but offers additional details and features.

Similar to land maps, charts vary by scale and level of detail. Large-scale charts offer a lot of detail but cover a small area. Harbor charts are typically between the scales of 1:2,001 and 1:20,000 and are used for navigation in harbors. Approach charts are used for approaching coasts and are available between the scales of 1:20,001 and 1:50,000. Coastal charts, used by fisheries, are a popular scale between 1:50,001 and 1:150,000 and show continuous extensive coverage with inshore detail to facilitate landfall sightings. Small-scale maps like general charts, with scales of 1:50,001 to 1:150,000, and sailing charts, with scales 1:500,001 and smaller, are used for offshore navigation beyond sight of land.

The two agencies in North America that publish, update, and distribute nautical charts are the Canadian Hydrographic Service (CHC), part of the Department of Fisheries and Oceans (http://www.charts.gc.ca/), and the Office of Coast Surveys, part of the National Oceanic and Atmospheric Association, or NOAA (http://www.nauticalcharts.noaa.gov/). CHC covers all three of Canada's coastlines plus major lakes and inland waterways. NOAA maintains nautical charts and publications for U.S. coasts and the Great Lakes. Over a thousand charts cover ninety-five thousand miles of shoreline and 3.4 million square nautical miles of waters. The following is a list of all the nautical charts that are currently available for download from either the CHC or NOAA, sorted by body of water.

## CANADA

### Atlantic North Central

Cape Freels to/à Exploits Islands
Scale: 1:150,000
Chart Number: 4820

White Bay and/et Notre Dame Bay
Scale: 1:150,000
Chart Number: 4821

Cape St John to/à St Anthony
Scale: 1:150,000
Chart Number: 4822

Cape Ray to/à Garia Bay
Scale: 1:75,000
Chart Number: 4823

Garia Bay to/à Burgeo
Scale: 1:75,000
Chart Number: 4824

Burgeo and/et Ramea Islands
Scale: 1:30,000
Chart Number: 4825

Burgeo to/à François
Scale: 1:75,000
Chart Number: 4826

Hare Bay to/à Fortune Head
Scale: 1:75,000
Chart Number: 4827

Fortune Bay: Northern Portion/Partie nord
Scale: 1:60,000
Chart Number: 4831

Fortune Bay: Southern Portion/Partie sud
Scale: 1:60,000
Chart Number: 4832

Plans, Conception Bay, Trinity Bay and/et Bonavista Harbour
Scale: 1:25,000
Chart Number: 4849

Cape St Francis to/à Baccalieu Island and/et Heart's Content
Scale: 1:60,000
Chart Number: 4850

Smith Sound and/et Random Sound
Scale: 1:40,000
Chart Number: 4852

Catalina Harbour to/à Inner Gooseberry Islands
Scale: 1:60,000
Chart Number: 4854

Bonavista Bay: Southern Portion/Partie sud
Scale: 1:60,000
Chart Number: 4855

Bonavista Bay: Western Portion/Partie ouest
Scale: 1:60,000
Chart Number: 4856

Indian Bay to/à Wadham Islands
Scale: 1:60,000
Chart Number: 4857

Greenspond Harbour to/à Pound Cove
Scale: 1:20,000
Chart Number: 4858

Carmanville to/à Bacalhoa Island and/et Fogo
Scale: 1:40,000
Chart Number: 4862

Black Island to/à Little Denier Island
Scale: 1:40,000
Chart Number: 4864

Approaches to/Approches à Lewisporte and/et Loon Bay
Scale: 1:30,000
Chart Number: 4865

Botwood and Approaches/et les approches
Scale: 1:30,000
Chart Number: 4866

Twillingate Harbours
Scale: 1:15,000
Chart Number: 4886

West Point à/to Baie de Tracadie
Scale: 1:100,000
Chart Number: 4906

Détroit de Northumberland/Northumberland Strait: Partie ouest/Western Portion: Ports/Harbours
Scale: 1:30,000
Chart Number: 4909

Caraquet Harbour, Baie de Shippegan and/et Miscou Harbour
Scale: 1:40,000
Chart Number: 4913

**Atlantic North East**

Gulf of Maine to/à Baffin Bay/Baie de Baffin
Scale: 1:4,500,000
Chart Number: 4000

Gulf of Maine to Strait of Belle Isle/au Detroit de Belle Isle
Scale: 1:3,500,000
Chart Number: 4001

Cape Breton to/à Cape Cod
Scale: 1:1,000,000
Chart Number: 4003

Newfoundland and Labrador/Terre-Neuve-et-Labrador to Bermuda/aux Bermudes
Scale: 1:3,500,000
Chart Number: 4006

Bay of Fundy/Baie de Fundy: Inner portion/Partie intérieure
Scale: 1:200,000
Chart Number: 4010

Approaches to/Approches à Bay of Fundy/Baie de Fundy
Scale: 1:300,000
Chart Number: 4011

Yarmouth to/à Halifax
Scale: 1:300,000
Chart Number: 4012

Halifax to/à Sydney
Scale: 1:350,000
Chart Number: 4013

Sydney to/à Saint-Pierre
Scale: 1:350,000
Chart Number: 4015

Saint-Pierre to/à St. John's
Scale: 1:350,000
Chart Number: 4016

Cape Race to/à Cape Freels
Scale: 1:350,000
Chart Number: 4017

Strait of Belle Isle/Détroit de Belle Isle
Scale: 1:150,000
Chart Number: 4020

Pointe Amour à/to Cape Whittle et/and Cape George
Scale: 1:150,000
Chart Number: 4021

Cabot Strait and approaches/Détroit de Cabot et les approches
Scale: 1:350,000
Chart Number: 4022

Northumberland Strait/Détroit de Northumberland
Scale: 1:300,000
Chart Number: 4023

Sable Island Bank/Banc de l'Île de Sable to/au St. Pierre Bank/Banc de Saint-Pierre
Scale: 1:400,000
Chart Number: 4045

St. Pierre Bank/Banc de Saint-Pierre to/au Whale Bank/Banc de la Baleine
Scale: 1:400,000
Chart Number: 4047

Grand Bank, Northern Portion/Grand Banc, Partie nord to/à Flemish Pass/Passe Flamande
Scale: 1:400,000
Chart Number: 4049

Sable Island/Île de Sable
Scale: 1:100,000
Chart Number: 4098

Sable Island/Île de Sable: Western Portion/Partie ouest
Scale: 1:100,000
Chart Number: 4099

Campobello Island
Scale: 1:20,000
Chart Number: 4114

Passamaquoddy Bay and/et St. Croix River
Scale: 1:50,000
Chart Number: 4115

Approaches to/Approches à Saint John
Scale: 1:60,000
Chart Number: 4116

Saint John Harbour and Approaches/et les approches
Scale: 1:15,000
Chart Number: 4117

St. Marys Bay
Scale: 1:60,000
Chart Number: 4118

Harbours in the Bay of Fundy/Ports dans la Baie de Fundy
Scale: 1:25,000
Chart Number: 4124

Petitcodiac River and/et Cumberland Basin
Scale: 1:50,000
Chart Number: 4130

Avon River and Approaches/et les approches
Scale: 1:37,500
Chart Number: 4140

Saint John to/à Evandale
Scale: 1:30,000
Chart Number: 4141

Evandale to/à Ross Island
Scale: 1:30,000
Chart Number: 4142

Mactaquac Lake—Saint John River/Rivière Saint-Jean
Scale: 1:30,000
Chart Number: 4145

Glace Bay Harbour
Scale: 1:6,000
Chart Number: 4170

Halifax Harbour: Bedford Basin
Scale: 1:10,000
Chart Number: 4201

Halifax Harbour: Point Pleasant to/à Bedford Basin
Scale: 1:10,000
Chart Number: 4202

Halifax Harbour: Black Point to/à Point Pleasant
Scale: 1:10,000
Chart Number: 4203

Lockeport Harbour and/et Shelburne Harbour
Scale: 1:20,000
Chart Number: 4209

Cape Sable to/à Pubnico Harbour
Scale: 1:30,000
Chart Number: 4210

Cape Lahave to/à Liverpool Bay
Scale: 1:37,500
Chart Number: 4211

Country Harbour to/au Ship Harbour
Scale: 1:150,000
Chart Number: 4227

Little Hope Island to/à Cape St Marys
Scale: 1:150,000
Chart Number: 4230

Cape Canso to/à Country Island
Scale: 1:60,000
Chart Number: 4233

Country Island to/à Barren Island
Scale: 1:60,000
Chart Number: 4234

Barren Island to/à Taylors Head
Scale: 1:60,000
Chart Number: 4235

Taylors Head to/à Shut-in Island
Scale: 1:60,000
Chart Number: 4236

Approaches to/Approches de Halifax Harbour
Scale: 1:40,000
Chart Number: 4237

Liverpool Harbour to/à Lockeport Harbour
Scale: 1:60,000
Chart Number: 4240

Lockeport to/à Cape Sable
Scale: 1:60,000
Chart Number: 4241

Cape Sable Island to/aux Tusket Islands
Scale: 1:60,000
Chart Number: 4242

Tusket Islands to/à Cape St Marys
Scale: 1:60,000
Chart Number: 4243

Wedgeport and Vicinity/et les abords
Scale: 1:30,000
Chart Number: 4244

Yarmouth Harbour and Approaches/et les approches
Scale: 1:10,000
Chart Number: 4245

Georges Bank/Banc de Georges: Eastern Portion/Partie est
Scale: 1:175,000
Chart Number: 4255

Sydney Harbour
Scale: 1:20,000
Chart Number: 4266

St. Peters Bay
Scale: 1:20,000
Chart Number: 4275

Little Bras D'or
Scale: 1:5,000
Chart Number: 4276

Great Bras D'Or, St. Andrews Channel and/et St. Anns Bay
Scale: 1:40,000
Chart Number: 4277

Great Bras D'Or and/et St. Patricks Channel
Scale: 1:40,000
Chart Number: 4278

Bras D'Or Lake
Scale: 1:60,000
Chart Number: 4279

Canso Harbour and Approaches/et les Approches
Scale: 1:15,000
Chart Number: 4281

Strait of Canso
Scale: 1:30,000
Chart Number: 4302

Canso Harbour to/au Strait of Canso
Scale: 1:37,500
Chart Number: 4307

St. Peters Bay to/à Strait of Canso
Scale: 1:37,500
Chart Number: 4308

Egg Island to/à West Ironbound Island
Scale: 1:145,000
Chart Number: 4320

Cape Canso to/à Liscomb Island
Scale: 1:108,836
Chart Number: 4321

Lunenburg Bay
Scale: 1:18,000
Chart Number: 4328

Strait of Canso and Approaches/et les approches
Scale: 1:75,000
Chart Number: 4335

Alma (and Approaches/et les Approaches)
Scale: 1:25,000
Chart Number: 4337

Grand Manan
Scale: 1:60,000
Chart Number: 4340

Grand Manan (Harbours/Havres)
Scale: 1:25,000
Chart Number: 4342

Cape Smoky to/à St Paul Island
Scale: 1:75,000
Chart Number: 4363

Ingonish Harbour and/et Dingwall Harbour
Scale: 1:18,000
Chart Number: 4365

Flint Island to/à Cape Smokey
Scale: 1:75,185
Chart Number: 4367

Red Point to/à Guyon Island
Scale: 1:75,000
Chart Number: 4374

Guyon Island to/à Flint Island
Scale: 1:75,733
Chart Number: 4375

Louisbourg Harbour
Scale: 1:9,600
Chart Number: 4376

Main-à-Dieu Passage
Scale: 1:18,000
Chart Number: 4377

Liverpool Harbour
Scale: 1:8,400
Chart Number: 4379

Mahone Bay
Scale: 1:38,900
Chart Number: 4381

Pearl Island to/à Cape La Have
Scale: 1:39,023
Chart Number: 4384

Chebucto Head to/à Betty Island
Scale: 1:39,000
Chart Number: 4385

St. Margaret's Bay
Scale: 1:39,416
Chart Number: 4386

LaHave River: Conquerall Bank to/à Bridgewater
Scale: 1:6,000
Chart Number: 4391

LaHave River: West Ironbound Island to/à Riverport
Scale: 1:12,150
Chart Number: 4394

LaHave River: Riverport to/à Conquerall Bank
Scale: 1:12,150
Chart Number: 4395

Annapolis Basin
Scale: 1:24,000
Chart Number: 4396

Parrsboro Harbour and Approaches/et les approches
Scale: 1:12,154
Chart Number: 4399

Wallace Harbour
Scale: 1:24,000
Chart Number: 4402

East Point to/à Cape Bear
Scale: 1:75,000
Chart Number: 4403

Cape George to/à Pictou
Scale: 1:75,888
Chart Number: 4404

Pictou Island to/aux Tryon Shoals
Scale: 1:75,730
Chart Number: 4405

Tryon Shoals to/à Cape Egmont
Scale: 1:75,574
Chart Number: 4406

Havre de Gaspé
Scale: 1:12,000
Chart Number: 4416

Souris Harbour and Approaches/et les approches
Scale: 1:12,000
Chart Number: 4419

Murray Harbour
Scale: 1:18,233
Chart Number: 4420

Boughton River
Scale: 1:18,000
Chart Number: 4421

Cardigan Bay
Scale: 1:24,311
Chart Number: 4422

Harbours on the North Shore/Hâvres sur la Côte Nord
Scale: 1:25,000
Chart Number: 4425

Rivière Ristigouche/Restigouche River
Scale: 1:36,360
Chart Number: 4426

Pictou Harbour
Scale: 1:12,000
Chart Number: 4437

East River of Pictou: Indian Cross Point to/à Trenton and New Glasgow
Scale: 1:7,200
Chart Number: 4443

Merigomish Harbour
Scale: 1:18,232
Chart Number: 4445

Antigonish Harbour
Scale: 1:12,000
Chart Number: 4446

Pomquet and Tracadie Harbours/Havres de Pomquet et Tracadie
Scale: 1:25,000
Chart Number: 4447

Port Hood and/et Mabou Harbour
Scale: 1:18,000
Chart Number: 4448

Chéticamp Harbour/Grand Étang Harbour/Margaree Harbours
Scale: 1:12,000
Chart Number: 4449

Saint Paul Island
Scale: 1:24,300
Chart Number: 4450

Summerside Harbour and Approaches/et les approches
Scale: 1:25,000
Chart Number: 4459

Charlottetown Harbour
Scale: 1:12,000
Chart Number: 4460

St George's Bay
Scale: 1:75,000
Chart Number: 4462

Chéticamp to/à Cape Mabou
Scale: 1:75,000
Chart Number: 4463

Chéticamp to/à Cape St. Lawrence
Scale: 1:74,488
Chart Number: 4464

Hillsborough Bay
Scale: 1:36,500
Chart Number: 4466

Rustico Bay and/et New London Bay
Scale: 1:15,000
Chart Number: 4467

Caribou Harbour
Scale: 1:18,000
Chart Number: 4483

Baie des Chaleurs/Chaleur Bay
Scale: 1:150,000
Chart Number: 4486

Malpeque Bay
Scale: 1:37,500
Chart Number: 4491

Cascumpeque Bay
Scale: 1:25,000
Chart Number: 4492

Pugwash Harbour and approaches/et les approches
Scale: 1:24,000
Chart Number: 4498

Great Cat Arms and/et Little Cat Arm
Scale: 1:20,000
Chart Number: 4504

Plans—East Coast of the Island of Newfoundland /Côte Est de l'Île de Terre Neuve
Scale: 1:15,000
Chart Number: 4505

Plans: Vicinity of Canada Bay/Environs de Canada Bay
Scale: 1:18,308
Chart Number: 4506

Plans: Northeast Coast/Côte Nord-Est Newfoundland/Terre-Neuve
Scale: 1:15,000
Chart Number: 4507

Pistolet Bay
Scale: 1:45,000
Chart Number: 4509

Sacred Bay
Scale: 1:15,000
Chart Number: 4511

Quirpon Harbour and Approaches/et les approches
Scale: 1:15,000
Chart Number: 4512

St. Anthony Bight and Harbour
Scale: 1:15,000
Chart Number: 4514

Hare Bay
Scale: 1:45,000
Chart Number: 4515

Harbours in/Havres dans Hare Bay
Scale: 1:15,000
Chart Number: 4516

Ariege Bay
Scale: 1:25,000
Chart Number: 4518

Maiden Arm, Big Spring Inlet and/et Little Spring Inlet and Approaches/et les approches
Scale: 1:15,000
Chart Number: 4519

Baie Verte
Scale: 1:25,000
Chart Number: 4521

Tilt Cove and/et La Scie Harbour
Scale: 1:25,000
Chart Number: 4522

Little Bay Arm and Approaches/et les approches
Scale: 1:6,000
Chart Number: 4523

Fogo Harbour/Seal Cove and Approaches/et les approches
Scale: 1:20,000
Chart Number: 4529

Hamilton Sound: Eastern Portion/Partie-est
Scale: 1:40,000
Chart Number: 4530

Canada Bay including/y compris Chimney Bay
Scale: 1:18,290
Chart Number: 4538

Anchorages in White Bay/Mouillages dans White Bay
Scale: 1:12,254
Chart Number: 4540

Sops Arm
Scale: 1:13,000
Chart Number: 4541

Hampden Bay
Scale: 1:12,160
Chart Number: 4542

Plans—Notre Dame Bay
Scale: 1:24,320
Chart Number: 4582

St. Julien Island to/à Hooping Harbour including/y compris Canada Bay
Scale: 1:73,500
Chart Number: 4583

White Bay: Southern Part/Partie sud
Scale: 1:73,500
Chart Number: 4584

Green Head to/à Little Bay Island
Scale: 1:24,540
Chart Number: 4585

Mortier Bay
Scale: 1:12,000
Chart Number: 4587

Pilley's Island Harbour—Halls Bay and/et Sunday Cove
Scale: 1:37,400
Chart Number: 4591

Little Bay Island to/à League Rock
Scale: 1:24,400
Chart Number: 4592

Sunday Cove Island to/à Thimble Tickles
Scale: 1:24,500
Chart Number: 4593

Harbours in Placentia Bay/Havres dams Placentia Bay: Petit forte to/à Broad Cove Head
Scale: 1:25,000
Chart Number: 4615

Burin Harbours and Approaches/et les approches
Scale: 1:20,000
Chart Number: 4616

Red Island to/à Pinchgut Point
Scale: 1:40,000
Chart Number: 4617

Presque Harbour to/à Bar Haven Island and/et Paradise Sound
Scale: 1:37,500
Chart Number: 4619

Cape St Mary's to/à Argentia Harbour and/et Jude Island
Scale: 1:80,000
Chart Number: 4622

Long Island to/à St. Lawrence Harbours
Scale: 1:80,000
Chart Number: 4624

Burin Peninsula to/à Saint-Pierre
Scale: 1:75,000
Chart Number: 4625

Wreck Island to/à Cinq Cerf Bay
Scale: 1:24,318
Chart Number: 4638

Garia Bay and/et Le Moine Bay
Scale: 1:25,000
Chart Number: 4639

Isle aux Morts and Approaches/et les approches
Scale: 1:10,000
Chart Number: 4640

Port aux Basques and Approaches/et les approches
Scale: 1:7,500
Chart Number: 4641

Great St. Lawrence Harbour and/et Lamaline Harbour
Scale: 1:20,000
Chart Number: 4642

Île Saint-Pierre (France)
Scale: 1:15,000
Chart Number: 4643

Bay D'Espoir and/et Hermitage Bay
Scale: 1:50,000
Chart Number: 4644

Humber Arm: Meadows Point to/à Humber River
Scale: 1:14,600
Chart Number: 4652

Bay of Islands
Scale: 1:50,000
Chart Number: 4653

Lark Harbour and/et York Harbour
Scale: 1:12,000
Chart Number: 4654

Bonne Bay
Scale: 1:40,000
Chart Number: 4658

Port au Port
Scale: 1:41,700
Chart Number: 4659

Bear Head to/à Cow Head
Scale: 1:147,300
Chart Number: 4661

Cow Head to/à Pointe Riche
Scale: 1:144,000
Chart Number: 4663

St. Margaret Bay and Approaches/et les approches
Scale: 1:18,295
Chart Number: 4665

St Barbe Point to/à Old Férolle Harbour
Scale: 1:25,000
Chart Number: 4666

Savage Cove to/à St. Barbe Bay
Scale: 1:25,000
Chart Number: 4667

Anchorages/Mouillages in the/dans le Strait of Belle Isle/Détroit de Belle Isle
Scale: 1:25,200
Chart Number: 4668

Red Bay
Scale: 1:12,000
Chart Number: 4669

Forteau Bay
Scale: 1:24,440
Chart Number: 4670

Hawkes Bay, Port Saunders, Back Arm
Scale: 1:25,000
Chart Number: 4679

Hawkes Bay to/à Ste Geneviève Bay including/y compris St. John Bay
Scale: 1:73,000
Chart Number: 4680

Larkin Point to/à Cape Anguille
Scale: 1:20,000
Chart Number: 4682

Belle Isle to/à Resolution Island
Scale: 1:1,000,000
Chart Number: 4700

Ship Harbour Head to/aux Camp Islands
Scale: 1:75,000
Chart Number: 4701

Corbett Island to/à Ship Harbour Head
Scale: 1:75,000
Chart Number: 4702

White Point to/à Corbet Island
Scale: 1:75,000
Chart Number: 4703

Plans on the Coast of Labrador/Plans sur la côte du Labrador
Scale: 1:30,000
Chart Number: 4712

Terrington Basin
Scale: 1:6,000
Chart Number: 4722

Ticoralak Island to/à Carrington Island
Scale: 1:50,000
Chart Number: 4724

Carrington Island to/à Etagaulet Bay
Scale: 1:50,000
Chart Number: 4725

Epinette Point to/à Terrington Basin
Scale: 1:50,000
Chart Number: 4728

Nain to/à Domino Point
Scale: 1:588,000
Chart Number: 4730

Forteau Bay to/à Domino Run
Scale: 1:250,000
Chart Number: 4731

Approaches to/Approches à Hamilton Inlet
Scale: 1:223,975
Chart Number: 4732

Approaches to/approches à Spotted Island Harbour
Scale: 1:30,000
Chart Number: 4744

White Point to/à Sandy Island
Scale: 1:30,000
Chart Number: 4745

Bay Bulls to/à St. Mary's Bay
Scale: 1:150,000
Chart Number: 4817

Bay Bulls to/à St. Mary's Bay
Scale: 1:30,000
Chart Number: 4830

Head of/Fond de Placentia Bay
Scale: 1:40,000
Chart Number: 4839

Cape St Mary's to/à Argentia
Scale: 1:60,000
Chart Number: 4841

Cape Pine to/au Cape St Mary's
Scale: 1:60,000
Chart Number: 4842

Head of/Fond de St Mary's Bay
Scale: 1:60,000
Chart Number: 4843

Cape Pine to/à Renews Harbour
Scale: 1:60,000
Chart Number: 4844

Renews Harbour to/à Motion Bay
Scale: 1:60,000
Chart Number: 4845

Motion Bay to/à Cape St Francis
Scale: 1:60,000
Chart Number: 4846

Conception Bay
Scale: 1:60,000
Chart Number: 4847

Holyrood and/et Long Pond
Scale: 1:15,000
Chart Number: 4848

Trinity Bay: Southern Portion/Partie sud
Scale: 1:60,000
Chart Number: 4851

Trinity Bay: Northern Portion/Partie nord
Scale: 1:60,000
Chart Number: 4853

Port Harmon and Approaches/et les approches
Scale: 1:40,000
Chart Number: 4885

Cape Tormentine à/to West Point
Scale: 1:100,000
Chart Number: 4905

Entrée à/Entrance to Miramichi River
Scale: 1:60,000
Chart Number: 4911

Miramichi
Scale: 1:25,000
Chart Number: 4912

Plans: Baie des Chaleurs/Chaleur Bay: Côte sud/South Shore
Scale: 1:20,000
Chart Number: 4920

Labrador Sea/Mer du Labrador
Scale: 1:3,500,000
Chart Number: 5001

Cape Harrison to/à Nunaksaluk Island
Scale: 1:200,000
Chart Number: 5023

Nunaksaluk Island to/à Cape Kiglapait
Scale: 1:200,000
Chart Number: 5024

Murphy Head to/aux Button Islands
Scale: 1:200,000
Chart Number: 5027

Green Bay to/à Double Island
Scale: 1:60,000
Chart Number: 5030

St. Lewis Sound and/et Inlet
Scale: 1:40,000
Chart Number: 5031

Approaches to/à White Bear Arm
Scale: 1:30,000
Chart Number: 5032

Hawke Bay and/et Squasho Run
Scale: 1:30,000
Chart Number: 5033

Hawke Bay and/et Squasho Run
Scale: 1:60,000
Chart Number: 5042

Quaker Hat to/à Cape Harrison
Scale: 1:60,000
Chart Number: 5043

Cape Harrison to/à Dog Islands
Scale: 1:60,000
Chart Number: 5044

Dog Islands to/à Cape Makkovik
Scale: 1:60,000
Chart Number: 5045

Kaipokok Bay and Cape Makkovik to/à Winsor Harbour Island
Scale: 1:60,000
Chart Number: 5046

Winsor Harbour Island to/aux Kikkertaksoak Islands
Scale: 1:60,000
Chart Number: 5047

Cape Harrigan to/aux Kidlit Islands
Scale: 1:60,000
Chart Number: 5048

Davis Inlet to/aux Seniartlit Islands
Scale: 1:60,000
Chart Number: 5049

Nunaksuk Island to/aux Calf Cow and/et Bull Islands
Scale: 1:60,000
Chart Number: 5051

Seniartlit Islands to/à Nain
Scale: 1:60,000
Chart Number: 5052

Domino Point to/à Cape North
Scale: 1:75,000
Chart Number: 5133

Approaches to/Approches à Cartwright: Black Island to/à Tumbledown Dick Island
Scale: 1:75,000
Chart Number: 5134

Approaches to/Approches à Hamilton Inlet
Scale: 1:75,000
Chart Number: 5135

Sandwich Bay
Scale: 1:50,000
Chart Number: 5138

South Green Island to/à Ticoralak Island
Scale: 1:50,000
Chart Number: 5140

Lake Melville
Scale: 1:100,000
Chart Number: 5143

Alexis Bay and/et Alexis River
Scale: 1:40,000
Chart Number: 5179

Central North Central
Rivière Koksoak
Scale: 1:30,000
Chart Number: 5338

Baker Lake
Scale: 1:80,000
Chart Number: 5626

Rankin Inlet Including/y compris Melvin Bay and/et Prairie Bay
Scale: 1:15,000
Chart Number: 5628

Marble Island to/à Rankin Inlet
Scale: 1:60,000
Chart Number: 5629

Dunne Foxe Island to/à Chesterfield Inlet
Scale: 1:150,000
Chart Number: 5630

Eskimo Point to/à Dunne Foxe Island
Scale: 1:150,000
Chart Number: 5631

Churchill Harbour
Scale: 1:12,000
Chart Number: 5640

Arviat and Approaches/et approches
Scale: 1:60,000
Chart Number: 5641

Whale Cove and Approaches/et approches
Scale: 1:60,000
Chart Number: 5642

Approches à/Approaches to Chisasibi
Scale: 1:30,000
Chart Number: 5720

Manitoulin Island Lakes/Lacs sur Manitoulin Island
Scale: 1:30,000
Chart Number: 6030

Seven Sisters Falls to/à Lac du Bonnet
Scale: 1:25,000
Chart Number: 6205

Kettle Island to/à Martin Point
Scale: 1:25,000
Chart Number: 6259

Martin Point to/à Wightman Point
Scale: 1:25,000
Chart Number: 6260

Playgreen Lake to/au Little Playgreen Lake
Scale: 1:15,000
Chart Number: 6263

East Channel to/au Little Playgreen Lake
Scale: 1:15,000
Chart Number: 6264

Eaglenest Lake to/à Whitedog Dam
Scale: 1:25,000
Chart Number: 6285

Lake Manitoba/Lac Manitoba (Southern Portion/Partie sud)
Scale: 1:300,000
Chart Number: 6505

Lake Manitoba/Lac Manitoba (Northern Portion/Partie nord)
Scale: 1:100,000
Chart Number: 6506

Robinson Bay and Approaches/et les approches
Scale: 1:300,000
Chart Number: 7134

Parry Bay to/au Navy Channel
Scale: 1:150,000
Chart Number: 7485

Navy Channel to/à Fury and Hecla Strait
Scale: 1:150,000
Chart Number: 7486

Fury and Hecla Strait
Scale: 1:150,000
Chart Number: 7487

Air Force Island to/au Longstaff Bluff
Scale: 1:150,000
Chart Number: 7488

Navy Channel to/à Longstaff Bluff
Scale: 1:150,000
Chart Number: 7489

Gulf of Boothia and/et Committee Bay
Scale: 1:500,000
Chart Number: 7502

Resolute Passage
Scale: 1:50,000
Chart Number: 7511

Strathcona Sound and/et Adams Sound
Scale: 1:80,000
Chart Number: 7512

Milne Inlet (Southern Portion/Partie sud)
Scale: 1:40,000
Chart Number: 7513

Prince of Wales Strait (Northern Portion/Partie nord)
Scale: 1:200,000
Chart Number: 7520

Prince of Wales Strait (Southern Portion/Partie sud)
Scale: 1:200,000
Chart Number: 7521

Clyde Inlet to/à Cape Jameson
Scale: 1:300,000
Chart Number: 7565

Cape Jameson to/au Cape Fanshawe
Scale: 1:300,000
Chart Number: 7566

Lancaster Sound and/et Admiralty Inlet
Scale: 1:300,000
Chart Number: 7568

Barrow Strait and/et Wellington Channel
Scale: 1:300,000
Chart Number: 7569

Barrow Strait and/et Viscount Melville Sound
Scale: 1:300,000
Chart Number: 7570

Viscount Melville Sound
Scale: 1:300,000
Chart Number: 7571

Viscount Melville Sound and/et M'clure Strait
Scale: 1:300,000
Chart Number: 7572

M'Clintock Channel, Larsen Sound and/et Franklin Strait
Scale: 1:300,000
Chart Number: 7573

Peel Sound and/et Prince Regent Inlet
Scale: 1:300,000
Chart Number: 7575

Pelly Bay
Scale: 1:125,000
Chart Number: 7578

Beaufort Sea/Mer de Beaufort
Scale: 1:1,000,000
Chart Number: 7600

Eskimo Lakes
Scale: 1:150,000
Chart Number: 7608

Demarcation Bay to/à Liverpool Bay
Scale: 1:500,000
Chart Number: 7620

Amundsen Gulf
Scale: 1:500,000
Chart Number: 7621

Putulik (Hat Island) and/et Wilkins Point
Scale: 1:15,000
Chart Number: 7646

Demarcation Bay to/à Philips Bay
Scale: 1:150,000
Chart Number: 7661

Mackenzie Bay
Scale: 1:150,000
Chart Number: 7662

Kugmallit Bay
Scale: 1:150,000
Chart Number: 7663

Liverpool Bay
Scale: 1:150,000
Chart Number: 7664

Franklin Bay and/et Darnley Bay
Scale: 1:150,000
Chart Number: 7665

Cape Lyon to/à Tinney Point
Scale: 1:150,000
Chart Number: 7666

Dolphin and Union Strait to/à Prince Albert Sound
Scale: 1:150,000
Chart Number: 7667

Prince Albert Sound (Western Portion/Partie ouest)
Scale: 1:150,000
Chart Number: 7668

Prince Albert Sound (Eastern Portion/Partie est)
Scale: 1:150,000
Chart Number: 7669

Tuktoyaktuk Harbour and Approaches/et les approches
Scale: 1:15,000
Chart Number: 7685

Police Point and Approaches/et les approches
Scale: 1:50,000
Chart Number: 7686

Police Point and Approaches/et les approches
Scale: 1:50,000
Chart Number: 7687

Lambert Channel and/et Cache Point Channel
Scale: 1:80,000
Chart Number: 7710

Requisite Channel and Approaches
Scale: 1:75,000
Chart Number: 7725

Storis Passage and Approaches
Scale: 1:75,000
Chart Number: 7731

**Central North East**

Île St-Régis to/à Croil Islands
Scale: 1:25,000
Chart Number: 1433

Croil Islands to/à Cardinal
Scale: 1:25,000
Chart Number: 1434

Cardinal to/à Whaleback Shoal
Scale: 1:25,000
Chart Number: 1435

Whaleback Shoal to/au Summerland Group
Scale: 1:25,000
Chart Number: 1436

Summerland Group to/à Grindstone Island
Scale: 1:25,000
Chart Number: 1437

Grindstone Island to/à Carleton Island
Scale: 1:25,000
Chart Number: 1438

Carleton Island to/au Charity Shoal
Scale: 1:30,000
Chart Number: 1439

Ottawa to/à Smiths Falls
Scale: 1:20,000
Chart Number: 1512

Smith Falls to/à Kingston including/y compris Tay River to/à Perth
Scale: 1:20,000
Chart Number: 1513

Carillon à/to Papineauville
Scale: 1:20,000
Chart Number: 1514

Britannia Bay à/to Chats Falls
Scale: 1:25,000
Chart Number: 1550

Chats Falls à/to Chenaux
Scale: 1:25,000
Chart Number: 1551

Portage-du-Fort à/to Île Marcotte
Scale: 1:25,000
Chart Number: 1552

Île Marcotte à/to Rapides-des-Joachims
Scale: 1:25,000
Chart Number: 1553

Rapides-des-Joachims au/to Lac la Cave
Scale: 1:25,000
Chart Number: 1554

Lac la Cave
Scale: 1:25,000
Chart Number: 1555

Lac Témiscamingue/Lake Timiskaming
Scale: 1:35,000
Chart Number: 1556

Lake Ontario/Lac Ontario
Scale: 1:400,000
Chart Number: 2000

Upper Gap to/à Telegraph Narrows
Scale: 1:30,000
Chart Number: 2006

Belleville to/à Telegraph Narrows
Scale: 1:30,000
Chart Number: 2007

Belleville Harbour
Scale: 1:6,000
Chart Number: 2011

Kingston Harbour and Approaches/et les approches
Scale: 1:15,000
Chart Number: 2017

Lower Gap to/à Adolphus Reach
Scale: 1:30,000
Chart Number: 2018

Murray Canal to Healey Falls Locks/Murray Canal aux Écluses de Healey Falls
Scale: 1:20,000
Chart Number: 2021

Healey Falls Locks to Peterborough/Écluses de Healey Falls à Peterborough
Scale: 1:20,000
Chart Number: 2022

Peterborough to/à Buckhorn including/y compris Stony Lake
Scale: 1:20,000
Chart Number: 2023

Buckhorn to/à Bobcaygeon including/y compris Chemong Lake
Scale: 1:20,000
Chart Number: 2024

Bobcaygeon to Lake Simcoe/Bobcaygeon au Lake Simcoe
Scale: 1:30,000
Chart Number: 2025

Lake Scugog and/et Scugog River
Scale: 1:30,000
Chart Number: 2026

Lakes Simcoe and Couchiching including the Holland River/Lacs Simcoe et Couchiching y compris Holland River
Scale: 1:80,000
Chart Number: 2028

Couchiching Lock to Port Severn/Écluse de Couchiching a Port Severn
Scale: 1:20,000
Chart Number: 2029

Welland Canal St.Catharines to/à Port Colborne
Scale: 1:15,000
Chart Number: 2042

Lower Niagara River and Approaches/et les approches
Scale: 1:20,000
Chart Number: 2043

Port Dalhousie
Scale: 1:5,000
Chart Number: 2044

Clarkson Harbour
Scale: 1:7,500
Chart Number: 2047

Port Credit
Scale: 1:5,000
Chart Number: 2048

Whitby Harbour
Scale: 1:5,000
Chart Number: 2049

Oshawa Harbour
Scale: 1:5,000
Chart Number: 2050

Port Hope Harbour
Scale: 1:5,000
Chart Number: 2053

Cobourg Harbour
Scale: 1:5,000
Chart Number: 2054

Frenchman's Bay
Scale: 1:10,000
Chart Number: 2055

Cobourg to/à Oshawa
Scale: 1:72,400
Chart Number: 2058

Scotch Bonnet Island to/à Cobourg
Scale: 1:70,400
Chart Number: 2059

Main Duck Island to/à Scotch Bonnet Island
Scale: 1:77,700
Chart Number: 2060

Kingston to/à False Duck Islands
Scale: 1:61,528
Chart Number: 2064

Hamilton Harbour
Scale: 1:10,000
Chart Number: 2067

Picton to/à Presqu'île Bay
Scale: 1:60,588
Chart Number: 2069

Harbours in Lake Ontario/Havres dans le lac Ontario
Scale: 1:10,596
Chart Number: 2070

Lake Ontario/Lac Ontario (Western Portion/Partie ouest)
Scale: 1:100,000
Chart Number: 2077

Toronto Harbour
Scale: 1:15,000
Chart Number: 2085

Toronto to/à Hamilton
Scale: 1:50,000
Chart Number: 2086

Lake Erie/Lac Érié
Scale: 1:400,000
Chart Number: 2100

Long Point Bay
Scale: 1:50,000
Chart Number: 2110

Niagara River to/à Long Point
Scale: 1:120,000
Chart Number: 2120

Long Point to/à Port Glasgow
Scale: 1:100,000
Chart Number: 2121

Pointe aux Pins to/à Point Pelee
Scale: 1:100,000
Chart Number: 2122

Pelee Passage to/à la Detroit River
Scale: 1:100,000
Chart Number: 2123

Port Maitland to/à Dunnville
Scale: 1:7,500
Chart Number: 2140

Harbours in Lake Erie/Havres dans le lac Érié
Scale: 1:5,000
Chart Number: 2181

Lake Huron/Lac Huron
Scale: 1:400,000
Chart Number: 2200

Georgian Bay/Baie Georgienne
Scale: 1:200,000
Chart Number: 2201

Port Severn to/à Parry Sound
Scale: 1:40,000
Chart Number: 2202

Carling Rock to/à Byng Inlet
Scale: 1:20,000
Chart Number: 2203

Byng Inlet to/à Killarney
Scale: 1:20,000
Chart Number: 2204

Killarney to/à Little Current
Scale: 1:40,000
Chart Number: 2205

McGregor Bay
Scale: 1:15,000
Chart Number: 2206

Baie des Chaleurs/Chaleur Bay aux/to Îles de la Madeleine
Scale: 1:350,000
Chart Number: 4024

Cap Whittle à/to Havre Saint-Pierre et/and Île d'Anticosti
Scale: 1:300,000
Chart Number: 4025

Havre Saint-Pierre et/and Cap des Rosiers à/to Pointe des Monts
Scale: 1:300,000
Chart Number: 4026

Baie D'Ungava/Ungava Bay
Scale: 1:500,000
Chart Number: 5300

Shaftesbury Inlet to/à Ashe Inlet
Scale: 1:75,000
Chart Number: 5316

Rivière George
Scale: 1:30,000
Chart Number: 5335

Approach to/Approches à Sorry Harbor
Scale: 1:25,000
Chart Number: 5340

Approches à/Approaches to Hopes Advance Bay
Scale: 1:75,000
Chart Number: 5348

Hopes Advance Bay
Scale: 1:24,000
Chart Number: 5349

Payne Bay and Approaches
Scale: 1:50,000
Chart Number: 5351

Payne Bay et/and Rivière Arnaud (Tuvalik Point à/to Ile Basking)
Scale: 1:30,000
Chart Number: 5352

Cap du Prince-De-Galles à/to Davies Island
Scale: 1:100,000
Chart Number: 5365

Approches à/Approaches to Rivière George
Scale: 1:60,000
Chart Number: 5373

Beacon Island à/to Qikirtaaluk Islands
Scale: 1:60,000
Chart Number: 5374

Qikirtaaluk Islands à/to Point Qirniraujaq
Scale: 1:60,000
Chart Number: 5375

Approches à/Approaches to Rivière Koksoak
Scale: 1:60,000
Chart Number: 5376

Wakeham Bay and Fisher Bay et les approches/and Approaches
Scale: 1:37,500
Chart Number: 5390

Douglas Harbour et les approches/and Approaches
Scale: 1:37,500
Chart Number: 5391

Egg Island to/à Eskimo Point
Scale: 1:150,000
Chart Number: 5399

Cape Churchill to/à Egg River
Scale: 1:146,198
Chart Number: 5400

Pritzler Harbour to/à Maniittur Cape
Scale: 1:136,179
Chart Number: 5403

Cape Tatnam to/à Port Nelson
Scale: 1:127,000
Chart Number: 5406

Coral Harbour and Approaches/et les approches
Scale: 1:50,000
Chart Number: 5410

Lower Savage Islands to/à Pritzler Harbour
Scale: 1:137,520
Chart Number: 5411

Erik Cove to/à Nuvuk Harbour including/y compris Digges
Scale: 1:75,000
Chart Number: 5412

Rupert Bay
Scale: 1:73,000
Chart Number: 5414

Plans du Détroit D'Hudson/Plans of Hudson Strait
Scale: 1:10,000
Chart Number: 5429

Wager Bay
Scale: 1:200,000
Chart Number: 5440

Hudson Bay Baie d'Hudson (Northern Portion/Partie nord)
Scale: 1:1,000,000
Chart Number: 5449

Hudson Strait/Détroit d'Hudson
Scale: 1:1,000,000
Chart Number: 5450

Cape Dorset and Approaches/et les approches
Scale: 1:25,000
Chart Number: 5451

Diana Bay
Scale: 1:50,000
Chart Number: 5452

Kimmirut and Approaches/et les approches
Scale: 1:25,000
Chart Number: 5455

Deception Bay
Scale: 1:25,000
Chart Number: 5457

Sugluk Inlet
Scale: 1:37,500
Chart Number: 5458

Resolution Harbour and/et Acadia Cove
Scale: 1:12,000
Chart Number: 5459

Diana Bay (Partie sud/Southern Portion)
Scale: 1:20,000
Chart Number: 5464

Baie aux feuilles/Leaf Bay et les approches/and Approaches
Scale: 1:50,000
Chart Number: 5467

Passage aux Feuilles
Scale: 1:30,000
Chart Number: 5468

Lac aux Feuilles
Scale: 1:30,000
Chart Number: 5469

Inukjuak et les approches/and Approaches
Scale: 1:25,000
Chart Number: 5471

Harbours and Anchorages Hudson Bay and James Bay/Ports et Mouillages Baie d'Hudson et Baie James
Scale: 1:75,000
Chart Number: 5476

Bélanger Island à/to Cotter Island
Scale: 1:250,000
Chart Number: 5505

Povungnituk et les approches/and Approaches
Scale: 1:60,000
Chart Number: 5510

Smith Island to/à Knight Harbour
Scale: 1:50,000
Chart Number: 5512

Roes Welcome Sound (Chesterfield Inlet to/à Cape Munn)
Scale: 1:500,000
Chart Number: 5533

Entrance to/Entrée à Chesterfield Inlet (Fairway Island to/à Ellis Island)
Scale: 1:40,000
Chart Number: 5620

Rockhouse Island to/à Centre Island
Scale: 1:40,000
Chart Number: 5621

Centre Island to/à Farther Hope Point
Scale: 1:40,000
Chart Number: 5622

Farther Hope Point to/à Terror Point
Scale: 1:40,000
Chart Number: 5623

Terror Point to/au Schooner Harbour
Scale: 1:40,000
Chart Number: 5624

Schooner Harbour to/à Baker Lake
Scale: 1:40,000
Chart Number: 5625

**North Central**

Little Current to/à Clapperton Island
Scale: 1:60,000
Chart Number: 2207

Penetang Harbour
Scale: 1:18,000
Chart Number: 2218

Midland Harbour
Scale: 1:8,000
Chart Number: 2221

Tiffin Harbour
Scale: 1:10,000
Chart Number: 2222

Port McNicoll and/et Victoria Harbour
Scale: 1:15,000
Chart Number: 2223

Rose Island to/à Parry Sound
Scale: 1:20,000
Chart Number: 2224

Approaches to/Approches à Parry Sound
Scale: 1:25,000
Chart Number: 2225

Lake Huron/Lac Huron (Southern Portion/Partie sud)
Scale: 1:120,000
Chart Number: 2228

Cape Hurd to/à Lonely Island
Scale: 1:60,000
Chart Number: 2235

Port Severn to/à Christian Island
Scale: 1:40,000
Chart Number: 2241

Giants Tomb Island to/à Franklin Island
Scale: 1:60,000
Chart Number: 2242

**North East**

Bateau Island to/à Byng Inlet
Scale: 1:60,000
Chart Number: 2243

Alexander Passage to/à Beaverstone Bay
Scale: 1:60,000
Chart Number: 2244

Beaverstone Bay to/à Lonely Island and/et McGregor Bay
Scale: 1:100,000
Chart Number: 2245

Bruce Mines to/à Sugar Island
Scale: 1:25,000
Chart Number: 2250

Meldrum Bay to/à St. Joseph Island
Scale: 1:60,000
Chart Number: 2251

Clapperton Island to/à John Island
Scale: 1:40,000
Chart Number: 2257

Bayfield Sound and Approaches/et les approches
Scale: 1:40,000
Chart Number: 2258

John Island to/à Blind River
Scale: 1:40,000
Chart Number: 2259

Sarnia to/à Bayfield
Scale: 1:80,000
Chart Number: 2260

Bayfield to/à Douglas Point
Scale: 1:80,000
Chart Number: 2261

Michael's Bay to/à Great Duck Island
Scale: 1:60,000
Chart Number: 2266

Great Duck Island to/à False Detour Passage
Scale: 1:60,000
Chart Number: 2267

Plans North Channel
Scale: 1:20,000
Chart Number: 2268

South Baymouth Harbour and Approaches
Scale: 1:6,000
Chart Number: 2273

Cape Hurd to/à Tobermory and/et Cove Island
Scale: 1:15,000
Chart Number: 2274

Owen Sound to/à Cabot Head
Scale: 1:80,000
Chart Number: 2282

Owen Sound to/à Giant's Tomb Island
Scale: 1:80,000
Chart Number: 2283

Point Clark to/à Southampton
Scale: 1:90,000
Chart Number: 2291

Chantry Island to Cove Island
Scale: 1:93,314
Chart Number: 2292

Byng Inlet and Approaches/et les approches
Scale: 1:12,000
Chart Number: 2293

Duck Islands to/à DeTour Passage
Scale: 1:91,085
Chart Number: 2297

Cove Island to/aux Duck Islands
Scale: 1:91,010
Chart Number: 2298

Clapperton Island to/à Meldrum Bay
Scale: 1:80,000
Chart Number: 2299

Lake Superior/Lac Supérieur
Scale: 1:600,000
Chart Number: 2300

Passage Island to/à Thunder Bay
Scale: 1:74,516
Chart Number: 2301

St. Ignace Island to/à Passage Island
Scale: 1:72,968
Chart Number: 2302

Jackfish Bay to/à St. Ignace Island
Scale: 1:70,000
Chart Number: 2303

Oiseau Bay to/à Jackfish Bay
Scale: 1:73,010
Chart Number: 2304

Jackfish Bay
Scale: 1:12,180
Chart Number: 2305

Peninsula Harbour and/et Port Munro
Scale: 1:12,172
Chart Number: 2306

Coppermine Point to/à Cape Gargantua
Scale: 1:74,486
Chart Number: 2307

Michipicoten Island to/à Oiseau Bay
Scale: 1:90,000
Chart Number: 2308

Cape Gargantua to/à Otter Head
Scale: 1:96,000
Chart Number: 2309

Caribou Island to/à Michipicoten Island
Scale: 1:97,280
Chart Number: 2310

Thunder Cape to/à Pigeon River
Scale: 1:72,880
Chart Number: 2311

Nipigon Bay and Approaches/et les approches
Scale: 1:60,800
Chart Number: 2312

Black Bay
Scale: 1:72,665
Chart Number: 2313

Port of Thunder Bay
Scale: 1:20,000
Chart Number: 2314

Harbours on the East Shore of Lake Superior/Ports sur la rive est du lac Supérieur
Scale: 1:12,000
Chart Number: 2315

Heron Bay
Scale: 1:12,161
Chart Number: 2318

Great Lakes/Grands Lacs
Scale: 1:1,584,000
Chart Number: 2400

Juan de Fuca Strait to/à Dixon Entrance
Scale: 1:1,250,000
Chart Number: 3000

Vancouver Island/Île de Vancouver, Juan de Fuca Strait to/à Queen Charlotte Sound
Scale: 1:525,000
Chart Number: 3001

Queen Charlotte Sound to/à Dixon Entrance
Scale: 1:525,000
Chart Number: 3002

Kootenay Lake and River
Scale: 1:75,000
Chart Number: 3050

Okanagan Lake
Scale: 1:50,000
Chart Number: 3052

Shuswap Lake
Scale: 1:50,000
Chart Number: 3053

Waneta to/à Hugh Keenleyside Dam
Scale: 1:20,000
Chart Number: 3055

Hugh Keenleyside Dam to/à Burton
Scale: 1:40,000
Chart Number: 3056

Burton to/à Arrowhead
Scale: 1:40,000
Chart Number: 3057

Arrowhead to/à Revelstoke
Scale: 1:20,000
Chart Number: 3058

Harrison Lake and/et Harrison River
Scale: 1:40,000
Chart Number: 3061

Pitt River and/et Pitt Lake
Scale: 1:25,000
Chart Number: 3062

Stuart Lake
Scale: 1:50,000
Chart Number: 3080

Sunshine Coast, Vancouver Harbour to/à Desolation Sound
Scale: 1:40,000
Chart Number: 3311

Jervis Inlet and Desolation Sound and Adjacent Waterways/ et les voies navigables adjacentes
Scale: 1:250,000
Chart Number: 3312

Gulf Islands and Adjacent Waterways/et les voies navigables adjacentes
Scale: 1:200,000
Chart Number: 3313

Sooke Inlet to/à Parry Bay
Scale: 1:20,000
Chart Number: 3410

Sooke
Scale: 1:12,000
Chart Number: 3411

Victoria Harbour
Scale: 1:12,000
Chart Number: 3412

Esquimalt Harbour
Scale: 1:5,000
Chart Number: 3419

Approaches to/Approches à Oak Bay
Scale: 1:10,000
Chart Number: 3424

Race Rocks to/à D'Arcy Island
Scale: 1:40,000
Chart Number: 3440

Haro Strait, Boundary Pass and/et Satellite Channel
Scale: 1:40,000
Chart Number: 3441

North Pender Island to/à Thetis Island
Scale: 1:40,000
Chart Number: 3442

Thetis Island to/à Nanaimo
Scale: 1:40,000
Chart Number: 3443

Nanaimo Harbour and/et Departure Bay
Scale: 1:10,000
Chart Number: 3447

Halibut Bank to/à Ballenas Channel
Scale: 1:30,000
Chart Number: 3456

Approaches to/Approches à Nanaimo Harbour
Scale: 1:20,000
Chart Number: 3458

Approaches to/Approches à Nanoose Harbour
Scale: 1:15,000
Chart Number: 3459

Juan de Fuca Strait (Eastern Portion/Partie est)
Scale: 1:80,000
Chart Number: 3461

Juan de Fuca Strait to/à Strait of Georgia
Scale: 1:80,000
Chart Number: 3462

Strait of Georgia (Southern Portion/Partie sud)
Scale: 1:80,000
Chart Number: 3463

Plans—Stuart Channel
Scale: 1:18,000
Chart Number: 3475

Plans—Gulf Islands
Scale: 1:15,000
Chart Number: 3477

Plans—Saltspring Island
Scale: 1:20,000
Chart Number: 3478

Approaches to/Approches à Sidney
Scale: 1:20,000
Chart Number: 3479

Approaches to/Approches à Vancouver Harbour
Scale: 1:25,000
Chart Number: 3481

Fraser River/Fleuve Fraser, Crescent Island to/à Harrison Mills
Scale: 1:20,000
Chart Number: 3488

Fraser River/Fleuve Fraser, Pattullo Bridge to/à Crescent Island
Scale: 1:20,000
Chart Number: 3489

Fraser River/Fleuve Fraser, Sand Heads to/à Douglas Island
Scale: 1:20,000
Chart Number: 3490

Fraser River/Fleuve Fraser, North Arm
Scale: 1:20,000
Chart Number: 3491

Roberts Bank
Scale: 1:20,000
Chart Number: 3492

Vancouver Harbour (Western Portion/Partie ouest)
Scale: 1:10,000
Chart Number: 3493

Vancouver Harbour (Central Portion/Partie centrale)
Scale: 1:10,000
Chart Number: 3494

Vancouver Harbour (Eastern Portion/Partie est)
Scale: 1:30,000
Chart Number: 3495

Strait of Georgia (Central Portion/Partie centrale)
Scale: 1:80,000
Chart Number: 3512

Strait of Georgia (Northern Portion/Partie nord)
Scale: 1:80,000
Chart Number: 3513

Jervis Inlet
Scale: 1:50,000
Chart Number: 3514

Knight Inlet
Scale: 1:80,000
Chart Number: 3515

Howe Sound
Scale: 1:40,000
Chart Number: 3526

Baynes Sound
Scale: 1:40,000
Chart Number: 3527

Plans—Howe Sound
Scale: 1:12,000
Chart Number: 3534

Malaspina Strait
Scale: 1:25,000
Chart Number: 3535

Plans—Strait of Georgia
Scale: 1:12,000
Chart Number: 3536

Okisollo Channel
Scale: 1:20,000
Chart Number: 3537

Desolation Sound and/et Sutil Channel
Scale: 1:40,000
Chart Number: 3538

Discovery Passage
Scale: 1:40,000
Chart Number: 3539

Approaches to/Approches à Campbell River
Scale: 1:10,000
Chart Number: 3540

Approaches to/Approches à Toba Inlet
Scale: 1:40,000
Chart Number: 3541

Bute Inlet
Scale: 1:40,000
Chart Number: 3542

Cordero Channel
Scale: 1:80,000
Chart Number: 3543

Johnstone Strait, Race Passage and/et Current Passage
Scale: 1:40,000
Chart Number: 3544

Johnstone Strait, Port Neville to/à Robson Bight
Scale: 1:40,000
Chart Number: 3545

Broughton Strait
Scale: 1:40,000
Chart Number: 3546

Queen Charlotte Strait (Eastern Portion/Partie est)
Scale: 1:40,000
Chart Number: 3547

Queen Charlotte Strait (Eastern Portion/Partie est)
Scale: 1:40,000
Chart Number: 3549

Approaches to/Approches à Seymour Inlet and/et Belize Inlet
Scale: 1:40,000
Chart Number: 3550

Seymour Inlet and/et Belize Inlet
Scale: 1:40,000
Chart Number: 3552

Plans Desolation Sound
Scale: 1:12,000
Chart Number: 3554

Plans—Vicinity of/Proximité de Redonda Islands and/et Loughborough Inlet
Scale: 1:18,000
Chart Number: 3555

Malaspina Inlet, Okeover Inlet and/et Lancelot Inlet
Scale: 1:12,000
Chart Number: 3559

Plans—Johnstone Strait
Scale: 1:20,000
Chart Number: 3564

Cape Scott to Cape Calvert
Scale: 1:74,490
Chart Number: 3598

Juan de Fuca Strait to/à Vancouver Harbour
Scale: 1:200,000
Chart Number: 3601

Approaches to/Approches à Juan de Fuca Strait
Scale: 1:150,000
Chart Number: 3602

Ucluelet Inlet to/à Nootka Sound
Scale: 1:150,000
Chart Number: 3603

Nootka Sound to/à Quatsino Sound
Scale: 1:150,000
Chart Number: 3604

Quatsino Sound to/à Queen Charlotte Strait
Scale: 1:150,000
Chart Number: 3605

Juan de Fuca Strait
Scale: 1:110,000
Chart Number: 3606

Kyuquot Sound to/à Cape Cook
Scale: 1:80,000
Chart Number: 3623

Cape Cook to Cape Scott
Scale: 1:90,000
Chart Number: 3624

Scott Islands
Scale: 1:80,000
Chart Number: 3625

Plans—Barkley Sound
Scale: 1:18,000
Chart Number: 3646

Port San Juan and/et Nitinat Narrows
Scale: 1:18,000
Chart Number: 3647

Scouler Entrance
Scale: 1:10,000
Chart Number: 3651

Alberni Inlet
Scale: 1:40,000
Chart Number: 3668

Broken Group
Scale: 1:20,000
Chart Number: 3670

Barkley Sound
Scale: 1:40,000
Chart Number: 3671

Clayoquot Sound, Tofino Inlet to/à Millar Channel
Scale: 1:40,000
Chart Number: 3673

Clayoquot Sound, Millar Channel to/à Estevan Point
Scale: 1:40,000
Chart Number: 3674

Nootka Sound
Scale: 1:40,000
Chart Number: 3675

Esperanza Inlet
Scale: 1:40,000
Chart Number: 3676

Kyuquot Sound
Scale: 1:40,000
Chart Number: 3677

Quatsino Sound
Scale: 1:50,000
Chart Number: 3679

Brooks Bay
Scale: 1:38,317
Chart Number: 3680

Plans—Quatsino Sound
Scale: 1:15,000
Chart Number: 3681

Checleset Bay
Scale: 1:36,493
Chart Number: 3683

Tofino
Scale: 1:20,000
Chart Number: 3685

Approaches to/Approches à Winter Harbour
Scale: 1:15,000
Chart Number: 3686

Plans Pitt Island
Scale: 1:18,337
Chart Number: 3721

Caamaño Sound and Approaches/et les approches
Scale: 1:71,594
Chart Number: 3724

Laredo Sound and Approaches
Scale: 1:72,217
Chart Number: 3726

Cape Calvert to Goose Island including Fitz Hugh Sound
Scale: 1:73,584
Chart Number: 3727

Milbanke Sound and Approaches/et les approches
Scale: 1:76,557
Chart Number: 3728

Catala Passage
Scale: 1:20,000
Chart Number: 3733

Laredo Channel including/y compris Laredo Inlet and/et Surf Inlet
Scale: 1:77,429
Chart Number: 3737

Otter Passage to Bonilla Island
Scale: 1:72,860
Chart Number: 3741

Otter Passage to/à McKay Reach
Scale: 1:70,920
Chart Number: 3742

Queen Charlotte Sound
Scale: 1:365,100
Chart Number: 3744

Stewart
Scale: 1:12,000
Chart Number: 3794

Langley Passage, Estevan Group
Scale: 1:12,000
Chart Number: 3795

Dixon Entrance
Scale: 1:200,000
Chart Number: 3800

Atli Inlet to/à Selwyn Inlet
Scale: 1:37,500
Chart Number: 3807

Juan Perez Sound
Scale: 1:37,500
Chart Number: 3808

Carpenter Bay to/à Burnaby Island
Scale: 1:37,500
Chart Number: 3809

Harbours in Queen Charlotte Islands/Havres dans Îles de la Reine-Charlotte
Scale: 1:18,315
Chart Number: 3811

Cape St. James to/à Houston Stewart Channel
Scale: 1:40,000
Chart Number: 3825

## Pacific North Central

Finlayson Channel and/et Tolmie Channel
Scale: 1:40,000
Chart Number: 3943

Princess Royal Channel
Scale: 1:40,000
Chart Number: 3944

Approaches to/Approches à Douglas Channel
Scale: 1:40,000
Chart Number: 3945

Grenville Channel
Scale: 1:40,000
Chart Number: 3946

Grenville Channel to/à Chatham Sound
Scale: 1:40,000
Chart Number: 3947

Gardner Canal
Scale: 1:40,000
Chart Number: 3948

Plans—Prince Rupert Harbour
Scale: 1:20,000
Chart Number: 3955

Malacca Passage to/à Bell Passage
Scale: 1:40,000
Chart Number: 3956

Approaches to/Approches à Prince Rupert Harbour
Scale: 1:40,000
Chart Number: 3957

Prince Rupert Harbour
Scale: 1:20,000
Chart Number: 3958

Hudson Bay Passage
Scale: 1:40,000
Chart Number: 3959

Approaches to/Approches à Portland Inlet
Scale: 1:40,000
Chart Number: 3960

Work Channel
Scale: 1:40,000
Chart Number: 3963

Tuck Inlet
Scale: 1:20,000
Chart Number: 3964

Dean Channel, Burke Channel and/et Bentinck Arms
Scale: 1:80,000
Chart Number: 3974

Douglas Channel
Scale: 1:80,000
Chart Number: 3977

Bonilla Island to/à Edye Passage
Scale: 1:80,000
Chart Number: 3978

Laredo Channel and/et Laredo Inlet
Scale: 1:40,000
Chart Number: 3981

Caamaño Sound to/à Whale Channel
Scale: 1:40,000
Chart Number: 3982

Principe Channel (Southern Portion/Partie sud)
Scale: 1:40,000
Chart Number: 3984

Principe Channel (Central Portion/Partie centrale) and/et Petrel Channel
Scale: 1:40,000
Chart Number: 3985

Browning Entrance
Scale: 1:40,000
Chart Number: 3986

Kitkatla Channel and/et Porcher Inlet
Scale: 1:40,000
Chart Number: 3987

Portland Inlet, Khutzeymateen Inlet and Pearse Canal
Scale: 1:40,000
Chart Number: 3994

## Pacific North East

Cape St. James to/à Cumshewa Inlet and/et Tasu Sound
Scale: 1:150,000
Chart Number: 3853

Tasu Sound to/à Port Louis
Scale: 1:141,935
Chart Number: 3854

Houston Stewart Channel
Scale: 1:20,000
Chart Number: 3855

Louscoone Inlet
Scale: 1:18,275
Chart Number: 3857

Flamingo Inlet
Scale: 1:18,247
Chart Number: 3858

Tasu Sound
Scale: 1:24,340
Chart Number: 3859

Harbours on the West Coast of/Havres sur la côte ouest de Graham Island
Scale: 1:36,513
Chart Number: 3860

Port Chanal
Scale: 1:18,258
Chart Number: 3863

Gowgaia Bay
Scale: 1:18,254
Chart Number: 3864

Vicinity of Englefield Bay
Scale: 1:36,600
Chart Number: 3865

Port Louis to/à Langara Island
Scale: 1:72,921
Chart Number: 3868

Skidegate Channel to/à Tian Rock
Scale: 1:70,620
Chart Number: 3869

Approaches to/Approches à Skidegate Inlet
Scale: 1:40,000
Chart Number: 3890

Skidegate Channel
Scale: 1:40,000
Chart Number: 3891

Masset Harbour and/et Naden Harbour
Scale: 1:40,000
Chart Number: 3892

Masset Inlet
Scale: 1:40,000
Chart Number: 3893

Selwyn Inlet to/à Lawn Point
Scale: 1:73,026
Chart Number: 3894

Plans—Dixon Entrance
Scale: 1:20,000
Chart Number: 3895

Plans—Hecate Strait
Scale: 1:250,000
Chart Number: 3902

Kitimat Harbour
Scale: 1:15,000
Chart Number: 3908

Plans Chatham Sound
Scale: 1:15,000
Chart Number: 3909

Plans—Milbanke Sound and/et Beauchemin Channel
Scale: 1:20,000
Chart Number: 3910

Plans Vicinity of/Proximité de Princess Royal Island
Scale: 1:25,000
Chart Number: 3911

Plans, Vicinity of/Proximité de Banks Island
Scale: 1:25,000
Chart Number: 3912

Nass Bay, Alice Arm and Approaches/et les approches
Scale: 1:40,000
Chart Number: 3920

Fish Egg Inlet and/et Allison Harbour
Scale: 1:20,000
Chart Number: 3921

Smith Inlet, Boswell Inlet and/et Draney Inlet
Scale: 1:40,000
Chart Number: 3931

Rivers Inlet
Scale: 1:40,000
Chart Number: 3932

Portland Canal and/et Observatory Inlet
Scale: 1:80,000
Chart Number: 3933

Approaches to/Approches à Smith Sound and/et Rivers Inlet
Scale: 1:40,000
Chart Number: 3934

Hakai Passage and Vicinity/et environs
Scale: 1:40,000
Chart Number: 3935

Fitz Hugh Sound to/à Lama Passage
Scale: 1:40,000
Chart Number: 3936

Queens Sound
Scale: 1:40,000
Chart Number: 3937

Queens Sound to/à Seaforth Channel
Scale: 1:40,000
Chart Number: 3938

Fisher Channel to/à Seaforth Channel and/et Dean Channel
Scale: 1:40,000
Chart Number: 3939

Spiller Channel and/et Roscoe Inlet
Scale: 1:40,000
Chart Number: 3940

Channels/Chenaux Vicinity of/Proximité de Milbanke Sound
Scale: 1:40,000
Chart Number: 3941

Cape Dufferin to/à Broughton Island
Scale: 1:250,000
Chart Number: 5705

Bélanger Island to/à Long Island
Scale: 1:250,000
Chart Number: 5707

Baie James/James Bay
Scale: 1:500,000
Chart Number: 5800

Long Island à/to Fort George
Scale: 1:150,000
Chart Number: 5801

Approaches to/Approches à Moose River
Scale: 1:24,000
Chart Number: 5860

Ship Sands Island to/à Moosonee
Scale: 1:24,000
Chart Number: 5861

Lake Muskoka
Scale: 1:25,000
Chart Number: 6021

Lake Rosseau and/et Lake Joseph
Scale: 1:25,000
Chart Number: 6022

Lake of Bays
Scale: 1:25,000
Chart Number: 6023

Wahwashkesh Lake
Scale: 1:12,000
Chart Number: 6026

Blind River and/et Lake Duborne
Scale: 1:12,000
Chart Number: 6028

Lake Nipissing/Lac Nipissing (Eastern Portion/Partie est)
Scale: 1:40,000
Chart Number: 6035

Iron Island to/à West Bay
Scale: 1:25,000
Chart Number: 6037

West Bay to/à West Arm
Scale: 1:25,000
Chart Number: 6038

Plans in Lake Nipigon/Plans dans le lac Nipigon
Scale: 1:36,000
Chart Number: 6050

Lac Saint-Jean
Scale: 1:100,000
Chart Number: 6100

Rainy Lake/Lac à la Pluie
Scale: 1:100,000
Chart Number: 6105

Northwest Bay to/à Ash Bay
Scale: 1:25,000
Chart Number: 6106

Hostess Island to/à Devils Cascade
Scale: 1:25,000
Chart Number: 6107

Fort Frances to/à Hostess Island and/et Sandpoint Island
Scale: 1:25,000
Chart Number: 6108

Sandpoint Island to/aux Anchor Islands
Scale: 1:25,000
Chart Number: 6109

Redgut Bay
Scale: 1:25,000
Chart Number: 6110

Rainy Lake/Lac à la Pluie (Eastern Portion/Partie est) Seine River Seine Bay to/à Sturgeon Falls
Scale: 1:25,000
Chart Number: 6111

Rainy Lake/Lac à la pluie (Southeast Portion/Partie sud-est) Anchor Islands to/à Oakpoint Island
Scale: 1:25,000
Chart Number: 6112

Lake of the Woods/Lac des Bois
Scale: 1:150,000
Chart Number: 6201

Seven Sisters Falls to/à Slave Falls
Scale: 1:25,000
Chart Number: 6206

Slave Falls to/à Eaglenest Lake
Scale: 1:25,000
Chart Number: 6207

Brereton Lake
Scale: 1:7,200
Chart Number: 6209

Big Traverse Bay
Scale: 1:80,000
Chart Number: 6211

Kenora to/à Aulneau Peninsula (Northern Portion/Partie nord)
Scale: 1:40,000
Chart Number: 6212

Whitefish Bay
Scale: 1:40,000
Chart Number: 6213

Sabaskong Bay
Scale: 1:40,000
Chart Number: 6214

Basil Channel to/à Sturgeon Channel
Scale: 1:40,000
Chart Number: 6215

Sturgeon Channel to/à Big Narrows Island
Scale: 1:40,000
Chart Number: 6216

Ptarmigan Bay and/et Shoal Lake
Scale: 1:40,000
Chart Number: 6217

Kenora, Rat Portage Bay
Scale: 1:40,000
Chart Number: 6218

Red River/Rivière Rouge to/à Berens River
Scale: 1:255,994
Chart Number: 6240

Berens River to/à Nelson River
Scale: 1:255,723
Chart Number: 6241

Winnipeg to/au Lake Winnipeg/Lac Winnipeg
Scale: 1:25,000
Chart Number: 6242

Winnipeg River/Rivière Winnipeg and Approaches/et les approches
Scale: 1:19,300
Chart Number: 6243

Wightman Point to/à Whiskey Jack Portage
Scale: 1:75,000
Chart Number: 6247

Observation Point to/à Grindstone Point
Scale: 1:48,000
Chart Number: 6248

Gull Harbour to/à Riverton
Scale: 1:48,000
Chart Number: 6249

Red River/Rivière Rouge to/à Gull Harbour
Scale: 1:100,000
Chart Number: 6251

Grindstone Point to/à Berens River
Scale: 1:125,000
Chart Number: 6267

Berens River and Approaches/et les approches
Scale: 1:24,000
Chart Number: 6268

Wanipigow River
Scale: 1:12,110
Chart Number: 6269

Lake Winnipegosis/Lac Winnipegosis
Scale: 1:200,000
Chart Number: 6270

Winnipegosis to/à Red Deer Point
Scale: 1:63,360
Chart Number: 6271

Red Deer Point to/à North Manitou Island
Scale: 1:63,360
Chart Number: 6272

North Manitou Island to/à Whiskey Jack Island
Scale: 1:63,360
Chart Number: 6273

Whiskey Jack Island to/à Red Deer River
Scale: 1:63,360
Chart Number: 6274

Lac La Ronge
Scale: 1:75,000
Chart Number: 6281

Whitedog Dam to/à Minaki
Scale: 1:25,000
Chart Number: 6286

Minaki to/à Kenora
Scale: 1:25,000
Chart Number: 6287

Lake Athabasca/Lac Athabasca
Scale: 1:25,000
Chart Number: 6310

Poplar Point to/à Stony Rapids
Scale: 1:50,000
Chart Number: 6311

Great Slave Lake/Grand lac des Esclaves (Eastern Portion/Partie est)
Scale: 1:250,000
Chart Number: 6341

McIver Point to/à Mirage Point
Scale: 1:31,680
Chart Number: 6354

Mirage Point to/à Hardisty Island
Scale: 1:31,680
Chart Number: 6355

Hardisty Island to/à North Head
Scale: 1:31,680
Chart Number: 6356

North Head to/à Moraine Point
Scale: 1:31,680
Chart Number: 6357

Northwest Point to/à Jones Point
Scale: 1:31,680
Chart Number: 6358

Jones Point to/à Burnt Point
Scale: 1:31,680
Chart Number: 6359

Windy Point to/à Slave Point
Scale: 1:31,680
Chart Number: 6360

Cabin Islands to/aux Pilot Islands
Scale: 1:31,680
Chart Number: 6368

Yellowknife Bay
Scale: 1:31,680
Chart Number: 6369

Great Slave Lake/Grand lac des Esclaves (Western Portion/Partie ouest)
Scale: 1:250,000
Chart Number: 6370

Harbours in Great Slave Lake/Havres dans le Grand Lacs
Scale: 1:7,200
Chart Number: 6371

Great Bear Lake
Scale: 1:350,000
Chart Number: 6390

Cache Island to/à Rabbitskin River Kilometre/Kilomètre 233–301
Scale: 1:50,000
Chart Number: 6408

Rabbitskin River to/à Fort Simpson Kilometre/Kilomètre 300–330
Scale: 1:25,000
Chart Number: 6409

Fort Simpson to/à Trail River Kilometre/Kilomètre 330–390
Scale: 1:50,000
Chart Number: 6410

Trail River to/à Camsell Bend Kilometre/Kilomètre 390–460
Scale: 1:50,000
Chart Number: 6411

Camsell Bend to/à McGern Island Kilometre/Kilomètre 510
Scale: 1:50,000
Chart Number: 6412

McGern Island to/à Wrigley River Kilometre/Kilomètre 510–580
Scale: 1:50,000
Chart Number: 6413

Wrigley River to/à Three Finger Creek Kilometre/Kilomètre 580–650
Scale: 1:50,000
Chart Number: 6414

Three Finger Creek to/à Saline Island Kilometre/Kilomètre 650–730
Scale: 1:50,000
Chart Number: 6415

Saline Island to/à Police Island Kilometre/Kilomètre 730–810
Scale: 1:50,000
Chart Number: 6416

Tulita (Fort Norman), Police Island to/aux Halfway Islands Kilometre/Kilomètre 810–860
Scale: 1:50,000
Chart Number: 6417

Norman Wells, Halfway Islands to/à Rader Island Kilometre/Kilomètre 850–920
Scale: 1:50,000
Chart Number: 6418

Norman Wells to/à Carcajou Ridge Kilometre/Kilomètre 910–980
Scale: 1:50,000
Chart Number: 6419

Carcajou Ridge to/à Hardie Island Kilometre/Kilomètre 980–1040
Scale: 1:50,000
Chart Number: 6420

Hardie Island to/à Fort Good Hope Kilometre/Kilomètre 1040–1100
Scale: 1:50,000
Chart Number: 6421

Fort Good Hope to à Askew Islands Kilometre 1100/Kilometre 1180
Scale: 1:50,000
Chart Number: 6422

Askew Islands to/à Bryan Island Kilometre/Kilomètre 1080–1240
Scale: 1:50,000
Chart Number: 6423

Bryan Island to/à Travaillant River Kilometre/Kilomètre 1240–1325
Scale: 1:50,000
Chart Number: 6424

Travaillant River to/à Adam Cabin Creek Kilometre/Kilomètre 1325–1400
Scale: 1:50,000
Chart Number: 6425

Adam Cabin Creek to/à Point Separation Kilometre/Kilomètre 1400–1480
Scale: 1:50,000
Chart Number: 6426

Point Separation to/au Aklavik Channel Kilometre/Kilomètre 1480–1540
Scale: 1:50,000
Chart Number: 6427

Aklavik Channel to/au Napoiak Channel including/y compris Aklavik Channel to/à Aklavik Kilometre/Kilomètre 1530–1597
Scale: 1:50,000
Chart Number: 6428

Kilometre/Kilomètre 1580–1645 including/y compris East Channel, Inuvik to/à Kilometre/Kilomètre 1645
Scale: 1:50,000
Chart Number: 6429

East Channel, Kilometre/Kilomètre 1645–1710
Scale: 1:50,000
Chart Number: 6430

East Channel, Lousy Point to/à Tuktoyaktuk Kilometre/Kilomètre 1710–1766
Scale: 1:50,000
Chart Number: 6431

Kilometre/Kilomètre 1500 to/à Inuvik East Channel
Scale: 1:50,000
Chart Number: 6432

West Channel, Aklavik to/à Shallow Bay
Scale: 1:50,000
Chart Number: 6433

Reindeer Channel, Tununuk Point to/à Shallow Bay
Scale: 1:50,000
Chart Number: 6434

Middle Channel, Tununuk Point to/à Mackenzie Bay Kilometre/Kilomètre 1670–1730
Scale: 1:50,000
Chart Number: 6435

Napoiak Channel, including/y compris Schooner and/et Taylor Channels Kilometre/Kilomètre 1590–1650
Scale: 1:50,000
Chart Number: 6436

Peel Channel including/y compris Husky Channel and/et Phillips Channel
Scale: 1:50,000
Chart Number: 6437

Peel River, Mackenzie River/Fleuve Mackenzie to/à Road Island
Scale: 1:50,000
Chart Number: 6438

West Channel including/y compris Anderton Channel, Ministicoog Channel and/et Moose Channel to/à Shoalwater Bay
Scale: 1:50,000
Chart Number: 6441

Sans Sault Rapids
Scale: 1:50,000
Chart Number: 6451

Mackenzie River/Fleuve Mackenzie (Kilometre/Kilomètre 0–58)
Scale: 1:50,000
Chart Number: 6452

Mackenzie River/Fleuve Mackenzie (Kilometre/Kilomètre 58–90)
Scale: 1:50,000
Chart Number: 6453

Mackenzie River/Fleuve Mackenzie (Kilometre/Kilomètre 90–147)
Scale: 1:50,000
Chart Number: 6454

Mackenzie River/Fleuve Mackenzie (Kilometre/Kilomètre 147–205)
Scale: 1:50,000
Chart Number: 6455

Arctic Archipelago/Archipel de l'Arctique
Scale: 1:5,000,000
Chart Number: 7000

Davis Strait and/et Baffin Bay
Scale: 1:2,000,000
Chart Number: 7010

Hudson Strait/Détroit D'Hudson to/à Groenland
Scale: 1:1,500,000
Chart Number: 7011

Resolution Island to/à Cape Mercy
Scale: 1:500,000
Chart Number: 7050

Cumberland Sound
Scale: 1:500,000
Chart Number: 7051

Cape Mercy to/à Kangeeak Point
Scale: 1:500,000
Chart Number: 7052

Paallavvik to/à Kangiqtugaapik
Scale: 1:500,000
Chart Number: 7053

Mill Island to/à Winter Island
Scale: 1:500,000
Chart Number: 7065

Cape Dorchester to/à Spicer Islands
Scale: 1:500,000
Chart Number: 7066

Spicer Islands to West Entrance of/à L'Entrée Ouest de Fury and/et Hecla Strait
Scale: 1:500,000
Chart Number: 7067

Cape Norton Shaw to/à Cape M'Clintock
Scale: 1:500,000
Chart Number: 7071

Kane Basin to/à Lincoln Sea
Scale: 1:500,000
Chart Number: 7072

Cape Baring to/à Cambridge Bay
Scale: 1:500,000
Chart Number: 7082

Cambridge Bay to/à Shepherd Bay
Scale: 1:500,000
Chart Number: 7083

Approaches to/Approches à Brevoort Harbour
Scale: 1:150,000
Chart Number: 7103

Cape Mills to/à Cape Rammelsberg
Scale: 1:75,000
Chart Number: 7121

Culbertson Island to/à Koojesse Inlet
Scale: 1:75,000
Chart Number: 7122

Pike-Resor Channel
Scale: 1:37,500
Chart Number: 7125

Culbertson Island to Frobisher's Farthest
Scale: 1:25,000
Chart Number: 7126

Koojesse Inlet and Approaches/et les approches
Scale: 1:40,000
Chart Number: 7127

Brevoort Harbour
Scale: 1:12,000
Chart Number: 7135

Cape Mercy and Approaches/et les approches
Scale: 1:30,000
Chart Number: 7136

Pangnirtung
Scale: 1:36,585
Chart Number: 7150

Exeter Bay and Approaches/et les approches
Scale: 1:50,000
Chart Number: 7170

Exeter Bay Landing Beach
Scale: 1:12,500
Chart Number: 7171

Padloping Island and Approaches/et les approches
Scale: 1:65,000
Chart Number: 7180

Durban Harbor
Scale: 1:12,500
Chart Number: 7181

Broughton Island and Approaches/et les approches
Scale: 1:50,000
Chart Number: 7184

Kangeeak Point and Approaches/et les approches
Scale: 1:50,000
Chart Number: 7185

Cape Hooper and Approaches/et les approches
Scale: 1:60,000
Chart Number: 7193

Cape Hooper to/à Arguyartu Point Including/y compris Ekalugad Fiord
Scale: 1:125,000
Chart Number: 7194

Kangok Fiord and Approaches/et les approches
Scale: 1:40,000
Chart Number: 7195

Bylot Island and Adjacent Channels
Scale: 1:250,000
Chart Number: 7212

Lancaster Sound, Eastern Approaches/approches est
Scale: 1:500,000
Chart Number: 7220

Dundas Harbour
Scale: 1:25,000
Chart Number: 7292

Lady Ann Strait to/à Smith Sound
Scale: 1:500,000
Chart Number: 7302

Lincoln Sea
Scale: 1:500,000
Chart Number: 7304

Jones Sound
Scale: 1:300,000
Chart Number: 7310

Alexandra Fiord
Scale: 1:25,000
Chart Number: 7371

Frozen Strait, Lyon Inlet and Approaches/et les approches
Scale: 1:200,000
Chart Number: 7404

Repulse Bay and Approaches/et les approches
Scale: 1:200,000
Chart Number: 7405

Spicer Islands to Longstaff Bluff
Scale: 1:200,000
Chart Number: 7411

Repulse Bay Harbours Islands to/à Talun Bay
Scale: 1:31,700
Chart Number: 7430

Approaches to/Approches à Cambridge Bay
Scale: 1:80,000
Chart Number: 7750

St. Roch and/et Rasmussen Basins
Scale: 1:200,000
Chart Number: 7760

Spence Bay and Approaches
Scale: 1:50,000
Chart Number: 7770

Dolphin and Union Strait
Scale: 1:150,000
Chart Number: 7776

Coronation Gulf (Western Portion/Partie ouest)
Scale: 1:150,000
Chart Number: 7777

Coronation Gulf (Eastern Portion/Partie est)
Scale: 1:150,000
Chart Number: 7778

Dease Strait
Scale: 1:150,000
Chart Number: 7779

Queen Maud Gulf (Western Portion/Partie ouest)
Scale: 1:150,000
Chart Number: 7782

Queen Maud Gulf (Eastern Portion/Partie est)
Scale: 1:150,000
Chart Number: 7783

Victoria Strait
Scale: 1:150,000
Chart Number: 7784

Melville Sound
Scale: 1:100,000
Chart Number: 7790

Bathurst Inlet—Northern Portion/Partie nord
Scale: 1:100,000
Chart Number: 7791

Bathurst Inlet—Central Portion
Scale: 1:60,000
Chart Number: 7792

Bathurst Inlet—Southern Portion/Partie sud
Scale: 1:60,000
Chart Number: 7793

Eglington Island to/à Cape Kellett
Scale: 1:500,000
Chart Number: 7832

Tanquary, Slidre and Glacier Fiords
Scale: 1:100,000
Chart Number: 7920

Hell Gate and/et Cardigan Strait
Scale: 1:75,000
Chart Number: 7930

Crozier Strait and/et Pullen Strait
Scale: 1:100,000
Chart Number: 7935

Eureka South and Southern Approaches/et les approches du sud Including/y Compris Baumann Fiord
Scale: 1:300,000
Chart Number: 7940

Nansen Sound and/et Greely Fiord
Scale: 1:300,000
Chart Number: 7941

Jones Sound, Norwegian Bay and Queens Channel
Scale: 1:500,000
Chart Number: 7950

Bathurst Island to/à Borden Island
Scale: 1:500,000
Chart Number: 7951

Cape Manning to/à Borden Island
Scale: 1:500,000
Chart Number: 7952

Borden Island to/à Cape Stallworthy
Scale: 1:500,000
Chart Number: 7953

Cape Stallworthy to/à Cape Discovery
Scale: 1:500,000
Chart Number: 7954

Byan Martin Channel to/au Maclean Strait
Scale: 1:300,000
Chart Number: 7980

Georges Bank
Scale: 1:300,000
Chart Number: 8005

Scotian Shelf/Plate-Forme Néo-Écossaise: Browns Bank to Emerald Bank/Banc de Brown au Banc D'Émeraude
Scale: 1:300,000
Chart Number: 8006

Halifax to/à Sable Island/Île de Sable, including/y compris Emerald Bank/Banc d'Émeraude and/et Sable Island Bank/Banc de l'Île de Sable

Scale: 1:300,000
Chart Number: 8007

Grand Bank/Grand Banc: Southern Portion/Partie sud
Scale: 1:350,000
Chart Number: 8010

Grand Bank/Grand Banc: Northern Portion/Partie nord
Scale: 1:350,000
Chart Number: 8011

Flemish Pass/Passe Flamande
Scale: 1:350,000
Chart Number: 8012

Flemish Cap/Bonnet Flamand
Scale: 1:350,000
Chart Number: 8013

Grand Bank/Grand Banc: Northeast Portion/Partie nord-est
Scale: 1:350,000
Chart Number: 8014

Funk Island and Approaches/et les approches
Scale: 1:350,000
Chart Number: 8015

Button Islands to/à Cod Island
Scale: 1:500,000
Chart Number: 8046

Cod Island to/à Cape Harrison
Scale: 1:500,000
Chart Number: 8047

Cape Harrison to/à St. Michael Bay
Scale: 1:500,000
Chart Number: 8048

St. Michael Bay to/aux Gray Islands
Scale: 1:500,000
Chart Number: 8049

**Quebec**

Saint-Fulgence à/to Saguenay
Scale: 1:15,000
Chart Number: 1201

Cap Éternité à/to Saint Fulgence
Scale: 1:37,500
Chart Number: 1202

Tadoussac à/to Cap Éternité
Scale: 1:37,500
Chart Number: 1203

Baie des Sept-Îles
Scale: 1:25,000
Chart Number: 1220

Pointe de Moisie à/to Île du Grand Caoui
Scale: 1:75,000
Chart Number: 1221

Chenal du Bic et les approches/and Approaches
Scale: 1:24,000
Chart Number: 1223

Mouillages et Installations Portuaires/Anchorages and Harbour Installations—Haute Côte-Nord
Scale: 1:50,000
Chart Number: 1226

Plans-Péninsule de la Gaspésie
Scale: 1:24,000
Chart Number: 1230

Cap aux Oies à/to Sault-au-Cochon
Scale: 1:50,000
Chart Number: 1233

Cap de la Tête au Chien au/to Cap aux Oies
Scale: 1:80,000
Chart Number: 1234

Pointe des Monts aux/to Escoumins
Scale: 1:200,000
Chart Number: 1236

Port de Montréal
Scale: 1:15,000
Chart Number: 1310

Sorel-Tracy à/to Varennes
Scale: 1:40,000
Chart Number: 1311

Lac Saint-Pierre
Scale: 1:40,000
Chart Number: 1312

Batiscan au/to Lac Saint-Pierre
Scale: 1:40,000
Chart Number: 1313

Donnacona à/to Batiscan
Scale: 1:40,000
Chart Number: 1314

Québec à/to Donnacona
Scale: 1:40,000
Chart Number: 1315

Port de Québec
Scale: 1:15,000
Chart Number: 1316

Sault-au-Cochon à/to Québec
Scale: 1:50,000
Chart Number: 1317

Île du Bic au/to Cap de la Tête au Chien
Scale: 1:80,000
Chart Number: 1320

Sorel-Tracy à/to Otterburn-Park
Scale: 1:15,000
Chart Number: 1350

Bassin de Chambly au lac/to Lake Champlain
Scale: 1:15,000
Chart Number: 1351

Lac Memphrémagog
Scale: 1:30,000
Chart Number: 1360

Montréal to/à Lake/Lac Ontario
Scale: 1:125,000
Chart Number: 1400

Canal de la Rive Sud
Scale: 1:20,000
Chart Number: 1429

Lac Saint-Louis
Scale: 1:25,000
Chart Number: 1430

Canal de Beauharnois
Scale: 1:25,000
Chart Number: 1431

Lac Saint-François/Lake St. Francis
Scale: 1:25,000
Chart Number: 1432

Lac Saint-François/Lake St. Francis
Scale: 1:25,000
Chart Number: 1432

Rivière des Prairies
Scale: 1:20,000
Chart Number: 1509

Lac des Deux Montagnes
Scale: 1:30,000
Chart Number: 1510

Carillon à/to Papineauville
Scale: 1:20,000
Chart Number: 1514

Papineauville à/to Ottawa
Scale: 1:20,000
Chart Number: 1515

**Quebec North East**

Havre de Natashquan et les approches/and Approaches
Scale: 1:10,000
Chart Number: 4428

Havre Saint-Pierre et les approches/and Approaches
Scale: 1:20,000
Chart Number: 4429

Plans—Île D'Anticosti
Scale: 1:72,000
Chart Number: 4430

Archipel de Mingan
Scale: 1:70,000
Chart Number: 4432

Îles Sainte-Marie à/to Île à la Brume
Scale: 1:75,000
Chart Number: 4440

Havres et Mouillages/Harbours and Anchorages—Côte-Nord/North Shore
Scale: 1:48,000
Chart Number: 4452

Île à la Brume à/to Pointe Curlew
Scale: 1:70,000
Chart Number: 4453

Pointe Curlew à/to Baie Washtawouka
Scale: 1:70,000
Chart Number: 4454

Baie Washtawouka à/to Baie Piashti
Scale: 1:69,950
Chart Number: 4455

Baie Piashti à/to Petite Île au Marteau
Scale: 1:70,000
Chart Number: 4456

Île du Petit Mécatina aux/to Îles Sainte-Marie
Scale: 1:75,000
Chart Number: 4468

Île Plate à/to Île du Petit Mécatina
Scale: 1:76,300
Chart Number: 4469

Blanc Sablon à/to Middle Bay
Scale: 1:36,643
Chart Number: 4470

Baie au Saumon à/to Baie des Homards
Scale: 1:36,500
Chart Number: 4471

Baie des Homards à/to Île de la Grande Passe
Scale: 1:36,550
Chart Number: 4472

Île de la Grande Passe aux/to Îles Bun
Scale: 1:36,600
Chart Number: 4473

Îles Bun à/to Baie des Moutons
Scale: 1:36,500
Chart Number: 4474

Cap des Rosiers à/to Chandler
Scale: 1:75,000
Chart Number: 4485

Plans, Baie des Chaleurs/Chaleur Bay (côte nord/North Shore)
Scale: 1:50,000
Chart Number: 4921

Îles de la Madeleine
Scale: 1:150,000
Chart Number: 4950

Chenal du Havre de la Grande Entrée
Scale: 1:15,000
Chart Number: 4954

Havre-aux-Maisons
Scale: 1:10,000
Chart Number: 4955

Cap-aux-Meules
Scale: 1:10,000
Chart Number: 4956

Havre-Aubert
Scale: 1:10,000
Chart Number: 4957

South Auliatsivik Island to/à Fenstone Tickle Island
Scale: 1:60,000
Chart Number: 5054

Cape Kiglapait to/à Khikkertarsoak North Island
Scale: 1:100,000
Chart Number: 5055

Khikkertarsoak North Island to/à Morhardt Point
Scale: 1:100,000
Chart Number: 5056

Hare Islands to/à North Head
Scale: 1:100,000
Chart Number: 5057

North Head to/à Murphy Head
Scale: 1:100,000
Chart Number: 5058

Saglek Bay
Scale: 1:40,000
Chart Number: 5059

Cape Daly to/à Amiktok Island
Scale: 1:40,000
Chart Number: 5060

Amiktok Island to/à Osborne Point
Scale: 1:40,000
Chart Number: 5061

Osborne Point to/à Cape Kakkiviak
Scale: 1:40,000
Chart Number: 5062

Cape Kakkiviak to/à Duck Islands
Scale: 1:40,000
Chart Number: 5063

McLelan Strait
Scale: 1:40,000
Chart Number: 5064

Gray Strait and/et Button Islands
Scale: 1:25,000
Chart Number: 5065

Satosoak Island to/à Akuliakatak Peninsula
Scale: 1:25,000
Chart Number: 5070

Punchbowl Inlet and Approaches/et les approches
Scale: 1:25,000
Chart Number: 5080

Punchbowl Inlet and Approaches/et les approches
Scale: 1:25,000
Chart Number: 5080

# UNITED STATES OF AMERICA

## Alaska

North Pacific Ocean (Eastern Part) Bering Sea Continuation
Scale: 1:10,000,000
NOAA Chart Number: 50

West Coast of North America Dixon Entrance to Unimak Pass
Scale: 1:3,500,000
NOAA Chart Number: 500

North Pacific Ocean, West Coast of North America, Mexican Border to Dixon Entrance
Scale: 1:3,500,000
NOAA Chart Number: 501

Bering Sea Southern Part
Scale: 1:3,500,000
NOAA Chart Number: 513

Bering Sea Northern Part
Scale: 1:3,500,000
NOAA Chart Number: 514

North America West Coast, San Diego to Aleutian Islands and Hawai'ian Islands
Scale: 1:4,860,700
NOAA Chart Number: 530

Gulf of Alaska Strait of Juan de Fuca to Kodiak Island
Scale: 1:2,100,000
NOAA Chart Number: 531

Point Barrow to Herschel Island
Scale: 1:700,000
NOAA Chart Number: 16004

Cape Prince of Wales to Point Barrow
Scale: 1:700,000
NOAA Chart Number: 16005

Bering Sea (Eastern Part); St. Matthew Island, Bering Sea; Cape Etolin, Achorage, Nunivak Island
Scale: 1:1,534,076
NOAA Chart Number: 16006

Alaska Peninsula and Aleutian Islands to Seguam Pass
Scale: 1:1,023,188
NOAA Chart Number: 16011

Aleutian Islands Amukta Island to Attu Island
Scale: 1:1,126,321
NOAA Chart Number: 16012

Cape St. Elias to Shumagin Islands; Semidi Islands
Scale: 1:969,761
NOAA Chart Number: 16013

Dixon Entrance to Cape St. Elias
Scale: 1:969,756
NOAA Chart Number: 16016

Demarcation Bay and Approaches
Scale: 1:51,639
NOAA Chart Number: 16041

Griffin Pt. and Approaches
Scale: 1:51,024
NOAA Chart Number: 16042

Barter Island and Approaches; Bernard Harbor
Scale: 1:50,819
NOAA Chart Number: 16043

Camden Bay and Approaches
Scale: 1:50,819
NOAA Chart Number: 16044

Bullen Pt. to Brownlow Pt.
Scale: 1:50,615
NOAA Chart Number: 16045

McClure and Stockton Islands and Vicinity
Scale: 1:50,204
NOAA Chart Number: 16046

Prudhoe Bay and Vicinity
Scale: 1:50,000
NOAA Chart Number: 16061

Jones Islands and Approaches
Scale: 1:49,794
NOAA Chart Number: 16062

Harrison Bay (Eastern Part)
Scale: 1:49,590
NOAA Chart Number: 16063

Harrison Bay (Western Part)
Scale: 1:49,794
NOAA Chart Number: 16064

Cape Halkett and Vicinity
Scale: 1:49,177
NOAA Chart Number: 16065

Pitt Pt. and Vicinity
Scale: 1:48,973
NOAA Chart Number: 16066

Approaches to Smith Bay
Scale: 1:48,767
NOAA Chart Number: 16067

Scott Pt. to Tangent Pt.
Scale: 1:48,149
NOAA Chart Number: 16081

Pt. Barrow and Vicinity
Scale: 1:47,943
NOAA Chart Number: 16082

Skull Cliff and Vicinity
Scale: 1:50,000
NOAA Chart Number: 16083

Peard Bay and Approaches
Scale: 1:50,000
NOAA Chart Number: 16084

Wainwright Inlet to Atainik
Scale: 1:50,000
NOAA Chart Number: 16085

Nakotlek Pt. to Wainwright
Scale: 1:50,000
NOAA Chart Number: 16086

Icy Cape to Nokotlek Pt.
Scale: 1:50,000
NOAA Chart Number: 16087

Utukok Pass to Blossom Shoals
Scale: 1:50,000
NOAA Chart Number: 16088

Pt. Lay and Approaches
Scale: 1:50,000
NOAA Chart Number: 16101

Kuchiak River to Kukpowruk Pass
Scale: 1:50,000
NOAA Chart Number: 16102

Cape Beaufort
Scale: 1:50,000
NOAA Chart Number: 16103

Cape Sabine (Metric)
Scale: 1:50,000
NOAA Chart Number: 16104

East of Cape Lisburne
Scale: 1:50,000
NOAA Chart Number: 16121

Cape Dyer to Cape Lisburne
Scale: 1:50,000
NOAA Chart Number: 16122

Point Hope to Cape Dyer
Scale: 1:50,000
NOAA Chart Number: 16123

Cape Thompson to Point Hope
Scale: 1:50,000
NOAA Chart Number: 16124

Alaska—West Coast, Delong Mountain Terminal
Scale: 1:40,000
NOAA Chart Number: 16145

Kotzebue Harbor
Scale: 1:50,000
NOAA Chart Number: 16161

Bering Strait North
Scale: 1:100,000
NOAA Chart Number: 16190

Norton Sound; Golovnin Bay
Scale: 1:400,000
NOAA Chart Number: 16200

Port Clarence and Approaches
Scale: 1:100,000
NOAA Chart Number: 16204

Nome Harbor and Approaches, Norton Sound; Nome Harbor
Scale: 1:20,000
NOAA Chart Number: 16206

Bering Sea—St. Lawrence Island to Bering Strait
Scale: 1:315,350
NOAA Chart Number: 16220

Cape Ramonzof to St. Michael; St. Michael Bay; Approaches to Cape Ramanzof
Scale: 1:300,000
NOAA Chart Number: 16240

Kuskokwim Bay; Goodnews Bay
Scale: 1:200,000
NOAA Chart Number: 16300

Kuskokwim Bay to Bethel
Scale: 1:100,000
NOAA Chart Number: 16304

Bristol Bay, Cape Newenham and Hagemeister Strait
Scale: 1:100,000
NOAA Chart Number: 16305

Bristol Bay, Togiak Bay and Walrus Islands
Scale: 1:100,000
NOAA Chart Number: 16315

Bristol Bay, Nushagak Bay and Approaches
Scale: 1:100,000
NOAA Chart Number: 16322

Bristol Bay, Kvichak Bay and Approaches
Scale: 1:100,000
NOAA Chart Number: 16323

Bristol Bay, Ugashik Bay to Egegik Bay
Scale: 1:100,000
NOAA Chart Number: 16338

Port Heiden
Scale: 1:80,000
NOAA Chart Number: 16343

Port Moller and Herendeen Bay
Scale: 1:80,000
NOAA Chart Number: 16363

Pribilof Islands
Scale: 1:200,000
NOAA Chart Number: 16380

St. George Island, Pribilof Islands
Scale: 1:50,000
NOAA Chart Number: 16381

St. Paul Island, Pribilof Islands
Scale: 1:50,000
NOAA Chart Number: 16382

Near Islands, Buldir Island to Attu Island
Scale: 1:300,000
NOAA Chart Number: 16420

Ingenstrem Rocks to Attu Island
Scale: 1:160,000
NOAA Chart Number: 16421

Shemya Island to Attu Island (Metric)
Scale: 1:100,000
NOAA Chart Number: 16423

Attu Island, Theodore Point to Cape Wrangell
Scale: 1:40,000
NOAA Chart Number: 16430

Temnac Bay
Scale: 1:20,000
NOAA Chart Number: 16431

Massacre Bay
Scale: 1:25,000
NOAA Chart Number: 16432

Sarana Bay to Holtz Bay; Chichagof Harbor
Scale: 1:20,000
NOAA Chart Number: 16433

Agattu Island
Scale: 1:40,000
NOAA Chart Number: 16434

Semichi Islands, Alaid and Nizki Islands
Scale: 1:20,000
NOAA Chart Number: 16435

Shemya Island; Alcan Harbor; Skoot Cove
Scale: 1:20,000
NOAA Chart Number: 16436

Rat Islands, Semisopochnoi Island to Buldir I.
Scale: 1:300,000
NOAA Chart Number: 16440

Kiska Island and Approaches
Scale: 1:80,000
NOAA Chart Number: 16441

Kiska Harbor and Approaches
Scale: 1:20,000
NOAA Chart Number: 16442

Constantine Harbor, Amchitka Island
Scale: 1:10,000
NOAA Chart Number: 16446

Amchitka Island and Approaches (Metric)
Scale: 1:100,000
NOAA Chart Number: 16450

Igitkin I. to Semisopochnoi Island
Scale: 1:300,000
NOAA Chart Number: 16460

Andrenof Islands, Tanga Bay and Approaches
Scale: 1:50,000
NOAA Chart Number: 16462

Kanaga Pass and Approaches
Scale: 1:50,000
NOAA Chart Number: 16463

Tanaga Island to Unalga Island (Metric)
Scale: 1:100,000
NOAA Chart Number: 16465

Adak Island to Tanaga Island (Metric)
Scale: 1:100,000
NOAA Chart Number: 16467

Atka Pass to Adak Strait; Three Arm Bay, Adak Island; Kanaga Bay, Kanaga Island; Chapel Roads and Chapel Cove, Adak Island
Scale: 1:120,000
NOAA Chart Number: 16471

Bay of Islands; Aranne Channel; Hell Gate
Scale: 1:12,000
NOAA Chart Number: 16474

Kuluk Bay and Approaches, including Little Tanaga and Kagalaska Strs.
Scale: 1:30,000
NOAA Chart Number: 16475

Sweeper Cove, Finger and Scabbard Bays
Scale: 1:10,000
NOAA Chart Number: 16476

Tagalak Island to Little Tanaga I.
Scale: 1:30,000
NOAA Chart Number: 16477

Tagalak Island to Great Sitkin Island; Sand Bay (Northeast Cove)
Scale: 1:30,000
NOAA Chart Number: 16478

Amkta Island to Igitkin Island; Finch Cove Seguam Island; Sviechnikof Harbor, Amilia Island
Scale: 1:300,000
NOAA Chart Number: 16480

Atka Island to Chugul Island, Atka Island
Scale: 1:30,000
NOAA Chart Number: 16484

Atka Island, Western Part
Scale: 1:40,000
NOAA Chart Number: 16486

Korovin Bay to Wall Bay, Atka Island; Martin Harbor
Scale: 1:40,000
NOAA Chart Number: 16487

Nazan Bay and Amilia Pass
Scale: 1:20,000
NOAA Chart Number: 16490

Unalaska I. to Amukta I.
Scale: 1:300,000
NOAA Chart Number: 16500

Islands of Four Mountains
Scale: 1:80,000
NOAA Chart Number: 16501

Inanudak Bay and Nikolski Bay, Umnak I.; River and Mueller Coves
Scale: 1:40,000
NOAA Chart Number: 16511

Unalaska Island Umnak Pass and Approaches
Scale: 1:40,000
NOAA Chart Number: 16513

Kulikak Bay and Surveyor Bay
Scale: 1:40,000
NOAA Chart Number: 16514

Chernofski Harbor to Skan Bay
Scale: 1:40,000
NOAA Chart Number: 16515

Chernofski Harbor
Scale: 1:10,000
NOAA Chart Number: 16516

Makushin Bay
Scale: 1:40,000
NOAA Chart Number: 16517

Cape Kavrizhka to Cape Cheerful
Scale: 1:40,000
NOAA Chart Number: 16518

Unimak and Akutan Passes and Approaches; Amak Island
Scale: 1:300,000
NOAA Chart Number: 16520

Unalaska Island, Protection Bay to Eagle Bay
Scale: 1:40,000
NOAA Chart Number: 16521

Beaver Inlet
Scale: 1:40,000
NOAA Chart Number: 16522

Unalaska Bay and Akutan Pass
Scale: 1:40,000
NOAA Chart Number: 16528

Dutch Harbor
Scale: 1:10,000
NOAA Chart Number: 16529

Captains Bay
Scale: 1:10,000
NOAA Chart Number: 16530

Krenitzan Islands
Scale: 1:80,000
NOAA Chart Number: 16531

Akutan Bay, Krenitzin Islands
Scale: 1:20,000
NOAA Chart Number: 16532

Morzhovoi Bay and Isanotski Strait
Scale: 1:80,000
NOAA Chart Number: 16535

Shumagin Islands to Sanak Islands; Mist Harbor
Scale: 1:300,000
NOAA Chart Number: 16540

Sanak Island and Sandman Reefs; Northeast Harbor; Peterson and Salmon Bays; Sanak Harbor
Scale: 1:81,326
NOAA Chart Number: 16547

Cold Bay and Approaches, Alaska Peninsula; King Cove Harbor
Scale: 1:80,000
NOAA Chart Number: 16549

Unga Island to Pavlof Bay, Alaska Peninsula
Scale: 1:80,000
NOAA Chart Number: 16551

Shumagin Islands, Nagai Island to Unga Island; Delarof Harbor; Popof Strait (Northern Part)
Scale: 1:80,000
NOAA Chart Number: 16553

Chiachi Island to Nagai Island; Chiachi Islands Anchorage
Scale: 1:80,000
NOAA Chart Number: 16556

Mitrofania Bay and Kuiukta Bay
Scale: 1:80,000
NOAA Chart Number: 16561

Chignik and Kujulik Bays, Alaska Peninsula; Anchorage and Mud Bays, Chignik Bay
Scale: 1:77,477
NOAA Chart Number: 16566

Wide Bay to Cape Kumlik, Alaska Peninsula
Scale: 1:106,600
NOAA Chart Number: 16568

Portage and Wide Bays, Alaska Peninsula
Scale: 1:50,000
NOAA Chart Number: 16570

Dakavak Bay to Cape Unalishagvak; Alinchak Bay
Scale: 1:80,000
NOAA Chart Number: 16575

Shelikof Strait, Cape Nukshak to Dakavak Bay
Scale: 1:80,000
NOAA Chart Number: 16576

Kodiak Island; Southwest Anchorage, Chirikof Island
Scale: 1:350,000
NOAA Chart Number: 16580

Semidi Islands and Vicinity
Scale: 1:135,000
NOAA Chart Number: 16587

Kodiak Island, Sitkinak Strait and Alitak Bay
Scale: 1:81,529
NOAA Chart Number: 16590

Alitak Bay, Cape Alitak to Moser Bay
Scale: 1:20,000
NOAA Chart Number: 16591

Kodiak Island, Gull Point to Kaguyak Bay; Sitkalidak Passage
Scale: 1:80,728
NOAA Chart Number: 16592

Chiniak Bay to Dangerous Cape
Scale: 1:80,000
NOAA Chart Number: 16593

Marmot Bay and Kupreanof Strait; Whale Passage; Ouzinkie Harbor
Scale: 1:78,900
NOAA Chart Number: 16594

Womens Bay
Scale: 1:10,000
NOAA Chart Number: 16596

Uganik and Uyak Bays
Scale: 1:80,000
NOAA Chart Number: 16597

Cape Ikolik to Cape Kuliuk
Scale: 1:80,000
NOAA Chart Number: 16598

Bays and Anchorages, Kodiak Island Karluk Anchorage; Larsen Bay; Uyak Anchorage
Scale: Various
NOAA Chart Number: 16599

Cape Alitak to Cape lkolik
Scale: 1:80,905
NOAA Chart Number: 16601

Kukak Bay, Alaska Peninsula
Scale: 1:30,000
NOAA Chart Number: 16603

Shuyak and Afagnak Islands and Adjacent Waters
Scale: 1:78,000
NOAA Chart Number: 16604

Shuyak Strait and Bluefox Bay
Scale: 1:20,000
NOAA Chart Number: 16605

Barren Islands
Scale: 1:77,062
NOAA Chart Number: 16606

Shelikof Strait, Cape Douglas to Cape Nukshak
Scale: 1:80,000
NOAA Chart Number: 16608

Cook Inlet, Southern Part
Scale: 1:200,000
NOAA Chart Number: 16640

Gore Point to Anchor Point
Scale: 1:82,662
NOAA Chart Number: 16645

Graham; Seldovia Bay; Seldovia Harbor; Approaches to Homer Harbor; Homer Harbor
Scale: Various
NOAA Chart Number: 16646

Cook Inlet, Cape Elizabeth to Anchor Point
Scale: 1:100,000
NOAA Chart Number: 16647

Kamishak Bay; lliamna Bay
Scale: 1:100,000
NOAA Chart Number: 16648

Cook Inlet, Northern Part
Scale: 1:194,154
NOAA Chart Number: 16660

Cook Inlet, Anchor Point to Kalgin Island; Ninilchik Harbor
Scale: 1:100,000
NOAA Chart Number: 16661

Cook Inlet, Kalgin Island to North Foreland; Drift River; Kasilof River to Kenai River; Rikiski
Scale: 1:100,000
NOAA Chart Number: 16662

Cook Inlet, East Foreland to Anchorage; North Foreland
Scale: 1:100,000
NOAA Chart Number: 16663

Cook Inlet, Approaches to Anchorage; Anchorage
Scale: 1:50,000
NOAA Chart Number: 16665

Point Elrington to East Chugach Island
Scale: 1:200,000
NOAA Chart Number: 16680

Seal Rocks to Gore Point
Scale: 1:83,074
NOAA Chart Number: 16681

Cape Resurrection to Two Arm Bay; Seaward
Scale: 1:81,847
NOAA Chart Number: 16682

Point Elrington to Cape Resurrection
Scale: 1:81,436
NOAA Chart Number: 16683

Prince William Sound
Scale: 1:200,000
NOAA Chart Number: 16700

Prince William Sound (Western Entrance)
Scale: 1:81,436
NOAA Chart Number: 16701

Latouche Passage to Whale Bay
Scale: 1:40,000
NOAA Chart Number: 16702

Drier Bay, Prince William Sound
Scale: 1:20,000
NOAA Chart Number: 16704

Prince William Sound (Western Part)
Scale: 1:80,000
NOAA Chart Number: 16705

Passage Canal incl. Port of Whittier; Port of Whittier
Scale: 1:20,000
NOAA Chart Number: 16706

Prince William Sound, Valdez Arm and Port Valdez; Valdez Narrows; Valdez and Valdez Marine Terminal
Scale: 1:40,000
NOAA Chart Number: 16707

Prince William Sound, Port Fidalgo and Valdez Arm; Tatitlek Narrows
Scale: 1:79,291
NOAA Chart Number: 16708

Prince William Sound (Eastern Entrance)
Scale: 1:80,000
NOAA Chart Number: 16709

Orca Bay and Inlet, Channel Islands to Cordova
Scale: 1:30,000
NOAA Chart Number: 16710

Port Wells, including College Fiord and Harriman Fiord
Scale: 1:50,000
NOAA Chart Number: 16711

Unakwik Inlet to Esther Passage and College Fiord
Scale: 1:50,000
NOAA Chart Number: 16712

Naked Island to Columbia Bay
Scale: 1:50,000
NOAA Chart Number: 16713

Controller Bay
Scale: 1:100,000
NOAA Chart Number: 16723

Icy Bay
Scale: 1:40,000
NOAA Chart Number: 16741

Cross Sound to Yakutat Bay
Scale: 1:300,000
NOAA Chart Number: 16760

Yakutat Bay; Yakutat Harbor
Scale: 1:80,000
NOAA Chart Number: 16761

Lituya Bay; Lituya Bay Entrance
Scale: 1:20,000
NOAA Chart Number: 16762

Stephens Passage to Cross Sound, including Lynn Canal
Scale: 1:209,978
NOAA Chart Number: 17300

Cape Spencer to Icy Point
Scale: 1:40,000
NOAA Chart Number: 17301

Icy Strait and Cross Sound; Inian Cove; Elfin Cove
Scale: 1:80,000
NOAA Chart Number: 17302

Yakobi Island and Lisianski Inlet; Pelican Harbor
Scale: 1:40,000
NOAA Chart Number: 17303

Holkham Bay and Tracy Arm—Stephens Passage
Scale: 1:40,000
NOAA Chart Number: 17311

Hawk Inlet, Charham Strait
Scale: 1:10,000
NOAA Chart Number: 17312

Port Snettisham
Scale: 1:40,000
NOAA Chart Number: 17313

Slocum and Limestone Inlets and Taku Harbor
Scale: 1:20,000
NOAA Chart Number: 17314

Gastineau Channel and Taku Inlet; Juneau Harbor
Scale: 1:40,000
NOAA Chart Number: 17315

Lynn Canal, Icy Strait to Point Sherman; Funter Bay; Chatham Strait
Scale: 1:80,000
NOAA Chart Number: 17316

Lynn Canal, Point Sherman to Skagway; Lutak Inlet; Skagway and Nahku Bay; Portage Cove, Chilkoot Inlet
Scale: 1:77,810
NOAA Chart Number: 17317

Glacier Bay; Bartlett Cove
Scale: 1:80,000
NOAA Chart Number: 17318

Coronation Island to Lisianski Strait
Scale: 1:217,828
NOAA Chart Number: 17320

Cape Edward to Lisianski Strait, Chichagof Island
Scale: 1:40,000
NOAA Chart Number: 17321

Khaz Bay, Chichagof Island Elbow Passage
Scale: 1:40,000
NOAA Chart Number: 17322

Salisbury Sound, Peril Strait and Hoonah Sound
Scale: 1:40,000
NOAA Chart Number: 17323

Sitka Sound to Salisbury Sound, Inside Passage; Neva Strait, Neva Point to Zeal Point
Scale: 1:40,000
NOAA Chart Number: 17324

South and West Coasts of Kruzof Island
Scale: 1:40,000
NOAA Chart Number: 17325

Crawfish Inlet to Sitka, Baranof Island; Sawmill Cove
Scale: 1:40,000
NOAA Chart Number: 17326

Sitka Harbor and Approaches; Sitka Harbor
Scale: 1:10,000
NOAA Chart Number: 17327

Snipe Bay to Crawfish Inlet, Baranof Island
Scale: 1:40,000
NOAA Chart Number: 17328

West Coast of Baranof Island, Cape Ommaney to Byron Bay
Scale: 1:20,000
NOAA Chart Number: 17330

Charham Strait, Ports Alexander, Conclusion, and Armstrong
Scale: 1:10,000
NOAA Chart Number: 17331

Ports Herbert, Walter, Lucy and Armstrong
Scale: 1:20,000
NOAA Chart Number: 17333

Patterson Bay and Deep Cove
Scale: 1:20,000
NOAA Chart Number: 17335

Harbors in Charham Strait and Vicinity, Gut Bay, Chatham Strait; Hoggatt Bay, Chatham Strait; Red Bluff Bay, Chatham Strait; Herring Bay and Chapin Bay, Frederick Sound; Surprise Harbor, and Murder Cove, Frederick Sound
Scale: Various
NOAA Chart Number: 17336

Harbors in Chatham Strait, Kelp Bay; Warm Spring Bay; Takatz and Kasnyku Bays
Scale: Various
NOAA Chart Number: 17337

Peril Strit, Hoonah Sound to Chatham Strait
Scale: 1:40,000
NOAA Chart Number: 17338

Hood Bay and Kootzmahoo Inlet, Chatham Strait; Killsnoo Harbor
Scale: 1:30,000
NOAA Chart Number: 17339

Whitewater Bay and Chaik Bay, Chatham Strait
Scale: 1:20,000
NOAA Chart Number: 17341

Etolin Island to Midway Islands, including Sumner Strait; Holkham Bay; Big Castle Island
Scale: 1:217,828
NOAA Chart Number: 17360

Gambier Bay, Stephens Passage
Scale: 1:40,000
NOAA Chart Number: 17362

Pybus Bay, Frederick Sound; Hobart and Windham Bays, Stephens Passage
Scale: 1:40,000
NOAA Chart Number: 17363

Woewodski and Eliza Harbors; Fanshaw Bay and Cleveland Passage
Scale: 1:20,000
NOAA Chart Number: 17365

Thomas, Farragut, and Portage Bays, Frederick Sound
Scale: 1:40,000
NOAA Chart Number: 17367

Keku Strait (Northern Part), including Saginaw and Security Bays and Port Camden; Kake Inset
Scale: 1:40,000
NOAA Chart Number: 17368

Bay of Pillars and Rowan Bay, Chatham Strait; Washington Bay, Chatham Strait
Scale: 1:20,000
NOAA Chart Number: 17370

Keku Strait-Monte Carlo Island to Entrance Island; The Summit; Devils Elbow

Scale: 1:20,000
NOAA Chart Number: 17372

Wrangell Narrows; Petersburg Harbor
Scale: 1:20,000
NOAA Chart Number: 17375

Tebenkof Bay and Port Malmesbury
Scale: 1:40,000
NOAA Chart Number: 17376

Le Conte Bay
Scale: 1:25,000
NOAA Chart Number: 17377

Port Protection, Prince of Wales Island
Scale: 1:20,000
NOAA Chart Number: 17378

Shaken Bay and Strait, Alaska
Scale: 1:10,000
NOAA Chart Number: 17379

Reb Bay, Prince of Wales Island
Scale: 1:20,000
NOAA Chart Number: 17381

Zarembo Island and Approaches; Burnett Inlet, Etolin Island; Steamer Bay
Scale: 1:80,000
NOAA Chart Number: 17382

Snow Passage, Alaska
Scale: 1:30,000
NOAA Chart Number: 17383

Wrangell Harbor and Approaches; Wrangell Harbor
Scale: 1:20,000
NOAA Chart Number: 17384

Ernest Sound, Eastern Passage and Zimovia Strait; Zimovia Strait
Scale: 1:80,000
NOAA Chart Number: 17385

Sumner Strait, Southern Part
Scale: 1:40,000
NOAA Chart Number: 17386

Shakan and Shipley Bays and Part of El Capitan Passage; El Capitan Pasage, Dry Pass to Shakan Strait
Scale: 1:40,000
NOAA Chart Number: 17387

Dixon Entrance to Charham Strait
Scale: 1:229,376
NOAA Chart Number: 17400

Lake Bay and Approaches, Clarence Strait
Scale: 1:10,000
NOAA Chart Number: 17401

Southern Entrances to Sumner Strait
Scale: 1:40,000
NOAA Chart Number: 17402

Davidson Inlet and Sea Otter Sound; Edna Bay
Scale: 1:40,000
NOAA Chart Number: 17403

San Christoval Channel to Cape Lynch
Scale: 1:40,000
NOAA Chart Number: 17404

Ulloa Channel to San Christoval Channel; North Entrance, Big Salt Lake; Shelter Cove, Craig
Scale: 1:40,000
NOAA Chart Number: 17405

Baker, Noyes, and Lulu Islands and Adjacent Waters
Scale: 1:40,000
NOAA Chart Number: 17406

Northern Part of Tlevak Strait and Uloa Channel
Scale: 1:40,000
NOAA Chart Number: 17407

Central Dall Island and Vicinity
Scale: 1:40,000
NOAA Chart Number: 17408

Southern Dall Island and Vicinity
Scale: 1:40,000
NOAA Chart Number: 17409

Hecate Strait to Etolin Island, including Behm and Portland Canals
Scale: 1:229,376
NOAA Chart Number: 17420

Behm Canal (Western Part); Yes Bay
Scale: 1:79,334
NOAA Chart Number: 17422

Harbor Charts, Clarence Strait and Behm Canal Dewey Anchorage, Etolin Island; Ratz Harbor, Prince of Wales Island; Naha Bay, Revillagigedo Island; Tolstoi and Thorne Bays, Prince of Wales Island; Union Bay, Cleveland Peninsula

Scale: Various
NOAA Chart Number: 17423

Behm Canal, Eastern Part
Scale: 1:80,000
NOAA Chart Number: 17424

Portland Canal, North of Hattie Island
Scale: 1:80,000
NOAA Chart Number: 17425

Kasaan Bay, Clarence Strait; Hollis Anchorage, Eastern Part; Lyman Anchorage
Scale: 1:40,000
NOAA Chart Number: 17426

Revillagigedo Channel, Nichols Passage, and Tongass Narrows; Seal Cove; Ward Cove
Scale: 1:80,000
NOAA Chart Number: 17427

Revillagigedo Channel, Nichols Passage, and Tongass Narrows; Seal Cove; Ward Cove
Scale: 1:40,000
NOAA Chart Number: 17428

Tongass Narrows
Scale: 1:10,000
NOAA Chart Number: 17430

North End of Cordova Bay and Hetta Inlet
Scale: 1:40,000
NOAA Chart Number: 17431

Clarence Strait and Moira Sound
Scale: 1:40,000
NOAA Chart Number: 17432

Kendrick Bay to Shipwreck Point, Prince of Wales Island
Scale: 1:40,000
NOAA Chart Number: 17433

Revillagigedo Channel; Ryus Bay; Foggy Bay
Scale: 1:80,000
NOAA Chart Number: 17434

Harbors in Clarence Strait Port Chester, Annette Island; Tamgas Harbor, Annette Island; Metlakatla Harbor
Scale: Various
NOAA Chart Number: 17435

Clarence Strait, Cholmondeley Sound and Skowl Arm
Scale: 1:40,000
NOAA Chart Number: 17436

Portland Inlet to Nakat Bay
Scale: 1:40,000
NOAA Chart Number: 17437

**Atlantic Coast**

Cape Hatteras to Straits of Florida
Scale: 1:1,200,000
NOAA Chart Number: 11009

Straits of Florida and Approaches
Scale: 1:1,200,000
NOAA Chart Number: 11013

Everglades National Park Shark River to Lostmans River
Scale: 1:50,000
NOAA Chart Number: 11432

Everglades National Park Whitewater Bay
Scale: 1:50,000
NOAA Chart Number: 11433

Florida Keys Sombrero Key to Dry Tortugas
Scale: 1:180,000
NOAA Chart Number: 11434

Dry Tortugas; Tortugas Harbor
Scale: 1:30,000
NOAA Chart Number: 11438

Sand Key to Rebecca Shoal
Scale: 1:80,000
NOAA Chart Number: 11439

Key West Harbor and Approaches
Scale: 1:30,000
NOAA Chart Number: 11441

Florida Keys Sombrero Key to Sand Key
Scale: 1:80,000
NOAA Chart Number: 11442

Intracoastal Waterway Bahia Honda Key to Sugarloaf Key
Scale: 1:40,000
NOAA Chart Number: 11445

Intracoastal Waterway Sugarloaf Key to Key West
Scale: 1:40,000
NOAA Chart Number: 11446

Key West Harbor
Scale: 1:10,000
NOAA Chart Number: 11447

Intracoastal Waterway, Big Spanish Channel to Johnston Key
Scale: 1:40,000
NOAA Chart Number: 11448

Intracoastal Waterway, Matecumbe to Grassy Key
Scale: 1:40,000
NOAA Chart Number: 11449

Fowey Rocks to American Shoal
Scale: 1:180,000
NOAA Chart Number: 11450

Intracoastal Waterway, Alligator Reef to Sombrero Key
Scale: 1:80,000
NOAA Chart Number: 11452

Florida Keys, Grassy Key to Bahia Honda Key
Scale: 1:40,000
NOAA Chart Number: 11453

Cape Canaveral to Key West
Scale: 1:466,940
NOAA Chart Number: 11460

Fowey Rocks to Alligator Reef
Scale: 1:80,000
NOAA Chart Number: 11462

Intracoastal Waterway, Sands Key to Blackwater Sound
Scale: 1:40,000
NOAA Chart Number: 11463

Intracoastal Waterway, Blackwater Sound to Matecumbe
Scale: 1:40,000
NOAA Chart Number: 11464

Intracoastal Waterway, Miami to Elliot Key
Scale: 1:40,000
NOAA Chart Number: 11465

Jupiter Inlet to Fowey Rocks; Lake Worth Inlet
Scale: 1:80,000
NOAA Chart Number: 11466

Intracoastal Waterway, West Palm Beach to Miami
Scale: 1:40,000
NOAA Chart Number: 11467

Miami Harbor
Scale: 1:10,000
NOAA Chart Number: 11468

Straits of Florida, Fowey Rocks, Hillsboro Inlet to Bimini Islands, Bahamas
Scale: 1:100,000
NOAA Chart Number: 11469

Fort Lauderdale, Port Everglades
Scale: 1:10,000
NOAA Chart Number: 11470

Intracoastal Waterway, Palm Shores to West Palm Beach; Loxahatchee River
Scale: 1:40,000
NOAA Chart Number: 11472

Bethel Shoal to Jupiter Inlet
Scale: 1:80,000
NOAA Chart Number: 11474

Fort Pierce Harbor
Scale: 1:10,000
NOAA Chart Number: 11475

Cape Canaveral to Bethel Shoal
Scale: 1:80,000
NOAA Chart Number: 11476

Port Canaveral; Canaveral Barge Canal Extension
Scale: 1:40,000
NOAA Chart Number: 11478

Charleston Light to Cape Canaveral
Scale: 1:449,659
NOAA Chart Number: 11480

Approaches to Port Canaveral
Scale: 1:25,000
NOAA Chart Number: 11481

Ponce de Leon Inlet to Cape Canaveral
Scale: 1:80,000
NOAA Chart Number: 11484

Intracoastal Waterway, Tolmato River to Palm Shores
Scale: 1:40,000
NOAA Chart Number: 11485

St. Augustine Light to Ponce de Leon Inlet
Scale: 1:80,000
NOAA Chart Number: 11486

St. Johns River, Racy Point to Crescent Lake
Scale: 1:40,000
NOAA Chart Number: 11487

Amelia Island to St. Augustine
Scale: 1:80,000
NOAA Chart Number: 11488

Intracoastal Waterway, St. Simons Sound to Tolmato River
Scale: 1:40,000
NOAA Chart Number: 11489

Approaches to St. Johns River; St. Johns River Entrance
Scale: 1:40,000
NOAA Chart Number: 11490

St. Johns River, Atlantic Ocean to Jacksonville
Scale: 1:20,000
NOAA Chart Number: 11491

St. John's River, Jacksonville to Racy Point
Scale: 1:40,000
NOAA Chart Number: 11492

St. Johns River, Dunns Creek to Lake Dexter
Scale: 1:40,000
NOAA Chart Number: 11495

St. Johns River, Lake Dexter to Lake Harney
Scale: 1:40,000
NOAA Chart Number: 11498

Doboy Sound to Fernadina
Scale: 1:80,000
NOAA Chart Number: 11502

St. Marys Entrance, Cumberland Sound and Kings Bay
Scale: 1:25,000
NOAA Chart Number: 11503

St. Andrew Sound and Satilla River
Scale: 1:40,000
NOAA Chart Number: 11504

Savannah River Approach
Scale: 1:40,000
NOAA Chart Number: 11505

St. Simons Sound, Brunswick Harbor and Turtle River
Scale: 1:40,000
NOAA Chart Number: 11506

Intracoastal Waterway, Beafort River to St. Simons Sound
Scale: 1:40,000
NOAA Chart Number: 11507

Altamaha Sound
Scale: 1:40,000
NOAA Chart Number: 11508

Tybee Island to Doboy Sound
Scale: 1:80,000
NOAA Chart Number: 11509

Sapelo and Doboy Sounds
Scale: 1:40,000
NOAA Chart Number: 11510

Ossabaw and St. Catherines Sounds
Scale: 1:40,000
NOAA Chart Number: 11511

Savannah River and Wassaw Sound
Scale: 1:40,000
NOAA Chart Number: 11512

St. Helena Sound to Savannah River
Scale: 1:80,000
NOAA Chart Number: 11513

Savannah River, Savannah to Brier Creek
Scale: 1:20,000
NOAA Chart Number: 11514

Savannah River, Brier Creek to Augusta
Scale: 1:20,000
NOAA Chart Number: 11515

Port Royal Sound and Inland Passages
Scale: 1:40,000
NOAA Chart Number: 11516

St. Helena Sound
Scale: 1:40,000
NOAA Chart Number: 11517

Intracoastal Waterway, Casino Creek to Beafort River
Scale: 1:40,000
NOAA Chart Number: 11518

Parts of Coosaw and Broad Rivers
Scale: 1:40,000
NOAA Chart Number: 11519

Cape Hatteras to Charleston
Scale: 1:432,720
NOAA Chart Number: 11520

Charleston Harbor and Approaches
Scale: 1:80,000
NOAA Chart Number: 11521

Stono and North Eclisto Rivers
Scale: 1:40,000
NOAA Chart Number: 11522

Charleston Harbor
Scale: 1:20,000
NOAA Chart Number: 11524

Wando River Upper Part
Scale: 1:20,000
NOAA Chart Number: 11526

Cooper River above Goose Creek
Scale: 1:20,000
NOAA Chart Number: 11527

Charleston Harbor Entrance and Approach
Scale: 1:40,000
NOAA Chart Number: 11528

Winyah Bay to Bulls Bay
Scale: 1:80,000
NOAA Chart Number: 11531

Winyah Bay
Scale: 1:40,000
NOAA Chart Number: 11532

Intracoastal Waterway, Myrtle Grove Sound and Cape Fear River to Casino Creek
Scale: 1:40,000
NOAA Chart Number: 11534

Little River Inlet to Winyah Bay Entrance
Scale: 1:80,000
NOAA Chart Number: 11535

Approaches to Cape Fear River
Scale: 1:80,000
NOAA Chart Number: 11536

Cape Fear River, Cape Fear to Wilmington
Scale: 1:40,000
NOAA Chart Number: 11537

New River Inlet to Cape Fear
Scale: 1:80,000
NOAA Chart Number: 11539

Intracoastal Waterway, Neuse River to Myrtle Grove Sound
Scale: 1:40,000
NOAA Chart Number: 11541

New River; Jacksonville
Scale: 1:40,000
NOAA Chart Number: 11542

Cape Lookout to New River
Scale: 1:80,000
NOAA Chart Number: 11543

Portsmouth Island to Beaufort, including Cape Lookout Shoals
Scale: 1:80,000
NOAA Chart Number: 11544

Beaufort Inlet and Part of Core Sound; Lookout Bight
Scale: 1:40,000
NOAA Chart Number: 11545

Morehead City Harbor
Scale: 1:15,000
NOAA Chart Number: 11547

Pamlico Sound, Western Part
Scale: 1:80,000
NOAA Chart Number: 11548

Ocracoke Inlet and Part of Core Sound
Scale: 1:40,000
NOAA Chart Number: 11550

Neuse River and Upper Part of Bay River
Scale: 1:40,000
NOAA Chart Number: 11552

Intracoastal Waterway, Albermarle Sound to Neuse River; Alligator River; Second Creek
Scale: 1:80,000
NOAA Chart Number: 11553

Pamlico River
Scale: 1:40,000
NOAA Chart Number: 11554

Cape Hatteras, Wimble Shoals to Ocracoke Inlet
Scale: 1:80,000
NOAA Chart Number: 11555

Cape May to Cape Hatteras
Scale: 1:419,706
NOAA Chart Number: 12200

Currituck Beach Light to Wimble Shoals
Scale: 1:80,000
NOAA Chart Number: 12204

Intracoastal Waterway, Norfolk to Albemarle Sound via North Landing River or Great Dismal Swamp Canal
Scale: 1:40,000
NOAA Chart Number: 12206

Cape Henry to Currituck Beach Light
Scale: 1:80,000
NOAA Chart Number: 12207

Approaches to Chesapeake Bay
Scale: 1:50,000
NOAA Chart Number: 12208

Chincoteague Inlet to Great Machipongo Inlet; Chincoteague Inlet
Scale: 1:80,000
NOAA Chart Number: 12210

Fenwick Inlet to Chincoteague Inlet; Ocean City Inlet
Scale: 1:80,000
NOAA Chart Number: 12211

Cape May to Fenwick Island
Scale: 1:80,000
NOAA Chart Number: 12214

Cape Henlopen to Indian River Inlet; Breakwater Harbor
Scale: 1:40,000
NOAA Chart Number: 12216

Chesapeake Bay Entrance
Scale: 1:80,000
NOAA Chart Number: 12221

Chesapeake Bay, Cape Charles to Norfolk Harbor
Scale: 1:40,000
NOAA Chart Number: 12222

Chesapeake Bay, Cape Charles to Wolf Trap
Scale: 1:40,000
NOAA Chart Number: 12224

Chesapeake Bay, Wolf Trap to Smith Point
Scale: 1:80,000
NOAA Chart Number: 12225

Chesapeake Bay, Wolf Trap to Pungoteague Creek
Scale: 1:40,000
NOAA Chart Number: 12226

Chesapeake Bay, Pocomoke and Tangier Sounds
Scale: 1:40,000
NOAA Chart Number: 12228

Chesapeake Bay, Smith Point to Cove Point
Scale: 1:80,000
NOAA Chart Number: 12230

Chesapeake Bay, Tangier Sound Northern Part
Scale: 1:40,000
NOAA Chart Number: 12231

Potomac River, Chesapeake Bay to Piney Point
Scale: 1:40,000
NOAA Chart Number: 12233

Chesapeake Bay, Rappahannock River Entrance, Piankatank and Great Wicomico Rivers
Scale: 1:40,000
NOAA Chart Number: 12235

Rappahannock River, Corrotoman River to Fredericksburg
Scale: 1:40,000
NOAA Chart Number: 12237

Chesapeake Bay, Mobjack Bay and York River Entrance
Scale: 1:40,000
NOAA Chart Number: 12238

York River, Yorktown and Vicinity
Scale: 1:20,000
NOAA Chart Number: 12241

York River, Yorktown to West Point
Scale: 1:40,000
NOAA Chart Number: 12243

Pamunkey and Mattaponi Rivers
Scale: 1:40,000
NOAA Chart Number: 12244

Hampton Roads
Scale: 1:20,000
NOAA Chart Number: 12245

James River, Newport News to Jamestown Island; Back River and College Creek
Scale: 1:40,000
NOAA Chart Number: 12248

James River, Jamestown Island to Jordan Point
Scale: 1:40,000
NOAA Chart Number: 12251

James River, Jordan Point to Richmond
Scale: 1:20,000
NOAA Chart Number: 12252

Norfolk Harbor and Elizabeth River
Scale: 1:20,000
NOAA Chart Number: 12253

Chesapeake Bay, Cape Henry to Thimble Shoal Light
Scale: 1:20,000
NOAA Chart Number: 12254

Little Creek Naval Amphibious Base
Scale: 1:5,000
NOAA Chart Number: 12255

Chesapeake Bay, Thimble Shoal Channel
Scale: 1:20,000
NOAA Chart Number: 12256

Chesapeake Bay, Honga, Nanticoke, Wicomico Rivers and Fishing Bay
Scale: 1:40,000
NOAA Chart Number: 12261

Chesapeake Bay, Cove Point to Sandy Point
Scale: 1:80,000
NOAA Chart Number: 12263

Chesapeake Bay, Patuxent River and Vicinity
Scale: 1:40,000
NOAA Chart Number: 12264

Chesapeake Bay, Choptank River and Herring Bay; Cambridge
Scale: 1:40,000
NOAA Chart Number: 12266

Choptank River, Cambridge to Greensboro
Scale: 1:40,000
NOAA Chart Number: 12268

Chesapeake Bay, Eastern Bay and South River; Selby Bay
Scale: 1:40,000
NOAA Chart Number: 12270

Chester River; Kent Island Narrows, Rock Hall Harbor and Swan Creek
Scale: 1:40,000
NOAA Chart Number: 12272

Chesapeake Bay, Sandy Point to Susquehanna River
Scale: 1:80,000
NOAA Chart Number: 12273

Head of Chesapeake Bay
Scale: 1:40,000
NOAA Chart Number: 12274

Chesapeake and Delaware Canal
Scale: 1:20,000
NOAA Chart Number: 12277

Chesapeake Bay, Approaches to Baltimore Harbor
Scale: 1:40,000
NOAA Chart Number: 12278

Chesapeake Bay
Scale: 1:200,000
NOAA Chart Number: 12280

Baltimore Harbor
Scale: 1:15,000
NOAA Chart Number: 12281

Chesapeake Bay, Severn and Magothy Rivers
Scale: 1:25,000
NOAA Chart Number: 12282

Annapolis Harbor
Scale: 1:10,000
NOAA Chart Number: 12283

Patuxent River, Solomons Island and Vicinity
Scale: 1:10,000
NOAA Chart Number: 12284

Potomac River, Piney Point to Lower Cedar Point
Scale: 1:40,000
NOAA Chart Number: 12286

Potomac River, Dahlgren and Vicinity
Scale: 1:20,000
NOAA Chart Number: 12287

Potomac River, Lower Cedar Point to Mattawoman Creek
Scale: 1:40,000
NOAA Chart Number: 12288

Potomac River, Mattawoman Creek to Georgetown; Washington Harbor
Scale: 1:40,000
NOAA Chart Number: 12289

Approaches to New York, Nantucket Shoals to Five Fathom Bank
Scale: 1:400,000
NOAA Chart Number: 12300

Delaware Bay
Scale: 1:80,000
NOAA Chart Number: 12304

Delaware River, Smyrna River to Wilmington
Scale: 1:40,000
NOAA Chart Number: 12311

Delaware River, Wilmington to Philadelphia
Scale: 1:40,000
NOAA Chart Number: 12312

Philadelphia and Camden Waterfronts
Scale: 1:15,000
NOAA Chart Number: 12313

Delaware River, Philadelphia to Trenton
Scale: 1:20,000
NOAA Chart Number: 12314

Intracoastal Waterway, Little Egg Harbor to Cape May; Atlantic City
Scale: 1:40,000
NOAA Chart Number: 12316

Cape May Harbor
Scale: 1:10,000
NOAA Chart Number: 12317

Little Egg Inlet to Hereford Inlet; Absecon Inlet
Scale: 1:80,000
NOAA Chart Number: 12318

Sea Girt to Little Egg Inlet
Scale: 1:80,000
NOAA Chart Number: 12323

Intracoastal Waterway, Sandy Hook to Little Egg Harbor
Scale: 1:80,000
NOAA Chart Number: 12324

Navesink and Shrewsbury Rivers
Scale: 1:15,000
NOAA Chart Number: 12325

Approaches to New York, Fire Island Light to Sea Girt
Scale: 1:80,000
NOAA Chart Number: 12326

New York Harbor
Scale: 1:40,000
NOAA Chart Number: 12327

Raritan Bay and Southern Part of Arthur Kill
Scale: 1:15,000
NOAA Chart Number: 12331

Raritan River, Raritan Bay to New Brunswick
Scale: 1:20,000
NOAA Chart Number: 12332

Kill Van Kull and Northern Part of Arthur Kill (Inset)
Scale: 1:15,000
NOAA Chart Number: 12333

New York Harbor Upper Bay and Narrows, Anchorage Chart
Scale: 1:10,000
NOAA Chart Number: 12334

Hudson and East Rivers, Governors Island to 67th Street
Scale: 1:10,000
NOAA Chart Number: 12335

Passaic and Hackensack Rivers
Scale: 1:20,000
NOAA Chart Number: 12337

East River, Newtown Creek
Scale: 1:5,000
NOAA Chart Number: 12338

East River, Tallman Island to Queensboro Bridge
Scale: 1:10,000
NOAA Chart Number: 12339

Hudson River, Days Point to George Washington Bridge
Scale: 1:10,000
NOAA Chart Number: 12341

Harlem River
Scale: 1:10,000
NOAA Chart Number: 12342

Hudson River, New York to Wappinger Creek
Scale: 1:40,000
NOAA Chart Number: 12343

Hudson River, George Washington Bridge to Yonkers
Scale: 1:10,000
NOAA Chart Number: 12345

Hudson River, Yonkers to Piermont
Scale: 1:10,000
NOAA Chart Number: 12346

Hudson River, Wappinger Creek to Hudson
Scale: 1:40,000
NOAA Chart Number: 12347

Hudson River, Coxsackie to Troy
Scale: 1:40,000
NOAA Chart Number: 12348

Jamaica Bay and Rockaway Inlet
Scale: 1:20,000
NOAA Chart Number: 12350

Shinnecock Light to Fire Island Light
Scale: 1:80,000
NOAA Chart Number: 12353

Long Island Sound, Eastern Part
Scale: 1:80,000
NOAA Chart Number: 12354

New York Long Island, Shelter Island Sound and Peconic Bays; Mattituck Inlet
Scale: 1:40,000
NOAA Chart Number: 12358

Port Jefferson and Mount Sinai Harbors
Scale: 1:10,000
NOAA Chart Number: 12362

Long Island Sound, Western Part
Scale: 1:80,000
NOAA Chart Number: 12363

South Shore of Long Island Sound, Oyster and Huntington Bays
Scale: 1:20,000
NOAA Chart Number: 12365

Long Island Sound and East River, Hempstead Harbor to Tallman Island
Scale: 1:20,000
NOAA Chart Number: 12366

North Shore of Long Island Sound, Greenwich Point to New Rochelle
Scale: 1:20,000
NOAA Chart Number: 12367

North Shore of Long Island Sound, Sherwood Point to Stamford Harbor
Scale: 1:20,000
NOAA Chart Number: 12368

North Shore of Long Island Sound, Stratford to Sherwood Point
Scale: 1:20,000
NOAA Chart Number: 12369

North Shore of Long Island Sound, Housatonic River and Milford Harbor
Scale: 1:20,000
NOAA Chart Number: 12370

New Haven Harbor; New Haven Harbor (Inset)
Scale: 1:20,000
NOAA Chart Number: 12371

North Shore of Long Island Sound, Guilford Harbor to Farm River
Scale: 1:20,000
NOAA Chart Number: 12373

North Shore of Long Island Sound, Duck Island to Madison Reef
Scale: 1:20,000
NOAA Chart Number: 12374

Connecticut River, Long Island Sound to Deep River
Scale: 1:20,000
NOAA Chart Number: 12375

Connecticut River, Deep River to Bodkin Rock
Scale: 1:20,000
NOAA Chart Number: 12377

Connecticut River, Bodkin Rock to Hartford
Scale: 1:20,000
NOAA Chart Number: 12378

New York Lower Bay, Southern Part
Scale: 1:15,000
NOAA Chart Number: 12401

New York Lower Bay Northern Part
Scale: 1:15,000
NOAA Chart Number: 12402

Cape Sable to Cape Hatteras
Scale: 1:200,000
NOAA Chart Number: 13003

West Quoddy Head to New York
Scale: 1:675,000
NOAA Chart Number: 13006

Gulf of Maine and Georges Bank
Scale: 1:500,000
NOAA Chart Number: 13009

Georges Bank and Nantucket Shoals
Scale: 1:400,000
NOAA Chart Number: 13200

Georges Bank, Western Part
Scale: 1:220,000
NOAA Chart Number: 13203

Georges Bank, Eastern Part
Scale: 1:220,000
NOAA Chart Number: 13204

Block Island Sound and Approaches
Scale: 1:80,000
NOAA Chart Number: 13205

Block Island Sound and Gardiners Bay; Montauk Harbor
Scale: 1:40,000
NOAA Chart Number: 13209

North Shore of Long Island Sound, Niantic Bay and Vicinity
Scale: 1:20,000
NOAA Chart Number: 13211

Approaches to New London Harbor
Scale: 1:20,000
NOAA Chart Number: 13212

New London Harbor and Vicinity; Bailey Point to Smith Cove
Scale: 1:10,000
NOAA Chart Number: 13213

Fishers Island Sound
Scale: 1:20,000
NOAA Chart Number: 13214

Block Island Sound, Point Judith to Montauk Point
Scale: 1:40,000
NOAA Chart Number: 13215

Block Island
Scale: 1:15,000
NOAA Chart Number: 13217

Martha's Vineyard to Block Island
Scale: 1:80,000
NOAA Chart Number: 13218

Point Judith Harbor
Scale: 1:15,000
NOAA Chart Number: 13219

Narragansett Bay
Scale: 1:40,000
NOAA Chart Number: 13221

Narragansett Bay, including Newport Harbor
Scale: 1:20,000
NOAA Chart Number: 13223

Providence River and Head of Narragansett Bay
Scale: 1:20,000
NOAA Chart Number: 13224

Providence Harbor
Scale: 1:10,000
NOAA Chart Number: 13225

Mount Hope Bay
Scale: 1:20,000
NOAA Chart Number: 13226

Fall River Harbor; State Pier
Scale: 1:2,000
NOAA Chart Number: 13227

Westport River and Approaches
Scale: 1:20,000
NOAA Chart Number: 13228

Buzzards Bay; Quicks Hole
Scale: 1:40,000
NOAA Chart Number: 13230

New Bedford Harbor and Approaches
Scale: 1:20,000
NOAA Chart Number: 13232

Martha's Vineyard; Menemsha Pond
Scale: 1:40,000
NOAA Chart Number: 13233

Woods Hole
Scale: 1:5,000
NOAA Chart Number: 13235

Cape Cod Canal and Approaches
Scale: 1:20,000
NOAA Chart Number: 13236

Nantucket Sound and Approaches
Scale: 1:80,000
NOAA Chart Number: 13237

Martha's Vineyard Eastern Part; Oak Bluffs Harbor; Vineyard Haven Harbor; Edgartown Harbor
Scale: 1:20,000
NOAA Chart Number: 13238

Nantucket Island
Scale: 1:40,000
NOAA Chart Number: 13241

Nantucket Harbor
Scale: 1:10,000
NOAA Chart Number: 13242

Eastern Entrance to Nantucket Sound
Scale: 1:40,000
NOAA Chart Number: 13244

Cape Cod Bay
Scale: 1:80,000
NOAA Chart Number: 13246

Chatham Harbor and Pleasant Bay
Scale: 1:20,000
NOAA Chart Number: 13248

Provincetown Harbor
Scale: 1:20,000
NOAA Chart Number: 13249

Wellfleet Harbor; Sesuit Harbor
Scale: 1:40,000
NOAA Chart Number: 13250

Barnstable Harbor
Scale: 1:20,000
NOAA Chart Number: 13251

Harbors of Plymouth, Kingston and Duxbury; Green Harbor
Scale: 1:20,000
NOAA Chart Number: 13253

Bay of Fundy to Cape Cod
Scale: 1:378,838
NOAA Chart Number: 13260

Massachusetts Bay; North River
Scale: 1:80,000
NOAA Chart Number: 13267

Cohasset and Scituate Harbors
Scale: 1:10,000
NOAA Chart Number: 13269

Boston Harbor
Scale: 1:25,000
NOAA Chart Number: 13270

Boston Inner Harbor
Scale: 1:10,000
NOAA Chart Number: 13272

Portsmouth Harbor to Boston Harbor; Merrimack River Extension
Scale: 1:80,000
NOAA Chart Number: 13274

Salem and Lynn Harbors; Manchester Harbor
Scale: 1:25,000
NOAA Chart Number: 13275

Salem, Marblehead and Beverly Harbors
Scale: 1:10,000
NOAA Chart Number: 13276

Portsmouth to Cape Ann; Hampton Harbor
Scale: 1:80,000
NOAA Chart Number: 13278

Ipswich Bay to Gloucester Harbor; Rockport Harbor
Scale: 1:20,000
NOAA Chart Number: 13279

Gloucester Harbor and Annisquam River
Scale: 1:10,000
NOAA Chart Number: 13281

Newburyport Harbor and Plum Island Sound
Scale: 1:20,000
NOAA Chart Number: 13282

Portsmouth Harbor, Cape Neddick Harbor to Isles of Shoals; Portsmouth Harbor
Scale: 1:20,000
NOAA Chart Number: 13283

Portsmouth to Dover and Exeter
Scale: 1:20,000
NOAA Chart Number: 13285

Cape Elizabeth to Portsmouth; Cape Porpoise Harbor; Wells Harbor; Kennebunk River; Perkins Cove
Scale: 1:80,000
NOAA Chart Number: 13286

Saco Bay and Vicinity
Scale: 1:20,000
NOAA Chart Number: 13287

Monhegan Island to Cape Elizabeth
Scale: 1:80,000
NOAA Chart Number: 13288

Casco Bay
Scale: 1:40,000
NOAA Chart Number: 13290

Portland Harbor and Vicinity
Scale: 1:20,000
NOAA Chart Number: 13292

Damariscotta, Sheepscot and Kennebec Rivers; South Bristol Harbor; Christmas Cove
Scale: 1:40,000
NOAA Chart Number: 13293

Kennebec and Sheepscot River Entrances
Scale: 1:15,000
NOAA Chart Number: 13295

Boothbay Harbor to Bath, including Kennebec River
Scale: 1:15,000
NOAA Chart Number: 13296

Kennebec River, Courthouse Point to Augusta
Scale: 1:15,000
NOAA Chart Number: 13297

Kennebec River, Bath to Courthouse Point
Scale: 1:15,000
NOAA Chart Number: 13298

Muscongus Bay; New Harbor; Thomaston
Scale: 1:40,000
NOAA Chart Number: 13301

Penobscot Bay and Approaches
Scale: 1:80,000
NOAA Chart Number: 13302

Approaches to Penobscot Bay
Scale: 1:40,000
NOAA Chart Number: 13303

Penobscot Bay; Carvers Harbor and Approaches
Scale: 1:40,000
NOAA Chart Number: 13305

Camden, Rockport and Rockland Harbors
Scale: 1:20,000
NOAA Chart Number: 13307

Fox Islands Thorofare
Scale: 1:15,000
NOAA Chart Number: 13308

Penobscot River; Belfast Harbor
Scale: 1:40,000
NOAA Chart Number: 13309

Frenchman and Blue Hill Bays and Approaches
Scale: 1:80,000
NOAA Chart Number: 13312

Approaches to Blue Hill Bay
Scale: 1:40,000
NOAA Chart Number: 13313

Deer Island Thorofare and Casco Passage
Scale: 1:20,000
NOAA Chart Number: 13315

Blue Hill Bay; Blue Hill Harbor
Scale: 1:40,000
NOAA Chart Number: 13316

Frenchman Bay and Mount Desert Island
Scale: 1:40,000
NOAA Chart Number: 13318

Southwest Harbor and Approaches
Scale: 1:10,000
NOAA Chart Number: 13321

Winter Harbor
Scale: 1:10,000
NOAA Chart Number: 13322

Bar Harbor Mount Desert Island
Scale: 1:10,000
NOAA Chart Number: 13323

Tibbett Narrows to Schoodic Island
Scale: 1:40,000
NOAA Chart Number: 13324

Quaddy Narrows to Petit Manan Island
Scale: 1:80,000
NOAA Chart Number: 13325

Machias Bay to Tibbett Narrows
Scale: 1:40,000
NOAA Chart Number: 13326

Grand Manan Channel, Southern Part (Metric)
Scale: 1:50,000
NOAA Chart Number: 13392

Grand Manan Channel, Northern Part (Metric); North Head and Flagg Cove
Scale: 1:50,000
NOAA Chart Number: 13394

Campobello Island (Metric); Eastport Harbor
Scale: 1:20,000
NOAA Chart Number: 13396

Passamaquoddy Bay and St. Croix River (Metric); Beaver Harbor; Saint Andrews; Todds Point
Scale: 1:50,000
NOAA Chart Number: 13398

Puerto Rico and Virgin Islands
Scale: 1:326,856
NOAA Chart Number: 25640

Virgin Islands, Virgin Gorda to St. Thomas and St. Croix; Krause Lagoon Channel
Scale: 1:100,000
NOAA Chart Number: 25641

Frederiksted Road; Frederiksted Pier
Scale: 1:20,000
NOAA Chart Number: 25644

Christiansted Harbor
Scale: 1:10,000
NOAA Chart Number: 25645

Pillsbury Sound
Scale: 1:15,000
NOAA Chart Number: 25647

Saint Thomas Harbor
Scale: 1:10,000
NOAA Chart Number: 25649

Virgin Passage and Sonda de Vieques
Scale: 1:100,000
NOAA Chart Number: 25650

**Great Lakes**

Hudson River, New York to Wappinger Creek
Scale: 1:40,000
NOAA Chart Number: 12343

Hudson River, George Washington Bridge to Yonkers
Scale: 1:10,000
NOAA Chart Number: 12345

Hudson River, Yonkers to Piermont
Scale: 1:10,000
NOAA Chart Number: 12346

Hudson River, Wappinger Creek to Hudson
Scale: 1:40,000
NOAA Chart Number: 12347

Hudson River, Coxsackie to Troy
Scale: 1:40,000
NOAA Chart Number: 12348

Great Lakes, Lake Champlain to Lake of the Woods
Scale: 1:1,500,000
NOAA Chart Number: 14500

St. Lawrence River, Morristown, N.Y. to Butternut Bay, Ont.
Scale: 1:15,000
NOAA Chart Number: 14770

St. Lawrence River, Butternut Bay, Ont. to Ironsides Island, N.Y.
Scale: 1:15,000
NOAA Chart Number: 14771

St. Lawrence River, Ironsides I., N.Y. to Bingham I., Ont.
Scale: 1:15,000
NOAA Chart Number: 14772

St. Lawrence River, Gananoque, Ont. to St. Lawrence Park, N.Y.
Scale: 1:15,000
NOAA Chart Number: 14773

St. Lawrence River, Round Island, New York and Gananoque, Ontario to Wolfe Island, Ont.
Scale: 1:15,000
NOAA Chart Number: 14774

Riviere Richelieu to South Hero Island
Scale: 1:40,000
NOAA Chart Number: 14781

Cumberland Head to Four Brothers Islands
Scale: 1:40,000
NOAA Chart Number: 14782

Four Brothers Islands to Barber Point
Scale: 1:40,000
NOAA Chart Number: 14783

Barber Point to Whitehall
Scale: 1:40,000
NOAA Chart Number: 14784

Burlington Harbor
Scale: 1:10,000
NOAA Chart Number: 14785

Oneida Lake—Lock 22 to Lock 23
Scale: 1:40,000
NOAA Chart Number: 14788

Cayuga and Seneca Lakes; Watkins Glen; Ithaca
Scale: 1:60,000
NOAA Chart Number: 14791

Lake Ontario (Metric)
Scale: 1:400,000
NOAA Chart Number: 14800

Clayton to False Ducks Is.
Scale: 1:80,000
NOAA Chart Number: 14802

Six Miles South of Stony Point to Port Bay; North Pond; Little Sodus Bay
Scale: 1:80,000
NOAA Chart Number: 14803

Port Bay to Long Pond; Port Bay Harbor; Irondequoit Bay
Scale: 1:80,000
NOAA Chart Number: 14804

Long Pond to Thirtymile Point; Point Breeze Harbor
Scale: 1:80,000
NOAA Chart Number: 14805

Thirtymile Point, N.Y., to Port Dalhousie, Ont.
Scale: 1:80,000
NOAA Chart Number: 14806

Olcott Harbor to Toronto (Metric); Olcott and Wilson Harbors
Scale: 1:100,000
NOAA Chart Number: 14810

Chaumont, Henderson and Black River Bays; Sackets Harbor; Henderson Harbor; Chaumont Harbor
Scale: 1:30,000
NOAA Chart Number: 14811

Oswego Harbor
Scale: 1:10,000
NOAA Chart Number: 14813

Sodus Bay
Scale: 1:10,000
NOAA Chart Number: 14814

Rochester Harbor, including Genessee River to Head of Navigation
Scale: 1:10,000
NOAA Chart Number: 14815

Lower Niagara River
Scale: 1:30,000
NOAA Chart Number: 14816

Lake Erie (includes Metric version)
Scale: 1:400,000
NOAA Chart Number: 14820

Approaches to Nigara River and Welland Canal
Scale: 1:80,000
NOAA Chart Number: 14822

Sturgeon Point to Twentymile Creek; Dunkirk Harbor; Barcelona Harbor
Scale: 1:80,000
NOAA Chart Number: 14823

Sixteenmile Creek to Conneaut; Conneaut Harbor
Scale: 1:80,000
NOAA Chart Number: 14824

Ashtabula to Chagrin River; Mentor Harbor; Chagrin River
Scale: 1:80,000
NOAA Chart Number: 14825

Moss Point to Vermilion; Beaver Creek; Vermilion Harbor; Rocky River
Scale: 1:80,000
NOAA Chart Number: 14826

Erie to Geneva (Metric)
Scale: 1:100,000
NOAA Chart Number: 14828

Geneva to Lorain (Metric); Beaver Creek; Rocky River; Mentor Harbor; Chagrin River
Scale: 1:100,000
NOAA Chart Number: 14829

West End of Lake Erie; Port Clinton Harbor; Monroe Harbor; Lorain to Detroit River (Metric); Vermilion
Scale: 1:100,000
NOAA Chart Number: 14830

Niagara Falls to Buffalo
Scale: 1:30,000
NOAA Chart Number: 14832

Buffalo Harbor
Scale: 1:15,000
NOAA Chart Number: 14833

Erie Harbor
Scale: 1:15,000
NOAA Chart Number: 14835

Ashtabula Harbor
Scale: 1:5,000
NOAA Chart Number: 14836

Fairport Harbor
Scale: 1:8,000
NOAA Chart Number: 14837

Buffalo to Erie; Dunkirk; Barcelone Harbor
Scale: 1:120,000
NOAA Chart Number: 14838

Cleveland Harbor, including Lower Cuyahoga River
Scale: 1:10,000
NOAA Chart Number: 14839

Lorain Harbor
Scale: 1:10,000
NOAA Chart Number: 14841

Huron Harbor
Scale: 1:5,000
NOAA Chart Number: 14843

Islands in Lake Erie; Put-In-Bay
Scale: 1:40,000
NOAA Chart Number: 14844

Sandusky Harbor
Scale: 1:10,000
NOAA Chart Number: 14845

Toledo Harbor; Entrance Channel to Harbor
Scale: 1:40,000
NOAA Chart Number: 14847

Detroit River
Scale: 1:30,000
NOAA Chart Number: 14848

Lake St. Clair
Scale: 1:60,000
NOAA Chart Number: 14850

St. Clair River; Head of St. Clair River
Scale: 1:40,000
NOAA Chart Number: 14852

Trenton Channel and River Rouge; River Rouge
Scale: 1:15,000
NOAA Chart Number: 14854

Lake Huron
Scale: 1:500,000
NOAA Chart Number: 14860

Port Huron to Pte aux Barques; Port Sanilac; Harbor Beach
Scale: 1:120,000
NOAA Chart Number: 14862

Saginaw Bay; Port Austin Harbor; Caseville Harbor; Entrance to Au Sable River; Sebewaing Harbor; Tawas Harbor
Scale: 1:120,000
NOAA Chart Number: 14863

Harrisville to Forty Mile Point; Harrisville Harbor; Alpena; Rogers City and Calcite
Scale: 1:120,000
NOAA Chart Number: 14864

South End of Lake Huron
Scale: 1:15,000
NOAA Chart Number: 14865

Saginaw River
Scale: 1:20,000
NOAA Chart Number: 14867

Thunder Bay Island to Presque Isle; Stoneport Harbor; Resque Isle Harbor
Scale: 1:60,000
NOAA Chart Number: 14869

Straits of Mackinac
Scale: 1:120,000
NOAA Chart Number: 14880

De Tour Passage to Waugoshance Point; Hammond Bay Harbor; Mackinac Island; Cheboygan; Mackinaw City; St. Ignace
Scale: 1:80,000
NOAA Chart Number: 14881

St. Marys River, De Tour Passage to Munuscong Lake; De Tour Passage
Scale: 1:40,000
NOAA Chart Number: 14882

St. Marys River, Munuscong Lake to Sault Ste. Marie
Scale: 1:40,000
NOAA Chart Number: 14883

St. Marys River, Head of Lake Nicolet to Whitefish Bay; Sault Ste. Marie
Scale: 1:40,000
NOAA Chart Number: 14884

Les Cheneaux Islands
Scale: 1:20,000
NOAA Chart Number: 14885

Lake Michigan (Mercator Projection)
Scale: 1:500,000
NOAA Chart Number: 14901

North End of Lake Michigan, including Green Bay
Scale: 1:240,000
NOAA Chart Number: 14902

Algoma to Sheboygan; Kewaunee; Two Rivers
Scale: 1:120,000
NOAA Chart Number: 14903

Port Washington to Waukegan; Kenosha; North Point Marina; Port Washington; Waukegan
Scale: 1:120,000
NOAA Chart Number: 14904

Waukegan to South Haven; Michigan City; Burns International Harbor; New Buffalo
Scale: 1:120,000
NOAA Chart Number: 14905

South Haven to Stony Lake; South Haven; Port Sheldon; Saugatuck Harbor
Scale: 1:120,000
NOAA Chart Number: 14906

Stony Lake to Point Betsie; Pentwater; Arcadia; Frankfort
Scale: 1:120,000
NOAA Chart Number: 14907

Dutch Johns Point to Fishery Point, including Big Bay de Noc and Little Bay de Noc; Manistique
Scale: 1:80,000
NOAA Chart Number: 14908

Upper Green Bay, Jackson Harbor and Detroit Harbor; Detroit Harbor; Jackson Harbor; Baileys Harbor
Scale: 1:80,000
NOAA Chart Number: 14909

Lower Green Bay; Oconto Harbor; Algoma
Scale: 1:80,000
NOAA Chart Number: 14910

Waugoshance Point to Seul Choix Point, including Beaver Island Group; Port Inland; Beaver Harbor
Scale: 1:80,000
NOAA Chart Number: 14911

Platte Bay to Leland; Leland; South Manitou Harbor
Scale: 1:80,000
NOAA Chart Number: 14912

Grand Traverse Bay to Little Traverse Bay; Harbor Springs; Petoskey; Elk Rapids; Suttons Bay; Northport; Traverse City
Scale: 1:80,000
NOAA Chart Number: 14913

Little Bay de Noc
Scale: 1:30,000
NOAA Chart Number: 14915

Menominee and Marinette Harbors
Scale: 1:15,000
NOAA Chart Number: 14917

Head of Green Bay, including Fox River below De Pere; Green Bay
Scale: 1:25,000
NOAA Chart Number: 14918

Sturgeon Bay and Canal; Sturgeon Bay
Scale: 1:30,000
NOAA Chart Number: 14919

Manitowoc and Sheboygan
Scale: 1:10,000
NOAA Chart Number: 14922

Milwaukee Harbor
Scale: 1:10,000
NOAA Chart Number: 14924

Racine Harbor
Scale: 1:10,000
NOAA Chart Number: 14925

Chicago Lake Front; Gary Harbor
Scale: 1:60,000
NOAA Chart Number: 14927

Chicago Harbor
Scale: 1:15,000
NOAA Chart Number: 14928

Calumet, Indiana and Buffington Harbors, and Lake Calumet
Scale: 1:15,000
NOAA Chart Number: 14929

St. Joseph and Benton Harbor
Scale: 1:10,000
NOAA Chart Number: 14930

Grand River from Dermo Bayou to Bass River
Scale: 1:15,000
NOAA Chart Number: 14931

Holland Harbor
Scale: 1:15,000
NOAA Chart Number: 14932

Grand Haven, including Spring Lake and Lower Grand River
Scale: 1:15,000
NOAA Chart Number: 14933

Muskegon Lake and Muskegon Harbor
Scale: 1:15,000
NOAA Chart Number: 14934

White Lake
Scale: 1:10,000
NOAA Chart Number: 14935

Ludington Harbor
Scale: 1:5,000
NOAA Chart Number: 14937

Manistee Harbor and Manistee Lake
Scale: 1:10,000
NOAA Chart Number: 14938

Portage Lake
Scale: 1:10,000
NOAA Chart Number: 14939

Lake Charlevoix; Charlevoix, South Point to Round Lake
Scale: 1:30,000
NOAA Chart Number: 14942

Lake Superior (Mercator Projection)
Scale: 1:600,000
NOAA Chart Number: 14961

St. Marys River to Au Sable Point; Whitefish Point; Little Lake Harbors; Grand Marais Harbor
Scale: 1:120,000
NOAA Chart Number: 14962

Grand Marais to Big Bay Point; Big Bay Harbor
Scale: 1:120,000
NOAA Chart Number: 14963

Big Bay Point to Redridge; Grand Traverse Bay Harbor; Lac La Belle Harbor; Copper and Eagle Harbors
Scale: 1:120,000
NOAA Chart Number: 14964

Redridge to Saxon Harbor; Ontonagon Harbor; Black River Harbor; Saxon Harbor
Scale: 1:120,000
NOAA Chart Number: 14965

Little Girls Point to Silver Bay, including Duluth and Apostle Islands; Cornucopia Harbor; Port Wing Harbor; Knife River Harbor; Two Harbors
Scale: 1:120,000
NOAA Chart Number: 14966

Beaver Bay to Pigeon Point; Silver Bay Harbor; Taconite Harbor; Grand Marais Harbor
Scale: 1:120,000
NOAA Chart Number: 14967

Grand Portage Bay, Minn. to Shesbeeb Point, Ont.
Scale: 1:120,000
NOAA Chart Number: 14968

Munising Harbor and Approaches; Munising Harbor
Scale: 1:30,000
NOAA Chart Number: 14969

Marquette and Presque Isle Harbors
Scale: 1:15,000
NOAA Chart Number: 14970

Keweenaw Bay; L'Anse and Baraga Harbors
Scale: 1:30,000
NOAA Chart Number: 14971

Keweenaw Waterway, including Torch Lake; Hancock and Houghton
Scale: 1:30,000
NOAA Chart Number: 14972

Apostle Islands, including Chequamegan Bay; Bayfield Harbor; Pikes Bay Harbor; La Pointe Harbor
Scale: 1:60,000
NOAA Chart Number: 14973

Ashland and Washburn Harbors
Scale: 1:15,000
NOAA Chart Number: 14974

Duluth-Superior Harbor; Uppers St. Louis River
Scale: 1:15,000
NOAA Chart Number: 14975

Isle Royale
Scale: 1:40,000
NOAA Chart Number: 14976

**Gulf Coast**

Gulf of Mexico
Scale: 1:2,160,000
NOAA Chart Number: 411

Gulf Coast, Key West to Mississippi River
Scale: 1:875,000
NOAA Chart Number: 11006

Straits of Florida and Approaches
Scale: 1:1,200,000
NOAA Chart Number: 11013

Galveston to Rio Grande
Scale: 1:460,732
NOAA Chart Number: 11300

Southern Part of Laguna Madre
Scale: 1:80,000
NOAA Chart Number: 11301

Intracoastal Waterway, Stover Point to Port Brownsville, including Brazos Santiago Pass
Scale: 1:40,000
NOAA Chart Number: 11302

Intracoastal Waterway, Laguna Madre—Chubby Island to Stover Point, including the Arroyo Colorado
Scale: 1:40,000
NOAA Chart Number: 11303

Northern Part of Laguna Madre
Scale: 1:80,000
NOAA Chart Number: 11304

Intracoastal Waterway, Laguna Madre—Middle Ground to Chubby Island
Scale: 1:40,000
NOAA Chart Number: 11306

Aransas Pass to Baffin Bay
Scale: 1:80,000
NOAA Chart Number: 11307

Intracoastal Waterway, Redfish Bay to Middle Ground
Scale: 1:40,000
NOAA Chart Number: 11308

Corpus Christi Bay
Scale: 1:40,000
NOAA Chart Number: 11309

Corpus Christi Harbor
Scale: 1:10,000
NOAA Chart Number: 11311

Corpus Christi Bay, Port Aransas to Port Ingleside
Scale: 1:20,000
NOAA Chart Number: 11312

Matagorda Light to Aransas Pass
Scale: 1:80,000
NOAA Chart Number: 11313

Intracoastal Waterway, Carlos Bay to Redfish Bay, including Copano Bay
Scale: 1:40,000
NOAA Chart Number: 11314

Intracoastal Waterway, Espiritu Santo Bay to Carlos Bay including San Antonio Bay and Victoria Barge Canal
Scale: 1:40,000
NOAA Chart Number: 11315

Matagorda Bay and Approaches
Scale: 1:80,000
NOAA Chart Number: 11316

Lavaca; Continuation of Lavaca River; Continuation of Tres Palacios Bay
Scale: 1:50,000
NOAA Chart Number: 11317

Intracoastal Waterway, Cedar Lakes to Espiritu Santo Bay
Scale: 1:40,000
NOAA Chart Number: 11319

San Luis Pass to East Matagorda Bay
Scale: 1:80,000
NOAA Chart Number: 11321

Intracoastal Waterway, Galveston Bay to Cedar Lakes
Scale: 1:40,000
NOAA Chart Number: 11322

Approaches to Galveston Bay
Scale: 1:80,000
NOAA Chart Number: 11323

Galveston Bay Entrance, Galveston and Texas City Harbors
Scale: 1:25,000
NOAA Chart Number: 11324

Houston Ship Channel, Carpenters Bayou to Houston
Scale: 1:10,000
NOAA Chart Number: 11325

Upper Galveston Bay, Houston Ship Channel, Dollar Pt. to Atkinson
Scale: 1:25,000
NOAA Chart Number: 11327

Houston Ship Channel, Atkinson Island to Alexander Island
Scale: 1:10,000
NOAA Chart Number: 11328

Houston Ship Channel, Alexander Island to Carpenters Bayou; San Jacinto and Old Rivers
Scale: 1:25,000
NOAA Chart Number: 11329

Mermentau River to Freeport
Scale: 1:250,000
NOAA Chart Number: 11330

Intracoastal Waterway, Ellender to Galveston Bay
Scale: 1:40,000
NOAA Chart Number: 11331

Sabine Bank
Scale: 1:80,000
NOAA Chart Number: 11332

Calcasieu River and Approaches
Scale: 1:50,000
NOAA Chart Number: 11339

Mississippi River to Galveston
Scale: 1:458,596
NOAA Chart Number: 11340

Calcasieu Pass to Sabine Pass
Scale: 1:80,000
NOAA Chart Number: 11341

Sabine Pass and Lake
Scale: 1:40,000
NOAA Chart Number: 11342

Sabine and Neches Rivers
Scale: 1:40,000
NOAA Chart Number: 11343

Rollover Bayou to Calcasieu Pass
Scale: 1:80,000
NOAA Chart Number: 11344

Intracoastal Waterway, New Orleans to Calcasieu River (West Section)
Scale: 1:175,000
NOAA Chart Number: 11345

Port Fourchon and Approaches
Scale: 1:20,000
NOAA Chart Number: 11346

Calcasieu River and Lake
Scale: 1:50,000
NOAA Chart Number: 11347

Intracoastal Waterway, Forked Island to Ellender, including the Mermantau River, Grand Lake and White Lake
Scale: 1:40,000
NOAA Chart Number: 11348

Vermilion Bay and Approaches
Scale: 1:80,000
NOAA Chart Number: 11349

Intracoastal Waterway, Wax Lake Outlet to Forked Island including Bayou Teche, Vermilion River, and Freshwater Bayou
Scale: 1:40,000
NOAA Chart Number: 11350

Point au Fer to Marsh Island
Scale: 1:80,000
NOAA Chart Number: 11351

Intracoastal Waterway, New Orleans to Calcasieu River (East Section)
Scale: 1:175,000
NOAA Chart Number: 11352

Baptiste Collette Bayou to Mississippi River Gulf Outlet; Baptiste Collette Bayou Extension
Scale: 1:40,000
NOAA Chart Number: 11353

Intracoastal Waterway, Morgan City to Port Allen, including the Atchafalaya River
Scale: 1:80,000
NOAA Chart Number: 11354

Intracoastal Waterway, Catahoula Bay to Wax Lake Outlet including the Houma Navigation canal
Scale: 1:40,000
NOAA Chart Number: 11355

Isles Dernieres to Point au Fer
Scale: 1:80,000
NOAA Chart Number: 11356

Timbalier and Terrebonne Bays
Scale: 1:80,000
NOAA Chart Number: 11357

Barataria Bay and Approaches
Scale: 1:80,000
NOAA Chart Number: 11358

Loop Deepwater Port, Louisiana Offshore Oil Port
Scale: 1:50,000
NOAA Chart Number: 11359

Cape St. George to Mississippi Passes
Scale: 1:456,394
NOAA Chart Number: 11360

Mississippi River Delta; Southwest Pass; South Pass; Head of Passes
Scale: 1:80,000
NOAA Chart Number: 11361

Chandeleur and Breton Sounds
Scale: 1:80,000
NOAA Chart Number: 11363

Mississippi River, Venice to New Orleans
Scale: 1:80,000
NOAA Chart Number: 11364

Barataria and Bayou Lafourche Waterways, Intracoastal Waterway to Gulf of Mexico
Scale: 1:50,000
NOAA Chart Number: 11365

Approaches to Mississippi River
Scale: 1:250,000
NOAA Chart Number: 11366

Intracoastal Waterway, Waveland to Catahoula Bay
Scale: 1:40,000
NOAA Chart Number: 11367

New Orleans Harbor, Chalmette Slip to Southport
Scale: 1:15,000
NOAA Chart Number: 11368

Lakes Pontchartrain and Maurepas
Scale: 1:80,000
NOAA Chart Number: 11369

Mississippi River, New Orleans to Baton Rouge
Scale: 1:40,000
NOAA Chart Number: 11370

Lake Borgne and Approaches, Cat Island to Point aux Herbes
Scale: 1:80,000
NOAA Chart Number: 11371

Intracoastal Waterway, Dog Keys Pass to Waveland
Scale: 1:40,000
NOAA Chart Number: 11372

Mississippi Sound and Approaches, Dauphin Island to Cat Island
Scale: 1:80,000
NOAA Chart Number: 11373

Intracoastal Waterway, Dauphin Island to Dog Keys Pass
Scale: 1:40,000
NOAA Chart Number: 11374

Pascagoula Harbor
Scale: 1:20,000
NOAA Chart Number: 11375

Mobile Bay Mobile Ship Channel (Northern End)
Scale: 1:80,000
NOAA Chart Number: 11376

Mobile Bay Approaches and Lower Half
Scale: 1:40,000
NOAA Chart Number: 11377

Intracoastal Waterway, Santa Rosa Sound to Dauphin Island
Scale: 1:40,000
NOAA Chart Number: 11378

Mobile Bay East, Fowl River to Deer River Pt.; Mobile Middle Bay Terminal
Scale: 1:20,000
NOAA Chart Number: 11380

Pensacola Bay and Approaches
Scale: 1:80,000
NOAA Chart Number: 11382

Pensacola Bay
Scale: 1:30,000
NOAA Chart Number: 11383

Pensacola Bay Entrance
Scale: 1:10,000
NOAA Chart Number: 11384

Intracoastal Waterway, West Bay to Santa Rosa Sound
Scale: 1:40,000
NOAA Chart Number: 11385

Choctawhatchee Bay
Scale: 1:80,000
NOAA Chart Number: 11388

St Joseph and St Andrew Bays
Scale: 1:80,000
NOAA Chart Number: 11389

Intracoastal Waterway, East Bay to West Bay
Scale: 1:40,000
NOAA Chart Number: 11390

St. Andrew Bay
Scale: 1:25,000
NOAA Chart Number: 11391

St. Andrew Bay, Bear Point to Sulphur Point
Scale: 1:5,000
NOAA Chart Number: 11392

Intracoastal Waterway, Lake Wimico to East Bay
Scale: 1:40,000
NOAA Chart Number: 11393

Tampa Bay to Cape San Blas
Scale: 1:456,394
NOAA Chart Number: 11400

Apalachicola Bay to Cape San Blas
Scale: 1:80,000
NOAA Chart Number: 11401

Intracoastal Waterway, Apalachicola Bay to Lake Wimico
Scale: 1:40,000
NOAA Chart Number: 11402

Intracoastal Waterway, Carrabelle to Apalachicola Bay; Carrabelle River
Scale: 1:40,000
NOAA Chart Number: 11404

Apalachee Bay
Scale: 1:80,000
NOAA Chart Number: 11405

St. Marks River and Approaches
Scale: 1:15,000
NOAA Chart Number: 11406

Horseshoe Point to Rock Islands; Horseshoe Beach
Scale: 1:80,000
NOAA Chart Number: 11407

Crystal River to Horseshoe Point; Suwannee River; Cedar Keys
Scale: 1:80,000
NOAA Chart Number: 11408

Anclote Keys to Crystal River
Scale: 1:80,000
NOAA Chart Number: 11409

Intracoastal Waterway, Tampa Bay to Port Richey
Scale: 1:40,000
NOAA Chart Number: 11411

Tampa Bay and St. Joseph Sound
Scale: 1:80,000
NOAA Chart Number: 11412

Tampa Bay Entrance; Manatee River Extension
Scale: 1:40,000
NOAA Chart Number: 11415

Tampa Bay; Safety Harbor; St. Petersburg; Tampa
Scale: 1:80,000
NOAA Chart Number: 11416

Havana to Tampa Bay
Scale: 1:470,940
NOAA Chart Number: 11420

Estero Bay to Lemon Bay, including Charlotte Harbor; Continuation of Peace River
Scale: 1:80,000
NOAA Chart Number: 11424

Intracoastal Waterway, Charlotte Harbor to Tampa Bay
Scale: 1:40,000
NOAA Chart Number: 11425

Estero Bay to Lemon Bay, including Charlotte Harbor; Continuation of Peace River
Scale: 1:80,000
NOAA Chart Number: 11426

Intracoastal Waterway, Fort Myers to Charlotte Harbor and Wiggins Pass
Scale: 1:40,000
NOAA Chart Number: 11427

Okeechobee Waterway, St. Lucie Inlet to Fort Myers; Lake Okeechobee
Scale: 1:40,000
NOAA Chart Number: 11428

Chatham River to Clam Pass; Naples Bay; Everglades Harbor
Scale: 1:40,000
NOAA Chart Number: 11429

Lostmans River to Wiggins Pass
Scale: 1:40,000
NOAA Chart Number: 11430

East Cape to Mormon Key
Scale: 1:80,000
NOAA Chart Number: 11431

Everglades National Park, Shark River to Lostmans River
Scale: 1:50,000
NOAA Chart Number: 11432

Everglades National Park, Whitewater Bay
Scale: 1:50,000
NOAA Chart Number: 11433

Florida Keys, Sombrero Key to Dry Tortugas
Scale: 1:180,000
NOAA Chart Number: 11434

Dry Tortugas; Tortugas Harbor
Scale: 1:30,000
NOAA Chart Number: 11438

Sand Key to Rebecca Shoal
Scale: 1:80,000
NOAA Chart Number: 11439

Key West Harbor and Approaches
Scale: 1:30,000
NOAA Chart Number: 11441

Florida Keys, Sombrero Key to Sand Key
Scale: 1:80,000
NOAA Chart Number: 11442

Intracoastal Waterway, Bahia Honda Key to Sugarloaf Key
Scale: 1:40,000
NOAA Chart Number: 11445

Intracoastal Waterway, Sugarloaf Key to Key West
Scale: 1:40,000
NOAA Chart Number: 11446

Key West Harbor
Scale: 1:10,000
NOAA Chart Number: 11447

Intracoastal Waterway, Big Spanish Channel to Johnston Key
Scale: 1:40,000
NOAA Chart Number: 11448

Intracoastal Waterway, Matecumbe to Grassy Key
Scale: 1:40,000
NOAA Chart Number: 11449

Fowey Rocks to American Shoal
Scale: 1:180,000
NOAA Chart Number: 11450

Intracoastal Waterway, Alligator Reef to Sombrero Key
Scale: 1:80,000
NOAA Chart Number: 11452

Florida Keys, Grassy Key to Bahia Honda Key
Scale: 1:40,000
NOAA Chart Number: 11453

Cape Canaveral to Key West
Scale: 1:466,940
NOAA Chart Number: 11460

Fowey Rocks to Alligator Reef
Scale: 1:80,000
NOAA Chart Number: 11462

Intracoastal Waterway, Sands Key to Blackwater Sound
Scale: 1:40,000
NOAA Chart Number: 11463

Intracoastal Waterway, Blackwater Sound to Matecumbe
Scale: 1:40,000
NOAA Chart Number: 11464

Intracoastal Waterway, Blackwater Sound to Matecumbe
Scale: 1:40,000
NOAA Chart Number: 11464

**Pacific Coast**

North Pacific Ocean, West Coast of North America, Mexican Border to Dixon Entrance
Scale: 1:3,500,000
NOAA Chart Number: 501

North America West Coast, San Diego to Aleutian Islands and Hawai'ian Islands
Scale: 1:4,860,700
NOAA Chart Number: 530

Gulf of Alaska, Strait of Juan de Fuca to Kodiak Island
Scale: 1:2,100,000
NOAA Chart Number: 531

Hawai'ian Islands
Scale: 1:3,121,170
NOAA Chart Number: 540

Cape Blanco to Cape Flattery
Scale: 1:736,560
NOAA Chart Number: 18003

San Francisco to Cape Flattery
Scale: 1:1,200,000
NOAA Chart Number: 18007

Monterey Bay to Coos Bay
Scale: 1:811,980
NOAA Chart Number: 18010

San Diego to Cape Mendocino
Scale: 1:1,444,000
NOAA Chart Number: 18020

San Diego to San Francisco Bay
Scale: 1:868,003
NOAA Chart Number: 18022

Strait of Georgia and Strait of Juan de Fuca
Scale: 1:200,000
NOAA Chart Number: 18400

Strait of Juan de Fuca to Strait of Georgia; Drayton Harbor
Scale: 1:80,000
NOAA Chart Number: 18421

Bellingham Bay; Bellingham Harbor
Scale: 1:40,000
NOAA Chart Number: 18424

Anacortes to Skagit Bay
Scale: 1:25,000
NOAA Chart Number: 18427

Oak and Crescent Harbors
Scale: 1:10,000
NOAA Chart Number: 18428

Rosario Strait (Southern Part)
Scale: 1:25,000
NOAA Chart Number: 18429

Rosario Strait (Northern Part)
Scale: 1:25,000
NOAA Chart Number: 18430

Rosario Strait to Cherry Point
Scale: 1:25,000
NOAA Chart Number: 18431

Boundary Pass
Scale: 1:25,000
NOAA Chart Number: 18432

Haro Strait, Middle Bank to Stuart Island
Scale: 1:25,000
NOAA Chart Number: 18433

San Juan Channel
Scale: 1:25,000
NOAA Chart Number: 18434

Puget Sound
Scale: 1:150,000
NOAA Chart Number: 18440

Puget Sound (Northern Part)
Scale: 1:80,000
NOAA Chart Number: 18441

Approaches to Everett
Scale: 1:40,000
NOAA Chart Number: 18443

Everett Harbor
Scale: 1:10,000
NOAA Chart Number: 18444

Puget Sound, Apple Cove Point to Keyport; Agate Passage
Scale: 1:25,000
NOAA Chart Number: 18446

Lake Washington Ship Canal and Lake Washington
Scale: 1:10,000
NOAA Chart Number: 18447

Puget Sound (Southern Part)
Scale: 1:80,000
NOAA Chart Number: 18448

Puget Sound, Seattle to Bremerton
Scale: 1:25,000
NOAA Chart Number: 18449

Seattle Harbor, Elliott Bay and Duwamish Waterway
Scale: 1:10,000
NOAA Chart Number: 18450

Sinclair Inlet
Scale: 1:10,000
NOAA Chart Number: 18452

Tacoma Harbor
Scale: 1:15,000
NOAA Chart Number: 18453

Olympia Harbor and Budd Inlet
Scale: 1:20,000
NOAA Chart Number: 18456

Puget Sound, Hammersley Inlet to Shelton
Scale: 1:10,000
NOAA Chart Number: 18457

Hood Canal, South Point to Quatsap Point including Dabob Bay
Scale: 1:25,000
NOAA Chart Number: 18458

Strait of Juan de Fuca Entrance (includes Metric Version)
Scale: 1:100,000
NOAA Chart Number: 18460

Port Townsend
Scale: 1:20,000
NOAA Chart Number: 18464

Strait of Juan de Fuca (Eastern Part)
Scale: 1:80,000
NOAA Chart Number: 18465

Port Angeles
Scale: 1:10,000
NOAA Chart Number: 18468

Approaches to Admiralty Inlet, Dungeness to Oak Bay
Scale: 1:40,000
NOAA Chart Number: 18471

Puget Sound, Oak Bay to Shilshole Bay
Scale: 1:40,000
NOAA Chart Number: 18473

Puget Sound, Shilshole Bay to Commencement Bay
Scale: 1:40,000
NOAA Chart Number: 18474

Puget Sound, Hood Canal and Dabob Bay
Scale: 1:40,000
NOAA Chart Number: 18476

Puget Sound, Entrance to Hood Canal
Scale: 1:25,000
NOAA Chart Number: 18477

Approaches to Strait of Juan de Fuca, Destruction Island to Amphitrite Point
Scale: 1:176,253
NOAA Chart Number: 18480

Neah Bay
Scale: 1:10,000
NOAA Chart Number: 18484

Cape Flattery
Scale: 1:40,000
NOAA Chart Number: 18485

Columbia River to Destruction Island
Scale: 1:180,789
NOAA Chart Number: 18500

Grays Harbor; Westhaven Cove
Scale: 1:40,000
NOAA Chart Number: 18502

Willapa Bay; Toke Pt.
Scale: 1:40,000
NOAA Chart Number: 18504

Yaquina Head to Columbia River; Netarts Bay
Scale: 1:185,238
NOAA Chart Number: 18520

Columbia River to Harrington Point; Ilwaco Harbor
Scale: 1:40,000
NOAA Chart Number: 18521

Columbia River, Harrington Point to Crims Island
Scale: 1:40,000
NOAA Chart Number: 18523

Columbia River, Crims Island to Saint Helens
Scale: 1:40,000
NOAA Chart Number: 18524

Columbia River, Saint Helens to Vancouver
Scale: 1:40,000
NOAA Chart Number: 18525

Port of Portland, including Vancouver; Multnomah Channel (Southern Part)
Scale: 1:20,000
NOAA Chart Number: 18526

Willamette River, Swan Island Basin
Scale: 1:5,000
NOAA Chart Number: 18527

Willamette River, Portland to Walnut Eddy
Scale: 1:15,000
NOAA Chart Number: 18528

Willamette River, Walnut Eddy to Newburg
Scale: 1:15,000
NOAA Chart Number: 18529

Columbia River Vancouver to Bonneville; Bonneville Dam
Scale: 1:40,000
NOAA Chart Number: 18531

Columbia River Bonneville to the Dalles; The Dalles; Hood River
Scale: 1:40,000
NOAA Chart Number: 18532

Columbia River, Lake Celilo
Scale: 1:20,000
NOAA Chart Number: 18533

Columbia River, John Day Dam to Blalock
Scale: 1:20,000
NOAA Chart Number: 18535

Columbia River, Sundale to Heppner Junction
Scale: 1:20,000
NOAA Chart Number: 18536

Columbia River, Alderdale to Blalock Islands
Scale: 1:20,000
NOAA Chart Number: 18537

Columbia River, Blalock Islands to McNary Dam
Scale: 1:20,000
NOAA Chart Number: 18539

Columbia River, McNary Dam to Juniper
Scale: 1:20,000
NOAA Chart Number: 18541

Columbia River, Juniper to Pasco
Scale: 1:20,000
NOAA Chart Number: 18542

Columbia River, Pasco to Richland
Scale: 1:20,000
NOAA Chart Number: 18543

Lake Sacajawea
Scale: 1:20,000
NOAA Chart Number: 18545

Snake River, Lake Herbert G. West
Scale: 1:20,000
NOAA Chart Number: 18546

Snake River, Lake Bryon
Scale: 1:20,000
NOAA Chart Number: 18547

Snake River, Lower Granite Lake
Scale: 1:20,000
NOAA Chart Number: 18548

Franklin D. Roosevelt Lake (Southern Part)
Scale: 1:50,000
NOAA Chart Number: 18551

Franklin D. Roosevelt Lake (Northern Part)
Scale: 1:50,000
NOAA Chart Number: 18553

Lake Pend Oreille
Scale: 1:50,000
NOAA Chart Number: 18554

Nehalem River
Scale: 1:20,000
NOAA Chart Number: 18556

Tillamook Bay
Scale: 1:50,000
NOAA Chart Number: 18558

Approaches to Yaquina Bay; Depoe Bay
Scale: 1:50,000
NOAA Chart Number: 18561

Cape Blanco to Yaquina Head
Scale: 1:191,730
NOAA Chart Number: 18580

Yaquina Bay and River; Continuation of Yaquina River
Scale: 1:10,000
NOAA Chart Number: 18581

Siuslaw River
Scale: 1:20,000
NOAA Chart Number: 18583

Umpqua River, Pacific Ocean to Reedsport
Scale: 1:20,000
NOAA Chart Number: 18584

Coos Bay
Scale: 1:20,000
NOAA Chart Number: 18587

Coquille River Entrance
Scale: 1:20,000
NOAA Chart Number: 18588

Port Orford to Cape Blanco; Port Orford
Scale: 1:40,000
NOAA Chart Number: 18589

Trinidad Head to Cape Blanco
Scale: 1:196,948
NOAA Chart Number: 18600

Cape Sebastian to Humbug Mountain
Scale: 1:40,000
NOAA Chart Number: 18601

Pyramid Point to Cape Sebastian; Chetco Cove; Hunters Cove
Scale: 1:40,000
NOAA Chart Number: 18602

St. George Reef and Crescent City Harbor; Crescent City Harbor
Scale: 1:40,000
NOAA Chart Number: 18603

Trinidad Harbor
Scale: 1:15,000
NOAA Chart Number: 18605

Point Arena to Trinidad Head; Rockport Landing; Shelter Cove
Scale: 1:200,000
NOAA Chart Number: 18620

Humboldt Bay
Scale: 1:25,000
NOAA Chart Number: 18622

Cape Mendocino and Vicinity
Scale: 1:40,000
NOAA Chart Number: 18623

Elk to Fort Bragg; Fort Bragg and Nayo Anchorage; Elk
Scale: 1:40,000
NOAA Chart Number: 18626

Albion to Caspar
Scale: 1:10,000
NOAA Chart Number: 18628

San Francisco to Point Arena
Scale: 1:207,840
NOAA Chart Number: 18640

Bodega and Tomales Bays; Bodega Harbor
Scale: 1:30,000
NOAA Chart Number: 18643

Gulf of the Farallones; Southeast Farallon
Scale: 1:100,000
NOAA Chart Number: 18645

Drakes Bay
Scale: 1:40,000
NOAA Chart Number: 18647

Entrance to San Francisco Bay
Scale: 1:40,000
NOAA Chart Number: 18649

San Francisco Bay, Candlestick Point to Angel Island
Scale: 1:20,000
NOAA Chart Number: 18650

San Francisco Bay, Southern Part; Redwood Creek; Oyster Point
Scale: 1:40,000
NOAA Chart Number: 18651

San Francisco Bay, Angel Island to Point San Pedro
Scale: 1:20,000
NOAA Chart Number: 18653

San Pablo Bay
Scale: 1:40,000
NOAA Chart Number: 18654

Mare Island Strait
Scale: 1:10,000
NOAA Chart Number: 18655

Suisun Bay
Scale: 1:40,000
NOAA Chart Number: 18656

Carquinez Strait
Scale: 1:10,000
NOAA Chart Number: 18657

Suissun Bay, Roe Island and Vicinity
Scale: 1:10,000
NOAA Chart Number: 18658

Suissun Bay, Mallard Island to Antioch
Scale: 1:10,000
NOAA Chart Number: 18659

San Joaquin River, Stockton Deep Water Channel, Antioch to Medford Island
Scale: 1:20,000
NOAA Chart Number: 18660

Sacramento and San Joaquin Rivers, Old River, Middle River and San Joaquin River Extension; Sherman Island
Scale: 1:80,000
NOAA Chart Number: 18661

Sacramento River, Andrus Island to Sacramento
Scale: 1:40,000
NOAA Chart Number: 18662

San Joaquin River, Stockton Deep Water Channel, Medford Island to Stockton
Scale: 1:20,000
NOAA Chart Number: 18663

Sacramento River, Sacramento to Fourmile Bend
Scale: 1:20,000
NOAA Chart Number: 18664

Lake Tahoe (Metric)
Scale: 1:40,000
NOAA Chart Number: 18665

Suisun Bay, Middle Ground to New York Slough
Scale: 1:10,000
NOAA Chart Number: 18666

Sacramento River, Fourmile Bend to Colusa
Scale: 1:20,000
NOAA Chart Number: 18667

Point Sur to San Francisco
Scale: 1:210,668
NOAA Chart Number: 18680

Half Moon Bay
Scale: 1:20,000
NOAA Chart Number: 18682

Monterey Bay; Monterey Harbor; Moss Landing Harbor; Santa Cruz Small Craft Harbor
Scale: 1:50,000
NOAA Chart Number: 18685

Pfeiffer Point to Cypress Point
Scale: 1:40,000
NOAA Chart Number: 18686

Lake Mead
Scale: 1:48,000
NOAA Chart Number: 18687

Point Conception to Point Sur
Scale: 1:216,116
NOAA Chart Number: 18700

Estero Bay; Morro Bay
Scale: 1:40,000
NOAA Chart Number: 18703

San Luis Obispo Bay, Port San Luis
Scale: 1:20,000
NOAA Chart Number: 18704

Point Dume to Purisma Point
Scale: 1:232,188
NOAA Chart Number: 18720

Santa Cruz Island to Purisima Point
Scale: 1:100,000
NOAA Chart Number: 18721

Port Hueneme and Approaches; Port Hueneme
Scale: 1:20,000
NOAA Chart Number: 18724

Port Hueneme to Santa Barbara; Santa Barbara; Channel Islands Harbor and Port Hueneme; Ventura
Scale: 1:50,000
NOAA Chart Number: 18725

San Miguel Passage; Cuyler Harbor
Scale: 1:40,000
NOAA Chart Number: 18727

Santa Cruz Channel
Scale: 1:40,000
NOAA Chart Number: 18728

Anacapa Passage; Prisoners Harbor
Scale: 1:40,000
NOAA Chart Number: 18729

San Diego to Santa Rosa Island
Scale: 1:234,270
NOAA Chart Number: 18740

Santa Monica Bay; King Harbor
Scale: 1:40,000
NOAA Chart Number: 18744

San Pedro Channel; Dana Point Harbor
Scale: 1:80,000
NOAA Chart Number: 18746

El Segundo and Approaches
Scale: 1:15,000
NOAA Chart Number: 18748

San Pedro Bay; Anaheim Bay, Huntington Harbor
Scale: 1:20,000
NOAA Chart Number: 18749

Los Angeles and Long Beach Harbors
Scale: 1:20,000
NOAA Chart Number: 18751

Newport Bay
Scale: 1:10,000
NOAA Chart Number: 18754

San Nicolas Island
Scale: 1:40,000
NOAA Chart Number: 18755

Santa Barbara Island
Scale: 1:20,000
NOAA Chart Number: 18756

Santa Catalina Island; Avalon Bay; Catalina Harbor; Isthmus Cove
Scale: 1:40,000
NOAA Chart Number: 18757

Del Mar Boat Basin
Scale: 1:5,000
NOAA Chart Number: 18758

San Clemente Island
Scale: 1:40,000
NOAA Chart Number: 18762

San Clemente Island, Northern Part; Wison Cove
Scale: 1:20,000
NOAA Chart Number: 18763

San Clemente Island, Pyramid Cove and Approaches
Scale: 1:15,000
NOAA Chart Number: 18764

Approaches to San Diego Bay; Mission Bay
Scale: 1:100,000
NOAA Chart Number: 18765

Approaches to San Diego Bay
Scale: 1:20,000
NOAA Chart Number: 18772

San Diego Bay
Scale: 1:12,000
NOAA Chart Number: 18773

Gulf of Santa Catalina; Delmar Boat Basin, Camp Pendleton
Scale: 1:100,000
NOAA Chart Number: 18774

Hawai'ian Islands
Scale: 1:600,000
NOAA Chart Number: 19004

Hawai'i to French Frigate Shoals
Scale: 1:1,650,000
NOAA Chart Number: 19007

Hawai'ian Islands (Southern Part)
Scale: 1:675,000
NOAA Chart Number: 19010

Hawai'ian Islands (Northern Part)
Scale: 1:675,000
NOAA Chart Number: 19013

Ni'ihau to French Frigate Shoals; Necker Island; Nihoa
Scale: 1:633,000
NOAA Chart Number: 19016

French Frigate Shoals to Laysan Island
Scale: 1:653,219
NOAA Chart Number: 19019

Laysan Island to Kure Atoll
Scale: 1:642,271
NOAA Chart Number: 19022

Island Of Hawai'i
Scale: 1:250,000
NOAA Chart Number: 19320

Harbors and Landings on the Northeast and Southeast Coasts of Hawai'i; Punalu'u Harbor; Honu'apo Bay; Honokaa Landing; Kukuihaele Landing
Scale: 1:2,500
NOAA Chart Number: 19322

Island of Hawai'i, Hilo Bay
Scale: 1:10,000
NOAA Chart Number: 19324

Pa'auhau Landing, Island of Hawai'i
Scale: 1:5,000
NOAA Chart Number: 19326

West Coast of Hawai'i, Cook Point to Upolu Point; Keauhou Bay; Honokohau Harbor
Scale: 1:80,000
NOAA Chart Number: 19327

Mähukona Harbor and Approaches, Island of Hawai'i
Scale: 1:5,000
NOAA Chart Number: 19329

Kawaihae Bay, Island of Hawai'i
Scale: 1:10,000
NOAA Chart Number: 19330

Kailua Bay, Island of Hawai'i
Scale: 1:5,000
NOAA Chart Number: 19331

Kealakekua Bay to Hönaunau Bay
Scale: 1:10,000
NOAA Chart Number: 19332

Hawai'i to O'ahu
Scale: 1:250,000
NOAA Chart Number: 19340

Häna Bay, Island of Maui
Scale: 1:5,000
NOAA Chart Number: 19341

Kahului Harbor and Approaches; Kahului Harbor
Scale: 1:30,000
NOAA Chart Number: 19342

Channels between Molokai, Maui, Läna'i and Kaho'olawe; Manele Bay
Scale: 1:80,000
NOAA Chart Number: 19347

Approaches to Lahaina, Island of Maui
Scale: 1:15,000
NOAA Chart Number: 19348

Island of Maui, Ma'alaea Bay
Scale: 1:10,000
NOAA Chart Number: 19350

Channels between O'ahu, Moloka'i and Läna'i; Kaumalapa'u Harbor
Scale: 1:80,000
NOAA Chart Number: 19351

Harbors of Moloka'i, Kaunakakai Harbor; Pükoo Harbor; Kamalö Harbor; Kolo Harbor; Lono Harbor
Scale: 1:5,000
NOAA Chart Number: 19353

Island of O'ahu; Barbers Point Harbor
Scale: 1:80,000
NOAA Chart Number: 19357

Southeast Coast of O'ahu, Waimänalo Bay to Diamond Head
Scale: 1:20,000
NOAA Chart Number: 19358

O'ahu East Coast, Käne'ohe Bay
Scale: 1:15,000
NOAA Chart Number: 19359

Port Wa'ianae, Island of O'ahu
Scale: 1:10,000
NOAA Chart Number: 19361

South Coast of O'ahu, Kalaeloa
Scale: 1:20,000
NOAA Chart Number: 19362

Pearl Harbor, O'ahu South Coast
Scale: 1:15,000
NOAA Chart Number: 19366

Island of O'ahu, Honolulu Harbor
Scale: 1:5,000
NOAA Chart Number: 19367

O'ahu South Coast, Approaches to Pearl Harbor
Scale: 1:20,000
NOAA Chart Number: 19369

O'ahu to Ni'ihau
Scale: 1:247,482
NOAA Chart Number: 19380

Island of Kaua'i
Scale: 1:80,000
NOAA Chart Number: 19381

Port Allen, Island of Kaua'i
Scale: 1:5,000
NOAA Chart Number: 19382

Kaua'i, Nawiliwili Bay
Scale: 1:5,000
NOAA Chart Number: 19383

Hanamaulu Bay, Island of Kaua'i
Scale: 1:2,000
NOAA Chart Number: 19384

North Coast of Kaua'i, Hä'ena Point to Kepuhi Point
Scale: 1:20,000
NOAA Chart Number: 19385

Kaua'i, Approaches to Waimea Bay
Scale: 1:1,000
NOAA Chart Number: 19386

French Frigate Shoals
Scale: 1:80,000
NOAA Chart Number: 19401

French Frigate Shoals, Anchorage
Scale: 1:25,000
NOAA Chart Number: 19402

Gardner Pinnacles and Approaches; Gardner Pinnacles
Scale: 1:100,000
NOAA Chart Number: 19421

Maro Reef
Scale: 1:80,000
NOAA Chart Number: 19441

Lisianski and Laysan Island; West Coast of Laysan Island
Scale: 1:40,000
NOAA Chart Number: 19442

Pearl and Hermes Atoll
Scale: 1:40,000
NOAA Chart Number: 19461

Gambia Shoal to Kure Atoll including Approaches to the Midway Islands
Scale: 1:180,000
NOAA Chart Number: 19480

Hawai'ian Islands, Midway Islands
Scale: 1:32,500
NOAA Chart Number: 19481

Hawai'ian Islands, Midway Islands
Scale: 1:10,000
NOAA Chart Number: 19482

Hawai'i, Kure Atoll
Scale: 1:20,000
NOAA Chart Number: 19483

# 8
# Fire Insurance Plans

A fire insurance plan is a large-scale map of a community that offers very detailed information about individual buildings that existed at the time of the map drawing. The original purpose of the plan was to have a detailed record of the potential fire risks of individual commercial, industrial, and residential buildings, including building material and the location of water and fire alarm boxes. Using a variety of symbols and colors, the plan reveals the construction type of each building exterior, as well as the building height, the use of the building (residential, commercial, and so on), and, for many nonresidential buildings, the organization and commercial names. Related structures such as fire hydrants and water or oil pipes are also available. Street information, such as street name, number, and street width, is printed on the map.

Due to its large-scale details, researchers find fire insurance plans incredibly valuable, offering information that cannot be obtained elsewhere. Reviewing and studying a fire insurance plan for a city or town will often help the researcher narrow the construction date for a building; will provide the historical street name and house number (often street names and house numbers change with future development); may highlight any building zoning changes, additions, or constructions to a house (such as a side garage); may highlight if a building has been demolished; and may indicate potential buried oil tanks or other contaminated resources.

In Canada, the most extensive coverage was produced between 1848 and 1910 by Charles E. Goad, who published around thirteen hundred plans of Canadian cities. In 1918, the Underwriters' Survey Bureau was established and acquired exclusive rights to update, revise, and reprint the Goad plans. Over the years, with amalgamation with other Canadian associations, the Canadian Underwriter's Association centralized the production of all plans, creating the Insurers Advisory Organization of Canada (IAO), and eventually ceased all plan production in 1975. Today, many Canadian libraries house Goad and Underwriters' Survey Bureau plans, published during the early to mid-twentieth century.

In the United States, the Sanborn Map Company has been creating and publishing fire insurance plans since 1866, offering users coverage of over twelve thousand U.S. cities and towns. The Library of Congress houses the largest collection of Sanborn maps, offering over twenty-five thousand sheets from over three thousand cities online.

Due to the large size of fire insurance collections, including a complete union list in this chapter is not possible. Many organizations and institutions have created union lists and partial lists of holdings, making them available online as well as publishing them in print. The library at the University of California, Berkeley, for example, has a very large collection, as reflected in their list (http://www.lib.berkeley.edu/EART/sanborn_union_list).

This chapter offers published union lists that should be used to determine if specific American or Canadian towns or cities have fire insurance plan coverage. Users can visit the Library of Congress website to determine if the plans have been digitized; the closest academic or public library can be consulted for access to the plans in print.

*Canadian Fire Insurance Plans in Ontario Collections, 1876–1973*, by M. Fortin, L. Dubreuil, and C. Woods. Ottawa, ON: Association of Canadian Map Libraries and Archives, 1995.

*Catalogue of Canadian Fire Insurance Plans, 1875–1975*, by L. Dubreuil and C. A. Woods. Ottawa, ON: Association of Canadian Map Libraries and Archives, 2002.

*Catalogue of Fire Insurance Plans of the Dominion of Canada, 1885–1973 = Catalogue des plans d'assurance du Canada, 1885–1973.* London, ON: Phelps, 1977.

*Catalogue of Fire Insurance Plans Published by Chas E. Goad Ltd. 1878–1970.* Old Hatfield, UK: Goad, 1984.

*Fire Insurance Maps in the Library of Congress: Plans of North American Cities and Towns Produced by the Sanborn Map Company: A Checklist.* Washington, DC: Library of Congress, 1981.

*Fire Insurance Plans in the National Map Collection*, by R. J. Hayward. Ottawa, ON: Public Archives, 1978.

*Fire Insurance Plans in the National Map Collection = Plans d'assurance-incendie de la Collection nationale de cartes de plans*, by R. J. Hayward. Ottawa, ON: National Map Collection, 1978.

*Fire Insurance Plans of Canadian Cities*. Toronto, ON: Chas E. Goad; Underwriters' Survey Bureau, 1959.

*Union List of Canadian Fire Insurance Plans Held in Ontario University Map Libraries*, by M. Fortin. London, ON: University of Western Ontario, 1992.

*Union List of Sanborn Fire Insurance Maps Held by Institutions in the United States and Canada: Vol. 1*, by R. P. Hoehn, W. S. Peterson-Hunt, and E. L. Woodruff. Santa Cruz, CA: Western Association of Map Libraries, 1976.

*Union List of Sanborn Fire Insurance Maps Held by Institutions in the United States and Canada: Vol. 2*, by R. P. Hoehn, W. S. Peterson-Hunt and E. L. Woodruff. Santa Cruz, CA: Western Association of Map Libraries, 1977.

# 9
# Air Photos

Air photos, or aerial photography, are photographs of the ground taken from an airplane. Some of the first air photos of towns in North America were taken in the 1920s. Today, photos continue to be captured but with digital cameras rather than on film. Photos taken with drones are also becoming more and more common. Digital air photos are easily viewed using online products like Google Earth and Google Maps; however, many municipalities are also offering digital air photos, or orthoimagery, from their websites. Older air photos are available from many academic map libraries. Because of the large interest in photos, many have been scanned and made available digitally as well.

Not all air photos have the same scale, so users looking for greater detail (to see individual homes, for instance) will want to search for larger-scale photos, such as 1:25,000 and larger. Small-scale air photos, with a scale of 1:50,000 and smaller, offer less detail but cover a larger area.

Air photos are available for all of Canada and the United States via the National Air Photo Library, or NAPL (http://www.nrcan.gc.ca/earth-sciences/geomatics/satellite-imagery-air-photos/9265), and the USGS EarthExplorer (https://earthexplorer.usgs.gov). This chapter lists air photo publishers available via EarthExplorer, organized by state. This list was compiled by obtaining information about the photography available for every state and then visiting each publisher's website for URL information. Unfortunately, a similar list for Canada is not available because of the sheer size of the collection (over six million photos) and a lack of search filters via the NAPL website. Users who are interested in obtaining Canadian aerial photography are encouraged either to visit the NAPL website to order prints or to visit local libraries and municipalities. Each item below includes the publisher, date range, and URL.

## Alabama

Army Map Service
1950–1958
https://www.nga.mil/ABOUT/HISTORY/NGAINHISTORY/Pages/ArmyMapService.aspx

NASA AMES Research Center
1979–1996
https://www.nasa.gov/centers/ames/home/index.html

NASA Johnson Space Center
1969–1975
https://www.nasa.gov/centers/johnson/home/index.html

National Park Service
1983
https://www.nps.gov/hfc/cfm/npsphoto2.cfm

U.S. Air Force
1960–1973
https://www.airforce.com

U.S. Geological Survey
1938–2011
https://pubs.usgs.gov/gip/AerialPhotos_SatImages/aerial.html

## Alaska

NASA AMES Research Center
1974–1996
https://www.nasa.gov/centers/ames/home/index.html

NASA Johnson Space Center
1970–1980
https://www.nasa.gov/centers/johnson/home/index.html

U.S. Air Force
1941–1977
https://www.airforce.com

U.S. Bureau of Reclamation
1951–1965
https://www.usbr.gov

U.S. Fish and Wildlife Services
1948–1976
https://www.fws.gov

U.S. Geological Survey
1962–1990
https://www.usgs.gov

U.S. Navy
1949–1955
http://www.navy.mil/

## Arizona

Army Map Service
1953–1958
https://www.nga.mil/ABOUT/HISTORY/NGAINHISTORY/Pages/ArmyMapService.aspx

McDonnell Douglas
1956–1975
http://mcdonnelldouglas.weebly.com

NASA AMES Research Center
1972–1996
URL: https://www.nasa.gov/centers/ames/home/index.html

NASA Johnson Space Center
1969–1977
https://www.nasa.gov/centers/johnson/home/index.html

National Park Service
1979–2003
https://www.nps.gov/hfc/cfm/npsphoto2.cfm

U.S. Air Force
1947–1978
https://www.airforce.com

U.S. Bureau of Reclamation
1947–1968
https://www.usbr.gov

U.S. Fish and Wildlife Services
1947–1977
https://www.fws.gov

U.S. Geological Survey
1940–1993
https://www.usgs.gov

## Arkansas

Army Map Service
1948–1957
https://www.nga.mil/ABOUT/HISTORY/NGAINHISTORY/Pages/ArmyMapService.aspx

McDonnell Douglas
1963
http://mcdonnelldouglas.weebly.com

NASA AMES Research Center
1973–1995
https://www.nasa.gov/centers/ames/home/index.html

NASA Johnson Space Center
1972–1975
https://www.nasa.gov/centers/johnson/home/index.html

National Park Service
2005–2006
https://www.nps.gov/hfc/cfm/npsphoto2.cfm

U.S. Air Force
1955–1969
https://www.airforce.com

U.S. Army Corps of Engineers
1978
http://www.usace.army.mil

U.S. Geological Survey
1941–1990
https://www.usgs.gov

## California

Army Map Service
1952–1956
https://www.nga.mil/ABOUT/HISTORY/NGAINHISTORY/Pages/ArmyMapService.aspx

McDonnell Douglas
1956–1975
http://mcdonnelldouglas.weebly.com

NASA AMES Research Center
1971–1996
https://www.nasa.gov/centers/ames/home/index.html

NASA Johnson Space Center
1969–1977
https://www.nasa.gov/centers/johnson/home/index.html

National Park Service
1979–1997
https://www.nps.gov/hfc/cfm/npsphoto2.cfm

U.S. Air Force
1947–1978
https://www.airforce.com

U.S. Bureau of Land Management
1967–2010
https://www.blm.gov

U.S. Bureau of Reclamation
1964
https://www.usbr.gov

U.S. Fish and Wildlife Services
1947–1976
https://www.fws.gov

U.S. Geological Survey
1940–1998
https://www.usgs.gov

## Colorado

Army Map Service
1953–1960
https://www.nga.mil/ABOUT/HISTORY/NGAINHISTORY/Pages/ArmyMapService.aspx

McDonnell Douglas
1956–1966
http://mcdonnelldouglas.weebly.com

NASA AMES Research Center
1974–1995
https://www.nasa.gov/centers/ames/home/index.html

NASA Johnson Space Center
1969–1976
https://www.nasa.gov/centers/johnson/home/index.html

National Park Service
1979–2004
https://www.nps.gov/hfc/cfm/npsphoto2.cfm

U.S. Air Force
1942–1972
https://www.airforce.com

U.S. Bureau of Land Management
1962–2002
https://www.blm.gov

U.S. Bureau of Reclamation
1939–1964
https://www.usbr.gov

U.S. Fish and Wildlife Services
1955
https://www.fws.gov

U.S. Geological Survey
1941–1991
https://www.usgs.gov

## Connecticut

Army Map Service
1958
https://www.nga.mil/ABOUT/HISTORY/NGAINHISTORY/Pages/ArmyMapService.aspx

NASA AMES Research Center
1972–1995
https://www.nasa.gov/centers/ames/home/index.html

NASA Johnson Space Center
1969–1973
https://www.nasa.gov/centers/johnson/home/index.html

National Park Service
1982–2004
https://www.nps.gov/hfc/cfm/npsphoto2.cfm

U.S. Air Force
1959–1960
https://www.airforce.com

U.S. Geological Survey
1938–1991
https://www.usgs.gov

## Delaware

Army Map Service
April 1957–May 1957
https://www.nga.mil/ABOUT/HISTORY/NGAINHISTORY/Pages/ArmyMapService.aspx

NASA AMES Research Center
1972–1995
https://www.nasa.gov/centers/ames/home/index.html

NASA Johnson Space Center
1969–1975
https://www.nasa.gov/centers/johnson/home/index.html

National Park Service
1982–1983
https://www.nps.gov/hfc/cfm/npsphoto2.cfm

U.S. Air Force
1955–1960
https://www.airforce.com

U.S. Geological Survey
1950–1995
https://www.usgs.gov

**Florida**

Army Map Service
1949–1956
https://www.nga.mil/ABOUT/HISTORY/NGAINHISTORY/Pages/ArmyMapService.aspx

FEMA
2004
https://www.fema.gov

NASA AMES Research Center
1975–1996
https://www.nasa.gov/centers/ames/home/index.html

NASA Johnson Space Center
1969–1977
https://www.nasa.gov/centers/johnson/home/index.html

National Park Service
1982–1987
https://www.nps.gov/hfc/cfm/npsphoto2.cfm

U.S. Air Force
1958–1968
https://www.airforce.com

U.S. Fish and Wildlife Services
July 1969
https://www.fws.gov

U.S. Geological Survey
1942–2012
https://www.usgs.gov

**Georgia**

Army Map Service
1950–1958
https://www.nga.mil/ABOUT/HISTORY/NGAINHISTORY/Pages/ArmyMapService.aspx

McDonnell Douglas
March 1969
http://mcdonnelldouglas.weebly.com

NASA AMES Research Center
1972–1992
https://www.nasa.gov/centers/ames/home/index.html

NASA Johnson Space Center
1969–1976
https://www.nasa.gov/centers/johnson/home/index.html

Marshall Space Flight Center
1972–1973
https://www.nasa.gov/centers/marshall/home/index.html

National Park Service
1982–1983
https://www.nps.gov/hfc/cfm/npsphoto2.cfm

U.S. Air Force
1952–1973
https://www.airforce.com

U.S. Fish and Wildlife Services
March 1949
https://www.fws.gov

**Hawaii**

NASA AMES Research Center
1974–1995
https://www.nasa.gov/centers/ames/home/index.html

U.S. Geological Survey
1949–1978
https://www.usgs.gov

U.S. Navy
March 1972
http://www.navy.mil

**Idaho**

Army Map Service
1953–1958
https://www.nga.mil/ABOUT/HISTORY/NGAINHISTORY/Pages/ArmyMapService.aspx

EROS
1976–1979
https://eros.usgs.gov

McDonnell Douglas
1966
http://mcdonnelldouglas.weebly.com

NASA AMES Research Center
1973–1995
https://www.nasa.gov/centers/ames/home/index.html

NASA Johnson Space Center
1969–1979
https://www.nasa.gov/centers/johnson/home/index.html

National Park Service
1969–1980
https://www.nps.gov/hfc/cfm/npsphoto2.cfm

U.S. Bureau of Land Management
1973–1999
https://www.blm.gov

U.S. Bureau of Reclamation
1940–1969
https://www.usbr.gov

U.S. Fish and Wildlife Services
1974–1976
https://www.fws.gov

U.S. Geological Survey
1939–1987
https://www.usgs.gov

## Illinois

Army Map Service
1950–1958
https://www.nga.mil/ABOUT/HISTORY/NGAINHISTORY/Pages/ArmyMapService.aspx

NASA AMES Research Center
1973–1999
https://www.nasa.gov/centers/ames/home/index.html

NASA Johnson Space Center
1969–1974
https://www.nasa.gov/centers/johnson/home/index.html

U.S. Air Force
1947–1969
https://www.airforce.com

U.S. Army Corps of Engineers
1974–1978
http://www.usace.army.mil

U.S. Fish and Wildlife Services
1964–1980
https://www.fws.gov

U.S. Geological Survey
1946–1995
https://www.usgs.gov

## Indiana

Army Map Service
1952–1956
https://www.nga.mil/ABOUT/HISTORY/NGAINHISTORY/Pages/ArmyMapService.aspx

NASA AMES Research Center
1973–1995
https://www.nasa.gov/centers/ames/home/index.html

NASA Johnson Space Center
1969–1977
https://www.nasa.gov/centers/johnson/home/index.html

National Park Service
July 1983
https://www.nps.gov/hfc/cfm/npsphoto2.cfm

U.S. Air Force
1948–1971
https://www.airforce.com

U.S. Army Corps of Engineers
1971–1974
http://www.usace.army.mil

U.S. Fish and Wildlife Services
August 1967
https://www.fws.gov

U.S. Geological Survey
1941–1990
https://www.usgs.gov

## Iowa

Army Map Service
1949–1958
https://www.nga.mil/ABOUT/HISTORY/NGAINHISTORY/Pages/ArmyMapService.aspx

EROS
July 1979
https://eros.usgs.gov

NASA AMES Research Center
1975–1999
https://www.nasa.gov/centers/ames/home/index.html

NASA Johnson Space Center
1970–1977
https://www.nasa.gov/centers/johnson/home/index.html

U.S. Army Corps of Engineers
1969–1978
http://www.usace.army.mil

U.S. Bureau of Reclamation
1951–1952
https://www.usbr.gov

U.S. Fish and Wildlife Services
1939–1999
https://www.fws.gov

U.S. Geological Survey
1945–1986
https://www.usgs.gov

**Kansas**

Army Map Service
1948–1957
https://www.nga.mil/ABOUT/HISTORY/NGAINHISTORY/Pages/ArmyMapService.aspx

McDonnell Douglas
May 1966
http://mcdonnelldouglas.weebly.com

NASA AMES Research Center
1972–1999
https://www.nasa.gov/centers/ames/home/index.html

NASA Johnson Space Center
1969–1977
https://www.nasa.gov/centers/johnson/home/index.html

U.S. Air Force
1948–1974
https://www.airforce.com

U.S. Army Corps of Engineers
1978
http://www.usace.army.mil

U.S. Bureau of Reclamation
1952–1963
https://www.usbr.gov

U.S. Geological Survey
1946–1995
https://www.usgs.gov

**Kentucky**

Army Map Service
1952–1956
https://www.nga.mil/ABOUT/HISTORY/NGAINHISTORY/Pages/ArmyMapService.aspx

NASA AMES Research Center
1973–1995
https://www.nasa.gov/centers/ames/home/index.html

NASA Johnson Space Center
1970–1973
https://www.nasa.gov/centers/johnson/home/index.html

National Park Service
March 1990
https://www.nps.gov/hfc/cfm/npsphoto2.cfm

U.S. Air Force
1953–1973
https://www.airforce.com

U.S. Army Corps of Engineers
1978
http://www.usace.army.mil

U.S. Geological Survey
1947–1995
https://www.usgs.gov

**Louisiana**

Army Map Service
1949–1956
https://www.nga.mil/ABOUT/HISTORY/NGAINHISTORY/Pages/ArmyMapService.aspx

EROS
January 1982
https://eros.usgs.gov

McDonnell Douglas
December 1963
http://mcdonnelldouglas.weebly.com

NASA AMES Research Center
1973–1996
https://www.nasa.gov/centers/ames/home/index.html

NASA Johnson Space Center
December 1973
https://www.nasa.gov/centers/johnson/home/index.html

NASA Stennis Space Center
November 1974
https://www.nasa.gov/centers/stennis/home/index.html

National Park Service
1982–1983
https://www.nps.gov/hfc/cfm/npsphoto2.cfm

U.S. Air Force
1968–1969
https://www.airforce.com

U.S. Army Corps of Engineers
April 1989
http://www.usace.army.mil

U.S. Geological Survey
1947–1987
https://www.usgs.gov

## Maine

Army Map Service
1951–1956
https://www.nga.mil/ABOUT/HISTORY/NGAINHISTORY/Pages/ArmyMapService.aspx

NASA AMES Research Center
1973–1994
https://www.nasa.gov/centers/ames/home/index.html

NASA Johnson Space Center
July 1975
https://www.nasa.gov/centers/johnson/home/index.html

National Park Service
1981–1997
https://www.nps.gov/hfc/cfm/npsphoto2.cfm

U.S. Air Force
1958–1960
https://www.airforce.com

U.S. Fish and Wildlife Services
1959–1965
https://www.fws.gov

U.S. Geological Survey
1939–2008
https://www.usgs.gov

## Maryland

Army Map Service
November 1955
https://www.nga.mil/ABOUT/HISTORY/NGAINHISTORY/Pages/ArmyMapService.aspx

NASA AMES Research Center
1972–1995
https://www.nasa.gov/centers/ames/home/index.html

NASA Johnson Space Center
1969–1975
https://www.nasa.gov/centers/johnson/home/index.html

National Park Service
1956–2002
https://www.nps.gov/hfc/cfm/npsphoto2.cfm

U.S. Air Force
1955–1968
https://www.airforce.com

U.S. Geological Survey
1946–1991
https://www.usgs.gov

## Massachusetts

Army Map Service
August 1958
https://www.nga.mil/ABOUT/HISTORY/NGAINHISTORY/Pages/ArmyMapService.aspx

NASA AMES Research Center
1973–1995
https://www.nasa.gov/centers/ames/home/index.html

NASA Johnson Space Center
1969–1975
https://www.nasa.gov/centers/johnson/home/index.html

National Park Service
1982–2003
https://www.nps.gov/hfc/cfm/npsphoto2.cfm

U.S. Air Force
1960–1971
https://www.airforce.com

U.S. Army Corps of Engineers
1977–1978
http://www.usace.army.mil

U.S. Geological Survey
1938–1991
https://www.usgs.gov

## Michigan

Army Map Service
1952–1957
https://www.nga.mil/ABOUT/HISTORY/NGAINHISTORY/Pages/ArmyMapService.aspx

NASA AMES Research Center
1973–1994
https://www.nasa.gov/centers/ames/home/index.html

NASA Johnson Space Center
1969–1975
https://www.nasa.gov/centers/johnson/home/index.html

National Park Service
1983–2004
https://www.nps.gov/hfc/cfm/npsphoto2.cfm

U.S. Air Force
1960–1971
https://www.airforce.com

U.S. Army Corps of Engineers
1969–1975
http://www.usace.army.mil

U.S. Fish and Wildlife Services
1977–2002
https://www.fws.gov

U.S. Geological Survey
1937–1990
https://www.usgs.gov

**Minnesota**

Army Map Service
1949–1958
https://www.nga.mil/ABOUT/HISTORY/NGAINHISTORY/Pages/ArmyMapService.aspx

EROS
1975–1979
https://eros.usgs.gov

McDonnell Douglas
1957
http://mcdonnelldouglas.weebly.com

NASA AMES Research Center
1975–1992
https://www.nasa.gov/centers/ames/home/index.html

NASA Johnson Space Center
1969–1977
https://www.nasa.gov/centers/johnson/home/index.html

National Park Service
1983–2003
https://www.nps.gov/hfc/cfm/npsphoto2.cfm

U.S. Air Force
1951–1970
https://www.airforce.com

U.S. Army Corps of Engineers
May 1974
http://www.usace.army.mil

U.S. Bureau of Reclamation
1952–1956
https://www.usbr.gov

U.S. Fish and Wildlife Services
1937–1994
https://www.fws.gov

U.S. Geological Survey
1945–1991
https://www.usgs.gov

**Mississippi**

Army Map Service
1949–1953
https://www.nga.mil/ABOUT/HISTORY/NGAINHISTORY/Pages/ArmyMapService.aspx

McDonnell Douglas
December 1963
http://mcdonnelldouglas.weebly.com

NASA AMES Research Center
1973–1996
https://www.nasa.gov/centers/ames/home/index.html

NASA Johnson Space Center
1969–1975
https://www.nasa.gov/centers/johnson/home/index.html

NASA Stennis Space Center
December 1973
https://www.nasa.gov/centers/stennis/home/index.html

National Park Service
1982–1983
https://www.nps.gov/hfc/cfm/npsphoto2.cfm

U.S. Air Force
1955–1969
https://www.airforce.com

U.S. Geological Survey
1940–1999
https://www.usgs.gov

## Missouri

Army Map Service
1948–1957
https://www.nga.mil/ABOUT/HISTORY/NGAINHISTORY/Pages/ArmyMapService.aspx

NASA AMES Research Center
1972–1994
https://www.nasa.gov/centers/ames/home/index.html

NASA Johnson Space Center
1969–1974
https://www.nasa.gov/centers/johnson/home/index.html

U.S. Air Force
1947–1971
https://www.airforce.com

U.S. Army Corps of Engineers
1978
http://www.usace.army.mil

U.S. Fish and Wildlife Services
1964–1981
https://www.fws.gov

U.S. Geological Survey
1941–1995
https://www.usgs.gov

## Montana

Army Map Service
1953–1959
https://www.nga.mil/ABOUT/HISTORY/NGAINHISTORY/Pages/ArmyMapService.aspx

NASA AMES Research Center
1973–1994
https://www.nasa.gov/centers/ames/home/index.html

NASA Johnson Space Center
1969–1975
https://www.nasa.gov/centers/johnson/home/index.html

National Park Service
1969–2002
https://www.nps.gov/hfc/cfm/npsphoto2.cfm

U.S. Air Force
1946–1952
https://www.airforce.com

U.S. Bureau of Land Management
May 1976
https://www.blm.gov

U.S. Fish and Wildlife Services
1970–1974
https://www.fws.gov

U.S. Geological Survey
1945–1986
https://www.usgs.gov

## Nebraska

Army Map Service
1950–1956
https://www.nga.mil/ABOUT/HISTORY/NGAINHISTORY/Pages/ArmyMapService.aspx

McDonnell Douglas
1956–1966
http://mcdonnelldouglas.weebly.com

NASA AMES Research Center
1975–1999
https://www.nasa.gov/centers/ames/home/index.html

NASA Johnson Space Center
1969–1977
https://www.nasa.gov/centers/johnson/home/index.html

National Park Service
July 1995
https://www.nps.gov/hfc/cfm/npsphoto2.cfm

U.S. Air Force
July 1974
https://www.airforce.com

U.S. Army Corps of Engineers
1969–1978
http://www.usace.army.mil

U.S. Bureau of Land Management
July 1976
https://www.blm.gov

U.S. Bureau of Reclamation
1950–1962
https://www.usbr.gov

U.S. Fish and Wildlife Services
1975–1997
https://www.fws.gov

U.S. Geological Survey
1946–1995
https://www.usgs.gov

### Nevada

Army Map Service
1952–1957
https://www.nga.mil/ABOUT/HISTORY/NGAINHISTORY/Pages/ArmyMapService.aspx

EROS
1979–1981
https://eros.usgs.gov

McDonnell Douglas
1956–1975
http://mcdonnelldouglas.weebly.com

NASA AMES Research Center
1972–1996
https://www.nasa.gov/centers/ames/home/index.html

NASA Johnson Space Center
1969–1975
https://www.nasa.gov/centers/johnson/home/index.html

National Park Service
1979–1981
https://www.nps.gov/hfc/cfm/npsphoto2.cfm

U.S. Air Force
1952–1980
https://www.airforce.com

U.S. Bureau of Land Management
1992–2004
https://www.blm.gov

U.S. Bureau of Reclamation
1953–1966
https://www.usbr.gov

U.S. Fish and Wildlife Services
1976–1978
https://www.fws.gov

U.S. Geological Survey
1941–1998
https://www.usgs.gov

### New Hampshire

NASA AMES Research Center
1973–1995
https://www.nasa.gov/centers/ames/home/index.html

NASA Johnson Space Center
1969–1976
https://www.nasa.gov/centers/johnson/home/index.html

National Park Service
April 2003
https://www.nps.gov/hfc/cfm/npsphoto2.cfm

U.S. Air Force
1958–1968
https://www.airforce.com

U.S. Army Corps of Engineers
1977–1978
http://www.usace.army.mil

U.S. Fish and Wildlife Services
August 1965
https://www.fws.gov

U.S. Geological Survey
1938–2007
https://www.usgs.gov

### New Jersey

Army Map Service
1956–1958
https://www.nga.mil/ABOUT/HISTORY/NGAINHISTORY/Pages/ArmyMapService.aspx

NASA AMES Research Center
1971–1995
https://www.nasa.gov/centers/ames/home/index.html

NASA Johnson Space Center
1969–1975
https://www.nasa.gov/centers/johnson/home/index.html

National Park Service
1981–1984
https://www.nps.gov/hfc/cfm/npsphoto2.cfm

U.S. Air Force
1948–1960
https://www.airforce.com

U.S. Geological Survey
1948–1995
https://www.usgs.gov

### New Mexico

Army Map Service
1942–1960

https://www.nga.mil/ABOUT/HISTORY/NGAINHISTORY/Pages/ArmyMapService.aspx

McDonnell Douglas
1956–1969
http://mcdonnelldouglas.weebly.com

NASA AMES Research Center
1973–1996
https://www.nasa.gov/centers/ames/home/index.html

NASA Johnson Space Center
1970–1975
https://www.nasa.gov/centers/johnson/home/index.html

National Park Service
1979–2003
https://www.nps.gov/hfc/cfm/npsphoto2.cfm

U.S. Air Force
1952–1978
https://www.airforce.com

U.S. Army Corps of Engineers
1978
http://www.usace.army.mil

U.S. Bureau of Reclamation
1946–1968
https://www.usbr.gov

U.S. Fish and Wildlife Services
1947–1974
https://www.fws.gov

U.S. Geological Survey
1942–1991
https://www.usgs.gov

**New York**

Army Map Service
1956–1958
https://www.nga.mil/ABOUT/HISTORY/NGAINHISTORY/Pages/ArmyMapService.aspx

NASA AMES Research Center
1972–1995
https://www.nasa.gov/centers/ames/home/index.html

NASA Johnson Space Center
1969–1977
https://www.nasa.gov/centers/johnson/home/index.html

National Park Service
1981–2004
https://www.nps.gov/hfc/cfm/npsphoto2.cfm

U.S. Air Force
1948–1970
https://www.airforce.com

U.S. Army Corps of Engineers
1970–1978
http://www.usace.army.mil

U.S. Fish and Wildlife Services
1955
https://www.fws.gov

U.S. Geological Survey
1939–1993
https://www.usgs.gov

**North Carolina**

Army Map Service
1950–1958
https://www.nga.mil/ABOUT/HISTORY/NGAINHISTORY/Pages/ArmyMapService.aspx

FEMA
September 1999
https://www.fema.gov

NASA AMES Research Center
1972–1995
https://www.nasa.gov/centers/ames/home/index.html

NASA Johnson Space Center
1969–1982
https://www.nasa.gov/centers/johnson/home/index.html

National Park Service
1982–1983
https://www.nps.gov/hfc/cfm/npsphoto2.cfm

U.S. Air Force
1940–1973
https://www.airforce.com

U.S. Geological Survey
1947–1990
https://www.usgs.gov

U.S. Navy
1957–1959
http://www.navy.mil

**North Dakota**

Army Map Service
1952

https://www.nga.mil/ABOUT/HISTORY/NGAINHISTORY/Pages/ArmyMapService.aspx

EROS
1975
https://eros.usgs.gov

McDonnell Douglas
1957
http://mcdonnelldouglas.weebly.com

NASA AMES Research Center
1978–1986
https://www.nasa.gov/centers/ames/home/index.html

NASA Johnson Space Center
1970–1977
https://www.nasa.gov/centers/johnson/home/index.html

National Park Service
2002
https://www.nps.gov/hfc/cfm/npsphoto2.cfm

U.S. Air Force
1951–1952
https://www.airforce.com

U.S. Bureau of Reclamation
1946–1966
https://www.usbr.gov

U.S. Fish and Wildlife Services
April 1997
https://www.fws.gov

U.S. Geological Survey
1946–1985
https://www.usgs.gov

## Ohio

Army Map Service
1952–1956
https://www.nga.mil/ABOUT/HISTORY/NGAINHISTORY/Pages/ArmyMapService.aspx

NASA AMES Research Center
1975–1994
https://www.nasa.gov/centers/ames/home/index.html

NASA Johnson Space Center
1969–1974
https://www.nasa.gov/centers/johnson/home/index.html

National Park Service
1979–2004
https://www.nps.gov/hfc/cfm/npsphoto2.cfm

U.S. Air Force
1948–1970
https://www.airforce.com

U.S. Army Corps of Engineers
1970–1974
http://www.usace.army.mil

U.S. Geological Survey
1947–1990
https://www.usgs.gov

## Oklahoma

Army Map Service
1949–1957
https://www.nga.mil/ABOUT/HISTORY/NGAINHISTORY/Pages/ArmyMapService.aspx

McDonnell Douglas
1966–1969
http://mcdonnelldouglas.weebly.com

NASA AMES Research Center
1973–1999
https://www.nasa.gov/centers/ames/home/index.html

NASA Johnson Space Center
1969–1977
https://www.nasa.gov/centers/johnson/home/index.html

National Park Service
August 2005
https://www.nps.gov/hfc/cfm/npsphoto2.cfm

U.S. Air Force
1955–1979
https://www.airforce.com

U.S. Army Corps of Engineers
June 1978
http://www.usace.army.mil

U.S. Bureau of Reclamation
1956–1963
https://www.usbr.gov

U.S. Geological Survey
1946–1987
https://www.usgs.gov

## Oregon

Army Map Service
1952–1957
https://www.nga.mil/ABOUT/HISTORY/NGAINHISTORY/Pages/ArmyMapService.aspx

NASA AMES Research Center
1971–1995
https://www.nasa.gov/centers/ames/home/index.html

NASA Johnson Space Center
1969–1979
https://www.nasa.gov/centers/johnson/home/index.html

National Park Service
1979–1986
https://www.nps.gov/hfc/cfm/npsphoto2.cfm

U.S. Air Force
1968
https://www.airforce.com

U.S. Bureau of Land Management
1960–2011
https://www.blm.gov

U.S. Bureau of Reclamation
1955–2002
https://www.usbr.gov

U.S. Fish and Wildlife Services
1943–1979
https://www.fws.gov

U.S. Geological Survey
1939–1987
https://www.usgs.gov

## Pennsylvania

Army Map Service
1955–1958
https://www.nga.mil/ABOUT/HISTORY/NGAINHISTORY/Pages/ArmyMapService.aspx

NASA AMES Research Center
1972–1993
https://www.nasa.gov/centers/ames/home/index.html

NASA Johnson Space Center
1969–1975
https://www.nasa.gov/centers/johnson/home/index.html

National Park Service
1981–2004
https://www.nps.gov/hfc/cfm/npsphoto2.cfm

## Rhode Island

Army Map Service
1953–1960
https://www.nga.mil/ABOUT/HISTORY/NGAINHISTORY/Pages/ArmyMapService.aspx

NASA AMES Research Center
1972–1993
https://www.nasa.gov/centers/ames/home/index.html

NASA Johnson Space Center
1969–1975
https://www.nasa.gov/centers/johnson/home/index.html

National Park Service
1982–1983
https://www.nps.gov/hfc/cfm/npsphoto2.cfm

U.S. Air Force
May 1960
https://www.airforce.com

U.S. Army Corps of Engineers
1970–1974
http://www.usace.army.mil

U.S. Geological Survey
1938–1990
https://www.usgs.gov

## South Carolina

Army Map Service
1950–1956
https://www.nga.mil/ABOUT/HISTORY/NGAINHISTORY/Pages/ArmyMapService.aspx

FEMA
September 1999
https://www.fema.gov

NASA AMES Research Center
1972–1992
https://www.nasa.gov/centers/ames/home/index.html

NASA Johnson Space Center
1969–1976
https://www.nasa.gov/centers/johnson/home/index.html

National Park Service
1982–1996
https://www.nps.gov/hfc/cfm/npsphoto2.cfm

U.S. Air Force
1960–1968
https://www.airforce.com

U.S. Geological Survey
1947–1990
https://www.usgs.gov

U.S. Navy
1957–1961
http://www.navy.mil

**South Dakota**

Army Map Service
1949–1956
https://www.nga.mil/ABOUT/HISTORY/NGAINHISTORY/Pages/ArmyMapService.aspx

EROS
July 1979
https://eros.usgs.gov

McDonnell Douglas
1956–1966
http://mcdonnelldouglas.weebly.com

NASA AMES Research Center
1973–1999
https://www.nasa.gov/centers/ames/home/index.html

NASA Johnson Space Center
1969–1977
https://www.nasa.gov/centers/johnson/home/index.html

National Park Service
1997–2000
https://www.nps.gov/hfc/cfm/npsphoto2.cfm

U.S. Air Force
October 1967
https://www.airforce.com

U.S. Army Corps of Engineers
April 1969
http://www.usace.army.mil

U.S. Bureau of Land Management
1976
https://www.blm.gov

U.S. Bureau of Reclamation
1946–1976
https://www.usbr.gov

U.S. Fish and Wildlife Services
1937–2002
https://www.fws.gov

U.S. Geological Survey
1946–1980
https://www.usgs.gov

**Tennessee**

Army Map Service
1950–1958
https://www.nga.mil/ABOUT/HISTORY/NGAINHISTORY/Pages/ArmyMapService.aspx

NASA AMES Research Center
1973–1994
https://www.nasa.gov/centers/ames/home/index.html

NASA Johnson Space Center
1969–1982
https://www.nasa.gov/centers/johnson/home/index.html

National Park Service
1982
https://www.nps.gov/hfc/cfm/npsphoto2.cfm

U.S. Air Force
1955–1973
https://www.airforce.com

U.S. Army Corps of Engineers
1978
http://www.usace.army.mil

U.S. Geological Survey
1947–1995
https://www.usgs.gov

**Texas**

Army Map Service
1942–1958
https://www.nga.mil/ABOUT/HISTORY/NGAINHISTORY/Pages/ArmyMapService.aspx

McDonnell Douglas
1956–1969
http://mcdonnelldouglas.weebly.com

NASA AMES Research Center
1973–1996
https://www.nasa.gov/centers/ames/home/index.html

NASA Johnson Space Center
1969–1980
https://www.nasa.gov/centers/johnson/home/index.html

National Park Service
1982–2002
https://www.nps.gov/hfc/cfm/npsphoto2.cfm

U.S. Air Force
1955–1978
https://www.airforce.com/

U.S. Army Corps of Engineers
1978
http://www.usace.army.mil

U.S. Bureau of Land Management
October 1971
https://www.blm.gov

U.S. Bureau of Reclamation
1948–1957
https://www.usbr.gov

U.S. Fish and Wildlife Services
1946–1977
https://www.fws.gov

U.S. Geological Survey
1937–1991
https://www.usgs.gov

## Vermont

NASA AMES Research Center
1975–1995
https://www.nasa.gov/centers/ames/home/index.html

NASA Johnson Space Center
1969–1973
https://www.nasa.gov/centers/johnson/home/index.html

National Park Service
April 2003
https://www.nps.gov/hfc/cfm/npsphoto2.cfm

U.S. Air Force
1958–1960
https://www.airforce.com

U.S. Army Corps of Engineers
1977–1978
http://www.usace.army.mil

U.S. Geological Survey
1939–1984
https://www.usgs.gov

## Virginia

Army Map Service
1950–1956
https://www.nga.mil/ABOUT/HISTORY/NGAINHISTORY/Pages/ArmyMapService.aspx

FEMA
September 1999
https://www.fema.gov

NASA AMES Research Center
1971–1995
https://www.nasa.gov/centers/ames/home/index.html

NASA Johnson Space Center
1969–1977
https://www.nasa.gov/centers/johnson/home/index.html

National Park Service
1957–2002
https://www.nps.gov/hfc/cfm/npsphoto2.cfm

U.S. Air Force
1955–1962
https://www.airforce.com

U.S. Geological Survey
1941–1981
https://www.usgs.gov

## Washington

Army Map Service
1952–1958
https://www.nga.mil/ABOUT/HISTORY/NGAINHISTORY/Pages/ArmyMapService.aspx

NASA AMES Research Center
1973–1995
https://www.nasa.gov/centers/ames/home/index.html

NASA Johnson Space Center
1969–1977
https://www.nasa.gov/centers/johnson/home/index.html

National Park Service
1979–1986
https://www.nps.gov/hfc/cfm/npsphoto2.cfm

U.S. Air Force
1942–1972
https://www.airforce.com

U.S. Bureau of Land Management
1973–2003
https://www.blm.gov

U.S. Bureau of Reclamation
1956–1972
https://www.usbr.gov

U.S. Fish and Wildlife Services
1968–1981
https://www.fws.gov

U.S. Geological Survey
1966–1992
https://www.usgs.gov

**Washington, D.C.**

NASA AMES Research Center
1972–1993
https://www.nasa.gov/centers/ames/home/index.html

NASA Johnson Space Center
1969–1975
https://www.nasa.gov/centers/johnson/home/index.html

National Park Service
1957–1984
https://www.nps.gov/hfc/cfm/npsphoto2.cfm

U.S. Air Force
1959–1968
https://www.airforce.com

U.S. Geological Survey
1949–1977
https://www.usgs.gov

**West Virginia**

Army Map Service
1956
https://www.nga.mil/ABOUT/HISTORY/NGAINHISTORY/Pages/ArmyMapService.aspx

NASA AMES Research Center
1972–1993
https://www.nasa.gov/centers/ames/home/index.html

NASA Johnson Space Center
1970–1973
https://www.nasa.gov/centers/johnson/home/index.html

National Park Service
1983–2004
https://www.nps.gov/hfc/cfm/npsphoto2.cfm

U.S. Air Force
1948–1960
https://www.airforce.com

U.S. Geological Survey
1941–1979
https://www.usgs.gov

**Wisconsin**

Army Map Service
1949–1958
https://www.nga.mil/ABOUT/HISTORY/NGAINHISTORY/Pages/ArmyMapService.aspx

NASA AMES Research Center
1980–1999
https://www.nasa.gov/centers/ames/home/index.html

NASA Johnson Space Center
1969–1974
https://www.nasa.gov/centers/johnson/home/index.html

National Park Service
1983–2004
https://www.nps.gov/hfc/cfm/npsphoto2.cfm

U.S. Air Force
1951–1970
https://www.airforce.com

U.S. Army Corps of Engineers
1970–1975
http://www.usace.army.mil

U.S. Fish and Wildlife Services
1976–2002
https://www.fws.gov

U.S. Geological Survey
1937–1991
https://www.usgs.gov

**Wyoming**

Army Map Service
1943–1954
https://www.nga.mil/ABOUT/HISTORY/NGAINHISTORY/Pages/ArmyMapService.aspx

EROS
September 1979
https://eros.usgs.gov

McDonnell Douglas
1956–1966
http://mcdonnelldouglas.weebly.com

NASA AMES Research Center
1972–1999
https://www.nasa.gov/centers/ames/home/index.html

NASA Johnson Space Center
1969–1974
https://www.nasa.gov/centers/johnson/home/index.html

National Park Service
1969–1995
https://www.nps.gov/hfc/cfm/npsphoto2.cfm

U.S. Air Force
1967–1972
https://www.airforce.com

U.S. Bureau of Land Management
1964–2003
https://www.blm.gov

U.S. Bureau of Reclamation
1939–1968
https://www.usbr.gov

U.S. Geological Survey
1943–1986
https://www.usgs.gov

# 10
# Bird's-Eye View Maps

A bird's-eye view map is an elevated oblique-angled view of a geographic area from above, often drawn using an aerial photograph. These maps, also known as panoramic maps, were a very popular cartographic form used to depict U.S. and Canadian cities and towns during the late nineteenth and early twentieth centuries. They are an artist's representation of what a specific geographic area looked like at a specific point in time and are generally not drawn to scale. Most maps showcase street patterns, individual buildings, and major landscapes. They may be one of the only visual resources that provides large-scale detailed drawings of residential and industrial buildings, trees, gardens, horse-drawn carriages, even ships and trains. The details seen in these drawings enable researchers to see whether homes had garages, additions, and trees, as well as to determine the street names at the time of production. Because of its easy readability, this map style was also sometimes used to depict areas for planned development. Land developers and real estate agents used these maps to demonstrate future growth possibilities to potential buyers. Advances in lithography and photolithography made the distribution of these prints possible, so they ended up becoming an extremely popular way to capture neighborhoods in the 1800s. Researchers today are very fortunate to have online access to the history of these towns.

The Library of Congress has the most impressive collection of over fifteen hundred panoramic maps, providing easily accessible free images of the maps. Most of the maps have been drawn by artists Albert Ruger, Thaddeus Mortimer Fowler, Lucien R. Burleigh, Henry Wellge, and Oakley H. Bailey. These five artists prepared more than 55 percent of the panoramic maps in the Library of Congress.

The following list represents U.S. maps available online from the Library of Congress, as well as Canadian maps available from Library and Archives Canada and university collections across the country.

## CANADA

### Alberta

*Bird's Eye View of Lethbridge, Alta.* Toronto, ON: Valentine & Son's, 1900.

*Bird's Eye View of the City of Calgary, 1910*, by H. M. Burton. Ottawa, ON: Association of Canadian Map Libraries and Archives = Association des cartothèques et archives cartographiques canadiennes, 2004.

*Calgary, 1910*, by H. M. Burton. Ottawa, ON: Association of Canadian Map Libraries and Archives = Association des cartothèques et archives cartographiques canadiennes, 2004.

### British Columbia

*Bird's Eye View of the City of North Vancouver B.C. from the Harbor 1907*, by C. H. Rawson. Vancouver, BC: Irwin & Billings, 1907.

*Bird's Eye View of the City of Vancouver, B.C., Canada, 1890.* Vancouver, BC: Elliott, 1890.

*Bird's Eye View of the City of Vancouver, B.C., Canada, 1908.* Vancouver, BC: Vancouver Tourist Association, 1908.

*Bird's Eye View of Vancouver from Shaughnessy Heights.* Vancouver, BC: C. S. Douglas, 1907.

*Bird's Eye View of Victoria, B.C., 1889.* Victoria, BC: Ellis, 1889.

*Fort St. James National Historic Park, British Columbia.* Ottawa, ON: Parks Canada, 1981.

*Panoramic View of the City of Vancouver, British Columbia, 1898*, by J. C. McLagan. Ottawa, ON: Association of Canadian Map Libraries and Archives = Association des cartothèques et archives cartographiques canadiennes, 1998.

*University of British Columbia Bird's Eye View: Prospective Plan of the University of British Columbia.* Vancouver: University of British Columbia, 1912.

*Vancouver & Fraser Valley, British Columbia, Canada.* Vancouver, BC: Vistascene Maps, 2012.

## Manitoba

*Bird's Eye View of the City of Winnipeg.* Winnipeg, MB: W. J. C. Bulman, 1900.

*Winnipeg, Incorporated in 1873, Manitoba, 1881, Population 12,000,* by A. Mortimer. Ottawa, ON: Association of Canadian Map Libraries and Archives = Association des cartothèques et archives cartographiques canadiennes, 2000.

## New Brunswick

*Bird's Eye View of City of St. John, New Brunswick.* Ottawa, ON: Public Archives of Canada, 1976.

## Newfoundland

*Panoramic View of St. John's, Newfoundland, 1879,* by A. Ruger. Ottawa, ON: Association of Canadian Map Libraries and Archives = Association des cartothèques et archives cartographiques canadiennes, 1996.

## Nova Scotia

*Bird's Eye View of Annapolis Royal & Granville, Nova Scotia, 1878,* by T. M. Fowler. Annapolis Royal, NS: Historic Restoration Society of Annapolis County, 1984.

*Bird's Eye View of Digby, Nova Scotia, 1878,* by T. M. Fowler. Digby, NS: Digby Historical Society, 1984.

*Bird's Eye View of Kentville, Nova Scotia, 1879,* by T. M. Fowler. Kentville, NS: Kentville Publishing, 1950.

*Bird's Eye View of Lunenburg, Lunenburg Co., Nova Scotia, 1879.* Lunenburg, NS: A. Ruger, 1879.

*Bird's Eye View of Windsor, Nova Scotia, 1878,* by T. M. Fowler. Morrisville, PA: T. M. Fowler, 1878.

## Ontario

*Bird's Eye View, London, Ontario, Canada, 1872.* Ottawa, ON: Association of Canadian Map Libraries and Archives = Association des cartothèques et archives cartographiques canadiennes, 1998.

*Bird's Eye View of Belleville, Ontario, Canada, 1874,* by J. J. Stoner. Belleville, ON: Hastings County Historical Society, 1977.

*Bird's Eye View of Brantford: Province of Ontario, Canada,* by H. Brosius. Chicago, IL: C. Shober & Co. Prop's Lith, 1875.

*Bird's Eye View of Brockville Province, Ontario, Canada,* by H. Brosius. Chicago, IL: Chicago Lithographing, 1874.

*Bird's Eye View of Cobourg in 1874, Province Ontario, Canada,* by H. Brosius. Ottawa, ON: Association of Canadian Map Libraries and Archives = Association des cartothèques et archives cartographiques canadiennes, 1985.

*Bird's Eye View of Lindsay, Ontario, Canada, 1875.* Ottawa, ON: National Archives of Canada, 1976.

*Bird's Eye View of Port Hope, Ontario, Canada, 1874.* Madison, WI: H. Brosius, 1874.

*Bird's Eye View of Stratford, Ontario, 1872,* by H. Brosius. Stratford, ON: Beacon Herald, 1972.

*Bird's Eye View of the City of London, Canada.* Toronto, ON: Toronto Lithographing 1893.

*Bird's Eye View of the City of Peterborough with Views of Principal Business Buildings.* Peterborough, ON: Howell Lith., 1900.

*Bird's Eye View of the City of Port Huron, Sarnia & Gratiot, St. Clair Co., Michigan, 1867 & Point Edwards, Lambton Co., Canada West,* by A. Ruger. Chicago, IL, Chicago Lithographing, 1867.

*Bird's-Eye View of the River Niagara from Lake Erie to Lake Ontario: Shewing the Situation and Extent of Navy Island, and the Towns and Villages on the Banks of the River in Canada and the United States with the Situation of the Caroline Steam Boat off Schlosser.* London: J. Robins, 1837.

*Bird's-Eye View of Toronto,* by P. A. Gross. Toronto, ON: P. A. Gross, 1876.

*Bird's Eye View of Windsor, Ontario, 1878.* Morrisville, PA: T. M. Fowler, 1878.

*City of Hamilton, Canada, with Views of Principal Business Buildings.* Ottawa, ON: Association of Canadian Map Libraries and Archives = Association des cartothèques et archives cartographiques canadiennes, 1894.

*City of Ottawa, Canada with Views of Principal Business Buildings.* Ottawa, ON: Association of Canadian Map Libraries and Archives = Association des cartothèques et archives cartographiques canadiennes, 1895.

*Kingston, Ontario, Canada 1875,* by H. Brosius. Ottawa, ON: Charles Shober, 2001.

*Metropolitan Toronto and Region Bird's-Eye View.* Don Mills, ON: Unique Media, 1985.

*Queen's Campus: Bird's Eye View.* Kingston, ON: Queen's University, 2006.

*Town of Waterloo, Canada: With Views of Principal Business Buildings.* Ottawa, ON: Association of Canadian Map Libraries and Archives = Association des cartothèques et archives cartographiques canadiennes, 1997.

*Your Bird's Eye View of Thunder Bay, Ontario, Canada: Come to Thunder Bay Just for the Fun of It!; Family Playground.* Thunder Bay, ON: City of Thunder Bay Visitors and Convention Bureau, 1970.

## Quebec

*Bird's Eye View of Coaticook, P.Q., 1881.* Ottawa, ON: National Map Collection, Public Archives of Canada, 1978.

*Bird's Eye View of Derby-Line, VT. & Rock Island, P.Q.* Ottawa, ON: Archives nationales du Canada, 1977.

*Bird's Eye View of Lennoxville, P.Q., 1881.* Ottawa, ON: National Map Collection, Public Archives of Canada, 1978.

*Bird's Eye View of Sherbrooke, P.Q., 1881.* Ottawa, ON: Archives publiques du Canada, 1881.

*Bird's Eye View of St. Johns, P.Q., Domin of Canada, 1881.* Ottawa, ON: National Map Collection, Public Archives of Canada, 1978.

*Bird's-Eye View of the City of Montreal, 1889.* Midland Park, NJ: MicroColor International, 1993.

*Bird's Eye View of Town of Farnham, P.Q., 1881.* Ottawa, ON: National Map Collection, Public Archives of Canada, 1978,

*Bird's Eye View of Waterloo, P.Q., 1881.* Ottawa, ON: National Map Collection, Public Archives of Canada, 1978.

*Canadian Cities, Bird's Eye Views: Montréal 1889.* Ottawa, ON: Association of Canadian Map Libraries and Archives = Association des cartothèques et archives cartographiques canadiennes, 2001.

*Canadian Cities, Bird's Eye Views: Québec 1905.* Ottawa, ON: Association of Canadian Map Libraries and Archives = Association des cartothèques et archives cartographiques canadiennes, 2000.

*Montreal 1889.* Ottawa, ON: Association of Canadian Map Libraries and Archives = Association des cartothèques et archives cartographiques canadiennes, 2001.

*Province of Quebec Canada: A Bird's-Eye View Showing Natural Resources, Mines, Water-Power Plants, Settled Districts.* Montreal, QC: Northland Studios, 1924.

*Vue à vol d'oiseau de Lévis, P.Q: Bird's Eye View of Levis, P.Q.* Ottawa, ON: Archives nationales du Canada, 1977.

### Yukon

*Bird's Eye View of Dawson, Yukon Ter., 1903.* Ottawa, ON: Association of Canadian Map Libraries and Archives = Association des cartothèques et archives cartographiques canadiennes, 1997.

*Bird's Eye View of White Pass & Yukon Ry. from Skaguay Alaska, to Whitehorse, Yukon Terr.* London: White Pass & Yukon Route, 1900.

## UNITED STATES

### Arkansas

*Bird's Eye View of Hot Springs, Ark.* Milwaukee, WI: Henry Wellge, 1888.

*Bird's Eye View of the City of Little Rock, the Capitol of Arkansas 1871*, by A. Ruger. Little Rock, AR: A. Ruger, 1871.

*Perspective Map of the City of Little Rock, Ark., State Capital of Arkansas, County Seat of Pulaski County. 1887.* Milwaukee, WI: Henry Wellge, 1887.

*Perspective Map of Texarkana, Texas and Arkansas.* Milwaukee, WI: Henry Wellge, 1888.

*Perspective Map of Van Buren, Ark. County Seat of Crawford County 1888.* Milwaukee, WI: Henry Wellge, 1888.

### California

*Auburn, Cal.*, by C. P. Cook. San Francisco: W. W. Elliott, 1887.

*Berkeley*, by C. Green. Berkeley, CA: C. Green, 1909.

*Bird's Eye View of Azusa, Los Angeles Co. Cal., 1887*, by E. S. Moore and J. S. Slauson. Los Angeles: Slauson, 1887.

*Bird's Eye View of Coronado Beach, San Diego Bay and City of San Diego, Cal. in Distance*, by E. S. Moore. San Diego, CA: Coronado Beach, 1880.

*Bird's Eye View of San Diego, California 1876*, by E. S. Glover. San Diego, CA: Schneider & Kueppers, 1876.

*Bird's Eye View of San Francisco*, by G. H. Goddard. San Francisco: Britton & Rey, 1868.

*Bird's Eye View of San Francisco and Surrounding Country*, by G. H. Goddard. San Francisco: Rey & Britton, 1876.

*Bird's Eye View of Santa Rosa, Sonoma County, Cal., 1876*, by E. S. Glover. San Francisco: Wm. M. Evans, 1876.

*Bird's Eye View of the City of San José, Cal.*, by C. B. Gifford, et al. San Jose, CA: Geo. H. Hare, 1869.

*Eureka, Humboldt County, California*, by G. R. Georgeson. Eureka, CA: Britton & Rey, 1902.

*Greater Los Angeles: The Wonder City of America*, by K. M. Leuschner. Los Angeles: Metropolitan Surveys, 1932.

*Lakeport, Lake County, California*, by S. Inchbold. San Francisco: Britton & Rey, 1888.

*Los Angeles as It Appeared in 1871.* Los Angeles: Women's University Club of L.A., 1871.

*Los Angeles, Cal., Population of City and Environs 65,000*, by H. B. Elliott. Los Angeles: Southern California Land, 1891.

*Los Angeles, California, 1894*, by B. W. Pierce. Los Angeles: Semi-Tropic Homestead, 1894.

*Oakland, California, 1900*, by F. Soderberg. Oakland, CA: Mutual L. & Lith., 1900.

*Sacramento.* Sacramento, CA: W. W. Elliott, 1890.

*San Jose, California.* San Francisco: Britton & Rey, 1901.

*View of City of Stockton, the Manufacturing City of California.* San Francisco: Dakin Publishing, 1895.

*View of Los Angeles from the East. Brooklyn Heights in the Foreground; Pacific Ocean and Santa Monica Mountains in the Background*, by E. S. Glover. Los Angeles: Brooklyn Land and Building, 1877.

### Colorado

*Bird's Eye View of Canon City, Colo. County Seat of Fremont County 1882*, by H. Wellge. Milwaukee, WI: American Publishing, 1882.

*Bird's-Eye-View of Colorado Springs, Colorado.* Denver: Benford-Bryan, 1909.

*Bird's Eye View of Gunnison, Colo. County Seat of Gunnison County, 1882.* Madison, WI: J. J. Stoner, 1882.

*Bird's Eye View of Maysville, Colo. Chaffee County 1882.* Madison, WI: J. J. Stoner, 1882.

*Bird's Eye View of the City of Denver, Colorado, 1881*, by J. H. Flett. Denver: J. H. Flett, 1881.

*Black Hawk, Colo. 1882.* Madison, WI: J. J. Stoner, 1882.

*Cripple Creek Mining District, the Great Gold Camp of Colorado*, by C. H. Amerine. Cripple Creek, CO: Denver Lith., 1895.

*Panoramic Bird's Eye View of Colorado Springs, Colorado City and Manitou, Colo. 1882.* Madison, WI: J. J. Stoner, 1882.

*Perspective Map of the City of Denver, Colo. 1889*, by H. Wellge. Milwaukee, WI: American Publishing, 1889.

*Trinidad, Colo. 1882, County Seat of Las Animas County.* Madison, WI: J. J. Stoner, 1882.

## Connecticut

*Aero View of Ansonia, Connecticut 1921.* Boston: O. H. Bailey, 1921.

*Aero View of Manchester, Connecticut 1914.* Boston: O. H. Bailey, 1914.

*Aero View of Meriden, Connecticut 1918.* New York: Hughes & Bailey, 1918.

*Aero View of Willimantic, Connecticut 1909.* Boston: O. H. Bailey, 1909.

*Aero View of Windsor Locks, Connecticut, 1913.* Boston: O. H. Bailey, 1913.

*Bird's-Eye View of Branford, Connecticut.* New York: Hughes & Bailey, 1905.

*Bird's Eye View of Bristol, Conn. 1907.* New York: Hughes & Bailey, 1907.

*Bird's-Eye View of Naugatuck, Connecticut 1906.* Boston: O. H. Bailey, 1906.

*Bird's-Eye View of Thompsonville now Enfield, Connecticut.* Boston: O. H. Bailey, 1880.

*Birmingham, Conn. 1876.* Boston: O. H. Bailey, 1876.

*City of New Haven, Conn. 1879.* Boston: O. H. Bailey, 1879.

*Derby, Shelton, and East Derby, Conn., 1898.* [New York: Landis and Hughes, 1898].

*Middletown, Conn. 1877.* Boston: O. H. Bailey, 1887.

*Norwalk, South Norwalk, and East Norwalk, Conn.* [New York: Landis & Hughes, 1899].

*Plainville, Connecticut 1878.* Boston: O. H. Bailey, 1878.

*View of Bethel, Conn.* Boston: O. H. Bailey, 1879.

*View of Cheshire, Connecticut 1882.* Boston: O. H. Bailey, 1882.

*View of Clinton, Connecticut 1881.* Boston: O. H. Bailey, 1881.

*View of Danbury, Conn. 1875.* Boston: O. H. Bailey, 1875.

*View of East Haddam, Connecticut, and Goodspeeds Landing: 1880.* Boston: O. H. Bailey, 1880.

*View of Higganum, Connecticut 1881.* Boston: O. H. Bailey, 1881.

*View of Mystic River & Mystic Bridge, Conn. 1879.* Boston: O. H. Bailey, 1879.

*View of New Britain, Conn. 1875.* Boston: O. H. Bailey, 1875.

*View of Unionville, Conn.* Boston: O. H. Bailey, 1878.

*Waterbury, Conn. 1899.* New York: Landis and Hughes, 1899

*Wolcottville, Conn.* Boston: D. Bremner and O. H. Bailey, 1875.

## Florida

*Bird's Eye View of Cedar-Key, Fla., Levy Co.* Madison, WI: J. J. Stoner, 1884.

*Bird's Eye View of De Land, Fla., Volusia County, 1884*, by H. Wellge. Madison, WI: J. J. Stoner, 1884.

*Bird's Eye View of Jacksonville, Fla.* Kansas City, MO: Alvord, Kellog & Campbell, 1876.

*Bird's Eye View of Key West, Fla., Key West Island, C. S. Monroe Co., 1884*, by H. Wellge. Madison, WI: J. J. Stoner, 1884.

*Cove Springs, County Seat of Clay County, Florida. 1885.* Milwaukee, WI: Wellge, 1885.

*Jacksonville, Florida.* Kansas City, MO: Hudson-Kimberly, 1893.

*Pensacola, Fla. County Seat of Escambia County 1885.* Milwaukee, WI: Wellge, 1885.

*View of the City of Tallahassee. State Capital of Florida, County Seat of Leon County 1885.* Milwaukee, WI: Wellge, 1885.

## Georgia

*Bird's Eye View Cordele, Georgia 1908.* Morrisville, PA: T. M. Fowler, 1908.

*Bird's Eye View of Atlanta, Fulton Co., State Capital, Georgia*, by A. Koch. Atlanta: Hughes Litho, 1892.

*Bird's Eye View of Fitzgerald, Georgia 1908.* Morrisville, PA: T. M. Fowler, 1908.

*Bird's Eye View of Ocilla, Georgia 1908.* Morrisville, PA: T. M. Fowler, 1908.

*Bird's Eye View of the City of Atlanta, the Capitol of Georgia 1871*, by A. Ruger. St. Louis, MO: A. Ruger, 1871.

*Macon, Ga. County Seat of Bibb County 1887.* Milwaukee, WI: Wellge, 1887.

*Perspective Map of Columbus, Ga., County Seat of Muscogee County, 1886.* Milwaukee, WI: Wellge, 1886.

*Panoramic View of Quitman, Ga. County-Seat of Brooks-County 1885.* Milwaukee, WI: Wellge, 1885.

*Panoramic View of Valdosta, Ga.: County Seat of Lowndes County.* Milwaukee, WI: Wellge, 1885.

*View of the City of Albany, Ga. (the Artesian City) County-Seat of Dougherty-County. 1885.* Milwaukee, WI: Wellge, 1885.

## Illinois

*Abingdon, Ill., Knox Co., 1874*, by D. D. Morse. Chicago: Chas. Shober, 1874.

*Alton, Madison Co., Illinois 1867*, by A. Ruger. Chicago: Chicago Lithographing, 1867.

*Aurora, Illinois 1867*, by A. Ruger. Chicago: Chicago Lithographing, 1867.

*Aurora, Illinois 1882*, by H. Brosius. Madison, WI: J. J. Stoner, 1882.

*Belleville, St. Clair Co., Illinois 1867*, by A. Ruger. Chicago: Chicago Lithographing, 1867.

*Bird's Eye View of Batavia, Kane County, Illinois, 1869*, by A. Ruger. Chicago: Chicago Lithographing, 1869.

*Bird's Eye View of Chenoa, McLean County, Illinois 1869*, by A. Ruger. Chicago: Chicago Lithographing, 1867.

*Bird's Eye View of Chicago, 1893*, by P. Roy. Chicago: P. Roy, 1893.

*Bird's-Eye-View of Chicago as It Was before the Great Fire, by T. R. Davis.* New York: T. R. Davis, 1871.

*Bird's Eye View of Loda, Iroquois Co., Illinois 1869*, by A. Ruger. Chicago: Chicago Lithographing, 1869.

*Bird's Eye View of Mattoon, Illinois 1884*, by J. W. Smith. Chicago: Shober & Carqueville, 1884.

*Bird's-Eye-View of the Business District of Chicago.* Chicago: Poole Brothers, 1898.

*Bird's Eye View of the City of Danville, Vermillion County, Illinois 1869*, by A. Ruger. Chicago: Chicago Lithographing, 1869.

*Bird's Eye View of the City of Kankakee, Kankakee County, Illinois 1869*, by A. Ruger. Chicago: Chicago Lithographing, 1869.

*Bird's Eye View of the City of Lincoln, Logan County, Illinois 1869*, by A. Ruger. Chicago: Chicago Lithographing, 1869.

*Bird's Eye View of the City of Moline, Rock Island County, Illinois 1869*, by A. Ruger. Chicago: Chicago Lithographing, 1869.

*Bird's Eye View of the City of Monmouth, Warren County, Illinois 1869*, by A. Ruger. Chicago: Chicago Lithographing, 1869.

*Bird's Eye View of the City of Princeton, Bureau County, Illinois 1870*, by A. Ruger. Chicago: Chicago Lithographing, 1870.

*Bird's Eye View of the City of Shelbyville, Shelby County, Illinois 1869*, by A. Ruger. Chicago: Chicago Lithographing, 1869.

*Bird's Eye View of the City of Urbana, Champaign County, Illinois 1869*, by A. Ruger. Chicago: Chicago Lithographing, 1869.

*Bloomington, Illinois 1867*, by A. Ruger. Chicago: Chicago Lithographing, 1867.

*Centralia, Marion Co., Illinois 1867*, by A. Ruger. Chicago: Chicago Lithographing, 1867.

*Chicago, Central Business Section.* Chicago: A. B. Reincke, 1916.

*City of Chicago, Showing the Burnt District.* New York: Currier & Ives, 1874.

*Elgin, Kane Co., Illinois 1880*, by A. B. Upham. Chicago: Shober & Carqueville, 1880.

*Mount Vernon, Jefferson County, Illinois 1881*, by H. Brosius. Madison, WI: J. J. Stoner, 1881.

*New Salem, Home of Abraham Lincoln 1831 to 1837*, by A. Brown. Mason City, IL: R. J. Onstott and Franks & Sons, 1837.

*Perspective Map of the City of Cairo, Ill. 1888.* Milwaukee, WI: Henry Wellge, 1888.

*Rock Island, Ill.*, by H. Wellge. Milwaukee, WI: American Publishing, 1889.

## Indiana

*Bend, Indiana 1866*, by A. Ruger. Chicago: Chicago Lithographing, 1866.

*Bird's Eye View of Michigan City, LaPorte County, Ind. 1869*, by A. Ruger. Chicago: Chicago Lithographing, 1869.

*Bird's Eye View of the City of Attica, Fountain County, Indiana 1869*, by A. Ruger. Chicago: Chicago Lithographing, 1869.

*Bird's Eye View of the City of Fort Wayne, Indiana 1868*, by A. Ruger. Chicago: Chicago Lithographing, 1868.

*Bird's Eye View of the City of Kokomo, Howard Co., Indiana 1868*, by A. Ruger. Chicago: Chicago Lithographing, 1868.

*Bird's Eye View of the City of Lafayette, Tippecanoe Co., Indiana 1868*, by A. Ruger. Chicago: Chicago Lithographing, 1868.

*Bird's Eye View of the City of Peru, Miami Co., Indiana 1868*, by A. Ruger. Chicago: Chicago Lithographing, 1869.

*City of Richmond, Indiana 1884*, by A. E. Downs. Boston: O. H. Bailey, 1884.

*Panoramic View of Evansville, Ind., 1880.* Milwaukee, WI: Beck & Pauli, 1880.

*Panoramic View of Terre Haute, Ind. 1880.* Milwaukee, WI: Beck & Pauli, 1880.

## Iowa

*1889 Perspective Map of the City of Burlington, Ia*, by H. Wellge. Milwaukee, WI: American Publishing, 1889.

*Bird's Eye View of Blairstown, Benton Co., Iowa 1868*, by A. Ruger. Chicago: Chicago Merchants Lithographing, 1868.

*Bird's Eye View of Pella, Marion County, Iowa, 1869*, by A. Koch. Chicago: Chicago Lithographing, 1869.

*Bird's Eye View of the City of Cedar Rapids and Kingston, Linn Co., Iowa 1868*, by A. Ruger. Chicago: Chicago Merchants Lithographing, 1868.

*Bird's Eye View of the City of Council Bluffs, Pottawattamie Co., Iowa 1868* / A. Ruger. Chicago: Chicago Merchants Lithographing, 1868.

*Bird's Eye View of the City of De Witt, Clinton Co., Iowa 1868*, by A. Ruger. Chicago: Chicago Merchants Lithographing, 1868.

*Bird's Eye View of the City of Marengo, Iowa Co., Iowa 1868*, by A. Ruger. Chicago: Chicago Merchants Lithographing, 1868.

*Bird's Eye View of the City of McGregor and North McGregor, Clayton County, Iowa 1869*, by A. Ruger. Chicago: Chicago Merchants Lithographing, 1869.

*Bird's Eye View of the City of Montana, Boone Co., Iowa 1868*, by A. Ruger. Chicago: Chicago Merchants Lithographing, 1868.

*Davenport, Ia. 1888*, by H. Wellge. Milwaukee, WI: American Publishing, 1888.

*Decorah, Winneshiek County, Iowa 1870*, by A. Ruger. Chicago: Chicago Merchants Lithographing, 1870.

*Perspective Map of Fort Madison, Ia. 1889*, by H. Wellge. Milwaukee, WI: American Publishing, 1889.

*Perspective Map of Sioux City, Iowa. 1888*. Milwaukee, WI: H. Wellge, 1888.

*Perspective Map of the City of Dubuque, Ia. 1889*, by H. Wellge. Milwaukee, WI: American Publishing, 1889.

## Kansas

*Bird's Eye View of Great Bend, Kans.: C.S. of Barton County, 1882*. Madison, WI: J. J. Stoner, 1882.

*Bird's-Eye View of Ottawa, the Largest City of Its Age in Kansas, 1872: Looking South-west*, by E. S. Glover. Cincinnati, OH: Strobridge & Co. Lith., 1872.

*Bird's Eye View of the City of Atchison, Atchison Co., Kansas 1869*, by A. Ruger. Chicago: Merchants Lithographing, 1869.

*Bird's Eye View of the City of Baxter Springs, Kansas, 1871*, by E. S. Glover. Chicago: Union Lith., 1871.

*Bird's Eye View of the City of Lawrence, Kansas 1869*, by A. Ruger. Chicago: Merchants Lithographing, 1869.

*Bird's Eye View of the City of Leavenworth, Kansas 1869*, by A. Ruger. Chicago: Merchants Lithographing, 1869.

*Bird's Eye View of the City of Topeka, the Capital of Kansas 1869*, by A. Ruger. Chicago: Merchants Lithographing, 1869.

*Bird's Eye View of Wyandotte, Wyandotte Co., Kansas 1869*, by A. Ruger. Chicago: Merchants Lithographing, 1869.

*Hiawatha, Kansas, 1879*. Morrisville, PA: T. M. Fowler, 1879.

## Kentucky

*Bird's-Eye View of Louisville from the River Front and Southern Exposition, 1883*, by W. F. Clarke. Cincinnati, OH: M. P. Levyeau, 1883.

*Bird's Eye View of Louisville, Kentucky 1876*, by A. Ruger. Chicago: Charles Shober, 1876.

*Bird's Eye View of Paducah, Kentucky 1889*, by J. Postlethwaite. Cincinnati, OH: Krebs Lith., 1889.

*Bird's Eye View of the City of Bowling Green, Warren County, Kentucky 1871*, by A. Ruger. Chicago: Chicago Lithographing, 1871.

*Bird's Eye View of the City of Frankfort, the Capital of Kentucky 1871*, by A. Ruger. Cincinnati, OH: Ehrgott & Krebs Lith., 1871.

*Bird's Eye View of the City of Lexington, Fayette County, Kentucky 1871*, by A. Ruger. Cincinnati, OH: Ehrgott & Krebs Lith., 1871.

*Bird's Eye View of the City of Paris, Bourbon County, Kentucky 1870*, by A. Ruger. Chicago: Chicago Lithographing, 1870.

## Louisiana

*Bird's Eye View of the City of Louisiana, Pike County, Mo.*, by A. Koch. Chicago: Chas. Shober, 1876.

*New Orleans, La. and Its Vicinity*, by J. Wells. New Orleans: Virtue, 1863.

## Maine

*Bar Harbor, Mt. Desert Island, Maine*, by G. W. Morris. Portland, ME: G. W. Morris, 1886.

*Bird's Eye View of Eastport, Washington Co., Maine*. Madison, WI: J. J. Stoner, 1879.

*Bird's Eye View of Great Falls, Strafford Co., New Hampshire & Berwick, York Co., Maine 1877*, by A. Ruger. Madison, WI: J. J. Stoner, 1878.

*Bird's Eye View of the City of Augusta, Maine, 1878*, by A. Ruger. Madison, WI: J. J. Stoner, 1878.

*Bird's Eye View of the City of Bangor, Penobscot County, Maine, 1875*, by A. Koch. Madison, WI: J. J. Stoner, 1875.

*Bird's Eye View of the City of Hallowell, Kennebec Co., Maine 1878*, by A. Ruger. Madison, WI: J. J. Stoner, 1878.

*Bird's Eye View of the City of Portland, Maine 1876*, by J. Warner. Madison, WI: J. J. Stoner, 1876.

*Bridgton, Maine, U.S.A.: 1888*. Troy, NY: Burleigh Lith., 1888.

*Cherryfield, Me., Washington Co., 1896*. Brockton, MA: Geo. E. Norris, 1896.

*Kennebunk, Me.* Brockton, MA: Geo. E. Norris, 1895.

*Kingfield, Me., Franklin Co., 1895*. Brockton, MA: Geo. E. Norris, 1895.

*Livermore Falls, Maine.* Brockton, MA: Geo. E. Norris, 1889.

*North Berwick, York Co., Maine 1877*, by A. Ruger. Madison, WI: J. J. Stoner, 1877.

*Peak's Island, Portland Harbor, Maine*, by G. W. Morris. Portland, ME: G. W. Morris, 1886.

*Pittsfield, Maine.* Brockton, MA: Geo. E. Norris, 1889.

*Sanford, Maine* Brockton, MA: Geo. E. Norris, 1889.

*Skowhegan, ME.* Brockton, MA: Geo. E. Norris, 1892.

*South Berwick, York Co., Maine, 1877*, by J. B. Richards. Madison, WI: Ruger & Stoner, 1877.

## Maryland

*Bird's Eye View of Chestertown, Kent Co., Maryland 1907*, by T. M. Fowler. Morrisville, PA: Fowler & Kelly, 1907.

*Bird's Eye View of Mountain Lake Park, Garrett Co., Maryland 1906*, by T. M. Fowler. Morrisville, PA: Fowler & Kelly, 1906.

*Bird's Eye View of Oakland, Maryland 1906*, by T. M. Fowler. Morrisville, PA: Fowler & Kelly, 1906.

*Bird's Eye View of Rising Sun, Cecil County, Maryland 1907*, by T. M. Fowler. Morrisville, PA: Fowler & Kelly, 1907.

*Bird's-Eye View of the Heart of Baltimore*, by E. W. Spofford. Baltimore, MD: Norman T. A. Munder, 1912.

*E. Sachse, & Co.'s Bird's Eye View of the City of Baltimore, 1869.* Baltimore, MD: E. Sachse, 1870.

*Lonaconing, Maryland*, by T. M. Fowler. Morrisville, PA: Fowler & Kelly, 1905.

*Midland, Maryland*, by T. M. Fowler. Morrisville, PA: T. M. Fowler, 1905.

*View of Elkton, Maryland 1907*, by T. M. Fowler. Morrisville, PA: Fowler & Kelly, 1907.

*View of Frostburg, Maryland 1905*, by T. M. Fowler. Morrisville, PA: T. M. Fowler, 1905.

## Massachusetts

*Aero-View of Amesbury, Massachusetts 1914.* New York: Hughes & Bailey, 1914.

*Amesbury, Mass.*, by G. E. Norris. Brockton, MA: G. E. Norris, 1890.

*Arlington, Massachusetts, 1884.* Boston: O. H. Bailey, 1884.

*Athol, Mass. 1887.* Troy, NY: L. R. Burleigh, 1887.

*Ayer, Mass.* Troy, NY: L. R. Burleigh, 1886.

*Baldwinville, Mass.* Troy, NY: L. R. Burleigh, 1886.

*Barre, Massachusetts.* Boston: O. H. Bailey, 1891.

*Bird's-Eye View of Boston, United States*, by T. Sulman. Boston: T. Sulman, 1870.

*Bird's Eye View of Boston*, by G. H. Walker. Boston: Geo. H. Walker, 1902.

*Bird's Eye View of Boston Harbor along the South Shore to Plymouth, Cape Cod Canal, and Provincetown: In Colors: Showing All Steamboat Routes.* Boston: Union News, 1920.

*Bird's Eye View of Boston Harbor and South Shore to Provincetown: Showing Steamboat Routes.* Boston: John F. Murphy, 1905.

*Bird's Eye View of the City of Brockton, Plymouth County, Mass. 1882*, by A. F. Poole. Madison, WI: J. J. Stoner, 1882.

*Bird's-Eye View of Village of Barre, Mass.*, by W. R. Spooner. Barre, MA: W. R. Spooner, 1890.

*Boston 1899*, by A. E. Downs. Boston: Geo H. Walker & Co., 1899.

*Boston Bird's-Eye View from the North*, by J. Bachmann. Boston: L. Prang, 1877.

*Boston Highlands, Massachusetts. Wards 19, 20, 21 & 22 of Boston.* Boston: O. H. Bailey, 1888.

*Chester, Mass., 1885.* Troy, NY: L. R. Burleigh, 1885.

*City of Boston.* New York: Currier & Ives, 1873.

*City of Boston 1879.* Boston: O. H. Bailey, 1879.

*City of Taunton, Mass. 1875*, by O. H. Bailey. Milwaukee, WI: American Oleograph, 1875.

*Dalton, Mass.* Troy, NY: L. R. Burleigh, 1884.

*East Douglas, Mass. 1886.* Troy, NY: L. R. Burleigh, 1886.

*Edgartown, Duke's County, Martha's Vineyard Id., Mass.*, by G. H. Walker. Boston: Geo. H. Walker, 1886.

*Fitchburg, Mass. 1882.* Troy, NY: L. R. Burleigh, 1882.

*Haverhill, Massachusetts.* Boston: O. H. Bailey, 1893.

*Hebronville, Massachusetts.* Boston: O. H. Bailey, 1891.

*Hopkinton, Mass.*, by J. C. Hazen. Boston: O. H. Bailey, 1893.

*Ipswich, Mass.: Incorporated 1634*, by G. E. Norris. Brockton, MA: G. E. Norris, 1893.

*North Abington, Plymouth County, Mass., Looking West*, by A. F. Pooie. Abington, MA: A. F. Pooie, 1974.

*North Attleborough, Mass. 1878*, by J. C. Hazen. Boston: O. H. Bailey, 1878.

*Pittsfield, Mass.*, by J. E. Rapp, and C. Fausel. South Schodack, NY: A. M. Van de Carr, 1899.

*Spencer, Massachusetts, 1892.* Boston: O. H. Bailey, 1892.

*Twentieth Century Boston*, by A. F. Poole. Boston: A. W. Elson, 1905.

*View of Ashland, Mass. 1878*, by J. C. Hazen. Boston: O. H. Bailey, 1878.

*View of Beverly, Massachusetts, 1886.* Boston: Greenough & Co., 1886.

*View of Boston, July 4th, 1870*, by F. Fuchs and J. Weik. Boston: New England Lith., 1870.

*View of East Boston, Mass. 1879.* Boston: O. H. Bailey, 1879.

*View of Palmer, Mass.* Boston: O. H. Bailey, 1879.

*View of the City of New Bedford, Mass., 1876*, by O. H. Bailey. New Bedford, MA: Leonard B. Ellis, 1876.

*View of Watertown, Mass., 1879*. Boston: O. H. Bailey, 1879.

*Webster, Massachusetts 1892*. Boston: O. H. Bailey, 1892.

*Woburn, Mass., 1883*. Syracuse, NY: D. Mason, 1883.

## Michigan

*Adrian, Michigan 1866*, by A. Ruger. Chicago: Chicago Lithographing, 1866.

*Benton Harbor, Mich. 1889*. Milwaukee, WI: C. J. Pauli, 1889.

*Bird's Eye View of Bay City, Portsmouth, Wenona & Salzburg, Bay Co., Michigan 1867*, by A. Ruger. Chicago: Chicago Lithographing, 1867.

*Bird's Eye View of Bessemer, Mich., Ontonagon County 1886*, by H. Wellge. Milwaukee, WI: Wellge, 1886.

*Bird's Eye View of Saint Clair, St. Clair Co., Michigan 1868*, by A. Ruger. Chicago: Chicago Lithographing, 1868.

*Bird's Eye View of the City of Albion, Calhoun Co., Michigan*, by A. Ruger. Chicago: Chicago Lithographing, 1880.

*Bird's Eye View of the City of Battle Creek, Calhoun Co., Mich.*, by A. Ruger. Battle Creek, MI: A. Ruger, 1870.

*Bird's Eye View of the City of Hillsdale, Hillsdale Co., Mich. 1866*, by A. Ruger. Chicago: Chicago Lithographing, 1866.

*Bird's Eye View of the City of Jackson, Michigan*, by A. Ruger. Chicago: Chicago Lithographing, 1868.

*Bird's Eye View of the City of Lansing, Michigan 1866*, by A. Ruger. Chicago: Chicago Lithographing, 1866.

*Bird's Eye View of the City of Muskegon, Muskegon Co., Michigan 1868*, by A. Ruger. Chicago: Chicago Lithographing, 1868.

*Bird's Eye View of the City of Port Huron, Sarnia & Gratiot, St. Clair Co., Michigan 1867 & Point Edwards, Lambton Co., Canada West*, by A. Ruger. Chicago: Chicago Lithographing, 1867.

*Bird's Eye View—Showing about Three Miles Square—of the Central Portion of the City of Detroit, Michigan*. Detroit, MI: Calvert Lithographing, 1889.

*Calumet, Hecla & Red Jacket, Mich.: 1881*, by H. Wellge. Madison, WI: J. J. Stoner, 1881.

*East Saginaw, Michigan, 1867*, by A. Ruger. Chicago: Chicago Lithographing, 1867.

*Grand Haven, Ottawa County, Michigan 1874*, by A. Ruger. Madison, WI: J. J. Stoner, 1874.

*Grand Rapids, Michigan 1868*, by A. Ruger. Chicago: Chicago Lithographing, 1868.

*Hudson, Lenawee Co., Michigan 1868*, by A. Ruger. Chicago: Chicago Lithographing, 1868.

*Kalamazoo, Michigan 1874*, by A. Ruger. Chicago: Chicago Lithographing, 1874.

*Mills & Manufacturing Establishments of the City of Battle Creek, Calhoun Co., Mich.*, by A. Ruger. Battle Creek, MI: A. Ruger, 1869.

*Negaunee, Mich. 1871*, by H. H. Bailey. Milwaukee, WI: Milwaukee Lith. & Engr., 1871.

*Niles, Berrien County, Michigan*, by A. Ruger. Chicago: Chicago Lithographing, 1868.

*Panoramic View of Mt. Clemens 1881, Macomb Co., Mich.*, by A. Ruger. Madison, WI: J. J. Stoner, 1881.

*Panoramic View of the City of Ann Arbor, Washtenaw Co., Michigan 1880*, by A. Ruger. Madison, WI: J. J. Stoner, 1880.

*Pontiac, Oakland Co., Michigan 1867*, by A. Ruger. Chicago: Chicago Lithographing, 1867.

*Saginaw City Michigan 1867*, by A. Ruger. Chicago: Chicago Lithographing, 1867.

*Saint Johns, Clinton Co., Michigan 1868*, by A. Ruger. Chicago: Chicago Lithographing, 1868.

*Ypsilanti, Washtenaw Co., Michigan*, by A. Ruger. Chicago: Chicago Lithographing, 1868.

## Minnesota

*Bird's Eye View of Anoka, Anoka County, Minnesota 1869*, by A. Ruger. Chicago: Merchant's Lithographing, 1869.

*Bird's Eye View of Austin, Mower County, Minnesota 1870*, by A. Ruger. Chicago: Merchant's Lithographing, 1870.

*Bird's Eye View of Luverne. County Seat of Rock County, Minnesota*, by H. Brosius. Madison, WI: J. J. Stoner; Rock Co. Herald, 1883.

*Bird's Eye View of Northfield, Rice County, Minnesota 1869*, by A. Ruger. Chicago: Merchant's Lithographing, 1869.

*Bird's Eye View of Owatonna, Steele County, Minnesota 1870*, by A. Ruger. Chicago: Merchant's Lithographing, 1870.

*Bird's Eye View of the City of Faribault, Rice County, Minnesota 1869*, by A. Ruger. Chicago: Merchant's Lithographing, 1869.

*Bird's Eye View of the City of Mankato, Blue Earth County, Minnesota 1870*, by A. Ruger. Chicago: Merchant's Lithographing, 1870.

*Bird's Eye View of the City of Minneapolis, Minn.*, by F. Pezolt. Minneapolis: A. M. Smith, 1891.

*Bird's Eye View of the City of Redwing, Coodhue Co., Minnesota 1868*, by A. Ruger. Chicago: Robert Teufel, 1868.

*Bird's Eye View of the City of Rochester, Olmsted County, Minnesota 1869*, by A. Ruger. Chicago: Merchant's Lithographing, 1869.

*Bird's Eye View of the City of Stillwater, Washington County, Minnesota 1870*, by A. Ruger. Chicago: Merchant's Lithographing, 1870.

*Brainerd, "City of Mines,"* by A. G. McCoy. Duluth, MN: Brainerd Townsite Company, 1914.

*Lake City, Wabasha Co., Minnesota 1867*, by A. Ruger. Chicago: Merchant's Lithographing, 1867.

*Minneapolis and Saint Anthony, Minnesota 1867*, by A. Ruger. Chicago: Merchant's Lithographing, 1867.

*Panoramic View of the City of Minneapolis, Minnesota, 1879*, by A. Ruger. Madison, WI: J. J. Stoner, 1879.

*Perspective Map of Duluth, Minn. 1887*, by H. Wellge. Duluth, MN: Duluth News, 1887.

*Saint Paul, Minnesota 1867*, by A. Ruger. Chicago: Merchant's Lithographing, 1867.

*Winona, Minn.*, by G. H. Ellsbury. Chicago: Chicago Lithographing, 1874.

## Missouri

*Aurora, Missouri 1891*, by T. M. Fowler and J. B. Moyer. Morrisville, PA: T. M. Fowler & James B. Moyer, 1891.

*Bird's Eye View of Brookfield, Linn Co., Missouri 1869*, by A. Ruger. Chicago: Merchant's Lithographing, 1869.

*Bird's Eye View of California, Moniteau Co., Missouri 1869*, by A. Ruger. Chicago: Merchant's Lithographing, 1869.

*Bird's Eye View of Jefferson City, the Capitol of Missouri 1869*, by A. Ruger. Chicago: Merchant's Lithographing, 1869.

*Bird's Eye View of Kansas City, Missouri. Jan'y. 1869*, by A. Ruger. Chicago: Merchant's Lithographing, 1869.

*Bird's Eye View of the City of Chillicothe, Livingston Co., Missouri 1869*, by A. Ruger. Chicago: Merchant's Lithographing, 1869.

*Bird's Eye View of the City of Columbia, Boone Co., Missouri 1869*, by A. Ruger. Chicago: Merchant's Lithographing, 1869.

*Bird's Eye View of the City of Hannibal, Marion Co., Missouri 1869*, by A. Ruger. Chicago: Merchant's Lithographing, 1869.

*Bird's Eye View of the City of Holden, Johnson Co., Missouri 1869*, by A. Ruger. Chicago: Merchant's Lithographing, 1869.

*Bird's Eye View of the City of Independence, Jackson Co., Missouri 1868*, by A. Ruger. Chicago: Merchant's Lithographing, 1868.

*Bird's Eye View of the City of Lexington, Lafayette Co., Missouri 1869*, by A. Ruger. Chicago: Merchant's Lithographing, 1869.

*Bird's Eye View of the City of Louisiana, Pike County, Mo.*, by A. Koch. Chicago: Chicago Lithographing, 1876.

*Bird's Eye View of the City of Mexico., Audrian Co., Missouri 1869*, by A. Ruger. Chicago: Merchant's Lithographing, 1869.

*Bird's Eye View of the City of Palmyra, Marion Co., Missouri, A.D. 1869*, by A. Ruger. Chicago: Merchant's Lithographing, 1869.

*Bird's Eye View of the City of Saint Charles, St. Charles Co., Missouri 1869*, by A. Ruger. Chicago: Merchant's Lithographing, 1869.

*Bird's Eye View of the City of Saint Joseph, Missouri 1868*, by A. Ruger. Chicago: Merchant's Lithographing, 1868.

*Bird's Eye View of the City of Sedalia, Pettis Co., Missouri 1869*, by A. Ruger. Chicago: Merchant's Lithographing, 1869.

*Pacific, Formerly Franklin, Franklin Co., Missouri 1869*, by A. Ruger. Chicago: Merchant's Lithographing, 1869.

*Panoramic View of the West Bottoms, Kansas City, Missouri & Kansas Showing Stock Yards, Packing & Wholesale Houses*, by A. Koch. Kansas City, MO: A. Koch, 1895.

## Montana

*Billings, Montana, County Seat of Yellowstone County 1904*. Milwaukee, WI: H. Wellge, 1904.

*Bird's Eye View of Butte-City, Montana, County Seat of Silver Bow Co., 1884*. Milwaukee, WI: H. Wellge, 1884.

*Bird's-Eye View of Helena, Montana 1875*, by E. S. Glover. Helena, MT: C. K. Wells, 1875.

*Bird's Eye View of Livingston, Mon., Gallatin County 1883*. Madison, WI: J. J. Stoner, 1883.

*Bird's Eye View of Miles City, C.S. of Custer County, Montana 1883*. Madison, WI: J. J. Stoner, 1883.

*Bird's Eye View of Missoula, Mon. County Seat of Missoula County 1884*. Milwaukee, WI: H. Wellge, 1884.

*Panoramic View of Butte, Montana, 1904: Population 60,000, Altitude Near Gov. Bldg. 6,000 Feet, Values Taken out of Butte Mines over 600 Million Dollars, the Largest Mining Camp on Earth, a Model City with All Modern Institutions and Conveniences.* Milwaukee, WI: H. Wellge, 1904.

*Perspective Map of Great Falls, Mont. 1891*. Milwaukee, WI: American Publishing, 1891.

## Nebraska

*1889 Perspective Map of Norfolk, Neb*, by H. Wellge. Milwaukee, WI: American Publishing, 1889.

*Lincoln, Neb., State Capitol of Nebraska, County Seat of Lancaster Co. 1889*. Milwaukee, WI: American Publishing, 1889.

*Perspective Map of the City of Kearney, Neb. County Seat of Buffalo Co. 1889*, by H. Wellge. Milwaukee, WI: American Publishing, 1889.

## New Hampshire

*Alton and Alton Bay, N.H. 1888*, by G. E. Norris. Brockton, MA: Burleigh Litho, 1888.

*Antrim, N.H. and Clinton Village*, by G. E. Norris. Brockton, MA: Burleigh Litho, 1887.

*Bird's Eye View, Newport, New Hampshire, 1877*, by A. Ruger. Chicago: Shober & Carqueville Litho., 1877.

*Bird's Eye View of Bethlehem, Grafton County, N.H. 1883*, by A. F. Poole. Brockton, MA: Poole & Norris, 1883.

*Bird's Eye View of Claremont, Sullivan County, N.H. 1877*, by A. Ruger. Chicago: Shober & Carqueville Litho., 1877.

*Bird's Eye View of Concord, N.H.: 1875*. Boston, MA: H. H. Bailey, 1875.

*Bird's Eye View of Dover, Strafford Co., New Hampshire 1877*, by A. Ruger. Milwaukee, WI: D. Bremner & Co. Lith., 1877.

*Bird's Eye View of Laconia, Belknap County, N.H. 1883*, by G. E. Norris. Brockton, MA: Burleigh Litho, 1883.

*Bird's Eye View of Littleton, Grafton County, N.H. 1883*, by G. E. Norris. Brockton, MA: Burleigh Litho, 1883.

*Bird's Eye View of Portsmouth, Rockingham Co., New Hampshire*, by A. Ruger. Madison, WI: J. J. Stoner, 1877.

*Bird's Eye View of Rochester, Strafford County, New Hampshire, 1877: From a Position, East of Town*. Madison, WI: J. J. Stoner, 1877.

*Bird's Eye View of the Village of Lancaster, Coos County, N.H. 1883*, by G. E. Norris. Brockton, MA: Burleigh Litho, 1883.

*Bird's Eye View of Whitefield, Coos County, N.H., 1883*, by G. E. Norris. Brockton, MA: Burleigh Litho, 1883.

*Exeter, N.H., County Seat of Rockingham County, 1884*. Brockton, MA: Norris & Wellge, 1884.

*Exeter, New Hampshire, 1896*. Boston, MA: Moore, 1896.

*Franklin and Franklin Falls, N.H., Merrimack County, 1884*. Brockton, MA: Norris & Wellge, 1884.

*Gorham, N.H. 1888*, by G. E. Norris. Brockton, MA: Burleigh Litho, 1888.

*Greenville, N.H. 1886*. Troy, NY: L. R. Burleigh, 1886.

*Henniker, N.H*, by G. E. Norris. Brockton, MA: Burleigh Litho, 1889.

*Hillsborough-Bridge, Hillsborough County, N.H. 1884*. Brockton, MA: Norris & Wellge, 1884.

*Lebanon, Grafton County, N.H., 1884*, by G. E. Norris. Brockton, MA: Burleigh Litho, 1884.

*Meredith Village, N.H.*, by G. E. Norris. Brockton, MA: Burleigh Litho, 1889.

*Milton, N.H., 1888*, by G. E. Norris. Brockton, MA: Burleigh Litho, 1888.

*Penacook, N.H*. Troy, NY: L. R. Burleigh, 1887.

*Peterborough, N.H. 1886*. Troy, NY: L. R. Burleigh, 1886.

*Pittsfield, Merrimack County, N.H. 1884*, by G. E. Norris. Brockton, MA: Burleigh Litho, 1884.

*Rochester, N.H., Gonic and East-Rochester, 1884*, by G. E. Norris. Brockton, MA: Burleigh Litho, 1884.

*South-New-Market, Rockingham County, N.H. 1884*, by G. E. Norris. Brockton, MA: Burleigh Litho, 1884.

*Wolfeborough, N.H., Lake Winnipesaukee*, by G. E. Norris. Brockton, MA: Burleigh Litho, 1889.

## New Jersey

*Aeroplane View of Asbury Park, N.J. Showing Location of "Asbury Park Estates" among the Hills on Asbury Ave.*, by H. M. Petit. New York: Barton & Spooner, 1910.

*Aero-View of Absecon, New Jersey 1924*, by R. Cinquin. New York: Hughes & Cinquin, 1924.

*Aero-View of Margate City, New Jersey 1925*, by R. Cinquin. New York: Hughes & Cinquin, 1925.

*Aero-View of New-Brunswick, New Jersey, 1910*. Boston: Fowler & Bailey, 1910.

*Aero-View of Somers-Point 1925, New Jersey*, by R. Cinquin. New York: Hughes & Cinquin, 1925.

*Aero-View of Westfield, N.J. 1929*, by R. Cinquin. New York: Hughes & Cinquin, 1929.

*Asbury Park, Ocean Grove and Vicinity, New Jersey 1897*. New York: Landis & Hughes, 1897.

*Atlantic City, N.J. 1880*, by T. J. Landis. Philadelphia: T. J. Shepherd Landis, 1880.

*Atlantic City, New Jersey*. Newark, NJ: Landis & Alsop, 1900.

*Bird's Eye View of Asbury Park, New Jersey, 1881*. Asbury Park, NJ: T. M. Fowler, 1881.

*Bird's Eye View of Egg Harbor City, N.J.*, by F. Scheu. Egg Harbor City, NJ: Herline & Hensel, 1865.

*Bird's Eye View of Garfield, New Jersey 1909*. Morrisville, PA: T. M. Fowler, 1909.

*Bird's-Eye-View of Maplewood, N.J.*, by H. S. Wyllie. Newark, NJ: H. S. Wyllie, 1910.

*Bird's Eye View of Morristown, Morris Co., New Jersey*. Milwaukee, WI: Fowler & Bulger, 1876.

*City of Bridgeton, New Jersey, 1886*. Boston: O. H. Bailey, 1886.

*City of Hoboken, New Jersey, 1881*, by A. Ward. Boston: O. H. Bailey, 1881.

*City of Woodbury, New Jersey, 1886*. Boston: O. H. Bailey, 1886.

*Dover, New Jersey 1903*. Boston: Fowler & Bailey, 1903.

*Elizabeth, N.J. 1898*. New York: Landis & Hughes, 1898.

*Hackensack, New Jersey*. Boston: O. H. Bailey, 1896.

*Properties of the Delaware River Improvement Company on Morrisville Island, Pa. Opposite Trenton, N.J*. Boston: Fowler & Bailey, 1900.

*Rockaway, New Jersey*. Boston: Fowler & Bailey, 1902.

## New York

*Aero-View of Amityville, Suffolk County, Long Island, N.Y. 1925*, by R. Cinquin. New York: Metropolitan Aero-View, 1925.

*Aero-View of Farmingdale, Nassau County, Long Island, N.Y. 1925*, by R. Cinquin. New York: Metropolitan Aero-View, 1925.

*Aero View of Freeport, Long Island, N.Y. 1909*. New York: Hughes & Bailey, 1909.

*Aero-View of Goshen, New York: 1922*. Brooklyn, NY: Hughes & Fowler, 1922.

*Aero-View of Hicksville, Long Island, Nassau County, New York*, by R. Cinquin. New York: Metropolitan Aero-View, 1925.

*Aero-View of Lindenhurst, Long Island, 1926, New York*, by R. Cinquin. New York: Metropolitan Aero-View, 1926.

*Aero-View of Monroe, New York 1923.* Milwaukee, WI: Hughes & Bailey, 1923.

*Aero-View of Pearl River, New York, 1924*, by R. Cinquin. New York: Hughes & Bailey, 1924.

*Albany, New York 1879.* Hartford, CT: H. H. Rowley, 1879.

*Altamont, N.Y.* Troy, NY: L. R. Burleigh, 1889.

*Amsterdam & Port Jackson, N.Y.* Hartford, CT: H. H. Rowley, 1881.

*Bainbridge, N.Y.* Troy, NY: L. R. Burleigh, 1889.

*Bay Side Park, 3d Ward, Borough of Queens, New York City.* New York: North Shore Realty, 1915.

*Binghamton, NY.* Troy, NY: L. R. Burleigh, 1882.

*Bird's-Eye View of Antwerp, N.Y.*, by C. Fausel. Troy, NY: L. R. Burleigh, 1888.

*Bird's-Eye-View of Brooklyn*, by A. R. Ohman. New York: A. R. Ohman, 1908.

*Bird's Eye View of Ilion, N.Y.: Population 4500.* Hartford, CT: H. H. Rowley, 1881.

*Bird's Eye View of Jamaica Park South: Queens, New York City, N.Y.*, by G. Wright. New York: Theo. J. Van Horen, 1908.

*Bird's Eye View of New York and Environs*, by J. Bachmann. New York: Kimmel & Forster, 1865.

*Bird's Eye View of Olean, Cattaraugus County, New York 1882*, by H. Brosius. Madison, WI: J. J. Stoner, 1882.

*Bird's Eye View of That Portion of the 23rd and 24th Wards of the City of New York, Lying Westerly of the New York and Harlem Railroad, and of the Grand Boulevard and Concourse Connecting Manhattan Island with the Park System North of the Harlem River*, by W. W. Klein and R. A Welcke. New York: Dept. of Street Improvements, 1897.

*Bird's-Eye View of the Borough of Brooklyn: Showing Parks, Cemeteries, Principal Buildings, and Suburbs*, by G. Welch. Brooklyn, NY: Brooklyn Daily Eagle, 1897.

*Bird's Eye View of the Village of Friendship, Allegany County, New York: 1882*, by H. Brosius and A. G. Poole. Madison, WI: J. J. Stoner, 1882.

*Brewster, N.Y.* Troy, NY: L. R. Burleigh, 1887.

*Caledonia, N.Y.* Troy, NY: L. R. Burleigh, 1892.

*Cambridge, N.Y. 1886.* Troy, NY: L. R. Burleigh, 1886.

*Camden, N.Y.* Troy, NY: L. R. Burleigh, 1885.

*Canajoharie & Palatine Bridge, N.Y., 1881.* Troy, NY: C. H. Vogt, 1881.

*Canastota, N.Y. 1885.* Cleveland, OH: C. H. Vogt & Son Lith., 1885.

*Carthage, N.Y. 1888.* Troy, NY: L. R. Burleigh, 1888.

*Catskill, N.Y.* Troy, NY: L. R. Burleigh, 1889.

*Cazenovia, N.Y.* Troy, NY: L. R. Burleigh, 1890.

*Chatham, N.Y. 1886.* Troy, NY: L. R. Burleigh, 1886.

*City of Brooklyn*, by C. R. Parsons. New York: Currier & Ives, 1879.

*City of Buffalo, N.Y. 1880*, by E. H. Hutchinson. Buffalo, NY: E. H. Hutchinson, 1880.

*City of New York.* New York: Currier & Ives, 1876.

*Clifton Springs, N.Y.* Troy, NY: L. R. Burleigh, 1892.

*Clinton, N.Y.* Troy, NY: L. R. Burleigh, 1885.

*Cobleskill, N.Y., 1883.* Albany, NY: J. McGregor, 1883.

*Cooperstown, N.Y.* Troy, NY: L. R. Burleigh, 1890.

*Corinth, N.Y. and Palmer Falls.* Troy, NY: L. R. Burleigh, 1888.

*Corning, N.Y., 1882: Junction of the New York, Lake Erie, and Western and the Syracuse, Geneva & Corning R. R.'s.* Philadelphia: Philadelphia Publication House, 1882.

*Cortland, N.Y.* Troy, NY: L. R. Burleigh, 1894.

*De Ruyter, N.Y.* Troy, NY: L. R. Burleigh, 1892.

*Delhi, N.Y. 1887.* Troy, NY: L. R. Burleigh, 1887.

*Depew, N.Y.* Buffalo, NY: Matthews-Northrup, 1898.

*Deposit, N.Y. 1887.* Troy, NY: L. R. Burleigh, 1887.

*Despatch, N.Y., Rochester's Great Industrial Suburb.* Boston: O. H. Bailey, 1890.

*Dolgeville, N.Y.*, by A. Dolge. Dolgeville, NY: Burleigh Litho, 1890.

*East Syracuse, N.Y.* Troy, NY: L. R. Burleigh, 1885.

*Elmira, N.Y.: 1873*, by H. H. Bailey. Cincinnati, OH: Strobridge & Co. Lith., 1873.

*Fairport, N.Y.* Troy, NY: L. R. Burleigh, 1885.

*Fishkill-on-the-Hudson, U.S.A. 1886.* Troy, NY: L. R. Burleigh, 1886.

*Fonda, N.Y.* Troy, NY: L.R. Burleigh, 1889.

*Fort Plain, N.Y. and Nelliston.* Troy, NY: L. R. Burleigh, 1891.

*Frankfort, N.Y. 1887.* Troy, NY: L. R. Burleigh, 1887.

*Freeport, L.I., 1925, N.Y.*, by R. Cinquin. New York: Metropolitan Aero-View, 1925.

*Fultonville, N.Y.* Troy, NY: L. R. Burleigh, 1889.

*Geneva, N.Y., 1893.* Troy, NY: L. R. Burleigh, 1836.

*Geneva & Corning R.R.'s.* Philadelphia: Philadelphia Pub. House, 1882.

*Glens Falls, N.Y.* Troy, NY: L. R. Burleigh, 1884.

*Gloversville, N.Y. 1875.* Albany, NY: H. H. Bailey, 1875.

*Gouverneur, N.Y.* Troy, NY: L.R. Burleigh, 1885.

*Gowanda, N.Y.* Troy, NY: L. R. Burleigh, 1892.

*Grand Bird's Eye View of the Great East River Suspension Bridge. Connecting the Cities of New York & Brooklyn Showing Also the Splendid Panorama of the Bay and Part of New York.* New York: Currier & Ives, 1885.

*Granville, N.Y.* Troy, NY: L. R. Burleigh, 1886.

*Greene, N.Y.* Troy, NY: L. R. Burleigh, 1890.

*Greenwich, N.Y.* Troy, NY: L. R. Burleigh, 1885.

*Hagamans Mills, N.Y.* Troy, NY: L. R. Burleigh, 1890.

*Hamilton, N.Y.* Troy, NY: L. R. Burleigh, 1885.

*Hoosick Falls, N.Y.* Troy, NY: L. R. Burleigh, 1889.

*Hunter, N.Y.* Troy, NY: L. R. Burleigh, 1890.

*Ithaca, N.Y.* Troy, NY: L. R. Burleigh, 1882.

*Johnsonville, N.Y. 1887.* Troy, NY: L. R. Burleigh, 1887.

*Keeseville, N.Y. 1887.* Troy, NY: L. R. Burleigh, 1887.

*Lancaster, N.Y.* Troy, NY: L. R. Burleigh, 1892.

*Le Roy, N.Y.* Troy, NY: L. R. Burleigh, 1892.

*Little Falls, N.Y.* Hartford, CT: H. H. Rowley, 1881.

*Lowville, N.Y.* Troy, NY: L. R. Burleigh, 1885.

*Luzerne, N.Y. and Hadley.* Troy, NY: L. R. Burleigh, 1888.
*Malone, N.Y.* Troy, NY: L. R. Burleigh, 1886.
*Map of Hamilton, Madison County, New York*, by B. A. Clark. Philadelphia: Benj. A. Clark, 1858.
*Marlborough, N.Y.* Troy, NY: L. R. Burleigh, 1891.
*Matteawan, N.Y.* Troy, NY: L. R. Burleigh, 1886.
*Mechanicville, N.Y.* Troy, NY: L. R. Burleigh, 1880.
*Middleville, N.Y.* Troy, NY: L. R. Burleigh, 1890.
*Millerton, N.Y.* Troy, NY: L. R. Burleigh, 1887.
*Mohawk, N.Y.* Troy, NY: L. R. Burleigh, 1893.
*Mount Morris, N.Y.* Troy, NY: L. R. Burleigh, 1893.
*New York.* New York: Currier & Ives, 1879.
*New York and Brooklyn.* New York: Currier & Ives, 1875.
*Newport, N.Y.* Troy, NY: L. R. Burleigh, 1890.
*Niagara-Falls, N.Y. 1882*, by H. Wellge. Madison, WI: J. J. Stoner, 1882.
*Oneonta, N.Y.* Troy, NY: L. R. Burleigh, 1884.
*Oxford, N.Y.* Troy, NY: L. R. Burleigh, 1888.
*Panoramic View of New York City and Vicinity*, by J. Ruppert. New York: Jacob Ruppert, 1912.
*Pawling, N.Y.*, by P. H. Smith and W. G. Tice. New York: Knickerbocker Litho, 1909.
*Proposed Site for World's Fair in 1883: Between 110th and 125th Streets, Morning Side and River Side Parks, N.Y. Area 300 Acres*, by W. Stranders. New York: Demorest's Illustrated Monthly Magazine, 1879.
*View of Albion, N.Y., 1880.* Hartford, CT: H. H. Rowley, 1880.
*View of Middletown, N.Y.: 1874.* Milwaukee, WI: American Oleograph, 1874.
*View of Oneida, N.Y.: 1874.* Milwaukee, WI: O. H. Bailey, 1874.
*View of the Borough of Larchmont, New York.* New York: Hughes & Bailey, 1904.

## North Carolina

*1891 Bird's-Eye View of the City of Asheville, North Carolina.* Madison, WI: Burleigh Litho, 1891.
*Aero View of High Point, North Carolina.* High Point, NC: J. J. Farris, 1913.
*Asheville, Buncombe Co. N.C. 1912.* Passaic, NJ: Charles Hart Litho, 1912.
*Bird's-Eye-View of Hickory, North Carolina*, by A. E. Downs. Boston: A. E. Downs, 1907.
*Bird's-Eye-View of Statesville, North Carolina*, by A. E. Downs. Boston: A. E. Downs, 1907.
*Bird's-Eye View of the City of Durham, North Carolina 1891.* Madison, WI: Burleigh Litho, 1891.
*Bird's Eye View of the City of Greensboro, North Carolina. 1891.* Madison, WI: Burleigh Litho, 1891.
*Bird's Eye View of the City of Raleigh, North Carolina 1872*, by C. N. Drie. Raleigh, NC: C. N. Drie, 1872.
*Bird's-Eye View of the Twin Cities, Winston-Salem, North Carolina 1891.* Madison, WI: Burleigh Litho, 1891.
*Bird's Eye View of Wilson, North Carolina 1908.* Morrisville, PA: T. M. Fowler, 1908.
*Black Mountain, N.C. 1912.* Asheville, NC: Fowler & Browning, 1912.
*Property of the South Rocky Mount Land Co. at South Rocky Mount, N.C.* Morrisville, PA: T. M. Fowler, 1900.

## Ohio

*Ashtabula Harbor, Ohio 1896*, by T. M. Fowler and J. B. Moyer. Morrisville, PA: T. M. Fowler & James B. Moyer, 1896.
*Barnesville, Ohio 1899*, by T. M. Fowler and J. B. Moyer. Morrisville, PA: T. M. Fowler & James B. Moyer, 1899.
*Bird's Eye View of Bellaire, Ohio 1882*, by H. Wellge. Madison, WI: J. J. Stoner, 1882.
*Bird's Eye View of Camp Chase near Columbus, Ohio*, by A. Ruger. Cincinnati, OH: Ehrgott, Forbriger, 1860.
*Bird's Eye View of Cleveland, Ohio 1877.* Madison, WI: Ruger & Stoner, 1877.
*Bird's Eye View of Massillon, Stark County, Ohio 1870.* Madison, WI: Ruger & Stoner, 1870.
*Bird's Eye View of the City of Akron, Summit County, Ohio 1870.* Madison, WI: Ruger & Stoner, 1870.
*Bird's Eye View of the City of Mount Vernon, Knox County, Ohio 1870.* Madison, WI: Ruger & Stoner, 1870.
*Bird's-Eye View of the City of Sandusky, Erie County, Ohio 1870.* Madison, WI: Ruger & Stoner, 1870.
*Bowling Green, Ohio 1888.* Troy, NY: Burleigh & Norris, 1888.
*Cambridge, Ohio 1899*, by T. M. Fowler and J. B. Moyer. Morrisville, PA: T. M. Fowler & James B. Moyer, 1899.
*Canal Dover, Tuscarawas County, Ohio 1899*, by A. E. Downs. Boston: T. M. Fowler, 1899.
*Conneaut, Ohio 1896*, by T. M. Fowler and J. B. Moyer. Morrisville, PA: T. M. Fowler & James B. Moyer, 1896.
*Dayton, Ohio 1870*, by A. Ruger. Chicago: Merchants Lithographing, 1870.
*Findlay, Ohio, the Gas City.* Troy, NY: Burleigh & Norris, 1889.
*Jefferson, Ohio 1901*, by T. M. Fowler and J. B. Moyer. Morrisville, PA: T. M. Fowler & James B. Moyer, 1901.
*Map of the City of Portsmouth and Wayne Township, Scioto County, Ohio*, by R. A. Bryan. Portsmouth, OH: Strobridge & Co. Lith., 1868.
*Martin's Ferry, Ohio 1899*, by T. M. Fowler and J. B. Moyer. Morrisville, PA: T. M. Fowler & James B. Moyer, 1899.
*Mingo Junction, Ohio 1899*, by T. M. Fowler and J. B. Moyer. Morrisville, PA: T. M. Fowler & James B. Moyer, 1899.
*Panoramic View, City of Cincinnati, U.S.A. 1900*, by J. L. Trout. Cincinnati, OH: Henderson Lithographing, 1900.
*Panoramic View of the City of Kent, Portage County, Ohio 1882.* Madison, WI: Ruger & Stoner, 1882.

*Panoramic View of the City of Niles, Trumbull Co., Ohio 1882.* Madison, WI: Ruger & Stoner, 1882.
*Panoramic View of the City of Youngstown, County Seat of Mahoning Co., Ohio 1882.* Madison, WI: Ruger & Stoner, 1882.
*Sixth Ward of Akron, Formerly Middlebury, Summit Co., Ohio 1882.* Madison, WI: Ruger & Stoner, 1882.
*Toledo, Ohio 1876.* Madison, WI: Ruger & Stoner, 1876.

## Oklahoma

*Aero View of Tulsa, Oklahoma, 1918.* Passaic, NJ: Fowler & Kelly, 1918.
*Bartlesville, Oklahoma 1917.* Passaic, NJ: Fowler & Kelly, 1917.
*Edmond, Oklahoma Territory, 1891.* Morrisville, PA: T. M. Fowler and James B. Moyer, 1891.
*El Reno, Oklahoma Territory 1891.* Morrisville, PA: T. M. Fowler and James B. Moyer, 1891.
*Fort Reno, Oklahoma Territory 1891.* Morrisville, PA: T. M. Fowler and James B. Moyer, 1891.

## Oregon

*Bird's Eye View of Jacksonville and the Rogue River Valley, Oregon 1883*, by F. A. Walpole. Milwaukee, WI: Beck & Pauli, 1883.
*Capital City of Oregon, Salem*, by E. Koppe. Portland, OR: E. Koppe, 1905.
*Panoramic View of Pendleton, Or., County Seat of Umatilla County 1884*, by H. Wellge. Madison, WI: J. J. Stoner, 1884.
*Bird's Eye View of Salem, Oregon 1876*, by E. S. Glover. Salem, OR: F. A. Smith, 1876.
*Portland, Oregon*, by E. S. Glover. San Francisco: E. S. Glover, 1879.
*Portland, Oregon 1890.* San Francisco: Clohessy & Strengele, 1890.

## Pennsylvania

*Aero View of Pen Argyl, Pennsylvania 1916*, by T. M. Fowler. New York: Hughes & Bailey, 1916.
*Alburtis and Lockridge, Lehigh County, Pennsylvania 1893*, by T. M. Fowler and J. B. Moyer. Morrisville, PA: Fowler & Moyer, 1893.
*Apollo, Armstrong County, Pennsylvania 1896*, by T. M. Fowler and J. B. Moyer. Morrisville, PA: Fowler & Moyer, 1896.
*Archbald, Lackawanna County, Pa. 1892*, by T. M. Fowler and J. B. Moyer. Morrisville, PA: Fowler & Moyer, 1892.
*Beaver, Pennsylvania 1900*, by T. M. Fowler and J. B. Moyer. Morrisville, PA: Fowler & Moyer, 1900.
*Belle Vernon, Pennsylvania*, by T. M. Fowler and J. B. Moyer. Morrisville, PA: Fowler & Moyer, 1902.

*Bird's Eye View of Burnham and Yeagertown, Mifflin Co., Pa. 1906*, by T. M. Fowler and J. B. Moyer. Morrisville, PA: Fowler & Moyer, 1906.
*Bird's Eye View of Orbisonia and Rock Hill, Pennsylvania 1906.* Morrisville, PA: Fowler & Kelly, 1906.
*Bird's Eye View of Oxford, Chester Co., Pennsylvania.* Morrisville, PA: Fowler & Kelly, 1907.
*Bird's Eye View of Philadelphia*, by J. Bachmann and J. Weik. Philadelphia: Duval & Son, 1857.
*Bird's Eye View of Sinking Spring, Pennsylvania*, by T. M. Fowler and J. B. Moyer. Morrisville, PA: Fowler & Moyer, 1898.
*Bird's Eye View of the City of Allentown, Pa.*, by O. H. Bailey. Boston: Fowler & Bailey, 1879.
*Bird's Eye View of the City of Erie, Erie County, Pennsylvania 1870.* Madison, WI: Ruger & Stoner, 1870.
*Bradford, McKean County, Pennsylvania, 1895*, by T. M. Fowler and J. B. Moyer. Morrisville, PA: Fowler & Moyer, 1895.
*Brookville, Jefferson County, Pennsylvania 1895*, by T. M. Fowler and J. B. Moyer. Morrisville, PA: Fowler & Moyer, 1895.
*Butler, Butler County, Pennsylvania, 1896*, by T. M. Fowler and J. B. Moyer. Morrisville, PA: Fowler & Moyer, 1896.
*Canonsburg, Washington County, Pennsylvania 1897*, by T. M. Fowler and J. B. Moyer. Morrisville, PA: Fowler & Moyer, 1897.
*Carbondale, Pennsylvania, 1890*, by T. M. Fowler and J. B. Moyer. Morrisville, PA: Fowler & Moyer, 1890.
*Carnegie, Allegheny County, Pennsylvania 1897*, by T. M. Fowler and J. B. Moyer. Morrisville, PA: Fowler & Moyer, 1897.
*Catasauqua, Pa.*, by T. M. Fowler and J. B. Moyer. Morrisville, PA: Fowler & Moyer, 1897.
*Clarion, Clarion County, Pennsylvania 1896*, by T. M. Fowler and J. B. Moyer. Morrisville, PA: Fowler & Moyer, 1896.
*Confluence, Pennsylvania 1905*, by T. M. Fowler and J. B. Moyer. Morrisville, PA: Fowler & Moyer, 1905.
*Connellsville, Fayette County, Pennsylvania 1897*, by T. M. Fowler and J. B. Moyer. Morrisville, PA: Fowler & Moyer, 1897.
*Dawson, Pennsylvania 1902*, by T. M. Fowler and J. B. Moyer. Morrisville, PA: Fowler & Moyer, 1902.
*Donora, Washington County, Pennsylvania, 1901*, by T. M. Fowler and J. B. Moyer. Morrisville, PA: Fowler & Moyer, 1897.
*Derry Station, Pennsylvania 1900*, by T. M. Fowler and J. B. Moyer. Morrisville, PA: Fowler & Moyer, 1900.
*Downingtown, Chester County, Pennsylvania 1893*, by T. M. Fowler and J. B. Moyer. Morrisville, PA: Fowler & Moyer, 1903.
*Du Bois, Clearfield County, Pennsylvania, 1895*, by T. M. Fowler and J. B. Moyer. Morrisville, PA: Fowler & Moyer, 1895.

*Dunmore, Pennsylvania*, by T. M. Fowler and J. B. Moyer. Morrisville, PA: Fowler & Moyer, 1892.

*Easton, Pa. and Phillipsburg, N.J.* Newark, NJ: Landis & Alsop, 1900.

*Edinboro, Pa. 1898*, by J. O'Brien. Erie, PA: John O'Brien, 1898.

*Elizabeth and West Elizabeth, Allegheny County, Pennsylvania 1897*, by T. M. Fowler and J. B. Moyer. Morrisville, PA: Fowler & Moyer, 1897.

*Emlenton, Venango County, Pennsylvania 1897*, by T. M. Fowler and J. B. Moyer. Morrisville, PA: Fowler & Moyer, 1897.

*Evans City, Pennsylvania 1900*, by T. M. Fowler and J. B. Moyer. Morrisville, PA: Fowler & Moyer, 1900.

*Forest City, Susquehanna County, Pa. 1889*, by T. M. Fowler and J. B. Moyer. Morrisville, PA: Fowler & Moyer, 1889.

*Frackville, Pennsylvania*, by T. M. Fowler and J. B. Moyer. Morrisville, PA: Fowler & Moyer, 1889.

*Glassport, Allegheny Co., Pennsylvania 1902*, by T. M. Fowler and J. B. Moyer. Morrisville, PA: Fowler & Moyer, 1902.

*Grand Bird's-Eye View of the Centennial Exhibition Grounds and the Surrounding Country: Ogontz Park, Ogontz, Montgomery Co., Penna.: Wm. T. B. Roberts, 410 Land Title Bldg., Philadelphia.* Boston: O. H. Bailey, 1880.

*Great Bend, Penn. 1887.* Troy, NY: L. R. Burleigh, 1887.

*Grove City, Mercer County, Pennsylvania 1901*, by T. M. Fowler and J. B. Moyer. Morrisville, PA: Fowler & Moyer, 1901.

*Hamburg, Berks Co., Penna. 1889*, by T. M. Fowler and J. B. Moyer. Morrisville, PA: Fowler & Moyer, 1889.

*Homestead, Pennsylvania, 1902*, by T. M. Fowler and J. B. Moyer. Morrisville, PA: Fowler & Moyer, 1902.

*Indiana, Pennsylvania, 1900*, by T. M. Fowler and J. B. Moyer. Morrisville, PA: Fowler & Moyer, 1897.

*Jeannette, Westmoreland County, Pennsylvania, 1897*, by T. M. Fowler and J. B. Moyer. Morrisville, PA: Fowler & Moyer, 1897.

*Johnsonburg, Elk County, Pennsylvania*, by T. M. Fowler and J. B. Moyer. Morrisville, PA: Fowler & Moyer, 1895.

*Kittanning, Armstrong County, Pennsylvania 1896*, by T. M. Fowler and J. B. Moyer. Morrisville, PA: Fowler & Moyer, 1896.

*League Island U.S. Navy Yard Philadelphia.* Philadelphia: Webster & Hunter, 1897.

*Lewisburgh.* Boston: O. H. Bailey, 1884.

*Mahanoy City, Pennsylvania*, by T. M. Fowler and J. B. Moyer. Morrisville, PA: Fowler & Moyer, 1889.

*Macungie, Lehigh County, Pennsylvania, 1893*, by T. M. Fowler and J. B. Moyer. Morrisville, PA: Fowler & Moyer, 1893.

*McKee's Rocks, Allegheny County, Pennsylvania 1901*, by T. M. Fowler and J. B. Moyer. Morrisville, PA: Fowler & Moyer, 1901.

*Mechanicsburg, Pa., 1903*, by T. M. Fowler and J. B. Moyer. Morrisville, PA: Fowler & Moyer, 1903.

*Mifflintown, Juniata County, Pennsylvania*, by T. M. Fowler and J. B. Moyer. Morrisville, PA: Fowler & Moyer, 1895.

*Millersville, Lancaster County, Pennsylvania 1894*, by T. M. Fowler and J. B. Moyer. Morrisville, PA: Fowler & Moyer, 1894.

*Miner's Mills and Mill Creek, Luzerne County, Pa. 1892*, by T. M. Fowler and J. B. Moyer. Morrisville, PA: Fowler & Moyer, 1892.

*Monaca, Pennsylvania 1900*, by T. M. Fowler and J. B. Moyer. Morrisville, PA: Fowler & Moyer, 1900.

*Montrose, Susquehanna County, Pa.*, by T. M. Fowler and J. B. Moyer. Morrisville, PA: Fowler & Moyer, 1890.

*Morrisville, Bucks County, Pennsylvania, 1893*, by T. M. Fowler and J. B. Moyer. Morrisville, PA: Fowler & Moyer, 1893.

*Moscow, Lackawanna County, Penn'a. 1891*, by T. M. Fowler and J. B. Moyer. Morrisville, PA: Fowler & Moyer, 1891.

*Mount Joy, Lancaster County, Pennsylvania, 1894*, by T. M. Fowler and J. B. Moyer. Morrisville, PA: Fowler & Moyer, 1894.

*Mount Union, Huntingdon Co., Pa. 1906*, by T. M. Fowler and J. B. Moyer. Morrisville, PA: Fowler & Moyer, 1906.

*New Brighton, Pennsylvania 1901*, by T. M. Fowler and J. B. Moyer. Morrisville, PA: Fowler & Moyer, 1901.

*New Castle, Pennsylvania 1896*, by T. M. Fowler and J. B. Moyer. Morrisville, PA: Fowler & Moyer, 1836.

*New Kensington, Pennsylvania, 1902*, by T. M. Fowler and J. B. Moyer. Morrisville, PA: Fowler & Moyer, 1902.

*Newport, Perry County, Pennsylvania, 1895*, by T. M. Fowler and J. B. Moyer. Morrisville, PA: Fowler & Moyer, 1895.

*Perkasie, Pennsylvania, Bucks County, 1894*, by T. M. Fowler and J. B. Moyer. Morrisville, PA: Fowler & Moyer, 1894.

*Pennsburgh, Montgomery County, Pennsylvania 1894*, by T. M. Fowler and J. B. Moyer. Morrisville, PA: Fowler & Moyer, 1894.

*Philadelphia in 1886.* Philadelphia: Burk & McFetridge, 1886.

*Pitcairn, Pa., Allegheny County 1901*, by T. M. Fowler and J. B. Moyer. Morrisville, PA: Fowler & Moyer, 1901.

*Plains, Luzerne County, Pa*, by T. M. Fowler and J. B. Moyer. Morrisville, PA: Fowler & Moyer, 1892.

*Point Marion, Pennsylvania 1902*, by T. M. Fowler. Morrisville, PA: T. M. Fowler, 1902.

*Providence, Pennsylvania 1892*, by A. E. Downs. Boston: A. E. Downs, 1892.

*Roscoe, Washington Co., Pennsylvania 1902*, by T. M. Fowler and J. B. Moyer. Morrisville, PA: Fowler & Moyer, 1902.

*Scranton, Penn. 1890*, by T. M. Fowler and J. B. Moyer. Morrisville, PA: Fowler & Moyer, 1890.

*Tidioute, Warren County, Pennsylvania 1896*, by T. M. Fowler and J. B. Moyer. Morrisville, PA: Fowler & Moyer, 1896.

*Tyrone, Blair County, Pennsylvania 1895*, by T. M. Fowler and J. B. Moyer. Morrisville, PA: Fowler & Moyer, 1895.

*View of Duncannon, Pennsylvania 1903*, by T. M. Fowler and J. B. Moyer. Morrisville, PA: Fowler & Moyer, 1903.

*View of Harrisburg, Penn*, by J. T. Williams. York, PA: Sachse, 1855.

*View of the City of Franklin, Pa., 1901*, by T. M. Fowler and J. B. Moyer. Morrisville, PA: Fowler & Moyer, 1901.

*Washington, Pennsylvania, 1897*. Morrisville, PA: T. M. Fowler & James B. Moyer, 1897.

*Waynesburg, Greene County, Pennsylvania, 1897*, by T. M. Fowler and J. B. Moyer. Morrisville, PA: Fowler & Moyer, 1897.

*West Newton, Pennsylvania 1900*, by T. M. Fowler and J. B. Moyer. Morrisville, PA: Fowler & Moyer, 1900.

*Zelienople, Butler County, Pennsylvania 1901*, by T. M. Fowler and J. B. Moyer. Morrisville, PA: Fowler & Moyer, 1901.

## South Dakota

*1883 Bird's Eye View of Madison, County Seat of Lake Co. Dakota*, by H. Brosius. Madison, WI: J. J. Stoner 1883.

*Bird's Eye View of Clark, Dakota, County Seat Clark Co. 1883*. Madison, WI: J. J. Stoner 1883.

*Bird's Eye View of Flandreau, County Seat of Moody Co., Dakota 1883*, by H. Brosius. Madison, WI: J. J. Stoner 1883.

*Bird's Eye View of Frederick, Dak. 1883*, by C. F. Campau. Frederick, SD: C. F. Campau, 1883.

*Bird's Eye View of Watertown, Dak. County-Seat of Codington Co. 1883*. Madison, WI: J. J. Stoner, 1883.

*Sioux Falls, Minnehaha County, Dakota*. Sioux Falls, SD: Chas. A. Carson, 1881.

## Tennessee

*Bird's Eye View of the City of Chattanooga, Hamilton County, Tennessee 1871*, by A. Ruger. St. Louis: A. Ruger, 1871.

*Bird's Eye View of the City of Clarksville, Montgomery County, Tennessee 1870*, by A. Ruger. Madison, WI: Stoner & Ruger, 1870.

*Bird's Eye View of the City of Jackson, Madison County, Tennessee 1870*, by A. Ruger. Chicago: Chicago Lithographing, 1870.

*Chattanooga, County Seat of Hamilton County, Tennessee 1886*. Milwaukee, WI: Wellge, 1886.

*Harriman, Tenn. 1892*, by G. E. Norris. Brockton, MA: G. E. Norris, 1892.

*Knoxville, Tenn. County Seat of Knox County 1886*. Milwaukee, WI: Wellge, 1886.

*Perspective Map of the City of Memphis, Tenn. 1887*. Milwaukee, WI: Wellge, 1887.

## Texas

*Aeroplane View of Business District Amarillo, Texas*, by E. E. Motter. Amarillo, TX: Panhandle Printing, 1912.

*Bird's Eye View of the City of Denison, Texas, 1873*, by J. W. Pearce. Denison, TX: J. W. Pearce, 1873.

*Childress, Texas 1890*. Morrisville, PA: T. M. Fowler, 1890.

*Clarendon, Texas, Donley Co. 1890*. Morrisville, PA: T. M. Fowler, 1890.

*Fort Worth, Tex., "The Queen of the Prairies," County Seat of Tarrant County 1886*. Milwaukee, WI; Wellge, 1886.

*Greenville, Tex., County Seat of Hunt County 1886*. Milwaukee, WI: Wellge, 1886.

*Honey Grove, Tex. Fannin County 1886*. Milwaukee, WI: Wellge, 1886.

*Houston—A Modern City*. Houston: Hopkins & Motter, 1912.

*Ladonia, Fannin County, Texas*. Morrisville, PA: T. M. Fowler and J. B. Moyer, 1890.

*Perspective Map of Fort Worth, Tex. 1891*. Milwaukee, WI: Wellge, 1891.

*Perspective Map of Texarkana, Texas and Arkansas*. Milwaukee, WI: Wellge, 1888.

*Wolfe City, Texas 1891*. Morrisville, PA: T. M. Fowler and J. B. Moyer, 1891.

## Utah

*Bird's-Eye View of Brigham City and Great Salt Lake, Utah, Ty. 1875*, by E. S. Glover. Salt Lake City: E. S. Glover, 1875.

*Bird's Eye View of Ogden City, Utah, Ty. 1875*, by E. S. Glover. Salt Lake City: E. S. Glover, 1875.

*Bird's Eye View of Salt Lake City, Utah Territory 1870*, by A. Koch. Chicago: Chicago Lithographing, 1870.

## Vermont

*Barre, Washington County, VT, 1884*, by G. E. Norris. Brockton, MA: Geo. E. Norris, 1884.

*Bellows Falls, Vt.* Troy, NY: L. R. Burleigh, 1886.

*Bennington, Vt. 1887*. Troy, NY: L. R. Burleigh, 1887.

*Bethel, Vt. 1886*. Troy, NY: L. R. Burleigh, 1886.

*Bird's Eye View of Bennington & Bennington Centre, Bennington Co., Vermont*, by J. J. Stoner. Chicago: Shober & Carqueville Litho., 1877.

*Brandon, Vt.* Troy, NY: L. R. Burleigh, 1890.

*Bristol, Vt.*, by G. E. Norris. Brockton, MA: G. E. Norris, 1889.

*Castleton, Vt. 1889*. Troy, NY: L. R. Burleigh, 1889.

*Enosburg Falls, Vt., Franklin Co., 1892*, by G. E. Norris. Brockton, MA: Geo. E. Norris, 1892.

*Fair Haven, Vt.* Troy, NY: L. R. Burleigh, 1886.
*Ludlow, Vt.* Troy, NY: L. R. Burleigh, 1885.
*Middlebury, Vt.* Troy, NY: L. R. Burleigh, 1886.
*Rutland, Vt.* Troy, NY: L. R. Burleigh, 1885.
*Springfield, Vt.* Troy, NY: L. R. Burleigh, 1886.
*West Randolph, Vt. 1886.* Troy, NY: L. R. Burleigh, 1886.
*Windsor, Vermont 1886.* Troy, NY: L. R. Burleigh, 1886.

## Virginia

*Aero View of Bristol, Va.=Tenn. 1912.* Passaic, NJ: T. M. Fowler, 1912.
*Bird's Eye Map of the City of Winchester*, by W. A. Ryan. Winchester, VA: W. A. Ryan, 1926.
*Bird's Eye View of Emporia, Virginia 1907.* Morrisville, PA: T. M. Fowler, 1907.
*Bird's Eye View of Suffolk, Nansemond Co., Va. 1907.* Morrisville, PA: Fowler & Kelly, 1907.
*Panorama of Norfolk and Surroundings 1892*, by H. Wellge. Milwaukee, WI: American Publishing, 1892.
*Perspective Map of Bedford City, Va., County Seat of Bedford Co. 1891*, by H. Wellge. Milwaukee, WI: American Publishing, 1891.
*Perspective Map of Buena Vista, Va. 1891.* Milwaukee, WI: American Publishing, 1891.
*Perspective Map of Newport News, Va., County Seat of Warwick County 1891.* Milwaukee, WI: American Publishing, 1891.
*Perspective Map of the City of Roanoke, Va. 1891.* Milwaukee, WI: American Publishing, 1891.
*Perspective Map of the City of Staunton, Va., County Seat of Augusta County, Virginia 1891.* Milwaukee, WI: American Publishing, 1891.
*Richmond, Va. and Its Vicinity*, by J. Wells and R. Hinshelwood. Richmond, VA: Virtue Yorston, 1863.
*View of Fredericksburg, Va. Nov. 1862.* Baltimore, MD: E. Sachse, 1863.

## Washington

*Bird's Eye View of Cheney, Wash. Ter., County Seat of Spokane County, 1884*, by H. Wellge. Madison, WI: J. J. Stoner, 1884.
*Bird's Eye View of Port Townsend, Puget Sound, Washington Territory 1878*, by E. S. Glover. Portland, OR: Bancroft, 1878.
*Bird's-Eye View of Seattle and Environs King County, Wash., 1891*, by A. Koch. Chicago: Hughes Litho, 1891.
*Bird's Eye View of Walla Walla, Washington Territory 1876*, by E. S. Glover. Walla Walla, WA: Everts & Able, 1876.
*Olympia, the Capital on Puget Sound, Washington*, by E. Lange. Olympia, WA: S. S. Churchill, 1903.
*Panoramic View of Dayton, W.T., County Seat of Columbia County 1884*, by H. Wellge. Madison, WI: J. J. Stoner, 1884.
*Seattle Birdseye View of Portion of City and Vicinity*, by E. C. Poland. Seattle, WA: Kroll Map Company, 1925.
*View of the City of Tacoma, W.T., Puget-Sound, County Seat of Pierce Cty. 1884*, by H. Wellge. Madison, WI: J. J. Stoner, 1884.

## Washington, D.C.

*B.H. Warner & Co.'s Map Showing a Bird's-Eye View of the City of Washington and Suburbs: Locating the Public Buildings and Places of Interest*, by A. G. Gedney. Washington, DC: B. H. Warner, 1886.
*Birdseye View of the National Capital, Including the Site of the Proposed World's Exposition of 1892 and Permanent Exposition of the Three Americas*, by J. E. Kurtz. Washington, DC: A. Hoen, 1888.
*Bird's Eye View of the National Zoological Park, Washington, D.C.* Washington, DC: National Zoological Park, 1996.
*Bird's-Eye-View of Washington City, D.C.* Washington, DC: W. H. & O. H. Morrison, 1872.
*Franklin Delano Roosevelt Memorial, Washington, D.C.* Washington, DC: U.S. National Park Service, 1997.
*National Capital, Washington, D.C. Sketched from Nature by Adolph Sachse, 1883 to 1884*, by A. Sachse. Baltimore: A. Sachse, 1884.
*View Looking Northwest from Anacostia: Washington D.C.*, by J. L. Trout. Washington, DC: J. L. Trout, 1901.
*View of Washington City.* Baltimore: E. Sachse, 1871.

## West Virginia

*Aero View of Bluefield, West Virginia 1911.* Flemington, NJ: Fowler & Basham, 1911.
*Aero View of Keystone, West Virginia 1911.* Flemington, NJ: T. M. Fowler, 1911.
*Aero View of North Fork and Town of Clark, West Virginia 1911.* Flemington, NJ: T. M. Fowler, 1911.
*Bird's Eye View of the City of Wheeling, West Virginia 1870*, by A. Ruger. Madison, WI: Ruger & Stoner, 1870.
*Buckhannon, West Virginia 1900.* Morrisville, PA: T. M. Fowler & James B. Moyer, 1900.
*Cairo, West Virginia 1899.* Morrisville, PA: T. M. Fowler & James B. Moyer, 1899.
*Cameron, West Virginia 1899.* Morrisville, PA: T. M. Fowler & James B. Moyer, 1899.
*Clarksburg, West Virginia 1898.* Morrisville, PA: T. M. Fowler & James B. Moyer, 1898.
*Elkins, Randolph County, W.Va. 1897.* Morrisville, PA: T. M. Fowler & James B. Moyer, 1897.

*Fairmont and Palatine, West Virginia 1897.* Morrisville, PA: T. M. Fowler & James B. Moyer, 1897.

*Grafton, West Virginia 1898.* Boston: Fowler & Downs, 1898.

*Mannington, West Virginia 1897.* Morrisville, PA: T. M. Fowler & James B. Moyer, 1897.

*Morgantown, West Virginia 1897.* Morrisville, PA: T. M. Fowler & James B. Moyer, 1897.

*Moundsville, West Virginia 1899*, by A. E. Downs. Myerstown, PA: James B. Moyer, 1899.

*Parkersburg, West Virginia 1899.* Morrisville, PA: T. M. Fowler & James B. Moyer, 1899.

*Salem, West Virginia 1899.* Morrisville, PA: T. M. Fowler & James B. Moyer, 1899.

*St. Mary's, West Virginia 1899.* Morrisville, PA: T. M. Fowler & James B. Moyer, 1899.

## Wisconsin

*Appleton, Outagamie County, Wisconsin 1867*, by A. Ruger. Chicago: Chicago Lithographing, 1867.

*Bird's Eye View of Bayfield, Wis., County Seat of Bayfield County 1886.* Milwaukee, WI: Wellge, 1886.

*Bird's Eye View of Boscobel, Grant Co., Wisconsin 1869*, by A. Ruger. Chicago: Chicago Lithographing, 1869.

*Bird's Eye View of Columbus, Columbia Co., Wisconsin 1868*, by A. Ruger. Chicago: Chicago Lithographing, 1868.

*Bird's Eye View of Fort Atkinson, Jefferson County, Wisconsin 1870*, by A. Ruger. Chicago: Chicago Lithographing, 1870.

*Bird's Eye View of Jefferson, Jefferson County, Wisconsin 1870*, by A. Ruger. Chicago: Chicago Lithographing, 1870.

*Bird's Eye View of Lake Geneva, Walworth Co., Wis. 1882.* Madison, WI: J. J. Stoner, 1882.

*Bird's Eye View of Merrill, Wis. County Seat Lincoln Co. 1883.* Madison, WI: J. J. Stoner, 1883.

*Bird's Eye View of Reedsburg, Sauk County, Wis. 1874*, by A. Ruger. Chicago: Chicago Lithographing, 1874.

*Bird's Eye View of River Falls, Pierce County, Wisconsin, 1880.* Madison, WI: J. J. Stoner, 1880.

*Bird's-Eye View of South Milwaukee.* Milwaukee, WI: Binner Engraving, 1906.

*Bird's Eye View of the City of Antigo, Wis. County Seat of Langlade County 1886.* Milwaukee, WI: Wellge, 1886.

*Bird's Eye View of the City of Ashland, Wis., County Seat of Ashland County 1886.* Milwaukee, WI: Wellge, 1886.

*Bird's Eye View of the City of Beaver Dam, Dodge Co., Wisconsin 1867*, by A. Ruger. Chicago: Chicago Lithographing, 1867.

*Bird's Eye View of the City of Hudson, St. Croix County, Wisconsin 1870*, by A. Ruger. Chicago: Chicago Lithographing, 1870.

*Bird's Eye View of the City of La Crosse, Wisconsin 1867*, by A. Ruger. Chicago: Chicago Lithographing, 1867.

*Bird's Eye View of the City of Portage, Columbia Co., Wisconsin 1868*, by A. Ruger. Chicago: Chicago Lithographing, 1868.

*Chippewa-Falls, Wis., County-Seat of Chippewa-County 1886.* Milwaukee: Wellge, 1886.

*Delavan, Walworth Co., Wisconsin 1884*, by H. J. Brosius. Madison, WI: J. J. Stoner, 1884.

*Fond du Lac, Wisconsin 1867*, by A. Ruger. Chicago: Chicago Lithographing, 1867.

*Green Bay and Fort Howard, Brown Co., Wisconsin 1867*, by A. Ruger. Chicago: Chicago Lithographing, 1867.

*Hurley, Wis., Ashland County 1886.* Milwaukee, WI: Wellge, 1886.

*Madison, State Capital of Wisconsin, County Seat of Dane County 1885.* Milwaukee, WI: Wellge, 1885.

*Medford, Wis., County Seat of Taylor County before the Great Fire May 28th 1885.* Milwaukee, WI: Wellge, 1885.

*Milwaukee, Wis.*, by H. H. Bailey. Milwaukee, WI: Holzapfel & Eskuche, 1872.

*Oshkosh, Winebago Co., Wisconsin 1867*, by A. Ruger. Chicago: Chicago Lithographing, 1867.

*Perspective Map of Beloit, Wis. 1890.* Milwaukee, WI: American Publishing, 1890.

*Racine, Wis. County Seat of Racine Co. 1883.* Madison, WI: J. J. Stoner, 1883.

*View of the City of Oconomowoc, Wis. Waukesha County 1885.* Milwaukee, WI: Wellge, 1885.

*View of the City of Whitewater, Wis. Walworth-County 1855.* Milwaukee, WI: Wellge, 1855.

# 11
# GIS Data Sources

Traditionally, cartographic materials have been made available in print format as globes, atlases, flat maps, folded maps, and air photos, to name a few. As discussed in previous chapters, these maps are static and cannot be manipulated in the way online maps can. Whether the electronic maps are in our cars, on our phones, or on our computer screens, they can be easily accessed, updated, and customized for our personal or professional uses. Popular online products such as Google Maps have been used for years to find locations, directions, and points of interest. Geographic information systems (GIS), the technology that fuels these easy-to-use maps, have been used by many organizations to develop digital and online cartographic products.

Many of the maps discussed in previous chapters can be discovered and accessed online. Instead of visiting a map library for a print topographic map, scholars can access websites that offer digital versions of updated topographic maps. Online maps have a variety of formats with varying features. Some sites offer static digitized maps, but more and more offer interactive mapping applications that allow users to build their own maps by selecting the features they are interested in seeing visualized. Scholars who know how to use GIS data with GIS software are able to access and download the online features (that is, the data layers) needed to create their own customized map or to perform spatial analysis using the data layers. Traditionally, GIS files were only available from map libraries and data centers, but, today, many government organizations freely offer their data online. Government data and open data may be used for research or personal interests at no cost. The federal, provincial, and state governments have recently introduced open data initiatives to make data transparent and easily available to the public.

Almost every province and state has a data clearinghouse, offering the public base or topographic datasets. Many counties and cities also offer their data online; these data tend to be more detailed, focusing on city or county services. Finding what is available online can be challenging, and a good place to start is a city's municipal website or a local university's GIS services website. All universities that offer GIS services list recommended online resources.

This chapter lists recommended GIS data sources available for Canada, the United States, and around the world. Data is sorted by country, province or state, and county or city. This is not an exhaustive list of all freely available datasets on the web, as such a list is not possible, but rather a list that has been collected using the recommendations of one or more academic institutions. The author has visited every Canadian and U.S. academic institution's GIS website and made note of the top resources recommended for their city or state.

Many of the municipal resources listed offer standard open datasets. GIS datasets offered by counties and cities or towns tend to include municipal information such as locations of airports, parks, buildings, bridges, cemeteries, elevation, streets, neighborhood boundaries, points of interests, schools, speed limits, storm sewers, storm ponds, trails, traffic signals, vacant land, and much more. Often the datasets can be searched by data categories such as facilities, land and property, local government, parks and recreation, public safety, tourism, and transportation. Many of the county and city resources include datasets similar to those just described. Annotations indicate if they deviate from the list or if the provincial and state resources focus on a specific topic.

## CANADIAN DATA RESOURCES

### National Data

Aboriginal Mapping Network
http://nativemaps.org/taxonomy/term/72
The Aboriginal Mapping Network (AMN) was established in 1998 as a joint initiative of the Gitxsan and Ahousaht First Nations and Ecotrust Canada, supporting aboriginal and indigenous peoples facing issues such as land claims, treaty negotiations, and resource development and offering

online tools such as traditional use studies, GIS mapping, and other information systems.

AgriMap
http://www.agr.gc.ca/atlas/agrimap/
Interactive agriculture-related maps, geospatial data, and tools can be accessed at AgriMap.

Canadian Climate Data Products
http://open.canada.ca/data/en/dataset/51dbf91e-509c-437b-ab9e-eca6d0bf6fa8
Climate Data Products at Environment Canada offers four different datasets: almanac averages and extremes, monthly climate summaries, Canadian climate normals, and Canadian historical weather radar. Data are available for stations with at least fifteen years of data between the periods of 1961 through 1990, 1971 through 2000, and 1981 through 2010.

Canadian Government GeoSpatial Data Search
http://www.data.gc.ca
A large number of datasets are available online from organizations related to agriculture, space, climate, natural resources, statistics, and more.

Canada's National Forest Inventory (Canfi)
https://nfi.nfis.org/data_and_tools.php?lang=en
The National Forest Inventory offers an interactive map of forestry resources, as well as data upon request.

Canadian Soil Information Service (CanSIS)
http://sis.agr.gc.ca/cansis/index.html
Canadian Soil Information Service offers a National Soil DataBase (NSDB) archive for soil and land resource information spatial datasets on ecological groupings, soil properties, and distribution. Soil survey reports and maps are also available.

Canadian Wildland Fire Information Systems
http://cwfis.cfs.nrcan.gc.ca/datamart
The National Fire Database is a collection of forest fire perimeters provided by Canadian fire management agencies including provinces, territories, and Parks Canada. The Large Fire Database (LFDB) is a collection of forest fire locations including only fires greater than two hundred hectares in final size for 1959 through 1999. An interactive map is also available.

Canadian Wind Energy Atlas
http://www.windatlas.ca/index-en.php
The Canadian Wind Energy Atlas provides maps and data on wind speed and energy.

CanMatrix
https://open.canada.ca/data/en/dataset?organization=nrcan-rncan
Although this product is no longer supported, it offers federal government topographic maps at scales of 1:25,000, 1:50,000, 1:125,000, 1:500,000 and 1:1,000,000.

CanVec
https://open.canada.ca/data/en/dataset?organization=nrcan-rncan
CanVec is a digital cartographic reference product of Natural Resources Canada offering topographical information in vector format. CanVec contains more than sixty topographic entities organized into eight distribution themes: transport features, administrative features, hydro features, land features, human-made features, elevation features, resource management features, and toponymic features.

Census Agriculture Boundary Files
http://www.statcan.gc.ca/pub/92-637-x/92-637-x2011001-eng.htm
The Census Agriculture Boundary Files for Canada contain the boundaries of all eighty-two census agricultural regions delineated for the Census of Agriculture. They were created to support the spatial analysis and thematic mapping of data from that census.

Census Boundary Files
http://www12.statcan.gc.ca/census-recensement/2011/geo/bound-limit/bound-limit-eng.cfm
Statistics Canada Census Boundary files, including cartographic and geographic boundary files, for 2016 and earlier.

Census Mapper
https://censusmapper.ca
Census Mapper offers an interactive map application using census data across all census aggregation levels.

Fisheries and Oceans Canada Interactive Maps
http://www.pac.dfo-mpo.gc.ca/fm-gp/maps-cartes/index-eng.htm
Fisheries and Oceans Canada Interactive Maps offers a large number of interactive maps, online maps, geoportals, and GIS data related to fish, conservation areas, and water.

Geography Network Canada
http://www.geographynetwork.ca
The Geography Network Canada is an online resource for finding and sharing geographic content, including maps and data, from many of the world's leading data publishers. The data collection includes online datasets that can be previewed online and downloaded immediately.

Geological Atlas of the Western Canada Sedimentary Basin
http://ags.aer.ca/publications/atlas-of-the-western-canada-sedimentary-basin.htm
The Geological Atlas of the Western Canada Sedimentary Basin is the online version of the *Geological Atlas of the Western Sedimentary Basin*, offering digital data and GIS files.

GeoScienceData
http://gdr.agg.nrcan.gc.ca/gdrdap/dap/search-eng.php
GeoScienceData is a geoscience data repository for geophysical data collections.

Global Forest Watch Canada
http://www.globalforestwatch.ca
Global Forest Watch monitors the state of Canada's forests to provide information on development activity and environmental impacts. The geospatial datasets include forestry concessions, mineral concessions, and petroleum/natural gas concessions.

Health Regions (HSDA) Boundary Files
http://www.statcan.gc.ca/pub/82-402-x/2015002/hrbf-flrs eng.htm
This site provides Statistics Canada Health Region boundary files for the country.

Inventory of Canada Lands
http://sis.agr.gc.ca/cansis/nsdb/cli/index.html
The Canada Land Inventory is a multidisciplinary land inventory of rural Canada, mapping land capability for agriculture, forestry, wildlife, recreation, and wildlife. This soil capability is available as a GIS dataset for download.

National Soil Database (NSDB)
http://sis.agr.gc.ca/cansis/nsdb/index.html
The National Soil Dataset offers datasets that contain soil, landscape, and climate data for Canada; the data were collected by federal and provincial field surveys or created by land data analysis projects.

National Topographic System
https://open.canada.ca/data/en/dataset?organization=nrcan-rncan
Topographic datasets are available for download from Geogratis. The National Topographic System covers Canada at various scales.

Natural Resources Canada Climate Change Data
https://www.nrcan.gc.ca/environment/resources/data/11017
Natural Resources Canada Climate Change Data offers for download several datasets related to climate change, including climate change, coastal erosion, glaciers, icefields, permafrost, sea-level rise, and temperature profiles.

Natural Resources Canada Land Survey Data
http://www.nrcan.gc.ca/earth-sciences/geomatics/canada-lands-surveys/11092#CLSR_Access_Method_Summary_Table
Natural Resources Canada Land Survey provides access to survey information and data for Canada Lands (that is, Indian reserves and national parks).

Open Government Data in Canada
http://open.canada.ca/en/open-data
The Open Government Data in Canada website lists a variety of open datasets offered by many sectors of the Canadian government.

Topographic Maps
http://atlas.gc.ca/toporama/en/index.html
Topographic Maps is an interactive map of topographic information for Canada.

Water Survey of Canada
http://www.ec.gc.ca/rhc-wsc/
The Water Survey of Canada offers standardized water resource data and information.

## Provincial, County, and City Data

*Alberta*

### Provincial Data

AgroClimatic Information Service (ACIS)
http://agriculture.alberta.ca/acis/
AgroClimatic Information Service (ACIS) provides interactive weather maps and data that describe Alberta's weather, climate, and related agricultural features.

Alberta Energy
http://www.energy.alberta.ca/OurBusiness/1072.asp
Alberta Energy consists of interactive maps that show the current disposition of mineral rights in Alberta for different mineral commodities.

Alberta Geological Survey Open Data
http://geology.ags-aer.opendata.arcgis.com
This open data catalog offers a large number of GIS datasets related to geology published by the Alberta Geological Survey.

Alberta Open Data—AltaLIS
http://www.altalis.com
Alberta Open Data offers current GIS data related to property, base, terrain, imagery, utility, and vegetation for the province of Alberta.

Alberta Open Government
http://open.alberta.ca
The Alberta Open Government offers interactive maps and visuals for several government organizations within the province of Alberta.

GeoDiscover Alberta Catalogue by Government of Alberta
https://open.alberta.ca/interact/geodiscover-alberta
The GeoDiscover Alberta Catalogue is an interactive map for viewing and downloading several GIS files for the province of Alberta.

### County and City Data

Banff Open Data
http://www.banffopendata.ca
Banff Open Data consists of a large number of downloadable municipal-related open data files for Banff.

Calgary City Open Data
https://data.calgary.ca

Calgary Region Open Data
http://www.calgaryregionopendata.ca
The Calgary Region Open Data catalog provides access to over one hundred datasets for the Calgary Region.

Edmonton Open Data
https://data.edmonton.ca
Edmonton Open Data offers many interactive maps and geospatial data for the city of Edmonton, including traffic, trails, bike routes, sandboxes, edible fruit trees, and much more.

Grand Prairie Open Data
https://data.cityofgp.com

Lethbridge Open Data
http://opendata.lethbridge.ca
Lethbridge Open Data includes geospatial open datasets such as aerial photography, transit schedules, facilities, land and property resources, parks and recreation, and public safety.

Red Deer Open Data
http://reddeer.cloudapp.net
Red Deer Open Data offers access to a small number of city-managed datasets such as building permits, cemetery locations, neighborhood boundaries, and public transit information.

Strathcona County Open Data
https://data.strathcona.ca/

## British Columbia

### Provincial Data

BC Ministry of Sustainable Resource Management—GIS Data FTP Site
http://www.empr.gov.bc.ca/Mining/Geoscience/MapPlace/GeoData/pages/default.aspx
GIS files relating to geology, minerals, and hydrology in British Columbia are available for download.

DataBC
http://catalogue.data.gov.bc.ca/dataset?download_audience=Public
The BC Data catalog offers a large number of datasets across several departments including Natural Resources, Economy, Justice, and Service.

Marine Conservation Analysis
http://bcmca.ca/maps-data/browse-or-search/

OpenDataBC
https://www.opendatabc.ca/dataset

Province of BC Geographic Data and Services
http://www2.gov.bc.ca/gov/content/governments/about-the-bc-government/databc/geographic-data-and-services
The Province of BC Geographic Data and Services includes a wide number of geographic services including iMapBC (offering hundreds of map data layers), B.C. Physical Address Geocoder (finding locations of addresses), and the Topographic Map Viewer (downloadable PDF topographic maps), to name a few.

Provincial Agricultural Land Commission—Maps and GIS
http://www.alc.gov.bc.ca/alc/content/alr-maps/maps-and-gis
The Provincial Agricultural Land Commission includes maps and GIS data related to agricultural parcels of land.

### County and City Data

Kamloops Open Data
http://www.kamloops.ca/downloads/maps/launch.htm
Kamloops Open Data offers a wide variety of data layers including cadastral, geology, hydrography, planimetric, telecommunications, orthophotos, and historical photos, as well as other county and city data.

Langley Township Open Data
https://data.tol.ca
Langley Township Open Data offers a large number of geospatial files and aerial photography for the Township of Langley, British Columbia.

Nanaimo Open Data
http://data.nanaimo.ca

North Okanagan Digital Data
http://www.rdno.ca/index.php/maps/digital-data
North Okanagan Digital Data includes typical county data as well as cadastral data and orthoimagery.

North Vancouver District Open Data
http://www.geoweb.dnv.org/data/
The North Vancouver District Open Data has over 170 datasets featuring biking routes, historical city boundaries, parcels, buildings, current and historical census, contours from 1 meter to 20 meters, and many more data layer files.

Prince George Open Data Catalogue
http://data.cityofpg.opendata.arcgis.com
The Prince George Open Data Catalogue has hundreds of city datasets including those related to infrastructure, property, transportation, parks and recreation, land use, government, topography, and more.

Surrey Open Data
http://surrey.ca/city-services/658.aspx

Vancouver GIS Data
http://data.vancouver.ca
Over 145 city datasets are available for the city of Vancouver, including data on street trees, parking tickets, crime, public washrooms, drinking fountains, and more.

Victoria Open Data
http://www.victoria.ca/EN/main/online-services/open-data-catalogue.html
Victoria Open Data provides GIS data layers for download and viewing from the interactive map application, VicMap (http://www.victoria.ca/EN/main/online-services/maps.html).

*Manitoba*

**Provincial Data**
Province of Manitoba Open Data
http://mli2.gov.mb.ca
The Manitoba Land Initiative offers a wide variety of maps and data related to cadastral boundaries, elevation, environment, forestry, geology, hydrography, land use, soil, and much more.

**County and City Data**
Winnipeg Open Data
http://www.winnipeg.ca/interhom/maps/
Winnipeg Open Data consists of a number of interactive maps including parks and amenities, e-CIS (address finder), and Navigo (Transit's online trip planner), as well as a large number of city-related open data files including aerial photography.

*New Brunswick*

**Provincial Data**
Digital Topographic Data Base (1998) New Brunswick
http://www2.gnb.ca/content/gnb/en/services/services_renderer.12755.Digital_Topographic_Data_Base_(1998).html
The Digital Topographic Data Base consists of topographic files available for download, including buildings, transportation, and other basic topographic layers.

GeoNB
http://www.snb.ca/geonb1/e/index-E.asp
GeoNB is New Brunswick's gateway to GIS information offering a wide variety of open data, including Lidar data.

**County and City Data**
Fredericton Open Data
http://data.fredericton.ca/en
Data layers are available for the city of Fredericton including city planning, recreation, public safety, transportation, finance, budgeting, and tourism.

St. John Open Data
http://www.moncton.ca/Government/Terms_of_use.htm

*Newfoundland*

GeoScience OnLine
http://gis.geosurv.gov.nl.ca

Government of Newfoundland Open Data
http://opendata.gov.nl.ca
Newfoundland Open Data offers open data on health, demographics, justice, and transportation.

*Northwest Territories*

Northwest Territories Discovery Portal by Northwest Territories Centre for Geomatics
http://www.geomatics.gov.nt.ca/nwtdp.aspx
The Northwest Territories Discovery Portal offers a large number of datasets for download, including applications; however, registration is necessary. It is a comprehensive source for environmental monitoring knowledge, offering opportunities for researchers to upload their own data.

NT GoMap by Northwest Territories Geoscience Office
http://ntgomap.nwtgeoscience.ca

*Nova Scotia*

**Provincial Data**
Forestry GIS Data by Natural Resources, Government of Nova Scotia
http://novascotia.ca/natr/forestry/gis/downloads.asp
The Forestry GIS site offers data related to forest inventory, wet areas, land capability, forest cover, and ecological land classification.

GeoNOVA
http://gov.ns.ca/geonova/home/default.asp
GeoNOVA is a gateway to GIS information in Nova Scotia, covering topics related to imagery, forestry, geology, location, and planning.

Nova Scotia Data Locator
http://gis8.nsgc.gov.ns.ca/datalocator
The Nova Scotia Data Locator offers downloadable files such as digital elevation models (DEM), forest inventory, Lidar, and topography.

Nova Scotia Geology
https://novascotia.ca/natr/meb/download/gis-data-maps.asp
Nova Scotia Geology consists of geological datasets available for download, including bedrock geology, surficial geology, hydrogeology, and mineral resource information.

**County and City Data**
Halifax Open Data
http://www.halifax.ca/opendata/

*Nunavut*

Canada-Nunavut Geoscience Office
http://cngo.ca
The Canada-Nunavut Geoscience Office offers resources related to Precambrian, Paleozoic, and Quaternary geology.

*Ontario*

### Provincial Data

AgMaps Geographic Information Portal (Ontario)
http://www.omafra.gov.on.ca/english/landuse/gis/portal.htm
AgMaps Geographic Information Portal offers interactive maps and GIS data related to land use, soil, and agriculture.

Land Information Ontario (LIO)
https://www.javacoeapp.lrc.gov.on.ca/geonetwork/srv/en/main.home
The Land Information Ontario's Discovering Ontario Data offers access to data layers related to elevation, geology, topography, imagery, and more.

OGS Earth
http://www.mndm.gov.on.ca/en/mines-and-minerals/applications/ogsearth
OGS Earth provides geological data collected by the Mines and Minerals division.

Ontario Basic Mapping (OBM)
http://www.geographynetwork.ca/website/obm/viewer.htm
The Ontario Base Mapping interactive map offers base map layers of dated topographic features in Ontario for viewing and download.

Ontario Government Data
https://www.ontario.ca/search/data-catalogue
Over two thousand government datasets are available for download, including arts and culture, business and economy, climate change, road conditions, and much more.

Ontario Road Network (Geography Network)
http://www.geographynetwork.ca/website/orn/viewer.htm
The Ontario Road Network offers street data for viewing and download.

Petroleum Well Data–Ontario
http://www.ogsrlibrary.com/data_free_petroleum_ontario
Petroleum Well Data offers data layers related to petroleum wells.

### County and City Data

Brampton Open Data
http://geohub.brampton.ca/pages/data
This site is the city of Brampton's public hub for exploring and downloading open data related to topics in culture and tourism, business, community, health and safety, land use, imagery, parks, and more.

Burlington Open Data
http://www.burlington.ca/en/services-for-you/Open-Data.asp
City of Burlington Open Data offers data on topics such as infrastructure, property, recreation and culture, land use, and transportation.

Essex Region Conservation Authority GIS and Interactive Mapping
http://ercamaps.countyofessex.on.ca

Grand River Information Network (GRIN)
http://maps.grandriver.ca
The Grand River Conservation Authority (GRCA) offers data for download through the Grand River Information Network. Data files include hydrological and climate resources.

Halton Region Data Centre
http://www.halton.ca/cms/One.aspx?portalId=8310&pageId=130441

Hamilton Open Data
https://www.hamilton.ca/city-initiatives/strategies-actions/open-data-program

London Open Data
https://www.london.ca/city-hall/open-data/Pages/Open-Data-Data-Catalogue.aspx
London Open Data offers a large number of municipal datasets related to recreation, such as tennis courts, parks, swing sets, trails, bicycle routes, and more.

Mississauga Open Data
http://www.mississauga.ca/portal/residents/mississaugadata
This resource offers open data on topics including population, growth forecasts, housing, employment, land use, vacant lands, and more for the city of Mississauga.

Niagara Falls Open Data
https://www.niagarafalls.ca/services/open/
This site provides open datasets for the city of Niagara Falls.

Niagara Region Open Data
http://www.niagararegion.ca/government/opendata/default.aspx?banner=1

Oakville Open Data
http://www.oakville.ca/data/

Ottawa-GeoOttawa
http://maps.ottawa.ca/geoottawa/
Ottawa-GeoOttawa includes an interactive map showing aerial maps, base maps, and a large number of municipal layers.

Ottawa Open Data
http://data.ottawa.ca
Open GIS data are available for a number of topics such as business and economy, city hall, demographics, environment, health and safety, planning, and development.

Peel Region Open Data
http://opendata.peelregion.ca
This site offers open data categories such as demographics, economics, environment, facilities and structures, and demographic forecasts.

Toronto Open Data
http://www.toronto.ca/open/catalogue.htm
Toronto Open Data offers a large number of datasets organized into fifteen categories.

Waterloo Open Data
http://www.waterloo.ca/en/opendata/index.asp
Waterloo Open Data includes datasets in a number of categories such as community, elections, environment, heritage, parks and recreation, points of interests, and transportation.

Waterloo Region Open Data
http://www.regionofwaterloo.ca/en/regionalGovernment/OpenDataCatalogue.asp

Welland Open Data
https://www.welland.ca/open/

Windsor Open Data
http://www.citywindsor.ca/opendata/Pages/Open-Data-Catalogue.aspx

York Region Open Data
https://ww6.yorkmaps.ca/YorkMaps/nindex.html

*Prince Edward Island*

GIS Data Catalogue by the Government of Prince Edward Island
http://www.gov.pe.ca/gis/
The GIS Data Catalogue consists of a small number of datasets including base maps, emergency data, and community-based data. Other datasets are available for a fee.

*Quebec*

### Provincial Data
Quebec Geographic
http://www.quebecgeographique.gouv.qc.ca/index.asp

### County and City Data
Montreal Open Data
http://donnees.ville.montreal.qc.ca/dataset
A large number of datasets are available within the categories of government and finances, transportation, tourism, culture, agriculture, economy, and much more.

*Saskatchewan*

### Provincial Data
Open Data SK by Northern Lights Data Lab
http://opendatask.ca/data/
This site lists a large number of geospatial resources available for Saskatchewan, including links to governmental organizations such as Finance, Energy, and Statistics.

### County and City Data
Regina Open Datasets
http://open.regina.ca
Open data for the city of Regina includes GIS files within several topics such as community and recreation, elections, heritage and history, land use, transportation, and more.

Saskatoon Open Datasets
http://opendata-saskatoon.cloudapp.net

*Yukon*

Environment Yukon
http://www.env.gov.yk.ca/publications-maps/geomatics/data.php
The Yukon Department of Environment maintains three broad categories of data: administrative (game management areas, outfitting concessions, registered trapping concessions), base map, and the Wildlife Key Area Inventory (locations used by wildlife for critical, seasonal life functions).

GeoYukon
http://www.geomaticsyukon.ca/data/datasets
GeoYukon offers access to the Yukon government's data and imagery. Users can view and download a number of base, topographic, and geologic data files.

## AMERICAN DATA RESOURCES

### National Data

American FactFinder
http://factfinder2.census.gov
American FactFinder provides census data (population, housing, economic) in the form of geospatial datasets, tables, and maps.

Bureau of Land Management Georeferenced PDF Maps
https://www.blm.gov/maps/georeferenced-PDFs
The Bureau of Land Management provides a series of georeferenced PDF maps pertaining to land management.

Bureau of Ocean Energy Management, Regulation, and Enforcement (BOEMRE): Maps and GIS Data
https://www.boem.gov/Maps-and-GIS-Data/
The Bureau of Ocean Energy Management, Regulation, and Enforcement (BOEMRE) provides administrative ocean boundary data. In addition, an interactive map displays land information, including legal, property ownership, physical, biological, ocean use, and cultural information.

Commission for Environmental Protection National Environmental Atlas
http://www.cec.org/sites/default/atlas/map/
The Commission for Environmental Protection National Environmental Atlas provides environmental data (ecosystems, climate, pollution, waste, and so forth) that pertain to North America.

CropScape
http://nassgeodata.gmu.edu/CropScape
CropScape provides data on cropland.

Data Basin
http://databasin.org
Data Basin provides biological, physical, and socioeconomic data.

Earth Resources Observation and Science Data Center
http://eros.usgs.gov/find-data
The Earth Resources Observation and Science (EROS) Center provides aerial photographs, digitized maps, and satellite imagery, in addition to elevation, land cover, topographic, and remotely sensed data.

Education Demographic and Geographic Estimates (EDGE)
http://nces.ed.gov/surveys/sdds/index.aspx
The Education Demographics and Geographic Estimates (EDGE) program provides data and information to help identify and understand the social and spatial characteristics of education in the United States.

Environmental Dataset Gateway
https://edg.epa.gov/metadata/catalog/main/home.page
The U.S. Environmental Protection Agency (EPA) provides data pertaining to the environment: climate change, forest conservation, ecoregions, environmental justice, and so on.

Federal Geographic Data Products
http://www.fgdc.gov
The Federal Geographic Data Committee provides geospatial data that cover a great number of topics. These data were collected from a number of federal agencies.

FedStats
http://fedstats.sites.usa.gov
FedStats provides federal statistical data and information that pertains to the United States.

FEMA Flood Map Service Center
https://msc.fema.gov
The FEMA Flood Map Service Center provides flood hazard data.

Food Access Research Atlas (USDA)
http://www.ers.usda.gov/data-products/food-access-research-atlas/go-to-the-atlas.aspx#.U1EryFf5O58
The Food Access Research Atlas, developed by the U.S. Department of Agriculture, provides data regarding food access for various communities in the United States.

Data.gov
https://www.data.gov/geospatial/
Data.gov provides a multitude of geospatial data and resources including an interactive map of submerged land information and maps of air quality and water resources.

GeoMAC Wildfire Viewer
http://www.geomac.gov
The GeoMAC Wildfire Viewer provides data regarding active and past fires that have occurred within the United States.

GeoPlatform
http://www.geoplatform.gov
GeoPlatform provides a great number of national datasets and web services.

Geospatial Data Gateway
http://datagateway.nrcs.usda.gov
The Geospatial Data Gateway provides natural resources data (soil, rural development, elevation, and so forth) as well as imagery for several years.

HydroSHEDS
http://hydrosheds.cr.usgs.gov/index.php
HydroSHEDS, a mapping product from the USGS, provides hydrological data and maps at a regional and global scale.

Industry Statistics Portal (ISP)
http://www.census.gov/econ/isp/
The Industry Statistics Portal (ISP) provides economic data for a number of industries in the United States.

Mineral Resources Online Spatial Data
http://mrdata.usgs.gov
The USGS Mineral Resources Online Spatial Data site provides regional and global datasets pertaining to geology, geochemistry, geophysics, and mineral resources.

National Agricultural Statistics Service (NASS)
http://www.nass.usda.gov

The National Agricultural Statistics Service (NASS) provides U.S. census data pertaining to agriculture.

National Archive of Criminal Justice Data (NACJD)
http://www.icpsr.umich.edu/icpsrweb/NACJD/index.jsp
The National Archive of Criminal Justice Data (NACID) provides U.S. crime and justice data.

National Center for Health Statistics (NCHS)
http://www.cdc.gov/nchs/index.htm
The National Center for Health Statistics (NCHS) provides statistical health data (diseases, injuries, disabilities, health care and insurance, and so on) pertaining to the United States.

National Centers for Environmental Information
https://maps.ngdc.noaa.gov
The National Centers for Environmental Information provide interactive maps and web services pertaining to the environment, including topics such as historical declination, thermal springs, and global natural hazards.

National Centers for Environmental Information
http://www.ngdc.noaa.gov/dem/
The National Centers for Environmental Information, part of the National Oceanic and Atmospheric Administration (NOAA), provides high-resolution coastal digital elevation models (DEMs) that integrate ocean bathymetry and land topography.

National Climatic Data Center
http://lwf.ncdc.noaa.gov/oa/ncdc.html
The National Climatic Data Center provides climate and historical weather data and information.

National Elevation Dataset
http://ned.usgs.gov
The USGS National Elevation Dataset provides elevation data in the form of contour lines and digital elevation models (DEMs).

National Historical Geographic Information System (NHGIS)
https://nhgis.org
The National Historical Geographic Information System (NHGIS) provides housing, agricultural, population, economic, and boundary data for the United States.

National Hydrography Dataset
http://nhd.usgs.gov
The National Hydrography Dataset provides data regarding drainage networks in the United States.

National Land Cover Database
http://www.mrlc.gov
The National Land Cover Database provides data pertaining to land cover at a national scale.

The National Map
https://nationalmap.gov
The National Map provides current and historical topographic maps of the United States in addition to datasets pertaining to elevation, hydrography, structures, geographic names, boundaries, orthoimagery, and more.

National Ocean Service Data Explorer
http://oceanservice.noaa.gov/dataexplorer/welcome.html
The National Ocean Service Data Explorer offers interactive mapping tools that allow users to locate NOS products in any area within the United States.

National Oceanographic Data Center
http://www.nodc.noaa.gov
The National Oceanographic Data Center provides oceanographic data for the United States.

National Park Service Data Store
http://www.nps.gov/GIS/data_info/
The National Park Service Data Store provides data pertaining to the national parks' natural and cultural resources. Examples of datasets include park boundaries, natural resource inventories, and park science reports.

National Park Service GIS Datasets
https://irma.nps.gov/App/Portal/Home/FeaturedContent
The National Park Service provides various geospatial data, web mapping tools, and reports pertaining to national parks within the United States.

National Resources Conservation Services (NRCS) Geospatial Data Gateway
https://gdg.sc.egov.usda.gov
The NRCS Geospatial Data Gateway provides data pertaining to natural resources in the United States (soil, elevation, conservation, rural development, and the like).

National Weather Service
http://www.weather.gov
The National Weather Service provides interactive weather maps showing current conditions.

Natural Earth
http://www.naturalearthdata.com/downloads/
Natural Earth provides cultural, physical, and raster data. Maps are available in three different scales: 1:10m, 1:50m, and 1:110m.

Nature Conservancy Core Datasets
http://maps.tnc.org/gis_data.html
The Nature Conservancy provides geospatial data and web maps pertaining to conservation.

NOAA Shoreline Data Explorer
http://www.ngs.noaa.gov/newsys_ims/shoreline/index.cfm
The National Oceanic and Atmospheric Administration (NOAA) Shoreline Data Explorer provides data pertaining to shorelines in the United States.

North American Transportation Atlas Data (NORTAD)
http://www.bts.gov/publications/north_american_transportation_atlas_data/
The North American Transportation Atlas Data (NORTAD) provides statistical transportation data.

Open Geoportal
http://opengeoportal.org
The Open Geoportal (OGP) provides a multitude of geospatial data. Examples of topics include land use, temperature, elevation, and population.

OpenTopography
http://www.opentopography.org/index.php
OpenTopography provides high-resolution topography data and tools. LIDAR point cloud, raster, and Google Earth data are provided.

Socioeconomic Data and Application Center (SEDAC)
http://sedac.ciesin.columbia.edu
The Socioeconomic Data and Application Center (SEDAC) provides a diverse range of datasets. Some of the topics covered include poverty, population, infrastructure, sustainability, and health.

Social Explorer
http://www.socialexplorer.com
Social Explorer provides a multitude of demographic, health, religion, crime, and economic data in the form of maps, reports, and data downloads.

Surface-Water Data for California
http://waterdata.usgs.gov/ca/nwis/sw/
The USGS Surface-Water Data website provides California data pertaining to stream levels, streamflow (discharge), reservoir and lake levels, surface-water quality, and rainfall.

USA Data
http://www.data.gov
This resource provides U.S. data pertaining to a multitude of different topics including agriculture, climate, ecosystems, education, energy, health, manufacturing, and science.

U.S. Census Bureau GIS Data
http://www.census.gov/geo/maps-data/
The U.S. Census Bureau GIS Data website provides a diverse range of geographic data and map products, including TIGER/Line Shapefiles, KMLs, TIGERweb, cartographic boundary files, thematic and reference maps, and geographic relationship files.

U.S. Department of Agriculture
http://www.ers.usda.gov/data-products.aspx
The U.S. Department of Agriculture provides data pertaining to agriculture, the environment, and natural resources. Examples of topics include food safety, rural economy and population, food choices and health, and crops.

U.S. Fish and Wildlife Service National GIS Datasets
http://www.fws.gov/gis/data/national/index.html
The U.S. Fish and Wildlife Service provides various data that pertain to conservation, wildlife, and ecosystems.

U.S. Geological Survey
http://earthexplorer.usgs.gov
With the USGS Earth Explorer, users can search, query, or purchase various satellite images, cartographic products, and aerial photographs from several different sources.

U.S. Geological Survey Disease Maps
https://diseasemaps.usgs.gov/mapviewer/
The U.S. Geological Survey Disease Maps provide information regarding disease and infection cases for humans, mosquitos, birds, sentinel animals, and veterinary animals.

U.S. Geological Survey Earthquake Hazard Program
http://earthquake.usgs.gov
The USGS Earthquake Hazard Program provides data about the significant earthquakes that have occurred within the United States.

U.S. Geological Survey Map Locator and Downloader
http://store.usgs.gov
The USGS Map Locator and Downloader allows users to search for historical topographic and topographic quadrangle maps based on their area of interest.

U.S. Geological Survey National Map
http://nationalmap.gov/viewers.html
The USGS Survey National Map provides a multitude of topographic maps. Additionally, data pertaining to boundaries, elevation, imagery, land cover, transportation, and woodlands are also provided.

U.S. Geological Survey Water Data
http://waterdata.usgs.gov/usa/nwis/nwis
The USGS Water Data site provides water-resources data for all fifty states, plus Puerto Rico, the Virgin Islands, Guam, American Samoa, the District of Columbia, and the Commonwealth of the Northern Mariana Islands. Water-resources data includes surface water, groundwater, water quality, and water use.

Web Soil Survey
http://websoilsurvey.nrcs.usda.gov/app/HomePage.htm
The United States Department of Agriculture provides soil data produced by the National Cooperative Soil Survey. State, County, and City Data

*Alabama*

**State Data**
Alabama GIS
http://www.alabamagis.com
Alabama GIS offers a comprehensive list of Alabama counties that offer free GIS data resources.

AlabamaView (Alabama GIS Data Portal)
http://www.alabamaview.org/index.php
AlabamaView is a consortium offering access to GIS layers and remotely sensed imagery. GIS shapefiles include roads, hydrography, urban areas, railway, and census tracts.

Geological Survey of Alabama (Oil and Gas Board)
http://www.gsa.state.al.us
The Geological Survey of Alabama offers maps and data pertaining to geology, groundwater assessment, ecosystem investigations, and topography.

University of Stanford Data Links
https://library.stanford.edu/research/stanford-geospatial-center/data
The Stanford Geospatial Center has a host of comprehensive base and thematic datasets. Their website lists many useful external links as well.

**County and City Data**
Alabama GIS
http://www.alabamagis.com
Alabama GIS offers a comprehensive list of Alabama counties that offer free GIS data resources.

Auburn—City of Auburn GIS
https://www.auburnalabama.org/maps/
The city of Auburn offers a large number of interactive maps; however, GIS files are available by request only and are not online.

*Alaska*

Alaska State Geospatial Data Clearinghouse
http://www.asgdc.state.ak.us
Alaska Department of Natural Resources offers spatial data sources with topics related to culture, boundaries, tax parcels, environment, monuments, parks and recreation, physical features, transportation, and more.

*Arizona*

**State Data**
Arizona Land Resource Information System
http://www.land.state.az.us/alris/
The Arizona Land Resource Information System includes interactive maps; GIS layers, although available, need to be ordered.

AZGEO Clearinghouse
https://azgeo.az.gov/azgeo/about-azgeo
The AZGEO Clearinghouse is an initiative of the Arizona Geographic Information Council and offers users access to interactive maps and downloadable geospatial data, including administrative boundaries, demographic, environment, hydrology, mining, transportation, imagery, and more.

**County and City Data**
Glendale Open Data Portal
http://data-cog-gis.opendata.arcgis.com

Maricopa County GIS Portal
http://gis.maricopa.gov/index.html
The Maricopa County GIS Portal includes GIS mapping applications that offer historical aerials, planning and zoning resources, and parks and trail information.

Mesa Open Data Portal
http://open.mesaaz.gov/home
The Mesa Open Data Portal offers city datasets such as energy and utilities, financial resources, recreation and culture, zoning and property, public safety, and much more.

Phoenix Open Data Portal
https://www.phoenix.gov/opendata
The Phoenix Open Data Portal offers a mapping portal where datasets can be viewed. Datasets that can be downloaded include census, parks, transportation, property, and development.

Pima County Geographic Data
http://gis.pima.gov/data/guide/

Scottsdale
https://lib.asu.edu/geo/data
The City of Scottsdale GIS Data and Applications site offers a select number of geospatial data resources for download as well as a handful of interactive maps related to city services, public amenities, property, and neighborhood information.

Tempe Open Data Portal
http://data-tempegov.opendata.arcgis.com
The City of Tempe Open Data Portal offers a platform for exploring and accessing geospatial data related to transportation, neighborhoods and historic preservation, land use, and public safety.

Tuscon GIS Data
http://gisdata.tucsonaz.gov

## Arkansas

### State Data

Arkansas GIS Office
http://gis.arkansas.gov
The Arkansas GIS Office has a large number of GIS files available for download including those related to climatology, elevation, environment, agriculture, health, society, cadastre, utilities, transportation, and more.

Arkansas Natural Resources Conservation Service
http://www.ar.nrcs.usda.gov
The Arkansas Natural Resources Conservation Service offers data related to soil, water, and climate.

Arkansas Watershed Data
http://watersheds.cast.uark.edu

Northwest Arkansas Regional Planning Commission—Regional Maps
http://nwarpc.org/planning/gis-and-mapping/
The Northwest Arkansas Regional Planning Commission has several map datasets available for download related to political boundaries, demographics, transportation, environment, and storm water.

### County and City Data

Benton County GIS Home Page
http://www.bentoncountyar.gov/BCHome.aspx?info=Agency/info/GISInfo.html&page=Agency/GIS/Default.aspx
The Benton County GIS department has several interactive and downloadable maps related to parcels, sex offenders, food inspections, and more.

Washington County GIS Portal
http://arcserv.co.washington.ar.us/flex/public/
The Washington County GIS Portal includes an interactive map that allows users to view base map, planning, and imagery data.

## California

### State Data

Association of Bay Area Governments (ABAG) GIS and Maps
http://gis.abag.ca.gov
The Association of Bay Area Governments (ABAG) GIS and Maps has a variety of data available including land use planning, hazard planning (earthquakes, flooding, wildfire, landslides, tsunamis), and trails.

Cal-Atlas
http://portal.gis.ca.gov/geoportal/
Cal-Atlas is California's geoportal, offering a very large number of data files related to public safety, natural resources, education, health, and government.

California Biogeographic Data
http://www.dfg.ca.gov/biogeodata/gis/clearinghouse.asp
The California Biogeographic Data site offers downloadable datasets including hydrology, hunting zones, wetlands, and vegetation.

California Climate Commons
http://climate.calcommons.org
California Climate Commons is a discovery tool for climate change data and related resources.

California Farmland Mapping and Monitoring Program
http://www.conservation.ca.gov/dlrp/fmmp/Pages/Index.aspx
The California Farmland Mapping and Monitoring Program offers datasets related to farmland mapping and monitoring.

California Geographic Data
http://data.ca.gov
The California Geographic Data site has several resources available pertaining to water, economy and demographics, fleet and transportation, recycling, and buildings.

California Geographical Survey
http://geogdata.csun.edu

California Geoportal
http://portal.gis.ca.gov/geoportal/catalog/main/home.page

California Geothermal Map Data
http://www.conservation.ca.gov/dog/geothermal/maps/Pages/Index.aspx
The California Geothermal Map Data site has geothermal-related data layers, showing geothermal fields, as well as those related to energy resources, showing sedimentary basins, oil, gas, and pipeline data.

California GIS Directory
http://www.coordinatedlegal.com/gis.html
The California GIS Directory is a comprehensive list of county GIS sites for California.

California Natural Diversity Database (CNDD)
http://www.dfg.ca.gov/biogeodata/cnddb/

California Vegetation Map Catalog
http://ice.ucdavis.edu/project/veg-map-catalog

CalTrans GIS Data
http://www.dot.ca.gov/hq/tsip/gis/datalibrary/gisdatalibrary.html

CalTrans GIS focuses on transportation-related data layers such as aviation, highway, and rail; the layers are available in downloadable GIS format.

Fire and Resource Assessment (FRAP) Data
http://frap.fire.ca.gov
The Fire and Resource Assessment Program offers datasets related to fire prevention and assessment, including protection areas, forestry boundaries, facilities, fire threats, fire hazards, and more.

GeoData@UC Berkeley
https://geodata.lib.berkeley.edu
GeoData@UC Berkeley is a resource that lists several different data resources for California.

National Assessment of Shoreline Change
http://pubs.usgs.gov/of/2006/1251/#gis
The Sea-Level Change GIS Data site includes datasets related to shoreline change analysis that are available for download.

Southern California Association of Governments Data Center
http://www.scag.ca.gov/resources.htm
The Southern California Association of Governments Data Center includes government resources related to housing and the economy.

Surface Water Data for California
http://waterdata.usgs.gov/ca/nwis/sw/

### County and City Data

Association of Bay Area Governments (ABAG) GIS and Maps
http://gis.abag.ca.gov/
The Association of Bay Area Governments offers datasets related to land use planning and hazard planning.

Davis City Mapping and Geographic Information Systems
http://gis.cityofdavis.org
The city of Davis has a large number of interactive maps available offering information about parcels, historic properties, government facilities, zoning, building permits, flood zones, street lights, street trees, and much more.

Humboldt County GIS Data
http://co.humboldt.ca.us/planning/maps/datainventory/gisdatalist.asp
Humboldt County GIS datasets include administrative boundaries, fish and wildlife, demographics, fire plan, geology, hazard disclosure, hydrography, imagery, and aerial photography.

Los Angeles County Department of Regional Planning
http://planning.lacounty.gov/gis/download
A large number of interactive maps and GIS data are available for download. Layers are related to planning and zoning, land use, coastal program, and airport land use, among others.

Los Angeles County GIS Data Portal
http://egis3.lacounty.gov/dataportal/

Los Angeles Data
https://data.lacity.org

Riverside County Open Data
https://data.countyofriverside.us

Russian River Watershed Data
http://bwc.lbl.gov/RussianRiver/
The Russian River Watershed Data offers resources related to hydrology and wastershed themes.

San Diego Association of Governments (SANDAG) Maps and GIS
http://www.sandag.org/index.asp?subclassid=70&fuseaction=home.subclasshome
The SANDAG Geographic Boundary Viewer displays a wide variety of geographic data and information for the San Diego region. Available data layers include census tracts and ZIP codes, community planning areas, school and legislative districts, major statistical and subregional areas, and jurisdictional boundaries.

San Diego Geographic Information Source (SanGIS)
http://www.sangis.org
The San Diego Geographic Information Source consists of an interactive map viewer that provides parcel information for addresses in San Diego. GIS datasets are also available for download, but users need to register first.

San Francisco Bay Area Regional Database (BARD)
http://bard.wr.usgs.gov
The San Francisco Bay Area Regional Database includes historical maps, elevation maps, and imagery available from the USGS San Francisco Bay Area USGS database.

*Colorado*

### State Data

Colorado Department of Local Affairs GIS Maps and Data
https://demography.dola.colorado.gov/gis/
The Colorado Department of Local Affairs GIS Maps and Data site offers administrative boundaries and census data for download.

Colorado Department of Transportation (CDOT)
http://dtdapps.coloradodot.info/otis
The Colorado Department of Transportation offers transportation-related interactive maps and datasets.

Colorado Geological Mapping
http://coloradogeologicalsurvey.org/geologic-mapping/gis-data/

The Colorado Geological Mapping database offers geological datasets for download.

### County and City Data

Boulder City Maps and GIS
https://bouldercolorado.gov/open-data

Boulder County Geospatial Open Data
http://gis.bouldercounty.opendata.arcgis.com
The Boulder County Geospatial Open Data site includes a wide variety of datasets such as buildings, watershed, elections, elevation, floodplain, hydrography, property, public safety, recreation, storm-water management, transportation, wildlife, and more.

Denver Maps
http://www.denvergov.org/maps
Denver Maps has interactive maps that include property, crime, marijuana licenses, zoning, outdoor warning sirens, floodplain, development services, and more.

Denver Regional Council of Governments (DRCOG) Regional Data Catalog
http://gis.drcog.org/datacatalog/

Open Colorado
http://data.opencolorado.org/dataset

*Connecticut*

Connecticut Census Data
http://ctsdc.uconn.edu/connecticut_census_data/

Connecticut GIS Data
http://www.ct.gov/dep/gis
Connecticut GIS Data has datasets related to parks and forests, boating, fishing, hunting, and recycling.

Connecticut State Data Center
http://ctsdc.uconn.edu
The Connecticut State Data Center has a wide variety of datasets for the state of Connecticut including census, education, election, hydrography, places, and transportation.

*Delaware*

### State Data

Delaware Census State Data Center
http://stateplanning.delaware.gov/census_data_center/

Delaware Department of Education (DEDOE) Mapping
http://doearcgis.doe.k12.de.us/doemapping/

Delaware Geological Survey
http://udspace.udel.edu/handle/19716/2519
The Delaware Geological Survey offers geological datasets for the state of Delaware.

Delaware Open Data
https://data.delaware.gov
Delaware Open Data offers datasets related to economic development, heath, education, energy and environment, government and finance, recreation, and public safety.

Delaware Spatial Data Clearinghouse
http://www1.udel.edu/dsacher/gis/nsdi/clearinghouse/

FirstMap
http://firstmap.gis.delaware.gov/
FirstMap is open data for the state of Delaware.

### County and City Data

Kent County Online Mapping
http://www.co.kent.de.us/gis-division/public-mapping-web.aspx

New Castle County Geographic Information Services Map Viewer
https://gis.nccde.org/gis_viewer/

*District of Columbia*

DC GIS Services
https://octo.dc.gov/service/dc-gis-services
The District of Columbia Geographic Information System (DC GIS) provides geospatial data and enterprise applications to district agencies and the public.

*Florida*

Florida Geographic Data Library
http://www.fgdl.org/metadataexplorer/explorer.jsp

Florida GIS Resources
http://www.library.fau.edu/depts/govdocs/federal/florida_links.htm
This site has a comprehensive list of geospatial resources available for the state of Florida.

*Georgia*

### State Data

Georgia Open Data
http://data-georgiagio.opendata.arcgis.com/

Georgia's GIS Clearinghouse
https://data.georgiaspatial.org/login.asp
Georgia's GIS Clearinghouse offers free downloads of post-1980s aerial photography and imagery of Georgia. GIS files include roads, utilities, education, geology, demographic, and more. Registration is required.

Georgia's Natural, Archaeological and Historic Resources GIS (GNAHRGIS)
https://www.gnahrgis.org/gnahrgis/index.do
Georgia's Natural, Archaeological and Historic Resources GIS is an interactive GIS designed to catalog information about the natural, archaeological, and historic resources of Georgia.

Georgia's Tax Assessors
http://www.gaassessors.com
Georgia's Tax Assessors is a one-stop portal to assessment, tax, parcel, and GIS data for Georgia counties.

University of Georgia Library Data Collection
http://guides.libs.uga.edu
A comprehensive list of geospatial resources available for the state of Georgia.

### County and City Data
Atlanta City GIS
http://gis.atlantaga.gov

Atlanta Regional Commission GIS Data
https://atlantaregional.org/browse/?browse=type&type=data-maps
The Atlanta Regional Commission provides access to various data and shapefiles for the Atlanta metropolitan region.

Fulton County GIS
http://www.fultoncountyga.gov/fcgis-home

Fulton County Open Data Portal
http://fultoncountyopendata-fulcogis.opendata.arcgis.com
The Fulton County Open Data Portal is a platform for sharing data resources for Fulton County, Georgia.

Sandy Springs GIS Data
http://data-coss.opendata.arcgis.com
Sandy Springs GIS Data includes a wide variety of municipal datasets for Sandy Springs.

## Hawaii

### State Data
Hawaii Statewide GIS Program
http://geoportal.hawaii.gov
The Hawaii Statewide GIS Program offers a large number of datasets related to business and economy, census, climate, coastal and marine environment, elevation, fresh water, hazards, geodetic control, land use, parcels and zoning, terrestrial environment, and transportation.

Office of Planning GIS Data
http://planning.hawaii.gov/gis/download-gis-data/

### County and City Data
Honolulu Land Information Systems
http://gis.hicentral.com
Honolulu Land Information Systems offers interactive GIS web mapping and data access.

Honolulu Open Geospatial Data
http://honolulu-cchnl.opendata.arcgis.com
Honolulu Open Geospatial Data provides access to a variety of datasets covering Honolulu.

Maui County GIS
http://www.mauigis.net/data/

## Idaho

### State Data
Boise City Open GIS Data
http://opendata.cityofboise.org
Boise City Open GIS Data includes parks and recreation, utilities, and parking.

Idaho Department of Water Resources Maps and GIS
https://www.idwr.idaho.gov/GIS/
The Idaho Department of Water Resources offers interactive maps and datasets that are hydrological in nature.

Idaho Open Data Portal
http://data.gis.idaho.gov

Inside Idaho
http://inside.uidaho.edu
Inside Idaho is Idaho's official geospatial data clearinghouse.

### County and City Data
Bonner County GIS
http://bonnercounty.us/technology/gis/

Jefferson County GIS
http://www.co.jefferson.id.us/gis_mapping.php

## Illinois

### State Data
Illinois Data Portal
https://data.illinois.gov
The Illinois Data Portal offers a wide variety of government data related to natural resources, transportation, human resources, government, and students.

Illinois Geospatial Data Clearinghouse
http://clearinghouse.isgs.illinois.edu
The Illinois Geospatial Data Clearinghouse offers USGS digital topographic maps, aerial photography, orthoimagery, orthophotography, and GIS files related to geology, land use, natural resources, and infrastructure.

### County and City Data

Chicago Data Portal
https://data.cityofchicago.org
The Chicago Data Portal offers datasets related to transportation, historic preservation, ethics, education, environment, sanitation, facilities, health and human resources, and more.

Cook County Open Data
https://datacatalog.cookcountyil.gov

DuPage County GIS Open Data
http://dupageilgis2016–12–02t201335972z-dupage.opendata.arcgis.com
The DuPage County GIS Open Data site has a number of datasets including parcel cadastral, LiDAR, hydrology, transportation, and schools.

Illinois County GIS Resources
http://www.gis2gps.com/GIS/illcounties/illcounties.html
Illinois County GIS Resources consists of links to GIS maps and data for most counties in Illinois.

Lake County GIS Products
http://www.lakecountyil.gov/636/GIS-Division
Lake County GIS Products include GIS datasets that are available online from the Lake County government.

Lake County Open Data
https://datacatalog.cookcountyil.gov

McLean County Regional GIS Consortium
http://www.mcgis.org
The McLean County Regional GIS Consortium has a number of resources including multiple interactive web mapping applications providing data on school districts, local elections, and county parks, as well as government GIS data.

Metro Chicago Data
https://www.metrochicagodata.org
The Metro Chicago Data site offers interactive information and data from the city of Chicago, Cook County, and the state of Illinois. Users can perform keyword searches or browse a unified set of topics and categories.

Shawnee National Forest
https://www.fs.usda.gov/main/shawnee/landmanagement/gis
The Shawnee National Forest site offers downloadable GIS datasets for the Shawnee National Forest.

Will County GIS Data
http://www.willcogis.org/website2014/gis/data.html
Will County GIS Data consists of topographic data, imagery, and a large number of municipal files for Will County.

*Indiana*

### State Data

IndianaMAP
http://indianamap.org
IndianaMAP is open data for the state of Indiana, including geology, demographics, hydrology, and imagery.

Indiana Spatial Data Portal
http://gis.iu.edu
The Indiana Spatial Data Portal (ISDP) offers statewide datasets including aerial photos, topographic maps, and elevation data. In addition, the ISDP hosts several local datasets for Allen, Bartholomew, Boone, Dearborn, Gibson, Hamilton, Hancock, Hendricks, Johnson, Marion, Monroe, Morgan, Posey, Shelby, and Wayne Counties.

IndianaView
http://www.indianaview.org
IndianaView offers Landsat images for several Indiana counties.

### County and City Data

City of Bloomington GIS
http://bloomington.in.gov/maps/
City of Bloomington GIS includes a map gallery, as well as an interactive map and GIS data downloads.

Indianapolis and Marion County GIS
http://maps.indy.gov/MapIndy/
The site provides interactive ArcIMS maps for the city of Indianapolis and Marion County.

LakeRim GIS
https://igs.indiana.edu/LakeRim/index.cfm
LakeRim GIS is a repository for GIS data for the Indiana counties adjacent to Lake Michigan.

Monroe County GIS
http://www.co.monroe.in.us/tsd/GIS.aspx
Monroe County GIS is an interactive map that includes parcel data, zoning, imagery, and more.

SAVI
http://www.savi.org/savi/
SAVI is a comprehensive electronic database of mapped and tabular data about the Indianapolis Metropolitan Statistical Area (MSA), which refers to Marion County and its eight contiguous counties: Boone, Hamilton, Madison, Hancock, Shelby, Johnson, Morgan, and Hendricks.

Wayne County GIS
http://www.co.wayne.in.us/gis/

*Iowa*

### State Data

Iowa Geographic Map Server
http://ortho.gis.iastate.edu

The Iowa Geographic Map Server provides access to Iowa geographic map data through an online map viewer and through Web Map Service (WMS) connections for GIS. Map layers include imagery, elevation, land use, and land cover.

### *County and City Data*
Boone County Iowa GIS
https://beacon.schneidercorp.com/Application.aspx?AppID=84&LayerID=795

### *Kansas*

#### *State Data*
State of Kansas GIS Data Access and Support Center
http://www.kansasgis.org/index.cfm
The State of Kansas GIS Data Catalog includes administrative boundaries, cadastral mapping, infrastructure, elevation, and imagery.

#### *County and City Data*
Hays City / Ellis County Geospatial Data Portal
http://www.geodataportal.net/index.html
The Geospatial Data Portal provides a one-stop Internet resource for locating GIS data related to the city of Hays and Ellis County, Kansas.

Johnson County Animated Mapping System
http://aims.jocogov.org/AIMSData/FreeData.aspx
The Johnson County Animated Mapping System includes GIS files related to administrative boundaries, economic development, elections, the environment, FEMA floods, parks and recreation, schools, and services.

Lawrence City GIS Data
https://lawrenceks.org/maps/#gis-data

Sedgwick County GIS Data and Maps
http://www.gis.sedgwick.gov/whatsavailable.asp
Municipal data layers are available for Sedgwick County, Kansas.

### *Kentucky*

#### *State Data*
Kentucky Geography Network (KyGeoNet)
http://kygeonet.ky.gov
The KyGeoNet is the geospatial data clearinghouse for the Commonwealth of Kentucky. A variety of datasets can be located and downloaded, and static map products can be viewed.

Kentucky Geoportal
http://kygisserver.ky.gov/geoportal/catalog/main/home.page
The Kentucky Geoportal is a data clearinghouse that provides access to a large number of imagery and GIS files.

Kentucky GovMaps
http://kygeonet.ky.gov/govmaps/
Kentucky GovMaps has several web mapping applications that feature airports, soils, elevation, electric service areas, legislative districts, school locations, geologic mapping, and more.

Kentucky Open GIS Data
http://kygovmaps-kygeonet.opendata.arcgis.com
Kentucky Open GIS Datasets featured on this site will be of interest to a broad range of audiences including those involved in research, community planning, economic development, education, decision support, engineering, and public policy.

Kentucky State Data Center
http://www.ksdc.louisville.edu
The Kentucky State Data Center (KSDC) is the state's lead agency in the U.S. Census Bureau's State Data Center Program and Kentucky's official clearinghouse for Census data. Datasets available for download are related to statistics and demographics.

#### *County and City Data*
Louisville Metro Open Data Portal
https://data.louisvilleky.gov

Louisville Open GeoSpatial Data
http://www.lojic.org/main/
The Louisville Open GeoSpatial Data provides users with datasets related to addresses, demographics, elevation, environment, flooding, planning, public safety, and more.

Murray City
http://murrayky.gov/planning/gis/index.htm
Murray City has a GIS website with an interactive mapping application that features zoning boundaries, sanitation route boundaries, school district boundaries, voting precincts, topographic information, city streets, and public buildings.

### *Louisiana*

Atlas Louisiana GIS
https://atlas.ga.lsu.edu
Atlas Louisiana is a data distribution system hosting a variety of publicly available GIS datasets for the state of Louisiana.

Louisiana Department of Transportation and Development
http://wwwsp.dotd.la.gov/Business/Pages/GIS_Maps.aspx
Louisiana Department of Transportation and Development offers users up-to-date maps for travel planning, GIS data, emergency/evacuation, relief funding eligibility, and more.

Louisiana FloodMaps Portal
http://maps.lsuagcenter.com/floodmaps/

Louisiana Geographic Information Center
http://lagic.lsu.edu

The Louisiana Geographic Information Center (LAGIC) contains a list of the available resources managed or used by LAGIC and the community, including data, interactive maps, and metadata.

*Maine*

Maine Department of Environmental Protection
http://www.maine.gov/dep/gis/datamaps/
The Maine Department of Environmental Protection offers a large number of interactive maps and datasets that include topics such as land resources, remediation and waste management, and air quality.

*Maryland*

**State Data**
Maryland's Mapping and GIS Data Portal (MD iMap)
http://imap.maryland.gov/Pages/default.aspx
Maryland's Mapping and GIS Data Portal (MD iMap) is the primary source for data and maps from Maryland state government agencies. Of particular value is the GIS Data Catalog and Data Downloads and Services, which offer a large number of datasets for download.

Maryland State Data Center
http://planning.maryland.gov/msdc/S5_Map_GIS.shtml
The Maryland State Data Center provides ZIP code boundary files (based on MdProperty View), Census 2010 boundary maps, historical census maps (1990 and 2000), and some other useful resources.

**County and City Data**
Baltimore City Open Data
https://data.baltimorecity.gov
Baltimore City Open Data includes city data as well as data about crime, citations, property taxes, parcel boundaries, building footprints, and more.

Baltimore County GIS Data Portal
http://gis.baltimorecountymd.gov/dataportal/

*Massachusetts*

**State Data**
Harvard Geospatial Library
http://hgl.harvard.edu:8080/opengeoportal/
The Harvard Geospatial Library offers a comprehensive list of selected geospatial resources for Massachusetts.

Massachusetts Office of Geographic Information (MassGIS)
http://www.mass.gov/anf/research-and-tech/it-serv-and-support/application-serv/office-of-geographic-information-massgis/
Through the Office of Geographic Information (MassGIS), Massachusetts has created a comprehensive, statewide database of spatial information for mapping and analysis supporting emergency response, environmental planning and management, transportation planning, economic development, and transparency in state government operations.

**County and City Data**
City of Cambridge GIS
http://www.cambridgema.gov/GIS

*Michigan*

**State Data**
Michigan GIS Open Data Portal
http://gis-michigan.opendata.arcgis.com
The Michigan GIS Open Data Portal provides users with GIS datasets that fall under the categories of geology, environment, demographics, elevation, fish and wildlife, hydro, transportation, and public health.

**County and City Data**
Ann Arbor GIS
http://www.a2gov.org/services/data/Pages/default.aspx

Detroit GIS
https://data.detroitmi.gov

Flint GIS
http://flint.maps.arcgis.com/home/index.html
Flint GIS includes street tree and demographic data.

Genesee County GIS
http://www.gc4me.com/departments/gis/
The Genesee County GIS provides interactive mapping applications that focus on a number of topics such as ecotourism, transportation hub, zoning, broadband Internet, and site development.

Kent County GIS
https://www.accesskent.com/Departments/GIS/
Kent County GIS includes parcel viewing (iMap) as well as a GIS data library with downloadable GIS data for administration, hydrography, property, and parcels.

Southeast Michigan Council of Governments (SEMCOG)
http://maps-semcog.opendata.arcgis.com
Southeast Michigan Council of Governments offers a few datasets including boundaries, economy, transportation, land use, and aerial photography.

Washtenaw County GIS
http://www.ewashtenaw.org/government/departments/gis
An interactive map viewer (MapWashtenaw) provides access to the latest parcel maps and property data available from Washtenaw County. The viewer shows a variety of property data for the entire county including parcels, aerial photos, topography, schools, and natural features.

*Minnesota*

**State Data**

Minnesota Geographic Data Clearinghouse
http://www.mngeo.state.mn.us/chouse/
The Minnesota Geographic Data Clearinghouse serves as a convenient source for geographic resources, ranging from simple state maps to complex geospatial data needed to power geographic information systems. Coordinated by MnGeo, the clearinghouse provides access to a wide variety of sources. Partners included the USGS, Bureau of the Census, Minnesota's DNR, DOT and PCA, the MetroGIS program, and many others.

Minnesota Geospatial Commons
https://gisdata.mn.gov
The Minnesota Geospatial Commons is a collaborative place for users and publishers of geospatial resources about Minnesota, including resources formerly distributed via the Minnesota Department of Resources' Data Deli and Metro-GIS DataFinder. The commons has replaced the GeoGateway search tool.

**County and City Data**

MetroGIS DataFinder
http://datafinder.org/index.asp
DataFinder is a one-stop shop for discovering geospatial data focused on and around the seven county Minneapolis–St. Paul Metropolitan Area. DataFinder provides metadata describing GIS data sets, many of which can be directly downloaded or used via map services.

Open Minneapolis
http://opendata.minneapolismn.gov

*Mississippi*

Mississippi Geospatial Clearinghouse
http://www.gis.ms.gov/Portal/home.aspx?x=1920&y=1105&browser=Netscape

*Missouri*

Missouri Spatial Data Information Service (MDIS)
http://msdis.missouri.edu

*Montana*

**State Data**

Montana Geographic Information Clearinghouse
http://geoinfo.msl.mt.gov
The Montana Geographic Information Clearinghouse links to Montana maps, the Montana GIS Portal, cadastral data, and many other data layers.

**County and City Data**

Flathead County GIS, Kalispell, Montana
http://maps.flathead.mt.gov/ims/Login.aspx?ReturnUrl=%2fims%2fdefault.aspx
The Flathead County Geographic Information Systems Interactive Map Site allows users to search for a particular parcel based on owner name, address, or assessor number and view a map (with or without aerial photography) of the area. Many different geographic datasets can be viewed, such as FEMA floodplains, census information, hunting districts, community wildfire protection plan, and so forth.

Missoula County Mapping/GIS Division
http://www.co.missoula.mt.us/gis/

*Nevada*

Virtual Clearinghouse for Nevada GIS
http://www.nbmg.unr.edu/Maps&Data/VirtualClearinghouse.html

*New Hampshire*

New Hampshire's Statewide GIS Clearinghouse
http://www.granit.unh.edu
New Hampshire's Statewide GIS Clearinghouse offers an array of geospatial services including data distribution, spatial analysis, online mapping, and image processing.

*New Jersey*

**State Data**

New Jersey Department of Environmental Protection Bureau of GIS
http://www.nj.gov/dep/gis/listall.html
New Jersey Department of Environmental Protection Bureau of GIS offers a large collection of data related to environmental themes and demographics.

New Jersey Geographic Information Network
https://njgin.state.nj.us/NJ_NJGINExplorer/jviewer.jsp?pg=about_njgin
The New Jersey Geographic Information Network offers a wide variety of datasets that can be browsed by topic such as agriculture and farming, biology and ecology, business and economics, land descriptions, demographics, facilities and structures, geology, and much more.

New Jersey Geographic Open Data Clearinghouse
http://njogis-newjersey.opendata.arcgis.com
The New Jersey Geographic Open Data Clearinghouse provides access to many of the state's spatial datasets, allowing users to search, query, and download filtered or complete records for each dataset. Themes include cadastral, environmental, governmental, infrastructure, transportation, and more.

### County and City Data

Burlington County GIS Data
http://www.co.burlington.nj.us/531/Access-to-Data-Maps
The Burlington County Department of Information Technology, GIS Section contains tax parcel, natural resource, transportation, demographic, planning, and environmental data.

Camden County Open Data Portal
http://www.camdencounty.com/service/public-works/gis-data/

Newark Open Data Portal
http://data.ci.newark.nj.us
The Newark Open Data Portal includes abandoned properties, wards, municipal utilities, zoning, land use, and more.

## New Mexico

New Mexico Resource GIS
http://rgis.unm.edu

## North Carolina

### State Data

North Carolina County GIS Data
https://www.lib.ncsu.edu/gis/counties.html
The NCSU Libraries have acquired a considerable amount of GIS data from North Carolina county governments. Key data resources from counties include orthophotography, land parcels, street centerlines, and other infrastructure or cultural data layers.

North Carolina Department of Transportation Data
https://www.lib.ncsu.edu/gis/datalist.html
The North Carolina Department of Transportation offers various transportation-related datasets and map products.

North Carolina GeoSpatial Portal—OneMap
http://data.nconemap.com/geoportal/catalog/main/home.page
The North Carolina GeoSpatial Portal (OneMap) provides a large collection of datasets within topics such as biology, ecology, business and economy, cadastral, facilities and structures, human health, military, inland water resources, transportation network, and more.

State Library of North Carolina Digital Collections
http://digital.ncdcr.gov/cdm/home/collections/gis-data
This online collection includes dataset categories such as local and state government datasets, orthoimagery, centralized datasets (Framework and non-Framework data), project files, and digitized maps. The GIS Data collection spans the years from 1947 through 2009, with the bulk of the items ranging from 1999 through 2009.

### County and City Data

North Carolina City GIS Data
https://www.lib.ncsu.edu/gis/cities.html
The North Carolina City GIS Data site offers an extensive list of links for individual city GIS webpages in North Carolina.

Wake County GIS Data
https://www.lib.ncsu.edu/gis/wake.html

## North Dakota

### State Data

North Dakota Department of Transportation (NDDOT) GIS
https://www.dot.nd.gov/business/gis-mapping.htm

North Dakota GIS Hub
https://www.nd.gov/itd/statewide-alliances/gis
The North Dakota GIS Hub offers a portal for finding data, maps, and applications, as well as a hub explorer, which is a mapping tool for exploring GIS hub datasets. The datasets available cover categories such as geoscientific, health, inland waters, location, transportation, earthquakes, railroads, abandoned mines, and much more.

### County and City Data

Ward County GIS
http://gis.wardnd.com
Ward County GIS includes flood, school, and parcel information, as well as planning and zoning layers.

## Ohio

### State Data

Ohio Department of Natural Resources GIS
http://www2.ohiodnr.gov/geospatial/map-services
The Ohio Department of Natural Resources offers a wide variety of datasets related to agriculture and farming, biology and ecology, demographics, elevation and bathymetry, environment and conservation, geology, imagery, and more.

Ohio Department of Transportation GIS
http://www.dot.state.oh.us/districts/D01/PlanningPrograms/gis/Pages/default.aspx
Shapefiles provided by ODOT include active rail, airports, cities, townships, counties, interstates, state highways, and U.S. highways.

Ohio Environmental Protection Agency
http://epa.ohio.gov/gis.aspx
The Ohio EPA GIS provides interactive maps that monitor and study water quality, air quality, ground water, biosolids, and cyanotoxins.

### County and City Data

Cincinnati Area GIS
http://cagisonline.hamilton-co.org/cagisonline/index.html
The Cincinnati Area GIS site includes aerials, street, topography, and property information.

Columbus Ohio Maps and GIS
https://www.columbus.gov/technology/gis/GIS-Maps---Applications-Gallery/

## Oklahoma

### State Data

Oklahoma Maps
https://okmaps.org/OGI/search.aspx
Oklahoma Maps offers an interactive mapping application to explore key resources in the state of Oklahoma including education, hydrology, transportation, weather, land parcels, structures, seismicity, imagery, and more.

### County and City Data

Bartlesville City GIS
http://www.cityofbartlesville.org/community-development/gis-mapping/
The Bartlesville City GIS site includes aerial imagery, topographic maps, property information, floodplains, park and school locations, and more.

Norman City GIS
http://www.normanok.gov/planning/gis
The Norman City GIS databases include wastewater collection system, water distribution system, storm-water collection system, zoning, Norman 2025 land use plan, city council wards, historic districts, recoupment districts, property ownership patterns, street centerlines, building footprints, paving, fences, trees, police beats, fire districts, sanitation routing, topography, hydrography, and digital aerial orthophotography.

Oklahoma City Open Data Portal
https://data.okc.gov/portal/page/start/
The Oklahoma City Open Data Portal offers the ability to browse and search for addresses, zoning, street, and other information for Oklahoma City.

## Oregon

### State Data

Oregon Spatial Data Library
http://spatialdata.oregonexplorer.info/geoportal/
The Oregon Spatial Data Library offers a large number of files available for download, including topics such as administrative boundaries, cadastral, elevation, hydrography, imagery, and transportation.

### County and City Data

Maps and GIS Resources for Cities in Oregon
https://library.uoregon.edu/map/or/gis_or_city
The Maps and GIS Resources for Cities in Oregon website was compiled by the University of Oregon, offering a list of Oregon cities that offer GIS data.

Oregon City GIS
https://www.orcity.org/maps

## Pennsylvania

### State Data

Open Data Pennsylvania
https://data.pa.gov

Pennsylvania Department of Transportation GIS Open Data Portal
http://data-pennshare.opendata.arcgis.com

Pennsylvania Open Data
http://data-dcnr.opendata.arcgis.com
Pennsylvania Open Data includes GIS data for Pennsylvania state parks, forests, trails, geology, and other relevant projects. An interactive maps gallery is also available that uses PA DCNR data.

Pennsylvania Spatial Data Access
http://www.pasda.psu.edu
PASDA is the official public access geospatial information clearinghouse for Pennsylvania. Through this arrangement, PASDA provides access to the geospatial data of most Pennsylvania state agencies.

### County and City Data

Centre County Open Data
http://gisdata-centrecountygov.opendata.arcgis.com
Centre County Open Data includes search and download options for open spatial data organized by data categories of environment, boundaries, locations, and infrastructure for Centre County, Pennsylvania.

## Rhode Island

### State Data

Rhode Island GIS
http://www.rigis.org
The Rhode Island Geographic Information System (RIGIS) distributes open geographically referenced datasets that represent a wide range of topics, including transportation, infrastructure, and the environment.

### County and City Data

Exeter Town GIS/Maps
http://www.town.exeter.ri.us/mapsgis.html
The town of Exeter allows users to view maps and data on property assessments, topography, soil types, wetlands, zoning, land use, and so on through an interactive mapping application.

Newport GIS Mapping and Data Services
http://cityofnewport.com/departments/civic-investment-planning/gis-mapping-services
Newport's online data mapping portal provides data such as property information (plat/lot information, zoning, land use, flood zones), maritime information (mooring locations, water access points), facilities, services and transportation (city facilities, parks, public restrooms), construction information (moratorium roads), and voter information (voting precincts, polling locations).

Providence's Open Data Portal
https://data.providenceri.gov
The Providence Open Data Portal provides data related to economy and finance, neighborhoods, and public safety.

Westerly Town GIS
http://gis.westerlyri.gov
The town of Westerly offers interactive mapping applications and GIS data for download such as imagery, inland water resources, utilities and communications, property parcels, tax assessment information, and much more.

## South Carolina

### State Data
South Carolina GIS
http://www.gis.sc.gov/data.html
This GIS portal offers a large variety of datasets including recreational waters, beach erosion, waste landfills, coastal zones, docks and boat ramps, public beach access, food facilities, funeral homes, and health facilities.

### County and City Data
Charleston City Open GIS Data
http://data-charleston-sc.opendata.arcgis.com

Columbia City Open GIS Data
http://coc-colacitygis.opendata.arcgis.com
The Columbia City Open GIS site provides interactive maps for community crime, parks, city landmarks, code violation properties, city information, and more.

## Tennessee

### State Data
Tennessee GIS Clearinghouse
http://www.tngis.org
The Tennessee GIS Clearinghouse offers datasets related to demographics, elevation, soil, land use and land cover, imagery, and more.

TNMap
http://tnmap.tn.gov
TNMap serves as the portal for accessing downloadable GIS datasets, web applications, data services, and the state's ArcGIS Online organization.

### County or City Data
Nashville Open Data Portal
https://data.nashville.gov

## Texas

### State Data
Texas Natural Resources Information System
https://www.tnris.org

### County and City Data
Brazos County GIS
http://www.brazoscountytx.gov/index.aspx?NID=268
The Brazos County GIS includes GIS files of county and city limits, cemeteries, rivers, streams, railroads, roads, and hundred-year floodplains.

Bryan City GIS
https://www.bryantx.gov/information-technology/gis-mapping-services/gis-maps/
The Bryan City GIS includes aerials, streets, new developments, school districts, capital improvements, cemetery information (ownership and grave location), census data, garage sales, and sex offender mapping services.

Dallas City GIS
http://gis.dallascityhall.com/EnterpriseGIS/

McKinney City GIS
http://www.mckinneytexas.org/index.aspx?NID=1174
The McKinney City GIS includes data for download for commonly requested Shapefiles, including address points, city limits, parks, schools, roads, and more.

## Utah

### State Data
Utah GIS Portal
https://gis.utah.gov/data/
The Utah GIS Portal offers data for download such as addresses, parcels, boundaries, land ownership, base maps, transportation, imagery, and more.

### County and City Data
Salt Lake City GIS Open Data
http://gis-slcgov.opendata.arcgis.com

Utah County GIS
http://www.utahcounty.gov/OnlineServices/maps/DataLayers.asp
The Utah County GIS includes layers such as address points, building outlines, tax parcels, roads, and more.

## Vermont

### State Data
Agency of Natural Resources Open Data Portal
http://anr.vermont.gov/maps/gis-data
This is the public platform for exploring, discovering, and downloading data from the Agency of Natural Resources. Files available include deer wintering areas, fishing access areas, waste facilities, lakes, flood hazard areas, and much more.

Vermont Center for Geographic Information
http://vcgi.vermont.gov

### County and City Data
City of Barre GIS Mapping
https://www.barrecity.org/gis.html
The city of Barre offers GIS layers such as parcel boundaries, old lot lines, roads, buildings, water, zoning districts, flood zone districts, contours, and imagery.

## Virginia

### State Data

Virginia GIS Clearinghouse
http://www.arcgis.com/home/item.html?id=c959e73b8d584ba581498be0ff7b1b31
Virginia's Statewide Clearinghouse for GIS data is hosted by the Virginia Geographic Information Network (VGIN) and the Virginia Information Technologies Agency (VITA). The clearinghouse is a repository of geospatial data produced and used by state agencies in Virginia, localities in Virginia, federal partnering agencies, nonprofits, and colleges and universities across Virginia.

### County and City Data

Virginia Beach Open GIS Datasets
https://www.vbgov.com
Virginia Beach Open GIS Datasets include geospatial categories such as public safety, business, finance and government, neighborhoods, and environment.

## Washington

### State Data

State of Washington Open Data
https://data.wa.gov

Washington Department of Natural Resources GIS
http://data-wadnr.opendata.arcgis.com
The Washington Department of Natural Resources GIS includes geospatial categories such as aquatics, cadastre, habitat conservation plan, soils, hydrography, natural heritage, transportation, wildlife and prevention, forest disturbance, forest practices, and geology.

Washington State Department of Ecology GIS
http://www.ecy.wa.gov/services/gis/data/data.htm
The Washington State Department of Ecology GIS has a large number of geospatial files available related to hydrology and environmental information monitoring.

Washington State Department of Transportation
http://www.wsdot.wa.gov/mapsdata/geodatacatalog/default.htm
The Washington State Department of Transportation offers geospatial datasets related to transportation such as bridges, ferry routes, fuel stations, noise walls, railroads, highways, and transit stops.

Washington State Geospatial Clearinghouse
http://wa-node.gis.washington.edu/geoportal/catalog/main/home.page
This geospatial clearinghouse provides a single point of access to geospatial information across Washington. The portal offers datasets across several themes and categories such as hazards, minerals, geology, soils, hydrology, topographic information, and much more.

Washington State Geospatial Data Archive (WAGDA)
https://wagda.lib.washington.edu

### County and City Data

King County GIS Center
http://www5.kingcounty.gov/gisdataportal/
Over seven hundred layers are available to explore, view, and download.

Pierce County Open Geospatial Data Portal
http://gisdata-piercecowa.opendata.arcgis.com
The Pierce County Open Geospatial Data Portal offers GIS datasets in a wide variety of topics including census, environment, fish and wildlife, hydro/topo, parcels, political, recreation, roads and rails, survey, and utilities.

Seattle Open Data GIS
https://data.seattle.gov
The city of Seattle offers GIS data related to business, community (organizations, neighborhoods), education, finance, public safety, and transportation.

## West Virginia

West Virginia Department of Environmental Protection GIS Server
https://tagis.dep.wv.gov/home/node/1
The West Virginia Department of Environmental Protection GIS Server offers a large number of shapefiles and geodatabases such as underground mining, valley fills, open dump cleanup sites, oil and gas wells, water and watershed resources, and much more.

West Virginia State Clearinghouse
https://www.mapwv.gov
West Virginia GIS offers its users a number of data layers related to farming, geoscience, biology, inland waters, economy, military intelligence, society and culture, transportation, and much more.

## Wisconsin

### State Data

Wisconsin Department of Natural Resources Maps and GIS
http://dnr.wi.gov/maps/
The DNR produces a wide range of publications and information for the public on hundreds of topics related to the environment, recreation, wildlife, and natural resources.

Wisconsin Open Data
https://data-wi-dnr.opendata.arcgis.com
The Wisconsin Department of Natural Resources GIS Open Data Portal offers interactive mapping viewers and GIS data related to surface water, public access lands,

burning restrictions and fire activity, fish and wildlife, forestry, transportation, managed lands, land cover and vegetation, parks and recreation, and more.

### *County and City Data*
Wisconsin Local Government Web Mapping Sites
https://gis.lic.wisc.edu/coastalweb/www/wisconsin-ims/wisconsin-ims.htm
This site provides a comprehensive list of county sites with interactive maps, GIS datasets for download, or both.

*Wyoming*

Wyoming Geospatial Hub
http://geospatialhub.org
The Wyoming Geospatial Hub is the primary site for the discovery and delivery of publicly accessible geospatial data produced, maintained, and shared by various partners in the state. This resource consists of three coupled web applications—*GeoHub Explorer*, *GeoHub Pathfinder,* and *GeoHub Imagery*—that provide access to downloadable data, tools, and techniques for finding and downloading data.

## INTERNATIONAL DATA RESOURCES

African Marine Atlas
http://www.africanmarineatlas.org
The African Marine Atlas provides rainfall climatology data (atmosphere, biosphere, geosphere, human environment, and hydrosphere data) for countries located around the African coast.

Antarctic Digital Database
http://www.add.scar.org
The Antarctic Digital Database includes Antarctic digital elevation models (DEMs), hillshade, coastline, topographic, moraine, water, human activity, and ice feature catchment data.

ArcGIS Open Data
http://opendata.arcgis.com
The ArcGIS Hub provides international datasets pertaining to a multitude of topics. Examples include bus routes, park services, and schools.

Australian Geoscience
https://ecat.ga.gov.au/geonetwork/srv/eng/search#fast=index&from=1&to=10&hitsperpage=10
Australian Geoscience provides a multitude of data, online tools, maps, videos, and publications pertaining to Australia.

Center for International Earth Science Network (CIESIN)
http://www.ciesin.org/data.html
The Center for International Earth Science Network (CIESIN) provides spatial and tabular data based on a diverse set of topics, including agriculture, ecosystems, climate change, environmental health, land use, and natural hazards.

China Historical Geographic Information System
http://www.fas.harvard.edu/~chgis/
This resource provides place-names and historical administrative unit data for China.

Diva-GIS
http://www.diva-gis.org/Data
This data is available at different scales (global level, country level, and so on). An example of a dataset provided at the global level is country boundaries. Datasets provided at the country level include roads, railroads, land cover, and population density. Diva-GIS also provides global climate data, species occurrence data, crop collection data, high-resolution LandSat satellite images, and global 90m resolution elevation data.

ESRI Data and Maps
http://www.arcgis.com/home/group.html?owner=esri&title=ESRI%20Data%20%26%20Maps&content=all
ESRI provides datasets pertaining to a multitude of topics. The datasets are available at the following scales: global, United States, North America, and Europe. Topics include population density, airports, highway exits, and coastal water bodies.

ESRI World Basemap Data
http://www.esri.com/data/free-data/index.html
ESRI provides a series of base maps at a global scale, including the following types: street, topographic, boundaries, gray base, ocean base, and imagery.

European Union Open Data Portal
https://data.europa.eu/euodp/en/data/
The European Union Open Data Portal provides datasets pertaining to a multitude of topics. Topics include economics, industry, finance, trade, science, and the environment.

Eurostat
http://ec.europa.eu/eurostat/web/gisco/overview
Eurostat offers European geographical data including administrative boundaries and thematic geospatial information, such as population grid data.

FAO GeoNetwork
http://www.fao.org/geonetwork/srv/en/main.home
FAO Geonetwork provides international interactive maps, satellite imagery, and more.

Free GIS Data
http://freegisdata.rtwilson.com

Free GIS Data provides datasets pertaining to a multitude of topics, including physical geography, land and ocean boundaries, hydrology, elevation, and climate.

## GADM
http://www.gadm.org
GADM provides global spatial administrative boundary data. Administrative areas in this particular database include country boundaries and lower-level subdivisions (that is, provinces, departments, counties, and the like).

## Gapminder
http://www.gapminder.org/world/#$majorMode=map
Gapminder provides a series of graphs pertaining to a number of topics (child health, climate, disasters, economy, education, family, global trends, poverty, and HIV).

## General Bathymetric Chart of Oceans (GEBCO) Gridded Bathymetric Datasets
https://www.bodc.ac.uk/data/online_delivery/gebco/
The General Bathymetric Chart of the Oceans (GEBCO) provides datasets of global terrain models for ocean and land.

## Geography Network
http://www.geographynetwork.com
Geography Network provides global spatial data and remote-sensing imagery.

## GEOnet World Place Names Server
http://geonames.nga.mil/gns/html/
The GEOnet Names Server (GNS) provides standard spellings for all foreign geographic names.

## German Historical GIS
http://www.hgis-germany.de

## Global Biodiversity Information Facility (GBIF)
http://www.gbif.org
The Global Biodiversity Information Facility (GBIF) provides global datasets pertaining to biodiversity.

## Global Irrigated Area and Rainfed Crops Areas
http://waterdata.iwmi.org/Applications/GIAM2000/
The International Water Management Institute (IWMI) provides datasets, maps, statistics, and publications pertaining to irrigation water sources, cropping intensity, and dominant crop types.

## Global Land Use Dataset
http://nelson.wisc.edu/sage/data-and-models/global-land-use/grid.php
The Center for Sustainability and the Global Environment provides datasets pertaining to a multitude of topics, including urban expansion, irrigated lands, agricultural lands, global river discharge, and global land use.

## Global Reservoir and Dam (GRandD) Database
http://atlas.gwsp.org/index.php?option=com_content&task=view&id=207&Itemid=63
The Global Reservoir and Dam (GRandD) database provides data displaying the location of large reservoirs and dams.

## Global Rural-Urban Mapping Project
http://sedac.ciesin.columbia.edu/data/collection/grump-v1/sets/browse
Global Rural-Urban Mapping Project provides several global datasets: population count grids, population density grids, national boundaries, national identifier grids, and coastlines.

## Harmonized Soil Database
http://webarchive.iiasa.ac.at/Research/LUC/External-World-soil-database/HTML/HWSD_Data.html?sb=4
The Harmonized World Soil Database provides soil data.

## Humanitarian Information Unit Data
https://hiu.state.gov
The Department of State Humanitarian Information Unit (HIU) provides data and products pertaining to humanitarian studies. Topics include food security, migration, and vulnerable populations.

## London Datastore
http://data.london.gov.uk
The City of London provides statistical data pertaining to London. Topics include jobs and economy, transportation, environment, community safety, housing, health, and GLA performance.

## Mapzen
https://mapzen.com/data/
Mapzen provides international datasets pertaining to transportation routes. Datasets include global street coverage and transit networks.

## Mexico Geospatial Data
https://www.diegovalle.net/projects.html#url=%23datasets
Diego Valle-Jones provides geospatial data, interactive maps, and web applications of Mexico based on a number of topics, including violence, divorce, elections, and crime.

## Mexico—Instituto Nacional de Estadistica y Geografia (INEGI)
http://www.inegi.org.mx

## NASA—Reverb/ECHO
http://reverb.echo.nasa.gov/reverb/
NASA provides earth sciences data.

## Open Flights
http://openflights.org/data.html

OpenFlights provides global datasets pertaining to airport locations and flight routes.

Open Portals around the World
https://www.opendatasoft.com/a-comprehensive-list-of-all-open-data-portals-around-the-world/
Open Data Soft provides a list of over twenty-six hundred open data portals.

OpenStreetMap
https://www.openstreetmap.org
OpenStreetMap offers a number of different datasets (railway stations, trails, roads, cafes, and the like) located all over the world.

Ordinance Survey
https://www.ordnancesurvey.co.uk/business-and-government/products/opendata-products-grid.html
Ordinance Survey provides data pertaining to a multitude of topics in Britain. Topics include roads, address points, rivers, and path networks.

PANGAEA
https://www.pangaea.de
Pangaea provides datasets pertaining to the environment. Topics include oceans, biological classification, ecology, and fisheries.

re3data
http://www.re3data.org
Re3data provides international datasets pertaining to a multitude of topics. Topics include humanities and social sciences, life sciences, natural sciences, and engineering sciences.

Socioeconomic Data and Applications Center (SEDAC)
http://sedac.ciesin.columbia.edu/data/set/groads-global-roads-open-access-v1
The Socioeconomic Data and Applications Center (SEDAC) provides global road data.

Southern Africa Regional Science Initiative (SAFARI)
http://daac.ornl.gov/S2K/safari.html
The Southern Africa Regional Science Initiative (SAFARI) provides data pertaining to land cover, soils, atmospheric and airborne studies, hydrology, remote sensing, and climate and meteorology.

Tanzania GIS User Group
http://www.tzgisug.org/wp/
The Tanzania GIS User Group lists a number of resources that provide GIS data for Tanzania.

United Nations Environment Programme
http://geodata.grid.unep.ch
The United Nations Environment Programme provides national, subregional, regional, and global statistics and geospatial datasets pertaining to a multitude of topics. Topics include population, freshwater, forests, emissions, climate, natural disasters, and health.

USGS—Earth Resources Observation and Science (EROS) Center Elevation Products
http://eros.usgs.gov/elevation-products
The Earth Resources Observation and Science (EROS) Center provides aerial photographs, digitized maps, and satellite imagery. In addition, the EROS also provides elevation, land cover, topographic, and remotely sensed data.

USGS Mineral Resources Data System (MRDS)
https://mrdata.usgs.gov/mrds/
The USGS Mineral Resources Data System (MRDS) provides a series of reports that describe metallic and nonmetallic mineral resources around the world.

Visible Earth
http://visibleearth.nasa.gov
Visible Earth provides a catalog of NASA images and animations of various parts of the world.

World Bank Data
http://data.worldbank.org
World Bank Data provides data pertaining to global development. Topics include population and GDP.

WorldClim—Global Climate Data
http://www.worldclim.org
WorldClim provides global climate data.

World Database of Large Urban Areas
https://nordpil.com/resources/world-database-of-large-cities/
The World Database of Large Urban Areas provides historic, current, and future estimates of the population for the largest urban areas in the world from 1950 to 2050.

World Factbook: Regional and World Maps
https://www.cia.gov/library/publications/the-world-factbook/docs/refmaps.html
The World Factbook provides a series of regional and world maps (available in JPG format and PDF).

World Geologic Maps
http://geodata.grid.unep.ch
The United Nations Environment Programme (UNEP) provides data that pertains to the environment. Topics include freshwater, population, forests, emissions, climate, health, and disasters.

# 12
# GIS Subject Guides

A GIS subject guide, also referred to as a GIS research guide, is a popular way to organize and share information about geospatial resources available in libraries and outside of them. Based on an online search of all North American postsecondary institutions that offer GIS services, about 75 percent of them use official GIS and other cartographically related subject guides. These guides typically offer resources for the new and expert GIS user, including resources that introduce visitors to GIS (GIS news, concepts, terminology), links to online mapping programs, lists of GIS software, and, very importantly, information about geospatial data or direct links to such data.

This chapter provides a directory of over 130 GIS subject guides, organized by province and state, listing important resources included in the guides. Users can use this list to determine which guides to visit for access to information about an institution's GIS teaching program, lists of GIS software, data, and other online resources.

The information here was obtained by visiting every North American library website that has a cartographic collection (see chapter 2 for the directory of websites) to determine if GIS services were offered as well. If they were, further exploration determined if subject guides were being used to disseminate information about their GIS resources. A thorough review and analysis of each subject guide revealed commonalities among many institutions, particularly as it relates to the services offered and the types of GIS software used.

Many subject guides, roughly 70 percent, provide GIS training resources, such as tutorials on GIS software and mapping methods, links to web tutorials (such as Esri), and embedded video tutorials. About 60 percent provide information about workshop sessions offered at the university. About half of all guides link to interactive online maps that display data or offer downloadable maps. Most guides offer additional GIS resources, such as books, articles, journals, atlases, air photos, gazetteers, citation information, reference materials, and job resources. All of these have been singled out in the following directory and listed under "Resources."

Almost all subject guides include a list of data resources, such as open data (data that is freely available, accessible, and can be downloaded by anybody), government data (data hosted by a government website or organization), statistical/census data, remote-sensing data (any imagery, LandSat, Lidar, radar, or spot data), and "other" data, such as links to other data repositories, tabular data, and the like). Almost all subject guides include open data, most offered government data and statistical/census data, but less than half offered remote-sensing data and other data. The most common online datasets listed in the subject guides are Data.gov, GeoCommons, GIS Data Depot, Natural Earth, DivaGIS, Geospatial Data Gateway, Open Geoportal, Clearinghouse, GeoNetwork, Data Basin, GeoPlatform, NationalAtlas.gov, the National Map, Earth Explorer, and ArcGIS Open Data. Information about data is listed in the directory under "Data."

The subject guides were explored to see whether they provide a map or data search engine to search the database for resources directly on the website or through a link on the website. Less than half of all subject guides offered search capabilities. This is denoted under "Data Search Capabilities" in the directory.

Lastly, the guides were analyzed to determine which GIS software programs were being used at the university. ArcGIS is by far the most popular GIS program, with seventy guides explicitly listing it as a resource available at the institution. Google Earth comes in second, with eighteen guides listing it. Other programs were listed less frequently: SPSS (listed by sixteen institutions), R (twelve), QGIS (eleven), SAS (ten), State (nine), AutoCAD (eight), ERDAS IMAGINE (eight), ENVI (six), FME (five), Sketchup (three), PCI Geomatica (two), SimplyMap (two), and LandScan (two). There were dozens of other software products listed by only one institution each. There may be more instances of these products, but they were not not mentioned in the guide. Many institutions recommend specific software, coining them as

"Popular Tools." The number one recommended software is QGIS (recommended by fifty-three institutions), followed by GRASS GIS (thirty-two), SimplyMap (twenty-three), Google Earth (seventeen), DIVA GIS (nine), Carto DB (seven), and Beyond 20/20 (three). Specific software used by each institution is denoted in the directory under "Software."

In the process of reviewing over 130 GIS subject guides, it became apparent that some guides had more useful information than others. A guide should be a one-stop shop for new and expert users, and many guides included many elements discussed above, such as training resources, links to useful resources, and access to data. The directory below includes, wherever applicable, a link to the library's cartographic guide. Cartographic resources often provide important historical information that geospatial data does not offer, so both spatial and printed resources should be consulted when conducting visual research. A separate list includes libraries that have cartographic guides but not GIS guides, since not all libraries offer GIS services. The list below offers the top twenty GIS subject guides (in alphabetical order), as elected by the authors, based on the richness of their information.

## TOP TWENTY GIS SUBJECT GUIDES

- Boston College Libraries
  http://libguides.bc.edu/gis
- Claremont University Consortium, Claremont Colleges Library
  http://libguides.libraries.claremont.edu/GIS
- Dartmouth College Library
  http://researchguides.dartmouth.edu/gis
- Massachusetts Institute of Technology (MIT) Libraries
  http://libguides.mit.edu/c.php?g=176295&p=1160692
- Ohio State University Libraries
  http://guides.osu.edu/maps-geospatial-data
- Purdue University Libraries
  http://guides.lib.purdue.edu/gis
- Syracuse University Libraries
  http://researchguides.library.syr.edu/GIS
- Texas A&M University Libraries
  http://guides.library.tamu.edu/content.php?pid=199766
- University of Arkansas Libraries
  http://uark.libguides.com/gis
- University of British Columbia
  http://guides.library.ubc.ca/gis
- University of Calgary
  http://library.ucalgary.ca/sands
- University of California, Davis Library
  http://guides.lib.ucdavis.edu/Maps-GIS
- University of California, Los Angeles Library
  http://guides.library.ucla.edu/gis
- University of California, San Diego Library
  http://ucsd.libguides.com/gis
- University of Hawaii at Manoa
  http://guides.library.manoa.hawaii.edu/magis
- University of Kansas Libraries
  http://guides.lib.ku.edu/gis
- University of Pittsburgh Library System
  http://pitt.libguides.com/gis
- University of Regina, Archer Library
  http://uregina.libguides.com/content.php?pid=181372
- University of Washington Libraries
  http://guides.lib.uw.edu/research/gis
- York University Libraries
  http://researchguides.library.yorku.ca/content.php?pid=245987&sid=2031504

## DIRECTORY OF GIS SUBJECT GUIDES

### Canada

*Alberta*

University of Alberta Libraries
Edmonton, Alberta
Subject Guide Name: Geospatial Data and Maps
Subject Guide URL: http://guides.library.ualberta.ca/c.php?g=301218&p=2009722
Resources: Training, Online Maps, Other
Data: Government, Open, Statistical/Census, Other
Data Search Capabilities: Yes

University of Calgary Libraries
Calgary, Alberta
Subject Guide Name: Spatial and Numeric Data Services
Subject Guide URL: http://library.ucalgary.ca/sands
Resources: Training, Online Maps, Other
Data: Government, Open, Remote Sensing, Statistical/Census, Other
Data Search Capabilities: No
Software: ArcGIS 10.3, Nvivo 10, PCI Geomatica 10.3, QGIS, R, SAS 9.3, SPSS 20, Stata 12
Other Related GIS Guides: http://libguides.ucalgary.ca/Data; http://libguides.ucalgary.ca/GIS
Cartographic Guide: Map Internet Resources, http://libguides.ucalgary.ca/c.php?g=255407&p=1703446

*British Columbia*

College of the Rockies
Cranbrook, British Columbia
Subject Guide Name: Geography, Geology, and GIS
Subject Guide URL: http://cotr.libguides.com/content.php?pid=609186&sid=5031776
Resources: Online Mapping, Other
Data: Government, Open, Statistical/Census
Data Search Capabilities: No
Additional Subject Guides: http://cotr.libguides.com/statistics

Simon Fraser University
Burnaby, British Columbia
Subject Guide Name: GIS and Maps
Subject Guide URL: http://www.lib.sfu.ca/find/other-materials/data-gis/gis
Resources: Training, Instruction/Workshop, Online Maps, Other
Data: Government, Open, Other
Data Search Capabilities: Yes
Software: ArcGIS, QGIS, Google Earth Pro, SimplyMap, LandScan Global Archive
Cartographic Guide: GIS and Maps Resources: Map Collection, http://www.lib.sfu.ca/find/other-materials/data-gis/gis/map-collection

University of British Columbia
Vancouver, British Columbia
Subject Guide Name: Geographic Information Systems (GIS)
Subject Guide URL: http://guides.library.ubc.ca/gis
Resources: Training, Instruction/Workshop, Online Maps, Other
Data: Government, Open, Remote Sensing, Statistical/Census, Other
Data Search Capabilities: Yes
Software: ArcGIS, QGIS, FME Desktop, Google Earth Pro, R, SPSS, Stata, StatTransfer
Additional Subject Guides: http://guides.library.ubc.ca/datastatistics
Cartographic Guide: Maps and Atlases, http://guides.library.ubc.ca/maps-atlases

*Manitoba*

University of Manitoba
Winnipeg, Manitoba
Subject Guide Name: GIS
Subject Guide URL: http://libguides.lib.umanitoba.ca/GIS
Resources: Training, Instruction/Workshop, Online Maps, Other
Data: Government, Open, Remote Sensing, Statistical/Census, Other
Data Search Capabilities: Yes
Software: ArcGIS 10.4, AutoCAD 15, Geomatica 2015, Google Earth Pro, GeoViewer 5.5
Additional Subject Guides: http://libguides.lib.umanitoba.ca/MapsImagery

*Ontario*

Carleton University Library
Ottawa, Ontario
Subject Guide Name: GIS
Subject Guide URL: https://library.carleton.ca/research/subject-guides/gis-quick-guide
Resources: Training, Online Maps, Other
Data: Government, Open, Remote Sensing, Statistical/Census, Other
Data Search Capabilities: Yes
Software: ArcGIS 10.4, AutoCAD Map 3D, CityEngine, FME Desktop, Global Mapper, Google Earth Pro, Lizardtech GeoViewer, QGIS
Additional Subject Guides: https://library.carleton.ca/research/subject-guides/geomatics
Cartographic Guide: Maps, https://library.carleton.ca/research/subject-guides/maps-detailed-guide

Lakehead University
Thunder Bay, Ontario
Subject Guide Name: Government Information/Data
Subject Guide URL: http://libguides.lakeheadu.ca/govinfo
Resources: Training, Online Maps, Other
Data: Government, Open, Remote Sensing, Other
Data Search Capabilities: Yes

McMaster University
Hamilton, Ontario
Subject Guide Name: GIS (Geographic Information Systems)
Subject Guide URL: https://library.mcmaster.ca/guides/gis
Data: Government, Open, Remote Sensing, Statistical/Census
Data Search Capabilities: Yes

Queen's University
Kingston, Ontario
Subject Guide Name: Geospatial Data
Subject Guide URL: http://guides.library.queensu.ca/c.php?g=501830&p=3436009
Resources: Training, Other
Data: Government, Open, Remote Sensing, Statistical/Census, Other
Data Search Capabilities: Yes

Ryerson University Library
Toronto, Ontario
Subject Guide Name: Geospatial Map and Data Centre
Subject Guide URL: http://library.ryerson.ca/gmdc/madar/geo-data/
Resources: Training, Instruction/Workshops, Online Maps, Other
Data: Government, Open, Statistical/Census, Remote Sensing, Other
Data Search Capabilities: Yes
Software: ArcGIS, QGIS, AutoDesk, MapInfo Professional 11.5, SPSS, Beyond 20/20

Trent University
Peterborough, Ontario
Subject Guide Name: MaDGIC

Subject Guide URL: https://www.trentu.ca/library/madgic
Resources: Training, Instruction/Workshops, Online Maps, Other
Data: Government, Open, Remote Sensing, Statistical/Census, Other
Data Search Capabilities: Yes

University of Ottawa
Ottawa, Ontario
Subject Guide Name: Geospatial Data
Subject Guide URL: http://uottawa.libguides.com/geospatial-data
Resources: Instruction/Workshops, Online Maps, Other
Data: Government, Open, Remote Sensing, Statistical/Census, Other
Data Search Capabilities: No
Cartographic Guide: Cartographic Resources, http://uottawa.libguides.com/friendly.php?s=carto-resources

University of Toronto Libraries
Toronto, Ontario
Subject Guide Name: Finding Geospatial Resources at the University of Toronto
Subject Guide URL: http://guides.library.utoronto.ca/geospatialdata
Resources: Training, Instruction/Workshops, Online Maps, Other
Data: Government, Open, Statistical/Census, Remote Sensing, Other
Data Search Capabilities: Yes
Cartographic Guide: Finding Geospatial Resources at the University of Toronto, http://guides.library.utoronto.ca/c.php?g=251122&p=1673131

University of Waterloo
Waterloo, Ontario
Subject Guide Name: GIS Subject Guide
Subject Guide URL: http://subjectguides.uwaterloo.ca/GIS
Resources: Training, Online Maps
Data: Government, Open, Statistical/Census, Remote Sensing, Other
Data Search Capabilities: No
Additional Subject Guides: http://subjectguides.uwaterloo.ca/content.php?pid=109421&sid=1112269

University of Windsor
Windsor, Ontario
Subject Guide Name: Geospatial Services
Subject Guide URL: http://leddy.uwindsor.ca/adc/geospatial
Resources: Training, Instruction/Workshops, Other
Data: Government, Open, Statistical/Census, Remote Sensing, Other
Data Search Capabilities: No

Western University Libraries
London, Ontario
Subject Guide Name: Geographic Information Systems (GIS)
Subject Guide URL: http://guides.lib.uwo.ca/gis/about
Resources: Training, Instruction/Workshop, Online Maps, Other
Data: Government, Open, Remote Sensing, Statistical/Census, Other
Data Search Capabilities: Yes
Software: ArcGIS, QGIS, Carto DB, Indie Mapper, Mango Map, MapBox, Scribble Maps, Target Map, World Map
Cartographic Guide: Maps and Atlases, https://www.lib.uwo.ca/madgic/mapsandatlases.html

York University
Toronto, Ontario
Subject Guide Name: Geospatial Data
Subject Guide URL: http://researchguides.library.yorku.ca/geospatial
Resources: Training, Online Maps, Other
Data: Government, Open, Statistical/Census, Remote Sensing, Other
Data Search Capabilities: No
Additional Subject Guides: http://researchguides.library.yorku.ca/data?hs=a
Cartographic Guide: Maps and Atlases, http://researchguides.library.yorku.ca/content.php?pid=461431

*Quebec*

McGill University
Montreal, Quebec
Subject Guide Name: GIS Guides
Subject Guide URL: http://libraryguides.mcgill.ca/gis_guides/home
Resources: Instruction/Workshops, Other
Data: Government, Open, Statistical/Census, Remote Sensing, Other
Data Search Capabilities: No

*Saskatchewan*

University of Regina
Regina, Saskatchewan
Subject Guide Name: GIS—Geographic Information Systems
Subject Guide URL: http://uregina.libguides.com/content.php?pid=181372
Resources: Training, Online Maps, Other
Data: Government, Open, Statistical/Census, Remote Sensing, Other
Data Search Capabilities: No
Software: ArcGIS, SimplyMap

**United States**

*Alabama*

Auburn University
Auburn, Alabama

Subject Guide Name: Geographic Information Systems (GIS)
Subject Guide URL: http://libguides.auburn.edu/gis
Resources: Training, Online Maps, Other
Data: Government, Open, Statistical/Census, Remote Sensing
Data Search Capabilities: Yes
Software: ArcGIS
Cartographic Guide: Maps, http://libguides.auburn.edu/maps

Samford University Library
Birmingham, Alabama
Subject Guide Name: Geography: GIS
Subject Guide URL: http://samford.libguides.com/c.php?g=185943&p=2427506
Resources: Online Maps, Other
Data: Government, Open
Data Search Capabilities: No

University of Alabama Libraries
Tuscaloosa, Alabama
Subject Guide Name: Esri ArcGIS
Subject Guide URL: http://guides.lib.ua.edu/software/arcgis
Resources: Training, Instruction/Workshops, Other
Data Search Capabilities: No
Software: ArcGIS

*Arizona*

Arizona State University Libraries
Tempe, Arizona
Subject Guide Name: GIS Resources
Subject Guide URL: http://libguides.asu.edu/GIS
Resources: Training, Other
Data: Government, Open, Statistical/Census, Remote Sensing, Other
Data Search Capabilities: No
Software: ArcGIS
Cartographic Guide: Map Collection, http://libguides.asu.edu/mapcollection

University of Arizona Libraries
Tucson, Arizona
Subject Guide Name: GIS and Geospatial Data
Subject Guide URL: http://libguides.library.arizona.edu/GIS
Resources: Training, Online Maps, Other
Data: Government, Open, Statistical/Census, Remote Sensing, Other
Data Search Capabilities: Yes
Software: ArcGIS

*Arkansas*

University of Arkansas Libraries
Fayetteville, Arkansas
Subject Guide Name: Geographic Information Systems (GIS)
Subject Guide URL: http://uark.libguides.com/c.php?g=79160&p=504418
Resources: Training, Online Maps, Other
Data: Government, Open, Statistical/Census, Remote Sensing, Other
Data Search Capabilities: No
Software: ArcGIS
Cartographic Guide: A Basic Guide to Maps and Geography, http://researchguides.ualr.edu/mapsandgeography

*California*

California Polytechnic State University Library
San Luis Obispo, California
Subject Guide Name: GIS
Subject Guide URL: http://guides.lib.calpoly.edu/gis
Resources: Training, Online Maps, Other
Data: Government, Open, Statistical/Census, Remote Sensing
Data Search Capabilities: No

California State University, Dominguez Hills
Carson, California
Subject Guide Name: GIS and Maps
Subject Guide URL: http://csudh.libguides.com/gis
Resources: Online Maps, Other
Data: Government, Open, Statistical/Census
Data Search Capabilities: No

California State University, Long Beach
Long Beach, California
Subject Guide Name: Maps and Geographic Information Systems
Subject Guide URL: http://csulb.libguides.com/c.php?g=39178&p=249846
Resources: Online Maps, Other
Data: Government, Open, Statistical/Census
Data Search Capabilities: No
Software: ArcGIS 10, ArcInfo 10, SPSS 21
Cartographic Guide: Maps and Geographic Information Systems, http://csulb.libguides.com/maps

Claremont Colleges Library
Claremont, California
Subject Guide Name: The Claremont Colleges Library GIS Services
Subject Guide URL: http://libguides.libraries.claremont.edu/c.php?g=317939&p=2120425
Resources: Training, Instruction/Workshops, Other
Data: Government, Open, Statistical/Census, Remote Sensing, Other
Data Search Capabilities: No
Software: ArcGIS 10.4, Google Earth Pro, Sketchup, MathLAB, SPSS, Visual Studio 2012

Humboldt State University
Arcata, California
Subject Guide Name: GIS and Digital Spatial Data
Subject Guide URL: http://libguides.humboldt.edu/GIS
Resources: Online Maps, Other
Data: Government, Open, Statistical/Census, Remote Sensing, Other
Data Search Capabilities: No
Software: ArcGIS, QGIS, FW Tools, GME, DNRGPS, 7-Zip, EasyGIS, GeoServer, NetLogo, R, R-Studio, Python, AutoCAD, PostgreSQL
Cartographic Guide: Maps and Atlases, http://libguides.humboldt.edu/mapsatlas

Stanford University Libraries
Stanford, California
Subject Guide Name: Stanford Geospatial Center
Subject Guide URL: http://library.stanford.edu/research/stanford-geospatial-center
Resources: Training, Instruction/Workshops, Online Maps, Other
Data: Government, Open, Remote Sensing
Data Search Capabilities: Yes
Software: ArcGIS 10.2.2, Basins 4.1, Diva GIS, GeoMapApp, Geospatial Modelling Environment (GME), Open Refine, Hydro Desktop, InVest, Envi/Idle, Google Earth Pro, GRASS, Marine
Cartographic Guide: Maps, http://library.stanford.edu/subjects/maps

University of California, Berkeley Library
Subject Guide Name: GIS (Geographic Information Systems)
Subject Guide URL: http://guides.lib.berkeley.edu/c.php?g=130718&p=854021
Resources: Training, Instruction/Workshops, Online Maps, Other
Data: Government, Open, Statistical/Census, Remote Sensing
Data Search Capabilities: Yes
Software: ArcGIS 10.4.1
Cartographic Guide: Maps and Air Photos: Finding Maps, http://guides.lib.berkeley.edu/maps

University of California, Davis Library
Davis, California
Subject Guide Name: Maps and GIS Subject Guide
Subject Guide URL: http://guides.lib.ucdavis.edu/Maps
Resources: Training, Online Maps, Other
Data: Government, Open, Statistical/Census, Remote Sensing, Other
Data Search Capabilities: Yes
Software: ArcGIS Suite, StreetMap, SurveyAnalyst, Tracking Analyst
Cartographic Guide: Map and GIS Subject Guide, http://guides.lib.ucdavis.edu/Maps

University of California, Irvine Libraries
Irvine, California
Subject Guide Name: Geographic Information Systems (GIS)
Subject Guide URL: http://guides.lib.uci.edu/gis/home
Resources: Training, Online Maps, Other
Data: Government, Open, Statistical/Census, Other
Data Search Capabilities: Yes
Software: ArcGIS 10.4.1
Additional Subject Guides: http://guides.lib.uci.edu/friendly.php?s=census

University of California, Los Angeles Library
Los Angeles, California
Subject Guide Name: Geographic Information Systems (GIS)
Subject Guide URL: http://guides.library.ucla.edu/c.php?g=180507&p=1187577
Resources: Training, Instruction/Workshops, Online Maps, Other
Data: Government, Open, Statistical/Census, Remote Sensing
Data Search Capabilities: Yes
Cartographic Guide: Maps, Atlases, Aerial Images, and Cartographic Resources, http://guides.library.ucla.edu/maps

University of California, San Diego
La Jolla, California
Subject Guide Name: GIS at UC San Diego
Subject Guide URL: http://ucsd.libguides.com/c.php?g=90732&p=585108
Resources: Training, Instruction/Workshops, Online Maps, Other
Data: Government, Open, Statistical/Census, Remote Sensing, Other
Data Search Capabilities: Yes

University of California, Santa Barbara Library
Santa Barbara, California
Subject Guide Name: Geospatial Data
Subject Guide URL: http://guides.library.ucsb.edu/c.php?g=546830&p=3751720
Resources: Training, Other
Data: Government, Open, Statistical/Census, Remote Sensing
Data Search Capabilities: Yes
Software: ArcMap, ArcGIS Pro, ArcGIS Earth
Cartographic Guide: Map Collection, http://www.library.ucsb.edu/map-imagery-lab/map-collection

*Colorado*

Colorado State University Libraries
Fort Collins, Colorado

Subject Guide Name: GIS (Geographic Information Systems)
Subject Guide URL: http://libguides.colostate.edu/c.php?g=64763&p=418702
Resources: Training, Online Maps, Other
Data: Government, Open, Statistical/Census, Remote Sensing
Data Search Capabilities: No
Cartographic Guide: Map Resources, http://libguides.colostate.edu/maps

University of Colorado, Boulder Libraries
Boulder, Colorado
Subject Guide Name: GIS (Geographic Information Systems)
Subject Guide URL: http://libguides.colorado.edu/gis
Resources: Training, Instruction/Workshops, Online Maps, Other
Data: Government, Open, Statistical/Census, Remote Sensing
Data Search Capabilities: Yes
Additional Subject Guides: http://libguides.colorado.edu/geospatialdata; http://libguides.colorado.edu/digitalmaps; http://libguides.colorado.edu/digitalmaps; http://libguides.colorado.edu/printmaps; http://libguides.colorado.edu/aerial
Cartographic Guide: Maps and GIS, http://libguides.colorado.edu/strategies/maps

*Connecticut*

Yale University
New Haven, Connecticut
Subject Guide Name: Geographic Information Systems at Yale
Subject Guide URL: http://guides.library.yale.edu/c.php?g=295854&p=1972661
Resources: Training, Instruction/Workshops, Online Maps, Other
Data: Government, Open, Statistical/Census, Remote Sensing
Data Search Capabilities: No
Cartographic Guide: South Asia: Maps and GIS Data, http://guides.library.yale.edu/SouthAsiaMapsGIS/Maps

*Delaware*

University of Delaware
Newark, Delaware
Subject Guide Name: Digital Mapping (GIS)
Subject Guide URL: http://guides.lib.udel.edu/c.php?g=85598&p=548103
Resources: Online Maps
Data: Government, Open, Statistical/Census, Remote Sensing, Other
Data Search Capabilities: Yes
Software: ArcGIS 10.4, Policy Map, Google Earth Pro

Cartographic Guide: Cartographic Resources, Maps, and Spatial Data, http://guides.lib.udel.edu/maps

*Florida*

University of Central Florida
Orlando, Florida
Subject Guide Name: Maps and Geographic Information System (GIS) Resources
Subject Guide URL: http://guides.ucf.edu/gis/home
Resources: Online Maps, Other
Data: Government, Open, Other
Data Search Capabilities: No
Cartographic Guide: Sanborn Fire Insurance Maps, http://guides.ucf.edu/sanborn

University of Miami
Miami, Florida
Subject Guide Name: GIS Resources at UM Libraries
Subject Guide URL: http://sp.library.miami.edu/subjects/gis
Resources: Training, Instruction/Workshops, Online Maps, Other
Data: Government, Open, Statistical/Census, Remote Sensing
Data Search Capabilities: No
Software: ArcGIS

*Georgia*

Georgia Institute of Technology
Atlanta, Georgia
Subject Guide Name: Maps, GIS, and Spatial Data
Subject Guide URL: http://libguides.gatech.edu/c.php?g=54023&p=349210
Resources: Training, Online Maps, Other
Data: Government, Open, Statistical/Census, Remote Sensing
Data Search Capabilities: No
Cartographic Guide: Maps, GIS, and Spatial Data, http://libguides.gatech.edu/c.php?g=54023&p=349210

Georgia State University Library
Atlanta, Georgia
Subject Guide Name: GIS Resources
Subject Guide URL: http://research.library.gsu.edu/GIS
Resources: Training, Online Maps
Data: Government, Open, Statistical/Census
Data Search Capabilities: No
Software: ArcGIS

University of Georgia Libraries
Athens, Georgia
Subject Guide Name: GIS Data
Subject Guide URL: http://guides.libs.uga.edu/gis
Resources: Training, Online Maps, Other

Data: Government, Open, Statistical/Census, Remote Sensing
Data Search Capabilities: No
Additional Subject Guides: http://guides.libs.uga.edu/airphoto; http://guides.libs.uga.edu/maps; http://guides.libs.uga.edu/map_reference
Cartographic Guide: Map and Government Information, http://www.libs.uga.edu/magil/

*Hawaii*

Brigham Young University Library
Laie, Hawaii
Cartographic Guide: Geography and Maps, http://libguides.byuh.edu/geography

University of Hawaii at Manoa
Honolulu, Hawaii
Subject Guide Name: Maps, Aerial Photographs, and GIS (MAGIS)
Subject Guide URL: http://guides.library.manoa.hawaii.edu/magis/home
Resources: Training, Online Maps, Other
Data: Government, Open, Statistical/Census, Remote Sensing
Data Search Capabilities: No
Software: ArcGIS 10.3, ENVI 4.7, Google Earth, Adobe Create Suite 4

*Illinois*

Eastern Illinois University
Charleston, Illinois
Subject Guide Name: GIS (Geographic Information System) Resources
Subject Guide URL: http://booth.library.eiu.edu/subjectsPlus/subjects/guide.php?subject=GIS
Resources: Training, Other
Data: Government, Open, Statistical/Census
Data Search Capabilities: No

Northwestern University Libraries
Evanston, Illinois
Subject Guide Name: GIS: Geographic Information System
Subject Guide URL: http://libguides.northwestern.edu/gis
Resources: Training, Instruction/Workshops, Other
Data: Government, Open, Statistical/Census, Remote Sensing, Other
Data Search Capabilities: No
Software: ArcGIS 10.2
Cartographic Guide: Map and Atlas Collections, http://libguides.northwestern.edu/NorthwesternMapLibrary

Southern Illinois University, Carbondale Library
Carbondale, Illinois
Subject Guide Name: Data/Statistics
Subject Guide URL: http://libguides.lib.siu.edu/content.php?pid=562578&sid=4637404
Resources: Training, Other
Data: Government, Open, Statistical/Census
Data Search Capabilities: No
Software: ArcView GIS 3.3, SAS, STATA, SPSS
Additional Subject Guides: http://libguides.lib.siu.edu/content.php?pid=500675&sid=4119544
Cartographic Guide: Map Resources Online, http://libguides.lib.siu.edu/MapsOnline

University of Chicago
Chicago, Illinois
Subject Guide Name: Geographic Information Systems
Subject Guide URL: https://gis.uchicago.edu/
Resources: Training, Other
Data: Government, Open, Statistical/Census, Remote Sensing
Data Search Capabilities: No
Software: ArcGIS
Cartographic Guide: How Do I Find Maps?, http://guides.lib.uchicago.edu/maps

University of Illinois, Chicago
Chicago, Illinois
Subject Guide Name: Geography and GIS
Subject Guide URL: http://researchguides.uic.edu/geography
Data: Government, Open, Statistical/Census, Remote Sensing
Data Search Capabilities: No
Cartographic Guide: Maps and GIS, http://library.uic.edu/collections/maps-gis

University of Illinois, Urbana-Champaign
Champaign, Illinois
Subject Guide Name: Geographic Information Systems (GIS)
Subject Guide URL: http://guides.library.illinois.edu/c.php?g=348425&p=2341023
Resources: Training, Instruction/Workshops, Other
Data: Government, Open, Statistical/Census, Other
Data Search Capabilities: No
Cartographic Guide: Map Library, http://guides.library.illinois.edu/c.php?g=348524

Western Illinois University
Macomb, Illinois
Subject Guide Name: Geography
Subject Guide URL: http://wiu.libguides.com/c.php?g=295318&p=1968783
Resources: Online Maps, Other
Data: Government, Open, Statistical/Census, Other
Data Search Capabilities: No
Cartographic Guide: Illinois Map Resources, http://www.wiu.edu/libraries/govpubs/map_collection/mapResources.php

*Indiana*

Ball State University
Muncie, Indiana
Subject Guide Name: GIS Research Area
Subject Guide URL: http://cms.bsu.edu/academics/libraries/collectionsanddept/gisandmaps/gisresearch
Resources: Training, Online Maps, Other
Data: Government, Open, Statistical/Census
Data Search Capabilities: No
Software: ArcGIS Desktop 10.4.1, ERDAS Imagine 2016

Earlham College
Richmond, Indiana
Subject Guide Name: GIS: Earlham College
Subject Guide URL: http://library.earlham.edu/c.php?g=82970&p=533153
Resources: Training, Online Maps, Other
Data: Government, Open, Statistical/Census, Other
Data Search Capabilities: No

Indiana University, Bloomington Libraries
Bloomington, Indiana
Subject Guide Name: GIS
Subject Guide URL: http://guides.libraries.indiana.edu/c.php?g=449146&p=3065527
Resources: Training, Instruction/Workshops, Online Maps, Other
Data: Government, Open, Statistical/Census, Remote Sensing
Data Search Capabilities: No
Software: ArcGIS
Additional Subject Guides: http://gisinfo.iu.edu/
Cartographic Guide: Maps and GIS, https://libraries.indiana.edu/maps-and-gis

Indiana University–Purdue University, Indianapolis
Indianapolis, Indiana
Subject Guide Name: IUPUI Department of Geography Library Guide
Subject Guide URL: http://iupui.campusguides.com/geography
Resources: Online Maps, Other
Data: Government, Open, Statistical/Census, Remote Sensing
Data Search Capabilities: No
Cartographic Guide: GIS, Maps and Atlases, http://iupui.campusguides.com/gismapsatlases#s-lg-box-5302265

Purdue University
Lafayette, Indiana
Subject Guide Name: GIS Data Guide
Subject Guide URL: http://guides.lib.purdue.edu/GIS
Resources: Training, Online Maps, Other
Data: Government, Open, Statistical/Census, Remote Sensing
Data Search Capabilities: No
Software: ArcGIS, IMAGINE, ENVI, eCognition, GeoSPHERIC, MultiSpec
Cartographic Guide: Map Resources, http://guides.lib.purdue.edu/mapresources

*Iowa*

Iowa State University
Ames, Iowa
Subject Guide Name: Geographic Information Systems
Subject Guide URL: http://www.gis.iastate.edu/
Resources: Training, Online Maps, Other
Data Search Capabilities: No

University of Iowa Libraries
Iowa City, Iowa
Subject Guide Name: Geographic Information Systems (GIS)
Subject Guide URL: http://guides.lib.uiowa.edu/gis
Resources: Online Maps, Other
Data: Government, Open, Statistical/Census, Remote Sensing
Data Search Capabilities: No
Cartographic Guide: Cartographic Information on the Web, http://guides.lib.uiowa.edu/maps/cartographic; Map Collection, http://www.lib.uiowa.edu/maps/

*Kansas*

Kansas State University
Manhattan, Kansas
Subject Guide Name: Data—Geospatial (GIS) Resources
Subject Guide URL: http://guides.lib.k-state.edu/c.php?g=181750&p=1196194
Resources: Training, Online Maps, Other
Data: Government, Open, Statistical/Census, Remote Sensing, Other
Data Search Capabilities: No
Software: ArcGIS 10.1, SPSS, FME Translator, SAS, R, GeoDa, Adobe PhotoShop
Cartographic Guide: Maps, http://guides.lib.k-state.edu/maps

University of Kansas Libraries
Lawrence, Kansas
Subject Guide Name: Geographic Information Systems (GIS)
Subject Guide URL: http://guides.lib.ku.edu/gis
Resources: Training, Instruction/Workshops, Online Maps, Other
Data: Government, Open, Statistical/Census, Remote Sensing, Other
Data Search Capabilities: No
Software: Adobe Suite Pro, ArcGIS, ATLAS, Business Analyst, Camtasia Studio 7, Colectica Express, Google Earth Pro, Maple, Minitab, Notepad++, NppToR, Nvivo, oXygen XML Editor, QGIS, R, SAS, SPSS

Cartographic Guide: Maps, Atlases, and Gazetteers, http://guides.lib.ku.edu/maps

Wichita State University
Wichita, Kansas
Subject Guide Name: GIS—Geographic Information Systems
Subject Guide URL: http://libresources.wichita.edu/gis
Resources: Training, Online Maps, Other
Data: Government, Open, Statistical/Census
Data Search Capabilities: No

*Kentucky*

University of Louisville
Subject Guide Name: Center for Geographic Information Sciences
Subject Guide URL: http://www.ulcgis.org/bldgs/
Data: Government, Open, Remote Sensing
Data Search Capabilities: No
Software: ArcGIS Desktop, ArcGIS Online, ArcGIS Server, Business Analyst, ESRI Maps and Data

*Maryland*

Johns Hopkins University
Baltimore, Maryland
Subject Guide Name: GIS and Maps
Subject Guide URL: http://guides.library.jhu.edu/gis
Resources: Training, Instruction/Workshops, Online Maps
Data: Government, Open, Statistical/Census, Remote Sensing, Other
Data Search Capabilities: No
Software: ArcGIS, Stata, SPSS, ERDAS, ENVI
Cartographic Guide: GIS and Maps, http://guides.library.jhu.edu/c.php?g=202579&p=1335290

University of Maryland
College Park, Maryland
Subject Guide Name: GIS and Geospatial Center
Subject Guide URL: http://www.lib.umd.edu/gis
Resources: Training, Instruction/Workshops, Online Maps
Data: Government, Open, Statistical/Census, Remote Sensing
Data Search Capabilities: No
Software: ArcGIS 10.3.1, ENVI 5.0, MATLAB 2015a, QGIS, Anaconda/Python, SPSS 23, SAS 9.4, Rstudio, Google Earth Pro, Weka, Panoply
Cartographic Guide: Maryland Maps, http://lib.guides.umd.edu/marylandmaps

*Massachusetts*

Boston College
Chestnut Hill, Massachusetts
Subject Guide Name: Finding Spatial Data for Mapping (GIS)
Subject Guide URL: http://libguides.bc.edu/gis/getinfo
Resources: Instruction/Workshops, Online Maps, Other
Data: Government, Open, Statistical/Census, Remote Sensing, Other
Data Search Capabilities: No
Software: ArcGIS 10

Harvard University
Cambridge, Massachusetts
Subject Guide Name: Center for Geographic Analysis
Subject Guide URL: http://gouldguides.carleton.edu/c.php?g=146933&p=963592
Resources: Training, Instruction/Workshops, Other
Data: Government, Open, Remote Sensing
Data Search Capabilities: No
Software: ArcGIS, ENVI, ERDAS, IDRISI
Cartographic Guide: Harvard Map Collection, http://hcl.harvard.edu/libraries/maps/collections/print.cfm

Massachusetts Institute of Technology
Cambridge, Massachusetts
Subject Guide Name: Geographic Information Systems (GIS)
Subject Guide URL: http://libguides.mit.edu/c.php?g=176295&p=1160692
Resources: Training, Instruction/Workshops, Online Maps, Other
Data: Government, Open, Statistical/Census, Remote Sensing, Other
Data Search Capabilities: No
Software: ArcGIS Desktop, Google Earth Pro, Geolytics, Adobe Creative Suite, AutoCAD
Cartographic Guide: Maps, http://libguides.mit.edu/maps

*Michigan*

Eastern Michigan University
Ypsilanti, Michigan
Subject Guide Name: GIS and GPS
Subject Guide URL: http://guides.emich.edu/c.php?g=187896&p=1242177
Data: Government, Statistical/Census
Data Search Capabilities: No
Cartographic Guide: Maps, http://guides.emich.edu/maps

Michigan State University
East Lansing, Michigan
Subject Guide Name: GIS Guide
Subject Guide URL: http://libguides.lib.msu.edu/gis
Resources: Training, Instruction/Workshops, Online Maps, Other
Data: Government, Open, Statistical/Census, Remote Sensing
Data Search Capabilities: No
Software: ArcGIS 10.1

Additional Subject Guides: http://libguides.lib.msu.edu/gisdata; http://libguides.lib.msu.edu/geocoding
Cartographic Guide: MSU Map Library, http://libguides.lib.msu.edu/maplinks

Michigan Technological University
Houghton, Michigan
Subject Guide Name: Geographic Information Science
Subject Guide URL: http://gis.mtu.edu/
Resources: Training, Online Maps, Other
Data: Government, Open, Statistical/Census, Remote Sensing
Data Search Capabilities: No
Software: ArcGIS Desktop 10.4.1, Google Earth Pro, DNRGPS
Cartographic Guide: Michigan Topographic Maps, http://libguides.lib.mtu.edu/topographicmaps

University of Michigan Library
Ann Arbor, Michigan
Subject Guide Name: Geospatial Data Resources
Subject Guide URL: http://guides.lib.umich.edu/gisdata
Resources: Training, Online Maps, Other
Data: Government, Open, Statistical/Census
Data Search Capabilities: No
Additional Subject Guides: http://guides.lib.umich.edu/govstatistics
Cartographic Guide: GIS, Map and Statistics, http://guides.lib.umich.edu/c.php?g=283152&p=1886396; http://guides.lib.umich.edu/c.php?g=283318&p=1886982; http://guides.lib.umich.edu/c.php?g=283110&p=1886161

*Minnesota*

Carleton College, Laurence McKinley Gould Library
Northfield, Minnesota
Subject Guide Name: GIS and Geospatial Data
Subject Guide URL: http://gouldguides.carleton.edu/c.php?g=146933&p=963592
Resources: Training, Online Maps, Other
Data: Government, Open, Statistical/Census, Remote Sensing
Data Search Capabilities: No
Cartographic Guide: Maps at Carleton, http://gouldguides.carleton.edu/maps

Minnesota State University, Mankato
Mankato, Minnesota
Subject Guide Name: GIS—Geographic Information Systems
Subject Guide URL: http://libguides.mnsu.edu/c.php?g=76753&p=497687
Data: Government, Open, Statistical/Census
Data Search Capabilities: No
Cartographic Guide: Maps Guide (Mary T Dooley Map Library), http://libguides.mnsu.edu/mapsarea

University of Minnesota
Minneapolis, Minnesota
Subject Guide Name: John R. Borchert Map Library
Subject Guide URL: https://www.lib.umn.edu/borchert/gis-data
Resources: Online Maps, Other
Data: Government, Open, Statistical/Census, Remote Sensing
Data Search Capabilities: No
Software: ArcGIS 10.2.2, Adobe CS5
Cartographic Guides: John R. Borchert Map Library, https://www.lib.umn.edu/borchert; Maps and GIS in Walter Library, https://www.lib.umn.edu/libdata/page.phtml?page_id=3042

*Missouri*

Saint Louis University
Saint Louis, Missouri
Subject Guide Name: Geographic Information Systems (GIS)
Subject Guide URL: http://libguides.slu.edu/c.php?g=185766&p=1227584
Resources: Training, Other
Data: Government, Open, Statistical/Census
Data Search Capabilities: No
Software: ArcGIS Suite

Washington University
Saint Louis, Missouri
Subject Guide Name: GIS (Geographic Information Systems)
Subject Guide URL: http://libguides.wustl.edu/c.php?g=46935&p=301616
Resources: Training, Online Maps, Other
Data: Government, Open, Statistical/Census, Remote Sensing
Data Search Capabilities: No
Software: ArcGIS, ERDAS Imagine, FME Pro

*Nebraska*

University of Nebraska, Lincoln
Lincoln, Nebraska
Subject Guide Name: GIS
Subject Guide URL: http://unl.libguides.com/gis
Resources: Training, Online Maps, Other
Data: Government, Open
Data Search Capabilities: Yes
Software: ArcGIS 10.2, Adobe Design Suite CS5, Google Earth Sketch Up, Google Earth Pro
Additional Subject Guides: http://unl.libguides.com/gistools
Cartographic Guide: Maps and Publications of the USGS, http://unl.libguides.com/usgs; Genealogy Research and Maps, http://unl.libguides.com/c.php?g=307485

## Nevada

University of Nevada, Reno
Reno, Nevada
Subject Guide Name: Geographic Information Systems (GIS)
Subject Guide URL: http://guides.library.unr.edu/gis
Resources: Training, Online Maps
Data: Government, Open, Statistical/Census, Remote Sensing
Data Search Capabilities: Yes
Software: ArcGIS 10.2
Additional Subject Guides: http://libguides.lib.msu.edu/gisdata; http://libguides.lib.msu.edu/maplinks; http://libguides.lib.msu.edu/geocoding
Cartographic Guide: Maps, http://guides.library.unr.edu/Maps

## New Hampshire

Dartmouth College
Hanover, New Hampshire
Subject Guide Name: Geographic Information Systems/Science: A Research Guide
Subject Guide URL: http://researchguides.dartmouth.edu/gis
Resources: Training, Online Maps, Other
Data: Government, Open, Statistical/Census, Remote Sensing
Data Search Capabilities: No
Software: ArcGIS
Additional Subject Guides: http://researchguides.dartmouth.edu/gisdata
Cartographic Guides: Maps and Atlases, http://researchguides.dartmouth.edu/maps; Maps Online, http://researchguides.dartmouth.edu/mapsonline; Cartography/History of Cartography, http://researchguides.dartmouth.edu/cartography

## New Jersey

Princeton University Library
Princeton, New Jersey
Subject Guide Name: Maps and Geospatial Information
Subject Guide URL: http://library.princeton.edu/collections/pumagic
Resources: Instruction/Workshops, Online Maps
Data: Open
Data Search Capabilities: Yes
Software: ArcGIS Desktop, ERDAS Imagine, Adobe Photoshop, Adobe Illustrator
Cartographic Guide: Maps and Geospatial Information, http://library.princeton.edu/collections/pumagic

Rutgers University, Camden Campus
New Brunswick, New Jersey
Subject Guide Name: GIS (Geographic Information Systems)
Subject Guide URL: http://libguides.rutgers.edu/c.php?g=336835&p=2267523
Resources: Training
Data: Government, Open, Statistical/Census, Remote Sensing
Data Search Capabilities: No
Cartographic Guide: Maps and Travel (Electronic Reference Sources), http://libguides.rutgers.edu/maps/geography; Scientific Maps, http://libguides.rutgers.edu/sci_maps

## New Mexico

University of New Mexico Library
Albuquerque, New Mexico
Subject Guide Name: Map and Geographic Information Center (MAGIC)
Subject Guide URL: http://libguides.unm.edu/magic
Resources: Online Maps, Other
Data: Government, Open
Data Search Capabilities: No
Software: ArcGIS 10.3

## New York

Cornell University Library
Ithaca, New York
Subject Guide Name: Finding Cartographic Resources Online
Subject Guide URL: http://guides.library.cornell.edu/c.php?g=31190&p=199261
Resources: Online Maps, Other
Data: Government, Open
Data Search Capabilities: No
Cartographic Guide: Map and Geospatial Information Collection, https://olinuris.library.cornell.edu/collections/maps

Pratt Institute Libraries
Brooklyn, New York
Subject Guide Name: Maps and GIS
Subject Guide URL: http://libguides.pratt.edu/content.php?pid=615559&sid=5088812
Resources: Online Maps, Other
Data: Government, Open
Data Search Capabilities: No
Cartographic Guide: Maps and GIS, http://libguides.pratt.edu/maps

Stony Brook University
Stony Brook, New York
Subject Guide Name: GIS Resources
Subject Guide URL: http://guides.library.stonybrook.edu/gis-resources
Resources: Online Maps, Other

Data: Government, Open, Statistical/Census, Remote Sensing, Other
Data Search Capabilities: No
Software: ArcGIS, Google Earth Pro
Cartographic Guide: Maps, http://guides.library.stonybrook.edu/Maps

Syracuse University Libraries
Syracuse, New York
Subject Guide Name: GIS, Statistics, Gaming, and Visualization
Subject Guide URL: http://researchguides.library.syr.edu/c.php?g=258118&p=1723812
Resources: Training, Online Maps, Other
Data: Government, Open, Statistical/Census, Remote Sensing
Data Search Capabilities: Yes
Software: ArcGIS ArcGIS 10.4 Suite, SAS, Stata, SPSS
Additional Subject Guides: http://libguides.lib.msu.edu/gisdata; http://libguides.lib.msu.edu/maplinks; http://libguides.lib.msu.edu/geocoding
Cartographic Guide: Maps, http://researchguides.library.syr.edu/Maps

University at Buffalo Libraries
Buffalo, New York
Subject Guide Name: Maps and Cartography
Subject Guide URL: http://research.lib.buffalo.edu/c.php?g=483023&p=3303034
Resources: Training, Online Maps, Other
Data: Government, Open, Statistical/Census, Remote Sensing, Other
Data Search Capabilities: No

*North Carolina*

Appalachian State University
Boone, North Carolina
Subject Guide Name: GIS (Geographic Information Systems)
Subject Guide URL: http://guides.library.appstate.edu/gis
Resources: Training, Instruction/Workshops, Online Maps
Data: Government, Open, Statistical/Census, Remote Sensing
Data Search Capabilities: No
Software: ArcGIS Suite

Duke University Libraries
Durham, North Carolina
Subject Guide Name: GIS Data
Subject Guide URL: http://guides.library.duke.edu/gisdata
Resources: Training, Instruction/Workshops
Data: Government, Open, Statistical/Census, Remote Sensing, Other
Data Search Capabilities: No

Additional Subject Guides: http://guides.library.duke.edu/webmap
Cartographic Guide: Maps, http://guides.library.duke.edu/maps

University of North Carolina, Chapel Hill University Libraries
Chapel Hill, North Carolina
Subject Guide Name: Spatial Data Online
Subject Guide URL: http://guides.lib.unc.edu/spatialdata/home
Resources: Training, Online Maps, Other
Data: Government, Open, Statistical/Census, Remote Sensing
Data Search Capabilities: Yes
Additional Subject Guides: http://guides.lib.unc.edu/software/home; http://guides.lib.unc.edu/mapsatUNC

Western Carolina University
Cullowhee, North Carolina
Subject Guide Name: Geographic Information Systems (GIS)
Subject Guide URL: http://researchguides.wcu.edu/GIS
Resources: Online Maps, Other
Data: Government, Open, Statistical/Census, Remote Sensing
Data Search Capabilities: No
Software: ArcGIS 10.1
Additional Subject Guides: http://researchguides.wcu.edu/geography; http://researchguides.wcu.edu/maps

*Ohio*

Case Western Reserve University Libraries
Cleveland, Ohio
Subject Guide Name: Geographic Information Systems (GIS)
Subject Guide URL: http://researchguides.case.edu/GIS
Resources: Training, Online Maps, Other
Data: Government, Open, Statistical/Census, Remote Sensing
Data Search Capabilities: No
Software: ArcGIS

Denison University Libraries
Granville, Ohio
Subject Guide Name: Denison University Libraries
Subject Guide URL: http://libguides.denison.edu/c.php?g=311693&p=2083604
Resources: Online Maps, Other
Data: Government, Remote Sensing, Other
Data Search Capabilities: No
Software: ArcGIS 10.1
Cartographic Guide: Cartographic Resources: Maps and Spatial Data, http://libguides.denison.edu/maps

Kent State University Libraries
Kent, Ohio
Subject Guide Name: GIS Data and Tools
Subject Guide URL: http://libguides.library.kent.edu/gis
Resources: Training, Online Maps, Other
Data: Government, Open, Statistical/Census, Remote Sensing
Data Search Capabilities: No
Software: Esri Suite
Cartographic Guide: Finding Paper Maps, http://libguides.library.kent.edu/findingmaps; Finding Online Map Resources, http://libguides.library.kent.edu/c.php?g=278156

Miami University Libraries
Oxford, Ohio
Subject Guide Name: Geographic Information Systems
Subject Guide URL: http://libguides.lib.miamioh.edu/GIS
Resources: Training, Other
Data Search Capabilities: No
Cartographic Guide: Mapping and Geospatial Imagery, http://libguides.lib.miamioh.edu/maps

Ohio State University Libraries
Columbus, Ohio
Subject Guide Name: Maps and Geospatial Data
Subject Guide URL: http://guides.osu.edu/maps-geospatial-data/getting-started
Resources: Training, Online Maps, Other
Data: Government, Open, Statistical/Census, Remote Sensing
Data Search Capabilities: No
Software: ArcGIS 10.2
Cartographic Guide: Maps and Geospatial Data, http://guides.osu.edu/maps-geospatial-data

Ohio University Libraries
Athens, Ohio
Subject Guide Name: Geography: Cartography, Remote Sensing and GIS
Subject Guide URL: http://libguides.library.ohiou.edu/cartography-gis
Resources: Online Maps, Other
Data: Government, Open
Data Search Capabilities: No
Cartographic Guide: Geography: Cartography, Remote Sensing and GIS, http://libguides.library.ohiou.edu/cartography-gis

University of Akron Libraries
Akron, Ohio
Subject Guide Name: Geosciences—Geology, Geography, and Environmental Science
Subject Guide URL: http://libguides.uakron.edu/c.php?g=226429&p=1500998
Resources: Training, Online Maps, Other
Data: Government, Open
Data Search Capabilities: No

University of Cincinnati Libraries
Cincinnati, Ohio
Subject Guide Name: GIS (Geographic Information Systems)
Subject Guide URL: http://guides.libraries.uc.edu/c.php?g=222807&p=1473492
Resources: Training Online Maps, Other
Data: Government, Open, Statistical/Census, Remote Sensing
Data Search Capabilities: No
Software: ArcGIS

University of Toledo Libraries
Toledo, Ohio
Subject Guide Name: Geography/Planning and Geographic Information Systems (GIS)
Subject Guide URL: http://libguides.utoledo.edu/gepl-gis/home
Resources: Online Maps, Other
Data: Government, Open, Statistical/Census, Remote Sensing
Data Search Capabilities: No
Cartographic Guide: Maps, http://libguides.utoledo.edu/findmaps

*Oregon*

Portland State University Library
Portland, Oregon
Subject Guide Name: Geography
Subject Guide URL: http://guides.library.pdx.edu/c.php?g=271169&p=1810460
Resources: Online Maps, Other
Data: Government, Open
Data Search Capabilities: No

*Pennsylvania*

Drexel University Libraries
Philadelphia, Pennsylvania
Subject Guide Name: Geographic Information Systems
Subject Guide URL: http://libguides.library.drexel.edu/GIS
Data: Government, Open, Statistical/Census
Data Search Capabilities: No
Software: ArcGIS

Kutztown University Rohrback Library
Kutztown, Pennsylvania
Subject Guide Name: GIS Resources
Subject Guide URL: http://library.kutztown.edu/c.php?g=229109&p=1518745
Cartographic Guide: Maps on the Web, http://library.kutztown.edu/maps

Pennsylvania State University Libraries
University Park, Pennsylvania
Subject Guide Name: GIS (Geographic Information Systems)

Subject Guide URL: http://guides.libraries.psu.edu/gis
Resources: Training, Online Maps, Other
Data: Government, Open, Statistical/Census, Remote Sensing
Data Search Capabilities: No
Software: ArcGIS
Additional Subject Guides: http://guides.libraries.psu.edu/friendly.php?s=remotesensing
Cartographic Guides: Maps and Geospatial: Maps, http://guides.libraries.psu.edu/maps; Maps and Geospatial: Newsmaps, http://guides.libraries.psu.edu/newsmaps; Maps and Geospatial: Pennsylvania Maps, http://guides.libraries.psu.edu/pamaps; Maps and Geospatial: Sanborn and Other Fire Insurance Maps; http://guides.libraries.psu.edu/sanbornmaps; Maps and Geospatial: Topographic Maps, http://guides.libraries.psu.edu/topomaps; Maps and Geospatial: Conflict/War Maps, http://guides.libraries.psu.edu/warmaps; Maps anbd Geospatial: Aerial Photographs (Historic), http://guides.libraries.psu.edu/aerialphotographs

TriCollege Libraries—Bryn Mawr, Haverford, and Swarthmore Colleges
Bryn Mawr, Pennsylvania
Subject Guide Name: Maps and Geography
Subject Guide URL: http://libguides.brynmawr.edu/maps
Resources: Online Maps
Data: Government, Open, Statistical/Census, Other
Data Search Capabilities: No

University of Pennsylvania
Philadelphia, Pennsylvania
Subject Guide Name: Data and GIS
Subject Guide URL: http://guides.library.upenn.edu/data
Resources: Training, Instruction/Workshops, Other
Data: Government, Open, Statistical/Census, Other
Data Search Capabilities: No
Software: SPSS, SAS, Stata, R, ArcGIS, QGIS

University of Pittsburgh Libraries
Pittsburgh, Pennsylvania
Subject Guide Name: Geographic Information System (GIS) Resources—Oakland Campus
Subject Guide URL: http://pitt.libguides.com/gis
Resources: Training, Online Maps, Other
Data: Government, Open, Statistical/Census, Remote Sensing, Other
Data Search Capabilities: No
Software: ArcGIS
Additional Subject Guides: http://pitt.libguides.com/uscensus
Cartographic Guides: Maps and Atlases @ Pitt Archives, http://pitt.libguides.com/ascmapsatlases; Finding Maps—Oakland Campus, http://pitt.libguides.com/hillman_maps; Finding Topographic Maps—Oakland Campus, http://pitt.libguides.com/topo

*Tennessee*

University of Tennessee, Knoxville University Libraries
Knoxville, Tennessee
Subject Guide Name: UT Geographic Information Systems (GIS) Services Hub
Subject Guide URL: http://libguides.utk.edu/gis
Resources: Training, Online Maps, Other
Data: Government, Open, Statistical/Census, Other
Data Search Capabilities: No
Software: ArcGIS
Cartographic Guide: Map Collection and Online Resources, http://libguides.utk.edu/map

Vanderbilt University Library
Nashville, Tennessee
Subject Guide Name: Geographic Information Systems (GIS)
Subject Guide URL: http://researchguides.library.vanderbilt.edu/gis
Resources: Training, Instruction/Workshops, Other
Data: Government, Open, Statistical/Census
Data Search Capabilities: No
Software: ArcGIS 10.4.1

*Texas*

Abilene Christian University Brown Library
Abilene, Texas
Subject Guide Name: Data Services
Subject Guide URL: http://guides.acu.edu/data
Resources: Training, Other
Data: Government, Open, Statistical/Census
Data Search Capabilities: No
Software: ArcGIS, SPSS

Rice University Fondren Library
Houston, Texas
Subject Guide Name: Geospatial/GIS Data
Subject Guide URL: http://libguides.rice.edu/GIS
Resources: Training, Other
Data: Government, Open
Data Search Capabilities: No
Software: ArcGIS 10 Suite, CommunityViz 4.1, ER Mapper, ERDAS IMAGINE, GME, Google Earth Pro 6, PASW Statistics 18, R 2.12, SAS 9.2, Adobe Design Premium CS5 Suite, Google SketchUp Pro 8, AutoCAD
Additional Subject Guides: http://libguides.rice.edu/gov_info_statistics_data
Cartographic Guide: Maps, http://libguides.rice.edu/maps

Texas A&M University Libraries
College Station, Texas
Subject Guide Name: GIS: Geographic Information Systems
Subject Guide URL: http://guides.library.tamu.edu/GIS

Resources: Training, Instruction/Workshops, Online Maps, Other
Data: Government, Open, Statistical/Census, Remote Sensing, Other
Data Search Capabilities: No
Software: ArcGIS 10.2, Google Earth
Additional Subject Guides: http://guides.library.tamu.edu/remotesensing; http://tamu.libguides.com/gisdata
Cartographic Guide: Map Resources, http://tamu.libguides.com/maps; Aerial and Satellite Imagery, http://tamu.libguides.com/aerialphotos

Texas Tech University Libraries
Lubbock, Texas
Subject Guide Name: Geographic Information Systems (GIS)
Subject Guide URL: http://guides.library.ttu.edu/c.php?g=543216
Data: Government, Open
Data Search Capabilities: No
Cartographic Guide: Maps, http://guides.library.ttu.edu/c.php?g=543218

University of Texas, Arlington UTA Libraries
Arlington, Texas
Subject Guide Name: GIS: Geospatial and Numeric Data
Subject Guide URL: http://libguides.uta.edu/gis/home
Resources: Instruction/Workshops, Online Maps
Data: Government, Open, Statistical/Census
Data Search Capabilities: No
Software: ArcGIS
Additional Subject Guides: http://libguides.lib.msu.edu/gisdata; http://libguides.lib.msu.edu/maplinks; http://libguides.lib.msu.edu/geocoding
Cartographic Guide: Map Guide, http://libguides.uta.edu/maps

University of Texas, Dallas Library
Richardson, Texas
Subject Guide Name: Geographic Information Systems
Subject Guide URL: http://libguides.utdallas.edu/gis
Data Search Capabilities: No

*Utah*

Brigham Young University Library
Provo, Utah
Subject Guide Name: Geospatial Technology
Subject Guide URL: http://guides.lib.byu.edu/geospatial
Resources: Training, Instruction/Workshops, Online Maps
Data: Open, Statistical/Census, Remote Sensing
Data Search Capabilities: No
Software: ArcGIS, AutoCAD Map 3D
Cartographic Guide: Maps, Atlases and Gazetteers, http://guides.lib.byu.edu/maps

University of Utah Libraries
Salt Lake City, Utah
Subject Guide Name: Geospatial Data and Resources
Subject Guide URL: http://campusguides.lib.utah.edu/c.php?g=160707
Data: Government, Open, Statistical/Census
Data Search Capabilities: No
Cartographic Guide: Map Collections in the Marriott Library, http://campusguides.lib.utah.edu/map_collections; Reconstructing the Past through Utah Sanborn Fire Insurance Maps, http://campusguides.lib.utah.edu/utahsanbornfireinsurancemaps

*Vermont*

University of Vermont Libraries
Burlington, Vermont
Subject Guide Name: Maps and GIS
Subject Guide URL: http://researchguides.uvm.edu/mapsandgis
Resources: Online Maps, Other
Data: Government, Open
Data Search Capabilities: No

*Virginia*

George Mason University Libraries
Fairfax, Virginia
Subject Guide Name: Geospatial Data and GIS
Subject Guide URL: http://infoguides.gmu.edu/geospatial
Resources: Training, Instruction/Workshops, Other
Data: Government, Open
Data Search Capabilities: Yes
Software: ArcGIS
Additional Subject Guides: http://libguides.lib.msu.edu/gisdata; http://libguides.lib.msu.edu/maplinks; http://libguides.lib.msu.edu/geocoding
Cartographic Guide: Atlases and Maps, http://infoguides.gmu.edu/maps

University of Mary Washington Libraries
Fredericksburg, Virginia
Subject Guide Name: GIS
Subject Guide URL: http://libguides.umw.edu/gis
Resources: Online Maps, Other
Data: Government, Open, Statistical/Census, Remote Sensing
Data Search Capabilities: No

University of Virginia Libraries
Charlottesville, Virginia
Subject Guide Name: Geographic Information Science
Subject Guide URL: http://guides.lib.virginia.edu/gis
Resources: Training, Instruction/Workshops, Other
Data: Government, Open

Data Search Capabilities: No
Cartographic Guide: Maps, http://guides.lib.virginia.edu/maps

Virginia Tech University Libraries
Blacksburg, Virginia
Subject Guide Name: Geospatial Data and Scanned Maps
Subject Guide URL: http://guides.lib.vt.edu/c.php?g=335141
Resources: Online Maps, Other
Data: Government, Open, Statistical/Census, Remote Sensing, Other
Data Search Capabilities: No
Software: ArcGIS Desktop Suite
Cartographic Guide: Maps, GIS and Cartographic Data, http://www.lib.vt.edu/subjects/maps/index.html

*Washington*

University of Washington Libraries
Seattle, Washington
Subject Guide Name: Geospatial Data Resources Guide
Subject Guide URL: http://guides.lib.uw.edu/research/gis
Resources: Training, Online Maps, Other
Data: Government, Open, Statistical/Census, Remote Sensing
Data Search Capabilities: No
Software: ArcGIS 10.4
Cartographic Guide: Maps and Cartographic Information, http://guides.lib.uw.edu/research/maps

Washington State University Libraries
Pullman, Washington
Subject Guide Name: GIS (Geographic Information Systems)
Subject Guide URL: http://libguides.libraries.wsu.edu/gis
Resources: Online Maps, Other
Data: Government, Open, Statistical/Census
Data Search Capabilities: No
Software: ArcGIS Suite, Autodesk Suite, R, Google Earth, Adobe Creative Cloud
Cartographic Guide: Maps and Atlases in Holland and Terrell Libraries, http://libguides.libraries.wsu.edu/mapsandatlases

*Wisconsin*

University of Wisconsin, Madison Libraries
Madison, Wisconsin
Subject Guide Name: Mapping and Geographic Information Systems (GIS)
Subject Guide URL: http://researchguides.library.wisc.edu/GIS
Resources: Online Maps, Other
Data: Government, Open, Statistical/Census, Remote Sensing
Data Search Capabilities: No

Software: ArcGIS
Cartographic Guide: Mapping and Geographic Information Systems (GIS), http://researchguides.library.wisc.edu/GIS

University of Wisconsin, Milwaukee
Milwaukee, Wisconsin
Subject Guide Name: Finding GIS Data
Subject Guide URL: http://guides.library.uwm.edu/c.php?g=567847
Data: Government, Open, Statistical/Census, Remote Sensing
Data Search Capabilities: No

*Wyoming*

University of Wyoming Libraries
Laramie, Wyoming
Subject Guide Name: GIS (Geographic Information Systems)
Subject Guide URL: http://libguides.uwyo.edu/GIS
Resources: Online Maps, Other
Data: Government, Open, Statistical/Census, Remote Sensing
Data Search Capabilities: No
Software: ArcGIS, Social Explorer
Cartographic Guide: Map Resources, http://libguides.uwyo.edu/maps

## DIRECTORY OF CARTOGRAPHIC GUIDES

This short list of cartographic guides stands alone. They were not included in the list above because the libraries below do not offer GIS subject guides.

### Canada

*British Columbia*

University of Victoria
Victoria, British Columbia
Cartographic Guide Name: Maps and GIS
Cartographic Guide URL: http://www.uvic.ca/library/locations/home/map/

*New Brunswick*

University of New Brunswick
Fredericton, New Brunswick
Cartographic Guide Name: Maps and Atlases
Cartographic Guide URL: https://lib.unb.ca/gddm/maps/maps-atlases.php

*Quebec*

Université de Sherbrooke
Sherbrooke, Quebec
Cartographic Guide Name: Maps and Aerial Photographs

Cartographic Guide URL: https://www.usherbrooke.ca/biblio/trouver-des/cartes-et-photographies-aeriennes/

Université du Québec à Montréal
Montreal, Quebec
Cartographic Guide Name: Maps on Paper
Cartographic Guide URL: http://www.bibliotheques.uqam.ca/cartotheque/cartes-sur-support-papier

Université Laval
Quebec City, Quebec
Cartographic Guide Name: Maps, Atlases and Aerial Photographs, Etc.
Cartographic Guide URL: http://www.bibl.ulaval.ca/services/centregeostat/cartes-atlas-et-photographies-aeriennes

## United States of America

*Alabama*

University of South Alabama
Mobile, Alabama
Cartographic Guide Name: Gov Docs: Maps/GIS Collection: Home
Cartographic Guide URL: http://libguides.southalabama.edu/govmaps

*Alaska*

University of Alaska, Anchorage
Anchorage, Alaska
Cartographic Guide Name: Maps at the Consortium Library
Cartographic Guide URL: http://libguides.consortiumlibrary.org/maps

*California*

California State University, Chico
Chico, California
Cartographic Guide Name: Fire Insurance Map Collection
Cartographic Guide URL: http://libguides.csuchico.edu/fireinsurancemaps

California State University, Fresno
Fresno, California
Cartographic Guide Name: Map and Aerial Photograph Collections
Cartographic Guide URL: http://guides.library.fresnostate.edu/mapcollections
Cartographic Guide Name: Geologic Maps
Cartographic Guide URL: http://guides.library.fresnostate.edu/geologicmaps

San Diego State University
San Diego, California
Cartographic Guide Name: Maps
Cartographic Guide URL: http://libguides.sdsu.edu/c.php?g=560610&p=3856716

San Jose State University
San Jose, California
Cartographic Guide Name: Map Collections of the King Library
Cartographic Guide URL: http://libguides.sjsu.edu/maps

University of California, Santa Cruz
Santa Cruz, California
Cartographic Guide Name: Maps
Cartographic Guide URL: http://guides.library.ucsc.edu/maps

*Colorado*

Colorado College
Colorado Springs, Colorado
Cartographic Guide Name: Maps
Cartographic Guide URL: http://coloradocollege.libguides.com/c.php?g=286907

Fort Lewis College
Fort Lewis, Colorado
Cartographic Guide Name: Maps at Reed Library
Cartographic Guide URL: http://subjectguides.fortlewis.edu/maps

*Connecticut*

University of Connecticut
Mansfield, Connecticut
Cartographic Guide Name: Historical Map Collection
Cartographic Guide URL: http://magic.lib.uconn.edu/historical_maps.htm

*District of Columbia*

Library of Congress
Washington, D.C.
Cartographic Guide Name: Geography and Maps
Cartographic Guide URL: http://www.loc.gov/rr/geogmap/guide/gmilltoc.html

*Florida*

University of Florida
Gainesville, Florida
Cartographic Guide Name: Map and Imagery Library
Cartographic Guide URL: http://cms.uflib.ufl.edu/maps/collections

*Georgia*

Emory University
Atlanta, Georgia
Cartographic Guide Name: Geography and Maps
Cartographic Guide URL: http://guides.main.library.emory.edu/c.php?g=49949&p=323922
Cartographic Guide Name: Topographic Maps
Cartographic Guide URL: http://guides.main.library.emory.edu/topographic

*Idaho*

Boise State University
Boise, Idaho
Cartographic Guide Name: Map Resources
Cartographic Guide URL: http://guides.boisestate.edu/map_resources

University of Idaho
Moscow, Idaho
Cartographic Guide Name: Map Resources
Cartographic Guide URL: http://libguides.uidaho.edu/Map_Resources

*Illinois*

Illinois State University
Normal, Illinois
Cartographic Guide Name: Maps at Milner Library
Cartographic Guide URL: http://guides.library.illinoisstate.edu/maps

Newberry Library
Chicago, Illinois
Cartographic Guide Name: Cartography—Special Map Collections and Strengths
Cartographic Guide URL: http://www.newberry.org/node/414/
Cartographic Guide Name: Fire Insurance Maps
Cartographic Guide URL: http://www.newberry.org/fire-insurance-maps

Northern Illinois University
DeKalb, Illinois
Cartographic Guide Name: Map Collection
Cartographic Guide URL: http://libguides.niu.edu/Maps

Southern Illinois University, Edwardsville
Edwardsville, Illinois
Cartographic Guide Name: Map Collection
Cartographic Guide URL: http://www.siue.edu/lovejoylibrary/govdocs/maps/index.shtml

*Indiana*

DePauw University
Greencastle, Indiana
Cartographic Guide Name: Maps and Atlases
Cartographic Guide URL: http://libguides.depauw.edu/maps

Indiana State Library
Terre Haute, Indiana
Cartographic Guide Name: Map Collection
Cartographic Guide URL: http://www.in.gov/library/3551.htm

*Iowa*

University of Northern Iowa
Cedar Falls, Iowa
Cartographic Guide Name: Maps
Cartographic Guide URL: http://guides.lib.uni.edu/maps

*Kansas*

Emporia State University
Emporia, Kansas
Cartographic Guide Name: Federal Depository and Earth Science Map Library
Cartographic Guide URL: http://academic.emporia.edu/abersusa/maplibrary.htm

*Kentucky*

Northern Kentucky University
Highland Heights, Kentucky
Cartographic Guide Name: Maps
Cartographic Guide URL: http://library.nku.edu/researchhelp/researchguides/maps.html

University of Kentucky
Lexington, Kentucky
Cartographic Guide Name: Maps
Cartographic Guide URL: http://libguides.uky.edu/c.php?g=222914&p=1476173

*Louisiana*

Louisiana State University
Baton Rouge, Louisiana
Cartographic Guide Name: Cartographic Information Center
Cartographic Guide URL: http://www.cic.lsu.edu/collections.htm

*Maine*

University of Maine
Orono, Maine
Cartographic Guide Name: Geography: Maps
Cartographic Guide URL: http://libguides.library.umaine.edu/c.php?g=103753&p=700651

*Massachusetts*

University of Massachusetts, Amherst
Amherst, Massachusetts
Cartographic Guide Name: Map Collection
Cartographic Guide URL: http://www.library.umass.edu/locations/du-bois-library/maps/

University of Massachusetts, Dartmouth
Dartmouth, Massachusetts
Cartographic Guide Name: Maps and Charts
Cartographic Guide URL: http://guides.lib.umassd.edu/maps

*Michigan*

Grand Valley State University
Grand Valley, Michigan
Cartographic Guide Name: Maps at GVSU
Cartographic Guide URL: http://libguides.gvsu.edu/maps

Western Michigan University
Kalamazoo, Michigan
Cartographic Guide Name: Maps
Cartographic Guide URL: http://libguides.wmich.edu/maps

*Minnesota*

Saint Cloud State University
Saint Cloud, Minnesota
Cartographic Guide Name: Atlases, Maps, and Gazetteers
Cartographic Guide URL: http://stcloud.lib.minnstate.edu/subjects/guide.php?subject=amg

*Missouri*

Missouri State University Libraries
Springfield, Missouri
Cartographic Guide Name: Maps and GIS
Cartographic Guide URL: http://guides.library.missouristate.edu/maps

*Montana*

Montana State University
Bozeman, Montana
Cartographic Guide Name: Maps and Mapping—Online Resources
Cartographic Guide URL: http://guides.lib.montana.edu/maps

Maureen and Mike Mansfield Library
Missoula, Montana
Cartographic Guide Name: Maps Research Guide
Cartographic Guide URL: http://guides.lib.montana.edu/maps

*New York*

University of Buffalo
Buffalo, New York
Cartographic Guide Name: Maps and Cartography
Cartographic Guide URL: http://research.lib.buffalo.edu/maps-cartography
Cartographic Guide Name: Maps and Mapping Resources from Government Agencies
Cartographic Guide URL: http://research.lib.buffalo.edu/government-resources-maps
Cartographic Guide Name: Sanborn Fire Insurance Maps
Cartographic Guide URL: http://research.lib.buffalo.edu/sanborn-maps
Cartographic Guide Name: Topographic Maps
Cartographic Guide URL: http://research.lib.buffalo.edu/topographic-maps
Cartographic Guide Name: Buffalo, New York in Maps, Charts, and Images
Cartographic Guide URL: http://research.lib.buffalo.edu/buffalo-maps
Cartographic Guide Name: Western New York in Maps, Charts, and Images
Cartographic Guide URL: http://research.lib.buffalo.edu/c.php?g=536188

*North Carolina*

East Carolina University
Greenville, North Carolina
Cartographic Guide Name: Map Series in the North Carolina Collections
Cartographic Guide URL: http://libguides.ecu.edu/ncmapseries
Cartographic Guide Name: Sanborn Fire Insurance Maps of North Carolina
Cartographic Guide URL: http://libguides.ecu.edu/ncsanbornmaps

Western Carolina University
Cullowhee, North Carolina
Cartographic Guide Name: Maps
Cartographic Guide URL: http://researchguides.wcu.edu/maps

*North Dakota*

North Dakota State University
Fargo, North Dakota
Cartographic Guide Name: Maps, Atlases, and Gazetteers

Cartographic Guide URL: https://library.ndsu.edu/guides/maps-atlases-gazetteers

*Oregon*

Lewis and Clark College
Portland, Oregon
Cartographic Guide Name: Maps
Cartographic Guide URL: http://library.lclark.edu/maps

Oregon State University
Corvallis, Oregon
Cartographic Guide Name: Maps
Cartographic Guide URL: http://guides.library.oregonstate.edu/subject-guide/551-Maps

Portland State University
Portland, Oregon
Cartographic Guide Name: Maps and Other Cartographic Information
Cartographic Guide URL: http://guides.library.pdx.edu/maps

*Pennsylvania*

State Library of Pennsylvania
Harrisburg, Pennsylvania
Cartographic Guide Name: Maps
Cartographic Guide URL: http://www.phmc.pa.gov/Archives/Research-Online/Pages/Maps.aspx

*Rhode Island*

University of Rhode Island
Kingstown, Rhode Island
Cartographic Guide Name: Maps and Mapping
Cartographic Guide URL: http://uri.libguides.com/maps

*South Carolina*

Citadel Military College
Charleston, South Carolina
Cartographic Guide Name: Maps
Cartographic Guide URL: http://library.citadel.edu/maps

University of South Carolina
Columbia, South Carolina
Cartographic Guide Name: World Maps and Data
Cartographic Guide URL: http://library.sc.edu/p/Collections/Government/maps/world

*South Dakota*

University of South Dakota
Vermillion, South Dakota
Cartographic Guide Name: Don Shorock Collection of Road Maps
Cartographic Guide URL: http://libguides.usd.edu/shorock

*Texas*

Saint Mary's University
San Antonio, Texas
Cartographic Guide Name: Maps
Cartographic Guide URL: http://lib.stmarytx.edu/maps

Southern Methodist University
Dallas, Texas
Cartographic Guide Name: Maps
Cartographic Guide URL: http://guides.smu.edu/maps

University of North Texas
Denton, Texas
Cartographic Guide Name: Mapping and GIS
Cartographic Guide URL: http://guides.library.unt.edu/mapping

University of Texas, Austin
Austin, Texas
Cartographic Guide Name: AP Summer Institute—PCL Map Collection
Cartographic Guide URL: http://guides.lib.utexas.edu/APSIMaps

University of Texas, Dallas
Dallas, Texas
Cartographic Guide Name: Maps
Cartographic Guide URL: http://libguides.utdallas.edu/c.php?g=217725

*Utah*

Utah State University
Logan, Utah
Cartographic Guide Name: Curriculum Maps
Cartographic Guide URL: http://libguides.usu.edu/curriculummaps

Weber State University
Ogden, Utah
Cartographic Guide Name: Maps and Atlases
Cartographic Guide URL: http://libguides.weber.edu/mapsatlases

*Washington*

Eastern Washington University
Cheney, Washington
Cartographic Guide Name: Maps and Atlases
Cartographic Guide URL: http://research.ewu.edu/maps_atlases

Western Washington University
Bellingham, Washington
Cartographic Guide Name: Map Collection
Cartographic Guide URL: http://libguides.wwu.edu/map-collection

*West Virginia*

West Virginia University
Morgantown, West Virginia
Cartographic Guide Name: Map Collection
Cartographic Guide URL: http://libguides.wvu.edu/map_collection

*Wisconsin*

University of Wisconsin, La Crosse
La Crosse, Wisconsin
Cartographic Guide Name: Maps and Atlases
Cartographic Guide URL: http://libguides.uwlax.edu/maps

# 13
# Handbooks and Manuals

This chapter provides readers with a list of handbooks, manuals, guides, and how-to resources for the fields of cartography and GIS. Handbooks and manuals are generally used to clarify guidelines, provide facts, or explain how to use specific products. Because cartography and GIS both have many guiding principles for the creation, interpretation, measurement, and analysis of maps and data, there are a large number of resources available to assist those who want to learn about cartography and map creation or want to use a specific GIS-related software program.

The following list covers instructions, principles, and techniques in cartography, GIS, and specific GIS software programs. Both older and more recent publications addressing topics such as map interpretation, map design, principles of cartography (that is, projections), cartographic tool techniques, air photo tools, cartographic citations, and many GIS-related topics, such as using GIS to solve problems and make decisions, are included. Readers may find the final section, "GIS Software," to be useful because it documents early GIS software programs and lists a variety of products, including open-source and industry-standard software such as Esri's ArcGIS, for the purposes of instruction and dissemination of technical procedures.

This list was compiled using a variety of resources, including WorldCat, Esri GIS bibliography (http://gis.library.esri.com), and Amazon.

## CARTOGRAPHY

This selection of titles includes manuals, guides, and handbooks for the field of cartography. Topics covered include map and air photo interpretation, cartographic projections, remote sensing, surveying, and general how-to skills for proper map design and creation. Some of the resources are older, covering traditional cartography and manual mapmaking techniques that are not being taught anymore. Modern advances in technology mean that geographic information systems (GIS), cartography-specific computer programs, are now used to create cartographic outputs. Resources related to GIS are listed in a separate section. For titles that require clarification, annotations have been included.

*Advanced Map and Aerial Photograph Reading*. Washington, DC: War Department, 1941.

This manual provides information on advanced map and aerial photograph reading for military personnel, covering conventional signs and military symbols, distances and map scales, directions and azimuths, proper use of the compass, orientation, coordinates, visualization of terrain, and interpretation of aerial photographs.

*Advances in Cartography*, by J. C. Müller. Oxford: Pergamon, 1991.

This guide provides an overview of research in the fields of cartography and GIS, focusing on design, development and use of spatial information, and their applications, including navigation, representation, and communication.

*Aerial Photo Interpretation in Classifying and Mapping Soils: Prep. by Soil Scientists and Cartographers, Soil Survey Staff, Soil Conservation Serv*. Washington, DC: U.S. Department of Agriculture, 1966.

*Aerial Photography for Planning and Development in Eastern North Carolina: A Handbook and Directory*, by S. Baker. Raleigh: UNC Sea Grant Program, North Carolina State University, 1976.

This directory lists freely available aerial photography of eastern North Carolina.

*Archiving Aerial Photography and Remote Sensing Data*, by R. Bewley. Oxford: Oxbow Books for the Arts and Humanities Data Service, 1999.

This guide to best practices for creating, maintaining, and archiving digital aerial photography, satellite imagery, and remote sensing also discusses air photo interpretation specifically for archaeological studies.

*Basic Cartography for Students and Technicians*, by R. W. Anson and F. J. Ormeling. London: Elsevier Applied Science, 2002.

This guide covers current practices for map generation including data collection, manipulation, information display, image processing, and cartographic principles such as projections, distortions, scale, coordinates, and so on.

*Basic Guide to Small-Format Hand-Held Oblique Aerial Photography*, by J. Fleming and R. G. Dison. Ottawa, ON: Canada Centre for Remote Sensing, 1981.

*Cartographer's Toolkit: Colors, Typography, Patterns*, by G. Peterson. London, UK: Petersongis, 2012.

This toolkit demonstrates how to produce highly visual, effective informative maps with a focus on using color palettes and color ramps, typography (with fifty typefaces showcased), and composition patterns.

*Cartographic Citations: A Style Guide*, by C. Kollen, M. L. Larsgaard, and W. Shawa. 2nd edition. Chicago: Map and Geography Round Table, American Library Association, 2010.

Written for authors and scholars, this style guide shows how to cite cartographic works, including printed maps, maps in periodicals and books, cross sections, facsimiles, relief models, globes, aerial photographs, and digital maps in the form of online maps and spatial data.

*Cartographic Information, Maps, and Spatial Data: A Guide to Internet Resources*. Newark, DE: University of Delaware Library, 1996. http://guides.lib.udel.edu/c.php?g=85444&p=548407.

*Cartographic Materials: A Manual of Interpretation for AACR2*, by V. Cartmell, V. Parker, and H. Stibbe, Chicago: American Library Association, 1982.

This manual is for library staff and other individuals who need to catalog and describe cartographic materials in detail. Best practices are outlined in this publication as they relate to titles, scales, accompanying keys and codes, standard descriptions, and agreed terms. It is a must for anyone creating final entries in catalog records.

*Cartographic Materials: A Manual of Interpretation for AACR2, 2002 Revision, 2004 Update*. Chicago: American Library Association, 2004.

This is an update to the previous listing, *Cartographic Materials: A Manual of Interpretation for AACR2*, including updated examples, applications, and policies in cartographic and digital cartographic materials. This manual has also been expanded to reflect current Anglo-American Cataloguing Rules terminology and additional forms of cartographic materials, including digital geospatial data.

*Cartographic Notebook: A Brief Guide to Some Aspects of Cartographic Design*, by S. P. Meszaros. Phoenix, AZ: U.S. Department of Interior, Bureau of Land Management, Arizona State Office, 2001.

*Cartographic Specifications and Symbols Handbook*. Washington, DC: United States Department of Agriculture, 1980.

*Cartographical Innovations: An International Handbook of Mapping Terms to 1900*, by H. Wallis and A. H. Robinson. Tring, UK: International Cartographic Association, 1987.

*Cartography: A Review and Guide for the Construction and Use of Maps and Charts*, by C. H. Deetz. Silver Spring, MD: U.S. Coast and Geodetic Survey, 1948.

*Cartography: Thematic Map Design*, by D. Borden. New York: McGraw-Hill Education, 2008.

This guide offers an introduction to concepts and theories needed for creating effective maps such as map projections, map design, and map production.

*Cartography: Visualization of Geospatial Data*, by M. J. Kraak and F. Ormeling. Harlow, UK: Prentice Hall, 2003.

This introduction to cartography and GIS focuses on the visualization of spatial data.

*Cartography: Visualization of Spatial Data*, by M. Kraak and F. Ormeling. Upper Saddle River, NJ: Pearson Education, 2003.

Principles and best practices for the visualization of spatial data, integrating cartography and GIS using theory and visuals in graphic format are described in this guide, which includes map design, copyright issues, and methods for usability testing.

*Cartography and Remote Sensing Imagery: Maps, Charts, Aerial Photographs, Satellite Images, Cartographic Literature, and Geographic Information Systems*, by R. E. Ehrenberg, Z. V. David, and A. K. Henrikson. Washington, DC: Smithsonian Institution Press, 1987.

This in-depth survey provides over two hundred collections of atlases, maps, air photos, and remote sensing images that can be found in the Washington, D.C., area, including collection description and facility information.

*Colour Handbook for GIS Users and Cartographers*, by A. Brown and W. F. Feringa. Enschede, Netherlands: ITC, Department of Geoinformatics, 1999.

This handbook covers the elements of proper color use in maps.

*Concise Guide to Map Projections, with Explanatory Notes*, by G. V. Gordon. Cambridge: W. Heffer, 1925.

*Consultants Guide to Geodesy, Navigation and Surveying*. Thornleigh, Australia: QCG and Associates 1995.

*Contours of Discovery: Printed Maps Delineating the Texas and Southwestern Chapters in the Cartographic History of North America, 1513–1930; a User's Guide*, by R. S. Martin, J. C. Martin, and J. C. Dunagan. Austin: Texas State Historical Association in cooperation with the Center for Studies in Texas History, University of Texas at Austin, 1982.

Sixty-six pages long, this user guide describes twenty-two maps of Texas and the Southwest, 1513–1900, covering the province's social and political history, history of art, and linguistics.

*Community Geography: GIS in Action; Teacher's Guide*, by L. Malone, A. M. Palmer, and C. L. Voigt. Redlands, CA: Esri Press, 2003.

This teacher's guide is a "how-to" for teachers seeking to use *Community Geography: GIS in Action* in their middle school or high school classrooms, with fifteen lesson plans covering geography and technology.

*Drone Camera Handbook: A Complete Step-by-Step Guide to Aerial Photography and Filmmaking; The Manual That Should Have Come in the Box*, by I. Marloh. London: Aurum, 2016.

This handbook provides an in-depth examination of current drones and the types of photography that can be taken with them.

*Elementary Map and Aerial Photography Reading.* Washington, DC: War Department, 1944.

This basic field manual covers elementary map reading, including military symbols, distances and scales, directions and azimuths, coordinates, relief, slopes, profiles and visibility, map reading, and air photo reading in the field.

*ESRI Guide to Cartography*, by B. Dent. Redlands, CA: Esri Press, 2002.

A guide to effective map design, this text includes data classification methods, map types, color and symbols, labeling, map scale and projection methods, layout design, and printing.

*Geography and Cartography: A Reference Handbook*, by C. B. M. Lock. 3rd edition, revised and enlarged. London, UK: Bingley, 1976.

Although an older publication, this guide offers a window into the way surveying and mapmaking were conducted without the technology we have today. Describes techniques in ground surveying, including chain surveying, traversing with a compass and chain, measuring slope, and surveying underwater features; in aerial photography, including types of cameras, stereoscopy and related instruments, and film type; and cartographic presentation by way of ink drawing, adhesive symbols, and scribing. Case studies are also presented in the field of mapping.

*Glossary of Cartographic Terms and Manual of Symbols and Abbreviations: Used on the Latest Navigational Charts of the Various Countries = Glossaire des termes cartographiques et manuel des signes conventionnels et abreviations employés sur les plus recentes cartes marines des divers pays.* Monaco: International Hydrographic Bureau, 1951.

*Guide to Aerial Photography of South Carolina.* Columbia, SC: Cartographic Information Center, South Carolina Land Resources Conservation Commission, 1989.

*Guide to Cartographic Records in the National Archives*, by C. M. Ashby. Washington, DC: National Archives, National Archives and Records Service, 1971.

This guide offers a comprehensive description of the cartographic holdings of the National Archives, one of the world's largest resources of cartographic material. Collections are described in detail, including collection size, geographical extent, dates, and descriptions of the map holdings (series information and themes).

*Guide to Commonly Used Map Projections Prepared for Use in Hyper Card*, by T. R. Alpha, J. F. Vigil, and L. Buchholz. Denver, CO: USGS, 1988.

This guide describes and illustrates seventeen map projections commonly used to present thematic data published by the U.S. Geological Survey, including planes (azimuthal), cones, cylinders, and others.

*Guide to GPS Positioning*, by D. Wells. Fredericton, NB: Department of Geodesy and Geomatics Engineering, University of New Brunswick, 1999.

This guide, designed for self-study or as lecture notes for postsecondary education courses, offers an introduction to the Global Positioning System. Ideal for those practicing or studying surveying, hydrography, engineering, geology, geography, forestry, agriculture, and more, this guide covers a description of GPS, GPS data collection and processing, GPS applications, and GPS receivers.

*Guide to Historical Cartography: A Selected Annotated List of References on the History of Maps and Map Making.* 2nd revision. Washington, DC: Library of Congress, 1962.

This guide was originally published in 1954 to assist the reference staff of the Library of Congress Map Division in satisfying requests related to historical cartography. The revised edition has been updated and revised to include historical cartography collections up to the end of the eighteenth century. This resource has been arranged in alphabetical order by author, as well as by subject. Library of Congress call numbers are included along with annotations.

*Guide to Map Projections*, by R. E. Bowyer and G. A. German. London, UK: Murray, 1959.

*Guide to the Cartographic Products of the Federal Depository Library Program*, by M. D. Schuler. Chicago: Map and Geography Round Table, American Library Association, 2005.

Designed as a collection development tool by map selectors of depository libraries, this guide assists them in selecting material from the Federal Depository Library Program.

*Guide to the Practical Use of Aerial Color-Infrared Photography in Agriculture*, by D. C. Rundquist and S. A. Samson. Lincoln: Conservation and Survey Division, Institute of Agriculture and Natural Resources, University of Nebraska–Lincoln, 1988.

This nontechnical guide offers users of infrared film with a reference on the physical properties and practices of CIR aerial photography as it relates to agricultural management.

*Handbook for Geologic Cartography: Guidelines and Graphic Techniques for the Production of Geologic Illustrations.* College, AK: Alaska Division of Geological and Geophysical Surveys, 1983.

*Handbook of Aerial Photography and Interpretation*, by K. K. Rampal. New Delhi, India: Concept, 1999.

This handbook covers the principles and applications of black-and-white aerial photography, including photo geometry, aerial cameras, stereoscopy, and image interpretation.

*Handbook of Geodesy*, by W. Jordan. Washington, DC: U.S. Army Corps of Engineers, Map Service, 1962.

This technical handbook of geodesy covers terrestrial ellipsoid, computation of the spherical triangle, spherical coordinates, and projection of the spherical surface on the plane.

*How to Identify a Mapmaker: An International Bibliographic Guide = Comment identifier un cartographe: guide bibliographique international*, by M. Pelletier, S. Rimbert, and L. Zögner. Paris: Comité français de cartographie, 1996.

*How to Read a Nautical Chart: A Complete Guide to Using and Understanding Electronic and Paper Charts*, by N. Calder. New York: McGraw Hill Professional, 2012.

This guide is written for mariners to teach nautical chart reading, including learning the system of signs, symbols and graphic elements found in nautical charts.

*Indiana Land Cover Classification Manual: A Guide to Mapping*. Indianapolis: State of Indiana, State Planning Services Agency, 1976.

*Information Sources in Cartography*, by C. R. Perkins. London, UK: Bowker-Saur, 1990.

This handbook of cartography is for practitioners, academics, and map librarians who are interested in learning about cartographic research resources, the history of cartography, map production, and map librarianship.

*Instruction Manual on World Magnetic Survey*, by E. H. Vestine. Paris: Impr. par L'institut géographique national, 1961.

The World Magnetic Survey is discussed in this mathematical and graphical manual with a focus on satellite surveys, instructing readers on the earth's main field and the calculations that are taken to obtain survey results.

*Interpretation of Geological Structures through Maps: An Introductory Practical Manual*, by D. Powell. Harlow, UK: Longman Scientific & Technical, 1992.

This manual addresses the basic problems and techniques in extracting geological information from an area presented in a map form, and it instructs readers how to read, interpret, and manipulate geological three-dimensional data.

*Interpretation of Landforms from Topographic Maps and Air Photographs Laboratory Manual*, by D. J. Easterbrook and D. J. Kovanen. Saddle River, NJ: Prentice Hall, 1999.

This is a manual on landform interpretation using maps and aerial photography, covering cartographic elements like map symbols, contour lines, topographic profiles, and geologic cross-sections.

*Managing Cartographic, Aerial Photographic, Architectural, and Engineering Records.*, by J. Young and N. G. Miller. Washington, DC: National Archives and Records Administration, Office of Records Administration, 1989.

*Manual for G.P.S. with Conventional Non-Electronic Maps*, by R. Kiser. Maitland, FL: Xulon, 2008.

This manual covers the basics of using GPS to find coordinates on a map and to locate positions on a paper map.

*Manual of AACR2 Examples for Cartographic Materials*, by B. N. Moore, E. Swanson, and M. H. McClaskey. Lake Crystal, MN: Soldier Creek, 1981.

*Manual of Aerial Photography*. Rochester, NY: U.S.A. School of Aerial Photography, 1910.

*Manual of Aerial Survey*, by R. Read and R. Graham. Boca Raton, FL: CRC, 2016.

This manual provides definitions and requirements for aerial surveying, including air camera instrumentation, photographic material and exposure for aerial photography, film processing, image quality, aircraft installations, and many aspects related to the actual navigation.

*Manual of Aerial Survey: Primary Data Acquisition*, by R. E. Read and R. Graham. Boca Raton, FL: CRC/Whittles, 2002.

*Manual of Cadastral Map Standards, Concepts, and Cartographic Procedures*, by R. A. Mead. Salem, OR: Department of Revenue, 1981.

Aimed to train the new cartographer, this manual discusses concepts and standards of the State Standard Cadastral Map System. It provides the technical fundamentals of cadastral cartography, offering reference resources that cover pertinent information such as conversion tables, geodetic tables, critical formulae, riparian charges, abbreviations, and a glossary of terms.

*Manual of Crime Analysis Map Production*, by M. Velasco and R. B. Santos. Washington, DC: U.S. Department of Justice, Office of Community Oriented Policing Services, Police Foundation, Crime Mapping Laboratory, 2009.

This manual provides guidelines for introductory-level crime analysis mapping for use in a law enforcement environment, covering basic mapmaking with attention to cartographic and design elements.

*Manual of Interpretation of Orbital Remote Sensing Satellite Photography and Imagery for Coastal and Offshore Environmental Features (Including Lagoons, Estuaries and Bays)*, by H. G. Gierloff-Emden. Munich: Inst. für Geographie d. Univ., 1976.

*Manual of Map Reading, Air Photo Reading and Field Sketching*. London: Her Majesty's Stationery Office, 1957.

This official War Office publication was intended for use by candidates for commissions in the Regular Army and in the Royal Air Force, providing instruction on map reading, air photo interpretation, and field sketching.

*Manual of Map Reading and Field Sketching, 1912*. London, UK: Her Majesty's Stationery Office, 1914.

*Manual of Map Reading and Land Navigation*. London, UK: Ministry of Defence, 1988.

Written for soldiers in the army, this field manual provides a standardized reference on map reading and land navigation, including training, and an introduction to orienteering.

*Manual of Map-Making and Mechanical Geography: Illustrated Sixty Engravings*, by A. Jamieson. London, UK: A. Fullarton, 2012. Originally published in 1846.

A technical manual for creating various projections.

*Map and Compass Manual*, by J. L. Carter and A. E. Carter. Estacada, OR: Carters Manual, 1877.

This brief guide introduces readers to the uses and interpretation of maps and compasses.

*Map Cataloging Manual*. Washington, DC: Cataloging Distribution Service, Library of Congress, 1991.

This manual covers practices and policies regarding the cataloging of maps.

*Map Indexing System User's Manual*. Fort Collins, CO: Department of the Interior, Fish and Wildlife Service, Office of Biological Services, Western Energy and Land Use Team, 1978.

This manual describes the map indexing system component of the U.S. Fish and Wildlfe Service's GIS, covering geospatial cataloging in a partially automated library system.

*Map Projections: A Reference Manual*, by L. M. Bugayevskiy and J. P. Snyder. London, UK: Taylor & Francis, 1995.

This manual focuses on the theory of map projections, using mathematical cartography to take geodetic measurements to develop graphical methods for solving projection-related problems.

*Map Projections: A Working Manual*, by J. P. Snyder. Washington, DC: U.S. Government Printing Office, 1987.

This manual discusses the several projections used in maps published by the U.S. Geological Survey maps and introduces readers to new projections conceived and designed by USGS, including the Space Oblique Mercator.

*Map Use: Reading, Analysis, Interpretation*, by A. J. Kimerling et al. 7th edition. Redlands, CA: Esri Press, 2016.

Aimed at college students, this book offers an introduction to map use and analysis, focusing on communicating with maps, with an extensive glossary and resources for instructors.

*Maxwell's Handbook for AACR2: Explaining and Illustrating the Anglo-American Cataloguing Rules through the 2003 Update*. Chicago: American Library Association, 2004.

*Modern Maps and Atlases: An Outline Guide to Twentieth Century Production*. C. B. M. Lock. London, UK: Bingley, 1969.

*National Cartographic Information Center User Guide*. Lexington: Kentucky Geological Survey, 1980.

This is a guide to the cartographic material available through the USGS National Cartographic Information Center.

*National Soils Handbook: Soil Survey, Cartography, Land Inventory and Monitoring*. Washington, DC: National Resources Conservation Service, 1983.

This handbook provides guidance for conducting soil surveys, map reading, and conducting soil land inventory for the Natural Resources Conservation Service.

*Nautical Chart Manual: Practical Guidance for Cartographers and Engineers Engaged in Constructing and Revising Nautical Charts*, by W. A. Bruder. Washington, DC: U.S. Government Printing Office, 1963.

A practical instruction guide for cartographers and engineers in the construction and revision of nautical charts, this book presents basic essentials of chart construction and details of current charting practices of the Coast and Geodetic Survey, U.S. Department of Commerce. The purpose of this manual is to ensure that the charts are accurate, complete, and uniform.

*New View of the World: A Handbook to the World Map, Peters Projection*, by W. L. Kaiser. New York: Friendship Press, 1987.

This is an introduction to the Peters projection map, including a general summary of what maps are and what messages they communicate. It includes teaching strategies, with activities and exercises for the classroom.

*Ordnance Survey Maps: A Descriptive Manual*, by J. B. Harley. Southampton, UK: Ordnance Survey, 1975.

This manual describes the history of the maps produced by the Ordnance Survey.

*Plane and Geodetic Surveying*, by A. Johnson. Boca Raton, FL: CRC, 2014.

This surveying manual offers students studying surveying with detailed guidance on surveying instruments (theodolites, Total Stations, levels, and PGS), as well as advice for conducting surveys.

*A Practical Guide to Aerial Photography with an Introduction to Surveying*, by J. A. Ciciarelli. Boston: Springer US, 1991.

An introduction to air photo interpretation and surveying, this book offers clear, short, simple, and well-illustrated instructions for those new in the field of aerial photography and surveying.

*Projection Handbook*, by J. A. Hilliard, U. Başoğlu, and P. Muehrcke. Madison: Cartographic Laboratory, University of Wisconsin, 1986.

*Remote Sensing: A Handbook for Archeologists and Cultural Resource Managers*, by T. R. Lyons et al. Washington, DC: Cultural Resources Management Division, U.S. Department of the Interior, 1981.

This handbook provides managers and archeologists with the basic information for applying remote sensing techniques to their needs.

*Remote Sensing: Aerial and Terrestrial Photography for Archeologists*. Washington, DC: Cultural Resources Management Division, U.S. Department of the Interior, 1981.

A thorough examination of how aerial and terrestrial photography has been used in the field of archeology, this text also offers examples of studies and discussions of archeological finds located through aerial photography.

*Remotely Sensed Imagery of South Carolina: A Guide to Aerial Photography, Skylab, and Landsat Imagery*, by D. A. Fairey. Columbia, SC: Cartographic Information Center, South Carolina Land Resources Conservation Commission, 1983.

This is a guide to imagery available for South Carolina.

*Routledge Handbook of Mapping and Cartography*, by A. Kent. New York: Routledge, 2016.

This handbook provides a comprehensive overview of cartographic research and practice within a multidisciplinary context, covering the history of cartography, visualization, cartographic design, and thematic uses, as well as topics like geodesy, GPS, and GIS.

*SAGE Handbook of Geomorphology*, by K. J. Gregory and A. Goudie. Los Angeles: SAGE, 2011.

This handbook provides the foundations, techniques, and processes of geomorphology, the study of the earth's physical land surface features. The history of the discipline is also included.

*Satellite Altimetry and Earth Sciences: A Handbook of Techniques and Applications*, by L. L. Fu and A. Cazenave. San Diego, CA: Academic Press, 2001.

Intended for researchers and students in geophysics, oceanography, and the space and earth sciences, this handbook demonstrates the techniques, missions, and accuracy of satellite altimetry.

*Scholars' Guide to Washington, D.C. for Cartography and Remote Sensing Imagery: Maps, Charts, Aerial Photographs, Satellite Images, Cartographic Literature, and Geographic Information Systems*, by R. E. Ehrenberg and Z. V. David. Washington, DC: Smithsonian Institution Press, 1987.

*South African Landscape: Exercise Manual for Map and Air Photo Interpretation*, by E. C. Liebenberg, P. J. Rootman, and M. K. R. Van Huyssteen. Durban: Butterworth, 1976.

*Soviet Topographic Map Symbols*. Washington, DC: Department of the Army, 1958.

This manual provides instructions on reading Soviet military topographic maps of various scales. General information about symbols is covered, and an explanation about the Soviet map numbering system is provided. The appendix includes abbreviations that appear on Soviet maps, arranged according to the Russian alphabet.

*Surveying and Mapping: A Manual of Simplified Techniques*, by R. F. G. Spier. New York: Holt, Rinehart, and Winston, 1970.

*Surveyors' Manual for the Exclusive Use and Guidance of Employees of the Sanborn Map Company*. New York: Sanborn Map, 1936.

This manual was published to ensure uniform standards of accuracy and presentation for Sanborn Fire Insurance Plan maps. It includes more than a hundred pages of precise instructions and sample maps, as well as a comprehensive symbol key.

*Training Manual in Topography, Map Reading and Reconnaissance*, by G. R. Spalding. Washington, DC: Government Printing Office, 1917.

*Who's Who in the History of Cartography: The International Guide to the Subject (D8)*, by M. A. Lowenthal. Tring, UK: Map Collector Publications for Imago Mundi, 1996.

This text is a directory of international cartographic researchers including phone numbers, email addresses, postal addresses, research interests, and publications.

## GIS

This GIS section offers readers a list of handbooks, manuals, and guides that offer overviews and instruction on general GIS topics that are not software specific. Among the topics are the use of GIS for analysis, remote sensing, and air photo interpretation; GIS for decision making; and the use of GIS for designing maps and geodatabases. Most of the publications are newer, published in the last fifteen years, but some older ones (published in the last thirty years) have been included to show the development and history of GIS, which is a relatively new topic. Bibliographical annotations have been included where necessary to help explain the subject.

*21st Century Complete Guide to the National Imagery and Mapping Agency (NIMA): Geospatial Intelligence for National Security; Geodesy for the Layman; Combat Support; Terrain Visualization*. Mount Laurel, NJ: Progressive Management, 2003.

This is a comprehensive guide of the work of the National Imagery and Mapping Agency, including geodesy (for lay readers), combat support, terrain visualization, and more.

*Advanced Land-Use Analysis for Regional Geodesign*, by P. D. Zwick. Redlands, CA: Esri Press, 2015.

Written for land-use planners, analysts, and advanced GIS students, this guide demonstrates how GIS is used to analyze land-use suitability, stakeholder preferences, and conflicts between competing land interests. Topics include land-use disasters, implications of sea-level rise, the interplay between transportation and land use, and the identification of urban mixed-use opportunities.

*Advanced Spatial Analysis: The CASA Book of GIS*, by P. Longley and M. Batty. Redlands, CA: Esri Press, 2003.

This handbook describes the Centre for Advanced Spatial Analysis (CASA), University College, London and their latest developments in GIS applications, drawing from archaeology, architecture, cartography, geography, remote sensing, planning, and other fields. Project highlights include Digital Egypt, a virtual-reality reconstruction for Egyptian archaeological finds.

*Application of Geographic Information Systems (GIS) and Remote Sensing: Training Manual for Managers*. Kathmandu, Nepal: International Centre for Integrated Mountain Development, 1996.

*Applying GIS and Spatial Data Technologies to Transportation: Participant Workbook and Reference Manual.*

Washington, DC: Federal Highway Administration, National Highway Institute, 2004.

The Federal Highway Administration, in collaboration with the Bureau of Transportation Statistics, offers a training manual for implementing transportation planning applications with the use of GIS technologies. Readers will evaluate and plan for the implementation of a variety of transportation planning applications that rely on GIS technology.

*Arts and Humanities Data Service (AHDS): Geographic Information System (GIS) Guide to Good Practice*, by M. Hillings and A. Wise. Essex, UK: Arts and Humanities Data Service, 1997.

Focusing on GIS for the arts and humanities, especially archaeology, this guide introduces readers to GIS and spatial data; it also includes definitions and references.

*Beyond Maps: GIS and Decision Making in Local Government*, by J. O'Looney. Redlands, CA: Environmental Systems Research Institute, 1997.

This guide uses case studies to demonstrate how local governments use GIS as a management tool.

*Biodiversity Characterization at Landscape Level Using Satellite Remote Sensing and GIS: Manual*. Dehradun, India: Indian Institute of Remote Sensing (National Remote Sensing Agency), Department of Space, Government of India, 2003.

This manual offers methodology for landscape-level characterization of biodiversity using remote sensing and GIS. Specific topics include classification scheme, multi-season data classification, and vegetation type mapping.

*Building a GIS: System Architecture Design Strategies for Managers*, by D. Peters. Redlands, CA: Environmental Systems Research Institute, 2008.

*Canadian Geographical Information Systems Source Book = Guide des sources Canadiennes sur les systèmes d'information géographique*. Ottawa, ON: Inter-Agency Committee on Geomatics, 1995.

*Concepts and Techniques in Geographic Information Systems*, by C. P. Lo, and A. K. W. Yeung. Upper Saddle River, NJ: Pearson Prentice Hall, 2003.

This guide, written for GIS specialists, GIS technologists, and urban planners, offers up-to-date coverage of concepts and techniques pertaining to the development life cycle of GIS.

*Confronting Catastrophe: A GIS Handbook*, by R. W. Greene. Redlands, CA: Esri Press, 2002.

This hands-on handbook is written to prepare those facing large-scale disaster threats, such as wildfires, earthquakes, hurricanes, or terrorist attacks. The guide includes case studies and lessons learned from real-world catastrophe responses in which GIS technology was used.

*Connecting Our World*, by W. Tang and J. Selwood. Redlands, CA: Esri, 2003.

This guide is written for managers in any enterprise who are interested in using spatial data and information in a web-service format. Many case studies that have used GIS web services are discussed, showing how GIS is imperative in the decision-making process.

*Data in Three Dimensions: A Guide to ArcGIS 3D Analyst*, by H. Kennedy. Independence, KY: Cengage Learning, 2004.

This self-study workbook provides over twenty-five exercises showcasing the manipulation of surface data in a 3-D environment using Esri's 3D Analyst ArcGIS extension. Readers will learn how to create TIN, raster, and 3-D vector data, set 3-D display properties, and render 2-D features such as rivers, roads, and buildings. The guide also covers the creation of 3-D animated films.

*Designing Better Map: A Guide for GIS Users*, by C. A. Brewer. Redlands, CA: Esri Press, 2008.

This elementary guide teaches readers to create effective maps, including proper layout design, scale, projection, color and color selection, and symbol placement.

*Designing Geodatabases for Transportation*, by J. A. Butler. Redlands, CA: Esri, 2008.

A guide for the transportation industry, discussing the use of GIS to manage data relating to transportation facilities.

*Empowering Electric and Gas Utilities with GIS*, by B. Meehan. Redlands, CA: Esri, 2007.

This guide offers case studies featuring electronic and gas utilities who use GIS in business related to electron and gas customers, shareholders, and government regulators.

*Essential Guide to GIS*, by E. Parsons. Cambridge: Longman GeoInformation, 1994.

*Exploration Geophysics, Remote Sensing and Geographic Information Systems*. Hobart, Australia: Centre for Ore Deposit and Exploration Studies, University of Tasmania, 1993.

*Florida Coastal Engineering and Bird Conservation Geographic Information System (GIS) Manual*, by C. A. Lott et al. Ft. Belvoir, FL: Defense Technical Information Center, 2009.

This manual provides an overview of coastal engineering activities in Florida, discussing and summarizing over five hundred GIS displays that present sand placement events over forty-five years. Some of these projects include renourishment, dune restoration, emergency berm placements, and beach-dredged material deposits.

*GCDB Handbook for GIS: Using the Geographic Coordinate Database as a Resource in a Geographic Information System*, by R. Zimmer. Helena, MT: Montana Technical Writing, 2013.

The GCDB (Geographic Coordinate Database) Handbook for GIS is a quick reference for GIS users and land surveyors, offering information on data and spatial accuracy.

*Geodesist's Handbook 2004*, by O. B. Andersen. Berlin, Germany: International Union of Geodesy and Geophysics, 2004.

The International Association of Geodesy (IAG) publishes the *Geodesist's Handbook* every four years, offering current organizational information regarding the IAG, as well as a listing of geodetic data centers and educational institutions offering geodetic instruction.

*Geographic Information System Reference Manual for Mapping: Current Land Use, Land Cover, Detailed Forest Inventory, Detailed Agricultural Inventory.* Olympia, WA: Washington State Department of Natural Resources, 2004.

*Geographic Information System Tools for Conservation Planning: User's Manual*, by T. J. Fox. Reston, VA: U.S. Department of the Interior, U.S. Geological Survey, 2003.

This technical manual offers managers and planners the ability to assess landscape attributes and link them with species-habitat information. A thorough discussion includes the tools available to analyze detailed land cover and identification of species and habitat. Users will be able to evaluate the merits of proposed landscape management scenarios and choose the scenario that best fits their needs.

*Geographic Information Systems: A Concise Handbook of Spatial Data Handling, Representation, and Computation*, by S. A. Roberts and C. Robertson. Toronto, ON: Oxford University Press, 2016.

Written for the advanced user, this handbook discusses common GIS research challenges.

*Geographic Information Systems: A Guide to the Technology*, by J. C. Antenucci. New York: Van Nostrand Reinhold, 1991.

This guide, written specifically for earth scientists, computer scientists, and information technologists, provides a general overview of GIS technology, introducing new users to concepts and terms, as well as addressing topics of interest for veteran GIS users such as database development, maintenance and access, cost-benefit analysis, selecting and upgrading hardware and software packages, and legal issues.

*Geographic Information Systems and Science*, by P. Longley et al. New York: John Wiley, 2005.

This guide provides examples of GIS used in the sciences, offering a primer on GIS methods and analysis and case studies of successful implementations of GIS in the field.

*Geographic Information Systems Demystified*, by S. R. Galati. Boston: Artech House, 2006.

This guide offers technical and nontechnical professionals with the fundamentals of GIS.

*Geographic Information Systems in Fisheries Management and Planning: Technical Manual*, by G. Graaf. Rome, Italy: Food and Agriculture Organization of the United Nations, 2003.

Written for fishery biologists, aquatic resource managers, and decision makers, this "do-it-yourself" manual advocates for using GIS and remote sensing in fisheries management, offering an introduction to GIS and its applications in fishery science. The manual showcases fishery managers who have successfully used GIS to foster the sustainable use of natural resources. The manual covers key GIS concepts such as geographic coordinate system and map projection, raster data and analysis, and regression analysis.

*Geographical Information Systems: A Beginner's Guide for Landscape Ecologists*, by N. Veitch. New York: Springer, 1995.

This guide examines how GIS can be used in the study of landscape ecology, providing an overview of the uses of GIS and case studies related to vegetation, animals, and hydrology.

*Geographical Information Systems: An Introduction*, by J. Delaney. South Melbourne, Australia: Oxford University Press, 2007.

Written for students with no prior GIS experience, this guide offers an introduction to GIS using real work examples. There is a strong pedagogical focus with examples of GIS applications used beyond academia.

*Geographical Information Systems (GIS) as a Tool for Analysis and Communication of Multidimensional Data*, by A. Sivertun. Umeå, Sweden: Department of Geography, University of Umeå, 1993.

This guide offers suggestions for how GIS can be used to support research in studies of agriculture and soil drainage. The flow of metals and nitrogen in soil moisture are specifically mentioned. A GIS model is presented that takes soil type, topography, vegetation, land use, agricultural drainage, and relative position in the watershed into account.

*Geospatial Data Infrastructure: Concepts, Cases, and Good Practice*, by R. Groot and J. D. McLaughlin. Oxford: Oxford University Press, 2000.

A guide to creating, managing, and maintaining datasets in a cost-effective way. Policies and standards are discussed, as well as recommendations for design, implementation, and management of geodatabases. It is written for academics and for practitioners at all levels of government and the private sector.

*Geospatial Research: Concepts, Methodologies, Tools, and Applications.* Washington, DC: IGI Global, 2016.

Written for academics, government agencies, and engineers, this multivolume publication explores multidisciplinary applications of GIS.

*Geospatial Tools for Urban Water Resources*, by P. L. Lawrence. Dordrecht, Netherlands: Springer, 2013.

This publication focuses on the development and implementation of GIS and remote sensing tools. Case studies are shared with a discussion of how GIS and data can be applied to manage water resources in urban areas.

*Geosurveillance: GIS-Based Exploratory Spatial Analysis Tools for Monitoring Spatial Patterns and Clusters*, by G. Lee, I. Yamada, and P. Rogerson. New York: Springer, 2010.

*Getting Started with GIS: A LITA Guide*, by E. Dodsworth. New York: Neal-Schuman, 2012.

This is a thorough guide to the use of GIS in libraries, introducing readers to online mapping, GIS software, GIS data, mapmaking, and GIS applications. It includes several step-by-step instructions on specific GIS software programs and tools.

*Getting to Know Web GIS*, by P. Fu. Redlands, CA: Esri Press, 2016.

This workbook offers step-by-step exercises for sharing GIS resources online as well as building web GIS applications. Some of the applications discussed include AppStudio, ArcGIS Pro, ArcGIS Online, Web AppBuilder for ArcGIS, and ArcGIS API for JavaScript.

*GIS 20*, by G. Clemmer. Redlands, CA: Esri Press, 2010.

Written for GIS beginners, this workbook demonstrates how to perform twenty essential GIS skills. Basic GIS functions are covered, and the workbook discusses how they can be applied in different analyses.

*GIS and Cartographic Modeling*, by D. C. Tomlin. Redlands, CA: Esri Press, 2013.

This book offers an introduction to GIS concepts and to the analytical use of raster-based GIS.

*GIS and Cultural Resource Management: A Manual for Heritage Managers*. Bangkok, Thailand: UNESCO, 1999.

*GIS and Spatial Analysis in Veterinary Science*, by P. A. Durr and A. C. Gatrell. Wallingford, UK: CABI, 2004.

This book introduces readers to GIS applications used in veterinary science. Topics include the application of GIS to epidemic disease response, animal epidemiology, and wildlife disease management.

*GIS Book: A Practitioner's Handbook*, by G. Korte. Santa Fe, NM: OnWord, 1992.

*GIS Cartography: A Guide to Effective Map Design*, by G. N. Peterson. Boca Raton, FL: CRC, 2015.

This guide offers readers the tools necessary to create maps using elements of effective design. Some of the topics discussed include projection theory, big data point density maps, scale-dependent map design, 3-D building modeling, and best practices in digital cartography.

*GIS for Decision Support and Public Policy Making*, by C. Thomas and N. Sappington. Redlands, CA: Esri Press, 2009.

This publication presents twenty-seven case studies and eight exercises that reveal the use of GIS in daily operations of the public sector, focusing on communication, collaboration, and decision making in organizations.

*GIS for Dummies*, by M. N. DeMers. New York: John Wiley, 2009.

This introductory text offers the basics of digital mapping, locating geographic features, analyzing patterns, and mapping terminology.

*GIS for Health Organizations*, by L. Lang. Redlands, CA: Esri Press, 2000.

Through the use of case studies, this guide explains the use of GIS in health care, demonstrating how it is used to improve patient care, monitor the spread of disease, and manage health facilities.

*GIS for the Curious: A Hands-On Guide to Geographic Information Systems*, by Y. C. Lee. Fredericton, NB: Geomatics Canada, 1995.

*GIS for the Geosciences: Shortcourse, GIS for the National Geological Society of America*, by R. L. Bedell. Sparks, NV: Exploration Department, Homestake Mining, 1995.

*GIS Fundamentals: A First Text of Geographic Information Systems*, by P. Bolstad. Ashland, OH: Bookmasters, 2002.

Written for introductory GIS classes, this guide provides an introduction to the theory and application of GIS. Topics covered include an introduction to GIS, spatial data models, map projections, data entry, image data, GPS, digital data, geodatabases, spatial analysis, metadata, and standards.

*GIS Guide for Elected Officials*, by C. Fleming. Redlands, CA: Esri Press, 2014.

Written for government officials, this guide describes how GIS can be used to help solve problems, focusing on building and maintaining a strong GIS program in light of changing technology and shrinking government budgets.

*GIS Guide for Local Government Officials*, by C. Fleming. Redlands, CA: Esri Press, 2005.

*GIS Guide to Good Practice*, by M. Gillings and A. Wise. Oxford: Oxbow Books, 1999.

This guide provides information on the creation, maintenance, use, and long-term preservation of GIS-based digital products, including long-term preservation and data archiving.

*GIS Guide to Public Domain Data*, by J. Kerski and J. Clark. Redlands, CA: Esri Press, 2012.

This guide offers GIS users information about public-domain spatial data, covering the essential elements of data acquisition, including finding, evaluating, and analyzing while taking into account issues such as copyright, cloud computing, volunteered geographic information, and international data.

*GIS in Organizations*, by H. Campbell and I. Masser. Boca Raton, FL: CRC, 1995.

*GIS in Schools*, by R. H. Audet and G. S. Ludwig. Redlands, CA: Esri Press, 2000.

This book presents the use of GIS software in the classroom, offering teachers practical ideas about bringing GIS into their classrooms.

*GIS Laboratory Exercises: Introduction to GIS*, by J. Taylor, J. Fletcher, and K. K. Kemp. Santa Barbara, CA: National Center for Geographic Information and Analysis, 1996.

*GIS Management Handbook: Concepts, Practices, and Tools for Planning, Implementing, and Managing Geographic Information System Projects and Programs*, by P. L. Croswell. Frankfort, KY: Kessey Dewitt. 2009.

*GIS Means Business*, by D. Boyles. Redlands, CA: Esri Press, 2002.

Of interest to business managers and GIS professionals, this guide demonstrates the ways a variety of businesses are using GIS for profitability and efficiency. Industries discussed include real estate firms, insurance companies, food distributors, and casinos.

*GIS Project Management Handbook*. Victoria, BC: Ministry of Forests, 1993.

*GIS Project Sustainability Handbook*. Manila, Philippines: Local Government Development Foundation, 2006.

*GIS Research Methods: Incorporating Spatial Perspectives*, by S. Steinberg and S. Steinberg. Redlands, CA: Esri Press, 2015.

This guide reflects on how researchers incorporate spatial thinking and GIS into research design and analysis, covering topics such as research design, digital data sources, volunteered geographic information, and spatial analysis.

*GIS Standards and Standardization: A Handbook*. New York: United Nations, 1998.

*GIS Survival Guide: An Introduction for First-Time GIS Users and Buyers Who Need More Than a Trade Leaflet but Less Than a Textbook*, by M. J. Clark. Southampton, UK: GeoData Institute, 1992.

Written for the IT professional or general manager, this guide examines GIS from a business perspective, offering an introduction to GIS in various industries. Some topics include project design, management implications, and investment decisions.

*GIS Tools for Decision Making*. Trenton, NJ: NJDEP, 1997.

*Guide to GIS Resources on the Internet*. Berkeley, CA: University of California Regents, 1998.

*Guide to the Environmental Visioning Process Utilizing a Geographic Information System (GIS)*. Cincinnati, OH: U.S. Environmental Protection Agency, 2000.

This publication covers new policies and discusses tactics used in addressing community-wide environmental issues. This guide is intended to help communities make decisions about land use and landscapes, using GIS to enhance the process.

*Handbook of Data Mining*, by N. Ye. Mahwah, NJ: Lawrence Erlbaum Associates, 2003.

This handbook, authored by some of the leading international authorities, offers concepts, methods, and tools for data mining. This book would be of interest to developers and those interested in data-mining methods and tools.

*Handbook of Geographic Information Systems and Archaeology*, by M. S. Aldenderfer. London, UK: Equinox, 2007.

This handbook was specifically written to address the fundamentals of GIS for archaeologists, covering spatial thinking, spatial data models and structures, projections and coordinate systems, sources of geographic data, geographic databases, and visualization of geographic data. Applications of GIS in archaeology are also discussed.

*Handbook of GIS*, by J. P. Wilson. New York: John Wiley, 2007.

This handbook and reference covers research and surveys conducted using GIS internationally. Intended for GIS students and others interested in using GIS in their own research.

*Handbook of GIS Standards and Procedures for the Lake Champlain Basin Program*. Montpelier, VT: Vermont Center for Geographic Information, 1993.

This handbook was written for those involved in collecting and analyzing GIS data in the Lake Champlain Basin. Topics include standards for digital data conversion, data layer documentation, coordinate systems, codes for land cover and land use, and geographic codes.

*Handbook of Management Information Systems: Applications in Remote Sensing and GIS*, by S. Keshmukh. London, UK: Koros, 2013.

*Handbook of Research on Geographic Information Systems Applications and Advancements*, by S. Faiz. Hershey, PA: IGI Global, 2017.

*Handbook of Satellite Applications*, by J. N. Pelton, S. Madry, and S. Camacho-Lara. New York: Springer, 2013.

This multidisciplinary reference handbook addresses system technologies and examines market dynamics, technical standards, and regulatory constraints as they relate to satellite telecommunications, remote sensing, and GIS.

*Handbook on Geospatial Infrastructure in Support of Census Activities*. New York: United Nations, 2009.

*Handbook of Applied Spatial Analysis: Software Tools, Methods and Applications*, by M. M. Fischer and A. Getis. Berlin: Springer, 2010.

This handbook was written both for newcomers and for those familiar with the field of spatial analysis. Step-by-step guides offer advanced instruction in spatial analysis.

*Handbook of Geographic Information Science*, by J. Wilson and A. S. Fotheringham. New York: John Wiley, 2007.

This handbook, written for students in geoinformation, geoinformatics, geocomputation, and geoprocessing, as well as for anyone else interested in applying GIS to their profession, offers an in-depth analysis of GIS and the full spectrum of ways GIS can be used in research.

*Harvesting Geospatial Safeguards Information with Open Source Tools*. Washington, DC: U.S. National Nuclear Security Administration, 2011.

*Illustrated Guide to Nonprofit GIS and Online Mapping*. Chicago: MapTogether, 2010.

This free guide was written for staff and volunteers of nonprofit organizations, nongovernmental organizations, and grassroots groups. It offers instruction on GIS technology and freely available public data, and it also includes reviews of free and low-cost tools for nonprofit mapping and GIS projects.

*Imagery and GIS: Best Practices for Extracting Information from Imagery*, by K. Green et al. Redlands, CA: Esri Press, 2017.

With more than two hundred full-color illustrations, this reference guide shows readers how to successfully incorporate imagery into maps and GIS projects. Imagery is used to enhance visualizations as well as to extract and analyze information.

*Implementing Geographic Information Systems.* Alexandria, VA: Water Environment Federation, 2004.

*Integrated Geospatial Technologies: A Guide to GPS, GIS, and Data Logging*, by J. Thurston, J. P. Moore, and T. K. Poiker. Hoboken, NJ: John Wiley, 2003.

This guide offers an integrative framework to present geotechnologies such as GIS, GPS, digital photogrammetry, and visualization. Through visualizations, photographs of hardware, and sidebars, users are introduced to core theories of GPS and GIS. Case studies share practices of real-world projects.

*Integrating Geographic Information Systems into Library Services: A Guide for Academic Libraries*, by J. Abresch. Hershey, PA: Information Science, 2008.

Written for library staff who work with cartographic and geospatial material, this guide offers resources and tools related to collection development, reference and research services, cataloging, and accessibility standards. Tools for integrating GIS into libraries are also provided.

*Integrating GIS and Spatial Statistical Tools for the Spatial Analysis of Health-Related Data*, by J. Ma. Sheffield, UK: University of Sheffield, 2000.

GIS and spatial statistical tools are brought together in the field of health research through the use of case studies. User guides and related publications are also included for reference.

*Interactive Spatial Data Analysis*, by T. C. Bailey and A. C. Gatrell. Harlow, UK: Longman Scientific & Technical, 1995.

This comprehensive introduction to spatial data analysis offers full explanations for a variety of methods, including several case studies and datasets.

*Introducing Geographic Information Systems to Community College Geography Instruction: A Resource Guide*, by M. Trembley. San Diego, CA: San Diego Mesa College, 1994.

*Introduction to Geographic Information System: A Lab Manual for STEM and Non-STEM Disciplines*, by S. Bhaskaran. Dubuque, IA: Kendall Hunt, 2015.

*Landscape Analysis Using Geospatial Tools*, by M. Madden. Berlin: Springer, 2009.

Geospatial tools, such as remote sensing, GIS, and GPS, are often used to analyze landscapes, to assess human impacts on the earth's resources, and to conduct research in landscape ecology. This guide integrates landscape ecology and GIS, offering case studies to highlight how these two disciplines can work together.

*Learning and Using Geographic Information Systems*, by W. L. Gorr and K. S. Kurland. Boston: Course Technology, 2006.

*Learning Progressions for Maps, Geospatial Technology, and Spatial Thinking: A Research Handbook*, by M. Solem, R. G. Boehm, and N. T. Huynh. Newcastle, UK: Cambridge Scholars, 2015.

*Local Government GIS Development Guides.* Albany, NY: Local Government Technology Services, State Archives and Records Administration, 1996.

This guide summarizes resources available for the state of New York related to available data, hardware, and software, and it also offers input regarding database planning and design, database construction, pilot studies, and benchmark tests.

*Local Government Guide to Geographic Information Systems: Planning and Implementation.* Washington, DC: PTI/Urban Consortium, 1991.

*Making Maps: A Visual Guide to Map Design for GIS*, by J. Krygier and D. Wood. New York: Guilford, 2011.

This guide offers students and professionals with instructions on creating effective maps, including good and poor design choices and the importance of locating and processing data. Map symbols and colors are discussed in depth as well.

*Making Spatial Decisions Using GIS and Remote Sensing*, by K. Keranen and R. Kolvoord. Redlands, CA: Esri Press, 2013.

This workbook highlights image processing and analysis using GIS software. Landsat imagery is also included.

*Manager's Guide to Geographic Information Systems*, by H. Fleet, W. P. Gregg, and J. Mathews. Denver, CO: Geographic Information Systems Field Unit in cooperation with Special Science Projects Division, 2006.

*Manual for Managing Geospatial Datasets in Service Centers: Version 3.0.* Washington, DC: U.S. Department of Agriculture, 2002.

*Manual of Canadian GIS-T Data Standards, Recommended Practices and Implementation Guidelines: Geographic Information Systems in Transportation (GIS-T).* Ottawa, ON: Transportation Association of Canada, 1995.

*Manual of Geographic Information Systems*, by M. Madden. Bethesda, MD: American Society for Photogrammetry and Remote Sensing, 2009.

*Manual of Geospatial Science and Technology*, by J. D. Bossler et al. Boca Raton, FL: CRC/Taylor & Francis, 2010.

This manual covers all aspects of the field of GIS, including the managing of GIS projects. Over fifty-three experts contributed to the manual in the areas of GIS, GPS, and remote sensing. The guide discusses many fundamentals such as area-wide mapping (with an introduction to math and physics), inventory, data conversion, and analysis. Case studies are also included.

*Manual on GIS for Planners and Decision Makers.* New York: United Nations, 1996.

*Map Librarianship: A Guide to Geoliteracy, Map and GIS Resources and Services*, by S. W. Aber and J. W. Aber: Amsterdam: Elsevier, 2017.

    Written for the map and/or GIS librarian, this guide discusses geoliteracy topics in map librarianship, such as reference and instruction, and related services that serve library clients, such as finding, downloading, delivering, and assessing maps and other GIS and cartographic resources. The guide also includes information about citations and copyright issues.

*Mapping Global Cities: GIS Methods in Urban Analysis*, by A. Pamuk. Redlands, CA: Esri Press, 2006.

    This guide focuses on how GIS technologies can be used to recognize residential patterns, shaped by population and employment, in cities.

*Mapping Hacks*, by E. Schuyler et al. Sebastopol, CA: O'Reilly Media, 2005.

    This collection of one hundred simple and mostly free techniques are made available to developers who are interested in creating digital maps, covering data access, interpreting and manipulating data, or incorporating personal photos into maps.

*Mapping It Out: Expository Cartography for the Humanities and Social Sciences*, by M. Monmonier. Chicago: University of Chicago Press, 1993.

    This book offers an introduction to the fundamental principles of graphic logic and design by showcasing a wide variety of maps covering subjects using many formats. Depending on the purpose of the map, mapmakers will use different patterns, typefaces, fonts, visibility of base maps, or frames to either clarify or obscure information. This guide aims to strengthen the mapmakers' use of design.

*Maps and Mapping: A Cartographic Manual*, by E. J. Hickin. Burnaby, BC: Department of Geography, Simon Fraser University, 1997.

*Maps and Mazes: A First Guide to Map Making*, by G. Chapman and P. Robson. South Melbourne: Macmillan Education Australia, 1993.

*Methods for Multilevel Analysis and Visualisation of Geographical Networks*, by C. Rozenblat and G. Melançon. Dordrecht, Netherlands: Springer, 2013.

*NCGIA Guide to GIS Laboratory Materials: 1991*, by R. F. Dodson, K. K. Kemp, and S. D. Palladino. Santa Barbara: National Center for Geographic Information and Analysis, University of California at Santa Barbara, 1991.

*Open Access: GIS in E-Government*, by R. W. Greene. Redlands, CA: Esri Press, 2001.

    This guide demonstrates how GIS is being used in e-government to increase efficiency in economic development and environmental protection.

*Oregon Geographic Information Systems: User Reference Guide.* Portland: Oregon's Geographic Information System Committee, 1988.

*Partnering with the Police to Prevent Crime Using Geographic Information Systems: A Guide for Housing Authorities and Other Community Stakeholders*, by H. R. Holtzman, W. D. Wheaton, and D. P. Chrest. Washington, DC: U.S. Department of Housing and Urban Development, Office of Policy Development and Research, 2003.

    This guidebook is for stakeholders such as police departments, housing authorities, school districts, neighborhood associations, and corporate managers interested in preventing crime using GIS.

*Pennsylvania GIS Directory: A Guide to GIS Users.* Harrisburg, PA: Department of Transportation, 1997.

*Pennsylvania GIS Resource and Implementation Guide: A Handbook for Understanding Geographic Information Systems*, by R. P. Sechrist. Harrisburg, PA: Center for Rural Pennsylvania, 1997.

*Place in History: A Guide to Using GIS in Historical Research*, by I. N. Gregory. Oxford: Oxbow Books, 2003.

    Written for historians who wish to use GIS in their research, this guide demonstrates how GIS can be used in history, describing how to create and use a GIS database. Case studies from a variety of historical projects have also been included.

*Place Matters: Geospatial Tools for Marine Science, Conservation, and Management in the Pacific Northwest*, by D. J. Wright and A. J. Scholz. Corvallis: Oregon State University Press, 2005.

    This guide explores how marine GIS is helping researchers understand and manage the conservation of the shores and ocean of the Pacific Northwest. GIS is used to analyze data from observatories, numerical models, simulations, and other sources, thus offering an understanding into the oceanographic, ecological, and socioeconomic conditions of the marine environment.

*Practical Handbook of Spatial Statistics*, by S. L. Arlinghaus and D. A. Griffith. Boca Raton, FL: CRC, 1996.

    This handbook offers proper spatial analysis techniques, including information on weighting, aggregation effects, sampling, spatial statistics, and GIS.

*Practical Manual on Basics of Remote Sensing Data Processing, GPS and GIS*, by R. N. Sahoo. New Delhi, India: Indian Agricultural Research Institute, 2012.

*Quantitative Methods and Applications in GIS*, by F. Wang. Boca Raton, FL: CRC/Taylor & Francis, 2006.

*Quick Guide to GIS Implementation and Management*, by R. Somers. Park Ridge, IL: Urban and Regional Information Systems Association, 2001.

*Reference Guide to Data Sources*, by J. Bauder. Chicago: American Library Association, 2014.

    This guide shows readers how to look for the best sources of data found freely online. Written for staff at public libraries, high school libraries, academic libraries, and other research institutions, the guide illustrates the variety of data sources available, how to find them, and how to cite them across a wide variety of subjects.

*Remote Sensing and GIS for Fisheries Management: A Course Manual*, by H. S. Mogalekar and C. Johnson. Saarbrücken, Germany: LAP Lambert Academic, 2015.

*Remote Sensing GIS/Imaging User's Manual*. Concord, NH: QW Communications, 1995.

*Remote Sensing Handbook*. Vol. 1. Boca Raton, FL: CRC/Taylor & Francis, 2016.

*Remote Sensing Handbook*. Vol. 2. Boca Raton, FL: CRC/Taylor & Francis, 2016.

*Remote Sensing Handbook*. Vol. 3. Boca Raton, FL: CRC/Taylor & Francis, 2016.

*SAGE Handbook of GIS and Society*, by T. Nyerges. Los Angeles: SAGE, 2011.

Written for academics and GIS practitioners, this handbook provides an overview of GIS and society research, highlighting ways GIS is applied to societal issues such as the value of community, fairness, and privacy.

*SAGE Handbook of Spatial Analysis*, by A. S. Fotheringham and P. Rogerson. London, UK: SAGE, 2009.

This handbook offers a comprehensive and authoritative discussion of issues and techniques in the field of spatial data analysis.

*SAGE Handbook of Visual Research Methods*, by E. Margolis and L. Pauwels. Los Angeles: SAGE, 2011.

*Self-Teaching Students' Manuals for Geographic Information Systems*, by G. Cho. Canberra, Australia: University of Canberra, 1995.

*Solutions Manual for Adjustment Computations: Statistics and Least Squares in Surveying and GIS*, by P. R. Wolf and C. D. Chilani. New York: John Wiley, 1997.

This technical manual, which was written for surveyors, photogrammetrists, and GIS professionals, introduces users to analyzing data, statistical testing, and using software to successfully adjust computations.

*Spatial Analysis: A Guide for Ecologists*, by M. J. Fortin and M. R. T. Dale. Cambridge: Cambridge University Press, 2005.

This GIS guide, written for ecologists, covers a number of topics that help professionals conduct spatial analysis. Ecological and statistical foundations are discussed, and advice for working with sample plot data is offered.

*Spatial Databases*, by P. Rigaux, M. O. Scholl, and A. Voisard. Amsterdam: Elsevier, 2002.

This book focuses on spatial databases, covering data models, algorithms, and indexing methods used to address features of spatial data. Spatial access methods are covered, including the R-tree and space-driven structures.

*Spatial Measurements and Statistics*, by A. Mitchell. Redlands, CA: Esri Press, 2005.

This guide covers the most commonly used spatial statistical tools, demonstrated in a range of disciplines. Topics include testing for statistical significance, defining spatial neighborhoods and weights, and using statistics with spatial data.

*Spatial Portals*, by W. Tang and J. Selwood. Redlands, CA: Esri Press, 2005.

*Teacher's Guide to GIS Operations: Using Geographic Information System (GIS) for Implementing Enquiry Learning in Geography*, by P. C. Lai, K. Chan, and S. A. Mak. Hong Kong: Personal, Social and Humanities Education Section, Education Bureau, 2009.

*Texas Rural Rail Transportation Districts: GIS Information Manual*, by J. E. Warner. College Station: Texas Transportation Institute, Texas A&M University System, 2001.

This ready reference was created for Texas Department of Transportation officials for the purposes of locating existing and abandoned rail lines within the state.

*Training Manual on Geographic Information Systems in Local/Regional Planning*, by M. Batty, D. F. Marble, and Y. A. Gar-On. Nagoya, Japan: United Nations Centre for Regional Development, 1995.

*Understanding Map Projections: GIS by ESRI*, by M. Kennedy and S. Kopp. Redlands, CA: Environmental Systems Research Institute, 2000.

This guide introduces geographic and projected coordinate systems, explaining terms and concepts of map projections, including spheres and spheroids, datums, and latitude and longitude.

*Understanding Place: GIS and Mapping across the Curriculum*, by D. S. Sinton and J. J. Lund. Redlands, CA: Environmental Systems Research Institute, 2007.

This collection of case studies describes how instructors have used GIS in their classrooms spanning a number of different subjects.

*Using Geographic Information Systems to Map Crime Victim Services: A Guide for State Victims of Crime Act Administrators and Victim Service Providers*, by D. A. Stoe. Washington, DC: U.S. Department of Justice, Office of Justice Programs, Office for Victims of Crime, 2003.

This guide introduces GIS to state Victims of Crime Act (VOCA) administrators and victim service providers, offering them increased awareness of using GIS as a crime-fighting tool.

*Using GIS for Oil and Gas Applications: A User's Guide*, by C. Wilson. Washington, DC: Bureau of Land Management, 1993.

This manual has been written to aid field personnel in using the Bureau of Land Management's GIS to store, retrieve, analyze, and model spatial data in oil and gas analysis.

*Watershed Soil Erosion, Runoff, and Sediment Yield Prediction Using Geographic Information Systems: A Manual of GIS Procedures*, by J. S. Blaszczynski. Denver: U.S. Department of the Interior, Bureau of Land Management, BLM Service Center, 1994.

This manual offers procedures for developing a soil erosion map, as well as calculating sediment yield from a storm event. The goal of this guide is to offer simple levels for modeling within a watershed soil erosion context.

*Web-Based Geospatial Tools to Address Hazard Mitigation, Natural Resource Management, and Other Societal Issues*, by P. P. Hearn. Reston, VA: U.S. Department of the Interior, U.S. Geological Survey, 2009.

This publication addresses the need for more accessible, manageable data tools by developing web-based GIS applications that incorporate USGS and other organization's data for decision making about critical issues. This guide offers a thorough discussion of all the tools that are available.

## GIS SOFTWARE

This selection of titles focuses on instructional user guides for GIS software programs and the tools available within them. Many are specific to software programs created by the Environmental Systems Research Institute (Esri), a supplier of GIS software used by most libraries and academic institutions in North America. These resources offer users step-by-step instructions for GIS-software tasks and tools related to mapmaking, spatial analysis, data creation, code development, and other popular tasks. Due to rapid changes in GIS technology, and advances in GIS software, many software programs that were used twenty-five years ago are obsolete or rarely used by institutions and professionals. Some of these, however, reveal the progression of GIS applications and may still be of interest to those studying the development of tools and interested in comparing GIS technologies throughout the years. This list, therefore, offers more recent publications (published within the last fifteen years) that specifically pertain to GIS software and also includes a few older resources that may still be useful. Where necessary, annotations have been included to provide more information about the title.

*Administering ArcGIS for Server*, by H. Nasser. Birmingham, UK: Packt, 2014.

Written for the advanced GIS user, analyst, or programmer, this step-by-step tutorial provides information for the setup, installation, and administration of ArcGIS Server technology.

*ArcCAD, the GIS for AutoCAD; User's Guide.* Redlands, CA: Esri Press, 2011.

ArcCAD software is the GIS engine for AutoCAD, a powerful graphic editing and geometric construction software. Combining the two provides the graphic software with spatial analysis tools and database management. The user guide provides step-by-step instructions for installing and using ArcCAD.

*ArcGIS and the Digital City: A Hands-On Approach for Local Government*, by W. Huxhold. Redlands, CA: Esri Press, 2004.

Written as a textbook for urban planning as well as a workbook for local governments, this book demonstrates how to use data to perform analysis and make city-themed layers such as displaying buildings and parcels, geocoding addresses, and creating geodatabase topology.

*ArcGIS Blueprints*, by E. Pimpler. Birmingham, UK: Packt, 2015.

This technical guide focuses on the use of Python, a scripting language, to create real-world ArcGIS applications. Readers will learn how to read and write JSON, CSV, and XML format data sources and write outputs to Google Earth Pro; in addition, they will be introduced to advanced ArcPy Mapping and ArcPy Data Access module techniques. The goal of this project-based guide is for readers to create their own GIS application program.

*ArcGIS by Example*, by H. Nasser. Birmingham, UK: Packt, 2015.

This advanced book guides application developers in building, designing, and running ArcGIS applications using ArcObjects SDK-Extend ArcGIS objects. Users will learn how to use ArcGIS code to query geodatabases and to design and develop three ArcGIS applications that will offer map browsing and searching, calculate costs, and work with reviews and ratings for tourism-themed maps.

*ArcGIS for JavaScript Developers by Example*, by J. Vijayaraghavan and Y. Dhanapal. Birmingham, UK: Packt, 2016.

Using ArcGIS JavaScript API, this guide leads users to create mapping applications in ArcGIS. The guide is written for JavaScript developers with an interest in learning how to develop mapping apps and to create code that meets best practices. Various data sources and open geospatial data sites are also covered.

*ArcGIS Imagery Book*, by C. Brown and C. Harder. Redlands, CA: Esri Press, 2016.

This is a guide to using imagery and remote sensing in ArcGIS to aid in mapmaking and storytelling.

*ArcGIS Introduction: An Easy Guide for Beginners*, by S. Casterson. Essex, UK: Conceptual Kings, 2016.

Created for those new to ArcGIS software and to GIS in general, this guide provides users with step-by-step instructions for using ArcGIS to create simple maps and to conduct some spatial analysis.

*ArcGIS 8 User Guides*, by B. Booth et al. Redlands, CA: Environmental Systems Research Institute, 2002.

*ArcGIS 9: Using ArcGIS Geostatistical Analyst*, by K. Johnston. Redlands, CA: Esri Press, 2005.

The ArcGIS Geostatistical Analyst is an extension to ArcGIS software, enabling users to fit models like kriging, cokriging, IDW, and others in order to take accurate continuous surface samples to make predictions about a landscape.

*ArcGIS 9 Documentation: Geoprocessing in ArcGIS.* Redlands, CA: Esri Press, 2004.

*ArcHydro: GIS for Water Resourcers*, by D. R. Maidment. Redlands, CA: Environmental Systems Research Institute, 2002.

This guide, written for hydrologists and GIS specialists, explores how hydrology projects are conducted using local, regional, national, and international data. The ArcGIS Hydro data model is the focus of this guide, demonstrating its success in solving water resource problems.

*ArcHydro Groundwater: GIS for hydrogeology*, by H. Strassberg, N. L. Jones, and D. R. Maidment. Redlands, CA: Esri Press, 2011.

This guide describes a new geodatabase model for representing groundwater systems in GIS—the groundwater data model. Similar to the ArcHydro model (listed above), this model provides an overview of water resources available and used for management, visualization, and analysis. Some of the topics covered include 3-D subsurface representation, geological mapping, aquifers, wells and boreholes, and time series and groundwater simulation modeling.

*Arc Marine: GIS for a Blue Planet*, by D. J. Wright. Redlands, CA: Esri, Inc., 2007.

Written for marine students and marine professionals, this book introduces readers to a specific Arc Marine data model that supports sea-floor mapping, fisheries management, marine mammal tracking, monitoring shoreline change, and water temperature analysis.

*ArcObjects Developer's Guide: GIS by ESRI; ArcInfo 8*. Redlands, CA: Environmental Systems Research Institute, 1999.

This guide is written for developers who wish to modify the ArcInfo 8 GIS program by using various ArcObjects models and Active X add-ins or by working with third-party CASE tools.

*ArcPy and ArcGIS: Geospatial Analysis with Python*, by S. Toms. Birmingham, UK: Packt, 2015.

This guide is written for GIS students and professionals who are interested in performing analysis faster and automating geospatial tasks by applying Python (ArcPy) programming.

*ArcSDE Administration Guide: ArcSDE 8*, by M. Harris and J. Clark. Redlands, CA: Environmental Systems Research Institute, 1999.

This guide provides administrators with resources for creating and managing ArcSDE services, allowing them to store and manage multiple databases in a variety of management systems (Oracle, SQL Server, DB2, Informix).

*ArcView: The Geographic Information System for Everyone; Quick Start Guide*. Redlands, CA: Environmental Systems Research Institute, 1995.

*ArcView GIS Developer's Guide: Programming with Avenue*, by A. H. Razavi. Albany, NY: Onword, 2002.

*Beginning ArcGIS for Desktop Development Using .NET*, by P. Amirian. Hoboken, NJ: John Wiley, 2013.

This guide introduces users to .NET (C# and VB.NET) programming basics, allowing them to customize and build their own commands, tools, and extensions. Topics include querying data, visualizing data, creating Desktop add-ins, and performing geoprocessing activities.

*Building a Geodatabase: GIS by ESRI*, by A. MacDonald. Redlands, CA: Environmental Systems Research Institute, 2001.

*Building Mapping Applications with QGIS*, by E. Westra. Birmingham, UK: Packt, 2014.

Using open-source software QGIS (Quantum GIS) and the built-in QGIS Python Console, readers will learn how to write Python code to build stand-alone mapping applications.

*Building Web and Mobile ArcGIS Server Applications with JavaScript*, by E. Pimpler. Birmingham, UK: Packt, 2014.

This hands-on guide offers developers the ability to create custom web and mobile ArcGIS Server applications using ArcGIS API for JavaScript. Users will learn HTML, CSS, and JavaScript in order to add layers, query and display spatial data, and use widgets in web and mobile applications.

*Building Web Applications with ArcGIS*, by H. Nasser. Birmingham, UK: Packt, 2014.

*Cartography and Site Analysis with Microcomputers: A Programming Guide for Physical Planning, Urban Design, and Landscape Architecture*, by N. B. Mutunayagam and A. Bahrami. New York: Van Nostrand Reinhold, 1997.

*Definitive Guide to MySQL*, by M. Kofler and D. Kramer. Berkeley, CA: Apress, 2010.

This guide offers a comprehensive overview of MySQL, a popular open-source database server, and how to use it, covering features such as stored procedures, triggers, spatial data types, administration, basic and advanced querying, and security.

*Desktop GIS: Mapping the Planet with Open Source Tools*, by G. E. Sherman. Raleigh, NC: Pragmatic Bookshelf, 2008.

This book guides the reader to explore open-source desktop GIS programs (such as GRASS, QGIS, uDIG, and GMT), providing strategies for selecting the ideal product, choosing the right tools, integrating the products, managing change, and acquiring support. The readers will also be introduced to scripting, using Python, shell, and other languages.

*Discovering GIS and ArcGIS*, by B. A. Shellito. New York: W. H. Freeman, 2014.

Written for new users, this guide introduces students to basic GIS concepts and practical ArcGIS software skills. With hands-on tutorials, students will create maps and perform simple spatial analysis.

*Editing in ArcMap: GIS by ESRI*, by J. Shaner and J. Wrightsell. Redlands, CA: Esri Press, 2000.

*ERDAS Field Guide.* Atlanta, GA: ERDAS, 2006.

Written for those who are both new to geoprocessing, as well as for expert users, this guide provides basic information about storing, publishing, and delivering various data types such as raster, vector, and LIDAR in the ERDAS Suite software program. A glossary of terms and a brief history of the field are also included.

*ERDAS Imagine 8.5 and 8.6 Training Manual: British Geological Survey Report IR/04/030*, by M. Hall, L. B. Bateson, and C. J. Jordon. London, UK: British Geological Survey, 2004.

*ESRI Guide to GIS Analysis: Modeling Suitability, Movement, and Interaction*, by A. Mitchell. Redlands, CA: Environmental Systems Research Institute, 2012.

Written for students interested in using GIS to evaluate locations and analyze movement, this guide demonstrates how users can explore spatial interaction, site selection, and routing and how they can best interpret analysis results.

*Extending ArcView GIS: With Network Analyst, Spatial Analyst and 3D Analyst*, by T. Ormsby and J. Alvi. Redlands, CA: Esri Press, 1999.

This guide offers detailed exercises leading users through the features of the three Analyst extensions used to solve route and modeling problems.

*Geocoding in ArcGIS: ArcGIS 9*, by S. Crosier. Redlands, CA: Esri Press, 2004.

This guide teaches users techniques and skills for geocoding address data. Topics include preparing data, determining the appropriate address locator style, building an address locator, creating a feature class, and distributing and customizing the address locator.

*GeoServer Beginner's Guide: Share and Edit Geospatial Data with This Open Source Software Server*, by S. Iacovella and B. Youngblood. Birmingham, UK: Packt, 2013.

This guide provides users with instructions for working with an open-source server-side Java-written software product to build custom interactive maps. Users will learn how to use backend data stores such as MySQL, PostGIS, MSSQL, and Oracle.

*Geospatial Analysis: A Comprehensive Guide to Principles, Techniques and Software Tools*, by M. J. de Smith, M. F. Goodchild, and P. Longley. Leicester, UK: Matador, 2007.

Written for those interested in geospatial analysis and modeling, this guide introduces users to Geospatial Analysis, a free web-based resource that offers concepts, methods, and tools for GIS.

*Geospatial Data Harmonization: Methods, Tools and Benefits*, by C. Higer. Boca Raton, FL: CRC, 2012.

*Geotours Workbook: A Guide for Exploring Geology and Creating Projects Using Google Earth*, by S. Marshak. New York: W. W. Norton, 2016.

This workbook introduces Google Earth's Geotour tool, enabling students to virtually visit locations all over the world to study plate tectonics, volcanoes, earthquakes, faults, folds, and other geological and environmental features. Students will learn how to interpret and analyze the sites and will learn about the other tools available in Google Earth.

*Getting Started with ArcGIS: GIS by ESRI*, by B. Booth and A. Mitchell. Redlands, CA: Environmental Systems Research Institute, 2001.

*Getting to Know ArcGIS Desktop*, by T. Ormsby. Redlands, CA: Esri Press, 2004.

*Getting to Know ArcGIS for Desktop*, by M. Law and A. K. Collins. Redlands, CA: Esri Press, 2013.

*Getting to Know ArcView GIS.* Redlands, CA: Esri Press, 1999.

*Getting to Know ArcObjects: Programming ArcGIS with VBA; for ESRI ArcView, ArcEditor, and ArcInfo*, by R. Burke. Redlands, CA: Esri Press, 2003.

ArcObjects are the building blocks of ArcGIS software, allowing GIS professionals to customize the interface. This guide describes basic object-oriented programming concepts as well as the use of Visual Basic.

*Getting to Know ESRI Business Analyst*, by F. L. Miller. Redlands, CA: Esri Press, 2010.

Written for business professionals and analysts, this workbook offers users an introduction to Esri Business Analyst, a suite of products that works with demographic and business data in business analysis, site selection, customer profiling, and segmentation to reveal patterns, trends, and business opportunities.

*GIS and Geocomputation*, by P. M. Atkinson and D. Martin. London, UK: Taylor & Francis, 2000.

*GIS Concepts and ArcGIS Methods*, by D. M. Theobad. Fort Collins, CO: Conservation Planning Technologies, 2003.

*GIS County User Guide: Laboratory Exercises in Urban Geographic Information Systems*, by W. E. Huxhold. New York: Oxford University Press, 1997.

*GIS Data Reviewer User Guide: Production Line Tool Set.* Redlands, CA: Environmental Systems Research Institute, 1998.

This guide offers instructions for using GIS Data Reviewer, an extension to ArcGIS for Desktop that provides a system to automate and simplify the data quality-control process, improving data integrity such as spatial, attribute, topology, connectivity, and database validation.

*GIS Tutorial 1: Basic Workbook*, by L. W. Gorr and K. S. Kurland. Redlands, CA: Esri Press, 2013.

This introductory text presents ArcGIS 10.3 for Desktop, offering an overview of GIS tools and functionality, including processes for collecting data, running geoprocessing tools, and analyzing street network data using ArcGIS Network Analysis.

*GIS Tutorial 2: Spatial Analysis Workbook*, by D. W. Allen and A. Mitchell. Redlands, CA: Esri Press, 2009.

This hands-on workbook offers intermediate GIS users resources for using spatial analysis methods such as location analysis, change-over-time comparisons, geographic distribution, pattern analysis, and cluster identification.

*GIS Tutorial 3: Advanced Workbook*, by D. W. Allen and J. M. Coffey. Redlands, CA: Esri Press, 2016.

This advanced workbook offers exercises that demonstrate the functionality of the ArcEditor and ArcInfo licenses of ArcGIS Desktop, covering geodatabase framework design, data creation and management, workflow optimization, and labeling and symbolizing.

*GIS Tutorial for Crime Analysis*, by W. Gorr and K. Kurland. Esri Press, 2012.

This workbook was written for analysts and criminology students, offering instructions for using GIS to prepare data, build templates, and produce maps.

*GIS Tutorial for Health*, by K. Kurland and W. Gorr. Redlands, CA: Esri Press, 2014.

This tutorial offers an introduction to using GIS for investigating patterns of specific population groups, analyzing environmental hazards and personal injuries, and more.

*GIS Tutorial for Homeland Security*, by R. L. Radke and E. Hanebuth. Redlands, CA: Esri Press, 2008.

*GIS Tutorial for Marketing*, by F. L. Miller. Redlands, CA: Esri, Inc., 2007.

*GIS Tutorial for Python Scripting*, by D. W. Allen. Redlands, CA: Esri Press, 2014.

*GIS Weasel User's Manual*, by R. J. Viger and G. H. Leavesley. Reston, VA: U.S. Department of the Interior, U.S. Geological Survey, 2007.

This is a user manual for the GIS Weasel software program, a GIS-based graphical user interface using the C programming language and external scripting languages.

*GISTRAN, Version 1.0: A Geographic Information System for Modelling Forest Products Transportation; User's Guide*, by D. C. Kapple and H. M. Hoganson. St. Paul: College of Natural Resources and the Agricultural Experiment Station, University of Minnesota, 1991.

This manual was written to facilitate the use of GISTRAN, a GIS for managing political and transportation spatial data in forest management planning models.

*Global Positioning System and ArcGIS*, by M. Kennedy. Boca Raton, FL: CRC, 2009.

*GRASS 4.1 Geographic Information System Beginner's Manual: Version 1.0*, by J. R. Hinthorne. Ellensburg, WA: Evergreen Software Services, 1996.

*GS-CAM: Geological Survey—Cartographic Automatic Mapping; User's Manual.* Reston, VA: USGS National Mapping Division, 1991.

*Hydrologic and Hydraulic Modeling Support*, by D. R. Maidment and D. Djokic. Redlands, CA: Esri Press, 2000.

This manual instructs users on the use of hydrologic and hydraulic modeling support with GIS for the purposes of watershed delineation, topographic characteristic extraction, and floodplain extent.

*HyPAS User's Manual: A Hydraulic Processes Analysis System, an Extension for ArcView GIS, Version 4.0.1*, by T. C. Pratt and D. S. Cook. Vicksburg, MS: U.S. Army Corps of Engineers, Engineer Research and Development Center, 2001.

HyPPAS, an extension for ArcView, was designed to analyze hydraulic information. The set of tools discussed in this manual demonstrate the extension's capability to visualize, analyze, reduce, and efficiently plot hydraulic data.

*Image Integration: Incorporating Images into Your GIS; 6.0.* Redlands, CA: Environmental Systems Research Institute, 1992.

*INSROP GIS v3.0a: User's Guide and System Documentation*, by S. M. Løvås and O. W. Brude. Lysaker, Norway: Fridtjof Nansen Institute, 1999.

*Instructional Guide for the Arcgis Book*, by K. Keranen and L. Malone. Redlands, CA: Esri Press, 2016.

*Introducing Geographic Information Systems with ArcGIS*, by M. D. Kennedy. New York: John Wiley, 2013.

*Introduction to ArcView GIS: Training Manual with Exercises.* Toronto, ON: Esri Canada, 1998.

*Knowledge Cartography: Software Tools and Mapping Techniques*, by A. Okada, S. S. J. Buckingham, and T. Sherborne. London, UK: Springer, 2008.

This book focuses on the process of using interactive maps to communicate and share information. Users will learn how to create knowledge maps for learning and teaching and for use in professional communities.

*Learning ArcGIS Geodatabases*, by H. Nasser. Birmingham, UK: Packt, 2014.

Written for geospatial developers, this guide showcases the use of ArcGIS geodatabases using real-world examples.

*Learning ArcGIS Runtime SDK for .NET*, by R. Vincent. Birmingham, UK: Packt, 2016.

This guide teaches developers how to build cross-platform mapping applications including ArcGIS Online and ArcGIS for Server. Some knowledge of .NET is required.

*Learning Geospatial Analysis with Python: An Effective Guide to Geographic Information System and Remote Sensing Analysis Using Python 3*, by J. Lawhead. Birmingham, UK: Packt, 2015.

This book guides the user in learning geospatial analysis with the Python 3 programming language.

*Learning GIS: A Lab Manual for Learning ArcGIS Desktop Version 10.2*, by S. E. Greco. Dubuque, IA: Kendall Hunt, 2014.

*Learning QGIS*, by A. Graser. Birmingham, UK: Packt, 2016.

This hands-on guide provides step-by-step exercises for creating maps in QGIS. Plugins and common geoprocessing and spatial analysis tasks are included.

*License Manager's Guide: ARC/INFO, ArcView GIS, and SDE on Windows NT and UNIX Workstations.* Redlands, CA: Environmental Systems Research Institute, 1998.

*Linear Referencing in ArcGIS*. Redlands, CA: Esri Press, 2004.

    This book guides readers through using linear referencing in ArcGIS, a feature that enables them to create, manage, display, query, and analyze data that has been modeled along a linear feature. Readers will learn how to create route data, display hatches on linear features, find and identify route locations, and perform a variety of geoprocessing operations with event data.

*Lining up Data in ArcGIS: A Guide to Map Projections*, by M. M. Maher. Redlands, CA: Esri Press, 2013.

    This reference guide helps users troubleshoot common problems of data misalignment. Topics include identifying projections, performing geographic transformations, adding x, y data, and making round buffers.

*Management Guide for Implementation of Geographic Information Systems (GIS) in State DOTs*, by A. P. Vonderohe. Washington, DC: Transportation Research Board, National Research Council, 1993.

*Manual for Analysis and Display of Lake Habitat and Fish Position Information Using GIS: Physical and Hydraulic Habitat, Draw Down, and Acoustic Telemetry*, by P. M. Cooley. Winnipeg, MB: Fisheries and Oceans Canada, 1999.

    This manual is specifically written for the novice Idrisi for Windows user, offering introductory information on using aquatic GIS techniques for modeling water levels and physical processes in lakes. Simple query tools are introduced as is modeling analysis.

*MAPPER, a Personal Computer Map Projection Tool*, by S. A. Bailey. Washington, DC: National Aeronautics and Space Administration, Office of Management, Scientific and Technical Information Program, 1993.

*Mastering ArcGIS*, by M. H. Price. New York: McGraw-Hill, 2014.

    This guide offers readers an introduction to GIS, data, and mapping in ArcGIS, covering data management, coordinate systems, data mapping and presentation, edits, queries and spatial joins, raster analysis, geodatabases, and metadata.

*Mastering ArcGIS Server Development with JavaScript*, by K. Donman. Birmingham, UK: Packt, 2015.

    Developers will be taught to create and share map applications for desktops, tablets, and mobile browsers. Users will create single-page mapping applications, search for geographic and tabular information, and customize maps and widgets. This guide is intended for developers who have experience with ArcGIS Server, HTML, CSS, and JavaScript.

*Mastering ArcGIS with Video Clips DVD-ROM*, by M. H. Price. New York: McGraw-Hill Education, 2013.

*Mastering QGIS*, by K. Menke. Birmingham, UK: Packt, 2015.

*Modeling Our World: The ESRI Guide to Geodatabase Concepts*, by M. Zeiler. Redlands, CA: Esri Press, 2010.

*NetEngine: A Programmer's Library for Network Analysis*. Redlands, CA: Environmental Systems Research Institute, 1998.

    This is a user guide for NetEngine, a programmer's library designed for network analysis.

*Open Source GIS: A GRASS GIS Approach*, by M. Neteler and H. Mitasova. Boston: Kluwer Academic, 2002.

    Aimed at researchers and practitioners in government and industry, this publication reviews code developed by the International GRASS Development Team.

*PostGIS in Action*, by R. O. Obe and L. S. Hsu. Cherry Hill, NJ: Manning, 2015.

    This guide teaches users to write spatial queries that solve real-world problems, covering basics in vector, raster, and topology-based GIS, as well as analyzing, viewing, and mapping data.

*Programming ArcGIS 10.1 with Python Cookbook*, by E. Pimpler. Birmingham, UK: Packt, 2013.

    This hands-on guide teaches Python users to create shortcuts, scripts, tools, and customizations in ArcGIS 10.1.

*Programming ASP.NET for ArcGIS Server*, by V. Zhuang. Albany, NY: OnWord, 2005.

    Programmers and GIS professionals are guided through developing ASP.NET applications for ArcGIS Server. Topics include GIS web applications, GIS web services, and wireless GIS applications for mobile devices such as PDAs and phones.

*PyQGIS Programmer's Guide: Extending QGIS 2.X with Python*, by G. Sherman. Chugiak, AK: Locate, 2014.

    PyGIS blends open-source QGIS and Python; it is designed to write scripts and plugins to implement new features and perform automated tasks. This guide offers an introduction to Python and PyQGIS, covering writing scripts and building plugins. A series of exercises are included as well.

*Python Geospatial Analysis Essentials: Process, Analyze, and Display Geospatial Data Using Python Libraries and Related Tools*, by E. Westra. Birmingham, UK: Packt, 2015.

    This guide offers resources for users to explore Python libraries and build their own geospatial applications using PostGIS and psycopg2 library. Users will also learn to use the Shapely and NetworkX libraries to create, analyze, and manipulate complex geometric objects.

*Python Primer for ArcGIS(r)*, by N. Jennings. Seattle, WA: Createspace, 2015.

*Python Scripting for ArcGIS* by P. A. Zandbergen. Redlands, CA: Esri Press, 2015.

*Quantum GIS Training Manual*, by R. Thiede et al. Chugiak, AK: Locate, 2013.

*QGIS 2.8 User Guide*. Cleveland, OH: Samurai Media, 2016.

*QGIS Python Programming Cookbook: Over 140 Recipes to Help You Turn QGIS from a Desktop GIS Tool into a Powerful Automated Geospatial Framework*, by J. Lawhead. Birmingham, UK: Packt, 2015.

This guide offers developers instructions for writing Python code to automate geoprocessing tasks in QGIS. Users will learn how to create and edit vector layers for faster queries, reproject vector layers, and convert raster layers to vector layers. Over 140 recipes are included to help users create customized maps and add specialized labels and annotations.

*S-Plus for ArcView GIS: User's Guide; Version 1.1*. Seattle WA: Insightful, 1998.

*SAS/GIS 9.2: Spatial Data and Procedure Guide*. Cary, NC: SAS, 2008.

*Secrets of Network Cartography: A Comprehensive Guide to Nmap*, by J. Messer. Tallahassee, FL: Professor Messer, 2007.

*Smart Land-Use Analysis, the LUCIS Model: Land Use Conflict Identification Strategy*, by M. H. Carr and P. D. Zwick. Esri Press, 2007.

Written for land-use planners, analysts, and scholars, this guide provides information needed to implement an innovative analysis model—the land-use conflict identification strategy (LUCIS). This model uses ArcGIS geoprocessing to analyze land-use categories in order to determine potential future conflict among the categories. Case study data and the LUCIS geoprocessing models are also provided.

*SML: Simple Macro Language for PC ARC/INFO*. Redlands, CA: Environmental Systems Research Institute, 1994.

*Spatial Analytics with ArcGIS*, by E. Pimpler. Redlands, CA: Esri, 2017.

*Spatial Analysis Workbook: For ArcGIS 10.1*, by D. W. Allen. Redlands, CA: Esri Press, 2013.

*Statistical Analysis of Geographic Information with ArcView GIS and ArcGIS*, by D. W. S. Wong and J. Lee. Hoboken, NJ: John Wiley, 2005.

*Teaching Introductory Geographical Data Analysis with GIS: A Laboratory Guide for an Integrated SpaceStat/ Idrisi Environment*, by R. F. Dodswon. Santa Barbara: National Center for Geographic Information and Analysis, University of California at Santa Barbara, 1993.

*Training Manual: Introduction to GIS; Using ARC/INFO on an UNIX Platform*, by T. Owens and D. Olsen. Onalaska, WI: U.S. Fish and Wildlife Service, Environmental Management Technical Center, 1993.

*TransCAD: Transportation GIS Software*. Newton, MA: Caliper, 2005.

TransCAD is a GIS application designed specifically for use by transportation professionals to store, display, manage, and analyze transportation data. This guide teaches users how to create and customize maps, as well as how to conduct transportation calculations such as network distances and travel times.

*Understanding GIS: An ArcGIS Project Workbook*, by C. Harder, T. Ormsby, and T. Balstrøm. Redlands, CA: Esri Press, 2013.

*User Manual Rock River Geographic Information System: ROCK-GIS*, by S. A. Tweddale. Champaign, IL: U.S. Army Corps of Engineers, Engineer Research and Development Center, Construction Engineering Research Laboratory, 2004.

*User's Guide for MapIMG 2: Map Image Re-Projection Software Package*, by M. P. Finn, J. R. Trent, and R. H. Buehler. Reston, VA: U.S. Geological Survey, 2006.

*Using ArcCatalog: GIS by ESRI*, by A. Vienneau. Redlands, CA: Environmental Systems Research Institute, 2001.

*Using ArcGIS 3D Analyst*. Redlands, CA: Environmental Systems Research Institute 2000.

This guide reviews the ArcGIS 3D Analyst, a three-dimensional visualization and analysis extension that enables users to create surface models. Some of the topics include querying and analyzing surfaces, determining the surface area and volume of parts of a surface, and animating local and worldwide three-dimensional images.

*Using ArcGIS Geostatistical Analyst*, by K. Johnston. Redlands, CA: Environmental Systems Research Institute 2001.

This guide offers instruction for using ArcGIS Geostatistical Analyst, a tool that creates a continuous surface or map from sample points stored in a point feature layer or a raster layer.

*Using ArcGIS Survey Analyst: GIS by ESRI*, by T. Hodson and K. L. Clark. Redlands, CA: Environmental Systems Research Institute, 2002.

ArcGIS Survey Analyst is one of the extensions available in ArcGIS Desktop, enabling users to store, manage, and analyze survey measurements and coordinates collected. Users will learn how to work with survey data, how to organize and visualize the data, and how to perform survey network analysis.

*Using ArcGIS Tracking Analyst: GIS by ESRI*. Redlands, CA: Environmental Systems Research Institute, 2002.

This guide focuses on the ArcGIS Tracking Analyst extension, enabling users to work with temporal data (data containing dates and times). Tracking layers allows users to view data that changes over time and lets them play back those changes at different speeds in forward and reverse. The data can also be animated and viewed in 3-D using ArcGlobe.

*Using ArcIMS 3*. Redlands, CA: Environmental Systems Research Institute, 2002.

*Using ArcMap* by M. Minami et al. Redlands, CA: Environmental Systems Research Institute, 2004.

*Using ArcScan for ArcGIS: GIS by ESRI*, by P. Sánchez. Redlands, CA: Environmental Systems Research Institute, 2002.

This guide steps readers through the capabilities of ArcScan, an extension in ArcGIS that provides tools for converting scanned images into vector-based feature layers.

*Using ArcToolbox: GIS by ESRI*, by C. Tucker. Redlands, CA: Environmental Systems Research Institute, 2000.

*Using Google Earth in Libraries: A Practical Guide for Librarians*, by E. Dodsworth and A. Nicholson. Lanham, MD: Roman & Littlefield, 2015.

This practical guide demonstrates how Google Earth is being used in research, in teaching, and in libraries. Through step-by-step exercises, users will learn how to use many of the tools available in Google Earth for mapmaking, research, analysis, and collecting data.

*Using Maplex for ArcGIS*. Redlands, CA: Environmental Systems Research Institute, 2004.

This guide shows how to use the Maplex Label Engine, a set of tools in ArcGIS designed to improve the quality of the labels on one's map.

*WaterGEMS for GIS: User's Guide. Geospatial Water Distribution Modeling Software*. Waterbury, CT: Haestad Methods, 2003.

*Watershed Assessment Model, GIS Application: User Guide*. Hinton, AB: Foothills Model Forest, 1997.

This is a user guide for the ArcInfo AML application used to generate watershed and stream characteristics from elevation, stream network, and plot location data.

*What Is ArcGIS? GIS by ESRI*. Redlands, CA: Esri Press, 2001.

# 14
# Cartographic and GIS Dictionaries

This chapter lists dictionaries and glossaries that offer definitions in alphabetical order of terms specific to cartography, GIS, GIS software, and related topics such as geoinformatics, geostatistics, computer cartography, geospatial databases, spatial and network analysis, geospatial data, geodesy, surveying, photogrammetry, aerial photography, and remote sensing. The dictionaries listed in this chapter vary in their focus; some cover only cartographic concepts and terms, whereas others emphasize terms commonly found in GIS software programs or GIS-related research. With the ever-expanding, multidisciplinary uses of GIS and the release of new GIS software products, dictionaries continue to be revised and updated. Several resources below offer online access to hundreds of terms that can be searched by category and keyword. Researchers will find this useful when, in the course of working with a software program or a document, they come across an unfamiliar term, a scenario that happens often to those new to GIS technology.

The dictionaries listed are written in English; the list has been compiled by researching library catalogs, online bookstores, and cartographic and GIS-related websites, including book publishers.

*A to Z GIS: An Illustrated Dictionary of Geographic Information Systems.* 2nd ed. Redlands, CA: Esri, 2006.

    This GIS dictionary contains over sixteen hundred terms found across multiple related fields such as cartography, computer science, surveying, geodesy, and remote sensing.

*Biotech's Dictionary of Geographical Information System,* by P. Kumar. Delhi, India: Biotech Books, 2007.

"Cartographic Terms." ProZ.com. http://www.proz.com/personal-glossary/123590?glossary=22960.

    A list of two hundred commonly used terms in cartography.

"DC GIS Glossary." DC.gov. https://octo.dc.gov/page/dc-gis-glossary.

    An online glossary for terms used in GIS data made available by DC GIS, the District of Columbia Geographic Information System.

*Dictionary of Abbreviations and Acronyms in Geographic Information Systems, Cartography, and Remote Sensing,* by P. Hoehn and M. L. Larsgaard. Berkeley: Earth Sciences and Map Library, University of California, Berkeley, Library, 2004. http://guides.lib.berkeley.edu/ld.php?content_id=22471089.

*Dictionary of Geographical Information System,* by K. Pradeep. New Delhi, India: Biotech Books, 2007.

*Dictionary of Geographical Information System,* by H. R. Singh and A. K. Jamwal. New Delhi, India: Jnanada Prakashan, in association with the Global Open University, Nagaland, 2010.

*Dictionary of Geographical Literacy: The Complete Geography Reference,* by K. O'Mahony. Seattle: EduCare, 1993.

    A dictionary defining more than one thousand geographical, meteorological, and geological terms.

*Dictionary of GIS Terminology,* by H. Kennedy. Redlands, CA: Esri, 2001.

    A list of over twelve hundred GIS terms for related areas like geography, cartography, GPS, remote sensing, and computing.

*Dictionary of Mapmakers including Cartographers, Geographers, Publishers, Engravers, etc. from Earliest Times to 1900,* by R. V. Tooley. London: Map Collectors' Circle, 1974.

*Dictionary of Remote Sensing,* by S. M. Rashid and M. M. A. Khan. Delhi, India: Manak Publications, 1993.

*Dictionary of Remote Sensing and Geoinformatics,* by A. Malik. Jawahar Nagar, India: Rawat Publications, 2016.

    A comprehensive compilation of almost four thousand entries defining terms in the subjects of remote sensing, geoinformatics, GIS, GPS, microwave system, aerial photography, geometry, and stereoscopy.

*Elsevier's Dictionary of Geographical Information Systems: In English, German, French and Russian*, by B. Delijska. Amsterdam: Elsevier, 2002.

A GIS dictionary containing over four thousand terms with seventeen hundred cross references used in the theory and practice of GIS, covering topics like geoinformatics, geostatistics, computer cartography, geospatial databases, computer graphics, geodesy, photogrammetry, remote sensing, and more.

"Esri Support GIS Dictionary." Esri.com. https://support.esri.com/other-resources/gis-dictionary.

An online searchable dictionary with terms related to GIS operations, cartography, and Esri technology.

"GIS Dictionary," by Caitlin Dempsey, October 29, 2012. https://www.gislounge.com/gis-dictionary/.

An online dictionary containing definitions of GIS, cartography, remote sensing, and geographic terms.

"GIS Dictionary." Compiled by the Association of Geographic Information and the University of Edinburgh Department of Geography. http://loi.sscc.ru/gis/defterm/agidict/welcome.html.

An online dictionary including definitions for 980 terms related to GIS. The dictionary can be browsed in alphabetical order, by category, or by acronym; it is also searchable by keyword.

*GIS Dictionary: An Alphabetical List of the Terms and Their Definitions Relating to Geographic Information and Its Associated Technologies*. London: Association for Geographic Information, Standards Committee, 2005.

"GIS Dictionary—Geospatial Definition Glossary." GIS-Geography.com. http://gisgeography.com/gis-dictionary-definition-glossary/.

An online comprehensive list of GIS definitions related to GIS technology and software.

"GIS Glossary: A-G definition." PCMag.com. www.pcmag.com/encyclopedia/term/43795/gis-glossary-a-g.

An online glossary of GIS terms reproduced with permission from "A Practitioner's Guide to GIS Terminology" by Stearns J. Wood, which contains more than ten thousand terms.

"GIS Glossary: H-T definition." PCMag.com. www.pcmag.com/encyclopedia/term/43796/gis-glossary-h-t.

An online glossary of GIS terms reproduced with permission from "A Practitioner's Guide to GIS Terminology" by Stearns J. Wood, which contains more than ten thousand terms.

"GIS Glossary of Terms." Missouri Department of Health and Senior Services. health.mo.gov/data/gis/pdf/GIS Glossary.pdf.

A glossary of GIS terms, including general GIS terms and GIS data and software terms.

"Glossary of Cartographic Terms." Alanpedia.com. http://www.alanpedia.com/Glossary_of_cartographic_terms/glossary_of_cartographic_terms_index_of_terms.html.

"Glossary of Cartographic Terms." Babylon.com. http://dictionary.babylon-software.com/science/geography/glossary-of-cartographic-terms/.

An online dictionary offering in-depth explanations and definitions of terms, abbreviations, and phrases in cartography.

"Glossary of Common GIS and GPS Terms." Natural Resources Conservation Services, U.S. Department of Agriculture, Cartographic and GIS Technical Note MT-1 (Rev. 1), August 2006. https://www.nrcs.usda.gov/Internet/FSE_DOCUMENTS/nrcs144p2_051844.pdf.

A glossary of common GIS and GPS terms.

"Glossary of Mapping and Surveying Terms." Business Queensland, Queensland Government, December 9, 2016. https://www.business.qld.gov.au/running-business/support-assistance/mapping-data-imagery/maps/mapping-glossary.

A glossary of mapping and surveying terms with a focus on cartography.

*Glossary of the Mapping Sciences*. Reston, VA: American Society of Civil Engineers, 1994.

A comprehensive 581-page glossary covering the fields of surveying, mapping, and remote sensing.

"A Glossary of Selected Terms Relevant to Geographic Information Systems." MassGIS (Bureau of Geographic Information). www.mass.gov/anf/docs/itd/services/massgis/gis-glossary.pdf. \

A subset of the glossary developed by the Urban and Regional Information Systems Association (URISA).

*Glossary of Terms in Computer Assisted Cartography*. London: International Cartographic Association, 1980.

*International GIS Dictionary*, by R. McDonnell and K. K. Kemp. Cambridge, UK: GeoInformation International, 1996.

A list of GIS terms used all over the world.

*Modern Dictionary of Remote Sensing*, by P. P. Singh. Delhi, India: Deep & Deep, 2006.

*Multilingual Dictionary of Geodesy*. Budapest: OMIKK Technoinform, 1989.

"Perry-Castañeda Library Map Collection Glossary of Cartographic Terms." University of Texas at Austin. http://www.lib.utexas.edu/maps/glossary.html.

A comprehensive glossary with hundreds of cartographic terms.

*Technical Memorandum: GIS Data Dictionary*, by E. Taylor and D. Ryder. Detroit, MI: Rouge River National Wet Weather Demonstration Project, 1993.

# 15
# Bibliographies

A bibliography is a list of resources, such as books, articles, maps, and other documents, that share a common principle such as subject, authorship, place of publication, or publisher. Bibliographies can play a large role in the research process, providing users with reading lists, threads for further information, sources of reliable material, and assistance for developing collections in new areas. For researchers who are studying specific subjects, a bibliography dedicated to that subject provides a comprehensive listing of related materials. Researchers studying cartography or GIS may benefit from a list of, for example, air photos published for a specific geographic region or a list of all historical topographic maps published within a given year. Not only is this comprehensive, specialized list an excellent way to begin the research process, but, for many, it is all that is needed to find the relevant resources.

The following work is the result of researching many large national libraries, such as the Library of Congress, as well as the online catalog WorldCat, to provide both subject and geographic bibliographies (also known as cartobibliographies) for cartographic and geospatial subjects such as general cartography, GIS, maps, aerial photography, and remote sensing. This list consists of bibliographies in book format and materials such as printed articles, manuscripts, maps, air photos, and other documents. We hope that readers find resources they did not know existed or ones that have been long forgotten.

## CARTOGRAPHY

*Annotated Bibliography of Articles on Cartography Appearing in Certain Journals*, compiled by M. L. Crawford. Kingston, RI: Geography Department, University of Rhode Island, 1965.

*Bibliography and Cartography of Maryland, Including Publications Relating to the Physiography, Geology and Mineral Resources*, compiled by E. B. Mathews. Baltimore, MD: Johns Hopkins Press, 1897.

*Bibliography of Bath County, Kentucky: Citations of Printed and Manuscript Sources Touching upon Its History, Cartography, Geology, Paleontology, and Mineral Resources with Annotations (1791–1961)*, compiled by W. R. Jillson. Frankfort, KY: Roberts Print., 1966.

*Bibliography of Cartography*. Boston: G. K. Hall, 1973.

*Bibliography of Cartography: Supplement 1*. Boston: G. K. Hall, 1980.

*Bibliography of Cartography or a Descriptive List of Books and Magazine Articles Relating to Maps, Map-Makers and Views*, compiled by P. L. Phillips. Washington, DC: Library of Congress, 1920.

*Bibliography of Elliott County, Kentucky: Citations of the Printed Sources Touching upon Its History, Geology, Cartography, Coal, Dikes, Ores, Oil, and Gas, with Annotations, 1861–1957*, compiled by W. R. Jillson. Frankfort, KY: Roberts Print., 1958.

*Bibliography of Estill County, Kentucky: Citations of Printed and Manuscript Sources Touching upon Its History, Geology, Cartography, Coal, Ores, Oil, and Gas, with Annotations, 1781–1956*, compiled by W. R. Jillson. Lawrenceburg, KY: Anderson, 1957.

*Bibliography of Green County, Kentucky: Citations of Printed and Manuscript Sources Touching upon Its History, Geology, Cartography, Onyx, Oil and Gas, with Annotations (1784–1955) / *, compiled by W. R. Jillson. Frankfort, KY: Roberts Print., 1955.

*Bibliography of Hart County, Kentucky: Citations of Printed and Manuscript Sources Touching upon Its History, Geology, Cartography, Onyx, Oil and Gas, with Annotations (1784–1955)*, compiled by W. R. Jillson. Frankfort, KY: Roberts Print., 1955.

*Bibliography of Jefferson County, Kentucky: Citations of Printed and manuscript Sources Touching upon Its History, Cartography, Geology, Paleontology, Oil and Gas,*

*1751–1960, with Annotations*, compiled by W. R. Jillson. Frankfort, KY: Roberts Print., 1964.

*Bibliography of Knox County, Kentucky: Citations of Printed and manuscript Sources Touching upon Its History, Geology, Cartography, Coal, Salt, Oil and Gas with Annotations (1750–1956)*, compiled by W. R. Jillson. Frankfort, KY: Roberts Print., 1958.

*Bibliography of Lawrence County, Kentucky: Citations of Printed and Manuscript Sources Touching upon Its History, Cartography, Geology, Paleontology, and Mineral Resources with Annotations (1789–1969)*, compiled by W. R. Jillson. Frankfort, KY: Roberts Print., 1969.

*Bibliography of Madison County, Kentucky: Citations of Printed and Manuscript Sources Touching upon Its History, Cartography, Geology, Calcite, Barite, Fluorite, Galenite, Sphalerite, Oil and Gas, with brief Annotations (1750–1963)*, compiled by W. R. Jillson. Frankfort, KY: Roberts Print., 1964.

*Bibliography of Map Projections*, compiled by J. P. Snyder and H. Steward. Washington, DC: Government Printing Office, 1988.

*Bibliography of Menifee County, Kentucky: Citations of Printed and Manuscript Sources Touching upon Its History, Cartography, Geology, Paleontology and Mineral Resources (1818–1964)*, compiled by W. R. Jillson. Frankfort, KY: Roberts Print., 1986.

*Bibliography of Morgan County, Kentucky: Citations of Printed and Manuscript Sources Touching upon Its History, Cartography, Geology, Paleontology and Mineral Resources (1822–1963)*, compiled by W. R. Jillson. Frankfort, KY: Roberts Print., 1968.

*Bibliography of Papers on the History of Cartography in American Periodicals of Bibliographical Interest Found in the Libraries of the University of Wisconsin*, compiled by D. Woodward. Madison: Department of Geography, University of Wisconsin, 1968.

*Bibliography of Statistical Cartography*, compiled by P. W. Porter. Minneapolis: Department of Geography, University of Minnesota, 1964.

*Bibliography of the Cumberland River Valley in Kentucky and Tennessee: Citations of Printed and Manuscript Sources Touching upon Its History, Geology, Cartography, Coal, Iron, Salt, Fluorspar, Phosphate, Clays, Oil and Gas, with Annotations*, compiled by W. R. Jillson. Frankfort, KY: Roberts Print., 1960.

*Bibliography of the Floyd County, Kentucky: Citations of Printed and Manuscript Sources Touching upon Its History, Geology, Cartography, Coal, Salt, Oil and Gas, with Annotations, 1750–1956*, compiled by W. R. Jillson. Frankfort, KY: Roberts Print., 1956.

*Bibliography of the History of Cartography at the Newberry Library*, compiled by R. W. Karrow, P. Hafner and P. A. Morris. Chicago: Newberry Library, 1998.

*Bibliography of the Licking River Valley in Kentucky: Annotated Citations of Printed and Manuscript Sources Touching upon Its History, Cartography, Geology, Paleontology, and Mineral Resources*, compiled by W. R. Jillson. Frankfort, KY: Roberts Print., 1968.

*Bibliography of the Paleontology of Kentucky: Citations of Printed Source Touching upon Its History, Cartography and Geology (1744–1969)*, compiled by W. R. Jillson. Frankfort, KY: Roberts Print., 1968.

*California Map List*. Reston, VA: U.S. Geological Survey, 1995.

*Cartographic Catalog, Software Sources*. Reston, VA: U.S. Geological Survey, 1993.

*Cartography of Rhode Island*, by H. M. Chapin. Providence, RI: Preston & Rounds, 1915.

*Cartography of the Northwest Coast of America to the Year 1800. [with Maps and a Bibliography]*, by H. R. Wagner. Berkeley: University of California Press, 1937.

*Catalog of Cartographic Data*. Reston, VA: Dept. of the Interior, U.S. Geological Survey, 1987.

*Computer Cartography: A Working Bibliography*. Toronto, ON: University of Toronto, 1972.

*Concise Bibliography of the History of Cartography: A Selected, Annotation List of Works on Old Maps and Their Makers, and on Their Coll., Cataloguing, Care, and Use*, compiled by R. W. Karrow. Chicago: Newberry Library, 1980.

*GIS, Cartography, and the Information Society: An Annotated Bibliography*, compiled by W. L. Dowdy. Santa Barbara, CA: National Center for Geographic Information and Analysis, 1994.

*Guide to Historical Cartography: A Selected, Annotated List of References on the History of Maps and Map Making*, compiled by W. W. Ristow and C. E. LeGear. Washington, DC: Library of Congress, 1960.

*Guide to the History of Cartography: An Annotated List of References on the History of Maps and Mapmaking*, compiled by W. W. Ristow. Washington, DC: Library of Congress, 1973.

*History of Cartography: A Bibliography, 1981–1992*, compiled by E. W. Wolf. Washington, DC: Washington Map Society in association with Fiat Lux, 1992.

*International Directory of Current Research in the History of Cartography and in Carto-bibliography*, by A. E. Clutton. Norwich, CT: Geo Books, 1985.

*Library of Congress: Geography and Map Division: The Bibliography of Cartography:* Boston: G. K. Hall, 1973.

*Maps: How to Make Them and Read Them; A Bibliography of General and Specialized Works on Cartography*, compiled by W. W. Ristow. New York: New York Public Library, 1943.

*Recent Trends in the History of Cartography: A Selective, Annotated Bibliography to the English-Language Literature*, compiled by E. Matthew. ALA Map and Geography Round Table, 2012.

*Select Bibliography of References on Cartography: Finding Aids to Maps, and the Classification and Cataloging of*

*Maps*, compiled by L. E. Kelsay. Washington, DC: National Archives, 1946.

*Urban Cartography: A Selected Bibliography*, compiled by A. G. White. Monticello, IL: Council of Planning Librarians, 1976.

*Who's Who in the History of Cartography: An International Directory of Current Research in the History of Cartography*, compiled by M. A. Lowenthal. Tring, UK: Map Collector Publications, 1992.

*William P. Cumming and the Study of Cartography: Two Brief Memoirs and a Bibliography*, edited by R. Cumming. Chapel Hill: North Caroliniana Society and the North Carolina Collection, 1998.

## GIS

*Accuracy of Spatial Databases: Annotated Bibliography*, compiled by H. Veregin. Santa Barbara, CA: National Center for Geographic Information and Analysis, 1989.

*ANCA Library Bibliography on GIS and Remote Sensing*. Canberra, Australia: ANCA Library, 1994.

*Annotated Bibliography: Concepts, Issues, and Developments on GIS-Related Standards*, compiled by P. L. Croswell. Washington, DC: URISA, 1994.

*Annotated Bibliography Developed for the Graduate Level Introductory GIS Classes Taught at EPA*. Washington, DC: U.S. Environmental Protection Agency, Office of Research and Development, 1999.

*Annotated Bibliography on GIS Related Standards*, compiled by P. L. Croswell. Washington, DC: URISA, 1993.

*Annotated Bibliography on Human Computer Interaction for GIS*. Santa Barbara, CA: National Center for Geographic Information and Analysis, 1991.

*Bibliographic Analysis of Statewide Geographic Information Systems, 1970–1990*, by R. T. Aangeenbrug, K. Schultz, and M. Hafen. Chicago: Council of Planning Librarians, 1992.

*Bibliography of Available Spatial Data for Southeast Coastal Texas between Galveston Bay and the Sabine River*, compiled by L. Handley and B. Hutchison. Washington, DC: U.S. Geological Survey, 1997.

*Bibliography of Available Spatial Data for St. Marks National Wildlife Refuge and Adjacent Areas*, compiled by L. Handley and B. Hutchison. Reston, VA: National Wetlands Research Center, 1997.

*Bibliography of Papers on Geographical Information Systems Published between 1984 and 1986*, compiled by N. Green. Swindon, UK: Natural Environment Research Council, 1986.

*Bibliography on Animation of Spatial Data*, compiled by B. P. Buttenfield. Buffalo: National Center for Geographic Information and Analysis, State University of New York, 1991.

*Categorized Bibliography of Coastal Applications of Geographic Information Systems*, compiled by T. L. Rickman and A. Miller. Madison: University of Wisconsin-Madison, Sea Grant Institute, 1995.

*Coastal Management: A Bibliography of Geographic Information System Applications*. Charleston, SC: NOAA Coastal Services Center, 1997.

*Concepts, Issues, and Developments on GIS-Related Standards: Annotated Bibliography*, compiled by P. Croswell. Washington, DC: Urban and Regional Information Systems Association, 1994.

*Emerging Ideas in Geographic Information Systems: An Annotated Bibliography (1980–1986)*, compiled by N. Chandra and J. Lakey. Cambridge, MA: Massachusetts Institute of Technology, 1986.

*Geographic Information Systems: A Partially Annotated Bibliography*, compiled by M. A. Wilson. Chicago: Council of Planning Librarians, 1990.

*Geographic Information Systems: A Subject Bibliography*. Washington, DC: NASA Scientific and Technical Information Program, 1993.

*Geographic Information Systems: Bibliography*. Cincinnati, OH: Environmental Research Center, 1988.

*Geographic Information Systems (GIS) and Agriculture: A Bibliography*, compiled by S. L. Beazley and K. B. Beesley. Truro: Rural Research Centre, Nova Scotia Agricultural College, 1996.

*Geographical Information System Map and Data Products Catalog 2001*. Boston: Massachusetts Highway Department, Bureau of Transportation Planning and Development, 2001.

*Geospatial Web Services, Open Standards, and Advances in Interoperability: A Selected, Annotated Bibliography*, compiled by C. Dietz. Chicago: ALA Map and Geography Round Table, 2012.

*GIS and Decision-Support: A Bibliography*, compiled by A. Rodrigues-Bachiller. Oxford: Oxford Brookes University, 1998.

*GIS and the Coastal Zone: An Annotated Bibliography*, compiled by D. Bartlett. Santa Barbara, CA: National Center for Geographic Information and Analysis, 1993.

*GIS Bibliography*, compiled by S. Vaught. Salem: Oregon State Forestry Department, Forestry Assistance Division, Resource Planning Section, 1981.

*GIS, Cartography, and the Information Society: An Annotated Bibliography*, compiled by W. Dowdy. Santa Barbara, CA: National Center for Geographic Information and Analysis, 1993.

*GIS/GPS in Law Enforcement Master Bibliography*, compiled by D. P. Albert. Washington, DC: Police Executive Research Forum, 2003.

*GIS in Archaeology: An Annotated Bibliography*, compiled by L. Petrie. Sydney, Australia: Archaeological Computing Laboratory, University of Sydney, 1995.

*GIS Videos: An Annotated Bibliography*, compiled by A. Ruggles. Santa Barbara, CA: NCGIA Publications, 1992.

*Multiple Roles for GIS in US Global Change Research: Annotated Bibliography*, compiled by A. Shortridge. Santa Barbara, CA: National Center for Geographic Information and Analysis, 1995.

*Natural Resources Information System's Bibliography for Geographic Information Systems: An Annotated Bibliography*, compiled by D. G. Ness. Edmonton: Alberta Energy and Natural Resources, Resource Evaluation and Planning Division, 1984.

*NCGIA Annual GIS Bibliography*. Santa Barbara, CA: National Center for Geographic Information and Analysis, 1986.

*NCGIA Annual GIS Bibliography for 1992*, compiled by S. Frank and H. J. Onsrud. Orono, ME: National Center for Geographic Information and Analysis, 1993.

*NCGIA Annual GIS Bibliography for 1993*, compiled by S. Frank et al. Orono, ME: University of Maine, 1994.

*Quantitative Resource Management Bibliography with Selected Papers in Geographic Information Systems, Spatial Statistics, Land and Water Evaluation and Remote Sensing*, compiled by L. M. Marciak and D. J. White. Edmonton: Land Use Branch, Alberta Agriculture, 1987.

*Selected Annotated Bibliography on Visualization of the Quality of Spatial Information: Research Initiative 7*, compiled by W. A. Mackaness, M. K. Buttenfield, and B. P. Buttenfield. Santa Barbara, CA: National Center for Geographic Information and Analysis, 1994.

*Selected Bibliography of Geographic Information Systems (GIS)*, compiled by J. Adams. Provo, UT: Brigham Young University, 2004.

*Selected Bibliography on the Applications of Geographic Information Systems for Resource Management*, compiled by W. J. Ripple. Corvallis: Environmental Remote Sensing Applications Laboratory, Oregon State University, 1986.

*Selected Bibliography Related to Geographic Information Systems*. Springfield, IL: Peter Ives, 1984.

*Selected Bibliography on Spatial Data Handling: Data Structures, Generalization and Three-Dimensional Mapping*, compiled by R. Sieber. Zurich, Switzerland: Department of Geography, University of Zurich, 1986.

*Selected GIS Bibliography*, compiled by C. C. Devonport. Canberra: Australia Government Publishing Service, 1992.

*Spatial Decision Support Systems: A Bibliography*, compiled by M. D. Gould and P. J. Densham. Santa Barbara, CA: National Center for Geographic Information and Analysis, 1991.

## MAPS

*10,000 Map and Travel Publications Reference Catalog.* Hollywood, CA: Travel Centers of the World, 1983.

*2002–2003 Map and List of Ohio's Bikeways.* Columbus: Ohio Department of Transportation, 2002.

*Alabama Catalog of Topographic and Other Published Maps.* Reston, VA: U.S. Geological Survey, 1989.

*Alabamiana 1823–1983: An Interesting Collection of Books, Pamphlets, Maps and Periodicals Related to Alabama and Incidentally to the Other Southeastern States; Catalog Thirty-Seven*, compiled by J. P. Cather and H. W. Brown. Birmingham, AL: Cather and Brown Books, 1983.

*Alaska and the Northwest Coast: Rare Books and Manuscripts, Early Maps, Prints, and Paintings of Outstanding Historical Importance.* New York: Edward Eberstadt, 1959.

*Alaska and the Northwest Part of North America 1588–1898: Maps in the Library of Congress*, compiled by P. L. Phillips. Washington, DC: Government Printing Office, 1898.

*Alaska: Catalog of Topographic and Other Published Maps.* Reston, VA: U.S. Geological Survey, 1991.

*Annotated Bibliography and Mapping Index of Precambrian of New Mexico*, compiled by J. M. Robertson. Socorro: New Mexico Bureau of Mines and Mineral Resources, 1976.

*Annotated Bibliography of Illinois County Landownership Maps and Atlases through 1929*, compiled by S. Gaetjens. Monticello, IL: Vance Bibliographies, 1991.

*Annotated Bibliography of Maps in the Wisconsin Legislative Reference Library.* Madison: Wisconsin Legislative Reference Library, 1954.

*Annotated Bibliography of Nearshore Fish Habitat Maps for the Strait of Georgia*, compiled by J. Lessard. West Vancouver, BC: Fisheries and Oceans Canada, 1996.

*Annotated Bibliography of Ontario Maps.* Ann Arbor, MI: R. B. Mancell, 1962.

*Annotated Bibliography of Surficial Geology (Regolith) Mapping, and Investigations of Glacial Geology in New Brunswick (Including Till and Soil Geochemistry Studies)*, compiled by A. A. Seaman. Fredericton: New Brunswick Natural Resources and Energy, Minerals and Energy, 2001.

*Arizona: Catalog of Topographic and Other Published Maps.* Reston, VA: U.S. Geological Survey, 1988.

*Arkansas: Catalog of Topographic and Other Published Maps.* Reston, VA: U.S. Geological Survey, 1986, 1988.

*Base Mapping Catalog: A Catalog of Maps Available to the General Public.* Wampsville, NY: Madison County Planning Board, 1972;

*Bathymetric Maps and Special Purpose Charts.* Rockville, MD: National Oceanic and Atmospheric Administration, National Ocean Survey, 1986.

*Bibliographic Guide to Maps and Atlases.* Boston: G. K. Hall, 1979.

*Bibliography for Geologic Map Index of Unpublished Mapping and Reports in California.* Sacramento: California Geological Survey, 1950.

*Bibliography for Map Reference Tools at the Illinois State Library.* Springfield: Illinois State Library, 1993.

*Bibliography for Surficial Mapping in Canada,* compiled by R. J. Fulton, L. Maurice, K. F. Bertrand. Ottawa, ON: Geological Survey of Canada, 1995.

*Bibliography of Arizona Landslide Maps and Reports,* compiled by J. W. Welty. Tucson: Arizona Geological Survey, 1988.

*Bibliography of Books, Pamphlets and Maps in the Colorado-Henkle Collection, Northern Illinois University, DeKalb, Illinois.* Denver, CO: Alan and Mary Culpin, 1981.

*Bibliography of Boston: A List of Maps and Views of Boston and Boston Harbor 1633–1899,* compiled by J. F. Carret and J. Murdoch. Boston: Boston Public Library, n.d.

*Bibliography of Delaware and New Jersey Maps and Atlases,* compiled by J. M. Moak. Philadelphia: Chestnut Hill Almanac, 1987.

*Bibliography of Early Books, Pamphlets, Articles and Maps Pertaining to the Geology, Paleontology and Seismology of Kentucky, 1744–1854,* compiled by W. R. Jillson. Frankfort, KY: Roberts Print.,1950.

*Bibliography of Geologic Literature of Nevada: And, Bibliography of Geologic Maps of Nevada Areas,* compiled by V. P. Gianella and R. W. B. N. Prince. Reno: Nevada State Geological Survey, 1945.

*Bibliography of Geologic Mapping for the Butte 1-degree by 2-degree Quadrangle, Western Montana,* by B. C. Sholes. Butte: Montana Bureau of Mines and Geology, 1984.

*Bibliography of Geologic Mapping for the Hardin 1 x 2 Quadrangle, Central Montana,* by B. C. Sholes. Butte: Montana Bureau of Mines and Geology, n.d.

*Bibliography of Geologic Mapping for the Hardin 1 x 2 Quadrangle, Eastern Montana,* by B. C. Sholes. Butte: Montana Bureau of Mines and Geology, n.d.

*Bibliography of Geologic Mapping for the Hardin 1 x 2 Quadrangle, East-Central Montana,* by B. C. Sholes. Butte: Montana Bureau of Mines and Geology, n.d.

*Bibliography of Geologic Mapping for the Hardin 1 x 2 Quadrangle, North-Central Montana,* by B. C. Sholes. Butte: Montana Bureau of Mines and Geology, n.d.

*Bibliography of Geologic Mapping for the Hardin 1 x 2 Quadrangle, North-Eastern Montana,* by B. C. Sholes. Butte: Montana Bureau of Mines and Geology, n.d.

*Bibliography of Geologic Mapping for the Hardin 1 x 2 Quadrangle, North-Western Montana,* by B. C. Sholes. Butte: Montana Bureau of Mines and Geology, n.d.

*Bibliography of Geologic Mapping for the Hardin 1 x 2 Quadrangle, South-Central Montana,* by B. C. Sholes. Butte: Montana Bureau of Mines and Geology, n.d.

*Bibliography of Geologic Mapping for the Hardin 1 x 2 Quadrangle, South-Eastern Montana,* by B. C. Sholes. Butte: Montana Bureau of Mines and Geology, n.d.

*Bibliography of Geologic Mapping for the Hardin 1 x 2 Quadrangle, South-Western Montana,* by B. C. Sholes. Butte: Montana Bureau of Mines and Geology, n.d.

*Bibliography of Geologic Mapping for the Hardin 1 x 2 Quadrangle, West-Central Montana,* by B. C. Sholes. Butte: Montana Bureau of Mines and Geology, n.d.

*Bibliography of Geologic Reports and Maps for Apache County, Arizona, South of Interstate 40,* compiled by R. A. Trapp. Tucson: Arizona Geological Survey, 1994.

*Bibliography of Map Projections,* compiled by J. P. Snyder and H. Steward. Washington, DC: Government Printing Office, 1988.

*Bibliography of Mapping from Space,* compiled by D. F. Woolnough. Fredericton: University of New Brunswick, 1973.

*Bibliography of North Carolina Geology, Mineralogy and Geography: With a List of Maps,* compiled by F. B. Laney. Raleigh, NC: E. M. Uzzell, 1909.

*Bibliography of Ohio County Atlases and Maps in the State Library of Ohio: Alphabetical by County and the Year of the Atlas or Map,* compiled by P. M. Immel. Columbus: State Library of Ohio, 2007.

*Bibliography of Pacific Area Maps: A Report in the International Research Series of the Institute of Pacific Relations,* compiled by C. H. MacFadden and R. B. Hall. San Francisco: American Council, Institute of Pacific Relations, 1941.

*Bibliography of Reports and Maps Related to Water Resources of Washington State as Published by the U.S. Geological Survey and Washington State Agencies.* Tacoma, WA: U.S. Geological Survey, 1982.

*Bibliography of Reports Containing Maps on South Dakota Geology Published before January 1, 1959,* compiled by M. J. Tipton, C. M. Christensen, and A. F. Agnes. Vermillion: Science Center, University of South Dakota, 1966.

*Bibliography of Southeastern Alberta with a Map and Air Photo Catalogue,* compiled by G. K. Willis. Edmonton: University of Alberta, 1977.

*Bibliography of the District of Columbia, Being a List of Books, Maps, and Newspapers, Including Articles in Magazines and Other Publications to 1898,* compiled by W. B. Bryan. Washington, DC: Columbia Historical Society, 1986.

*Bibliography of the Greater Lansing Area: A List of Books, Pamphlets, Maps, Newspapers and Other Material in the Michigan State Library Relating to the History and Present of Lansing, East Lansing and the Tri-County Area,* compiled by J. Janego. Lansing: Michigan Department of Education, Bureau of Library Services, 1970.

*Bibliography of the Printed Maps of Michigan, 1804–1880: With a Series of over One Hundred Reproductions of Maps Constituting an Historical Atlas of the Great Lakes and Michigan,* compiled by L. C. Karpinski and W. L. Jenks. Lansing: Michigan Historical Commission, 1931.

*Bibliography of United States Landslide Maps and Reports,* compiled by C. S. Alger and E. E. Brabb. Reston, VA: U.S. Department of the Interior, Geological Survey, 1985.

*Bibliography of Vegetation Maps of North America*, compiled by A. W. Kuchler and J. McCormick. Frankfurt: G. Fischer, 1986.

*Biking Ohio: Map and List of Bikeways in Ohio.* Columbus: Ohio Department of Transportation, 1998.

*Birmingham and Surrounding Communities within Jefferson County, Alabama: A Bibliography of Maps Held by the Linn-Henley Research Library*, compiled by J. R. Keeton. Birmingham, AL: Birmingham Public Library, 1983.

*California: Catalog of Topographic and Other Published Maps.* Reston, VA: U.S. Geological Survey, 1986.

*Carto-Bibliography of Pre-1986 Maps in the Map Collection of Hamilton Library, University of Hawai'i*, compiled by C. Nishimura and C. Tachihata. Honolulu: University of Hawaii, 1966.

*Catalog of Books, Maps, and Charts Belonging to the Library of the Two Houses of Congress: April, 1802.* Washington, DC: Library of Congress, 1802.

*Catalog of CIA Maps.* Washington, DC: Central Intelligence Agency, OIR/Map Services Center, 1990.

*Catalog of Map Sources.* Charleston, WV: Governor's Office, 1974.

*Catalog of Maps.* Reston, VA: U.S. Geological Survey, 1991.

*Catalog of Maps and Charts Available as of January 1, 1951.* Detroit, MI: Detroit Metropolitan Area Regional Planning Commission, 1951.

*Catalog of Maps, Charts, and Related Products: Part 1.* Washington, DC: Defense Mapping Agency Combat Support Center, 1990, 1993.

*Catalog of Maps, Charts, and Related Products: Part 2.* Washington, DC: Defense Mapping Agency Combat Support Center, 1997.

*Catalog of Maps of the North Central States Project*, compiled by P. A. Moore. Chicago: Hermon Dunlap Smith Center for the History of Cartography, Newberry Library, 1978.

*Catalog of Maps, Ships' Papers and Logbooks.* Boston: G. K. Hall, 1964.

*Catalog of Maps: Topographic Maps, Topographic-Bathymetric Maps, Antarctic Maps, Photoimage Maps, Satellite Image Maps, Geologic Maps, Hydrologic Maps, Land Use Maps, Maps of the Planets and Moons, National Atlas Separates, Special Maps.* Washington, DC: U.S. Geological Survey, 1987.

*Catalog of Nautical Charts.* Washington, DC: Defense Mapping Agency, 1970, 1986.

*Catalog of Publications Available through the Utah Geological and Mineralogical Survey.* Salt Lake City: Utah Geological and Mineralogical Survey, 1986.

*Catalog of Published Pennsylvania Maps: Topographic Map Series, Orthophotoquads, Orthophotomaps, State Maps, County Maps, United States Maps, National Parks Maps, National Atlas Maps, Special Maps.* Reston, VA: National Cartographic Information Center (NCIC), U.S. Geological Survey, 1985.

*Catalog of Published Virginia Maps: Topographic Map Series, Orthophotoquads, Orthophotomaps, State Maps, County Maps, United States Maps, National Park Maps, National Atlas Maps, Special Maps.* Reston, VA: U.S. Geological Survey, 1978.

*Catalog of Rand McNally Commercial Maps and Map Cases.* Chicago: Rand, McNally, 1920.

*Catalog of 17th and 18th Century Maps in the William Andrews Clark Memorial Library, University of California at Los Angeles*, compiled by M. D. Fleming. Los Angeles: University of Southern California, University Library, 1979.

*Catalog of the Map Collection of the Montana Historical Society.* Helena: Montana Historical Society, 1983.

*Catalog of Topographic and Other Published Maps.* Reston, VA: U.S. Geological Survey, 1983.

*Catalog of Travel, Topographical and Other Antiquarian Books, Maps and Atlases.* London: Bloomsbury Auctions, 2011.

*Catalog of U.S. Coast and Geodetic Survey Nautical and Aeronautical Charts, Coast Pilots, Tide Tables, Current Tables, Tidal Current Charts.* Washington, DC: Coast and Geodetic Survey, 1936.

*Catalog of U.S. Coast and Geodetic Survey Nautical Charts, Coast Pilots, Tide Tables, Current Tables, Tidal Current Charts, Airway Maps.* Washington, DC: Coast and Geodetic Survey, 1932.

*Catalog of West Virginia Geological Maps, 1752–1992*, compiled by P. Lessing. Charleston: West Virginia Geological and Economic Survey, 1993.

*Catalog of West Virginia Maps*, compiled by S. A. Kasales and P. Lessing. Morgantown: West Virginia Geological and Economic Survey in cooperation with the United States Geological Survey, 1983.

*Catalogue of Books, Maps, Plates on America, and of a Remarkable Collection of Early Voyages, Offered for Sale by Frederik Muller at Amsterdam: Including a Large Number of Books in All Languages with Bibliographical and Historical Notes and Presenting an Essay towards a Dutch-American Bibliography.* Amsterdam: Muller, 1872.

*Checklists and Catalogs of Map and Print Collections Containing North American Town Plans and Views: A Bibliography of Guides to the Location of Graphic Records of Urban Development Prior to 1986*, compiled by J. W. Reps. Ithaca, NY: Department of City and Regional Planning, Cornell University, 1971.

*Civil War Maps: An Annotated List of Maps and Atlases in the Library of Congress*, compiled by R. W. Stephenson. Washington, DC: Library of Congress, 1961, 1989.

*Colorado Catalog of Topographic and Other Published Maps.* Reston, VA: U.S. Geological Survey, 1986.

*Colorado Map List.* Reston, VA: U.S. Geological Survey, 1995.

*Contribution to Bibliography of Mineral Resources: Annotated Bibliography and Index Map of Barite Deposits in*

*United States*, compiled by D. A. Brobst and B. G. Dean. Reston, VA: U.S. Geological Survey, 1955.

*Defense Mapping Agency Catalog of Maps, Charts, and Related Products. Part 3, Topographic Products. Semiannual Bulletin Digest.* Washington, DC: Defense Mapping Agency Combat Support Center, 1986.

*Descriptive List of the Map Collection in the Pennsylvania State Archives: Catalogue of Maps in the Principal Map Collection (MG 11)*, compiled by M. L. Simonetti, D. H. Kent, and H. E. Whipkey. Harrisburg: Pennsylvania Historical and Museum Commission, 1976.

*Dictionary Catalog of the Map Division.* Boston: G. K. Hall & Company, 1971.

"Early Soil Maps of California, 1986–1940, a Bibliography with Indexes," by R. Soares. *Information Bulletin Western Association of Map Libraries* 32 (2001): 189–203.

*Earth Science Maps of Wisconsin, 1818–1974: A Bibliography and Index*, compiled by C. Reinhard. Madison, WI: State Cartographer's Office, 1975.

*Eastern Oregon Books and Print: An Annotated and Historical Bibliography (with Some Emphasis on Maps)*, compiled by W. Kee. Prineville, OR: Paunina, 2005.

*Edmonton District Directory for the Year 1895: Containing Full and Authentic Information, Statistics, Tables, Maps and Guide to Northern Alberta.* Edmonton, AB: J. B. Spurr, 1895.

*Explorers' Maps of the Canadian Arctic, 1818–1860: [A Bibliography]*, compiled by C. Verner and F. Woodward. Toronto: B. V. Gutsell, Department of Geography, York University, 1972.

*Fire Insurance Maps in the Library of Congress: Plans of North American Cities and Towns Produced by the Sanborn Map Company: A Checklist Compiled by the Reference and Bibliography Section, Geography and Map Division; Introd. by Walter W. Ristow.* Washington, DC: Library of Congress and Sanborn Map, 1981.

*Florida: Catalog of Topographic and Other Published Maps.* Reston, VA: U.S. Geological Survey, 1986, 1988.

*Geologic Bibliography and Index Maps of the Ocean Floor off Oregon and the Adjacent Continental Margin*, compiled by C. P. Peterson et al. Portland: State of Oregon, Department of Geology and Mineral Industries, 1985.

*Geological Map of the United States: A Narrative Outline and Annotated Bibliography (1752–1946)*, compiled by W. R. Jillson. Frankfort, KY: Roberts Print., 1950.

*Georgia Catalog of Topographic and Other Published Maps.* Reston, VA: U.S. Geological Survey, 1986, 1988.

*Glacial Map of North America: Part 2, Bibliography and Explanatory Notes*, by R. F. Flint. New York: Geological Society of America, 1945.

*Hawaii: Catalog of Topographic and Other Published Maps.* Reston, VA: U.S. Geological Survey, 1990.

*Historical Maps of Louisiana: An Annotated Bibliography*, compiled by J. N. Rolston and A. G. Stanton. Baton Rouge: Louisiana State University, 1999.

*Hotchkiss Map Collection: A List of Manuscript Maps, Many of the Civil War Period*, compiled by J. Hotchkiss and C. E. LeGear. Washington, DC: Library of Congress, 1951.

*Idaho: Catalog of Topographic and Other Published Maps.* Reston, VA: U.S. Geological Survey, 1989.

*Illinois Catalog of Topographic and Other Published Maps.* Reston, VA: U.S. Geological Survey, 1990.

*Illinois County Landownership Map and Atlas, Bibliography and Union List*, compiled by M. P. Conzen, J. R. Akerman, and D. T. Thackery. Springfield: Illinois Cooperative Collection Management Coordinating Committee, Illinois Board of Higher Education, 1991.

*Index to Maps in the Catalog of the Everett D. Graff Collection of Western Americana*, compiled by B. Berkman, R. W. Karrow, and C. Storm. Chicago: Hermon Dunlap Smith Center for the History of Cartography, Newberry Library, 1972.

*Indiana: Catalog of Topographic and Other Published Maps.* Reston, VA: U.S. Geological Survey, 1989.

*Iowa: Catalog of Topographic and Other Published Maps.* Reston, VA: U.S. Geological Survey, 1987.

*International Bibliography of Maps and Atlases.* Munich: K. G. Saur, 1998.

*International Bibliography of Vegetation Maps: Volume 1, North America*, compiled by A. W. Kuchler. Lawrence: University of Kansas Libraries, 1965.

*Kansas: Catalog of Topographic and Other Published Maps.* Reston, VA: U.S. Geological Survey, 1983.

*Kansas Map List.* Reston, VA: U.S. Geological Survey, 1996.

*Kentucky: Catalog of Topographic and Other Published Maps.* Reston, VA: U.S. Geological Survey, 1989.

*Library of Congress Catalog: A Cumulative List of Works Represented by Library of Congress Printed Cards. Maps and Atlases.* Washington, DC: Library of Congress, 1953.

*List of Arctic and Subarctic Maps in the McGill University Map Collection*, compiled by G. Shields and L. Dubreuil. Montreal, QC: McGill University, Centre for Northern Studies and Research, University Map Collection, 1975.

*List of Maps and Views of Washington and District of Columbia in the Library of Congress*, compiled by P. L. Phillips. Washington, DC: Government Printing Office, 1986.

*List of Maps of America in the Library of Congress.* Washington, DC: Library of Congress, 1901.

*List of Maps of America in the Library of Congress: Preceded by a List of Works Relating to Cartography*, compiled by P. Phillips. New York: B. Franklin, 1986.

*List of Printed Maps Contained in the Map Department.* Sacramento: California State Library, 1899.

*Louisiana Catalog of Topographic and Other Published Maps.* Reston, VA: U.S. Geological Survey, 1986.

*Maine: Catalog of Topographic and Other Published Maps.* Reston, VA: U.S. Geological Survey, 1987.

*Map Catalog.* Santa Barbara, CA: Pacific Travellers Supply, 1986.

*Map Catalog.* Santa Cruz: University of California Library, 1974.

*Map Catalog: Every Kind of Map and Chart on Earth and Even above It*, compiled by J. Makower. New York: Vintage, 1992.

*Map Catalog 5, United States: Topographic/Bathymetric, Bathymetric and Fishing Maps.* Rockville, MD: U.S. Dept. of Commerce, National Oceanic and Atmospheric Administration, National Ocean Service, 1986.

*Map Collections in Midwestern Universities and Colleges: Survey Tables and Bibliography*, by J. V. Bergen. Macomb: Western Illinois University, 1973.

*Map Depository Catalog.* Washington, 1950.

*Map, Description and Bibliography of the Mineralized Areas of the Basin and Range Province in Arizona*, by S. B. Keith. Denver, CO: U.S. Geological Survey, 1983.

*Map Index to Topographic Quadrangles of the United States, 1882–1940: A Graphic Bibliography of Out-of-Print Topographic Maps Published from 1882 to 1940 in the 15-, 30-, and 60-Minute Series for All States except Alaska*, compiled by R. M. Moffat. Santa Cruz, CA: Western Association of Map Libraries, 1985.

*Mapping Texas History: A Select Bibliography of Material Available at the Daughters of the Republic of Texas Library at the Alamo*, compiled by L. Zelnick. San Antonio, TX: Daughters of the Republic of Texas, 1995.

*Mapping Upper Canada, 1780–1867: An Annotated Bibliography of Manuscript and Printed Maps*, compiled by Joan Winearls. Toronto, ON: University of Toronto Press, 2016.

*Maps and Charts Published in America before 1800: A Bibliography*, compiled by J. C. Wheat and C. Brun. New Haven, CT: Yale University Press, 1969.

*Maps and Mapping Agencies in Washington State: A Selective and Analytical Bibliography*, compiled by R. E. Black and H. E. Vogel. Seattle: University of Washington, 1956.

*Maps of Famous Cartographers Depicting North America: A Historical Atlas of the Great Lakes and Michigan to 1880*, compiled by L. C. Karpinski. Amsterdam: Meridan Publishing, 1977.

*Maps of Famous Cartographers Depicting North America: An Historical Atlas of the Great Lakes and Maryland, Delaware, and District of Columbia Catalog of Topographic and Other Published Maps.* Reston, VA: U.S. Geological Survey, 1986.

*Maps of Hudson and James Bay, 1612–1969: A Bibliography of Related Holdings of the Public Archives of Canada, National Map Collection*, compiled by J. Murdoch. Ottawa, ON: Public Archives of Canada, 1975.

*Maps of the 16th to 19th Centuries in the University of Kansas Libraries: An Analytical Carto-Bibliography*, compiled by T. R. Smith and B. L. Thomas. Lawrence: University of Kansas Libraries, 1963.

*Maps of the State of Maine: A Bibliography of the Maps of the State of Maine*, compiled by E. C. Smith. Bangor, ME: Priv. Print, 1903.

*Maps, Plans, and Sketches of Herman Ehrenberg: A Carto-Bibliography*, compiled by D. M. T. North. Los Angeles: California Map Society, 1988.

*Marine Surveys of James Cook in North America 1758–1868, Particularly the Survey of Newfoundland: A Bibliography of Printed Charts and Sailing-Directions*, compiled by R. A. Skelton and R. V. Tooley. London: Map Collectors' Circle, 1986.

*Marketing Maps of the United States, an Annotated Bibliography*, compiled by Walter W. Ristow. Washington, DC: Library of Congress, 1958.

*Massachusetts, Rhode Island, and Connecticut Catalog of Topographic and Other Published Maps.* Reston, VA: U.S. Geological Survey, 1986.

*Michigan Bibliography: A Partial Catalogue of Books, Maps, Manuscripts and Miscellaneous Materials Relating to the Resources, Development and History of Michigan from Earliest Times to July 1, 1917; Together with Citation of Libraries in Which the Materials May Be Consulted, and a Complete Analytical Index by Subject and Author.* Lansing: Michigan Historical Commission, 1921.

*Michigan Carto-Bibliography: An Annotated Guide to Sources Pertaining to Michigan Maps, Atlases, and Related Cartographic Materials*, compiled by J. M. Walsh. Washington, DC: ERIC Clearinghouse, 1983.

*Michigan: Catalog of Topographic and Other Published Maps.* Reston, VA: U.S. Geological Survey, 1986.

*Minnesota: Catalog of Topographic and Other Published Maps.* Reston, VA: U.S. Geological Survey, 1987.

*Mississippi: Catalog of Topographic and Other Published Maps.* Reston, VA: U.S. Geological Survey, 1986.

*Missouri: Catalog of Topographic and Other Published Maps.* Reston, VA: U.S. Geological Survey, 1986.

*Montana: Catalog of Topographic and Other Published Maps.* Reston, VA: U.S. Geological Survey, 1987.

*Montana Map List.* Reston, VA: U.S. Geological Survey, 1995.

*Nebraska: Catalog of Topographic and Other Published Maps.* Reston, VA: U.S. Geological Survey, 1986.

*Nevada: Catalog of Topographic and Other Published Maps.* Reston, VA: U.S. Geological Survey, 1986.

*New Hampshire and Vermont: Catalog of Topographic and Other Published Maps.* Reston, VA: U.S. Geological Survey, 1991.

*New Jersey: Catalog of Topographic and Other Published Maps.* Reston, VA: U.S. Geological Survey, 1986.

*New Mexico: Catalog of Topographic and Other Published Maps.* Reston, VA: U.S. Geological Survey.

*New York: Catalog of Topographic and Other Published Maps.* Reston, VA: U.S. Geological Survey, 1986.

*North Carolina: Catalog of Topographic and Other Published Maps.* Reston, VA: U.S. Geological Survey, 1986.

*North Dakota: Catalog of Topographic and Other Published Maps.* Reston, VA: U.S. Geological Survey, 1986.

*Official Western North America Map and Chart Index Catalog.* Neenah, WI: U.S./Canadian Map Service Bureau, 1975.

*Ohio: Catalog of Topographic and Other Published Maps.* Reston, VA: U.S. Geological Survey, 1986.

*Oklahoma: Catalog of Topographic and Other Published Maps.* Reston, VA: U.S. Geological Survey, 1986.

*Old Maps of Canada: A Checklist of Pre-Twentieth Century Maps of Canada in the Map Section, Trent University Library.* Peterborough, ON: Trent University, 1976.

*Oregon: Catalog of Topographic and Other Published Maps.* Reston, VA: U.S. Geological Survey, 1986.

*Pacific Islands: Catalog of Topographic and Other Published Maps.* Reston, VA: U.S. Geological Survey, 1991.

*Pennsylvania: Catalog of Topographic and Other Published Maps.* Reston, VA: U.S. Geological Survey, 1991.

*Potentiometric Surfaces on Long Island, New York: A Bibliography of Maps*, compiled by D. A. Smolensky. Syosset, NY: U.S. Department of the Interior, Geological Survey, 1983.

*Preliminary Bibliography of Alaskan Geologic Maps by 1:250,000 Quadrangle*, compiled by J. F. Morrone. Denver, CO: U.S. Geological Survey, 1982.

*Preliminary Geologic Map, Index to Geologic Mapping, and Annotated Bibliography of the Hartford, Connecticut, New York, New Jersey, Massachusetts 2° sheet*, compiled by M. H. Pease. Reston, VA: U.S. Geological Survey, 1978.

*Pre-Nineteenth Century Maps in the Collection of the Georgia Surveyor General Department*, compiled by J. G. Blake. Atlanta: State Printers Office, 1975.

*Printed Maps of Upper Canada 1800–1864: Select Bibliography*, compiled by C. F. J. Whebell. Toronto: Ontario Historical Society, 1957.

*Publications and Maps Available from Montana Bureau of Mines and Geology.* Butte: Montana Bureau of Mines and Geology, 1981.

*Publications Directory: A Bibliography of Periodicals, Publications, Legislation and Maps Available from the New Jersey Department of Community Affairs.* Trenton: New Jersey Department of Community Affairs, 1984.

*Published Maps of Montana: An Annotated Bibliography*, compiled by V. K. Shaudys. Missoula: Bureau of Business and Economic Research, Montana State University, 1958.

*Puerto Rico and the Virgin Islands: Catalog of Topographic and Other Published Maps.* Reston, VA: U.S. Geological Survey, 1991.

*Railroad Maps of the United States: A Selective Annotated Bibliography of Original 19th-Century Maps in the Geography and Map Division of the Library of Congress*, compiled by A. M. Modelski. Washington, DC: The Library, 1975.

*Research Catalog of Maps of America to 1860 in the William L. Clements Library*, compiled by D. W. Marshall. Boston: G. K. Hall, 1972.

*Revised Bibliography for Geologic Map Index of Unpublished Mapping and Reports in California, January 1937 to January 1954.* Sacramento: California Geological Survey, 1954.

*Selected Bibliography and Index of Earth Science Reports and Maps Relating to Land-Resource Planning and Management Published by the U.S. Geological Survey through October 1976*, compiled by M. F. Eister. Reston, VA: U.S. Geological Survey, 1977.

*Selected Bibliography of Available Maps of Some of the Larger Cities of the United States.* Salt Lake City, UT: E. Kay Kirkham, 1955.

*Selected Bibliography of Map Books and Periodicals, Many of Which Relate to Virginia*, compiled by M. M. McKee. Richmond: Library of Virginia, Archival and Information Services, 1999.

*Selected Bibliography of Reports and Maps Relating to the Geology of Massachusetts: July 1, 1951.* Boston: USGS, 1951.

*Selected Bibliography of Southern California Maps*, compiled by E. L. Chapin. Berkeley: University of California Press, 1953.

*Sixteenth-Century Maps Relating to Canada: A Check-List and Bibliography.* Ottawa, ON: Public Archives of Canada, 1956.

*South Carolina: Catalog of Topographic and Other Published Maps.* Reston, VA: U.S. Geological Survey, 1987.

*South Dakota: Catalog of Topographic and Other Published Maps.* Reston, VA: U.S. Geological Survey, 1987.

*Status of Bedrock and Surficial Mapping in the NWT: Bibliography and Summary Maps*, compiled by K. Pollock. Yellowknife: Resources, Wildlife and Economic Development, Government of the Northwest Territories, 1997.

*Status of Geologic Mapping in Alaska: A Digital Bibliography*, compiled by J. P. Galloway and J. Laney. Menlo Park, CA: U.S. Geological Survey, 1994.

*Status of Quaternary Geology Mapping in Canada with Bibliography: États des travaux cartographique dans le domaine de la géologie du quarternaire au Canada et bibliographie.* Ottawa, ON: Energy, Mines and Resources, Canada, 1988.

*Tennessee: Catalog of Topographic and Other Published Maps.* Reston, VA: U.S. Geological Survey, 1989.

*Texas: Catalog of Topographic and Other Published Maps.* Reston, VA: U.S. Geological Survey, 1985.

*Topographic/Bathymetric, Bathymetric and Fishing Maps.* Rockville, MD: National Oceanic and Atmospheric Administration, National Ocean Service, 1986.

*Travel and Exploration, Polar Exploration, Maps, California and the West, Western Bibliography.* San Francisco: California Book Auction Galleries, 1983.

*Union Catalog of Maps.* Berkeley, CA: Berkeley Documentation Center, 1986.

*Union List of Foreign Topographic Map Series in Canadian Map Collections.* Ottawa, ON: Public Archives Canada, 1986.

"U.S./Canada Soil Map Bibliography," by J. R. Brunner. *Bulletin of Special Libraries Association Geography and Map Division* 160 (January 1990): 1–23.

*U.S. Geological Survey Map Sales Catalog.* Reston, VA: Department of the Interior, U.S. Geological Survey, National Mapping Program, 1987.

*Use of Maps as Historical Sources: A Bibliography of Materials in the Libraries of the University of Michigan for the Study of British History through Maps,* compiled by J. Attig. Ann Arbor: University of Michigan, 1974.

*Utah: Catalog of Topographic and Other Published Maps.* Reston, VA: U.S. Geological Survey, 1986.

*Virginia: Catalog of Topographic and Other Published Maps.* Reston, VA: U.S. Geological Survey, 1988.

*Washington: Catalog of Topographic and Other Published Maps.* Reston, VA: U.S. Geological Survey, 1986.

*Washington Map List.* Reston, VA: U.S. Geological Survey, 2000.

*Washington State Coal Mine Map Collection: A Catalog, Index, and User's Guide,* by H. W. Schasse. Olympia: Washington Division of Geology and Earth Resources, 1994.

*West Virginia: Catalog of Topographic and Other Published Maps.* Reston, VA: U.S. Geological Survey, 1986.

*Western Expansion, Early Americana, Maps of the Far East.* New York: Cohen & Taliaferro, 2000.

*Wisconsin: Catalog of Topographic and Other Published Maps.* Reston, VA: U.S. Geological Survey, 1986.

*Wyoming: Catalog of Topographic and Other Published Maps.* Reston, VA: U.S. Geological Survey, 1988.

*Wyoming Highway Department Catalog of Maps.* Cheyenne, WY: State Highway Department, Planning Division, 1972.

## REMOTE SENSING AND AERIAL PHOTOGRAPHY

*Aerial Photography: Bibliography of Available Material Relating to the Means, Methods, Experiments and Results of Aerial Photography,* compiled by H. E. Haferkorn. Washington, DC: Press of the Engineer School, 1918.

*Aerial Photography; Crops and Vegetation; 1947–1968; Annotated Bibliography.* Maidenhead, UK: Commonwealth Bureau of Pastures and Field Crops, 1986.

*Aerial Remote Sensing: A Bibliography,* compiled by D. B. Stafford. Washington, DC: Office of Water Resources, 1973.

*Agricultural Meteorology and Remote Sensing: January 1980–February 1991,* by S. Whitmore. Beltsville, MD: National Agricultural Library, 1991.

*Annotated Bibliography: 1268. Bibliography on Aerial Photography, 1930–1968.* Harpenden, UK: Commonwealth Agricultural Bureaux, 1968.

*Annotated Bibliography and Evaluation of Remote Sensing Publications Relating to Military Geography of Arid Lands,* by W. G. McGinnies. Natick, MA: U.S. Army Natick Laboratories, 1970.

*Annotated Bibliography of Aerial Remote Sensing in Coastal Engineering,* compiled by D. Stafford, R. Bruno, and H. Goldstein. Washington, DC: Army Coastal Engineering Research Center, 1973.

"Annotated Bibliography of Bibliographies on Photo Interpretation and Remote Sensing," compiled by D. Steiner. *Photogrammetria* 26, no. 4 (January 1970): 143–61.

*Annotated Bibliography of Remote Sensing at the University of Wisconsin.* Madison: University of Wisconsin, Remote Sensing Program, Environmental Studies Institute, 1970.

*Annotated Bibliography of Remote Sensing Methods for Monitoring Desertification,* compiled by A. S. Walker and C. J. Robinove. Alexandria, VA: U.S. Geological Survey, 1981.

*Annotated Bibliography of Remote Sensing of Air and Water Pollution,* compiled by P. Brooks and G. Thomson. Washington, DC: Army Topographic Command, 1971.

*Applications of Imaging Radar: A Bibliography; Final Edition,* compiled by L. F. Dellwig, N. E. Hardy, and R. K. Moore. Lawrence: University of Kansas Space Technology Center, 1974.

*Bibliography: Applications of Remote Sensing, Aerial Photography, and Instrumented Imagery Interpretation to Urban Area Studies,* by J. B. Kracht and W. A. Howard. Washington, DC: U.S. Geological Survey, Geographic Applications Program, 1970.

*Bibliography of Aeronautics: Part 30, Aerial Photography.* New York: Institute of the Aeronautical Sciences, 1939.

*Bibliography of Photogrammetry.* Washington, DC: American Society of Photogrammetry, 1936.

*Bibliography of Remote Sensing Applications to Urban Studies,* compiled by V. Singhroy. Halifax, NS: Canadian Institute of Planners, 1979.

*Bibliography of Remote Sensing in Forestry 1950–1978,* compiled by B. J. Myers and I. E. Craig. Canberra, Australia: CSIRO, 1980.

*Bibliography of Remote Sensing Techniques Used in Wetland Research: Final Report,* by J. Lampman. Vicksburg, MS: Army Engineer Waterways Experiment Station, 1992.

*Bibliography of Snow Mapping with Remote Sensing Techniques,* compiled by H. Haefner. Zurich, Switzerland: University of Zurich, 1979.

*Bibliography on Aerial Photo Volume Tables.* Blacksburg: School of Forestry and Wildlife Resources, Virginia Polytechnic Institute and State University, 1980.

*Bibliography on Application of Remote Sensing and Aerial Photography to Agricultural Crops, Soil Resources, and Land Use.* Ottawa, ON: Agriculture Canada, Research Branch, 1973.

*Bibliography on Application of Remote Sensing and Aerial Photography to Agricultural Crops, Soil Resources and Land Use (Supplement to February 1981)*, compiled by P. C. Geib, P. H. Crown, and A. R. Mack. Switzerland: Wild Heerbrugg, 1982.

*Bibliography on Interpretation of Vegetation from Aerial Photography*, compiled by T. Hart. Washington, DC: U.S. Naval Photographic Interpretation Center, U.S. Naval Receiving Station, 1950.

*Bibliography on the Interpretation of Aerial Photographs and Recent Bibliographies on Aerial Photography and Related Subjects*, compiled by G. C. Cobb. New York: Geological Society of America, 1943.

*Catalog of Aerial Photography: In the Map Division, Wilson Library, University of Minnesota, Minneapolis*. St. Paul: Minnesota State Planning Agency, 1977.

*Catalog of Aerial Photography of Iowa, 1937–1986: A List of Holdings at the University of Iowa Libraries Map Collection, 1994*. Iowa City: University of Iowa Libraries, 1994.

*Catalog of Aerial Photos by Fairchild Aerial Surveys, Inc. Now in the Collections of the Department of Geography, University of California at Los Angeles*. Santa Cruz: University of California, 1982.

*Catalog of Aerial Photos in the Map Collection of the University Library, University of California, Santa Cruz*, compiled by S. D. Stevens. Santa Cruz: Dean E. McHenry Library, University of California, 1979.

*Cross-Referenced Bibliography of Aerial Photography in Forestry*, compiled by B. J. Myers. Canberra, Australia: Distributed by Forestry and Timber Bureau, 1975.

*Forest Photogrammetry and Aerial Mapping: A Bibliography, 1887–1955*, compiled by S. H. Spurr. Ann Arbor: School of Natural Resources, University of Michigan, 1956.

*GIS and Remote Sensing Applications in Developing Countries: An Annotated Bibliography A–L and M–Z*, compiled by M. Barkhof, S. M. Jong, and P. Teeffelen. Utrecht, Netherlands: University of Utrecht, 1988.

*Infrared Photography and Remote Sensing Thermal Techniques for Mapping Saline Soils and Vegetation, Annotated Bibliography*, compiled by G. Sethi and M. Abdel-Hady. Stillwater: School of Civil Engineering, Oklahoma State University, 1968.

*Integration of GIS, Remote Sensing and Image Processing Systems: An Annotated Bibliography*, compiled by A. Oosterhoff. Perth, Australia: Department of GIS, University of Technology, 1993.

*Quantitative Resource Management Bibliography with Selected Papers in Geographic Information Systems, Spatial Statistics, Land and Water Evaluation and Remote Sensing*, edited by L. M. Marciak and D. J. White. Edmonton: Land Use Branch, Alberta Agriculture, 1987.

*Radar Remote Sensing for Geosciences: An Annotated and Tutorial Bibliography*, compiled by M. L. Bryan. Ann Arbor: Environmental Research Institute of Michigan, 1973.

*Regional Remote Sensing Bibliography*. Bangkok, Thailand: ESCAP/UNDP Regional Remote Sensing Programme, 1988.

*Remote Sensing and Geoinformation Systems as Related to the Regional Planning of Health Services: A Bibliography*, compiled by L. D. Miller, M. L. Mattews, and C. L. Walthall. Chicago: CPL Bibliographies, 1981.

*Remote Sensing and Highway Transportation Planning: An Annotated Bibliography*, compiled by B. N. Haack. Ann Arbor: Environmental Research Institute of Michigan, 1975.

*Remote Sensing Applications for Urban and Regional Planning: An Annotated Bibliography of Literature, 1973–May 1975*, compiled by M. E. Carberry. College Station: Texas A&M University, Remote Sensing Center, 1975.

*Remote Sensing Bibliography for Archaeology and History*, compiled by A. G. Hahn. Denver, CO: Office of the State Archaeologist, 1978.

*Remote Sensing Bibliography for Earth Resources*, compiled by R. K. Llaverias. Springfield, VA: National Technical Information Service, 1970.

*Remote Sensing for Natural Resource, Environmental, and Regional Planning: A Bibliography with Abstracts*, compiled by A. S. Hundermann. Springfield, VA: National Technical Information Service, 1976.

*Remote Sensing/Global Change: A Special Bibliography*. Washington, DC: National Aeronautics and Space Administration, 1994.

*Remote Sensing in Geophysics: A Bibliography*, compiled by T. C. O'Callaghan. Falls Church, VA: General Pub. Services, 1977.

*Remote Sensing of Terrestrial Vegetation: A Comprehensive Bibliography*, compiled by P. F. Krumpe. Knoxville: University of Tennessee, 1972.

*Remote Sensing of the Ocean: A Bibliography with Abstracts*, compiled by R. J. Brown. Springfield, VA: NTIS, U.S. Department of Commerce, 1978.

*Remote Sensing of the Ocean Surface: A Bibliography with Abstracts*, compiled by E. A. Harrison. Springfield, VA: Reproduced by National Technical Information Service, 1973.

*RESORSfiche: A Bibliography and Index of Remote Sensing*. Ottawa, ON: Canada Centre for Remote Sensing, 1984.

*Selected Bibliography: General Remote Sensing References*. Sioux Falls, SD: EROS Data Center, 1978.

*Selective Bibliography: Remote-Sensing Applications in Land-Use and Land-Cover Inventory Tasks*, compiled by W. J. Todd. Sioux Falls, SD: Technicolor Graphic Services, 1978.

*Selected Bibliography: Remote Sensing Applications in Wildlife*, compiled by D. M. Carneggie, D. O. Ohlen, and L. R. Pettinger. Sioux Falls, SD: U.S. Geological Survey, 1980.

*Selected Bibliography of Remote Sensing in Meteorology*, compiled by J. Kingwell and R. Ward. Melbourne, Australia: Bureau of Meteorology, 1986.

*Selected Bibliography of the Interpretation of Aerial Photographs and on Aerial Photography and Related Subjects*, compiled by M. Parvis. Lafayette, IN: Purdue University, 1949.

*Selected Bibliography on Maps, Mapping, and Remote Sensing*. Reston, VA: U.S. Department of the Interior, Geological Survey, 1981.

*Small Format Aerial Photography: A Selected Bibliography*, compiled by W. H. Anderson and F. X. Wallner. Sioux Falls, SD: EROS Data Center, 1978.

*Space Based Remote Sensing for Agricultural Applications: January 1983–December 1989*, compiled by E. A. Brownlee. Beltsville, MD: National Agricultural Library, 1990.

*Specialists Involved in Remote Sensing in Alberta and a Bibliography of Remote Sensing in Alberta*. Edmonton: Alberta Remote Sensing Center, Alberta Environment, 1979.

*Techniques and Application of Aerial Photography to Anthropology: A Bibliography*, compiled by L. Kruckman. Monticello, IL: Council of Planning Librarians, 1972.

*Thermal Infrared Remote Sensing: A Bibliography*, compiled by J. Cihlar. Ottawa, ON: Canada Centre for Remote Sensing, 1976.

*University of Illinois Air Photo Repository Catalog*. Champaign: Committee on Aerial Photography of the University of Illinois, Urbana-Champaign, 1970.

*Urban and Regional Planning Utilization of Remote Sensing Data: A Bibliography and Review of Pertinent Literature*. Springfield, VA: Reproduced by National Technical Information Service, U.S. Dept. of Commerce, 1972.

*Use of Remote Sensing in Detecting and Analyzing Natural Hazards and Disasters, 1972–1998: A Partially Annotated Bibliography*, compiled by P. S. Showalter. San Marcos: James and Marilyn Lovell Center for Environmental Geography and Hazards Research, Department of Geography, Southwest Texas State University, 1999.

*Uses of Conventional Aerial Photography in Urban Areas: Review and Bibliography*, by A. S. Manji. Evanston, IL: Northwestern University, Department of Geography, 1968.

*World Remote Sensing Bibliographic Index: A Comprehensive Geographic Index Bibliography to Remote Sensing Site Investigations of Natural and Agricultural Resources throughout the World*, compiled by P. F. Krumpe. Fairfax, VA: Tensor Industries, 1976.

# 16
# Gazetteers

The gazetteer, also known as a geographical index, is an alphabetical listing of place names, or toponyms, and geographical features. Gazetteers are commonly used to locate the areas associated with the names. When place names have changed several times, an up-to-date gazetteer is the ideal resource for identifying locations. There are three styles of gazetteers: alphabetical list, dictionary, and encyclopedic.

The alphabetical list consists of place names and locations, often including geographic coordinates. The dictionary-style gazetteer includes geographic coordinates and descriptions of places as they relate to other places spatially. Sometimes demographic information is included, as well as a pronunciation guide, similar to a word dictionary. Encyclopedic gazetteers include similar information but also offer considerably more geographic detail, often in the form of articles. The latter two styles may concentrate on specific themes, such as county or state history, business, hydrology (that is, a gazetteer of rivers and lakes), parks, land ownership, and so forth. All gazetteers cover a specific geographic area—some are specific to counties, whereas others may cover the entire country. This chapter focuses on Canadian and American gazetteers at the national and regional levels.

The first gazetteer that covered Canada was the *North American and the West Indian Gazetteer*, initially published in London in 1759 with a second edition following in 1778; the guide included information on climate, soil, produce, and trade. Over the years, several gazetteers were published covering Upper Canada and Lower Canada, but it was not until 1851 that the firm John Lovell issued the *Dominion of Canada and Newfoundland Gazetteer and Classified Business Directory*, a publication updated many times over and still referred to today. Throughout the years, as cities developed and expanded, individual provincial gazetteers were published, followed by county gazetteers. Several Ontario county gazetteers were published in the mid- to late nineteenth century, such as the *Gazetteer and Directory of the County of Grey* (Toronto, 1865), *Gazetteer and Directory of the Counties of Kent, Lambton, and Essex* (Toronto, 1866), *Gazetteer and Directory of the County of Simcoe* (Toronto, 1866), and the *Gazetteer and Directory of the City of Brantford and County of Brant* (Hamilton, 1883). In the 1890s, gazetteers were published for British Columbia, Manitoba, and the Northwest Territories, followed by a gazetteer for the Maritime Provinces in 1911. Since then, gazetteers have been published for many other counties and areas in Canada, and they still prove to be popular resources for place names today.

An excellent print series resource is the *Concise Gazetteer of Canada*, which covers the entire country and offers correct names and specific locations of geographical areas using latitude and longitude coordinates. First produced in 1952 by the Canadian Permanent Committee on Geographical Names, Natural Resources Canada, it contains an alphabetical list of populated places, rivers, lakes, mountains, parks, and native reserves. The gazetteer is also available in individual provincial editions in softcover format in both English and French.

Canada's naming authority is the Geographical Names Board of Canada (GNBC), and the GNBC Secretariat is located at Natural Resources Canada. Today it offers online access to Canada's official names via the Canadian Geographical Names Data Base (http://geonames.nrcan.gc.ca/search/search_e.php). This online query tool provides many ways to search for a current or former official name in either English or French and to find names for physical features, populated places, and undersea features. One can search by geographical coordinates as well. Other popular Canadian online gazetteers include the following:

- BC Geographical Names Search
  http://apps.gov.bc.ca/pub/bcgnws/
- Canadian Islands
  http://www.acsu.buffalo.edu/~dbertuca/maps/cat/canada-islands.html
- Noms et lieux du Québec
  http://www.toponymie.gouv.qc.ca/ct/accueil.aspx
- Nova Scotia Gazetteer
  http://geonova.novascotia.ca/place-names

- NWT Place Names Search
  http://www.pwnhc.ca/cultural-places/geographic-names/database-of-nwt-geographic-names/
- Place Names in Nunavut
  http://ihti.ca/eng/place-names/pn-agreementform.html
- Ontario Locator
  http://www.geneofun.on.ca/ontariolocator
- Ontario Provincial Parks
  http://www.acsu.buffalo.edu/~dbertuca/maps/cat/ontario-prov-parks-list.html

The first significant American gazetteer, *The American Gazetteer*, was published in 1797 by Jedidiah Morse and consisted of over seven thousand articles describing the U.S. population. National gazetteers were infrequently published in the first three decades of the nineteenth century, and it was not until 1854 that the 1,364-page *A New and Complete Gazetteer of the United States* was printed by Baldwin and Thomas. In 1890 U.S. place names became standardized with the creation of the U.S. Board on Geographic Names. The board is comprised of federal agencies concerned with geographic information, population, ecology, and management of public lands, and it serves as a central authority for all name-related inquiries. Today, the U.S. Board on Geographic Names, in collaboration with the U.S. Geological Survey (USGS), offers access two almost two million physical features and populated places in the United States via the online Geographic Names Information System, or GNIS (https://geonames.usgs.gov). This database offers search results on feature type, elevation, estimated population, county, and geographic coordinates. Other popular online U.S. gazetteers include the following:

- United States Board on Geographic Names
  https://geonames.usgs.gov
- U.S. Gazetteer (1990, 2000, and 2010 Censuses)
  http://www.census.gov/geo/maps-data/data/gazetteer.html
- U.S. Home Town Locator
  http://www.HomeTownLocator.com/

The following selected list of national and regional gazetteers has been extracted from several library databases. Many are historical in nature, having been reproduced from microfilm. Because gazetteers represent the places that existed during a specific time, they are often used for historical studies. Therefore, the selection of resources below includes some of the very first gazetteers published for the regions. Readers may notice that, because settlements at the time were quite small, the historical gazetteers focus on describing the counties and their people. Throughout the years, as populations increased, gazetteers began leaving the town sketches out and only listed point-blank facts, like place names and geographic coordinates. The gazetteers that are published in print, listed below, are not equivalent to those made available online today. Whether at the national or regional levels, most printed gazetteers paint a picture of the geographical area at that time—the landowners, the businesses, and transportation routes that all play a role in defining that place. This is why it is important to list all gazetteers printed, especially those transcribed from microfilm, as those offer some of the richest sources of geographical information. In many cases, rich historical information can be derived from the printed advertisements alone. Since the depth of the publication is not always obvious from the title, some of the items listed include annotations to highlight those that offer more than a list of places.

In the compiled list below, readers will find many titles that include directories and business information. Some gazetteers focus on business and industry, listing information vital to the business community. A publication is often dedicated to a specific industry such as shipping and transportation, so some of the titles include shipper's directories; lake and river routes with proximity to railroad stations; sea, lake, and river ports; and so on. In some instances, the publications include agricultural information (listing vegetable growers) and locations of post, telegraph, and express offices. Some of the directories include not only the names of businesses but also the names of businesspeople, interest tables, census returns, and other statistical information that paints a picture of the times.

## CANADIAN NATIONAL GAZETTEERS

*Cambridge Gazetteer of the United States and Canada: A Dictionary of Places*, edited by A. Hobson. Cambridge, UK: Cambridge University Press, 2011.

    A comprehensive listing of over twelve thousand entries based on census data regarding economic, cultural, historic, and topographic sources. Includes coverage of urban neighborhoods, rural communities, lakes, rivers, ocean areas, mountains, swamps, historical sites, and more.

*Canadian Guide: Canada's Most Up-to-Date Gazetteer and Shipper's Directory: 1986 Annual, 121st Year, No. 1387*. Burlington, ON: Interguide, 1985.

*Gazetteer and Classified Business Directory of Canada Including Newfoundland, 1930*. Manotick, ON: Archive CD Books Canada, 2004.

    The gazetteer takes up seventy pages in the two-thousand-page book, with the rest devoted to a complete directory of Canadian businesses and professionals and their advertising. The business directory is indexed by trade and professional, making it easy to search for specific businesses.

*Gazetteer of Canada*. Ottawa, ON: Canadian Government Publication Centre, Department of Supply and Services, 1993.

*Gazetteer of Canada: Répertoire géographique du Canada. Colombie-Britannique*. Ottawa, ON: Geographical Services Division, Surveys and Mapping Branch, Department of Energy, Mines and Resources Canada, 1985.

*Lovell's Gazetteer of British North America: Containing the Latest and Most Authentic Descriptions of over 7,500 Cities, Towns, Villages and Places, in the Provinces of Ontario, Quebec, Nova Scotia, New Brunswick, Prince Edward Island, Manitoba, British Columbia, the North West Territories, and Newfoundland*, edited by P. A. Crossby. Montreal, QC: J. Lovell, 1881.

*Lovell's Gazetteer of British North America, 1873*, edited by P. A. Crossby. Milton, ON: Global, 1988.

This historical gazetteer contains descriptions of over six thousand cities, towns, and villages in the provinces of Ontario, Quebec, Nova Scotia, New Brunswick, Newfoundland, Prince Edward Island, Manitoba, British Columbia, and the Northwest Territories, as well as general information drawn from official sources. Also included is a table of routes, showing the proximity of the railroad stations and water ports to the cities and towns.

*McAlpines' Gazetteer and Guide of the Maritime Provinces, Nova Scotia, New Brunswick and Prince Edward Island: Containing General Information from Official and Other Sources of the Latest Date, an Historical Sketch of Each County and the Principal Cities . . . Routes of Steamers and Railways*. St. John, NB: D. McAlpine, 1985.

This 660-page historical gazetteer has fifty-five pages devoted to advertisements that includes hotels, stables, plumbers, druggists, typing colleges, notaries, and more.

*Settlements of Northern Canada: A Gazetteer and Index*, edited by R. J. Fletcher. Edmonton, AB: University of Alberta, Boreal Institute for Northern Studies, 1975.

A listing of settlements of the North West Territories, Yukon, Labrador, Nouveau Quebec, and the Hudson Bay/James Bay coast, including unofficially named English or Eskimo settlements, trading posts, and mining exploration camps.

*Smith's Canadian Gazetteer: Canada West (Ontario), 1846*, edited by W. H. Smith. Milton, ON: Global Heritage Press, 1999, 2013.

This gazetteer covers western Canada and includes information on first settlements, climate, and productions. It lists post offices, magistrates, ministers of various denominations, table of distances, list of hotels, boarding houses, banks, and government offices. This historical gazetteer includes place names listed both by its old name and its new name, or cross reference from one to the other.

## CANADIAN PROVINCIAL AND COUNTY GAZETTEERS

### Alberta

*Repertoire geographique du Canada: Alberta = Gazetteer of Canada: Alberta*. Ottawa, ON: Geographical Services Division, Canada Centre for Mapping, Department of Energy, Mines and Resources Canada, 1988.

### British Columbia

*British Columbia: Colombie-Britannique*. Ottawa, ON: Department of Energy, Mines and Resources, 1985.

*Gazetteer of Canada: British Columbia = Répertoire géographique du Canada: Colombie Britannique*. Ottawa, ON: Geographical Services Division, Surveys and Mapping Branch, Department of Energy, Mines and Resources, 1985.

### Manitoba

*Gazetteer of Canada: Cumulative Supplement to June 30, 1990 = Répertoire góographique du Canada. Manitoba (1981): supplément cumulatif au 30 juin, 1990*. Manitoba: Manitoba Geographical Names Program Surveys and Mapping Branch, 1990.

*Gazetteer of Canada: Manitoba*. Ottawa, ON: Canadian Government Publication Centre, Department of Supply and Services, 1993.

*Henderson's Manitoba and North West Territories Gazetteer and Directory*. Winnipeg, MB: Henderson Directory, 1884.

This gazetteer and directory lists addresses of citizens and businesses dating back to 1905, and includes information about their occupation, business address, and residential address. The directory provides information on banks, public buildings, clubs, schools, government offices, law courts, police departments, cemeteries, churches, and places of recreation and leisure.

### New Brunswick

*Gazetteer of Canada: New Brunswick*. 3rd ed. Ottawa, ON: Canada Centre for Mapping, 1994.

*Geographic Place Names of New Brunswick*, by P. Fulton. Fredericton, NB: Non-Entity, 1992.

### Nova Scotia

*Gazetteer of Canada: Nova Scotia*. Ottawa, ON: Canadian Government Publication Centre, Department of Supply and Services, 1993.

*Gazetteer with Indexed Map of the Province of Nova Scotia: Rand McNally & Co.'s Series, Showing the Railroads and the Express Company Doing Business over Each, Also Counties, Islands, Lakes & Rivers, Together with Every Post Office, Railroad Station or Town, Carefully Indexed, Referring to the Exact Location Where Each May Be Found on the Map*. Montreal, QC: Rand McNally, 1986.

### Ontario

*1861 Census Index and Gazetteer of Hasting County, Ontario*. Bolton, ON: L. Corupe, 2005.

*1865 Mitchell & Co.'s Gazetteer of the County of Lincoln*. St. Catharines: Ontario Genealogical Society, Niagara Peninsula Branch, 1994.

*1865 Mitchell & Co.'s Gazetteer of the County of Welland.* St. Catharines: Ontario Genealogical Society, Niagara Peninsula Branch, 1994.

*Arthur Township, Wellington County Directory & Gazetteer, 1867,* by D. Pike. Kitchener: Waterloo Wellington Branch, Ontario Genealogical Society, 1995.

Gazetteer and directory for Arthur Township, giving names, occupation, lot, concession, and post offices.

*Bruce County Gazetteer and Business Directory for 1880–81,* by W. W. Evans. Port Elgin, ON: Bruce County Genealogical Society, 2002.

*Charlton & Co's County of Kent Gazetteer, General & Business Directory for 1874–5.* Toronto: Kent County Branch, Ontario Genealogical Society, 1995.

*Charlton's Gazetteer, Business, Street & General Directory of the Town of Guelph for 1875–76.* Ottawa, ON: Canadian Institute for Historical Micro Reproductions, 1989.

*County of Lambton Gazetteer, Commercial Advertiser and Business Directory, 1864–5.* Sarnia: Ontario Genealogical Society, 1994.

*Directory of the Province of Ontario, 1857: with a Gazetteer,* by T. B. Wilson and E. S. Wilson. Lambertville, NJ: Hunterdon House, 1987.

The directory lists business and professional people, as well as tradespeople. A list of Native North Americans and a list of counties of Ontario and their seats is also included.

*Directory of Waterloo County 1864: Compiled from the County of Waterloo Gazetteer and General Business Directory for 1864,* by N. Huber. Kitchener: Waterloo Region Branch, Ontario Genealogical Society, 2003.

*Gazetteer and Directory of the County of Grey for 1865–6.* Toronto, ON: W. W. Smith, 1985.

*Gazetteer and Directory of the County of Simcoe: Including the District of Muskoka and the Townships of Mono and Mulmur for 1872–3.* Elmvale, ON: East Georgian Bay Historical Foundation, 1985.

*Gazetteer and Directory of the County of Simcoe (Ontario) 1872–1873,* by W. H. Irwin. Campbellville, ON: Global Heritage, 2006.

*Gazetteer of Canada: Ontario = Répertoire géographique du Canada. Ontario.* Ottawa, ON: Geographical Services Division, Canada Centre for Mapping (Ottawa), Department of Energy, Mines, and Resources Canada, 1988.

*Gazetteer of Ottawa County: A Work in Progress.* Holland, MI: P. Trap, 1997.

*Gazetteer with Indexed Map of Ontario . . . , Showing the Railroads and the Express Company Doing Business over Each: Also, Counties, Islands, Lakes & Rivers, Together with Every Post Office, Railroad Station or Town, Carefully Indexed, Referring to the Exact Location Where Each May Be Found on the Map.* Montreal, QC: Rand McNally, 1994.

*Master Index Wellington County Directory and Gazetteer 1867, Ontario, Canada: Townships of Amaranth, Arthur, Eramosa, Erin, Garafraxa, Guelph, Luther, Maryborough, Minto, Nichol, Peel, Pilkington and Oxford Gazetteer: Containing an Abstract of Each Census of the County of Oxford and of The Townships Comprising It; Carefully Compiled from the Original Abstracts: to Which Is Added a Large Map of the Count,* by T. S. Shenston. Hamilton, ON: C.W, 1852.

*Mitchell & Co.'s Gazetteer of the County of Welland, 1865.* St Catharines: Niagara Peninsula Branch, Ontario Genealogical Society, 1994.

*Ontario Central Places in 1871: A Gazetteer Compiled from Contemporary Sources,* by G. T. Bloomfield, E. Bloomfield, and N. B. Van. Guelph, ON: Department of Geography, University of Guelph, 1990.

*Perth County Gazetteer,* by L. Manktelow. Stratford: Ontario Genealogical Society, Perth County Branch, 1992.

*Puslinch.* Kitchener: Waterloo-Wellington Branch, Ontario Genealogical Society, 1997.

*Smith's Canadian Gazetteer,* by W. H. Smith. Toronto, ON: Book On Demand, 2013.

*Wellington County Directory and Gazetteer 1867, Ontario, Canada: Township of Eramosa.* Kitchener: Waterloo-Wellington Branch, Ontario Genealogical Society, 1997.

*Western Ontario Gazetteer and Directory, 1898–99: Containing That Portion of the Province West of, and Including, the City of Toronto and South of Georgian Bay.* Ingersoll: Ontario Pub. & Advertising, 1986.

### Prince Edward Island

*Répertoire géographique du Canada: Ile-du-Prince-Édouard = Gazetteer of Canada: Prince Edward Island.* Ottawa, ON: Geographical Services Division, Canada Centre for Mapping, Department of Energy, Mines and Resources Canada, 1988.

### Quebec

*Gazetteer, with Indexed Map of the Province of Quebec, Canada: Showing the Railroads, and the Express Company Doing Business over Each: Also, Counties, Islands, Lakes & Rivers, Together with Every Post Office, Railroad Station or Town, Carefully Indexed, Referring to the Exact Location Where Each May Be Found on the Map.* Montreal, QC: Rand McNally, 1994.

*Inuttitut Nunait Atingitta Katirsutauningit Nunavimi, Kupaimmi, Kanatami = Gazetteer of Inuit Place Names in Nunavik, Quebec, Canada = Répertoire toponymique inuit du Nunavik, Québec, Canada,* by L. Müller-Wille. Inukjuak, QC: Avataq Cultural Institut, 1987.

### Saskatchewan

*Gazetteer of Canada: Saskatchewan = Repertoire toponymique du Canada: Saskatchewan.* Ottawa, ON: Minister of Public Works and Government Services Canada, 1985, 1998.

## Yukon

*Gazetteer of Canada: Yukon = Répertoire géographique du Canada. Territoire du Yukon.* Ottawa, ON: Geographical Services Division, Canada Centre for Mapping, Department of Energy, Mines and Resources Canada, 1988.

## U.S. NATIONAL GAZETTEERS

*American Gazetteer, Exhibiting a Full and Accurate Account of the States*, by J. Morse. London: British Library, 2011.

*Cambridge Gazetteer of the United States and Canada: A Dictionary of Places*, by A. Hobson. Cambridge: Cambridge University Press, 1995, 2011.

This provides a comprehensive listing of over twelve thousand entries based on census data regarding economic, cultural, historic, and topographic sources. Includes coverage of urban neighborhoods, rural communities, lakes, rivers, ocean areas, mountains, swamps, historical sites, and more.

*Centennial Gazetteer of the United States: A Geographical and Statistical Encyclopedia of the States, Territories, Counties, Townships . . . Etc., in the American Union*, by A. Steinwehr. Ann Arbor, MI: Making of America, 2011.

*Columbia Gazetteer of North America*, edited by S. B. Cohen. New York: Columbia University Press, 2000.

This book offers fifty thousand entries on places in North America, including information on the continent's physical geography, political boundaries, economic descriptions, natural and agricultural resources, trade and industry activities, transportation routes, points of interests, population from the last census, and more.

*Complete Reference Gazetteer of the United States of North America*, by W. Chapin. Cleveland, OH: Book On Demand, 2013.

*Fanning's Illustrated Gazetteer of the United States: With the Population and Other Statistics from the Census of 1850: Illustrated with Seals and Thirty-One State Maps in Counties, and Fourteen Maps of Cities.* Westminster, MD: Heritage Books, 2004.

*Gazetteer of the United States (1854)*, by T. Baldwin and J. Thomas. Cinderford, UK: Archive CD Books Project, 2001.

This book contains statistics and information obtained from the 1850 Census, including the population of states, counties, and townships, as well as agricultural and other statistics on the counties.

*Gazetteer of the United States of America: With the Governments and Literary and Other Public Institutions of the Country; Also, Its Mineral Springs, Waterfalls, Caves, Beaches, and Other Fashionable Resorts*, by J. Hayward. Ann Arbor, MI: Making of America, 2000.

*Geographic Names Information System (GNIS): Concise Gazetteer*. Reston, VA: National Cartographic Information Center, 1985.

*Historical Gazetteer of the United States*, by P. T. Hellmann. New York: Taylor & Francis, 2004, 2013.

This book consists of a place-by-place chronology of U.S. history, offering the most important events that have occurred at any locality in the country.

*National Gazetteer: A Geographical Dictionary of the United States, Compiled from the Latest Official Authorities and Original Sources, Embracing a Comprehensive Account of Every State, Territory, County, City, Town and Village throughout the Union, with Populations from the Last National Census (1884)*, by L. Colange. Salt Lake City: Genealogical Society of Utah, 1990.

*National Gazetteer of the United States of America.* Washington, DC: U.S. Geological Survey, 1984, 1987, 1989, 1990, 1993.

*National Municipal Gazetteer*. Plattsburgh, NY: Target Exchange, 2000.

*New and Complete Statistical Gazetteer of the United States of America*, by R. S. Fisher. London: British Library, 2000, 2011.

*New Gazetteer of the United States of America*, by W. S. A. D. Darby. London: British Library, 2011.

*New Gazetteer or Geographical Dictionary of North America and the West*, by B. Davenport. London: British Library, 2011.

Includes a general description of North America and the United States, including all of the states, counties, cities, towns, villages, forts, seas, harbors, capes, rivers, canals, railroads, and mountains. Tables relating to commerce, population, revenue, debt, and various institutions are also included.

*New Universal Gazetteer or Geographical Dictionary (1823)*, by J. Morse. Cleveland, OH: Book On Demand, 2013.

*Omni Gazetteer of the United States of America: Volumes 1–11*. Detroit, MI: Omnigraphics, 1991.

This resource covers nearly 1,500,000 populated places, structures, facilities, historical places, and named geographic features in the United States. It is the largest repository of place names available in print. Every entry includes name, type of feature, population, ZIP code, county, USGS topographic map name, latitude and longitude coordinates, elevation, and sources of information.

*Postal Route Gazetteer*, by R. D. Harris. Fishkill, NY: Printer's Stone, 1992.

*Traveller's Guide, or, Pocket Gazetteer of the United States: Extracted from the Latest Edition of Morse's Universal Gazetteer: with an Appendix, Containing Tables of*

*Distances, Longitude and Latitude of Important Towns, and of the Population, Commerce, Revenue, Debt, and Various Institutions of the United States*, by J. Morse and R. C. Morse. New Haven, CT: N. Whiting, 1823.

*Universal Gazetteer: or, a New Geographical Dictionary*, by W. Darby. Salt Lake City: Genealogical Society of Utah, 2009.

## GAZETTEERS BY U.S. STATE AND REGION

### Alabama

*National Gazetteer: Interim Products; Alabama*. Reston, VA: U.S. Geological Survey, 1985.

### Arizona

*Arizona Gazetteer and Distance Guide*. Phoenix, AZ: N to the 4th Power, Ltd, 1992, 1996.

*National Gazetteer of the United States of America—Arizona*. Washington, DC: U.S. Geological Survey, 1987.

### Arkansas

*N4 Gazetteer and Distance Guide of Arkansas*. Phoenix, AZ: N to the 4th Power, 1996.

### California

*California Gazetteer*. Wilmington, DE: American Historical Publications, 1985.

*California's Geographic Names: A Gazetteer of Historic and Modern Names of the State*, by D. L. Durham. Clovis, CA: Word Dancer, 1998.

This book offers over fifty thousand geographical features including topographical features like ridges, peaks, canyons, and valleys; water features like streams, lakes, and waterfalls; and administrative features like cities and towns.

*Durham's Place Names of Greater Los Angeles: Includes Los Angeles, Orange and Ventura Counties*, by D. L. Durham. Clovis, CA: Word Dancer, 2000.

This gazetteer differs from the others in that it lists feature names but also provides information about who named the feature, when, and why. Alternate or obsolete names are also given. Although factual, it is also an entertaining read used by tourists, travelers, genealogists, historians, and geographers.

*N4 Gazetteer and Distance Guide of California*. Phoenix, AZ: N to the 4th Power, 1996.

*National Gazetteer of the United States of America—California*. Reston, VA: National Cartographic Information Center, 1987.

*State of California: Alphabetic Listing of Geographic Names*. Reston, VA: U.S. Geological Survey, 1991.

*Western Shore Gazetteer and Commercial Directory for the State of California: Yolo County: One Volume Being Devoted to Each County of the State, Giving a Brief History of Each County*, by C. P. Sprague and H. W Atwell. Salem, MA: Higginson Book, 1990.

### Colorado

*N4 Gazetteer and Distance Guide of Colorado*. Phoenix, AZ: N to the 4th Power, 1996.

*People and Places: Historical Gazetteer, Dictionary of Place Names, Weld County, Colorado, Pre-Historic Indians to 1992*, edited by C. R. Shwayder. Greeley, CO: Unicorn Ventures, 1992.

### Connecticut

*Gazetteer of the States of Connecticut and Rhode Island: Written with Care and Impartiality from Original and Authentic Materials, Consisting of Two Parts . . . with an Accurate and Improved Map of Each State*, by J. C. Pease and J. M. Niles. New Haven, CT: Yale University Microfilming Unit, 1989.

### Delaware

*National Gazetteer of the United States of America, Delaware 1983: USGS Professional Paper*. Reston, VA: U.S. Geological Survey, 1983, 2013.

*State of Delaware: Alphabetic Listing of Geographic Names*. Reston, VA: U.S. Geological Survey, 1990.

### Florida

*1887 Orange County, Florida Gazetteer and Business Directory*, by B. J. Stockton. Orlando: Central Florida Genealogical Society, 1999.

*Florida Gazetteer, Containing Also a Guide to and through the State: Complete Official Business Directory; State and National Statistics*, by J. M. Hawks. Salem, MA: Higginson Book, 1998.

*Florida Native: A Gazetteer*, by J. L. Saunders. Panama City, FL: Garage Band Books, 2003.

*Historical Gazetteer of Imperial Polk County, Florida*, by F. Nerod. Lake Alfred, FL: F. Nerod, 1986.

*Historical Gazetteer of Polk County, Florida*, by O. H. Wright. Bartow, FL: Polk County Historical Association, 2003.

*Key Names: A Gazetteer of the Islands of the Florida Keys*, by J. Clupper. Key West, FL: Monroe County Public Library, 2002.

This book is an alphabetical compilation of every place name for an island or key in the Florida Keys and includes a description of the evolution of each name based on Spanish sailing charts from the 1600s, sea captains' logs from the 1700s, and survey and expedition logs from the 1800s. The author additionally traced local lore and diaries to paint the history of the Key names.

*National Gazetteer of the United States of America—Florida 1992.* Washington, DC: USGPO, 1992.

*Orange County Gazetteer and Business Directory.* Orlando: University of Central Florida Libraries, John R. Richards, 2006.

*Orange County Gazetteer and Business Directory: Embracing a Resident Directory of Orlando, Sanford, Eustis, Kissimmee, Apopka, Longwood, Tavares, and Winter Park Together with a Classified Business Directory of the Entire County with a Sketch of the Various Towns and Other Useful Information, to Which Is Added a Complete List of the Orange and Vegetable Growers, a Post Office, Telegraph and Express Office Directory of Florida,* by B. J. Stockton. Orlando: Central Florida Genealogical Society, 1999.

## Georgia

*Gazetteer of the State of Georgia,* by A. Sherwood. Philadelphia: J. W. Martin and W. K. Boden, 1829, 2007.

*Gazetteer of the State of Georgia: Embracing a Particular Description of the Counties, Towns, Villages, Rivers, &c., and Whatsoever Is Usual in Geographies, and Minute Statistical Works, Together with a New Map of the State,* by A. Sherwood. Baltimore, MD: Clearfield, 2002.

## Hawaii

*National Gazetteer: Interim Products; Hawaii.* Reston, VA: U.S. Geological Survey, 1985.

*National Gazetteer of the United States of America—Hawaii.* Reston, VA: U.S. Geological Survey, 1987.

## Illinois

*All Name Index, Kane County Gazetteer, 1867: Geneva Township Only,* by J. N. Klinkey and J. C. W. Bailey. West Chicago, IL: Micro Data, 1989.

*County and Township Gazetteer,* by S. Kelly and L. Lovett. Springfield: Illinois State Archives, 1988.

*Gazetteer of Illinois in Three Parts, Containing a General View of the State, a General View of Each County,* by J. M. Peck. Bowie, MD: Heritage Books, 1993.

*Illinois Gazetteer.* Wilmington, DE: American Historical Publications, 1985.

*Illinois State Almanac, Diary and Gazetteer, 1884.* Burlington, VT: Wells, Richardson, 1884.

*Knox County Gazetteer and Farmer's and Land Owner's Directory: to Which Is Added a Post Office, Telegraph and Express Office Directory of Illinois.* Galesburg, IL: Knox County Genealogical Society, 1993.

## Indiana

*1850 Indiana Gazetteer: or, Topographical Dictionary of the State of Indiana,* by E. Chamberlain. West Lafayette, IN: HAR, 2005.

*Gazetteer of Limestone Mills of Owen, Monroe, and Lawrence Counties to 1950,* by C. W. Stuckey. Bedford, IN: Lawrence County Historical Genealogical Society, 2004.

*Gazetteer of the St. Joseph Valley, Michigan and Indiana, with a View of Its Hydraulic and Business Capacities,* by T. G. Turner. Salt Lake City: Genealogical Society of Utah, 1990.

*G.W. Hawes' Indiana State Gazetteer and Business Directory,* by J. Sutherland and G. W. Hawes. Salt Lake City: Genealogical Society of Utah, 1990.

*Indiana Gazetteer: or, Topographical Dictionary of the State of Indiana,* by E. Chamberlain. Salem, MA: Higginson Book, 1993.

*Indiana State Gazetteer and Shipper's Guide for 1866–7,* by M. V. B. Cowen. Salt Lake City: Genealogical Society of Utah, 1990.

*Monroe County, Indiana,* by R. Ryle. Evansville, IN: Ryle Publications, 2008.

*N4 Gazetteer and Distance Guide of Indiana.* Phoenix, AZ: N to the 4th Power, 1996.

*National Gazetteer of the United States of America—Indiana, 1987.* Washington, DC: U.S. Geological Survey, 1987.

*State of Indiana: Alphabetic Listing of Geographic Names.* Reston, VA: U.S. Geological Survey, 1990.

## Iowa

*N4 Gazetteer and Distance Guide, Iowa.* Phoenix, AZ: N to the 4th Power, 1995.

## Kansas

*Great Bend City Directory and Business Gazetteer,* by W. E. Stoke and J. D. Welch. Great Bend, KS: Barton County Historical Society, 1990.

*N4 Gazetteer and Distance Guide of Kansas.* Phoenix, AZ: N to the 4th Power, 1996.

*National Gazetteer of the United States of America: Kansas, 1984.* Washington, DC: U.S. Geological Survey, 1985.

*State of Kansas: Alphabetic Listing of Geographic Names.* Reston, VA: U.S. Geological Survey, 1990.

## Kentucky

*George W. Hawes' Kentucky State Gazetteer and Business Directory: for 1859 and 1860,* by G. W. Hawes. Vine Grove, KY: Ancestral Trails Historical Society, 2006.

*Kentucky Gazetteer: Identifying over 6,500 Towns Which Have Existed within the Present State of Kentucky,* by J. C. Gioe. Indianapolis, IN: Researchers, 1993.

*Kentucky State Gazetteer and Business Directory: 1895*, by R. L. Polk. Vine Grove, KY: Ancestral Trails Historical Society, 2006.

*Kentucky State Gazetteer and Business Directory: 1896, Vol. 7*, by R. L. Polk. Vine Grove, KY: Ancestral Trails Historical Society, 2000.

## Louisiana

*N4 Gazetteer and Distance Guide of Louisiana*. Phoenix, AZ: N to the 4th Power, 1996.

*National Gazetteer: Interim Products; Louisiana*. Reston, VA: U.S. Geological Survey, 1985.

## Maine

*Gazetteer of the State of Maine with Numerous Illustrations*, by G. J. Varney. Boston: B. B. Russell, 1882, 1991.

*Maine State Almanac, 1884: Diary and Gazetteer*. Burlington, VT: Wells, Richardson, 1884.

*National Gazetteer: Interim Products; Maine*. Reston, VA: U.S. Geological Survey, 1985.

## Maryland

*Gazetteer of Maryland*. Washington, DC: Library of Congress Photoduplication Service, 1988.

*Gazetteer of Old, Odd & Obscure: Place Names of Frederick County, Maryland*, by L. B. O'Donoghue. Frederick, MD: Historical Society of Frederick County, 2008.

*Gazetteer of the State of Maryland*, by R. S. Fisher. Reston, VA: Book On Demand, 2013.

## Massachusetts

*Gazetteer of Berkshire County, Mass., 1725–1885*, by H. Child. Salt Lake City: Genealogical Society of Utah, 1985.

*Gazetteer of Hampshire County, Mass., 1654–1887*, by W. B. Gay. Ann Arbor, MI: University Microfilms International, 1998.

*Gazetteer of Hydrologic Characteristics of Streams in Massachusetts: Coastal River Basins of the South Shore and Buzzards Bay*, by S. W. Wandle and M. A. Morgan. Boston: U.S. Department of the Interior, Geological Survey, 1985.

*Gazetteer of Massachusetts: Containing Descriptions of All the Counties, Towns and Districts in the Commonwealth; Also, of Its Principal Mountains, Rivers, Capes, Bays, Harbors, Islands, and Fashionable Resorts*, by J. Hayward. Washington, DC: Library of Congress Photoduplication Service, 1989.

*Gazetteer of the State of Massachusetts: With Numerous Illustrations*, by E. Nason and G. J. Varney. Bowie, MD: Heritage Books, 1998.

*Gazetteer of the Town of Dennis, Massachusetts: Incorporated from the East Precinct of Old Yarmouth, 1793*. South Dennis, MA: Dennis Historical Society, 2013.

*Historical and Statistical Gazetteer of Massachusetts: with Sketches of the Principal Events from Its Settlement; a Catalogue of Prominent Characters, and Historical and Statistical Notices of the Several Cities and Towns, Alphabetically Arranged. With a New Map of the State*, by J. Spofford. Washington, DC: Library of Congress Photoduplication Service, 1989.

*Massachusetts Gazetteer*. Wilmington, DE: American Historical Publications, 1985.

## Michigan

*Allegan County Historical Atlas and Gazetteer*, edited by K. Lane. Douglas, MI: Pavilion, 1998.

*Gazetteer of the St. Joseph Valley, Michigan and Indiana*, by T. G. Turner. Reston, VA: Book On Demand, 2013.

*Gazetteer of the St. Joseph Valley, Michigan and Indiana, with a View of Its Hydraulic and Business Capacities*, by T. G. Turner. Salt Lake, UT: Genealogical Society of Utah, 1990.

*Gazetteer of the State of Michigan*, by J. T. Blois. New York: Book On Demand, 2013.

*Gazetteer of the State of Michigan: In Three Parts; with a Succinct History of the State, from the Earliest Period to the Present Time; with an Appendix, Containing the Usual Statistical Tables and a Directory for Emigrants*, by J. T. Blois. Ann Arbor, MI: University Microfilms International, 1988.

*Historical Gazetteer of Newaygo County, Michigan*, by R. S. Taylor. Newaygo, MI: R. S. Taylor, 1985.

*Michigan Gazetteer*. Wilmington, DE: American Historical Publications, 1992.

*Michigan State Gazetteer and Business Directory of Windsor, Walkerville and Ford, Ontario, 1921–22*. Salt Lake City: Genealogical Society of Utah, 1987.

*N4 Gazetteer and Distance Guide of Michigan*. Phoenix, AZ: N to the 4th Power, 1996.

*National Gazetteer: Interim Products; Michigan*. Reston, VA: U.S. Geological Survey, 1985.

## Minnesota

*Gazetteer of Minnesota Railroad Towns, 1861–1997*, by H. Leighton. Roseville, MN: Park Genealogical Books, 1998.

*Minnesota Gazetteer and Distance Guide: Geography Book of Minnesota*. Phoenix, AZ: N to the 4th Power, 1998.

*N4 Gazetteer and Distance Guide of Minnesota*. Phoenix, AZ: N to the 4th Power, 1996.

*National Gazetteer: Interim Products; Minnesota*. Reston, VA: U.S. Geological Survey, 1985.

*State of Minnesota: Alphabetic Listing of Geographic Names*. Reston, VA: U.S. Geological Survey, 1990.

## Mississippi

*N4 Gazetteer and Distance Guide of Mississippi.* Phoenix, AZ: N to the 4th Power, 1996.

*National Gazetteer: Interim Products; Mississippi.* Reston, VA: U.S. Geological Survey, 1985.

## Missouri

*1890–91 Springfield City Directory & Greene County Gazetteer and 1896 Map of Springfield, Missouri.* Springfield, MO: Greene County Archives and Records Center, 1999.

*Discovering Historic Clay County, Missouri: An Illustrated Gazetteer.* Liberty, MO: Clay County Archives and Historical Library, 2014.

*Gazetteer of Missouri,* by R. A. Campbell. Reston, VA: Book On Demand, 2014.

*Gazetteer of the State of Missouri,* by A. Wetmore. Baltimore, MD: Clearfield, 2013.

*Gazetteer of the State of Missouri: With a Map of the State from the Office of the Surveyor-General, Including the Latest Additions and Surveys: To Which Is Added, an Appendix, Containing Frontier Sketches and Illustrations of Indian Character,* by A. Wetmore. Santa Maria, CA: Janaway, 2012.

*Missouri as It Is in 1867: An Illustrated Historical Gazetteer of Missouri, Embracing the Geography, History, Resources and Prospects, the Mineralogical and Agricultural Wealth and Advantages, the Population, Business Statistics, Public Institutions, Etc. of Each County in the State. The New Constitution, the Emancipation Ordinance, and Important Facts Concerning "Free Missouri." An Original Article on Geology, Mineralogy, Soils, Etc. by Prof. G.C. Swallow. Also Special Articles on Climate, Grape Culture, Hemp, and Tobacco: Illustrated with Numerous Original Engravings,* by H. H. McEvoy and J. Sutherland. New York: Milward, 2009.

*Missouri Geological Survey Gazetteer Listing by County.* Rolla: Missouri Geological Survey, 1991.

*N4 Gazetteer and Distance Guide of Missouri.* Phoenix, AZ: N to the 4th Power, 1996.

*National Gazetteer: Interim Products; Missouri.* Reston, VA: U.S. Geological Survey, 1985.

*Polk's Missouri State Gazetteer and Business Directory 1893–1894.* Salt Lake City: Genealogical Society of Utah, 1987.

*State of Missouri: Alphabetic Listing of Geographic Names.* Reston, VA: U.S. Geological Survey, 1991.

## Nebraska

*N4 Gazetteer and Distance Guide of Nebraska.* Phoenix, AZ: N to the 4th Power, 1996.

## Nevada

*National Gazetteer of the United States of America—Nevada.* Reston, VA: National Cartographic Information Center, 1987.

*Nevada State Gazetteer and Business Directory, 1907–08.* Salt Lake City: Genealogical Society of Utah, 2001.

*U.S. Geological Survey Geographic Names Information System for Nevada, Also Known as the Nevada Gazetteer.* Reston, VA: United States Geological Survey, 1998.

## New England

*New England Gazetteer: Containing Descriptions of All the States, Counties and Towns in New England,* by J. Hayward. Whitefish, MT: Kessinger, 2008.

*New England Gazetteer: Containing Descriptions of All the States, Counties and Towns in New England: Also Descriptions of the Principal Mountains, Rivers, Lakes, Capes, Bays, Harbors, Islands, and Fashionable Resorts within That Territory: Alphabetically Arranged,* by J. Hayward. Bowie, MD: Heritage Books, 1988, 1997.

## New Hampshire

*Gazetteer of Cheshire County, N.H., 1736–1885,* by H. Child. Salem, MA: Higginson Book, 1997.

*Gazetteer of Grafton County, N.H. 1709–1886,* by H. Child. Salem, MA: Higginson Book, 1990.

*Gazetteer of New Hampshire: Containing Descriptions of All the Counties, Towns, and Districts in the State, Also of Its Principal Mountains, Rivers, Waterfalls, Harbors, Islands, and Fashionable Resorts: To Which Are Added Statistical Accounts of its Agriculture, Commerce and Manufactures, with a Great Variety of Other Useful Information,* edited by J. Hayward. Bowie, MD: Heritage Books, 1993.

*Gazetteer of the State of New-Hampshire,* by J. Farmer and J. B. Moore. Concord, NH: D. L. Guernsey, 1997, 2014.

*Gazetteer of the State of New-Hampshire in Three Parts,* by E. Merrill and P. Merrill. Bowie, MD: Heritage Books, 1987.

*Statistics and Gazetteer of New-Hampshire,* by A. J. Fogg. Concord, NH: D. L. Guernsey, 2014.

## New Jersey

*Gazetteer of the State of New Jersey,* by T. F. Gordon. New York: Heritage Books, 2013.

*Gazetteer of the State of New Jersey: Comprehending a General View of Its Physical and Moral Condition, Together with a Topographical and Statistical Account of Its Counties, Towns, Villages, Canals, Rail Roads, Etc.,* by T. F. Gordon. Bowie, MD: Heritage Books, 1997.

*National Gazetteer of the United States of America, New Jersey 1982: USGS Professional Paper 1200-nj.* Reston, VA: U.S. Geological Survey, 2013.

*State of New Jersey: Alphabetic Listing of Geographic Names*. Reston, VA: U.S. Geological Survey, 1990.

*Sussex County ... a Gazetteer*, by W. T. McCabe. Newton, NJ: Historic Preservation Alternatives, Inc, 2009.

*Waterways of Camden County: A Historical Gazetteer*, by W. R. Farr. Camden, NJ: Camden County Historical Society, 2002.

### New Mexico

*N4 Gazetteer and Distance Guide of New Mexico*. Phoenix, AZ: N to the 4th Power, 1996.

*National Gazetteer: Interim Products; New Mexico*. Reston, VA: U.S. Geological Survey, 1985.

*National Gazetteer of the United States of America—New Mexico*. Reston, VA: National Cartographic Information Center, 1987.

*State of New Mexico: Alphabetic Listing of Geographic Names*. Reston, VA: U.S. Geological Survey, 1990.

### New York

*Characteristics of New York State Lakes, Gazetteer of Lakes, Ponds, and Reservoirs*, by J. M. Swart and J. A. Bloomfield. Albany: Lakes Assessment Section, Division of Water, New York State Department of Environmental Conservation, 1985.

*Fredonia Block Directory 1873 Based on Child's Gazetteer*, by H. Child. Fredonia, NY: D. H. Shepard, 1987.

*Gazetteer and Biographical Record of Genesee County, N.Y., 1788–1890*. Ann Arbor, MI: University Microfilms International, 1987.

*Gazetteer and Business Directory of Albany & Schenectady Co., N.Y. for 1870–71*, by H. Child. Ann Arbor, MI: University Microfilms International, 1989.

*Gazetteer and Business Directory of Allegany County, N.Y. for 1875*, by H. Child. Ann Arbor, MI: University Microfilms International, 1985.

*Gazetteer and Business Directory of Broome and Tioga Counties, N.Y. for 1872–3*, by H. Child. Ann Arbor, MI: University Microfilms International, 1985.

*Gazetteer and Business Directory of Cattaraugus County, 1874–75*, by H. Child. Ann Arbor, MI: University Microfilms, 1987.

*Gazetteer and Business Directory of Cayuga County, N.Y. for 1867–8*, by H. Child. Bowie, MD: Heritage Books, 1989.

*Gazetteer and Business Directory of Columbia County*, by C. Hamilton. Syracuse, NY: Book On Demand, 2013.

*Gazetteer and Business Directory of Cortland County, N.Y. for 1869*, by C. Hamilton. New York: Book On Demand, 2013.

*Gazetteer and Business Directory of Genesee County, N.Y. for 1869–70*, by H. Child. Washington, DC: Library of Congress Photoduplication Service, 1990.

*Gazetteer and Business Directory of Lewis County, N.Y. for 1872–3*, by H. Child. Ann Arbor, MI: University Microfilms International, 1987.

*Gazetteer and Business Directory of Madison County, N.Y. for 1868–9*, by H. Child. Ann Arbor, MI: University Microfilms International, 1987.

*Gazetteer and Business Directory of Monroe County, N.Y. for 1869–70*, by H. Child. Ann Arbor, MI: University Microfilms International, 1985.

*Gazetteer and Business Directory of Montgomery and Fulton Counties, N.Y. for 1869–70*, by H. Child. Ann Arbor, MI: University Microfilms International, 1987.

*Gazetteer and Business Directory of Niagara County for 1869*, by H. Child. Ann Arbor, MI: University Microfilms International, 1985.

*Gazetteer and Business Directory of Oneida County, N.Y. for 1869*, by H. Child. Arbor, MI: University Microfilms International, 1987.

*Gazetteer and Business Directory of Onondaga County, N.Y. for 1868–9*. Ann Arbor, MI: University Microfilms International, 1986.

*Gazetteer and Business Directory of Ontario County, N.Y. for 1867–8*, by H. Child. Ann Arbor, MI: University Microfilms International, 1985.

*Gazetteer and Business Directory of Orleans County, N.Y. for 1869*, by H. Child. Ann Arbor, MI: University Microfilms International, 1987.

*Gazetteer and Business Directory of Oswego County, N.Y. for 1866–7*, by H. Child. Ann Arbor, MI: University Microfilms International, 1987.

*Gazetteer and Business Directory of Rensselaer County N.Y. for 1870–71*, by C. Hamilton. New York: Book On Demand, 2013.

*Gazetteer and Business Directory of Saratoga County, NY and Queensbury, Warren County for 1871*, by H. Child. Bowie, MD: Heritage Books, 2001.

*Gazetteer and Business Directory of Seneca County, N.Y. for 1867–8*, by H. Child. Ann Arbor, MI: University Microfilms International, 1987.

*Gazetteer and Business Directory of St. Lawrence County, N.Y. for 1873–4*, by H. Child. Ann Arbor, MI: University Microfilms International, 1987.

*Gazetteer and Business Directory of Steuben County, N.Y. for 1868–9*, by H. Child. Ann Arbor, MI: University Microfilms International, 1985.

*Gazetteer and Business Directory of Tompkins County, N.Y. for 1868*, by H. Child. Ann Arbor, MI: University Microfilms International, 1985.

*Gazetteer and Business Directory of Ulster County, N.Y. for 1871–2*, by H. Child. Bowie, MD: Heritage Books, Inc., 2001.

*Gazetteer and Business Directory of Wyoming County, New York, 1870–71*, by H. Child. Interlaken, NY: Heart of the Lakes, 1994.

*Gazetteer and Directory of Franklin and Clinton Counties: With an Almanac for 1862–3, Embracing the Names of Business Men, County Officers, Distances, Interest Tables, Census Returns, and Much Other Valuable Statistical Information*, by H. Child. Ann Arbor, MI: University Microfilms International, 1987.

*Gazetteer of City Property 1991*. New York: Department of City Planning, 1991.

*Gazetteer of Monroe County, N.Y., for 1869–70: Reprint of 1869 Edition of the Gazetteer Section of Child's Gazetteer and Business Directory*, by H. Child and C. S. Harmon. Sarasota, FL: Aceto Bookmen, 2001.

*Gazetteer of the County of Washington, NY: Comprising a Correct Statistical and Miscellaneous History of the County and Several Towns, from Their Organization to the Present Time*, by A. Corey. Salt Lake City: Genealogical Society of Utah, 1993.

*Gazetteer of the Eighth (1860) Census of New York State*, by D. P. Davenport. Syracuse: Central New York Genealogical Society, 1986.

*Gazetteer of the State of New York*, by J. H. French. Berwyn Heights, MD: Heritage Books, 2007.

*Gazetteer of the State of New-York*, by H. G. Spafford. London: British Library, 2011.

*Gazetteer of the State of New York: Comprehending Its Colonial History, General Geography, Geology, and Internal Improvements, Its Political State, a Minute Description of Its Several Counties, Towns, and Villages, Statistical Tables, Exhibiting the Area, Improved Lands, Population, Stock, Taxes, Manufactures, Schools, and Cost of Public Instruction, in Each Town: with a Map of the State, and a Map of each County, and Plans of the Cities and Principal Villages*, by T. F. Gordon. Salem, MA: Higginson Book, 1990.

*Gazetteer of the State of New York: Comprehending Its Colonial History; General Geography, Geology, and Internal Improvements; Its Political State; a Minute Description of Its Several Counties, Towns and Villages, with a Map of the State, and Plans of the Cities and Principal Villages*. Salt Lake City: Genealogical Society of Utah, 1987.

*Gazetteer of the State of New York: Embracing a Comprehensive View of the Geography, Geology, and General History of the State, and a Complete History and Description of Every County, City, Town, Village and Locality with Full Table of Statistics*, by J. H. French. Ann Arbor, MI: University Microfilms International, 1988.

*Gazetteer of the State of New York: Reprinted with an Index of Names Compiled by Frank Place; Two Volumes in One*, by J. H. French. Baltimore, MD: Genealogical Pub., 1994.

*Historical Gazetteer 1785–1888 and Directory, 1887–1888 of Tioga County New York: Volume One*, by W. B. Gay. Tioga, NY: On Demand, 2008.

*Historical Gazetteer and Biographical Memorial of Cattaraugus County, N.Y*, by W. Adams. Boston: New England Historic Genealogical Society, 1988, 1999.

*Historical Gazetteer of Steuben County, New York: With Memoirs and Illustrations*, by M. F. Roberts. Salem, MA: Higginson Book, 1994.

*Historical Gazetteer of Tioga County, New York*, by W. B. Gay. Tioga, NY: On Demand, 2013.

*Historical Gazetteer of Tioga County, New York, 1785–1888: Followed by a Directory of Tioga County, New York, 1887–1888. All Having Been Published Previously by W.B. Gay & Co., Syracuse, New York, 1887. With Added Index to Historical Gazetteer*, by W. B. Gay. Owego, NY: Tioga County Historical Society, 1985.

*New Gazetteer and Business Directory for Livingston County, N.Y. for 1868*, by G. E. Stetson. New York: Book On Demand, 2013.

*New York Gazetteer*. Wilmington, DE: American Historical Publications, 1985.

*New York State Almanac, Diary and Gazetteer*, by C. E. Allen. Burlington, VT: Wells, Richardson, 1884.

## North Carolina

*Brunswick County Gazetteer*, by C. M. Paty. Southport, NC: Southport Historical Society, 1994.

*Gazetteer of Buncombe County: Contains Full Line of Churches, High Schools, Societies, &c*. Asheville, NC: Old Buncombe County Genealogical Society, 2000.

*Granville County, North Carolina Gazetteer*, by L. F. Dean. Raleigh, NC: Leonard F. Dean, 2011.

*National Gazetteer: Interim Products; North Carolina*. Reston, VA: U.S. Geological Survey, 1985.

*North Carolina Gazetteer*, by W. S. Powell. Chapel Hill: University of North Carolina Press, 1987.

*North Carolina Gazetteer, 2nd ed: Dictionary of Tar Heel Places and Their History*, by W. S. Powell and M. Hill. Chapel Hill: University of North Carolina Press, 2010.

## Ohio

*Gazetteer and Directory of Clermont County, Ohio, 1882*, by A. P. Bancroft. Batavia, OH: Clermont County Historical Society, 2004.

*Gazetteer of Montgomery County, Ohio*, by J. M. Overton. Dayton: Ohio Genealogical Society, 1998.

*Gazetteer of Ohio Streams*, by J. C. Krolczyk and V. Childress. Columbus, OH: Department of Natural Resources, Division of Water, 2001.

*Historical Gazetteer of Wood County, Ohio*, by L. R. Fletcher. Evansville, IN: Whipporwill Publications, 1998.

*National Gazetteer: Interim Products; Ohio*. Reston, VA: U.S. Geological Survey, 1985.

*Ohio Gazetteer.* Wilmington, DE: American Historical Publications, 1985.

*Ohio Gazetteer,* by W. Jenkins. New York: Book On Demand, 2013.

*Ohio Gazetteer and Traveler's Guide: Containing a Description of the Several Towns, Townships and Counties, with Their Water Courses, Roads, Improvements, Mineral Productions,* by W. Jenkins. Salt Lake City: Genealogical Society of Utah, 1990.

*Ohio State Almanac, Diary and Gazetteer, 1884.* Burlington, VT: Wells, Richardson, 1884.

*Ohio State Gazetteer and Business Directory for 1860–61,* by G. W. Hawes. Ann Arbor, MI: Making of America, 1996.

*State of Ohio: Alphabetic Listing of Geographic Names.* Reston, VA: U.S. Geological Survey, 1991.

## Oklahoma

*N4 Gazetteer and Distance Guide of Oklahoma.* Phoenix, AZ: N to the 4th Power, 1996.

## Oregon

*From Abbott Butte to Zimmerman Burn: A Geographic-Names History and Gazetteer of the Rogue River National Forest,* by J. M. LaLande. Medford, OR: Rogue River-Siskiyou National Forest, 2007.

*N4 Gazetteer and Distance Guide of Oregon.* Phoenix, AZ: N to the 4th Power, 1996.

*Oregon and Washington Gazetteer and Business Directories, 1901–1904.* Portland, OR: R. L. Polk, 2007.

*Oregon Geographic Names Information System Alphabetical Listing.* Reston, VA: U.S. National Cartographic Information Center, 1987.

## Pennsylvania

*Gazetteer and Business Directory of Erie County, PA., 1873–1874,* by H. Child. Bowie, MD: Heritage Books, 2002.

*Gazetteer of Lebanon County, Pennsylvania, 2007,* by D. J. Bachman. Lebanon, PA: Lebanon County Historical Society, 2007.

*Gazetteer of the State of Pennsylvania,* by T. F. Gordon. Apollo, PA: Closson, 1999.

*Gazetteer of the State of Pennsylvania, etc.,* by T. F. Gordon. London: British Library, 2011.

*Gazetteer of York and Adams Counties, Pennsylvania.* York: South Central Pennsylvania Genealogical Society, 2004.

*Historical Gazetteer of Butler County, Pennsylvania,* by L. R. Eisler, G. C. McKnight, and J. Smith. Butler, PA: Butler Area Public Library, 2006.

*Pennsylvania Gazetteer.* Wilmington, DE: American Historical Publications, 1989.

*Pennsylvania Gazetteer of Streams.* Harrisburg, PA: U.S. Geological Survey, 1989, 2001.

*Pennsylvania State Almanac, Diary and Gazetteer.* Burlington, VT: Wells, Richardson, 1880.

*State of Pennsylvania: Alphabetic Listing of Geographic Names.* Reston, VA: U.S. Geological Survey, 1990.

## Rhode Island

*Gazetteer of the States of Connecticut and Rhode Island: Written with Care and Impartiality from Original and Authentic Materials, Consisting of Two Parts . . . with an Accurate and Improved Map of Each State,* by J. C. Pease and J. M. Niles. New Haven, CT: Yale University Microfilming Unit, 1989.

*Official Gazetteer of Rhode Island.* Ann Arbor, MI: University Microfilms International, 1988.

## South Carolina

*National Gazetteer: Interim Products; South Carolina.* Reston, VA: U.S. Geological Survey, 1985.

*South Carolina in the 1880s: A Gazetteer,* by J. H. Moore. Orangeburg, SC: Sandlapper, 1989.

*State of South Carolina: Alphabetic Listing of Geographic Names.* Reston, VA: U.S. Geological Survey, 1990.

## South Dakota

*Gazetteer of Lake County, South Dakota: Compiled in October and November, 1892,* by G. L. Houghton. Madison, SD: Lake County Historical Society, 2001.

*National Gazetteer of the United States of America, South Dakota 1989: USGS Professional Paper.* Reston, VA: USGS Geological Survey, 2013.

## Tennessee

*National Gazetteer: Interim Products; Tennessee.* Reston, VA: U.S. Geological Survey, 1985.

*State of Tennessee: Alphabetic Listing of Geographic Names.* Reston, VA: U.S. Geological Survey, 1991.

*Tennessee Gazetteer, or Topographical Dictionary: Containing a Description of the Several Counties, Towns, Villages, Post Offices, Rivers, Creeks, Mountains, Valleys, &c. in the State of Tennessee, Alphabetically Arranged: to Which Is Prefixed a General Description of the State, Its Civil Divisions, Resources, Population, &c. and a Condensed History from the Earliest Settlements down to the Rise of the Convention in the Year 1834: With an Appendix, Containing a List of the Practising Attorneys at Law in Each County, Principal Officers of the General and State Governments, Times of Holding Courts, and Other Valuable Tables,* by E. Morris. Santa Maria, CA: Janaway, 2007.

*Tennessee State Gazetteer and Business Directory for 1876–7*. Knoxville, TN: R. L. Polk, 1987.

## Texas

*N4 Gazetteer and Distance Guide of Texas*. Phoenix, AZ: N to the 4th Power, 1996.

*Biographical Gazetteer of Texas*. Austin, TX: Morrison, 1985.

*Gazetteer of Texas: USGS Bulletin 190*, by H. Gannett. Reston, VA: U.S. Geological Survey, 2013.

*National Gazetteer: Interim Products; Texas*. Reston, VA: U.S. Geological Survey, 1985.

*R.L. Polk & Co's Texas State Gazetteer and Business Directory, 1914–1915*. Galveston, TX: Galveston County Genealogical Society, 1993.

*Texas Gazetteer*. Wilmington, DE: American Historical Publications, 1985.

## Utah

*Counties & Communities in Utah: A Descriptive Gazetteer of Cities, Towns, Villages, Hamlets, Railroad Sidings and Resorts*. Salt Lake City, UT: G. W. Thornblom, 2000.

*Gazetteer of Utah, and Salt Lake City Directory*, by E. L. Sloan. Salt Lake City, UT: Salt Lake Herald Pub., 1869.

*N4 Gazetteer and Distance Guide of Utah*. Phoenix, AZ: N to the 4th Power, 1996.

*Wayne County Gazetteer Containing Directories of the Various Cities: Historical and Descriptive Sketches of the Several Townships of the County and a Directory of the Names, Occupations and Post Office Addresses of the Merchants, Manufacturers and Farmers throughout Wayne County to Which Are Added Complete Business Registers and Registers of City and County Organizations, Societies, Public Buildings, Etc.*, by C. W. Bailey. Salt Lake City: Genealogical Society of Utah, 1989.

## Vermont

*Alphabetical Atlas, or Gazetteer of Vermont: Affording a Summary Description of the State, Its Several Counties, Towns, and Rivers; Calculated to Supply, in Some Measure, the Place of a Map; and Designed for the Use of Offices, Travellers, Men of Business*, by J. Dean. Washington, DC: Library of Congress Photoduplication Service, 1989.

*Clarendon, Vermont from Child's Rutland County Gazetteer and Directory*, by H. Child. Fair Haven, VT: Reprinted by Sleeper Books, 2003.

*Gazetteer and Business Directory of Addison County, Vt., for 1881–82*, by H. Child. Salt Lake City: Filmed by the Genealogical Society of Utah, 1985.

*Gazetteer and Business Directory of Bennington County, Vt. for 1880–81*, by H. Child. Salem, MA: Higginson Book, 1997, 2003.

*Gazetteer and Business Directory of Rutland County, Vt., for 1881–82*, by H. Child. Boston: New England Historic Genealogical Society, 1986.

*Gazetteer and Business Directory of Windham County, Vt., 1724–1884*, by H. Child. Bowie, MD: Heritage Books, 2001.

*Gazetteer of Vermont: Containing Descriptions of All the Counties, Towns, and Districts in the State, and of Its Principal Mountains, Rivers, Waterfalls, Harbors, Islands, and Curious Places*, by J. Hayward. Bowie, MD: Heritage Books, 1990.

*Gazetteer of Washington County, Vt., 1783–1889*, by H. Child and W. Adams. Salem, MA: Higginson Book Company, 2009.

*History of the Town of Athens, Vermont: As Originally Published in Vermont Historical Gazetteer*, by A. M. Hemenway. Brandon, VT: Carrie E. H. Page, 1987.

*National Gazetteer: Interim Products; Vermont*. Reston, VA: U.S. Geological Survey, 1985.

*State of Vermont: Alphabetic Listing of Geographic Names*. Reston, VA: U.S. Geological Survey, 1990.

*Vermont Historical Gazetteer: A Local History of All the Towns in the State*, by A. M. Hemenway. Boston: New England Historic Genealogical Society, 2003.

*Vermont Historical Gazetteer: Embracing a History of Each Town, Civil, Ecclesiastical, Biographical and Military*, by A. M. Hemenway. Saco, ME: Toni W. Feeney, 1996, 2004.

## Virginia

*1815 Directory of Virginia Landowners (and Gazetteer): Albemarie, Amelia, Amherst, Buckingham, Charles City, Chesterfield, Cumberland, Dinwiddle, Fluvanna, Goochland, Hanover, Henrico, Louisa, Nelson, New Kent, Nottoway, Powhatan, Prince George, & the Independent Cities of Petersburg and Richmond*, by R. G. Ward. Athens, GA: Iberian Pub., 1997.

*1815 Directory of Virginia Landowners (and Gazetteer): Comprising the Counties of Alexandria County, Culpeper County, Fairfax County, Fauquier County, Frederick County, Independent City of Alexandria, Independent City of Fredericksburg, Independent City of Winchester, Loudoun County, Madison County, Orange County, Prince William County, Rockingham County, Shenandoah County, Spotsylvania County and Stafford County*, by R. G. Ward. Athens, GA: Iberian Pub., 1999.

*1815 Directory of Virginia Landowners (and Gazetteer): Comprising the Counties of Bedford, Brunswick, Campbell, Charlotte, Franklin, Greensville, Halifax, Henry, Lunenburg, Mecklenburg, Patrick, Pittsylvania, Prince Edward, Southampton, and Sussex*, by R. G. Ward. Athens, GA: Iberian Pub., 1997.

*1815 Directory of Virginia Landowners (and Gazetteer): Comprising the Following: Berkeley County, Brooke*

*County, Cabell County, Hampshire County, Hardy County, Harrison County, Jefferson County, Kanawha County, Mason County, Monongalia County, Ohio County, Pendleton County, Randolph County, Tyler County and Wood County*, by R. G. Ward. Athens, GA: Iberian Pub., 1999.

*City Directories of the United States, Virginia Gazetteer, 1906*. Woodbridge, CT: Primary Source Microfilm, 2002.

*Gazetteer of Virginia and West Virginia*, by H. Gannett. Baltimore, MD: Clearfield, 1998.

*Hill's Virginia Gazetteer and Classified Business Directory, 1911, for Grayson County, Virginia*. Saltville, VA: New River Notes Books, 2005.

*New and Comprehensive Gazetteer of Virginia, and the District of Columbia: . . . Collected and Compiled from the Most Respectable and Chiefly from Original Sources*, by J. Martin and W. H. Brockenbrough. Westminster, MD: Willow Bend Books, 2000.

*On the Frontier of Virginia & North & South Carolina: A Gazetteer of the First "Old West."* Chesapeake, OH: C. Eldridge, 1999.

*Statistical Gazetteer of the State of Virginia: Embracing Important Topographical and Historical Information from Recent and Original Sources, Together with the Results of the Last Census Population, in Most Cases to 1854*, by R. Edwards. Saltville, VA: J. Weaver/New River Notes, 2004.

## Washington

*Disturnell's Business Directory and Gazetteer of the West Coast of North America, 1882–1883*, by J. Disturnell. Salt Lake City: Genealogical Society of Utah, 2007.

*Oregon and Washington Gazetteer and Business Directories, 1901–1904*, by R. L. Polk. Salt Lake City: Genealogical Society of Utah, 2007.

## West Virginia

*Gazetteer of Communities in Harrison County, West Virginia with Extinct Towns*, by C. Eldridge. Athens, GA: Iberian Pub., 1995.

*Gazetteer of Virginia and West Virginia*, by H. Gannett. Baltimore, MD: Clearfield, 1998.

*Gazetteer of West Virginia*, by H. Gannett. Santa Maria, CA: Janaway, 2006.

*West Virginia Gazetteer of Physical and Cultural Place Names*. Morgantown: West Virginia Geological and Economic Survey, 1987.

*West Virginia Historical Almanac and Gazetteer*, by S. Clagg. Huntington, WV: John Deaver Drinko Academy, Marshall University, 2002.

## Wisconsin

*Gazetteer of the Communities, Churches & Cemeteries of Washington County, WI, 1840–2001*, by G. Wendelborn. West Bend, WI: Washington County Historical Society, 2002.

*N4 Gazetteer and Distance Guide of Wisconsin*. Phoenix, AZ: N to the 4th Power Ltd, 1996.

*Wisconsin Gazetteer*, by J. W. Hunt. Amherst Junction, WI: Bill Handrich Historicals, 2001.

*Wisconsin Gazetteer: Containing the Names, Location, and Advantages of the Counties, Cities, Towns, Villages, Post Offices and Settlements Together with a Description of the Lakes, Water Courses, Prairies and Public Localities in the State of Wisconsin*, by J. W. Brown and B. Brown. West Lafayette, IN: HAR, 2005.

*Wisconsin State Gazetteer and Business Directory (with Latest Map of the State) 1921–1922*. Salt Lake City: Filmed by the Genealogical Society of Utah, 1987.

# 17
# Citing Cartographic Material

Cartographic materials that have been used in research, referenced in a paper, or included in a publication or website need to be cited, crediting the author of the map, the data, and the GIS program, if applicable. There are varying citation guidelines depending on the type of cartographic material, whether it is a single work or part of series, whether it is part of a book or atlas, and whether it is a static map, interactive map, found online, or created in a GIS program. Often, in order to discover the essential elements the map itself will need to be scanned for information, such as title and scale, publisher, and date. When citing cartographic work, include the author of the map, the title of the map, the format (print map, software, Geospatial data), the edition of the map (if applicable), the scale of the map, the place of publication, the publisher, and the date. When citing online map and GIS data sources, include the URL and the date of the citation along with the publication date of the map.

There are many online resources and books in print that provide detailed examples of how maps, geospatial data, and GIS software should be cited. Many either follow or have adapted their recommended practices from the second edition of *Cartographic Citations: A Style Guide*, by Christine Kollen, Shawa Wangyal, and Mary Lynette Larsgaard, published by the Map and Geography Round Table, American Library Association in 2010, or from the Canadian version, *ACMLA Recommended Best Practices in Citation of Cartographic Materials*, by Alberta Auringer Wood and the Association of Canadian Map Libraries and Archives, published in 2012 (https://acmla-acacc.ca/docs/ACMLA_BestPracticesCitations.pdf). The Arthur H. Robinson Map Library of the University of Wisconsin-Madison (https://geography.wisc.edu/maplibrary/2016/09/23/guide-to-citing-geospatial-data-maps-or-atlases/) has a detailed list of map, GIS data, and GIS software citation examples. Other notable library guides worth a visit include the following:

- Brock University, "How to Reference Geospatial Data, Maps, Atlases, Air Photos" https://brocku.ca/library/wp-content/uploads/sites/51/MDG-How-to-Reference.pdf/
- Library of Congress, "Chicago: Maps and Charts" http://www.loc.gov/teachers/usingprimarysources/chicago.html
- MADGIC Library Queen's University, "Citation Guide for Maps" http://library.queensu.ca/webdoc/maps/citation.htm
- McMaster University, "Guide to Citing Maps and Atlases" https://library.mcmaster.ca/maps/cite
- Ohio Wesleyan University, "Citing Maps" http://library.owu.edu/friendly.php?s=citing-maps
- RMIT University, "Citing Printed Maps" http://rmit.libguides.com/c.php?g=336211&p=2262349
- University of Ottawa, "Citation Examples" https://library.carleton.ca/sites/default/files/help/writing-citing/citation-e.pdf
- University of Texas UTA Libraries, "Map Guide: Citing Maps" http://libguides.uta.edu/maps/citingmaps
- Western Washington University, "Citation Quick Guide and Style Manuals: Citing Maps" http://libguides.wwu.edu/c.php?g=308303&p=2063297

Because the citation style varies for each map type, having one place to look up citation guidelines and examples may be helpful to researchers. This chapter summarizes possible map formats and provides citation examples for each one.

## SINGLE SHEET MAP

### Format

Author. *Title* [format]. Edition. Scale. Place of publication: Publisher, Date.

## Examples

Goddard, Ives. *Native Languages and Language Families of North America* [map] Rev. and enlarged ed., with additions and corrections. 1:7,500,000. Washington, DC: Smithsonian Institute, 1996.

Manitoba Natural Resources, Surveys and Mapping Branch. *Manitoba: Municipalities Local Government Districts 1988* [map]. 3rd edition. 1:1,000,000. Winnipeg: Manitoba Natural Resources, Surveys and Mapping Branch, 1987.

National Geographic Society. *South Asia, with Afghanistan and Myanmar* [map]. 1:7,345,000, 1" = 116 miles. Washington, DC: National Geographic Society, May 1997.

U.S. Department of the Air Force. *U.S. Army Forces in WWII, 1941–1945* [map]. Scale not given. Washington, DC: Department of the Air Force, 1993.

## MAP IN A TOPOGRAPHIC SERIES

### Format

Author. *Sheet Title* [format]. Edition. Scale. Series title. Place of publication: Publisher, Date.

### Examples

Canada Department of Energy, Mines and Resources, Canada Centre for Mapping. *Brantford, Ontario* [map]. Edition 7. Canada 1:50,000, sheet 40P/1. Ottawa, ON: Canada Centre for Mapping, 1994.

Ontario Ministry of Natural Resources. *Ontario Base Maps* [map]. 1:10,000, sheet 10 17 5850 47900. Toronto, ON: Ministry of Natural Resources, 1983. U.S. Geological Survey. *Puzzle Mountain, Maine* [map]. 1:24,000. 7.5 Minute Series (Topographic). Reston, VA: USGS, 1977.

U.S. Geological Survey. *Raleigh West Quadrangle, North Carolina* [map]. Photo revised 1993. 1:24,000. 7.5 Minute Series. Reston, VA: United States Department of the Interior, USGS, 1999.

## MAP IN A SERIES (NONTOPOGRAPHIC)

### Format

Author. *Sheet Title* [format]. Edition. Scale. Series title. Place of publication: Publisher, Date.

### Examples

Easterbrook, Don J. *Geologic Map of Western Whatcom County, Washington* [map]. 1:62,500. Miscellaneous Investigations Series, map I–854-B. Reston, VA: U.S. Geological Survey, 1976.

Ontario Institute of Pedology. *Soils of Waterloo County, Regional Municipality of Waterloo, Ontario* [map]. Scale 1:25,000. Soils of the Regional Municipality of Waterloo, Ontario, Sheet 1. Waterloo, ON: The Institute, 1989.

Sado, E. V., and B. F. Carswell. *Surficial Geology of Northern Ontario* [map]. 1:1,200,000. Ontario Geological Survey Map 2518. Toronto: Ontario Geological Survey, 1987.

U.S. Geological Survey. *The North America Tapestry of Time and Terrain* [map]. 1:8,000,000. Geologic Investigations Series, I-2781. Reston, VA: U.S. Department of the Interior, USGS, 2003.

## FACSIMILE OR REPRODUCTION MAPS

### Format

Author. *Title* [format]. Scale. Place of Publication: Publisher, Date. As reproduced by Publisher, Date.

### Examples

Delisle, Guillaume. *Carte du Canada ou de la Nouvelle France et des découvertes qui y ont été faites* [facsimile]. 1:8,875,000. Paris: Guillaume Delisle, 1703. As reproduced by Association of Canadian Map Libraries, 1981.

Melish, John. *Map of the Seat of War in North America* [facsimile]. 1:9,875,100. Toronto, Ontario, 1813–1815. As reproduced by Association of Canadian Map Libraries and Archives, ACML Facsimile Map Series 144, 1993.

Reichard, C. G. *Nord-America, 1818* [facsimile]. 1:21,000,000. Gotha: Justus Perthes, 1831. As reproduced by Map Society of British Columbia, 1986.

Spencer, Jonathon. *18th Century North America* [facsimile]. 1 inch to 50 miles. London, England, 1721. As reproduced by Mark Nowak, 2017.

## AERIAL PHOTOGRAPHS

### Format

Source. *Title* [format]. Scale. Line/roll number. Photo number. Place of publication: Publisher, Date.

### Examples

Airborne Sensing Corp. *Hamilton Airport* [air photo]. 1:5,000. Line 4257/roll 15. Photos 98 and 99. Hamilton, ON: Hamilton-Wentworth, 1995.

UCLA Department of Geography. *Malibu* [air photo]. 1:30,000. Photo 17a. Los Angeles: University of California, Los Angeles, 1947.

United States Department of Agriculture Aerial Photography Field Office. *QT-3N-42* [air photo]. 1:21,000. Salt Lake City: USDA-FSA-APFO Aerial Photography Field Office, June 23, 1954.

## BIRD'S-EYE VIEW

**Format**

Author. *Title*. Scale. Place of publication: Publisher, Date.

**Examples**

Brosius, H. *Bird's Eye View of the City of Ottawa, Province, Ontario, Canada, 1876*. Scale not given. Chicago: Chas. Shober, 1876.

Gross, Peter Alfred. *Bird's-Eye View of Toronto, 1876*. Scale not given. Toronto, ON: Gross, 1876.

Penthouse Studios Inc. *Montreal in 3D: A Balloon's Eye View of North America's Most Exciting City, Montreal = Montréal en 3D: la ville la plus excitante d'Amérique du Nord, vue d'un ballon, Montréal*. 1:3,750. Montreal, QC: Penthouse Studios, 1966.

## SATELLITE IMAGERY

**Format**

Author. *Title*. Scale. Place of publication: Publisher, Date of image collection.

**Examples**

Google Inc. *Bloomington, Indiana*. Scale not given. Mountain View, CA: Google, 2014.

*Georgian Bay/Muskoka and Lake Simcoe from Space, Landsat 5*. 1:169 000. Missaussauga, ON: WorldSat International, July 21, 1986.

*Landsat-4 Thematic Mapper Image of Santa Barbara, California and Channel Islands, November 19, 1983: Spectral Bands .48um, .57um, .66um*. Scale not given. Santa Barbara, CA: EOSAT Santa Barbara Research Center, 1983.

Manitoba, Surveys and Mapping Branch. *Manitoba Landsat-1 Mosaic*. 1:000,000. Manitoba: Surveys and Mapping Branch, 1973–1974.

## ATLAS

**Format**

Author. *Title*. Edition. Place of publication: Publisher, Date.

**Examples**

Brown, Jeffery. *Atlas of the Grand River*. Ottawa: Geological Survey of Canada, 2017. Canada, Department of Energy, Mines and Resources. Surveys and Mapping Branch. *Atlas and Gazetteer of Canada*. Ottawa: The Queen's Printer, 1969.

Goode, J. Paul. *Goode's World Atlas*. 27th ed. Skokie, IL: Rand McNally, 2016.

National Geographic Society. *Atlas of the World*. Washington, DC: National Geographic Society, 1999.

## MAP IN A BOOK

**Format**

Map Author. *Map Title* [format]. Scale. In *Book Title*, by Book Author, page. Edition. Place of Publication: Publisher, Date.

**Examples**

Baum, Frank L. *The Yellow Brick Road* [map]. Scale not given. In *The Wizard of Oz*, by Frank L. Baum, 32. Kansas City: Munchkin, 1938.

Hulbert, Archer Butler. *Map of French Forts in America, 1750–60* [map]. Scale not given. In *History of the Niagara River*, by Archer Butler Hulbert, 165. Harrison, NY: Harbor Hill Books, 1978.

Smith, Dianna. *Land Use in York County* [map]. 1 cm = 1 m. In *Development of a Sleepy Town, by Dana Burford, 20*. Toronto, ON: Treetop Scholars Publishing, 2017.

## MAP IN A JOURNAL OR PERIODICAL

**Format**

Map Author, if known. *Map Title* [format]. Scale if known. In Article Author, "Article Title," *Journal Title* volume (year): page.

**Examples**

Crimpson, Josh. *Regional Cuisine of Poland* [map]. Scale unknown. In Jan Brick, "More Than Beets and Mushrooms," *Ethnic Bites* 24 (2014): 14.

*The Distribution of Canadian Multinational Headquarters in Ontario, 1992* [map]. 3.5 cm = 50 km. In Stephen P. Meyer, "Canadian Multinational Headquarters: The Importance of Toronto's Inner City," *Great Lakes Geographer* 3, no. 1 (1996): 7.

Jefferson, Louise. *Africa: A Friendship Map* [map]. Scale not given. In Glen Creason, "A Smile of Understanding," *Mercator's World* 4, no. 5 (September/October 1999): 13.

*Figure 1. Map of Landfills in San Diego* [map]. 1:400,000. In Jonathan Manes et al., "Online Map Design for Public-Health Decision Makers," *Cartographica* 66, no. 4 (Winter 2015): 96.

Verne, Jules. *The Bottom of the Sea* [map]. ½" = 20 leagues. In Jules Verne, "Fantastic Voyage," *Travel and Leisure 186 (1992)*: 127.

## GLOBE

### Format

Author. *Title*. Edition. Scale. Place of publication: Publisher, Date.

### Examples

George F. Cram & Co. *7 Inch Terrestrial Globe*. 1:71,723,520. "1 in. equals 1132 statute miles." Indianapolis: George F. Cram, 1934.

*Lunar Globe*. 1:41,800,000. Santa Clara, CA: Explore Technologies, 1997.

Tolman, LeRoy M. *Replogle 12 Inch Diameter Globe World Classic Series*. 1:41,849,600. Chicago: Replogle Globes, 1993.

## MAP IN AN ATLAS

### Format

Map Author. *Map Title* [format]. Scale. In *Atlas Title*, by Atlas Author, page. Edition. Place of publication: Publisher, Date. Page.

### Examples

Canada, Department of Energy, Mines and Resources, Surveys and Mapping Branch. *Relief* [map]. 1:2,000,000. In *The National Atlas of Canada*, by Canada Surveys and Mapping Branch, 1–2. 4th edition. Toronto: Macmillan Co. of Canada; Ottawa: Department of Energy, Mines and Resources, Information Canada, 1974, 1–2.

*Canada Population* [map]. 1:45,000,000. In *Canadian Oxford World Atlas*, 26. 4th Edition. Toronto: Oxford University Press, 1998.

*Hillsborough* [map]. Scale not given. In *Street Atlas of Raleigh, Durham, Chapel Hill and Vicinity: North Carolina*, by Universal Map, 26. Williamston, MI: Universal Map, 1995.

Küchler, A. W. *Natural Vegetation* [map]. 1:75,000,000. In *Goode's World Atlas*, 18. 21st ed. Skokie, IL: Rand McNally, 2005.

*Military Service* [map]. Scale not given. In *Women in the World: An International Atlas*, by Joni Seeger and Ann Olson, plate 32. London: Pluto, 1986.

*Sin City* [map]. 1:62,500. In *The Under-World Atlas*, by Dante Alighieri, 13. 2nd ed. Hades: Firestorm Press, 1298.

## MAP IN A THESIS OR DISSERTATION

### Format

Map Author if different from thesis or dissertation author. *Map Title*. Scale. In Author, "Thesis or Dissertation Title." Type of paper, Name of University, Date of issuance, page or map number if applicable.

### Examples

*Carte de localisation de la région à l'étude*. 1:2,500,000. In Martine Rocheleau, "Sédimentologie des paléoplages de la plaine d'Old Crow, territoire du Yukon, Canada," 6, Fig. 1.1. Masters thesis, University of Ottawa, 1997.

*Geology of the Hermitage Peninsula, Southern Newfoundland*. 1:50,000. In Cyril F. O'Driscoll, "Geology, Petrology and Geochemistry of the Hermitage Peninsula, Southern Newfoundland," Fig 1.2. M.Sc. thesis, Memorial University of Newfoundland, 1977.

*Map of U.S. Topography*. 1:63,360. In Donald S. Brown, "The Changing Elevation of Low Lying Areas," map 4. M.A. thesis, University of Santa Barbara, 2016.

## MAP PRODUCED USING GIS SOFTWARE

### Format

*Map Title*. [format]. Scale. Database name [type of medium]. Place of Publication: Publisher, Year. Using Title of Software [type of software].

### Examples

*California Railway Network* [map]. 1:25,000. National Transportation Atlas Databases [GIS]. Washington, DC: U.S. Department of Transportation, 2000. Using ArcGIS Version 10.0 [GIS software].

*Great Lakes Bathymetry* [map]. 1:15,000. North American Bathymetric Data [GIS]. Washington, DC: NOAA, 2015. Using ArcGIS Version 10.1 [GIS software].

*Percentage of Total Population over 40 Years Old* [map]. 1" = 15 miles approx. Census of Canada 2016 [GIS]. Ottawa, ON, 2016. Using ArcGIS Version 10.1 [GIS software].

*Wetlands and Floodplain in GRCA* [map] 1:10,000. GRCA GIS Data [GIS]. Cambridge, ON: Grand River Conservation Authority, 2011. St. Catharines, ON: John Doe, March 2008. Using ArcView Version 3.2 [GIS software].

## STATIC MAP ON THE WEB

### Format

Author if known. *Map Title* [format]. Date of map creation if known. Scale. "Title of the Web Site." Date posted if known. URL (Date accessed).

### Examples

*Delaware, Ohio* [map]. 1885. Scale not given. "Sanborn Fire Insurance Maps, 1867–1970—Ohio." OhioLINK Digital Media Center. April 19, 1016. http://dmc.ohiolink.edu

/mrsid/bin/viewmap.pl?client=Sanborn& image (May 2, 2016).

North Carolina Dept. of Agriculture. *Agriculture Overview* [map]. 1:50,000. "North Carolina Department of Agriculture and Consumer Services." Last updated September 2015. http://www.ncagr.com/stats/general/general1.htm (December 20, 2017).

*Russia* [map]. 1:500,000. "The World Factbook 2: Afghanistan." October 7, 2015. http://www.odci.gov/cia/publications/factbook/geos/af.html (September 5, 2016)

## INTERACTIVE MAP ON THE WEB

### Format

Author or statement of responsibility. *Map Title* [format]. Data date, if known. Scale; Name of person who generated map; Name of software used to generate the map or "Title of the Web Site." URL (Date generated).

### Examples

*Kitchener-Waterloo: 1955 to Present* [map]. 2016. Scale not known; generated by S. Kirkland; "Esri ArcGIS Online." https://uwaterloo.ca/library/geospatial/collections/digital-projects/kitchener-waterloo-1955-present (October 2017).

*Map of Electric Vehicle Charging Stations* [map]. 2017. Scale not known; generated by D. D. Winslett; "Google Maps." https://www.google.com/maps (October 2017).

Region of Waterloo. *Map of Landfills in Waterloo Region* [map]. 2011. Scale not known; generated by Eva Dodsworth; "Region of Waterloo Locator." http://gis.region.waterloo.on.ca/ (October 2017).

## REAL-TIME ONLINE MAP

### Format

Author. *Title* [format]. Date and time produced, if known. Scale. "Title of Document or Site." URL (Date accessed).

### Examples

Ohio Wesleyan University. *Ohio Wesleyan University* [map]. April 21, 2016. Scale not given. "The JAYwalk CAM." https://www.owu.edu/ex/liveCampusViews/JAYwalk.html (April 21, 2016).

North Carolina Department of Transportation. *Current Wake County Traffic Conditions* [map]. March 10, 2004, 15:07:20. Scale not given. "North Carolina Department of Transportation." http://apps.dot.state.nc.us/tims (March 10, 2004).

United States Geological Survey. *1 Day, Magnitude 2.5+ Worldwide* [map]. January 31, 2014, 12:51:47 UTC-05:00. Scale not given. "Real-Time Earthquake Map." https://www.earthquakes.usgs.gov/earthquakes/map/ (January 31, 2014).

The Weather Channel. *Interactive Weather Map* [map]. January 19, 2011, 15:40. Scale not given. "The Weather Channel: Maps." http://www.weather.com/weather/map/interactive/ (January 19, 2011).

The Weather Network. *Lightning* [map]. January 19, 2011, 13:30. Scale not given. "Lightning Maps: North America." http://www.theweathernetwork.com/lightning/ (January 19, 2011).

## GIS SOFTWARE

### Format

Author. *Title*. Edition or version. Place of production: Producer, Date of copyright or production.

### Example

Environmental Systems Research Institute (ESRI). *ArcGIS*. Version 10.3. Redlands, CA: Environmental Systems Research Institute, 2016.

# 18
# Professional Map and GIS Associations

Professional map and GIS associations offer a wealth of resources to individuals wishing to learn, grow, and contribute in their fields of interest. The following list provides a directory of such associations in Canada and the United States, compiled using a large number of resources related to cartography, geomatics, surveying, data, and GIS.

## CANADIAN MAP AND GIS ASSOCIATIONS

http://www.albertageomaticsgroup.ca/
200, 6 Crowfoot Circle NW
Calgary, AB T3G 2T3
The Alberta Geomatics Group facilitates geomatics innovations among the sectors and promotes business development opportunities.

Association of Canadian Map Libraries and Archives (ACMLA)
http://www.acmla-acacc.ca/
P.O. Box 60095
University of Alberta Postal Outlet
Edmonton, AB T6G 2S4
ACMLA is the representative professional group for Canadian map librarians, cartographic archivists, and others interested in geographic information.

Canadian Association of Geographers (CAG)
http://www.cag-acg.ca/
Department of Geography, Environment and Geomatics
University of Ottawa, Simard Hall, Room 031
60 University Private
Ottawa, ON K1N 6N5
The Canadian Association of Geographers (CAG) is committed to disseminating geographic research and promoting geography as a key discipline in education and research and in the public and private sectors nationally and internationally.

Canadian Association of Geographers (ACAG), Atlantic Division
http://community.smu.ca/acag/
Department of Geography, McGill University
425-805 Sherbrooke Street W
Montreal, QC H3A 2K6

Canadian Cartographic Association (CCA)
http://cca-acc.org/
The Canadian Cartographic Association promotes interest in cartographic materials and encourages research in cartography.

Canadian Geophysical Union (CGU)
http://people.ucalgary.ca/~geodesy/
National Resources Canada, Geodetic Survey Division
9860 West Saanich Road
Sidney, BC V8L4B2
The Canadian Geophysical Union (CGU) serves as a national group for geophysical sciences, with annual meetings, an awards program, and significant student involvement. The CGU also carries on the traditional responsibility of representing Canada in the International Union of Geodesy and Geophysics through a Canadian National Committee (CNC/IUGG).

Canadian Institute of Geomatics (CIG)
http://www.cig-acsg.ca/
900 Dynes Road, Bureau 100D
Ottawa, ON K2C 3L6
The Canadian Institute of Geomatics represents the interests of the geomatics community and is the Canadian member to the International Federation of Surveying (FIG), the International Society of Photogrammetry and Remote Sensing (ISPRS), and the International Cartographic Association (ICA). The organization is also a founding member of GeoAlliance.

Canadian Institute of Geomatics (CIG), British Columbia Branch
http://www.cig-acsg.ca/British-Columbia

Canadian Institute of Geomatics (CIG), Champlain Branch
http://www.acsg-champlain.ca/

Canadian Institute of Geomatics (CIG), Montreal Branch
http://acsg-montreal.ca/

Canadian Institute of Geomatics (CIG), New Brunswick Branch
http://www.cig-acsg.ca/New-Brunswick

Canadian Institute of Geomatics (CIG), Newfoundland and Labrador Branch
http://www.cig-acsg.ca/Newfoundland-Labrador

Canadian Institute of Geomatics (CIG), Nova Scotia Branch
http://www.cig-acsg.ca/Nova-Scotia

Canadian Institute of Geomatics (CIG), Ottawa Branch
http://cigottawabranch.wixsite.com/cig-ottawa/about-us

Canadian Remote Sensing Society (CRSS)
http://www.crss-sct.ca/
2202 Hanover Crescent
Regina, SK S4V 0Z6
The Canadian Remote Sensing Society (CRSS) is a fully independent, nonprofit professional society that provides leadership and excellence to advance the field of remote sensing.

Geomatics Association of Nova Scotia (GANS)
http://www.gans.ca/
P.O. Box 961
Halifax, NS B3J 2V9
The Geomatics Association of Nova Scotia (GANS) is a nonprofit association created to promote the development of the geomatics industry in Nova Scotia.

Ontario Association of Remote Sensing (OARS)
http://www.oars.on.ca/
225 Lloyd Avenue
Newmarket, ON L3Y 5L4
The Ontario Association of Remote Sensing (OARS) educates, communicates, and facilitates on a wide variety of remote sensing and geomatics topics. The group is composed of small business, industrial, and educational members.

## AMERICAN MAP AND GIS ASSOCIATIONS

American Association for Geodetic Surveying (AAGS)
https://www.aagsmo.org/
The American Association for Geodetic Surveying (AAGS) aims to lead the community of geodetic, surveying, and land information data users, focusing on educational programs.

American Association of Geographers (AAG)
http://www.aag.org/
1710 16th Street NW
Washington, DC 20009-3198
The American Association of Geographers (AAG) is a nonprofit scientific and educational society that contributes to the advancement of geography. It consists of members from nearly one hundred countries that share their interests in the theory, methods, and practice of geography.

American Association of Geographers (AAG), Cartography Specialty Group
https://aagcartography.wordpress.com/
The mission of the Cartography Specialty Group is to encourage cartographic research and promote education in cartography and map use.

American Association of Geographers (AAG), GIS Specialty Group
http://aag-giss.org/
The mission of the GIS Speciality Group is to promote the exchange of ideas and information relating to GIS.

American Association of Geographers (AAG), Remote Sensing Specialty Group
http://www.aagrssg.org/index.html
The mission of the Remote Sensing Specialty Group is to promote the exchange of ideas and information relating to remote sensing.

American Association of Geographers (AAG), Spatial Analysis and Modeling Specialty Group
http://sam-aag.org/
The mission of the Spatial Analysis and Modeling Specialty Group is to promote the exchange of ideas and information relating to the analysis of geo-referenced data, modeling of spatial-temporal processes, and the use of analytical and computational techniques in solving geographic problems.

American Congress on Surveying and Mapping (ACMS)
http://landsurveyorsunited.com/acsm
The American Congress on Surveying and Mapping (ACSM) is a nonprofit association dedicated to serving the public interest and advancing the profession of surveying and mapping.

American Geophysical Union (AGU)
http://sites.agu.org/leadership/strategic-plan/mission/
2000 Florida Avenue NW
Washington, DC 20009-1227
The mission of the American Geophysical Union is to promote discovery in earth and space science for the benefit of humanity.

American Society for Photogrammetry and Remote Sensing (ASPRS)
http://www.asprs.org/
425 Barlow Place, Suite 210
Bethesda, MD 20814-2160
The American Society for Photogrammetry and Remote Sensing (ASPRS) is a scientific association serving over seven thousand professional members around the world, with a mission to advance knowledge and improve understanding of mapping sciences to promote applications of photogrammetry, remote sensing, GIS, and supporting technologies.

California Geographic Information Association (CGIA)
http://cgia.org/
The California Geographic Information Association (CGIA) is a nonprofit, statewide association that facilitates the coordination, collaboration, and advocacy for California's GIS community.

Cartography and Geographic Information Society (CAGIS)
http://www.cartogis.org/
The Cartography and Geographic Information Society (CAGIS) is composed of educators, researchers, and practitioners involved in the design, creation, use, and dissemination of geographic information.

Federal Geographic Data Committee (FGDC)
https://www.fgdc.gov/
12201 Sunrise Valley Drive
Reston, VA 20192
The Federal Geographic Data Committee (FGDC) is an organized structure of federal geospatial professionals that provides executive, managerial, and advisory direction and oversight for geospatial decisions and initiatives across the federal government.

Geographic and Land Information Society (GLIS)
http://www.g-lis.org/
6300 Ocean Drive, Unit 5868
Corpus Christi, TX 78412-5868
The mission of the Geographic and Land Information Society is to encourage the appropriate use of surveying and mapping technologies in the development and use of geographic and land information systems.

Geospatial Information and Technology Association (GITA)
http://www.gita.org/
1360 University Avenue West, Suite 455
St. Paul, MN 55104-4086
The Geospatial Information and Technology Association (GITA) is a member-led nonprofit professional association for users of geospatial technology to help operate, maintain, and protect infrastructure, which includes utilities, telecommunication companies, and the public sector, among others.

Management Association for Private Photogrammetric Surveyors (MAPPS)
http://www.mapps.org/
1856 Old Reston Avenue, Suite 205
Reston, VA 20190
The Management Association for Private Photogrammetric Surveyors (MAPPS) consists of members who are interested in surveying, photogrammetry, satellite and airborne remote sensing, aerial photography, hydrography, aerial and satellite image processing, and GPS and GIS data collection and conversion services.

Map and Geospatial Information Round Table, American Library Association (MAGIRT)
http://www.ala.org/magirt/
1615 New Hampshire Avenue NW, 1st Floor
Washington, DC 20009-2520
The Map and Geospatial Information Round Table (MAGIRT) leads and inspires information professionals in their work with map and geospatial information resources, collections, and technologies in all formats through community, education, and advocacy.

Mid-America GIS Consortium (MAGIC)
http://www.magicgis.org/
P.O. Box 4533
Lawrence, KS 66046
The Mid-America GIS Consortium (MAGIC) is a network of leaders in the fields of mapmaking, location services, and data development.

National Alliance for Public Safety GIS (NAPSG)
https://www.napsgfoundation.org/
5335 Wisconsin Avenue NW, Suite 440
Washington, DC 20015
The National Alliance for Public Safety GIS (NAPSG) Foundation is a nonprofit organization that educates GIS users about information sharing and data interoperability associated with GIS and advanced technologies used by the public safety and homeland security communities.

National States Geographic Information Council (NSGIC)
https://www.nsgic.org/
3436 Magazine Street, #449
New Orleans, LA 70115
The National States Geographic Information Council (NSGIC) promotes the development and management of location-based information resources and advocates for innovative, strategic use of these assets to advance the interests of states, tribes, regions, local governments, and the nation.

North American Cartographic Information Society (NACIS)
http://nacis.org/
2311 E. Hartford Avenue
Milwaukee, WI 53211

The North American Cartographic Information Society (NACIS) is composed of specialists from private, academic, and government organizations along with people whose common interest lies in facilitating communication in the map information community.

Society for Conservation GIS (SCGIS)
https://www.scgis.org/

The Society for Conservation GIS (SCGIS) is a nonprofit organization that assists conservationists worldwide in using GIS through communication, networking, scholarships, and training. Their mission is to build community, provide knowledge, and support individuals using GIS and science for the conservation of natural resources and cultural heritage.

# Appendix A
## *Example of a Map Library Collection: University of Waterloo*

Millions of maps are published and made available via North American libraries. Many maps are not catalogued in library systems because of the sheer quantity of map sheets available. For any given map series, there may be tens of thousands of maps that constitute a collection covering an entire country. Producing a list that captures the entire collection is often not possible. Researchers may wonder, however, what a typical map collection looks like. How many maps cover a city or a lake? How many air photos does it take to capture coverage for a specific town? Who publishes the maps? How often are they revised? This appendix offers a snapshot of a southwest Ontario academic map library print collection. Although not a very large collection, it offers a representation of maps of the local area, North America, as well as maps of the world, varying in themes and topics. Many of the maps housed in the collection have been discussed throughout this book and offer such details as map name, scale, and year published. The following list is organized by geography, followed by map topic. Map themes include air photos, fire insurance plans, geology, road maps, historical maps, and maps emphasizing human and physical geography, land use, and meteorology. Although the focus of this book is on North American maps, a short list of world maps has been included to show the general collection. The library also has a collection of maps for other continents and regions, covering Europe, Asia, South America, the Caribbean, and Africa; however, these maps were not included in the list.

## CANADA

### Aerial Photographs

The Geospatial Centre has a large collection of over forty thousand aerial photographs that cover many areas of Ontario.

Date of coverage: 1930 (April–May)
Geographical area covered: Northern Waterloo County and part of Wellington County
Scale: 1:18,500, 1:11,500
Publisher: National Air Photo Library

Date of coverage: 1930 (May)
Geographical area covered: Part of Kitchener-Waterloo and part of Breslau
Scale: flying height 1,500 feet
Publisher: National Air Photo Library

Date of coverage: 1934 (September–November)
Geographical area covered: Parts of Hamilton-Wentworth and Niagara Counties
Scale: 1:15,840
Publisher: National Air Photo Library

Date of coverage: 1939 (June)
Geographical area covered: Toronto
Scale: 1:17,000
Publisher: National Air Photo Library

Date of coverage: 1945 (September)
Geographical area covered: Eastern Waterloo County (North Dumfries Township, most of Waterloo Township)
Scale: 1:20,600
Publisher: National Air Photo Library

Date of coverage: 1946 (June)
Geographical area covered: Western Waterloo County (Wilmot Township, parts of adjacent townships)
Scale: 1:15,500
Publisher: National Air Photo Library

Date of coverage: 1947 (May)
Geographical area covered: Kitchener, Waterloo, Preston, and Galt
Scale: varies, 1:11,700–1:12,300
Publisher: L & F (negatives Ontario Archives)

Date of coverage: 1948
Geographical area covered: St. Catherine's and most of Thorold
Scale: 1:4,800
Publisher: Unknown

Date of coverage: 1949
Geographical area covered: Waterloo, Kitchener, Preston, and Galt
Scale: 1:12,500
Publisher: Ontario Ministry of Natural Resources

Date of coverage: 1952
Geographical area covered: Southern Kitchener, Preston, and Hespeler
Scale: 1:11,500
Publisher: Unknown

Date of coverage: 1953–1955 (May–October)
Geographical area covered: Parts of midwestern and central Ontario (all of NTS 40P; most of NTS 30M; parts of NTS 31C, 40I, 40J, 41A, and 41H)
Scale: 1:15,840
Publisher: L & F (negatives Ontario Archives)

Date of coverage: 1959–1960 (September)
Geographical area covered: North of Metropolitan Toronto (including Caledon Hills, Newmarket, Uxbridge, Richmond Hill, Oshawa, and Pickering)
    Index on two sheets
Scale: 1:28,500
Publisher: National Air Photo Library

Date of coverage: 1960 (April)
Geographical area covered: North of Metropolitan Toronto (including Aurora, Brampton, and Richmond Hill)
Scale: 1:25,000, 1:27,000
Publisher: National Air Photo Library

Date of coverage: 1962
Geographical area covered: Parts of Essex County and north of Peel
Scale: 1:20,000
Publisher: National Air Photo Library

Date of coverage: 1963
Geographical area covered: Kitchener-Waterloo and adjacent areas
Scale: 1:11,950
Publisher: Spartan Air Services

Date of coverage: 1963
Geographical area covered: Parts of Huron, Perth, Lambton, Middlesex, Elgin, and Oxford Counties
Scale: 1:20,000
Publisher: National Air Photo Library

Date of coverage: 1964–1965
Geographical area covered: Kitchener-Waterloo
Scale: 1:25,000
Publisher: National Air Photo Library

Date of coverage: 1964 (August)
Geographical area covered: Sauble Beach, Wiarton, Lake Charles, and Owen Sound
Scale: 1:30,000
Publisher: National Air Photo Library

Date of coverage: 1965 (October)
Geographical area covered: Parts of Hamilton-Wentworth, Haldimand-Norfolk, and Niagara Counties
Scale: 1:20,000
Publisher: National Air Photo Library

Date of coverage: 1966 (Spring)
Geographical area covered: Northern Metropolitan Toronto, Markham, and Pickering
Scale: 1:12,000
Publisher: Lockwood (Northway)

Date of coverage: 1966 (June)
Geographical area covered: Waterloo County, part of Perth County, and part of Wellington County
Scale: 1:37,560
Publisher: National Air Photo Library

Date of coverage: 1966 (August)
Geographical area covered: Northern Waterloo County (Wellesley, Woolwich, and part of Waterloo Township), part of Wellington County (Elora, Fergus, Orangeville), and Bruce Peninsula
Scale: 1:15,840
Publisher: L & F (negatives MNR)

Date of coverage: 1967 (April)
Geographical area covered: Waterloo County
Scale: 1:18,000
Publisher: Lockwood (Northway)

Date of coverage: 1968 (Spring)
Geographical area covered: Waterloo County (Wilmot, Woolwich Township; part of North Dumfries, Waterloo, and Wellesley Townships), part of Oxford County (Plattsville)
Scale: 1:24,000
Publisher: International Mapping Services

Date of coverage: 1969 (May)
Geographical area covered: Beaver Valley, Grey County
Scale: 1:36,500
Publisher: National Air Photo Library

Date of coverage: 1969 (May 26)
Geographical area covered: Bruce Nuclear Station, Kincardine, Rocky Saugeen, Lueck Mill
Scale: 1:40,000
Publisher: National Air Photo Library

Date of coverage: 1969 (May 27)
Geographical area covered: Southhampton, MacGregor Provincial Park, Rockford, Arnott
Scale: 1:40,000
Publisher: National Air Photo Library

Date of coverage: 1969 (June 16)
Geographical area covered: Bruce Peninsula, Sauble Beach, Meaford
Scale: 1:40,000
Publisher: National Air Photo Library

Date of coverage: 1969–1973
Geographical area covered: Port Loring and Restoule areas, Parry Sound District
Scale: varies, 1:36,000–1:38,000
Publisher: National Air Photo Library

Date of coverage: 1970
Geographical area covered: St. Thomas Expressway, Kincardine—Underwood—Highway 21, Highway 407—Highway 27—Whitevale, Highway 8—Galt—Preston Bypass, St. Thomas, Duffs Corners to Brantford Highway 403, Welland Canal
Scale: varies, 1:3,000, 1:9,000, 1:12,000, 1:24,000
Publisher: MTO (Ontario Ministry of Transportation)

Date of coverage: 1970–1987
Geographical area covered: Parts of southwestern Ontario.
  Photos flown for the purposes of highway route planning and construction. Flight lines aligned along proposed or existing highways.
Scale: varies, 1:3,000–1:40,000
Publisher: Ontario Ministry of Transportation/National Air Photo Library

Date of coverage: 1970 (July)
Geographical area covered: Orangeville, Brantford, Toronto, Hamilton
  High altitude, color infrared transparencies
Scale: 1:120,000
Publisher: NASA/National Air Photo Library

Date of coverage: 1971
Geographical area covered: Highway 3—Jarvis to Delhi, Go Transit—Hamilton to Oshawa, Canadian National Railway (CNR) Toronto Area
Scale: varies, 1:5,160, 1:6,000, 1:12,000
Publisher: MTO (Ontario Ministry of Transportation)

Date of coverage: 1971 (May)
Geographical area covered: Waterloo, Kitchener, Breslau, and Cambridge
Scale: 1:15,840
Publisher: Northway-Photomap

Date of coverage: 1971 (June–August)
Geographical area covered: Georgetown, Caledon, Tottenham, Alliston, Newmarket, Uxbridge, Richmond Hill, Pickering, Milton
  Although the photos are dated 1971, some were flown in 1972.
Scale: 1:15,840
Publisher: MNR

Date of coverage: 1971 (October)
Geographical area covered: London through Woodstock, and Tillsonburg to Brantford
  High-altitude, color transparencies
Scale: 1:140,000
Publisher: NASA/National Air Photo Library

Date of coverage: 1971 and 1972
Geographical area covered: Golden Horseshoe to Hamilton to Niagara Falls; south of Hamilton to north of Burlington; Saltfleet to Niagara River; North Grimsby to Gainsborough; North Grimsby to Grimsby Centre; Mono to Flamborough; Stoney Creek to Woodburn; Castlederg to Trafalgar; Eversley to Maple; Lake Ontario (Long Branch); Bewdley to Port Hope; Pontypool to Port Hope; Thickson Point to Scugog Island; Claremont to Liverpool; Niagara on the Lake to Stamford; Leskard to New Castle; Teunton to Port Whitney; Fingerboard to Ghirley; Niagara Falls to Mulgrave; Stoney Creek to Hwy 20.
  High-altitude, color infrared transparencies
Scale: 1:60,000, 1:120,000, and 1:140,000
Publisher: NASA/NAPL/CCRS

Date of coverage: 1972
Geographical area covered: Conestoga Lake Area—St. Jacobs, Highway 24 and Highway 53—Brantford Expressway, Highway 7—New St. Mary's Bypass—New Hamburg, Mountain Freeway—Hamilton, Highway 402—Highway 81 London, Highway 7—London, Highway 403—Hamilton—Duffs Corners, Highway 7—Markham—Brooklin
Scale: varies, 1:3,000, 1:12,000, 1:40,000, 1:50,000
Publisher: MTO (Ontario Ministry of Transportation)

Date of coverage: 1972 (June–July)
Geographical area covered: Southern Waterloo County (Wilmot and Waterloo Townships; part of North Dumfries and Wellesley Townships)
Scale: 1:25,200
Publisher: National Air Photo Library

Date of coverage: 1972 (June–July)
Geographical area covered: Southern Waterloo County (Wilmot, Waterloo, and North Dumfries Townships; part of Wellesley Township; Long Point)
Scale: 1:15,840
Publisher: MNR

Date of coverage: July 1972
Geographical area covered: St. Mary's, Wilderwood and adjacent
Scale: 1:13,600
Publisher: National Air Photo Library

Date of coverage: 1973
Geographical area covered: Highway 401 and Highway 8—Kitchener—Preston, Highway 8—Kitchener—Galt, Highway 401—interchange 14—Windsor, Highway 402 Strathroy—401, QEW Guelph line—Highway 20, Welland Canal—Highway 406, Welland
Scale: varies, 1:3,000, 1:6,000, 1:7,200, 1,12,000, 1:20,568
Publisher: MTO (Ontario Ministry of Transportation)

Date of coverage: 1973 (March 23)
Geographical area covered: Tillsonburg to Durham; Cayuga to Angus (near Alliston).
  High-altitude, color infrared photos
Scale: 1:129,000
Publisher: NASA/National Air Photo Library

Date of coverage: 1973 (April)
Geographical area covered: Point Pelee
Scale: 1:35,000
Publisher: National Air Photo Library

Date of coverage: 1973
Geographical area covered: UWaterloo Campus, west to Fischer-Hallman Road
Scale: 1:6,000
Publisher: General Photogrammertric Services

Date of coverage: 1974
Geographical area covered: Pukaskwa National Park
Scale: 1:17,000
Publisher: Ontario Ministry of Natural Resources

Date of coverage: 1974
Geographical area covered: Hwy 407—Hwy 48 and Hwy 35, Hwy 8 and Hwy 7—Hwy 401—New Hamburg—Hwy 85, Hwy 402—Sarnia—Lambton—Middlesex County Line, Hwy 401 Chatham—Kent/Elgin, Hwy 8, Hwy 7—401
Scale: varies, 1:3,000, 1:4,200, 1:12,000, 1:24,000
Publisher: MTO (Ontario Ministry of Transportation)

Date of coverage: 1975
Geographical area covered: Strathroy Area—Environmental Impact of Highway 402, Highway 48—Mt. Albert—Highway 47, Highway 3—St. Thomas—Aylmer, Conestoga River—St. Jacobs, Highway 401—Elgin County, Grand River and Irvine River
Scale: varies, 1:3,000, 1:4,000. 1:5,000, 1:8,000, 1:10,000
Publisher: MTO (Ontario Ministry of Transportation)

Date of coverage: 1975
Geographical area covered: Kitchener, Cambridge, and part of North Dumfries Township
Scale: 1:11,900
Publisher: Northway

Date of coverage: 1976
Geographical area covered: Bruce Peninsula
Scale: 1:50,000
Publisher: Department of Energy, Mines and Resources

Date of coverage: 1976
Geographical area covered: Highway 4—Highway 22 and Highway 9—London—Teeswater, Highway 8—Cambridge Bypass, Highway 7—Kitchener—Guelph, Highway 40—New Sarnia Area, Highway 403—Highway 10—QEW
Scale: varies, 1:4,000, 1:5,000, 1:10,000, 1:12,000
Publisher: MTO (Ontario Ministry of Transportation)

Date of coverage: 1977
Geographical area covered: Highway 58—Thoroldstone Road—Highway 20, QEW—McLeod Road Interchange, Highway 403—Ancaster—Brantford, Highway 3—Jarvis—Dunnville, QEW—Highway 20—Fifty Road
Scale: varies, 1:4,000, 1:10,000, 1:12,000
Publisher: MTO (Ontario Ministry of Transportation)

Date of coverage: 1977 (July–August)
Geographical area covered: Georgian Bay shoreline (Port Severn to Pointe au Baril); Muskoka Lakes; southern Algonquin Park (south of Highway 60). Although photos are dated 1977, some were flown in June 1978
Scale: 1:15,840
Publisher: MNR

Date of coverage: 1978 (May 29)
Geographical area covered: Strathroy Area—Environmental Impact of Highway 402
Scale: 1:4,900
Publisher: MTO (Ontario Ministry of Transportation)

Date of coverage: 1978 (June–October)
Geographical area covered: Parts of midwestern and central Ontario (east half of NTS 40P, most of NTS 30M, and parts of NTS 30N, 31D, 31G, 40G, 40J, 41A, 41H)
  Index on seven sheets

Scale: 1:10,000
Publisher: MNR

Date of Coverage: 1978 (September)
Geographical area covered: Hamilton to Guelph
 Synthetic aperture radar images
Scale: 1:41,250
Publisher: National Air Photo Library

Date of coverage: 1978
Geographical area covered: Highway 403—QEW—Hamilton—Burlington—Oakville, Highway 8 and Highway 24—Preston—Freeport, Highway 59—Woodstock NW Bypass, Highway 406—QEW Interchange at Louth
Scale: varies, 1,3,000, 1,4,000, 1:10,000
Publisher: MTO (Ontario Ministry of Transportation)

Date of coverage: 1979 (September 20)
Geographical area covered: Strathroy Area—Environmental Impact of Highway 402
Scale: 1:5,000
Publisher: MTO (Ontario Ministry of Transportation)

Date of coverage: 1979
Geographical area covered: Bruce Peninsula (Lake Huron Coast)
Scale: 1:11,000
Publisher: National Air Photo Library

Date of coverage: 1980 (April–July)
Geographical area covered: Lake Huron shoreline to Acton; London to Mount Forest
Scale: 1:50,000
Publisher: National Air Photo Library

Date of coverage: 1980
Geographical area covered: Lake Huron shoreline Dunlop south to Grand Bend, east to St. Mary's
Scale: 1:55,650
Publisher: Northway

Date of coverage: 1980 (May 5)
Geographical area covered: West of Strathroy—Highway 402
Scale: 1:5,000
Publisher: MTO (Ontario Ministry of Transportation)

Date of coverage: 1980 (September 6)
Geographical area covered: West of Strathroy—Highway 402
Scale: 1:5,000
Publisher: MTO (Ontario Ministry of Transportation)

Date of coverage: 1980
Geographical area covered: Highway 7 and Highway 8—Stratford—New Hamburg, Highway 401, Highway 410, and Highway 403—Toronto, Highway 402—Lambeth area, Highway 3—Fort Erie—Jarvis
Scale: varies, 1:3,000, 1:5,000, 1:10,000, 1:12,000
Publisher: MTO (Ontario Ministry of Transportation)

Date of Coverage: 1981
Geographical area covered: St. Thomas Expressway and Highway 3, Cayuga, E. C. Rowe Expressway—Windsor, Highway 40—Sarnia, Highway 85—Waterloo Area—Northfield Drive
Scale: varies, 1:3,000, 1:8,000, 1:12,000, 1:20,000
Publisher: MTO (Ontario Ministry of Transportation)

Date of coverage: 1982 (October)
Geographical area covered: Waterloo Region (except for western part of Wilmot and Wellesley Twp.); Guelph and parts of Wellington County; Milton to Orangeville; Newmarket-Aurora area
Scale: 1:30,000
Publisher: Ontario Basic Mapping/MNR

Date of Coverage: 1982
Geographical area covered: Highway 402—London—Strathroy Area, Highway 403—Woodstock—Brantford, Highway 18—Canard River—Turkey Creek—Windsor, Highway 404—Highway 401 to Gormley
Scale: varies, 1:3,000, 1:5,000, 1:14,400
Publisher: MTO (Ontario Ministry of Transportation)

Date of coverage: 1982–1983
Geographical area covered: QEW East of Burlington—Oakville
Scale: 1:3,300
Publisher: MTO (Ontario Ministry of Transportation)

Date of coverage: 1983 (May)
Geographical area covered: Durham and Walkerton areas; Dundalk, Shelburne, Alliston
Scale: 1:30,000
Publisher: Ontario Basic Mapping/MNR

Date of coverage: 1983
Geographical area covered: Parts of Mississauga, Scarborough, Pickering, and Ajax
Scale: 1:11,700
Publisher: Kenting Earth Sciences

Date of coverage: 1984
Geographical area covered: Highway 407—Woodbine—Ninth Line, Highway 410—Brampton—Victoria
Scale: 1:10,000
Publisher: MTO (Ontario Ministry of Transportation)

Date of coverage: 1985 (July)
Geographical area covered: Pinery Provincial Park shoreline, Grand Bend and Wallaceburg area

Black and white infrared photos
Index on two sheets
Scale: 1:8,000
Publisher: MNR

Date of coverage: 1985
Geographical area covered: QEW—Burlington to Beamsville, QEW—Niagara Road 24 to Highway 405
Scale: 1:10,000
Publisher: MTO (Ontario Ministry of Transportation)

Date of coverage: 1986
Geographical area covered: Highway 406—Port Robinson—Thorold, Highway 420—Thoroldstone Road, QEW—Niagara Falls—Fort Erie
Scale: 1:10,000
Publisher: MTO (Ontario Ministry of Transportation)

Date of coverage: 1987
Geographical area covered: QEW—Freeman to Royal Windsor Drive; Burlington—Oakville
Scale: 1:3,000
Publisher: MTO (Ontario Ministry of Transportation)

Date of coverage: 1990 (April)
Geographical area covered: Waterloo Region
Scale: 1:20,000
Publisher: Northway Map Technology

Date of coverage: 1990 (July)
Geographical area covered: Brampton
Scale: 1:30,000
Publisher: Aquarius Flight

Date of coverage: 1992 (March)
Geographical area covered: Parts of Mississauga, Oakville, Meadowvale, and Erin Mills
Scale: 1:8,000
Publisher: Aquarius Flight

Date of coverage: 1995 (May)
Geographical area covered: Waterloo Region
Scale: 1:20,000
Publisher: Northway Map Technology

**Fire Insurance Plans**

The Geospatial Centre has a collection of fire insurance plans of many towns in Southwestern Ontario, with a focus on the region of Waterloo.

Title: Ayr
Date: December 1884 (revised 1904)
Publisher: Chas. E. Goad

Title: Baden
Date: July 1894 (revised 1904)
Publisher: Chas. E. Goad

Title: Baden Linseed Oil Mills
Date: August 1953
Publisher: Toronto Elevators

Title: Berlin (see also Kitchener)
Date: August 1894 (revised 1904)
Publisher: Chas. E. Goad

Title: Blyth
Date: October 1926
Publisher: Underwriters' Survey Bureau

Title: Burlington
Date: May 1924
Publisher: Underwriters' Survey Bureau

Title: Chesterville
Date: May 1928
Publisher: Underwriters' Survey Bureau

Title: Clarksburg
Date: August 1926
Publisher: Underwriters' Survey Bureau

Title: Creemore
Date: June 1929
Publisher: Underwriters' Survey Bureau

Title: Delhi
Date: January 1949
Publisher: Underwriters' Survey Bureau

Title: Dundas
Date: July 1951
Publisher: Underwriters' Survey Bureau

Title: Elmira
Date: August 1894 (revised July 1904)
Publisher: Chas. E. Goad

Title: Elmira
Date: April 1923
Publisher: Underwriters' Survey Bureau

Title: Elmira
Date: August 1949
Publisher: C. N. Lloyd

Title: Elora
Date: October 1890
Publisher: Chas. E. Goad

Title: Exeter
Date: September 1926
Publisher: Underwriters' Survey Bureau

Title: Fergus
Date: October 1921
Publisher: Underwriters' Survey Bureau

Title: Fergus
Date: October 1921 (revised August 1935)
Publisher: Underwriters' Survey Bureau

Title: Galt
Date: August 1910 (revised December 1913)
Publisher: Chas. E. Goad

Title: Galt
Date: August 1929
Publisher: Underwriters' Survey Bureau

Title: Guelph
Date: February 1897 (revised November 1907, revised 1916)
Publisher: Chas. E. Goad

Title: Guelph
Date: March 1922 (revised October 1929)
Publisher: Underwriters' Survey Bureau

Title: Guelph
Date: March 1922 (revised April 1946)
Publisher: Underwriters' Survey Bureau

Title: Guelph
Date: 1960
Publisher: Underwriters' Survey Bureau

Title: Hanover
Date: October 1925
Publisher: Underwriters' Survey Bureau

Title: Harriston
Date: May 1894 (reprinted June 1920)
Publisher: Underwriters' Survey Bureau

Title: Hespeler
Date: June 1885
Publisher: Chas. E. Goad

Title: Hespeler
Date: June 1885 (revised 1904)
Publisher: Chas. E. Goad

Title: Hespeler
Date: September 1910 (contains some notes with date 1917)
Publisher: Chas. E. Goad

Title: Hespeler
Date: September 1910 (revised 1917)
Publisher: Chas. E. Goad

Title: Hespeler
Date: Febuary 1947
Publisher: Underwriters' Survey Bureau

Title: Jarvis
Date: May 1909
Publisher: Chas. E. Goad

Title: Kitchener
Date: February 1908 (revised March 1917)
Publisher: Chas. E. Goad

Title: Kitchener (including the village of Bridgeport)
Date: February 1908 (revised and reprinted January 1947)
Publisher: Underwriters' Survey Bureau

Title: Linwood
Date: April 1937
Publisher: Provincial Insurance Surveys

Title: Meaford
Date: December 1925
Publisher: Underwriters' Survey Bureau

Title: Mitchell
Date: September, October 1925
Publisher: Underwriters' Survey Bureau

Title: Mount Forest
Date: October 1904 (revised October 1926)
Publisher: Underwriters' Survey Bureau

Title: New Hamburg
Date: 1885 (revised 1904)
Publisher: Chas. E. Goad

Title: Norwich
Date: December 1928
Publisher: Underwriters' Survey Bureau

Title: Orangeville
Date: December 1907
Publisher: Chas. E. Goad

Title: Orangeville
Date: December 1907 (revised December 1935)
Publisher: Underwriters' Survey Bureau

Title: Palmerston
Date: May 1914
Publisher: Chas. E. Goad

Title: Paris
Date: November 1913
Publisher: Chas. E. Goad

Title: Port Stanley
Date: February 1925
Publisher: Underwriters' Survey Bureau

Title: Preston
Date: 1910 (revised 1917)
Publisher: Chas. E. Goad

Title: Preston
Date: September 1910 (revised and reprinted March 1924)
Publisher: Underwriters' Survey Bureau

Title: Rockwood
Date: October 1904
Publisher: Chas. E. Goad

Title: Seaforth
Date: August 1923
Publisher: Underwriters' Survey Bureau

Title: Shakespeare
Date: April 1901
Publisher: Chas. E. Goad

Title: Simcoe
Date: April 1949
Publisher: Underwriters' Survey Bureau

Title: Stayner
Date: December 1924
Publisher: Underwriters' Survey Bureau

Title: St. Jacobs
Date: July 1904
Publisher: Chas. E. Goad

Title: St. Marys
Date: December 1919
Publisher: Underwriters' Survey Bureau

Title: Stratford
Date: December 1949
Publisher: Underwriters' Survey Bureau

Title: Tara
Date: August 1926
Publisher: Underwriters' Survey Bureau

Title: Tavistock
Date: February 1926
Publisher: Underwriters' Survey Bureau

Title: Thornbury
Date: July 1925
Publisher: Underwriters' Survey Bureau

Title: Toronto
Date: 1883 and 1890, atlas
Publisher: Chas. E. Goad

Title: Tottenham
Date: December 1928
Publisher: Underwriters' Survey Bureau

Title: Wasaga Beach
Date: April 1936
Publisher: Underwriters' Survey Bureau

Title: Waterford
Date: May 1926
Publisher: Underwriters' Survey Bureau

Title: Waterloo
Date: March 1908
Publisher: Chas. E. Goad

Title: Waterloo
Date: March 1908 (revised November 1913)
Publisher: Chas. E. Goad

Title: Waterloo
Date: June 1920 (revised and reprinted December 1946)
Publisher: Underwriters' Survey Bureau

Title: Waterloo
Date: April 1942
Publisher: Provincial Insurance Surveys

Title: Wellesley
Date: July 1894
Publisher: Chas. E. Goad

Title: Wellesley
Date: 1894 (revised 1904)
Publisher: Chas. E. Goad

Title: Wiarton
Date: July 1923
Publisher: Underwriters' Survey Bureau

Title: Wingham
Date: August 1904 (revised and reprinted October 1928)
Publisher: Underwriters' Survey Bureau

Title: Woodstock
Date: July 1913 (revised September 1932)
Publisher: Underwriters' Survey Bureau

Title: Woodstock
Date: July 1913 (revised April 1949)
Publisher: Underwriters' Survey Bureau

## Geological Maps

*Canadian Geological Survey*

### British Columbia
Map Number: 1339A
Map Name: Athabasca River, Alberta–British Columbia
Year: 1979
Scale: 1:1,000,000

Map Number: 1385A
Map Name: Skeena River, British Columbia–Alaska
Year: 1979
Scale: 1:1,000,000

Map Number: 1386A
Map Name: Fraser River, British Columbia–Washington
Year: 1979
Scale: 1:1,000,000

Map Number: 1418A
Map Name: Iskut River, British Columbia–Alaska
Year: 1979
Scale: 1:1,000,000

Map Number: 1424A
Map Name: Parsnip River, British Columbia
Year: 1979
Scale: 1:1,000,000

### Ontario
Map Number: 584A
Map Name: Toronto-Hamilton
Year: 1941
Scale: 1:253,400

Map Number: 624A
Map Name: Waterloo
Year: 1941
Scale: 1:253,400

Map Number: 1062A
Map Name: Southern Ontario—Oil and Natural Gas
Year: 1958
Scale: 1:380,160

Map Number: 1194A
Map Name: Bruce Peninsula Area
Year: 1969
Scale: 1:253,440

Map Number: 1263A
Map Name: Toronto-Windsor Area
Year: 1969
Scale: 1:250,000

Map Number: 1334A
Map Name: Riviere Gatineau, Quebec-Ontario
Year: 1978
Scale: 1:1,000,000

Map Number: 1335A
Map Name: Southern Ontario
Year: 1983
Scale: 1:1,000,000

Map Number: 1970-26
Map Name: Stratford (East Half)
Year: 1970
Scale: 1:50,000

Map Number: 1970-27
Map Name: Stratford (West Half)
Year: 1970
Scale: 1:50,000

Map Number: 1970-28
Map Name: Conestogo (East Half)
Year: 1970
Scale: 1:50,000

Map Number: 1970-29
Map Name: Conestogo (West Half)
Year: 1970
Scale: 1:50,000

### Yukon and the Arctic
Map Number: 900A
Map Name: Principal Mineral Areas of Canada
Year: 2009
Scale: 1:6,000,000

Map Number: 1398A
Map Name: MacMillan River, Yukon-Alaska
Year: 1980
Scale: 1:1,000,000

Map Number: 1765A
Map Name: Circumpolar Geological Map of the Arctic
Year: 1989
Scale: 1:6,000,000

Map Number: 1818A
Map Name: Circumpolar Map of Quaternary Deposits of the Arctic
Year: 1991
Scale: 1:6,000,000

### Canada-Wide

Map Number: 1246A
Map Name: Permafrost in Canada
Year: 1967
Scale: 1:7,603,200

Map Number: 1250A
Map Name: Geological Map of Canada
Year: 1969
Scale: 1:5,000,000

Map Number: 1251A
Map Name: Tectonic Map of Canada
Year: 1970
Scale: 1:5,000,000

Map Number: 1252A
Map Name: Mineral Deposits of Canada
Year: 1969
Scale: 1:5,000,000

Map Number: 1253A
Map Name: Glacial Map of Canada
Year: 1968
Scale: 1:5,000,000

Map Number: 1254A
Map Name: Physiographic Regions of Canada
Year: 1969
Scale: 1:5,000,000

Map Number: 1255A
Map Name: Magnetic Anomaly Map of Canada
Year: 1987
Scale: 1:5,000,000

Map Number: 1257A
Map Name: Retreat of Wisconsin and Recent Ice in North America
Year: 1969
Scale: 1:5,000,000

Map Number: 1399A
Map Name: Physiography of Eastern Canada and Adjacent Areas
Year: 1977
Scale: 1:2,000,000

Map Number: 1401A
Map Name: Geology of Eastern Canada and Adjacent Areas
Year: 1979
Scale: 1:2,000,000

Map Number: 1702A
Map Name: Late Wisconsinan and Holocene Retreat of the Laurentide Ice Sheet
Year: 1987
Scale: 1:5,000,000

Map Number: 1703A
Map Name: Paleogeography of Northern North America 18000–5000 Years Ago
Year: 1987
Scale: 1:12,500,000

Map Number: 1712A
Map Name: Tectonic Assemblage Map of the Canadian Cordillera and Adjacent Parts of the United States of America
Year: 1991
Scale: 1:2,000,000

Map Number: 1860A
Map Name: Geological Map of Canada
Year: 1996
Scale: 1:5,000,000

Map Number: 1880A
Map Name: Surficial Materials of Canada
Year: 1995
Scale: 1:5,000,000

### Ontario Geological Survey

The Geospatial Centre has complete coverage of Ontario at the 1:250,000 scale.

### Road Maps—Historical

The Geospatial Centre has a small collection of historical road maps that cover cities and counties in Ontario, between 1928 and 1976.

County: Brant, Elgin, Middlesex, Oxford, and Norfolk
Year: 1961
Scale: 1" = 4 miles
Publisher: Department of Highways Ontario

County: Carleton
Year: 1928
Scale: 1" = 4 miles
Publisher: Ontario Department of Public Highways

County: Dufferin
Year: 1928

Scale: 1" = 4 miles
Publisher: Ontario Department of Public Highways

County: Dufferin, Huron, Perth, Waterloo, and Wellington
Year: 1972
Scale: 1" = 4 miles
Publisher: Ontario Department of Transportation and Communications

County: Dundas, Stormont, Glengarry
Year: 1928
Scale: 1" = 4 miles
Publisher: Ontario Department of Public Highways

County: Elgin
Year: 1928
Scale: 1" = 4 miles
Publisher: Ontario Department of Public Highways

County: Essex
Year: 1928
Scale: 1" = 4 miles
Publisher: Ontario Department of Public Highways

County: Frontenac
Year: 1935
Scale: 1" = 4 miles
Publisher: Ontario Department of Highways

County: Frontenac, Lennox, Addington
Year: 1961
Scale: 1" = 4 miles
Publisher: Department of Highways Ontario

County: Grey and Bruce
Year: 1961
Scale: 1" = 3 miles
Publisher: Fleming Publishing

County: Haldimand
Year: 1928, 1937
Scale: 1" = 4 miles
Publisher: Ontario Department of Public Highways

County: Haliburton, District Municipality of Muskoka
Year: 1972
Scale: 1" = 4 miles
Publisher: Ontario Deptartment of Transportation and Communications

County: Halton
Year: 1928
Scale: 1" = 4 miles
Publisher: Ontario Department of Public Highways

County: Hastings
Year: 1932
Scale: 1" = 4 miles
Publisher: Ontario Department of Highways

County: Huron, Perth
Year: 1957
Scale: 1" = 4 miles
Publisher: Ontario Department of Highways

County: Huron, Wellington, Dufferin, Perth, and Waterloo
Year: 1976
Scale: 1" = 4 miles
Publisher: Ontario Ministry of Transportation and Communications

County: Kent
Year: 1928
Scale: 1" = 4 miles
Publisher: Ontario Department of Public Highways

County: Lambton
Year: 1928
Scale: 1" = 4 miles
Publisher: Ontario Department of Public Highways

County: Lanark
Year: 1928
Scale: 1" = 4 miles
Publisher: Ontario Department of Public Highways

County: Leeds and Grenville
Year: 1928
Scale: 1" = 4 miles
Publisher: Ontario Department of Public Highways

County: Lennox and Addington
Year: 1935
Scale: 1" = 4 miles
Publisher: Ontario Department of Highways

County: Lincoln
Year: 1932
Scale: 1" = 4 miles
Publisher: Ontario Department of Highways

County: Manitoulin
Year: 1972
Scale: 1" = 4 miles
Publisher: Ontario Department of Transportation and Communications

County: Middlesex
Year: 1928
Scale: 1" = 4 miles
Publisher: Ontario Department of Public Highways

County: Norfolk
Year: 1928
Scale: 1" = 4 miles
Publisher: Ontario Department of Public Highways

County: Northumberland and Durham
Year: 1928
Scale: 1" = 4 miles
Publisher: Ontario Department of Public Highways

County: Northumberland, Durham, Peterborough, and Victoria
Year: 1961
Scale: 1" = 4 miles
Publisher: Department of Highways Ontario

County: Ontario
Year: 1932
Scale: 1" = 4 miles
Publisher: Ontario Department of Highways

County: Ontario and York
Year: 1963
Scale: 1" = 4 miles
Publisher: Ontario Department of Public Highways

County: Parry Sound (District)
Year: 1972
Scale: 1" = 4 miles
Publisher: Ontario Department of Transportation and Communication

County: Peel
Year: 1932
Scale: 1" = 4 miles
Publisher: Ontario Department of Highways

County: Peel, Halton, Dufferin, Wellington, and Waterloo
Year: 1951, 1954, 1959
Scale: 1" = 4 miles
Publisher: Ontario Department of Highways

County: Prescott and Russell
Year: 1928
Scale: 1" = 4 miles
Publisher: Ontario Department of Public Highways

County: Prescott and Russell, Stormont, Dundas, and Glengarry
Year: 1972
Scale: 1" = 4 miles
Publisher: Ontario Department of Transportation and Communications

County: Prince Edward
Year: 1928
Scale: 1" = 4 miles
Publisher: Ontario Department of Public Highways

County: Renfrew
Year: 1935
Scale: 1" = 4 miles
Publisher: Ontario Department of Highways

County: Simcoe
Year: 1932
Scale: 1" = 4 miles
Publisher: Ontario Department of Highways

County: Victoria
Year: 1932
Scale: 1" = 4 miles
Publisher: Ontario Department of Highways

County: Waterloo
Year: 1928
Scale: 1" = 4 miles
Publisher: Ontario Department of Highways

County: Welland
Year: 1928
Scale: 1" = 4 miles
Publisher: Ontario Department of Highways

County: Wentworth
Year: 1928
Scale: 1" = 4 miles
Publisher: Ontario Department of Highways

County: York
Year: 1935
Scale: 1" = 4 miles
Publisher: Ontario Department of Highways

**Base Maps**

Map Name: 911 Municipal Address Ranges City of Nanticoke
Year: 1997
Scale: 1:50,000

Map Name: Alberta Provincial Base Map
Year: 1994
Scale: 1:1,000,000

Map Name: Bechtel Park, City of Waterloo
Year: 1982
Scale: 1:2,400

Map Name: British Columbia
Year: 1973
Scale: 1:2,000,000

Map Name: British Columbia
Year: 1990
Scale: 1:2,000,000

Map Name: City of Barrie Municipal Works Department
Year: 1992
Scale: 1:20,000

Map Name: City of Brantford
Year: 1993
Scale: 1:20,000

Map Name: City of Cambridge
Year: 1986—2003
Scale: 1:2,000–1:18,000

Map Name: City of Cornwall
Year: 1997
Scale: 1:10,000

Map Name: City of Gloucester
Year: 1992
Scale: 1:25,000

Map Name: City of Guelph
Year: 1986
Scale: 1:10,000

Map Name: City of Hamilton, Map and Street Guide
Year: 1995
Scale: 1:19,200

Map Name: City of Kitchener
Year: 1988
Scale: 1:10,000

Map Name: City of Kitchener, Downtown Area
Year: 1984
Scale: 1:1,200

Map Name: City of Niagara Falls, Ontario, Canada
Year: 1996
Scale: 1:25,000

Map Name: City of North Bay, Rural Area
Year: 1996
Scale: 1:126,720

Map Name: City of North Bay, Urban Area
Year: 1996
Scale: 1:20,000

Map Name: City of Owen Sound
Year: 1991
Scale: 1:6,000

Map Name: City of Scarborough
Year: 1993
Scale: 1:18,000

Map Name: City of St. Catharines
Year: 1996
Scale: 1:20,000

Map Name: City of Vancouver, British Columbia
Year: 1992
Scale: 1:15,000

Map Name: City of Waterloo
Year: 1988
Scale: 1:10,000

Map Name: City of Waterloo (4 sheets)
Year: 1988
Scale: 1:2,000

Map Name: Close-up, Canada, British Columbia, Alberta, and the Yukon
Year: 1978
Scale: 1:3,500,000

Map Name: Counties of Bruce, Grey
Year: 1977
Scale: 1:250,000

Map Name: County of Brant, Country Seat—Brantford
Year: 1979
Scale: 1:100,000

Map Name: County of Bruce
Year: 1980
Scale: 1:100,000

Map Name: County of Dufferin
Year: 1976, 1979
Scale: 1:100,000

Map Name: County of Elgin
Year: 1981
Scale: 1:100,000

Map Name: County of Essex
Year: 1979
Scale: 1:100,000

Map Name: County of Frontenac
Year: 1980
Scale: 1:100,000

Map Name: County of Grey
Year: 1980
Scale: 1:100,000

Map Name: County of Hastings
Year: 1980
Scale: 1:100,000

Map Name: County of Huron
Year: 1979
Scale: 1:100,000

Map Name: County of Kent
Year: 1979
Scale: 1:100,000

Map Name: County of Kent: County Seat—Chatham
  Map Subject: Base Map
Year: 1980
Scale: 1:126,720

Map Name: County of Lambton
Year: 1981
Scale: 1:100,000

Map Name: County of Lanark
Year: 1980
Scale: 1:100,000

Map Name: County of Lennox and Addington North Portion
Year: 1980
Scale: 1:100,000

Map Name: County of Lennox and Addington South Portion
Year: 1980
Scale: 1:100,000

Map Name: County of Middlesex
Year: 1978, 1981
Scale: 1:52,800–1:100,000

Map Name: County of Northumberland
Year: 1979
Scale: 1:100,000

Map Name: County of Oxford
Year: 1970, 1981
Scale: 1:100,000–1:142,857

Map Name: County of Perth
Year: 1979
Scale: 1:100,000

Map Name: County of Peterborough
Year: 1979
Scale: 1:100,000

Map Name: County of Prince Edward
Year: 1980
Scale: 1:100,000

Map Name: County of Renfrew
Year: 1979
Scale: 1:250,000

Map Name: County of Renfrew: County Seat—Pembroke
Year: 1980
Scale: 1:100,000

Map Name: County of Simcoe
Year: 1974, 1980
Scale: 1:100,000–1:250,000

Map Name: County of Victoria
Year: 1979
Scale: 1:100,000

Map Name: County of Wellington
Year: 1963, 1981
Scale: 1:100,000–1:126,720

Map Name: District of Manitoulin
Year: 1977
Scale: 1:250,000

Map Name: District of Muskoka
Year: 1971, 1980
Scale: 1:100,000

Map Name: District of Nipissing
Year: 1978
Scale: 1:100,000–1:250,000

Map Name: District of Parry Sound
Year: 1979
Scale: 1:250,000

Map Name: District of Rainy River
Year: 1947
Scale: 1:126,720

Map Name: District of Sudbury—Southwest Portion
Year: 1972
Scale: 1:126,720

Map Name: District of Timiskaming, Ontario
Year: 1972
Scale: 1:126,720

Map Name: Edmonton, Military City Map
Year: 1979
Scale: 1:25,000

Map Name: Flood Risk Map: Laurel Creek (6 sheets, 1 index)
Year: 1984
Scale: 1:2,000

Map Name: Great Lakes Region Canada
Year: 1969
Scale: 1:2,000,000

Map Name: Labrador
Year: 1992
Scale: 1:1,000,000

Map Name: Manitoba
Year: 1964
Scale: 1:1,760,320

Map Name: Metropolitan Windsor
Year: 1995
Scale: 1:2,000

Map Name: Military City Map 1:25,000. Whitehorse
Year: 1980
Scale: 1:25,000

Map Name: Northern Canada
Year: 2012
Scale: 1:4,000,000

Map Name: Northwest Territories and Yukon Territory
Year: 1994
Scale: 1:4,000,000

Map Name: Ontario
Year: 1991
Scale: 1:1,500,000

Map Name: Port Perry/Port Albert, Township of Scugog
Year: 1993
Scale: 1:5,000

Map Name: Prairie Region, Canada
Year: 1964
Scale: 1:2,000,000

Map Name: Region Municipality of Niagara, Town of Lincoln
Year: 1970
Scale: 1:48,000

Map Name: Regional Base Map Series, Ontario
Year: 1992
Scale: 1:2,000,000

Map Name: Regional Municipality of Durham
Year: 1974, 1981, 1991
Scale: 1:100,000

Map Name: Regional Municipality of Haldimand-Norfolk
Year: 1973
Scale: 1:126,720

Map Name: Regional Municipality of Haldimand-Norfolk
Year: 1979, 1992
Scale: 1:125,000

Map Name: Regional Municipality of Halton
Year: 1974
Scale: 1:31,680

Map Name: Regional Municipality of Halton
Year: 1981
Scale: 1:63,360

Map Name: Regional Municipality of Hamilton-Wentworth
Year: 1974, 1980
Scale: 1:120,000–1:126,720

Map Name: Regional Municipality of Niagara
Year: 1976, 1981
Scale: 1:100,000–1:125,000

Map Name: Regional Municipality of Niagara
Year: 1976, 1988
Scale: 1:100,000–1:200,000

Map Name: Regional Municipality of Niagara Area Municipality, Town of Fort Erie
Year: 1970
Scale: 1:24,000

Map Name: Regional Municipality of Niagara Municipality (Niagara on the Lake)
Year: 1970
Scale: 1:24,000

Map Name: Regional Municipality of Ottawa-Carleton
Year: 1974, 1979
Scale: 1:100,000–1:126,720

Map Name: Regional Municipality of Peel
Year: 1974, 1979
Scale: 1:63,360–1:100,000

Map Name: Regional Municipality of Sudbury
Year: 1980
Scale: 1:126,720

Map Name: Regional Municipality of Waterloo
Year: 1973, 1975, 1982, 1988
Scale: 1:50,000–1:100,000

Map Name: Regional Municipality of Waterloo—Township of Wilmot
Year: 1998
Scale: 1:7,500

Map Name: Regional Municipality of York
Year: 1971, 1991
Scale: 1:10,000–1:100,000

Map Name: Saskatchewan
Year: 1963
Scale: 1:760,320

Map Name: Southern Ontario
Year: 1977, 1980
Scale: 1:600,000–1:1,000,000

Map Name: Thunder Bay, Military City Map
Year: 1981
Scale: 1:25,000

Map Name: Town of Ancaster
Year: 1994
Scale: 1:40,000

Map Name: Town of Brighton
Year: 1993
Scale: 1:5,000

Map Name: Town of Caledon
Year: 1993
Scale: 1:75,000

Map Name: Town of Grimsby
Year: 1970
Scale: 1:24,000

Map Name: Town of Midland
Year: 1998
Scale: 1:10,000

Map Name: Town of Milton, Base Map
Year: 1992
Scale: 1:25,000

Map Name: Town of Newmarket
Year: 1998
Scale: 1:10,000

Map Name: Town of Pelham
Year: 1989
Scale: 1:18,000

Map Name: Town of Port Hope
Year: 1991
Scale: 1:10,000

Map Name: Town of Renfrew
Year: 1993
Scale: 1:4,800

Map Name: Town of Simcoe
Year: 1990
Scale: 1:12,000

Map Name: Town of St. Marys
Year: 1990
Scale: 1:5,000

Map Name: Town of Ville de Rockland
Year: 1993
Scale: 1:5,000

Map Name: Township of Brantford
Year: 1993
Scale: 1:30,000

Map Name: Township of Wellesley
Year: 1970, 1979
Scale: 1:20,000–1:31,680

Map Name: Township of Wilmot
Year: 1976, 1979, 1988
Scale: 1:20,000–1:180,000

Map Name: Township of Woolwich
Year: 1970, 1979
Scale: 1:20,000–1:60,000

Map Name: United Counties of Leeds and Grenville
Year: 1979, 1992
Scale: 1:100,000–1:126,720

Map Name: United Counties of Prescott and Russell
Year: 1979
Scale: 1:100,000

Map Name: United Counties of Stormont, Dundas, and Glengarry
Year: 1979, 1993
Scale: 1:100,000

Map Name: University of Waterloo
Year: 1990, 2003, 2004
Scale: 1:2,000

Map Name: Waterloo Park
Year: 1978
Scale: 1:1,200

Map Name: Waterloo Park
Year: 1988
Scale: 1:1,000

Map Name: Western Canada
Year: 1966
Scale: 1:5,068,800

Map Name: Yukon Territory, Northwest Territories, and Nunavut
Year: 1974
Scale: 1:4,000,000

**Biogeography**

Map Name: Canada, Plant Hardiness Zones
Year: 2001
Scale: 1:10,000,000

Map Name: Canada, Vegetation Cover
Year: 1993
Scale: 1:7,500,000

Map Name: Canada's Land Cover
Year: 2004
Scale: 1:6,000,000

Map Name: Ecoclimatic Regions of Canada
Year: 1989
Scale: 1:7,500,000

Map Name: Ecoregions of Alberta
Year: 1992
Scale: 1:1,500,000

Map Name: Ecoregions of Saskatchewan
Year: 1994
Scale: 1:2,000,000

Map Name: Ecozones, Ecoregions and Ecodistricts of Quebec
Year: 1995
Scale: 1:3,500,000

Map Name: Fish and Wildlife Resources, Ontario
Year: 1986
Scale: 1:50,000

Map Name: Forest Map of Manitoba
Year: 1988
Scale: 1:1,000,000

Map Name: Forest Regions of Canada
Year: 1972
Scale: 1:6,336,000

Map Name: Forest Stand Map, Ontario
Year: 1978
Scale: 1:10,000

Map Name: Forestry Resources, Ontario
Year: 1980
Scale: 1:50,000

Map Name: Grape Climatic Zones in Niagara
Year: 1976
Scale: 1:63,360

Map Name: Greening Plan (Niagara Falls)
Year: 1993
Scale: 1:22,500

Map Name: Land Cover Map of Manitoba
Year: 1989
Scale: 1:1,000,000

Map Name: Land Cover Map of the Maritimes
Year: 1991
Scale: 1:1,000,000

Map Name: Land Cover of Canada
Year: 1999
Scale: 1:5,000,000

Map Name: Natural Areas, Cambridge District, Ontario
Year: 1986
Scale: 1:50,000

Map Name: Natural History Map of Nova Scotia
Year: 1987
Scale: 1:625,000

Map Name: Niagara Escarpment Study, Fruit Belt Report, Orchards and Vineyards, 1965
Year: 1965
Scale: 1:63,360

Map Name: Northern Perspectives. Map No. 2: Caribou of Canada's Lands
Year: 1987
Scale: 1:5,000,000

Map Name: Northern Perspectives. Map No. 3: Native Hunting and Trapping Areas in Canada's Northern Lands 1970–1985: An Overview
Year: 1987
Scale: 1:5,000,000

Map Name: Ottawa Valley
Year: 2005
Scale: 1:500,000

Map Name: Plant Hardiness Zones in Canada
Year: 1967
Scale: 1:4,750,000

Map Name: Provincially Significant Wetland Buffer (Niagara Falls)
Year: 1993
Scale: 1:22,500

Map Name: Provisional Forest Map of Canada
Year: 1979
Scale: 1:5,000,000

Map Name: Soil Landscapes of Canada, Yukon Territory
Year: 1992
Scale: 1:1,000,000

Map Name: Terrestrial Ecozones, Ecoregions, and Ecodistricts: Alberta, Saskatchewan and Manitoba
Year: 1995
Scale: 1:2,000,000

Map Name: Terrestrial Ecozones, Ecoregions, and Ecodistricts: Alberta, Saskatchewan and Manitoba
Year: 1995
Scale: 1:3,500,000

Map Name: Terrestrial Ecozones, Ecoregions, and Ecodistricts: British Columbia and the Yukon
Year: 1995
Scale: 1:3,500,000

Map Name: Terrestrial Ecozones, Ecoregions and Ecodistricts: New Brunswick, Nova Scotia, Prince Edward Island and Newfoundland
Year: 1995
Scale: 1:7,500,000

Map Name: Terrestrial Ecozones, Ecoregions, and Ecodistricts of Northwest Territories
Year: 1995
Scale: 1:3,500,000

Map Name: Terrestrial Ecozones, Ecoregions and Ecodistricts: Province of Ontario
Year: 1995
Scale: 1:3,500,000

Map Name: Urban Wooded and Treed Site Locations (Niagara Falls)
Year: 1993
Scale: 1:22,500

Map Name: Vegetation Patterns of the Hudson Bay Lowlands
Year: 1969
Scale: 1:633,600

Map Name: Water Erosion Risk Map, Alberta
Year: 1993
Scale: 1:1,000,000

Map Name: Water Erosion Risk Map, Manitoba
Year: 1986
Scale: 1:1,000,000

Map Name: Water Erosion Risk Map, Saskatchewan
Year: 1988
Scale: 1:1,000,000

Map Name: Wildland Fire Occurrence in Canada
Year: 1975
Scale: 1:6,336,000

Map Name: Wind Erosion Risk Map, Alberta
Year: 1989
Scale: 1:1,000,000

Map Name: Wind Erosion Risk Map, Manitoba
Year: 1988
Scale: 1:1,000,000

Map Name: Wind Erosion Risk Map, Saskatchewan
Year: 1987
Scale: 1:1,000,000

Map Name: Woodlots Oxford County
    Map Subject: Biogeography
Year: 1999
Scale: 1:50,000

**Historical**

Map Name: Berlin
Year: 1879
Scale: 15 chains per inch

Map Name: Bird's Eye View of London, Ontario, Canada, 1872
Year: 1872
Scale: N/A

Map Name: Bird's Eye View of the City of Hamilton: Province Ontario, Canada
Year: 1876
Scale: 1:2,933,000

Map Name: Bird's Eye View of the City of Ottawa, Province Ontario, Canada
Year: 1876
Scale: 1:16,200

Map Name: Bird's Eye View of Toronto, 1876
Year: 1876
Scale: N/A

Map Name: Canada Divided into Counties and Ridings as per Union Bill
Year: 1840
Scale: 1:2,600,000

Map Name: Canada West, Formerly Upper Canada
Year: 1850
Scale: 1:2,111,789

Map Name: Carleton and Russell with a Correct Map of the City of Ottawa
Year: 1856
Scale: 1:137,000

Map Name: City of Hamilton, Canada with Views of Principal Business Buildings
Year: 1894
Scale: 1:4,300

Map Name: City of Kitchener
Year: 1937
Scale: 1:3,375

Map Name: City of Kitchener and Town of Waterloo
Year: 1923
Scale: 1:4,000

Map Name: City of Ottawa, Canada with Views of Principal Business Buildings
Year: 1893
Scale: N/A

Map Name: City of Toronto
Year: 1860
Scale: N/A

Map Name: City of Toronto and Liberties
Year: 1834
Scale: 1:15,840

Map Name: City Plan for Greater Berlin
Year: 1912
Scale: 0.597222222

Map Name: City Plan for Greater Berlin, Canada, Showing Waterloo
Year: 1914
Scale: 0.597222222

Map Name: County of Essex
Year: 1879
Scale: 1:126,720

Map Name: County of Haldimand, Ontario
Year: 1879
Scale: 1:190,000

Map Name: County of Huron
Year: 1879
Scale: N/A

Map Name: County of Kent
Year: 1879
Scale: 1:158,400

Map Name: County of Norfolk
Year: 1879
Scale: N/A

Map Name: County of Waterloo
Year: 1879
Scale: 1:126,720

Map Name: County of Waterloo
Year: 1908
Scale: 1:150,000

Map Name: Hudson's Bay Company Historic Trading Posts and Territories of the Governor and Company of Adventurers of England Trading into Hudson's Bay
Year: 1976
Scale: 1:8,500,000

Map Name: Lower Canada in Counties as Previous to Act of 1829; Lower Canada as Divided by Act of 1829
Year: 1829
Scale: N/A

Map Name: Making of Canada, Ontario
Year: 1996
Scale: 1:2,667,000

Map Name: Manitoba Natural Resources Historical Map
Year: 1990
Scale: 1:2,000,000

Map Name: Map of Part of the Proposed Canal through the District of Niagara and Gore to Form a Junction of Lakes and Ontario by the Grand River
Year: 1823
Scale: 1:3,068

Map Name: Map of Part of the Province of Upper Canada
Year: 1825
Scale: 1:506,880

Map Name: Map of Part of the Town of Berlin, Capital of the County of Waterloo
Year: 1982
Scale: 1:6,336

Map Name: Map of the City of Ottawa
Year: 1874
Scale: 1:20,100

Map Name: Map of the Counties of Wentworth Part of Brant and Lincoln, Haldimand, Welland
Year: 1866
Scale: 1:2,600,000

Map Name: Map of the County of Elgin, Ont.
Year: 1879
Scale: 1:4,800

Map Name: Map of the County of Lambton, Ontario
Year: 1879
Scale: 1:158,400

Map Name: Map of the County of Middlesex, Ontario
Year: 1879
Scale: 1:158,500

Map Name: Map of the County of Oxford
Year: 1879
Scale: 1:150,000

-Map Name: Map of the County of Perth
Year: 1879
Scale: 1:2,880,979

Map Name: Map of the Island of Montreal
Year: 1892
Scale: N/A

Map Name: Map of the Province of Upper Canada, Describing All the New Settlements, Townships
Year: 1800
Scale: N/A

Map Name: Map of the Province of Upper Canada and the Adjacent Territories in North America
Year: 1826
Scale: N/A

Map Name: Map of the Town of Berlin
Year: 1855
Scale: 8 chains to the inch

Map Name: Map of the Town of Berlin
Year: 1879
Scale: 2.241666667

Map Name: Map of the Town of Berlin, Capital of the County of Waterloo C.W.
Year: 1855
Scale: 1:13,000

Map Name: Map of the Town of Berlin, Waterloo Co., Ontario, from Actual Surveys and Records
Year: 1879
Scale: 4 chains per inch

Map Name: Map of the Town of Brantford, County of Brant, Canada West
Year: 1852
Scale: 1:4,300

Map Name: Map of the Town of Dundas in the Counties of Wentworth and Halton, Canada West
Year: 1851
Scale: N/A

Map Name: Map of the Town of Galt
Year: 1867
Scale: 2.708333333

Map Name: Map of the Townships of Waterloo and Woolwich U.C.
Year: 1815
Scale: N/A

Map Name: Map of the Village of Waterloo
Year: 1855
Scale: 2.241666667

Map Name: Map of Toronto
Year: 1878
Scale: 1:12,000

Map Name: Map of Upper Canada Engraved for Statistical Account
Year: 1873
Scale: N/A

Map Name: Maps Shewing Mounted Police Stations & Patrols 1888
Year: 1888
Scale: 1:160,934

Map Name: Montreal (1859)
Year: 1859
Scale: N/A

Map Name: Montreal (1889)
Year: 1889
Scale: N/A

Map Name: New Map of Upper & Lower Canada from the Latest Authorities
Year: 1807
Scale: 1:6,200,000

Map Name: Niagara Frontier
Year: 1865
Scale: 1:63,360

Map Name: Plan of John Hoffman's Survey of Parts of Lots no. 14 and 15 in Waterloo, C. W.

Year: 1855
Scale: N/A

Map Name: Plan of Quebec
Year: 1759
Scale: 1:3,600

Map Name: Plan of the City and Environs of Quebec, with Its Siege and Blockade by the Americans
Year: 1776
Scale: N/A

Map Name: Plan of the City of Quebec the Capital of Canada
Year: 1759
Scale: 1:3,430

Map Name: Plan of the City of Toronto
Year: 1885
Scale: N/A

Map Name: Plan of the City of Toronto
Year: 1898
Scale: 1:20,000

Map Name: Plan of the City of Toronto
Year: 1902
Scale: 1:20,000

Map Name: Plan of the City of Toronto
Year: 1912
Scale: 1:24,000

Map Name: Plan of the City of Toronto, Canada West
Year: 1857
Scale: 1:4,752

Map Name: Plan of the Harbour, Fort and Town of York, the capital of Upper Canada
Year: 1816
Scale: 1:150,000

Map Name: Plan of the Line of the Rideau Canal
Year: 1829
Scale: 1:14,400

Map Name: Plan of the Ordnance Reserve at Toronto, Canada
Year: 1862
Scale: 1:4,000

Map Name: Plan of the Town and Fortifications of Montreal or Ville Marie in Canada
Year: 1758
Scale: 600 yards

Map Name: Plan of the Town and Harbour of York
Year: 1814
Scale: 1:11,880

Map Name: Plan of the Town and Harbour of York, Upper Canada and also Military Reserve
Year: 1833
Scale: 1:126,720

Map Name: Plan of the Town of Hamilton District of Gore
Year: 1842
Scale: 1:1,500,000

Map Name: Plan of the Town of Niagara
Year: 1966
Scale: 1:3,068

Map Name: Plan of York
Year: 1823
Scale: 1:14,400

Map Name: Plan of York Harbour
Year: 1793
Scale: 1:20,000

Map Name: Plan of York Harbour and Humber Bay
Year: 1815
Scale: N/A

Map Name: Plan Shewing the Survey of the Land Reserved for Government Buildings, East End of the Town of York
Year: 1810
Scale: 1:9,130

Map Name: Proposed Rural Mail Route (Waterloo and Return)
Year: 1895
Scale: 40 chains to 1 inch

Map Name: Province of Canada (West Sheet)
Year: 1865
Scale: 1:15,840

Map Name: Quebec A.D. MDCCCCV
Year: 1905
Scale: N/A

Map Name: Renvoy des chifres qui se trouvent dans la Ville de Quebec
Year: 1694
Scale: N/A

Map Name: Sketch Map of Upper Canada, Showing the Routes Lt. Gov. J.G. Simcoe Took on Trips between March, 1792, and September, 1795
Year: 1795
Scale: 1:250,000

Map Name: Sketch of the Proposed District of Wellington
Year: 1837
Scale: N/A

Map Name: Southwestern Ontario—Origins—the Long Point Settlement
Year: 1990
Scale: 1:84,480

Map Name: St. Catharines, 1875, Province Ontario, Canada
Year: 1875
Scale: 1:6,336

Map Name: Topographical Plan of the City of Toronto
Year: 1851
Scale: 1:18,500

Map Name: Toronto, Canada West
Year: 1857
Scale: 1:12,950

Map Name: Town of Berlin
Year: 1853
Scale: 4.441666667

Map Name: Town of Berlin
Year: 1856
Scale: 1.141666667

Map Name: Town of Goderich
Year: 1879
Scale: 1:7,920

Map Name: Town of Waterloo, Canada
Year: 1891
Scale: N/A

Map Name: Town of Waterloo, Canada. With Views of Principal Business Buildings
Year: 1997
Scale: N/A

Map Name: Tremaine's Map of the County of Waterloo Canada West (Part A)
Year: 1861
Scale: N/A

Map Name: Tremaine's Map of the County of Waterloo Canada West (Part B)
Year: 1861
Scale: N/A

Map Name: Tremaine's Map of the County of York, Canada West, Compiled and Drawn by Geo R. Tremain
Year: 1860
Scale: 60 chains to 1 inch

Map Name: Upper Canada in Counties and Riding as at Present Divided
Year: 1840
Scale: 1:253,440

Map Name: Wellesley
Year: 1843
Scale: 40 chains per inch

Map Name: Wilmot
Year: 1828
Scale: 40 chains per inch

Map Name: Woolwich
Year: 1807
Scale: N/A

**Human Geography**

Map Name: Aboriginal Communities and Minerals and Metals Activities
Year: 2001
Scale: 1:2,000,000

Map Name: Ajax—Plan of Roads
Year: 1993
Scale: 1:12,500

Map Name: Alberta Air Facilities Map
Year: 1984
Scale: 1:1,000,000

Map Name: Atlantic Provinces Aboriginal Communities and Minerals and Metals Activities
Year: 2001
Scale: 1:2,000,000

Map Name: British Columbia Aboriginal Communities and Minerals and Metals Activities
Year: 1996
Scale: 1:2,000,000

Map Name: Canada Air Transportation Network
Year: 1983
Scale: 1:7,500,000

Map Name: Canada Distribution of Population, 1961
Year: 1961
Scale: 1:7,500,000

Map Name: Canada, Electricity, 1987
Year: 1987
Scale: 1:7,500,000

Map Name: Canada Employment Growth
Year: 1993
Scale: 1:7,500,000

Map Name: Canada Employment Variability
Year: 1990
Scale: 1:7,500,000

Map Name: Canada, Ethnic Diversity
Year: 1993
Scale: 1:7,500,000

Map Name: Canada Exploration 1497–1650
Year: 1991
Scale: 1:7,500,000

Map Name: Canada Exploration 1651–1760
Year: 1991
Scale: 1:7,500,000

Map Name: Canada, Farm Operators
Year: 1985
Scale: 1:7,500,000

Map Name: Canada, Fisheries Resources
Year: 1987
Scale: 1:7,500,000

Map Name: Canada—Heritage Areas
Year: 1994
Scale: 1:5,000,000

Map Name: Canada, Indian and Inuit Communities and Languages
Year: 1980
Scale: 1:7,500,000

Map Name: Canada-Indian and Inuit Communities Atlantic Provinces
Year: 1983
Scale: 1:2,000,000

Map Name: Canada-Indian and Inuit Communities, British Columbia
Year: 1984
Scale: 1:2,000,000

Map Name: Canada, Indian and Inuit Communities, Northwest Territories and Yukon
Year: 1984
Scale: 1:4,000,000

Map Name: Canada-Indian and Inuit Communities Quebec
Year: 1984
Scale: 1:2,000,000

Map Name: Canada-Indian and Inuit Communities within the Prairies
Year: 1984
Scale: 1:2,000,000

Map Name: Canada, Indian and Inuit Population Distribution
Year: 1981
Scale: 1:7,500,000

Map Name: Canada Indian Treaties
Year: 1991
Scale: 1:7,500,000

Map Name: Canada Manpower Centres
Year: 1976
Scale: 1:2,000,000

Map Name: Canada, Manufacturing
Year: 1981
Scale: 1:7,500,000

Map Name: Canada, Manufacturing Key Sectors
Year: 1986
Scale: 1:25,000,000

Map Name: Canada, Manufacturing Productivity
Year: 1986
Scale: 1:7,500,000

Map Name: Canada—National Parks
Year: 1985
Scale: 1:5,000,000

Map Name: Canada, Native Peoples 1630
Year: 1988
Scale: 1:7,500,000

Map Name: Canada, Native Peoples, 1740
Year: 1988
Scale: 1:7,500,000

Map Name: Canada, Native Peoples, 1823
Year: 1988
Scale: 1:7,500,000

Map Name: Canada Oil Pipelines
Year: 1980
Scale: 1:7,500,000

Map Name: Canada, Population Density 1976
Year: 1976
Scale: 1:7,500,000

Map Name: Canada, Population Density, 1871
Year: 1871
Scale: 1:7,500,000

Map Name: Canada, Population Density, 1976
Year: 1976
Scale: 1:7,500,000

Map Name: Canada, Public Fish Hatcheries
Year: 1983
Scale: 1:7,500,000

Map Name: Canada, Pulp and Paper Mills
Year: 1986
Scale: 1:7,500,000

Map Name: Canada Railway Transportation Network
Year: 1984
Scale: 1:7,500,000

Map Name: Canada, Road Transportation Network
Year: 1984
Scale: 1:7,500,000

Map Name: Canada, Settlement Pattern
Year: 1981
Scale: 1:7,500,000

Map Name: Canada, Showing Location of Indian Bands with Linguistic Affiliations 1965
Year: 1965
Scale: 1:6,336,000

Map Name: Canada Telecommunications Systems
Year: 1984
Scale: 1:7,500,000

Map Name: Canada, the Aging Population
Year: 1993
Scale: 1:7,500,000

Map Name: Canada, the Urban System
Year: 1991
Scale: 1:7,500,000

Map Name: Canada Transportation Routes
Year: 1983
Scale: 1:7,500,000

Map Name: Canada Water Transportation Infrastructure
Year: 1986
Scale: 1:7,500,000

Map Name: Canada's Sawmills
Year: 2003
Scale: 1:6,000,000

Map Name: Centrales et lignes de transport d'Hydro-Quebec
Year: 1987
Scale: 1:1,000,000

Map Name: Circonscriptions electorales federales—1987
Year: 1987
Scale: 1:5,000,000

Map Name: Cities of Waterloo RISC Boundaries, Population
Year: 1985
Scale: 1:500

Map Name: City of Belleville Street Map
Year: 1996
Scale: 1:10,000

Map Name: City of Brampton, Proposed Highway 407
Year: 1996
Scale: 1:25,000

Map Name: City of Burlington—Small Street Map
Year: 1994
Scale: 1:25,000

Map Name: City of Burlington Planning Department Large Street Map
Year: 1987
Scale: 1:12,500

Map Name: City of Cambridge, Official Plan
Year: 1981
Scale: 1:25,000

Map Name: City of Cambridge, Wards and Population
Year: 1991
Scale: 1:18,000

Map Name: City of Guelph—City Street Map
Year: 2001
Scale: 1:15,000

Map Name: City of Guelph Department of Planning and Development City Street Map
Year: 1990
Scale: 1:15,000

Map Name: City of Hamilton, Map and Street Guide
Year: 1995
Scale: 1:15,000

Map Name: City of Kingston, Consolidated Street Map
Year: 1995
Scale: 1:10,000

Map Name: City of Sudbury Street Map
Year: 1992
Scale: 1:25,000

Map Name: City of Toronto, Building Construction Dates
Year: 2003
Scale: 1:30,000

Map Name: City Street Map—City of Guelph
Year: 1994
Scale: 1:10,000

Map Name: Completing Canada's National Park System
Year: 1998
Scale: 1:6,000,000

Map Name: Conservation Areas in Ontario
Year: 2001
Scale: 1:800,000

Map Name: County of Lambton Road System
Year: 1970
Scale: 1:144,822

Map Name: County Road System—County of Essex
Year: 1980
Scale: 1:63,360

Map Name: Distribution des terres des communautes et groupes culturels autochtones
Year: 1989
Scale: 1:4,000,000

Map Name: Distribution of First Nations by Language Groupings
Year: 1996
Scale: 1:2,600,000

Map Name: Ecclesiastical Divisions of Canada
Year: 1963
Scale: 1:6,336,000

Map Name: Energy Resources of British Columbia
Year: 1991
Scale: 1:2,000,000

Map Name: First Generations: An Ontario Heritage Map
Year: 1986
Scale: 1:1,000,000

Map Name: First Nations Ontario
Year: 2008
Scale: N/A

Map Name: Greater Toronto Area Retail Density
Year: 2008
Scale: N/A

Map Name: Hydroelectric Development and Native Communities of Northern Quebec
Year: 1991
Scale: 1:4,000,000

Map Name: Inuit Owned Lands in Nunavut
Year: 2000
Scale: 1:3,100,000

Map Name: Kootenay Park, Alberta–British Columbia
Year: 1974
Scale: 1:126,720

Map Name: Les nations autochtones au Quebec
Year: 1988
Scale: 1:2,500,000

Map Name: Les sites de dechets toxiques Au Quebec
Year: 1986
Scale: 1:1,000,000

Map Name: Metropolitan Toronto (Parklands and Selected Open Spaces)
Year: 1998
Scale: 1:38,000

Map Name: Midwestern Ontario, Outdoor Recreation
Year: 1981
Scale: 1:300,000

Map Name: Midwestern Ontario: Outdoor Recreation: 25,000 square kilometers
Year: 1981
Scale: 1:300,000

Map Name: Northwest Territories: Aboriginal Communities and Mineral and Metal Activities
Year: 1996
Scale: 1:4,000,000

Map Name: Official Plan for the Town of Halton Hills, Schedule 5: Transportation
Year: 1994
Scale: 1:25,000

Map Name: Official Plan of Richmond Hill, Schedule 4: Transportation
Year: 1991
Scale: 1:20,000

Map Name: Official Plan of Richmond Hill, Schedule 5: Services
Year: 1991
Scale: 1:20,000

Map Name: Official Plan of the City of Brantford, Schedule 4: Transportation Plan
Year: 1995
Scale: 1:20,000

Map Name: Ontario Aboriginal Communities and Minerals and Metals Activities
Year: 2001
Scale: 1:2,000,000

Map Name: Ontario Hydro Principal Power Facilities and Municipal Systems Served
Year: 1986
Scale: 1:1,000,000

Map Name: Ontario Parks, a World-Renowned Park System
Year: 1999
Scale: 1:2,200,000

Map Name: Ontario Transportation Map Series
Year: 1983
Scale: 1:250,000

Map Name: Ontario's Living Legacy, Land Use Strategy
Year: 1999
Scale: 1:1,200,000

Map Name: Provincial Electric System of Alberta
Year: 1984
Scale: 1:1,500,000

Map Name: Quetico Provincial Park, Canoe Map
Year: 1999
Scale: 1:73,000

Map Name: Railway Map of Southern Ontario
Year: 1985
Scale: 1:1,000,000

Map Name: Regional Municipality of Niagara, Pollution Control Facilities
Year: 1993
Scale: 1:100,000

Map Name: Regional Municipality of Sudbury Regional Road System
Year: 1990
Scale: 1:125,000

Map Name: Riding Mountain National Park, Manitoba
Year: 1976
Scale: 1:250,000

Map Name: Road Map of the Counties of Bruce Grey
Year: 1972
Scale: 1:253,440

Map Name: Road Maps of the Counties of Dufferin, Huron, Perth, Waterloo, Wellington
Year: 1972
Scale: 1:253,440

Map Name: Street Map, Town of Richmond Hill
Year: 1993
Scale: 1:20,000

Map Name: Terra Nova National Park
Year: 1966
Scale: 1:50,000

Map Name: Town of Fort Erie, Street Map
Year: 1996
Scale: 1:24,000

Map Name: Town of Richmond Hill, Community Facilities
Year: 1992
Scale: 1:20,000

Map Name: Township of Halton Hills, Municipal Street Numbering
Year: 1991
Scale: 1:40,000

Map Name: Traditional Territories of British Columbia First Nations
Year: 1994
Scale: 1:2,000,000

Map Name: Treaties and Comprehensive Land Claims in Canada
Year: 2001
Scale: 1:7,500,000

Map Name: Yoho Park, Alberta–British Columbia
Year: 1961
Scale: 1:126,720

**Land Use**

Map Name: 1975 Municipal Division County of Oxford
Year: 1975
Scale: 1:100,000

Map Name: 1982 Land Use, City of Kingston
Year: 1982
Scale: 1:10,000

Map Name: 1983 Existing Land Use, Metropolitan Toronto and Immediate Region
Year: 1983
Scale: 1:120,000

Map Name: 1985 Rural Land Use—Regional Municipality of Ottawa-Carleton
Year: 1985
Scale: 1:100,000

Map Name: 1985 Urban Land Use Map for the Region Municipality of Ottawa
Year: 1985
Scale: 1:25,000

Map Name: Agricultural Land Use Systems of the Regional Municipality of Niagara Ontario
Year: 1980
Scale: 1:25,000

Map Name: Agricultural Land Use Systems of the Regional Municipality of Ottawa-Carleton
Year: 1977
Scale: 1:50,000

Map Name: Ajax—Designated Land Uses
Year: 1993
Scale: 1:12,000

Map Name: Amendment to the City of Kitchener Official Plan—Map 1
Year: 1993
Scale: 1:24,000

Map Name: Appendix A, Zoning By-law, Township of Thorold
Year: 1991
Scale: 1:25,000

Map Name: Ayr Settlement Area—Township of North Dumfries
Year: 1988
Scale: 1:2,400

Map Name: Background Story—Town of Grimsby—Existing Land Use, Transportation and Energy Corridors
Year: 1984
Scale: 1:14,400

Map Name: Borough of East York—Consolidated Zoning Map
Year: 1967, 1985, 1987, 1992
Scale: 1:12,000

Map Name: Canada, Electricity Generation and Transmission
Year: 1983
Scale: 1:5,000,000

Map Name: Canada, Farm Types
Year: 1988
Scale: 1:7,500,000

Map Name: City of Barrie—Schedule A—Land Use Plan
Year: 1993
Scale: 1:10,000

Map Name: City of Barrie Zoning By-law 85–95
Year: 1994
Scale: 1:10,000

Map Name: City of Brampton, General Land Use Designations
Year: 1993
Scale: 1:120,000

Map Name: City of Brantford—School and Park Sites
Year: 1992, 1996
Scale: 1:10,000

Map Name: City of Brantford—Generalized Existing Land Use
Year: 1996
Scale: 1:15,000

Map Name: City of Brockville Planning Department—Plate A to Zoning By-law
Year: 1987
Scale: 1:4,800

Map Name: City of Brockville Planning Department—Schedule A Future Land Use
Year: 1994
Scale: 1:5,000

Map Name: City of Burlington Planning Department By-law 4000–3
Year: 1993
Scale: 1:12,500

Map Name: City of Cambridge General Zoning Map
Year: 2003
Scale: 1:20,000

Map Name: City of Cambridge Official Plan General City Plan—Map 15
Year: 2002
Scale: 1:41,600

Map Name: City of Cambridge Recreational Facilities
Year: 1992, 2002
Scale: 1:24,400

Map Name: City of Chatham Land Use Plan Schedule "A"
Year: 1988
Scale: 1:15,000

Map Name: City of Edmonton, Land Use
Year: 1983
Scale: 1:30,000

Map Name: City of Etobicoke Official Plan Map 4—Land Uses
Year: 1992
Scale: 1:30,000

Map Name: City of Gloucester, Land Use Map
Year: 1988
Scale: 1:50,000

Map Name: City of Guelph—Land Use Plan
Year: 1990
Scale: 1:20,000

Map Name: City of Guelph Zoning By-law (1971)- 7666 as Amended December 1989
Year: 1989
Scale: 1:100,000

Map Name: City of Guelph—Facility and Open Space Inventory 1986
Year: 1986
Scale: 1:13,200

Map Name: City of Guelph—Schedule 2 Development Constraints
Year: 1990
Scale: 1:20,000

Map Name: City of Hamilton, Parks and Open Space Locations
Year: 1996
Scale: 1:20,000

Map Name: City of Kanata, Official Plan
Year: 1993
Scale: 1:10,000

Map Name: City of Kanata, Urban Area
Year: 1993
Scale: 1:9,909

Map Name: City of Kingston, Consolidated Zoning Map
Year: 1996
Scale: 1:12,000

Map Name: City of Kingston Zoning By-law 8499
Year: 1993
Scale: 1:10,000

Map Name: City of Kitchener Municipal Plan. Map 5, Land Use Plan
Year: 2002
Scale: 1:24,000

Map Name: City of Kitchener Official Plan. Map 2, Flood Plain and Environmental Areas
Year: 1990
Scale: 1:24,000

Map Name: City of Kitchener Official Plan. Map 3, Primary Aggregate Resource Areas
Year: 1990
Scale: 1:24,000

Map Name: City of Kitchener, Official Plan Map 4
Year: 1990
Scale: 1:24,000

Map Name: City of Kitchener, Planning Communities
Year: 1992
Scale: 1:24,000

Map Name: City of Nepean Official Plan, Map Schedule 1
Year: 1993
Scale: 1:25,000

Map Name: City of Niagara Falls, Canada. Schedule A to the Official Plan
Year: 1997
Scale: 1:25,000

Map Name: City of Niagara Falls, Canada. Schedule E to the Official Plan
Year: 1993
Scale: 1:22,500

Map Name: City of North York, Land Use Plan
Year: 1993
Scale: 1:20,000

Map Name: City of Oshawa, Existing Land Use
Year: 1989
Scale: 1:10,000

Map Name: City of Oshawa. Zoning Map
Year: 1992
Scale: 1:10,000

Map Name: City of Pembroke, Zoning By-law
Year: 1993
Scale: 1:3,000

Map Name: City of Scarborough Communities and Industrial Districts
Year: 1991
Scale: 1:27,000

Map Name: City of Scarborough Existing Land Use
Year: 1996
Scale: 1:20,000

Map Name: City of Sudbury, Schedule A to By-law 62-192, Zoning Map
Year: 1993
Scale: 1:12,000

Map Name: City of Thorold Zone Map, Schedule A
Year: 1991
Scale: 1:7,000

Map Name: City of Toronto 1979, Schools and Parks
Year: 1979
Scale: 1:12,069

Map Name: City of Toronto 2005 Zoning Map
Year: 2005
Scale: 1:10,000

Map Name: City of Waterloo Land Use
Year: 2006
Scale: 1:12,500

Map Name: City of Waterloo Zoning Map
Year: 2006
Scale: 1:6,500

Map Name: City of Welland Consolidated Zoning Map
Year: 1994, 1996
Scale: 1:9,600

Map Name: City of Welland Land Use
Year: 1991
Scale: 1:15,000

Map Name: City of Windsor Existing Land Use
Year: 1980
Scale: 1:1,000

Map Name: City of Windsor Planning Area. Schedule B, Land Use Plan
Year: 1994
Scale: 1:22,000

Map Name: City of Winnipeg, Industrial Development Map
Year: 2000
Scale: 1:250,000

Map Name: City of Woodstock Zoning Map
Year: 1996
Scale: 1:10,000

Map Name: Commission's Recommended Niagara Escarpment Plan
Year: 2001
Scale: 1:50,000

Map Name: Community Trails Location Map, City of Waterloo
Year: 1995
Scale: 1:16,300

Map Name: Composite Map of Selected Proposed Activities in and Contiguous to Environmentally Sensitive Policy Areas
Year: 1976
Scale: 1:100,000

Map Name: Conestoga River Watershed—Areas Recommended for Authority Forest, Areas for Private Reforestation and Existing Woodland
Year: 1954
Scale: 1:63,360

Map Name: Consolidated Zoning By-law I-83 Map—City of York
Year: 1985
Scale: 1:9,525

Map Name: Corporation of the City of Cambridge—Neighbourhoods
Year: 2003
Scale: 1:15,300

Map Name: Corporation of the City of Cambridge—Status of Plans of Subdivisions
Year: 1999
Scale: 1:19,200

Map Name: Corporation of the Town of Campbellford—Schedule "A"—Transportation and Land Use Plan
Year: 1993
Scale: 1:3,600

Map Name: Designated Wetlands, Town of Perth Planning Area
Year: 1993
Scale: 1:5,000

Map Name: Developments Pending for the Town of Newmarket
Year: 1995
Scale: 1:11,905

Map Name: Draft Plan Index, City of Kitchener
Year: 1990
Scale: 1:24,000

Map Name: Environmental Management Official Plan, Schedule C, Town of Whitby
Year: 1996
Scale: 1:31,200

Map Name: Existing Vacant Land Inventory, Town of Perth Planning Area
Year: 1993
Scale: 1:5,000

Map Name: Future Predominant Land Use—Burlington Urban Area
Year: 1993
Scale: 1:12,500

Map Name: Generalised Agricultural Land Use Systems of the Regional Municipality of Niagara
Year: 1985
Scale: 1:100,000

Map Name: Generalized Land Use: Greater Vancouver Regional District
Year: 1979
Scale: 1:50,000

Map Name: Georgian Bay Development Region, Predominant Land Use 1971
Year: 1971
Scale: 1:253,440

Map Name: Halton Region Conservation Authority Watershed Area
Year: 1995
Scale: 1:100,000

Map Name: Heritage Structures Kitchener Cambridge Waterloo, Regional Municipality of Waterloo
Year: 1983
Scale: 1:50,000

Map Name: Identification of a Regional Greenlands System—Alliston
Year: 1990
Scale: 1:50,000

Map Name: Identification of a Regional Greenlands System—Beaverton
Year: 1990
Scale: 1:50,000

Map Name: Identification of a Regional Greenlands System—Bolton
Year: 1990
Scale: 1:50,000

Map Name: Identification of a Regional Greenlands System—Brampton
Year: 1990
Scale: 1:50,000

Map Name: Identification of a Regional Greenlands System—Greater Toronto Greenland Strategy
Year: 1990
Scale: 1:50,000

Map Name: Identification of a Regional Greenlands System—Guelph
Year: 1990
Scale: 1:50,000

Map Name: Identification of a Regional Greenlands System—Hamilton—Burlington
Year: 1990
Scale: 1:50,000

Map Name: Identification of a Regional Greenlands System—Lindsay
Year: 1990
Scale: 1:50,000

Map Name: Identification of a Regional Greenlands System—Markham
Year: 1990
Scale: 1:50,000

Map Name: Identification of a Regional Greenlands System—Newmarket
Year: 1990
Scale: 1:50,000

Map Name: Identification of a Regional Greenlands System—Orangeville
Year: 1990
Scale: 1:50,000

Map Name: Identification of a Regional Greenlands System—Oshawa
Year: 1990
Scale: 1:50,000

Map Name: Identification of a Regional Greenlands System—Scugog
Year: 1990
Scale: 1:50,000

Map Name: Identification of a Regional Greenlands System—Toronto
Year: 1990
Scale: 1:50,000

Map Name: Industrial Lands and Access: The Regional Municipality of Waterloo
Year: 1979
Scale: 1:100,000

Map Name: Kitchener Planning Communities and Ward Map
Year: 2002
Scale: 1:21,000

Map Name: Lake Erie Development Region, Predominant Land Use 1971
Year: 1971
Scale: 1:253,440

Map Name: Land Cover Map of Alberta
Year: 1989
Scale: 1:1,000,000

Map Name: Land Cover Map of British Columbia
Year: 1989
Scale: 1:1,000,000

Map Name: Land Cover Map of Saskatchewan
Year: 1989
Scale: 1:1,000,000

Map Name: Land Cover of the Maritimes
Year: 1989
Scale: 1:1,000,000

Map Name: Land Cover Map of the Yukon and N.W.T.
Year: 1991
Scale: 1:1,000,000

Map Name: Land Use in Gloucester and Nepean Townships, Carleton County, Ontario
Year: 1976
Scale: 1:25,000

Map Name: Land Use Map, City of Scarborough Official Plan
Year: 1986
Scale: 1:18,000

Map Name: Land Use Official Plan, Schedule A, Town of Whitby
Year: 1996
Scale: 1:31,200

Map Name: Land Use Plan, Official Plan of the Belleville Section of the Belleville and Suburbs Land
Year: 1993
Scale: 1:9,600

Map Name: Land Use Plan, Official Plan of the City of Sarnia Planning Area, Schedule A
Year: 1978
Scale: 1:25,000

Map Name: Laurel Creek University of Waterloo Channel Modifications
Year: 1996
Scale: 1:6,000

Map Name: Licensed Pits and Quarries Ottawa Carleton
Year: 1993
Scale: 1:100,000

Map Name: Map No.1: Planning Districts, District Centers and Planning Units of the Official Plan for the City of London Planning Area
Year: 1982
Scale: 1:40,000

Map Name: Map No. 2: Land Use of the Official Plan for the City of London Planning Area
Year: 1982
Scale: 1:40,000

Map Name: Midwestern Ontario Development Region, Generalized Existing Land Use
Year: 1980
Scale: 1:250,000

Map Name: Midwestern Ontario Development Region, Predominant Land Use
Year: 1971
Scale: 1:250,000

Map Name: Niagara Development Region, Predominant Land Use 1971
Year: 1972
Scale: 1:253,440

Map Name: Niagara Escarpment Plan
Year: 1994
Scale: 1:50,000

Map Name: Niagara Falls, Zoning By-law
Year: 1990
Scale: 1:14,620

Map Name: North Bay Official Plan, Schedule B, Land Use Plan
Year: 1979
Scale: 1:12,000

Map Name: Nottswaga Valley Watershed
Year: 1997
Scale: 1:100,000

Map Name: Official Consolidation. Georgetown Urban Area
Year: 1995
Scale: 1:5,000

Map Name: Official Plan, City of Kingston. Schedule B
Year: 1993
Scale: 1:10,000

Map Name: Official Plan for the Kitchener Planning Area. Map 1, Plan for Land Use
Year: 1982
Scale: 1:2,000

Map Name: Official Plan for the Town of Halton Hills. Schedule 1: Land Use
Year: 1994
Scale: 1:25,000

Map Name: Official Plan for the Town of Halton Hills. Schedule 4: Environment
Year: 1994
Scale: 1:25,000

Map Name: Official Plan of Richmond Hill, Appendix 4, Concept Plan
Year: 1991
Scale: 1:20,000

Map Name: Official Plan of Richmond Hill, Schedule 1, Land Use
Year: 1991
Scale: 1:20,000

Map Name: Official Plan of Richmond Hill, Schedule 6
Year: 1991
Scale: 1:20,000

Map Name: Official Plan of the City of Brantford: Schedule 1, Land Use Plan
Year: 1995
Scale: 1:20,000

Map Name: Official Plan of the City of Brantford: Schedule 8, Special Policy Area
Year: 1992
Scale: 1:20,000

Map Name: Official Plan of the City of Thorold Planning Area: Schedule B, Land Use Plan
Year: 1986
Scale: 1:17,000

Map Name: Official Plan of the City of Waterloo Existing Land Use
Year: 1965
Scale: 1:1500

Map Name: Official Plan of the City of Waterloo Planning Area, Land Use Plan
Year: 1986
Scale: 1:9,600

Map Name: Official Plan of the County of Prince Edward: Schedule A, Environmentally Sensitive Areas
Year: 1993
Scale: 1:63,360

Map Name: Official Plan of the County of Prince Edward: Schedule B, Environmental Constraints
Year: 1993
Scale: 1:63,360

Map Name: Official Plan of the County of Prince Edward: Schedule E, Land Use Designations
Year: 1993
Scale: 1:63,360

Map Name: Official Plan of the Town of Ancaster
Year: 1978
Scale: 1:12,000

Map Name: Official Plan of the Town of Rockland. Schedule A, Land Use
Year: 1993
Scale: 1:5,000

Map Name: Official Plan of the Township of Brantford Planning Area. Schedule "B," Land Use Plan
Year: 1992
Scale: 1:31, 680

Map Name: Official Plan Ottawa-Carleton Planning Area Consolidation and Amendments
Year: 1993
Scale: 1:100,000

Map Name: Ontario Parks Northeast Zone Protected Areas Compilation Map
Year: 2004
Scale: 1:760,000

Map Name: Ownership, City of Kitchener, Ontario
Year: 1983
Scale: 1:24,000

Map Name: Plan of the City of Toronto
Year: 1988
Scale: 1:18,000

Map Name: Potential Townhouse and Apartment Sites—City of Guelph
Year: 1993
Scale: 1:15,000

Map Name: Proposed Commercial Designation Revisions, Map No. 2: Land Use of the Official Plan for the City of London Planning Area
Year: 1980
Scale: 1:40,000

Map Name: Proposed Land Use Map of the City of Owen Sound
Year: 1981
Scale: 1:9,600

Map Name: Recreational Facilities Located on Park and School Sites
Year: 1990
Scale: 1:20,000

Map Name: Regional Municipality of Waterloo—Municipal Addressing of Township of Woolwich, Wilmont, Wellesley and North Dumfries
Year: 2001
Scale: 1:54,000

Map Name: Regional Municipality of Waterloo—Official Policies Plan—Institutional Facilities and Open Space Resources
Year: 1976
Scale: 1:100,000

Map Name: Regional Municipality of Waterloo—Official Policies Plan—Land Use 1974
Year: 1976
Scale: 1:100,000

Map Name: Registered Plans 1940–Present: City Street Map—City of Guelph
Year: 1990
Scale: 1:15,000

Map Name: Residential Community Structure Official Plan Schedule B, Town of Whitby
Year: 1996
Scale: 1:31,200

Map Name: Schedule "A" Land Use and Roads Plan—Town of Brighton
Year: 1999
Scale: 1:20,000

Map Name: Schedule A—Land Use and Transportation Plan Town of Collingwood
Year: 1990
Scale: 1:7,200

Map Name: Schedule A, Land Use Plan for the City of Waterloo
Year: 1987
Scale: 1:25,000

Map Name: Schedule A, Zone Map, Town of Milton
Year: 1991
Scale: 1:25,000

Map Name: Schedule B, Land Use Plan, City of Windsor Planning Area
Year: 1996
Scale: 1:22,000

Map Name: Schedule B, Land Use Plan, Town of Goderich
Year: 1990
Scale: 1:7,200

Map Name: Schedule D to the Official Plan Community Planning Districts
Year: 1993
Scale: 1: 22,500

Map Name: Schedule 8 Special Policy Area Official Plan of the City of Brantford
Year: 1992
Scale: 1:20,000

Map Name: St. Clair Development Region Predominant Land Use
Year: 1971
Scale: 1:253,440

Map Name: Staging of Development (1993–1994), Kitchener
Year: 1993
Scale: 1:24,000

Map Name: Stouffville, Town of Whitchurch-Stoufville
Year: 1990
Scale: 1:4,800

Map Name: The Corporation of the City of Cambridge—Consolidated Zoning Map
Year: 1986
Scale: 1:12,678

Map Name: The Official Plan of the Township of South Dumfries Schedule "A" Land Use Plan
Year: 1988
Scale: 1:40,000

Map Name: The Regional Municipality of Waterloo—Heritage Houses
Year: 1980
Scale: 1:100,000

Map Name: Toronto and Region, Conservation for the Living City
Year: 2005
Scale: 1:1,250,000

Map Name: Toronto-Centred Region, Predominant Land Use 1971
Year: 1971
Scale: 1:253,440

Map Name: Town of Arnprior Schedule A—Land Use
Year: 1993
Scale: 1:4,800

Map Name: Town of Caledon Land Use Plan Schedule A
Year: 1991
Scale: 1:50,000

Map Name: Town of Caledon Official Plan Schedule B—Environmental Constraint Policy Areas
Year: 1991
Scale: 1:50,000

Map Name: Town of Fort Erie, Land Use Plan
Year: 1992
Scale: 1:24,000

Map Name: Town of Grimsby—Official Plan
Year: 1991
Scale: 1:14,400

Map Name: Town of Newmarket, Assessment Subdivision Map
Year: 1991
Scale: 1:11,905

Map Name: Town of Orangeville Street Plan
Year: 1982
Scale: 1:45,000

Map Name: Town of Paris, Key Plan
Year: 1994
Scale: 1:50,000

Map Name: Town of Paris, Schedule A, Land Use Plan
Year: 1994
Scale: 1:3,000

Map Name: Town of Perth Planning Area, Land Use, Schedule B
Year: 1993
Scale: 1:5,000

Map Name: Town of Perth Planning Area, Official Plan, Schedule A
Year: 1993
Scale: 1:5,000

Map Name: Town of Perth Planning Area, Zoning By-law, Schedule B
Year: 1993
Scale: 1:5,000

Map Name: Town of Richmond Hill, Composite Zoning Map
Year: 1993
Scale: 1:12,192

Map Name: Town of Rockland, Schedule A, Zoning By-law
Year: 1993
Scale: 1:5,000

Map Name: Town of Whitby Restricted Area (Zoning) By-law No. 2585, Schedule A
Year: 1996
Scale: 1:4,800

Map Name: Town of Whitchurch-Stoufville, 1997 Ward Boundaries and Electoral Polls
Year: 1997
Scale: 1:20,000

Map Name: Town of Whitchurch-Stoufville, Property Map
Year: 1993
Scale: 1:24,000

Map Name: Township of Brantford Existing Land Uses
Year: 1984
Scale: 1:25,000

Map Name: Township of North Dumfries Land Use
Year: 1989
Scale: 1:12,000

Map Name: Township of Onondaga: Schedule A, Land Use
Year: 1993
Scale: 1:25,000

Map Name: Township of South Dumfries Zoning By-law Number Schedule "A," Map 1
Year: 1988
Scale: 1:25,000

Map Name: Transportation Official Plan D, Town of Whitby
Year: 1996
Scale: 1:31,200

Map Name: Village of Blyth Land Use Plan Schedule B
Year: 1980
Scale: 1:2,400

Map Name: Waterloo County Area Development Plan Phase I: Adopted Strategy Landuse 1971–1981
Year: 1971
Scale: 1:126,720

Map Name: Wellington County Restructuring Proposal—Map 1
Year: 1998
Scale: 1:12,700

Map Name: Woodlot Inventory, City of Waterloo
Year: 1990
Scale: 1:26,660

Map Name: Zone Map 17 Schedule A North Bolton Area Town of Caledon
Year: 1990
Scale: 1:5,000

Map Name: Zone Map Town of Collingwood
Year: 1987
Scale: 1:7,200

Map Name: Zoning—City of Etobicoke
Year: 1992
Scale: 1:20,000

Map Name: Zoning Map, City of Chatham
Year: 1988
Scale: 1:8,000

Map Name: Zoning Map, City of Kitchener
Year: 1993
Scale: 1:9,000

Map Name: Zoning map, City of Vancouver, British Columbia, March 1993
Year: 1993
Scale: 1:63,790

Map Name: Zoning Map, City of Vancouver, British Columbia, March 1995
Year: 1995
Scale: 1:163,790

Map Name: Zoning Map of the City of Waterloo
Year: 1976
Scale: 1:3,000

## Meteorology

Map Name: Bouguer Gravity Anomaly Map of Canada
Year: 1969, 1987
Scale: 1:5,000,000–1:10,000,000

Map Name: Canada, Climatic Regions Moisture Regimes
Year: 1990
Scale: 1:7,500,000

Map Name: Canada, Heating Degree Days
Year: 1981
Scale: 1:7,500,000

Map Name: Canada, Precipitation
Year: 1991
Scale: 1:12,500,000 and 1:25,000,000

Map Name: Canada Snowfall
Year: 1991
Scale: 1:12,000,000 and 1:25,000,000

Map Name: Canada Solar Radiation
Year: 1984
Scale: 1:12,500,000 and 1:25,000,000

Map Name: Canada Temperature
Year: 1970, 1984
Scale: 1:12,500,000 and 1:25,000,000

Map Name: Forest Fire Weather Zones of Canada
Year: 1973
Scale: 1:10,000,000

Map Name: Free Air Gravity Anomaly Map of Canada
Year: 1987
Scale: 1:10,000,000

Map Name: Gravimetric Geoid Map of Canada
Year: 1989
Scale: 1:10,000,000

Map Name: Gravity Map of Canada
Year: 1980
Scale: 1:5,000,000

Map Name: Isostatic Gravity Anomaly Map of Canada
Year: 1987
Scale: 1:10,000,000

Map Name: Observed Gravity Values of Canada
Year: 1987
Scale: 1:10,000,000

Map Name: Principal Climatological Station Network
Year: 1988
Scale: 1:7,500,000

Map Name: Seismicity Map of Canada
Year: 1988
Scale: 1:10,000,000

**Physical Geography**

Map Name: Alberta, Canada, Relief
Year: 1978
Scale: 1:1,000,000

Map Name: Alliston Aquifer Complex
Year: 1977
Scale: 1:100,000

Map Name: Amenagements et potentiel hydroelectrique du Quebec
Year: 1982
Scale: 1:2,000,000

Map Name: Ausable-Bayfield Conservation Authority
Year: 1972
Scale: 1:125,000

Map Name: Bathymetric Chart, Oceanic Regions Adjacent to Canada
Year: 1971
Scale: 1:6,750,000

Map Name: Bedrock Formation Rocheuse Ottawa-Carleton
Year: 1993
Scale: 1:100,000

Map Name: Biogeoclimatic Zones of British Columbia
Year: 1971
Scale: 1:1,900,800

Map Name: Biogeoclimatic Zones of British Columbia
Year: 1992
Scale: 1:2,000,000

Map Name: Biogeoclimatic Zones of Vancouver Island and the Adjacent Mainland Based on Climax Vegetation
Year: 1974
Scale: 1:380,160

Map Name: British Columbia, Physical
Year: 1964
Scale: 1:1,900,000

Map Name: British Columbia, Relief Map
Year: 1960
Scale: 1:1,900,000

Map Name: Canada, Distribution of Wetlands
Year: 1986
Scale: 1:7,500,000

Map Name: Canada, Drainage Basins
Year: 1985
Scale: 1:7,500,000

Map Name: Canada, Drainage Patterns
Year: 1988
Scale: 1:7,500,000

Map Name: Canada, Glaciers
Year: 1985
Scale: 1:7,500,000

Map Name: Canada, Land Cover Associations
Year: 1989
Scale: 1:7,500,000

Map Name: Canada, Land Suitability for Forestry
Year: 1988
Scale: 1:7,500,000

Map Name: Canada, Permafrost
Year: 1995
Scale: 1:7,500,000

Map Name: Canada, Relief
Year: 1986
Scale: 1:7,500,000

Map Name: Canada, Sea Ice
Year: 1993
Scale: 1:12,500,000

Map Name: Canada, Seismicity
Year: 1994
Scale: 1:7,500,000

Map Name: Canada, Soil Capability for Agriculture
Year: 1980
Scale: 1:7,500,000

Map Name: Canada, Streamflow
Year: 1993
Scale: 1:7,500,000

Map Name: Canada, Terrestrial Ecoregions
Year: 1993
Scale: 1:7,500,000

Map Name: Canada, Wetland Regions
Year: 1986
Scale: 1:7,500,000

Map Name: Canadian Oil and Gas Map: Showing Oil Fields, Gas Fields, Pipelines, Refineries
Year: 1985
Scale: 1:2,217,500

Map Name: Classified Wetlands Ottawa-Carleton
Year: 1993
Scale: 1:100,000

Map Name: Contour Map of Kitchener/Waterloo
Year: 197-
Scale: 1:12,000

Map Name: County of Brant—Ground Water Probability
Year: 1977
Scale: 1:100,000

Map Name: County of Grey—Groundwater Probability Map 3111
Year: 1980
Scale: 1:100,000

Map Name: County of Haldimand—Ground Water Probability
Year: 1974
Scale: 1:100,000

Map Name: County of Kent—Water Probability Map
Year: 1970
Scale: 1:100,000

Map Name: County of Simcoe Ground-Water Probability: Ground Water Quality
Year: 1978
Scale: 1:300,000

Map Name: County of Simcoe Ground-Water Probability: Water Supplies in Bedrock
Year: 1978
Scale: 1:150,000 and 1:300,000

Map Name: County of Simcoe Ground-Water Probability: Water Supplies in Deep Overburden
Year: 1978
Scale: 1:150,000 and 1:300,000

Map Name: County of Simcoe Ground-Water Probability: Water Supplies in Shallow Overburden
Year: 1978
Scale: 1:150,000 and 1:300,000

Map Name: Discover Canada's Watersheds
Year: 2006
Scale: 1:5,000,000

Map Name: Generalized Soil Map—Regional Municipality of Haldimand-Norfolk
Year: 1985
Scale: 1:100,000

Map Name: Geological Highway Map of Alberta
Year: 1975
Scale: 1:1,584,000

Map Name: Geological Map of Alberta
Year: 1970, 1999
Scale: 1:1,000,000–1:1,267,000

Map Name: Geological Map of Saskatchewan
Year: 1980
Scale: 1:250,000

Map Name: Geological Map of the Province of Nova Scotia
Year: 1979
Scale: 1:500,000

Map Name: Geology of Nunavut
Year: 1999
Scale: 1:3,500,000

Map Name: Geology of the Island of Newfoundland
Year: 1990
Scale: 1:1,000,000

Map Name: Geothermal Potential Map of British Columbia
Year: 1983
Scale: 1:2,000,000

Map Name: Glacier Map of Southern British Columbia and Alberta
Year: 1965
Scale: 1:1,000,000

Map Name: Grand River Basin Bedrock Geology and Topography
Year: 1980
Scale: 1:250,000

Map Name: Grand River Basin Bedrock Water Quality
Year: 1980
Scale: 1:250,000

Map Name: Grand River Basin Brantford Paris Area
Year: 1980
Scale: 1:50,000

Map Name: Grand River Basin Drift Thickness
Year: 1980
Scale: 1:250,000

Map Name: Grand River Basin Dry and Flowing Wells
Year: 1980
Scale: 1:250,000

Map Name: Grand River Basin Elora-Fergus Area
Year: 1980
Scale: 1:33,000

Map Name: Grand River Basin Ground-Water Yields from Bedrock
Year: 1980
Scale: N/A

Map Name: Grand River Basin Ground-Water Yields from Overburden
Year: 1980
Scale: N/A

Map Name: Grand River Basin Hydrogeologic Cross Sections
Year: 1980
Scale: N/A

Map Name: Grand River Basin Kitchener-Waterloo-Cambridge-Guelph Area
Year: 1980
Scale: 1:600,000

Map Name: Grand River Basin Overburden Water Quality
Year: 1980
Scale: 1:250,000

Map Name: Grand River Basin Piezometric Surface Elevation in Bedrock
Year: 1980
Scale: 1:250,000

Map Name: Grand River Basin Poor Natural Water Quality
Year: 1980
Scale: 1:250,000

Map Name: Grand River Basin Surficial Geology
Year: 1980
Scale: N/A

Map Name: Grand River Basin Susceptibility of Ground Water to Contamination
Year: 1980
Scale: 1:250,000

Map Name: Grand River Basin Water Table Configuration
Year: 1980
Scale: 1:250,000

Map Name: Grand River Watershed
Year: 199-
Scale: 1:253,440

Map Name: Grey Sauble Conservation Authority Watershed Map
Year: 1990
Scale: 1:100,000

Map Name: Ground Water Probability Map—County of Elgin
Year: 1972
Scale: 1:100,000

Map Name: Ground Water Probability Map—County of Huron
Year: 1986
Scale: 1:100,000

Map Name: Ground Water Probability Map—County of Lambton
Year: 1969
Scale: 1:100,000

Map Name: Ground-Water Probability of the Regional Municipality of Durham
Year: 1986
Scale: 1:100,000

Map Name: Guelph-Amabel Aquifier, Hamilton to Orangeville
Year: 1978
Scale: 1:100,000

Map Name: Guelph-Amabel Aquifer, Markdale to Owen Sound
Year: 1978
Scale: 1:100,000

Map Name: Guelph-Amabel Aquifer, Orangeville to Markdale
Year: 1978
Scale: 1:100,000

Map Name: Guelph-Lockport Aquifer
Year: 1978
Scale: 1:100,000

Map Name: Industrial Mineral Commodities Map of the Province of Nova Scotia
Year: 1985
Scale: 1:500,000

Map Name: Lakes, Rivers and Glaciers
Year: 1969
Scale: 1:7,500,000

Map Name: Landform Map of Canada
Year: 1950
Scale: 1:6,500,000

Map Name: Landscapes of Southern Saskatchewan
Year: 1977
Scale: 1:1,000,000

Map Name: Laurel Creek Channel Modifications
Year: 1996
Scale: 1:13,000

Map Name: Low-Flow Characteristics of Streams in Southeastern Ontario
Year: 1977
Scale: 1:500,000

Map Name: Low-Flow Characteristics of Streams in Southwestern Ontario
Year: 1974
Scale: 1:300,000

Map Name: Low-Flow Characteristics of Streams in Toronto-Centred Region
Year: 1973
Scale: 1:250,000

Map Name: Manitoba, Geological Highway Map
Year: 1987
Scale: 1:1,000,000

Map Name: Manitoba, Relief
Year: 1979
Scale: 1:1,000,000

Map Name: Map of the Greater River St. John and Waters
Year: 1981
Scale: 1:253,440

Map Name: Mineral Map of Manitoba
Year: 1980
Scale: 1:1,000,000

Map Name: Mineral Occurrence Map of New Brunswick
Year: 1978
Scale: 1:500,000

Map Name: Mount Revelstoke and Glacier National Parks
Year: 1986
Scale: 1:70,000

Map Name: Native Prairie in the Habitat Subregions of Prairie Canada
Year: 1988
Scale: 1:42.6064

Map Name: Oak Ridges Aquifer Complex
Year: 1977
Scale: 1:100,000

Map Name: Oak Ridges Moraine Area within the Greater Toronto Area
Year: 1991
Scale: 1:100,000

Map Name: Parcourir le Relief du Quebec
Year: 2001
Scale: 1:2,500,000

Map Name: Peatland Areas of Nova Scotia
Year: 1986
Scale: 1:250,000

Map Name: Physiography of Southern Ontario
Year: 1965, 1972
Scale: 1:1,250,000

Map Name: Polar Continental Shelf Project
Year: 1988
Scale: 1:7,500,000

Map Name: Prince Albert National Park, Saskatchewan
Year: 1972
Scale: 1:125,000

Map Name: Quaternary Geology, Central Alberta
Year: 1986
Scale: 1:500,000

Map Name: Quaternary Geology, Southern Alberta
Year: 1983
Scale: 1:500,000

Map Name: Quebec—Aboriginal Communities and Minerals and Metals Activities
Year: 2001
Scale: 1:2,000,000

Map Name: Regional Municipality of Haldimand-Norfolk (Western Portion)—Ground Water Probability Sheet 1—Water Supplies in Shallow Overburden
Year: 1979
Scale: 1:250,000

Map Name: Regional Municipality of Haldimand-Norfolk (Western Portion)—Ground Water Probability Sheet 2—Water Supplies in Deep Overburden
Year: 1979
Scale: 1:250,000

Map Name: Regional Municipality of Haldimand-Norfolk (Western Portion)—Ground Water Probability Sheet 3—Water Supplies in Bedrock
Year: 1979
Scale: 1:250,000

Map Name: Regional Municipality of Haldimand-Norfolk (Western Portion)—Ground Water Probability Sheet 4—Ground Water Quality
Year: 1979
Scale: 1:250,000

Map Name: Regional Municipality of Niagara, Wetlands
Year: 1993
Scale: 1:100,000

Map Name: Regional Municipality of Peel—Water Supplies in Bedrock
Year: 1977
Scale: 1:250,000

Map Name: Regional Municipality of Peel—Water Supplies in Deep Overburden
Year: 1977
Scale: 1:100,000

Map Name: Regional Municipality of Peel—Water Supplies in Shallow Overburden
Year: 1977
Scale: 1:100,000

Map Name: Regional Municipality of Peel Ground Water Probability
Year: 1977
Scale: 1:250,000

Map Name: Regional Municipality of Waterloo—Official Policies Plan—Flood Plain and Environmentally Sensitive Policy Areas
Year: 1976
Scale: 1:63,360

Map Name: Relief Map of Canada
Year: 1998
Scale: 1:6,000,000

Map Name: Saugeen Valley Watershed
Year: 1978
Scale: 1:175,000

Map Name: Soil Associations Map—Waterloo County
Year: 1971
Scale: 1:100,000

Map Name: Soil Associations of Southern Ontario
Year: 1960
Scale: 1:633,600

Map Name: Soil Capability for Agriculture in Gloucester and Nepean Townships, Carleton County
Year: 1976
Scale: 1:25,000

Map Name: Soil Classification Oxford County
Year: 1990
Scale: 1:100,000

Map Name: Soil Landscapes of Canada, Alberta
Year: 1986
Scale: 1:1,000,000

Map Name: Soil Landscapes of Canada, British Columbia
Year: 1993
Scale: 1:1,000,000

Map Name: Soil Landscapes of Canada, British Columbia
Year: 1994
Scale: 1:1,000,000

Map Name: Soil Landscapes of Canada, Manitoba
Year: 1989
Scale: 1:1,000,000

Map Name: Soil Landscapes of Canada, Maritime Provinces
Year: 1988
Scale: 1:1,000,000

Map Name: Soil Landscapes of Canada, Ontario North
Year: 1994
Scale: 1:1,000,000

Map Name: Soil Landscapes of Canada, Ontario South
Year: 1988
Scale: 1:1,000,000

Map Name: Soil Landscapes of Canada, Quebec Southeast
Year: 1994
Scale: 1:1,000,000

Map Name: Soil Landscapes of Canada, Quebec Southwest
Year: 1990
Scale: 1:1,000,000

Map Name: Soil Landscapes of Canada, Saskatchewan
Year: 1990
Scale: 1:1,000,000

Map Name: Soil Map—County of Elgin
Year: 1929
Scale: 1:126,720

Map Name: Soil Map—County of Kent, Province of Ontario
Year: 1936
Scale: 1:126,720

Map Name: Soil Map, County of Norfolk
Year: 1928
Scale: 1:126,720

*Appendix A*

Map Name: Soil Map for Lincoln County, Ontario
Year: 1962
Scale: 1:63,360

Map Name: Soil Map of Dufferin County, Ontario
Year: 1963
Scale: 1:63,360

Map Name: Soil Map of Dundas County, Ontario
Year: 1951
Scale: 1:63,360

Map Name: Soil Map of Durham County
Year: 1945
Scale: 1:125,000

Map Name: Soil Map of Essex County
Year: 1947
Scale: 1:63,360

Map Name: Soil Map of Frontenac County
Year: 1965
Scale: 1:63,360

Map Name: Soil Map of Grenville County, Ontario
Year: 1949
Scale: 1:63,360

Map Name: Soil Map of Halton County, Ontario
Year: 1971
Scale: 1:63,360

Map Name: Soil Map of Hastings County
Year: 1958
Scale: 1:63,360

Map Name: Soil Map of Kent
Year: 1936
Scale: 1:126,720

Map Name: Soil Map of Lambton County
Year: 1956
Scale: 1:63,360

Map Name: Soil Map of Lanark County, Ontario
Year: 1966
Scale: 1:63,360

Map Name: Soil Map of Leeds County
Year: 1968
Scale: 1:63,360

Map Name: Soil Map of Lennox and Addington County, Ontario
Year: 1961
Scale: 1:63,360

Map Name: Soil Map of Manitoulin Island, Ontario
Year: 1957
Scale: 1:63, 360

Map Name: Soil Map of Ontario County
Year: 1979
Scale: 1:63,360

Map Name: Soil Map of Oxford County, Ontario
Year: 1958
Scale: 1:63,360

Map Name: Soil Map of Part of the Parry Sound District
Year: 1960
Scale: 1:126,720

Map Name: Soil Map of Peel County, Ontario, Soil Survey Report No. 18
Year: 1953
Scale: 1:63,360

Map Name: Soil Map of Perth County, Ontario
Year: 1977
Scale: 1:63,360

Map Name: Soil Map of Prescott County, Ontario
Year: 1961
Scale: 1:63,360

Map Name: Soil Map of Prince Edward County, Ontario
Year: 1947
Scale: 1:63,360

Map Name: Soil Map of Renfrew County
Year: 1964
Scale: 1:63,360

Map Name: Soil Map of Russell County
Year: 1961
Scale: 1:63,360

Map Name: Soil Map of Simcoe County
Year: 1959
Scale: 1:63,360

Map Name: Soil Map of the New Liskeard-Englehart Area
Year: 1995
Scale: 1:63,360

Map Name: Soil Map of Victoria County Ontario
Year: 1956
Scale: 1:63,360

Map Name: Soil Map of Wentworth County
Year: 1967
Scale: 1:63,360

Map Name: Soil Map of York County
Year: 1977
Scale: 1:63,360

Map Name: Soil Survey Map of County of Haldimand, Province of Ontario
Year: 1935
Scale: 1:125,000

Map Name: Soil Survey Map of Middlesex
Year: 1931
Scale: 1:125,720

Map Name: Soil Survey Map of the County of Welland, Ontario
Year: 1935
Scale: 1:126,720

Map Name: Soils of Blind River–Sault Ste Marie Area Ontario: Algoma
Year: 1983
Scale: 1:50,000

Map Name: Soils of Blind River–Sault Ste Marie Area Ontario
Year: 1983
Scale: 1:250,000

Map Name: Soils of Blind River–Sault Ste Marie Area Ontario: Bruce Mines
Year: 1983
Scale: 1:50,000

Map Name: Soils of Blind River–Sault Ste Marie Area Ontario: Dean Lake
Year: 1983
Scale: 1:50,000

Map Name: Soils of Blind River–Sault Ste Marie Area Ontario: Ile Parisienne
Year: 1983
Scale: 1:50,000

Map Name: Soils of Blind River–Sault Ste Marie Area Ontario: Iron Bridge
Year: 1983
Scale: 1:50,000

Map Name: Soils of Blind River–Sault Ste Marie Area Ontario: Madawanson
Year: 1983
Scale: 1:50,000

Map Name: Soils of Blind River–Sault Ste Marie Area Ontario: Pancake Bay
Year: 1983
Scale: 1:50,000

Map Name: Soils of Blind River–Sault Ste Marie Area Ontario: Sault Ste. Marie
Year: 1983
Scale: 1:50,000

Map Name: Soils of Blind River–Sault Ste Marie Area Ontario: Searchmount
Year: 1983
Scale: 1:50,000

Map Name: Soils of Blind River–Sault Ste Marie Area Ontario: Spanish
Year: 1983
Scale: 1:50,000

Map Name: Soils of Blind River–Sault Ste Marie Area Ontario: St. Joseph Island
Year: 1983
Scale: 1:50,000

Map Name: Soils of Blind River–Sault Ste Marie Area Ontario: Whiskey Lake
Year: 1983
Scale: 1:50,000

Map Name: Soils of Brant County
Year: 1994
Scale: 1:25,000

Map Name: Soils of Bruce County
Year: 1983
Scale: 1:63,360

Map Name: Soils of Canada
Year: 1972
Scale: 1:5,000,000

Map Name: Soils of Chapleau-Foleyet Area
Year: 1985
Scale: 1:250,000

Map Name: Soils of Dryden-Kenora Area Ontario
Year: 1987
Scale: 1:50,000

Map Name: Soils of Elgin County
Year: 1989
Scale: 1:50,000

Map Name: Soils of Fort Erie–Port Colborne: Regional Municipality Ontario
Year: 1980
Scale: 1:25,000

Map Name: Soils of Fort Frances–Rainy River Area: Arbour Vitae
Year: 1984
Scale: 1:50,000

Map Name: Soils of Fort Frances–Rainy River Area: Emo
Year: 1984
Scale: 1:50,000

Map Name: Soils of Fort Frances–Rainy River Area: Fort Frances
Year: 1984
Scale: 1:50,000

Map Name: Soils of Fort Frances–Rainy River Area: International Falls
Year: 1984
Scale: 1:250,000

Map Name: Soils of Fort Frances–Rainy River Area: Northwest Bay
Year: 1984
Scale: 1:50,000

Map Name: Soils of Fort Frances–Rainy River Area: Rainy River
Year: 1984
Scale: 1:50,000

Map Name: Soils of Frontenac County
Year: 1965
Scale: 1:63,360

Map Name: Soils of Glengarry County, Ontario
Year: 1956
Scale: 1:63,360

Map Name: Soils of Gloucester and Nepean Townships, Carleton County, Ontario
Year: 1976
Scale: 1:25,000

Map Name: Soils of Gogma Area, Ontario
Year: 1986
Scale: 1:250,000

Map Name: Soils of Gogma Area, Ontario: Charlton Station
Year: 1986
Scale: 1:50,000

Map Name: Soils of Gogma Area, Ontario: Elk Lake
Year: 1986
Scale: 1:50,000

Map Name: Soils of Grey County
Year: 1981
Scale: 1:63,360

Map Name: Soils of Grimsby-Lincoln: Regional Municipality of Niagara Ontario
Year: 1980
Scale: 1:25,000

Map Name: Soils of Haldimand-Norfolk Regional Municipality
Year: 1979
Scale: 1:25,000

Map Name: Soils of Huron County Ontario
Year: 1979
Scale: 1:63,360

Map Name: Soils of Middlesex County, Ontario
Year: 1991
Scale: 1:50,000

Map Name: Soils of Niagara Falls: Regional Municipality of Niagara Ontario
Year: 1980
Scale: 1:25,000

Map Name: Soils of North Bay Area Ontario
Year: 1961
Scale: 1:250,000

Map Name: Soils of North Bay Area Ontario—Kiosk
Year: 1986
Scale: 1:50,000

Map Name: Soils of North Bay Area Ontario—Marten Lake
Year: 1986
Scale: 1:50,000

Map Name: Soils of North Bay Area Ontario—Mattawa
Year: 1986
Scale: 1:50,000

Map Name: Soils of North Bay Area Ontario—North Bay
Year: 1986
Scale: 1:50,000

Map Name: Soils of North Bay Area Ontario—Powassan
Year: 1986
Scale: 1:50,000

Map Name: Soils of North Bay Area Ontario—Sturgeon Falls
Year: 1986
Scale: 1:50,000

Map Name: Soils of North Bay Area Ontario—Temiscaming
Year: 1986
Scale: 1:50,000

Map Name: Soils of Pelham-Thorold-Welland: Regional Municipality of Niagara Ontario
Year: 1980
Scale: 1:25,000

Map Name: Soils of Peterborough County
Year: 1979
Scale: 1:63,360

Map Name: Soils of Prince Edward Island, Kings County
Year: 1985
Scale: 1:75,000

Map Name: Soils of Prince Edward Island, Prince County
Year: 1985
Scale: 1:75,000

Map Name: Soils of Prince Edward Island, Queens County
Year: 1985
Scale: 1:75,000

Map Name: Soils of Pukaskawa National Park Ontario: Soil Survey Report No. 53
Year: 1985
Scale: 1:100,000

Map Name: Soils of St. Catherines–Niagara-on-the-Lake: Regional Municipality of Niagara Ontario
Year: 1980
Scale: 1:25,000

Map Name: Soils of Sudbury Area
Year: 1983
Scale: 1:250,000

Map Name: Soils of Sudbury Area—Capreol
Year: 1983
Scale: 1:50,000

Map Name: Soils of Sudbury Area—Chelmsford
Year: 1983
Scale: 1:50,000

Map Name: Soils of Sudbury Area—Coniston
Year: 1983
Scale: 1:50,000

Map Name: Soils of Sudbury Area—Copper Cliff
Year: 1983
Scale: 1:50,000

Map Name: Soils of Sudbury Area—Espanola
Year: 1983
Scale: 1:50,000

Map Name: Soils of Sudbury Area—Lake Temagami
Year: 1983
Scale: 1:50,000

Map Name: Soils of Sudbury Area—Milnet
Year: 1983
Scale: 1:50,000

Map Name: Soils of Sudbury Area—Noelville
Year: 1983
Scale: 1:50,000

Map Name: Soils of Sudbury Area—Verner
Year: 1983
Scale: 1:50,000

Map Name: Soils of Sudbury Area—Whitefish Falls
Year: 1983
Scale: 1:50,000

Map Name: Soils of the Regional Municipality of Ottawa-Carleton (Goulborn Township, West Carleton Township, and City of Kanata)
Year: 1981
Scale: 1:50,000

Map Name: Soils of the Regional Municipality of Ottawa-Carleton (Rideau and Osgoode Townships)
Year: 1981
Scale: 1:50,000

Map Name: Soils of the Thunder Bay Area—Jarvis River
Year: 1981
Scale: 1:50,000

Map Name: Soils of the Thunder Bay Area—Kakabeka
Year: 1981
Scale: 1:50,000

Map Name: Soils of the Thunder Bay Area—Loon
Year: 1981
Scale: 1:50,000

Map Name: Soils of the Thunder Bay Area—Onion Lake
Year: 1981
Scale: 1:50,000

Map Name: Soils of the Thunder Bay Area—Parth
Year: 1981
Scale: 1:50,000

Map Name: Soils of the Thunder Bay Area—Pigeon River
Year: 1981
Scale: 1:50,000

Map Name: Soils of the Thunder Bay Area—Sunshine
Year: 1981
Scale: 1:50,000

Map Name: Soils of the Thunder Bay Area—Thunder Bay
Year: 1981
Scale: 1:50,000

Map Name: Soils of Timmins-Noranda—Rouyn Area Ontario
Year: 1978
Scale: 1:250,000

Map Name: Soils of Timmins-Noranda—Rouyn Area Ontario—Iroquois Falls Sheet
Year: 1978
Scale: 1:50,000

Map Name: Soils of Timmins-Noranda—Rouyn Area Ontario—Kirkland Lake Sheet
Year: 1978
Scale: 1:50,000

Map Name: Soils of Timmins-Noranda—Rouyn Area Ontario—Matheson
Year: 1978
Scale: 1:50,000

Map Name: Soils of Timmins-Noranda—Rouyn Area Ontario—Pamour Sheet
Year: 1978
Scale: 1:50,000

Map Name: Soils of Timmins-Noranda—Rouyn Area Ontario—Porquis Junction Sheet
Year: 1978
Scale: 1:50,000

Map Name: Soils of Timmins-Noranda—Rouyn Area Ontario—Timmins Sheet
Year: 1978
Scale: 1:50,000

Map Name: Soils of Wainfleet: Regional Municipality of Niagara Ontario
Year: 1980
Scale: 1:25,000

Map Name: Soils of West Lincoln: Regional Municipality of Niagara Ontario
Year: 1980
Scale: 1:25,000

Map Name: Soil Salinity Map, Saskatchewan
Year: 1988
Scale: 1:1,000,000

Map Name: Soil Water Reserves for Spring Wheat in the Prairie Provinces
Year: 1970
Scale: 1:5,000,000

Map Name: Southern Ontario Drainage Basins
Year: 1973
Scale: 1:500,000

Map Name: Southern Part of the Province of Ontario 1954 Mosaics
Year: 1957
Scale: 1:15,840

Map Name: Speed River Watershed
Year: 1951
Scale: 1:60,000

Map Name: Surficial Geological Map of Manitoba
Year: 1981
Scale: 1:1,000,000

Map Name: Surficial Geological Map of Saskatchewan
Year: 1997
Scale: 1:1,000,000

Map Name: Surficial Geology, Alberta Foothills and Rocky Mountains
Year: 1974/1975
Scale: 1:250,000

Map Name: Surficial Geology of Insular Newfoundland
Year: 1990
Scale: 1:500,000

Map Name: Surficial Geology of the Province of Nova Scotia
Year: 1992
Scale: 1:500,000

Map Name: Susceptibility of Ground Water to Contamination, Goderich
Year: 1982
Scale: 1:50,000

Map Name: Susceptibility of Ground Water to Contamination, Grand Bend

Year: 1983
Scale: 1:50,000

Map Name: Susceptibility of Ground Water to Contamination, Palmerston
Year: 1986
Scale: 1:50,000

Map Name: Susceptibility of Ground Water to Contamination, Parkhill Sheet
Year: 1983
Scale: 1:50,000

Map Name: Susceptibility of Ground Water to Contamination, Seaforth
Year: 1982
Scale: 1:50,000

Map Name: Susceptibility of Ground Water to Contamination, St. Mary's
Year: 1985
Scale: 1:50,000

Map Name: Susceptibility of Ground Water to Contamination, Tillsonburg
Year: 1982
Scale: 1:50,000

Map Name: Susceptibility of Ground Water to Contamination, Walkerton
Year: 1986
Scale: 1:50,000

Map Name: Susceptibility of Ground Water to Contamination, Woodstock
Year: 1983
Scale: 1:50,000

Map Name: Tectonic Map of the Province of Nova Scotia
Year: 1982
Scale: 1:500,000

Map Name: Terrestrial Ecozones, Ecoregions and Ecodistricts, New Brunswick, Nova Scotia, Prince Edward Island, and Newfoundland, Canada
Year: 1995
Scale: 1:3,500,000

Map Name: Terrestrial Ecozones, Ecoregions and Ecodistricts, Province of Ontario
Year: 1995
Scale: 1:3,500,000

Map Name: Terrestrial Ecozones and Ecoregions of Canada, 1995
Year: 1995
Scale: 1:7,500,000

Map Name: Thunder Bay Surficial Geology
Year: 1965
Scale: 1:506,880

Map Name: Toponymie des principaux reliefs du Quebec
Year: 1971
Scale: 1:2,500,000

Map Name: Usable Bedrock Formations with 1st Level Constraint Removed, Ottawa-Carleton
Year: 1993
Scale: 1:100,000

Map Name: Ville-Marie, Quebec, Ontario
Year: 1990
Scale: 1:250,000

Map Name: Water Deficits for the Prairie Provinces
Year: 1970
Scale: 1:5,000,000

Map Name: Water Erosion Risk Map Maritime Provinces
Year: 1991
Scale: 1:1,000,000

Map Name: Water Erosion Risk Map Ontario
Year: 1991
Scale: 1:1,000,000

Map Name: Water Systems 1993 the Regional Municipality of Niagara
Year: 1993
Scale: 1:100,000

Map Name: Water Use, Strait of Georgia–Puget Sound Basin
Year: 1972
Scale: 1:500,000

Map Name: Waterloo Region—Elevation Map
Year: 1981
Scale: 1:100,000

Map Name: Watershed Map—Upper Thames River Conservation Authority
Year: 1994
Scale: 1:190,000

Map Name: Wetland Mapping Series, Third Approximation, Ontario
Year: 1985
Scale: 1:50,000

Map Name: Wetlands of Canada Distribution of Wetlands
Year: 1981
Scale: 1:7,500,000

Map Name: Yukon and Northwest Territories
Year: 1989
Scale: 1:2,000,000

Map Name: Yukon Territory
Year: 1976, 1993
Scale: 1:2,000,000

**Political Geography**

Map Name: Administrative Regions, Districts and Areas 1993
Year: 1993
Scale: 1:1,584,000

Map Name: Annexation Map Corporation Boundary from 1840 to present April 1993
Year: 1993
Scale: 1:15,000

Map Name: Annexations, 1948–1967, City of Waterloo
Year: 1967
Scale: N/A

Map Name: Atlantic Provinces
Year: 1992
Scale: 1:2,000,000

Map Name: Cambridge—Military City Map
Year: 1982
Scale: 1:25,000

Map Name: Cambridge Administrative District
Year: 1987
Scale: 1:190,080

Map Name: Canada—Confederation
Year: 1982
Scale: 1:7,500,000

Map Name: Canada—Political Divisions
Year: 1987
Scale: 1:7,500,000

Map Name: Canada—Territorial Evolution
Year: 1982
Scale: 1:7,500,000

Map Name: Canada, 31st Parliament
Year: 1979
Scale: 1:7,500,000

Map Name: Canada, 32nd Parliament
Year: 1980
Scale: 1:7,500,000

Map Name: Canada, 33rd Parliament
Year: 1984
Scale: 1:7,500,000

Map Name: Canada, 34th Parliament
Year: 1988
Scale: 1:7,500,000

Map Name: Canada, 35th Parliament
Year: 1993
Scale: 1:7,500,000

Map Name: Canada, 36th Parliament
Year: 1997
Scale: 1:7,500,000

Map Name: Canada, 37th Parliament
Year: 2000
Scale: 1:7,500,000

Map Name: Canada, Referendum 1992
Year: 1992
Scale: 1:5,000,000

Map Name: Canada's Federal Lands
Year: 1984
Scale: 1:5,000,000

Map Name: Carte ecclesiastique du Canada
Year: 1963
Scale: 1:6,500,000

Map Name: Charlottetown
Year: 1980
Scale: 1:25,000

Map Name: City of Brantford
Year: 1995, 2001
Scale: 1:20,000

Map Name: City of Brantford Annexations
Year: 1996
Scale: 1:20,000

Map Name: City of Cambridge 2003 Wards Map
Year: 2003
Scale: 1:25,000

Map Name: City of Guelph—Corporation Boundaries from 1840 to Date
Year: 1990
Scale: 1:15,000

Map Name: City of Hamilton, Ward Boundaries 1–8
Year: 2001
Scale: 1:17,000

Map Name: City of Scarborough Community Map
Year: 1996
Scale: 1:20,000

Map Name: City of Scarborough Ward Map
Year: 1996
Scale: 1:20,000

Map Name: City of St. Catharines Composite Zoning Map
Year: 1995
Scale: 1:12,000

Map Name: City of Waterloo Ward Map
Year: 2003
Scale: 1:20,000

Map Name: City of Waterloo Zoning Map
Year: 2002
Scale: 1:12,500

Map Name: Close-up: Quebec and Newfoundland
Year: 1980
Scale: 1:3,700,000

Map Name: Cornwall—Military City Map
Year: 1979
Scale: 1:25,000

Map Name: District of Cochrane
Year: 1977
Scale: 1:600,000

Map Name: District of Kenora Patricia Portion Northeast
Year: 1978
Scale: 1:600,000

Map Name: District of Thunder Bay
Year: 1979
Scale: 1:600,000

Map Name: Districts of Algoma, Sudbury and Timiskaming
Year: 1972, 1977
Scale: 1:600,000

Map Name: Districts of Rainy River, Kenora and Part of Kenora Patricia Portion
Year: 1978
Scale: 1:600,000

Map Name: Elections Ontario—Waterloo North Map A—Polling Divisions
Year: 1994
Scale: 1:12,000

Map Name: Elections Ontario—Waterloo North Map B—Polling Divisions
Year: 1994
Scale: 1:12,000

Map Name: Elections Ontario—Waterloo North Map C—Polling Divisions
Year: 1994
Scale: 1:12,000

Map Name: Elections Ontario—Waterloo North Map D—Polling Divisions
Year: 1994
Scale: 1:12,000

Map Name: Elections Ontario, Kitchener/Waterloo 038 (Map A)
Year: 1999
Scale: 1:18,000

Map Name: Electoral District of Brant Ontario
Year: 1976
Scale: 1:63,360

Map Name: Electoral District of Kitchener, Ontario
Year: 1966, 1976, 1987
Scale: 1:12,000–1:25,344

Map Name: Electoral District of Waterloo Ontario
Year: 1976
Scale: 1:63,360

Map Name: Evolution du territoire du Quebec
Year: 1971
Scale: 1:12,500,000

Map Name: Federal Electoral District of Kitchener, Ontario
Year: 1987
Scale: 1:25,300

Map Name: Federal Electoral District of Waterloo, Ontario
Year: 1986
Scale: 1:50,000

Map Name: Federal Electoral Districts, 1987
Year: 1987
Scale: 1:5,000,000

Map Name: Geographic Townships in the Province of Ontario
Year: 1992
Scale: 1:1,584,000

Map Name: Guelph—Military City Map
Year: 1981
Scale: 1:25,000

*Appendix A* 395

Map Name: Halifax
Year: 1985
Scale: 1:25,000

Map Name: Kingston, Military City Map
Year: 1975, 1985
Scale: 1:25,000

Map Name: Kitchener-Waterloo Military City Map
Year: 1984
Scale: 1:25,000

Map Name: le Quebec
Year: 1987
Scale: 1:2,000,000

Map Name: Les Municipalites regionales de comte
Year: 1983, 1992
Scale: 1:1,250,000

Map Name: Les regions administratives
Year: 1988
Scale: 1:1,250,000

Map Name: London, Military City Map
Year: 1985
Scale: 1:25,000

Map Name: Maritime Provinces
Year: 1988
Scale: 1:633,600

Map Name: Military City Map—Brantford
Year: 1982
Scale: 1:25,000

Map Name: Military City Map—Hamilton
Year: 1979
Scale: 1:25,000

Map Name: Military City Map, Chicoutimi
Year: 1979
Scale: 1:25,000

Map Name: Military City Map, Fredericton
Year: 1981
Scale: 1:25,000

Map Name: Military City Map, Moncton
Year: 1976
Scale: 1:25,000

Map Name: Military City Map, Montreal
Year: 1983
Scale: 1:25,000

Map Name: Military City Map, North Bay
Year: 1979
Scale: 1:25,000

Map Name: Military City Map, Oakville
Year: 1981
Scale: 1:25,000

Map Name: Military City Map, Ottawa-Hull
Year: 1972
Scale: 1:25,000

Map Name: Military City Map, Ottawa-Hull
Year: 1980
Scale: 1:25,000

Map Name: Military City Map, Quebec
Year: 1978
Scale: 1:25,000

Map Name: Military City Map, Saint John
Year: 1980
Scale: 1:25,000

Map Name: Military City Map, Shawinigan
Year: 1974
Scale: 1:25,000

Map Name: Military City Map, Sherbrooke
Year: 1979
Scale: 1:25,000

Map Name: Military City Map, Sudbury
Year: 1978
Scale: 1:25,000

Map Name: Military City Map, Trois-Rivières
Year: 1981
Scale: 1:25,000

Map Name: Military Map of Oshawa
Year: 1984
Scale: 1:25,000

Map Name: Mississauga, 1994 Municipal Wards and Polling Subdivisions
Year: 1994
Scale: 1:20,000

Map Name: New Brunswick
Year: 1972
Scale: 1:500,000

Map Name: Newfoundland and Labrador
Year: 1995
Scale: 1:1,500,000

Map Name: Northeastern Ontario Municipalities 1999
Year: 1999
Scale: 1:1,000,000

Map Name: Northeastern Ontario Municipalities 2000
Year: 2000
Scale: 1:1,000,000

Map Name: Northwestern Ontario Municipalities 2000
Year: 2000
Scale: 1:1,000,000

Map Name: Ontario Federal Electoral Districts Pursuant to the Electoral Boundaries Readjustment Act
Year: 1986
Scale: 1:633,600

Map Name: Ontario First Nation Treaty Map
Year: 2002
Scale: 1:2,000,000

Map Name: Peterborough Military City Map
Year: 1978
Scale: 1:25,000

Map Name: Prince Edward Island
Year: 1974
Scale: 1:250,000

Map Name: Property Boundaries Map Showing Polling Subdivisions
Year: 1990
Scale: 1:12,000

Map Name: Provisional County of Haliburton
Year: 1981
Scale: 1:100,000

Map Name: Quebec
Year: 1973, 1991
Scale: 1:2,000,000

Map Name: Results of Federal Election
Year: 1968, 1972
Scale: 1:7,500,000

Map Name: Richmond Hill, Military City Map
Year: 1983
Scale: 1:25,000

Map Name: Southern Ontario
Year: 1980
Scale: 1:600,000

Map Name: Southern Ontario Municipalities 1999
Year: 1999
Scale: 1:1,000,000

Map Name: Southern Ontario Municipalities 2000
Year: 2000
Scale: 1:800,000

Map Name: St. John's Military City Map
Year: 1976
Scale: 1:25,000

Map Name: Territorial Evolution of Canada
Year: 1967
Scale: 1:50,500,000

Map Name: The Provisional County of Haliburton and the District Municipality of Muskoka
Year: 1979
Scale: 1:250,000

Map Name: Thunder Bay, Military Map
Year: 1981
Scale: 1:25,000

Map Name: Toronto, Military City Map
Year: 1983
Scale: 1:25,000

Map Name: Town of Blenheim—House Numbering Plan
Year: 1978
Scale: 1:2,400

Map Name: Town of Halton Hills (Rural Area)
Year: 1986
Scale: 1:14,000

Map Name: Trenton-Belleville, Military City Map
Year: 1981
Scale: 1:25,000

Map Name: Upper Thames River Conservation Authority, Member Municipalities
Year: 1975
Scale: 1:100,000

Map Name: Waterloo Annexations to 1979
Year: 1979
Scale: 1:12,000

Map Name: Winnipeg, Provincial Electoral Division Boundaries
Year: 1989
Scale: 1:30,000

## Recreational Geography

Map Name: Alberta's Protected Areas Map
Year: 2000
Scale: 1:1,000,000

Map Name: Banff, Kootenay, and Yoho National Parks 1:200,000
Year: 1984
Scale: 1:200,000

Map Name: City of Waterloo, Parks and Community Trail
Year: 1999
Scale: 1:13,000

Map Name: Jasper National Park
Year: 1984
Scale: 1:200,000

Map Name: Les parcs du Quebec
Year: 1974
Scale: 1:1,250,000

Map Name: Quttinirpaaq National Park of Canada
Year: 2007
Scale: 1:435,000

Map Name: Sirmilik National Park of Canada
Year: 2006
Scale: 1:400,000

## Collections in a Series

*Ontario Geological Survey*

Map Name: Alliston
Map Type: Bedrock Topography
Map Number: P3213

## UNITED STATES OF AMERICA

### Base Map

Map Name: Acadia National Park and Vicinity, Hancock Co., Maine
Year: 1976
Scale: 1:50,000

Map Name: Alaska
Year: 1980, 1983, 1994, 1996
Scale: 1:2,500,000–1:5,000,000

Map Name: Bouguer Gravity Map of Arkansas
Year: 1981
Scale: 1:500,000

Map Name: Central Plains
Year: 1985
Scale: 1:2,301,000

Map Name: Chicago
Year: 1982
Scale: not drawn to scale

Map Name: Close-up: U.S.A., Alaska
Year: 1975
Scale: 1:3,295,000

Map Name: Close-up: U.S.A., the South Central States
Year: 1974
Scale: 1:2,350,000

Map Name: Close-up: U.S.A., Western New England
Year: 1975
Scale: 1:614,600

Map Name: Colorado, Base Map with Highways and Contours
Year: 1980
Scale: 1:500,000

Map Name: Deep South
Year: 1983
Scale: 1:2,566,000

Map Name: Hawaii
Year: 1995
Scale: 1:850,000

Map Name: New England
Year: 1987
Scale: 1:1,056,000

Map Name: New Orleans
Year: 1983
Scale: 1:4,000

Map Name: Northeastern United States
Year: 1959
Scale: 1:2,851,200

Map Name: Oklahoma
Year: 1963
Scale: 1:500,000

Map Name: Oregon
Year: 1982
Scale: 1:500,000

Map Name: Pennsylvania, County Divisions and Places
Year: 1970
Scale: 1:4,828,032

Map Name: Philadelphia
Year: 1984
Scale: 1:4,800

Map Name: Relief Map of Missouri
Year: 1974
Scale: 1:1,000,000

Map Name: Rhode Island: County Subdivisions, Towns and Places
Year: 1971
Scale: 1:500,000

Map Name: South Central United States
Year: 1961
Scale: 1:2,851,200

Map Name: State of Alabama
Year: 1989
Scale: 1:500,000

Map Name: State of Arizona
Year: 1981, 1990
Scale: 1:1,000,000

Map Name: State of California, Base Map with Highways and Contours
Year: 1981
Scale: 1:500,000

Map Name: State of California, Northern Half
Year: 1981
Scale: 1:500,000

Map Name: State of Connecticut
Year: 1974
Scale: 1:125,000

Map Name: State of Delaware Base Map
Year: 1993
Scale: 1:175,000

Map Name: State of Florida
Year: 1986
Scale: 1:500,000

Map Name: State of Florida
Year: 1989
Scale: 1:750,000

Map Name: State of Hawaii, Principal Islands
Year: 1971
Scale: 1:500,000

Map Name: State of Idaho
Year: 1976
Scale: 1:1,000,000

Map Name: State of Illinois: Base Map with Highways
Year: 1987
Scale: 1:500,000

Map Name: State of Indiana
Year: 1961
Scale: 1:500,000

Map Name: State of Indiana, Base Map with Highways and Contours
Year: 1973
Scale: 1:500,000

Map Name: State of Iowa
Year: 1984
Scale: 1:500,000

Map Name: State of Kansas
Year: 1961
Scale: 1:500,000

Map Name: State of Kentucky
Year: 1973
Scale: 1:500,000

Map Name: State of Louisiana
Year: 1968, 1990
Scale: 1:1,000,000

Map Name: State of Maine
Year: 1973
Scale: 1:500,000

Map Name: State of Michigan, Base Map with Highways and Contours (1 map on 2 sheets)
Year: 1970
Scale: 1:500,000

Map Name: State of Minnesota
Year: 1963
Scale: 1:500,000

Map Name: State of Mississippi
Year: 1972
Scale: 1:1,000,000

Map Name: State of Missouri
Year: 1973
Scale: 1:500,000

Map Name: State of Montana
Year: 1983
Scale: 1:500,000

Map Name: State of Nebraska
Year: 1984
Scale: 1:500,000

Map Name: State of Nevada
Year: 1984
Scale: 1:500,000

Map Name: State of New Jersey
Year: 1974
Scale: 1:500,000

Map Name: State of New Jersey: Base Map with Highways and Contours
Year: 1975
Scale: 1:500,000

Map Name: State of New Mexico
Year: 1968
Scale: 1:1,000,000

Map Name: State of New Mexico, Base Map with Highways and Contours
Year: 1967
Scale: 1:500,000

Map Name: State of North Carolina
Year: 1957
Scale: 1:500,000

Map Name: State of North Dakota
Year: 1983
Scale: 1:500,000

Map Name: State of Oklahoma
Year: 1975
Scale: 1:500,000

Map Name: State of Oregon
Year: 1982
Scale: 1:1,000,000

Map Name: State of Pennsylvania
Year: 1977
Scale: 1:500,000

Map Name: State of South Carolina: Base Map with Highways and Contours
Year: 1970
Scale: 1:500,000

Map Name: State of South Dakota
Year: 1984
Scale: 1:500,000

Map Name: State of Tennessee
Year: 1973
Scale: 1:1,000,000

Map Name: State of Texas
Year: 1982
Scale: 1:500,000

Map Name: State of Utah
Year: 1988
Scale: 1:500,000

Map Name: State of Virginia: Base Map with Highways and Contours
Year: 1980
Scale: 1:500,000

Map Name: State of Washington: Base Map with Highways and Contours
Year: 1982
Scale: 1:500,000

Map Name: State of West Virginia
Year: 1984
Scale: 1:500,000

Map Name: State of Wisconsin
Year: 1968
Scale: 1:500,000

Map Name: State of Wisconsin: Base Map with Highways and Contours
Year: 1984
Scale: 1:500,000

Map Name: State of Wyoming
Year: 1980
Scale: 1:1,000,000

Map Name: States of Maryland and Delaware
Year: 1977, 1990
Scale: 1:500,000

Map Name: States of Massachusetts, Rhode Island and Connecticut
Year: 1973
Scale: 1:500,000

Map Name: States of New Hampshire and Vermont
Year: 1976
Scale: 1:500,000

Map Name: The Heart of the Grand Canyon
Year: 1978
Scale: 1:24,000

Map Name: The North Central States
Year: 1974
Scale: 1:2,350,000

Map Name: Tidewater and Environs
Year: 1988
Scale: 1:1,260,000

Map Name: United States
Year: 1957, 1959, 1972, 1975, 1978, 1982, 1987
Scale: 1:4,560,000–1:10,000,000

Map Name: United States Contour Map
Year: 1957
Scale: 1:7,000,000

Map Name: United States General Reference
Year: 1970
Scale: 1:7,500,000

Map Name: United States General Reference
Year: 1973
Scale: 1:7,500,000

Map Name: U.S.A Florida, Close-up
Year: 1973
Scale: 1:1,331,000

Map Name: Washington D.C.
Year: 1985
Scale: Not drawn to scale

Map Name: Wyoming: Basemap with Highways and Contours
Year: 1980
Scale: 1:500,000

## Biogeography

Map Name: Arizona General Soil Map
Year: 1975
Scale: 1:1,000,000

Map Name: Audubon's America, the Nature Map of the National Audubon Society
Year: 1979
Scale: 1:5,300,000

Map Name: Biotic Communities of the Southwest
Year: 1980
Scale: 1:1,000,000

Map Name: Classes of Land-Surface Form
Year: 1966
Scale: 1:7,500,000

Map Name: Classes of Land-Surface Form in the Forty Eight States, U.S.A.
Year: 1963
Scale: 1:1,500,000

Map Name: Climate Zones for Cropland
Year: 1979
Scale: 1:1,000,000

Map Name: General Landcover of South Carolina 1980
Year: 1980
Scale: 1:500,000

Map Name: Iowa Major Forest Types
Year: 1990
Scale: 1:1,000,000

Map Name: Major Forest Types—Minnesota
Year: 1977
Scale: 1:1,000,000

Map Name: Major Michigan Soil Associations
Year: 1957
Scale: 1:600,000

Map Name: Minnesota Major Forest Types
Year: 1990
Scale: 1:1,000,000

Map Name: Original Vegetation of Minnesota
Year: 1930
Scale: 1:500,000

Map Name: Productivity Potential for Cropland
Year: 1979
Scale: 1:1,000,000

Map Name: San Francisco Bay Habitats
Year: 1985
Scale: 1:80,000

Map Name: Seismicity Map of the State of Vermont
Year: 1980
Scale: 1:100,000

Map Name: Soil Groups for Cropland
Year: 1979
Scale: 1:1,000,000

Map Name: South San Francisco Bay Habitat Change
Year: 1985
Scale: 1:70,000

## Geology

Map Name: Active Mines and Oil Fields in Nevada
Year: 1976
Scale: 1:1,000,000

Map Name: Aeromagnetic Map of Georgia
Year: 1980
Scale: 1:1,000,000

Map Name: Aeromagnetic Map of North Carolina
Year: 1977
Scale: 1:1,000,000

Map Name: Aeromagnetic Map of South Carolina
Year: 1982
Scale: 1:1,000,000

Map Name: Aeromagnetic Map of Tennessee and Kentucky
Year: 1984
Scale: 1:1,000,000

Map Name: Arizona Highway Geologic Map
Year: 1967
Scale: 1:1,000,000

Map Name: Badlands National Monument and Vicinity
Year: 1970
Scale: 1:62,500

Map Name: Badlands National Park
Year: 1981
Scale: 1:50,000

Map Name: Bedrock Geologic Map of Indiana
Year: 1987
Scale: 1:500,000

Map Name: Bedrock Geologic Map of Maine
Year: 1985
Scale: 1:500,000

Map Name: Bedrock Geologic Map of Massachusetts
Year: 1983
Scale: 1:160,934

Map Name: Bedrock Geologic Map of Wisconsin
Year: 1982
Scale: 1:1,000,000

Map Name: California, Nevada, and the Pacific Ocean Floor: The Present, Five Million Years Ago, and Five Million Years in the Future
Year: 1986
Scale: not drawn to scale

Map Name: Classes of Land-Surface Form in the Forty Eight States, U.S.A.
Year: 1963
Scale: 1:5,000,000

Map Name: Coal Resources and Distribution
Year: 1976
Scale: 1:7,500,000

Map Name: Depth to Bedrock in Wisconsin
Year: 1973
Scale: 1:1,000,000

Map Name: Earthquake Hazards Map
Year: 1995
Scale: 1:2,000,000

Map Name: Energy Resources Map of Colorado
Year: 1977
Scale: 1:500,000

Map Name: Energy Resources of Texas
Year: 1976
Scale: 1:1,000,000

Map Name: Fault Activity Map of California and Adjacent Areas, with Locations and Ages of Recent Volcanic Eruptions
Year: 1994
Scale: 1:750,000

Map Name: Geologic Map of Washington
Year: 1961
Scale: 1:500,000

Map Name: General Soil Map of Nebraska
Year: 1990
Scale: 1:1,000,000

Map Name: General Soil Map, South Carolina
Year: 1979
Scale: 1:750,000

Map Name: Generalized Pre-Pleistocene Geologic Map of the Northern United States Atlantic Continental Margin
Year: 1974
Scale: 1:1,000,000

Map Name: Generalized Tectonic-Metamorphic Map of New York
Year: 1971
Scale: 1:1,584,000

Map Name: Geologic Hazards Map of Tennessee
Year: 1977
Scale: 1:633,600

Map Name: Geologic Map of Arizona
Year: 1969
Scale: 1:500,000

Map Name: Geologic Map of Alabama
Year: 1989
Scale: 1:500,000

Map Name: Geologic Map of Alaska
Year: 1980
Scale: 1:2,500,000

Map Name: Geologic Map of Arkansas
Year: 1993
Scale: 1:500,000

Map Name: Geologic Map of California
Year: 1973
Scale: 1:750,000

Map Name: Geologic Map of Colorado
Year: 1979
Scale: 1:500,000

Map Name: Geologic Map of Florida
Year: 1981
Scale: 1:500,000

Map Name: Geologic Map of Georgia
Year: 1976
Scale: 1:500,000

Map Name: Geologic Map of Idaho
Year: 1978
Scale: 1:500,000

Map Name: Geologic Map of Illinois
Year: 1967
Scale: 1:500,000

Map Name: Geologic Map of Iowa
Year: 1969
Scale: 1:500,000

Map Name: Geologic Map of Kansas
Year: 1991
Scale: 1:500,000

Map Name: Geologic Map of Kentucky
Year: 1988
Scale: 1:500,000

Map Name: Geologic Map of Louisiana
Year: 1984
Scale: 1:500,000

Map Name: Geologic Map of Maryland
Year: 1968
Scale: 1:250,000

Map Name: Geologic Map of Minnesota
Year: 1970
Scale: 1:1,000,000

Map Name: Geologic Map of Minnesota, Quaternary Geology
Year: 1982
Scale: 1:500,000

Map Name: Geologic Map of Mississippi
Year: 1969
Scale: 1:500,000

Map Name: Geologic Map of Missouri
Year: 1979
Scale: 1:500,000

Map Name: Geologic Map of Montana
Year: 1955
Scale: 1:500,000

Map Name: Geologic Map of Nevada
Year: 1977
Scale: 1:500,000

Map Name: Geologic Map of New Hampshire
Year: 1955
Scale: 1:250,000

Map Name: Geologic Map of New Mexico
Year: 1965
Scale: 1:500,000

Map Name: Geologic Map of North Carolina
Year: 1985
Scale: 1:500,000

Map Name: Geologic Map of North Dakota
Year: 1980
Scale: 1:500,000

Map Name: Geologic Map of Ohio
Year: 1965
Scale: 1:500,000

Map Name: Geologic Map of Oklahoma
Year: 1954
Scale: 1:500,000

Map Name: Geologic Map of Oregon
Year: 1991
Scale: 1:500,000

Map Name: Geologic Map of South Carolina
Year: 1995
Scale: 1:1,000,000

Map Name: Geologic Map of Texas
Year: 1992
Scale: 1:500,000

Map Name: Geologic Map of the United States (exclusive of Alaska and Hawaii)
Year: 1974
Scale: 1:2,500,000

Map Name: Geologic Map of Utah
Year: 1980
Scale: 1:500,000

Map Name: Geologic Map of Washington
Year: 1961
Scale: 1:500,000

Map Name: Geologic Map of West Virginia
Year: 1986
Scale: 1:250,000

Map Name: Geologic Map of Wyoming
Year: 1985
Scale: 1:500,000

Map Name: Geological Highway Map of Texas
Year: 1973
Scale: 1:1,900,800

Map Name: Geological Map of New Mexico
Year: 1965
Scale: 1:500,000

Map Name: Geological Map of Pennsylvania
Year: 1960
Scale: 1:250,000

Map Name: Geothermal Energy in Alaska and Hawaii
Year: 1979
Scale: 1:5,000,000

Map Name: Geothermal Energy in the Western United States
Year: 1978
Scale: 1:2,500,000

Map Name: Glacial Deposits of Wisconsin, Sand and Gravel Resource Potential
Year: 1976
Scale: 1:500,000

Map Name: Map of Indiana Showing Thickness of Unconsolidated Deposits
Year: 1983
Scale: 1:500,000

Map Name: Map of Indiana Showing Topography of Bedrock Surface
Year: 1982
Scale: 1:500,000

Map Name: Map of Sedimentary Basins in the Conterminous United States
Year: 1988
Scale: 1:5,000,000

Map Name: Map of Surface Deposits of Illinois
Year: 1960
Scale: 1:1,013,760

Map Name: Map of Yosemite Valley
Year: 1970
Scale: 1:24,000

Map Name: Map Showing Extent of Glaciations in Alaska
Year: 1965
Scale: 1:2,500,000

Map Name: Maps, the Landscapes, and Fundamental Themes in Geography
Year: 1986
Scale: 1:8,340,000

Map Name: Mineral Resource Map, Georgia
Year: 1969
Scale: 1:500,000

Map Name: Mineral Resources of Iowa
Year: 1970
Scale: 1:500,000

Map Name: Mineral Resources of San Francisco Bay
Year: 1975
Scale: 1:250,000

Map Name: Mineral Resources of Virginia
Year: 1971
Scale: 1:500,000

Map Name: Mineral Resources of West Virginia
Year: 1994
Scale: 1:500,000

Map Name: New York State Geological Highway Map
Year: 1990
Scale: 1:1,000,000

Map Name: Ohio
Year: 1977
Scale: 1:500,000

Map Name: Oil and Gas Map of Wyoming
Year: 1980
Scale: 1:500,000

Map Name: Physiographic Diagram of the United States
Year: 1957
Scale: 1:9,000,000

Map Name: Quaternary Deposits of Illinois
Year: 1979
Scale: 1:500,000

Map Name: Quaternary Geologic Map of Indiana
Year: 1989
Scale: 1:500,000

Map Name: Quaternary Geologic Map of San Francisco
Year: 1993
Scale: 1:1,000,000

Map Name: Quaternary Geologic Map of the Dakotas Quadrangle
Year: 1995
Scale: 1:1,000,000

Map Name: Quaternary Geologic Map of the Des Moines 4 Degree by 6 Degree Quadrangle, United States
Year: 1991
Scale: 1:1,000,000

Map Name: Quaternary Geology of Michigan
Year: 1982
Scale: 1:500,000

Map Name: Seismicity Map of the Conterminous United States and Adjacent Areas, 1965–1974
Year: 1977
Scale: 1:5,000,000

Map Name: Seismicity Map of the Conterminous United States and Adjacent Areas, 1975–1984
Year: 1986
Scale: 1:5,000,000

Map Name: Seismicity Map of the State of Alabama
Year: 1979
Scale: 1:1,000,000

Map Name: Seismicity Map of the State of Arizona
Year: 1986
Scale: 1:1,000,000

Map Name: Seismicity Map of the State of Arkansas
Year: 1979
Scale: 1:1,000,000

Map Name: Seismicity Map of the State of Colorado
Year: 1984
Scale: 1:1,000,000

Map Name: Seismicity Map of the State of Florida
Year: 1987
Scale: 1:1,000,000

Map Name: Seismicity Map of the State of Georgia
Year: 1991
Scale: 1:1,000,000

Map Name: Seismicity Map of the State of Idaho
Year: 1986
Scale: 1:1,000,000

Map Name: Seismicity Map of the State of Illinois
Year: 1979
Scale: 1:1,000,000

Map Name: Seismicity Map of the State of Indiana
Year: 1974
Scale: 1:1,000,000

Map Name: Seismicity Map of the State of Kansas
Year: 1981
Scale: 1:1,000,000

Map Name: Seismicity Map of the State of Kentucky
Year: 1979
Scale: 1:1,000,000

Map Name: Seismicity Map of the State of Louisiana
Year: 1987
Scale: 1:1,000,000

Map Name: Seismicity Map of the State of Maine
Year: 1981
Scale: 1:1,000,000

Map Name: Seismicity Map of the State of Massachusetts
Year: 1980
Scale: 1:100,000

Map Name: Seismicity Map of the State of Michigan
Year: 1980
Scale: 1:1,000,000

Map Name: Seismicity Map of the State of Minnesota
Year: 1981
Scale: 1:1,000,000

Map Name: Seismicity Map of the State of Missouri
Year: 1979
Scale: 1:1,000,000

Map Name: Seismicity Map of the State of Montana
Year: 1985
Scale: 1:1,000,000

Map Name: Seismicity Map of the State of Nebraska
Year: 1981
Scale: 1:1,000,000

Map Name: Seismicity Map of the State of New Hampshire
Year: 1980
Scale: 1:1,000,000

Map Name: Seismicity Map of the State of New Jersey
Year: 1980
Scale: 1:1,000,000

Map Name: Seismicity Map of the State of New Mexico
Year: 1983
Scale: 1:1,000,000

Map Name: Seismicity Map of the State of New York
Year: 1981
Scale: 1:1,000,000

Map Name: Seismicity Map of the State of North Carolina
Year: 1980
Scale: 1:1,000,000

Map Name: Seismicity Map of the State of North Dakota
Year: 1981
Scale: 1:1,000,000

Map Name: Seismicity Map of the State of Ohio
Year: 1979
Scale: 1:1,000,000

Map Name: Seismicity Map of the State of Oklahoma
Year: 1981
Scale: 1:1,000,000

Map Name: Seismicity Map of the State of Pennsylvania
Year: 1981
Scale: 1:1,000,000

Map Name: Seismicity Map of the State of South Carolina
Year: 1980
Scale: 1:1,000,000

Map Name: Seismicity Map of the State of South Dakota
Year: 1981
Scale: 1:1,000,000

Map Name: Seismicity Map of the State of Tennessee
Year: 1979
Scale: 1:1,000,000

Map Name: Seismicity Map of the State of Texas
Year: 1982
Scale: 1:1,000,000

Map Name: Seismicity Map of the State of Virginia
Year: 1980
Scale: 1:1,000,000

Map Name: Seismicity Map of the State of West Virginia
Year: 1980
Scale: 1:1,000,000

Map Name: Seismicity Map of the State of Wisconsin
Year: 1980
Scale: 1:1,000,000

Map Name: Seismicity Map of the State of Wyoming
Year: 1985
Scale: 1:1,000,000

Map Name: Seismicity Map of the States of Connecticut and Rhode Island
Year: 1981
Scale: 1:1,000,000

Map Name: Seismicity Map of the States of Delaware and Maryland
Year: 1981
Scale: 1:1,000,000

Map Name: Sequoia and Kings Canyon National Parks and Vicinity, California
Year: 1967
Scale: 1:125,000

Map Name: Shaded Relief (of the United States)
Year: 1972
Scale: 1:7,500,000

Map Name: Soil Amplification/Liquefaction Potential Map
Year: 1999
Scale: 1:2,000,000

Map Name: Soils in the Tennessee Valley
Year: 1968
Scale: 1:633,300

Map Name: Soils of the Great Plains
Year: 1972
Scale: 1:2,500,000

Map Name: Solution Mining and Subsidence in Evaporite Rocks in the United States
Year: 1981
Scale: 1:5,000,000

Map Name: State of Georgia Geology
Year: 1963
Scale: 1:500,000

Map Name: State of Idaho Geology
Year: 1969
Scale: 1:500,000

Map Name: State of Nebraska Geology
Year: 1972
Scale: 1:500,000

Map Name: State of New York Geology
Year: 1974
Scale: 1:1,000,000

Map Name: State of North Carolina Geology
Year: 1972
Scale: 1:500,000

Map Name: State of Ohio Geology
Year: 1977
Scale: 1:500,000

Map Name: State of Pennsylvania Geology
Year: 1975
Scale: 1:1,000,000

Map Name: Surficial Geologic Map of Maine
Year: 1985
Scale: 1:500,000

Map Name: Surficial Geologic Map of Vermont
Year: 1969
Scale: 1:250,000

Map Name: Tectonic Features
Year: 1967
Scale: 1:7,500,000

Map Name: Tectonic Lithofacies Map of the Appalachian Orogen
Year: 1978
Scale: 1:1,000,000

Map Name: Tectonic Map of the United States, Exclusive of Alaska and Hawaii
Year: 1968
Scale: 1:2,500,000

Map Name: Volcanic Hazards and Energy Infrastructure, United States
Year: 1995
Scale: 1:4,000,000

Map Name: Wyoming Mines and Minerals
Year: 1979
Scale: 1:500,000

## Historical

Map Name: Atlantic Gateways
Year: 1983
Scale: 1:1,318,000

Map Name: Battles of the Civil War
Year: 2005
Scale: 1:2,950,000

Map Name: Boston to Washington, circa 1830/Boston to Washington Megalopolis
Year: 1994
Scale: 1:1,000,000

Map Name: Historical Map of the Conterminous United States
Year: 1967
Scale: 1:5,000,000

Map Name: Mapping an American Frontier, Oregon in 1850
Year: 1975
Scale: 1:710,000

Map Name: Native American Heritage
Year: 1991
Scale: 1:7,747,000

Map Name: Northern Approaches, Maine and the Maritimes
Year: 1985
Scale: 1:735,000

Map Name: Southwestern United States
Year: 1940
Scale: 1:2,500,000

Map Name: The United States, History of the Land
Year: 2006
Scale: 1:5,200,000

## Human Geography

Map Name: 1980 Population Distribution in the United States
Year: 1980
Scale: 1:5,000,000

Map Name: 1980 Population, Urban and Rural, United States
Year: 1980
Scale: 1:5,000,000

Map Name: Aids in LA 1983–1989
Year: 1989
Scale: 1:580,833

Map Name: Are You Safe Where You Live?
Year: 1993
Scale: 1:7,420,000

Map Name: Arkansas Population Distribution: With Shaded Relief Features of the Physical Landscape
Year: 1984
Scale: 1:1,000,000

Map Name: California Population Distribution in 1970
Year: 1970
Scale: 1:1,000,000

Map Name: Changing Population of Georgia
Year: 1970
Scale: 1:5,068,800

Map Name: Coal Fields of the United States
Year: 1979
Scale: 1:5,000,000

Map Name: Coal Movement by Highways
Year: 1974
Scale: 1:7,500,000

Map Name: Coal Movement by Railroads
Year: 1974
Scale: 1:7,500,000

Map Name: Coal Movement by Water
Year: 1974
Scale: 1:7,500,000

Map Name: Crude Oil Movement by Pipeline
Year: 1974
Scale: 1:7,500,000

Map Name: Detroit Area Ethnic Groups, 1988
Year: 1988
Scale: 1:150,000

Map Name: Electric Power Transmission
Year: 1974
Scale: 1:7,500,000

Map Name: Ethnic Patterns in Los Angeles
Year: 1989
Scale: 1:225,000

Map Name: Generalized Types of Farming in the United States
Year: 1949
Scale: 1:10,000,000

Map Name: Great Lakes, St. Lawrence Seaway Intermodal Transportation System
Year: 1975
Scale: 1:7,300,000

Map Name: Indian Lands, 1992
Year: 1992
Scale: 1:5,000,000

Map Name: Indian Lands in the United States
Year: 2000
Scale: 1:5,000,000

Map Name: Intercity Rail Passenger Routes
Year: 1976
Scale: 1:4,000,000

Map Name: Major Army, Navy, and Air Force Installations in the United States
Year: 1985
Scale: 1:3,500,000

Map Name: Natural Gas Movement by Pipeline
Year: 1974
Scale: 1:7,500,000

Map Name: New York, County Subdivisions, Towns, Indian Reservations, and Places
Year: 1970
Scale: 1:750,000

Map Name: Nightime Lights of the USA
Year: 1995
Scale: N/A

Map Name: Nuclear Fuel Materials Movement by Highways
Year: 1975
Scale: 1:7,500,000

Map Name: Petroleum Products Movement by Pipelines
Year: 1974
Scale: 1:7,500,000

Map Name: Petroleum Products Movement by Water
Year: 1976
Scale: 1:7,500,000

Map Name: Pipelines Transportation Systems
Year: 1974
Scale: 1:7,500,000

Map Name: Population Density by Counties of the United States
Year: 1971
Scale: 1:5,000,000

Map Name: Population Distribution, Urban and Rural in the United States
Year: 1960, 1970
Scale: 1:5,000,000

Map Name: Population Distribution, Washington State
Year: 1970
Scale: 1:760,000

Map Name: Population Origin Groups in Rural Texas
Year: 1970
Scale: 1:1,500,000

Map Name: Railroad, Highway, and Water Transportation Systems
Year: 1974
Scale: 1:7,500,000

Map Name: Railroads of the Continental United States
Year: 1996
Scale: 1:4,250,000

Map Name: Total Coal Movement
Year: 1974
Scale: 1:7,500,000

Map Name: Total Crude Oil Movement
Year: 1974
Scale: 1:7,500,000

Map Name: Total Interstate Energy Movement
Year: 1974
Scale: 1:7,500,000

Map Name: Total Petroleum Movement
Year: 1974
Scale: 1:7,500,000

Map Name: Transportation/Planning Map, New York State
Year: 1974
Scale: 1:250,000

Map Name: West Virginia Transportation System: Railroads, Major Highways, Airports and Navigable Waterways, as of November 1, 1986
Year: 1986
Scale: 1:500,000

Map Name: Western Migration: Dreams of Gold and a Better Life Drive Mass Movement
Year: 2000
Scale: 1:4,499,000

**Land Use**

Map Name: America's Federal Lands, or Federal Property, a Third of the Nation
Year: 1982
Scale: 1:5,889,000

Map Name: At Home Downtown: The Residential Transformation of Chicago's New Global-Era Core, 1985–2005
Year: 2005
Scale: 1:10,000

Map Name: Colorado Land Use Map Folio
Year: 1974
Scale: 1:500,000

Map Name: Downtown Manhattan
Year: 1982
Scale: not drawn to scale

Map Name: Forest Regions of the United States
Year: 1960
Scale: 1:12,000,000

Map Name: General Land Use in Nebraska
Year: 1973
Scale: 1:1,000,000

Map Name: Kansas Land-Use Patterns
Year: 1973
Scale: 1:1,000,000

Map Name: Land Use and Land Cover, 1973 and 1969, Allegheny County, Pennsylvania
Year: 1980
Scale: 1:125,000

Map Name: Land-Use in Iowa 1976
Year: 1976
Scale: 1:500,000

Map Name: Major Land Uses in the United States
Year: 1958
Scale: 1:5,000,000

Map Name: NYC Midtown Core
Year: 1982
Scale: Not drawn to scale

Map Name: South Dakota Land Use Patterns
Year: 1973
Scale: 1:1,000,000

Map Name: United States of America, Showing the Extent of Public Land Surveys, Remaining Public Land, Historical Boundaries, National Forests, Indian Reservations, Wildlife Refuges, National Parks, and Monuments
Year: 1965
Scale: 1:2,500,000

## Meteorology

Map Name: Climatic Regions of Arizona
Year: 1976
Scale: 1:3,027,200

Map Name: Magnetic Field in the United States
Year: 1992
Scale: 1:5,000,000

Map Name: Map of Solar Energy in the United States and Southern Canada
Year: 1980
Scale: 1:5,000,000

Map Name: U.S. Tornadoes 1930–74
Year: 1976
Scale: 1:9,500,000

## Physical Geography

Map Name: Aeromagnetic Map of the Northeastern United States
Year: 1981
Scale: 1:2,000,000

Map Name: Appalachian Region, as Designated by the Appalachia Regional Commission
Year: 1968
Scale: 1:2,500,000

Map Name: Average Elevation Map of the Conterminous United States
Year: 1981
Scale: 1:2,500,000

Map Name: Canyonlands National Park and Vicinity, Utah
Year: 1968
Scale: 1:62,500

Map Name: Digital Terrain Map of the United States
Year: 1981
Scale: 1:7,500,000

Map Name: Ecoregions of the United States
Year: 1976
Scale: 1:7,500,000

Map Name: Estimated Proportion of Land in Floodplain
Year: 1985
Scale: 1:8,500,000

Map Name: Federal Lands in the Fifty States
Year: 1996
Scale: 1:6,766,000

Map Name: Generalized Soil Erosion Hazard
Year: 1975
Scale: 1:500,000

Map Name: Geothermal Energy Resources of the Western United States
Year: 1977
Scale: 1:2,500,000

Map Name: Geothermal Gradient Map of the United States, Exclusive of Alaska and Hawaii
Year: 1982
Scale: 1:2,500,000

Map Name: Glacial Map of Ohio
Year: 1967
Scale: 1:500,000

Map Name: Glacial Map of the United States East of the Rocky Mountains
Year: 1959
Scale: 1:1,750,000

Map Name: Glacier National Park
Year: 1968
Scale: 1:100,000

Map Name: Great Salt Lake, Utah
Year: 1971
Scale: 1:125,000

Map Name: Heat Flow Map of the Eastern United States
Year: 1988
Scale: 1:2,500,000

Map Name: Illinois Drainage Map
Year: 1971
Scale: 1:506,880

Map Name: Land Resource Regions and Major Land Resource Areas of the United States
Year: 1963
Scale: 1:10,000,000

Map Name: Landform Map of Alaska
Year: 1966
Scale: 1:3,150,000

Map Name: Landforms of Ohio
Year: 1967
Scale: 1:1,000,000

Map Name: Landforms of the Northwestern States
Year: 1965
Scale: 1:1,350,000

Map Name: Landforms of Utah in Proportional Relief
Year: 1963
Scale: 1:650,000

Map Name: Major Areas of Potential Flood Hazard and Steep Slope
Year: 1975
Scale: 1:500,000

Map Name: Maps of Physiographic Divisions of Alaska
Year: 1965
Scale: 1:2,500,000

Map Name: National Atlas of the United States of America, Hydrologic Units
Year: 1998
Scale: 1:3,500,000

Map Name: National Atlas of the United States of America, Principal Aquifers
Year: 1998
Scale: 1:5,000,000

Map Name: National Atlas, Coastal Erosion and Accretion
Year: 1988
Scale: 1:7,500,000

Map Name: National Wild and Scenic River System
Year: 1991
Scale: 1:5,000,000

Map Name: Native Vegetation of Nebraska
Year: 1993
Scale: 1:1,000,000

Map Name: Northeastern United States Showing Relation of Land and Submarine Topography
Year: 1939
Scale: 1:1,000,000

Map Name: Physical Map, New York State and Adjacent Areas
Year: 1961
Scale: 1:750,000

Map Name: Physiographic Diagram of New Mexico
Year: 1982
Scale: 1:1,300,000

Map Name: Pikes Peak and Vicinity
Year: 1957
Scale: 1:62,500

Map Name: Potential Natural Vegetation of the Conterminous United States
Year: 1964
Scale: 1:3,168,000

Map Name: Potential Natural Vegetation of the Conterminous United States
Year: 1967
Scale: 1:7,500,000

Map Name: Prime Farmlands of Oklahoma
Year: 1982
Scale: 1:1,584,000

Map Name: Seasonal Land Cover Regions
Year: 1993
Scale: 1:7,500,000

Map Name: Shaded Elevation Map of Ohio
Year: 2003
Scale: 1:500,000

Map Name: Shaded Relief Map of Nevada
Year: 1981
Scale: 1:1,000,000

Map Name: Soils of Wisconsin
Year: 1968
Scale: 1:710,000

Map Name: State of New Mexico
Year: 1967
Scale: 1:500,000

Map Name: Surface-Water and Related-Land Resources Development in the United States and Puerto Rico
Year: 1983
Scale: 1:3,168,000

Map Name: United States; Precious Resource Water
Year: 1993
Scale: 1:4,560,000

Map Name: Vegetation of Nebraska
Year: 1850
Scale: 1:1,000,000

Map Name: Wild and Scenic Rivers of the United States
Year: 1977
Scale: 1:7,500,000

**Political Geography**

Map Name: Alabama; County Subdivisions, Census Divisions, and Places
Year: 1971
Scale: 1:750,000

Map Name: Arkansas; County Subdivisions, Townships and Places
Year: 1976
Scale: 1:750,000

Map Name: Colorado: County Subdivisions, Census County Divisions and Places
Year: 1970
Scale: 1:750,000

Map Name: Congressional Districts for the 92nd Congress
Year: 1970
Scale: 1:5,000,000

Map Name: Congressional Districts for the 94th Congress
Year: 1975
Scale: 1:7,500,000

Map Name: Congressional Districts for the 100th Congress
Year: 1987
Scale: 1:7,500,000

Map Name: Congressional Districts of the 98th Congress
Year: 1985
Scale: 1:5,000,000

Map Name: Delaware; County, Subdivisions, Census County Divisions and Places
Year: 1971
Scale: 1:510,000

Map Name: Florida; County Subdivisions, Census County Divisions and Places
Year: 1970
Scale: 1:775,000

Map Name: Georgia; County Subdivisions, Census County Divisions and Places
Year: 1970
Scale: 1:750,000

Map Name: Idaho; County Subdivisions, Census County Divisions and Places
Year: 1970
Scale: 1:750,000

Map Name: Illinois; County Subdivisions, Townships, Election Precincts, and Places
Year: 1976
Scale: 1:750,000

Map Name: Indian Land Areas Judicially Established
Year: 1978
Scale: 1:4,000,000

Map Name: Indian Land Cessions
Year: 1972
Scale: 1:10,000,000

Map Name: Indiana; County Subdivisions, Townships and Places
Year: 1976
Scale: 1:500,000

Map Name: Iowa; County Subdivisions, Townships and Places
Year: 1970
Scale: 1:750,000

Map Name: Kansas; County Subdivisions, Townships and Places
Year: 1970
Scale: 1:775,000

Map Name: Kentucky; County Subdivisions, Census County Divisions and Places
Year: 1970
Scale: 1:570,000

Map Name: Maryland; County Subdivisions, Election Districts and Places
Year: 1970
Scale: 1:500,000

Map Name: Massachusetts; County, Subdivisions, Towns and Places
Year: 1970
Scale: 1:160,934

Map Name: Metropolitan Statistical Area (CMSAs, PMSAs, and MSAs)
Year: 1986
Scale: 1:5,000,000

Map Name: Michigan; County Subdivisions, Townships and Places
Year: 1977
Scale: 1:1,000,000

Map Name: Minnesota; County Subdivisions, Townships, Organized Territories, and Places
Year: 1977
Scale: 1:750,000

Map Name: Mississippi; County Subdivisions, Supervisors Districts and Places
Year: 1970
Scale: 1:800,000

Map Name: Missouri; County Subdivisions, Townships and Places
Year: 1970
Scale: 1:750,000

Map Name: Montana; County Subdivisions, Census County Divisions, and Places
Year: 1970
Scale: 1:1,000,000

Map Name: Nebraska; County Subdivisions, Election Precincts, Townships, and Places
Year: 1970
Scale: 1:800,000

Map Name: Nevada; County Divisions, Townships and Places
Year: 1970
Scale: 1:750,000

Map Name: New Hampshire; County Subdivisions, Towns, Grants, Townships, Purchases, Locations, and Places
Year: 1970
Scale: 1:510,000

Map Name: New Jersey; County Subdivisions, Townships and Places
Year: 1970
Scale: 1:250,000

Map Name: New Mexico; County Subdivisions, Census County Divisions and Places
Year: 1970
Scale: 1:750,000

Map Name: North Carolina; County Subdivisions, Census County Divisions and Places
Year: 1976
Scale: 1:750,000

Map Name: Ohio; County Subdivisions, Townships and Places
Year: 1971
Scale: 1:500,000

Map Name: Oklahoma; County Subdivisions, Census Boundary Divisions, and Places
Year: 1970
Scale: 1:750,000

Map Name: Oregon; County Subdivisions, Census County Divisions and Places
Year: 1970
Scale: 1:750,000

Map Name: Principal Meridians and Base Lines Governing the United States Public Land Surveys
Year: 1963
Scale: 1:7,500,000

Map Name: South Carolina; County Subdivisions, Census County Divisions and Places
Year: 1970
Scale: 1:750,000

Map Name: Tennessee; County Subdivisions, Census County Divisions and Places
Year: 1970
Scale: 1:800,000

Map Name: Texas; County Subdivisions, Census County Divisions and Places
Year: 1976
Scale: 1:1,500,000

Map Name: Utah; County Subdivisions, Census County Divisions and Places
Year: 1970
Scale: 1:750,000

Map Name: Vermont; County Subdivisions, Census County Divisions and Places
Year: 1970
Scale: 1:510,000

Map Name: West Virginia; County Subdivisions, Census County Divisions and Places
Year: 1976
Scale: 1:500,000

Map Name: Wisconsin; Minor Civil Divisions, Towns, Cities and Villages
Year: 1961
Scale: 1:750,000

Map Name: Wyoming; County Subdivisions, Census County Divisions and Places
Year: 1970
Scale: 1:750,000

## Recreational Geography

Map Name: 1:100,000-Scale Metric Topographic Map of Olympic National Park and Vicinity, Washington
Year: 1987
Scale: 1:100,000

Map Name: Aviators' Reference Map of the United States
Year: 1988
Scale: 1:5,780,000

Map Name: Explorers Map and Directory of the New England Coast
Year: 1981
Scale: 1:330,000

Map Name: Invitation to Enjoyment, Federal Recreation Lands of the United States
Year: 1992
Scale: 1:7,729,920

Map Name: Isle Royale National Park
Year: 1987
Scale: 1:62,500

Map Name: Mount St. Helens and Vicinity, Washington-Oregon
Year: 1980
Scale: 1:1,000,000

Map Name: Mt. Rainier National Park, Washington
Year: 1971
Scale: 1:50,000

Map Name: National Park System
Year: 1980
Scale: 1:7,700,000

Map Name: National Wilderness Preservation System
Year: 1989
Scale: N/A

Map Name: National Wilderness Preservation System
Year: 1987
Scale: 1:5,000,000

Map Name: North Cascades National Park and Lake Chelan and Ross Lake National Recreation Area, Washington
Year: 1974
Scale: 1:100,000

Map Name: Vacationlands of the United States and Canada
Year: 1967
Scale: 1:5,132,160

Map Name: Yellowstone National Park, Wyoming-Montana-Idaho
Year: 1974
Scale: 1:125,000

## WORLD

### Base Maps

Map Name: Canada, United States of America, Estados Unidos Mexicanos
Year: 2004
Scale: 1:10,000,000

Map Name: Generalized Tectonic Map of North America
Year: 1972
Scale: 1:15,000,000

Map Name: Geologic Map of North America
Year: 1965
Scale: 1:13,823,998

Map Name: Isodemographic Map of North America
Year: 1976
Scale: N/A

Map Name: Isodemographic Map of North America
Year: 1981
Scale: N/A

Map Name: Nordamerika
Year: 1948
Scale: 1:20,000,000

Map Name: North America
Year: 1971, 1979, 1984, 1994, 2005
Scale: 1:13,825,000

### Geology

Map Name: Quaternary Geologic Map of the Lake Erie
Year: 1991
Scale: 1:1,000,000

Map Name: Quaternary Map of Greenland
Year: 1969
Scale: 1:2,500,000

Map Name: Tectonic/Geological Map of Greenland
Year: 1970
Scale: 1:2,500,001

Map Name: Terrestrial Heat Flow Data
Year: 1976
Scale: 1: 30,000,000

## Historical

Map Name: Americae sive novi orbis nova descriptio
Year: 1570
Scale: 1:39,000,000

Map Name: Carte des cinq grands lacs du Canada
Year: 1764
Scale: 1:4,500,000

Map Name: Colonization and Trade in the New World
Year: 1977
Scale: N/A

Map Name: Dawn of Humans
Year: 1997
Scale: 1:35,660,000

Map Name: North America in the Age of the Dinosaurs
Year: 1993
Scale: 1:19,112,000

Map Name: Planisfero del mondo nuovo
Year: 1695
Scale: 1:36,000,000

Map Name: Western Hemisphere from the Sanches Planisphere
Year: 1623
Scale: N/A

## Human Geography

Map Name: Indians of North America
Year: 1982
Scale: 1:10,610,000

Map Name: Map of North American Indian languages
Year: 1966
Scale: 1:6,336,000

Map Name: North American Indian Cultures, a Legacy of Language and Inspired Ideas; Indian Country
Year: 2004
Scale: N/A

Map Name: Toxic Great Lakes Hotspots
Year: 1984
Scale: 1:2,275,000

## Physical Geography

Map Name: Magnetic Anomaly Map of North America
Year: 2002
Scale: 1:10,000,000

Map Name: Natural Hazards of North America
Year: 1998
Scale: 1:15,750,000

Map Name: North America
Year: 1995
Scale: 1:10,000,000

Map Name: Physiographic Provinces of North America; Physiographic Diagram of North America
Year: 1948
Scale: 1:12,000,000

Map Name: Seismicity Map of North America
Year: 1988
Scale: 1:5,000,000

Map Name: Tectonic Map of North America
Year: 1996
Scale: 1:5,000,000

Map Name: USDA Plant Hardiness Zone Map
Year: 1990
Scale: 1:6,000,000

Map Name: Watersheds
Year: 2006
Scale: 1:10,000,000

## Physical Geography

Map Name: A New World of Understanding
Year: 1982
Scale: N/A

Map Name: International Permafrost Association Circum-Arctic Map of Permafrost and Ground Ice Conditions
Year: 1997
Scale: 1:10,000,000

Map Name: North Circumpolar Region
Year: 2000
Scale: 1:10,000,000

Map Name: North Polar Region
Year: 1966
Scale: 1:20,000,000

Map Name: North Pole International Polar Year
Year: 2007
Scale: 1:11,000,000

Map Name: Prospective Maritime Jurisdictions in the Polar Seas
Year: 1983
Scale: 1:25,000,000

Map Name: Top of the World
Year: 1965
Scale: 1:20,021,760

# Appendix B
## *Depository Library Programs*

Both Canada and the United States have a depository map program created by the government to offer maps and other publications to the public at no cost. In the United States, the Federal Depository Library Program (FDLP) offers over one thousand libraries U.S. federal government publications. In Canada, the Depository Services Program (DSP) offers Canadian government publications to over one hundred libraries. In Canada, there are two types of depositories: full depositories receive publications automatically, and selective depositories receive publications upon request. In the United States, there are also two types of depository libraries. One or two regional depository libraries in each state receive a copy of all government publications, and selective depository libraries select individual publications of interest. Each congressional district is permitted to have only two selective depositories. In Canada, as of January 1, 2014, the DSP program was modified and transitioned to an electronic-only model, offering libraries access to digital versions of maps.

Libraries with depository status develop their government map collections at no cost; however, they are responsible for processing the maps and making them available to the general public. Knowing which cities and libraries offer government map publications will aid researchers in search of federal maps. Below is a list of Canadian and U.S. libraries that are a part of the FDLP or DSP.

### CANADIAN LIBRARY MAP DEPOSITORIES

**Alberta**

*Calgary*

Mount Royal University
Riddell Library and Learning Centre

University of Calgary
Taylor Family Digital Library—Spatial and Numeric Data Services

*Edmonton*

University of Alberta
Cameron Library

*Lethbridge*

University of Lethbridge
University of Lethbridge Library

**British Columbia**

*Burnaby*

Simon Fraser University
Bennet Library

*Prince George*

College of New Caledonia
College of New Caledonia Library

University of Northern British Columbia
Geoffrey R. Weller Library

*Richmond*

Richmond Public Library

*Vancouver*

University of British Columbia
Walter C. Koerner Library

Vancouver Public Library
Central Library

*Victoria*

Greater Victoria Public Library

University of Victoria
McPherson Library—William C. Mearns Centre for Learning

## Manitoba

*Brandon*

Brandon University
John E. Robbins Library—Map Library

*Winnipeg*

University of Manitoba
Elizabeth Dafoe Library

University of Winnipeg
Centennial Hall Mezzanine—Map Library

Winnipeg Public Library

## New Brunswick

*Fredericton*

University of New Brunswick
Harriet Irving Library

*Moncton*

Université de Moncton
Champlain Library

*Sackville*

Mount Allison University
R. P. Bell Library

## Nova Scotia

*Halifax*

Dalhousie University
Killam Memorial Library

St. Mary's University
Patrick Power Library

*Lawrencetown*

Centre of Geographic Sciences
Library—Centre of Geographic Sciences

*Wolfville*

Acadia University
Vaughan Memorial Library

## Ontario

*Brampton*

Brampton Library

*Guelph*

University of Guelph
University of Guelph Library—Data Resource Centre

*Hamilton*

Hamilton Public Library

McMaster University
Lloyd Reeds Map Collection

*Kingston*

Queen's University
W. D. Jordan Rare Books and Special Collections

*London*

Western University
D. B. Weldon Library—Map and Data Centre

*Mississauga*

University of Toronto at Mississauga
Hazel McCallion Academic Centre—AstraZeneca Canada Centre for Information and Technological Literacy

*Orillia*

Lakehead University
Map Library—Department of Geography

*Oshawa*

Oshawa Public Library

*Ottawa*

Carleton University
MacOdrum Library

Library and Archives Canada

Ottawa Public Library

University of Ottawa
Geographic, Statistical and Government Information Centre

*Peterborough*

Trent University
Library and Archives—Maps, Data and Government Information Centre

*Scarborough*

University of Toronto at Scarborough
Bladen Library

*St. Catharines*

Brock University
James A. Gibson Library

*Sudbury*

Laurentian University
Parker Building—Map Library

*Toronto*

Legislative Assembly of Ontario

Ryerson University
R. D. Besse Information and Learning Commons

Toronto Public Library

University of Toronto
Robarts Library—Map and Data Library

York University
Scott Library—Map Library

*Waterloo*

University of Waterloo
Dana Porter Library—Geospatial Centre

Wilfrid Laurier University
Wilfrid Laurier University Library

*Windsor*

University of Windsor
Leddy Library

**Prince Edward Island**

*Charlottetown*

University of Prince Edward Island
Robertson Library

**Quebec**

*Montreal*

Bibliothèque et Archives nationales du Québec

Cégep de Saint-Laurent
La Bibliothèque

Concordia University
Webster Library

McGill University
McGill Library—Map and Geospatial Data

Université de Montréal
Cartothèque

Université du Québec à Montréal
Bibliothèques

*Quebec City*

Université Laval
Pavillon Jean-Charles-Bonenfant—GeoStat Center

*Rimouski*

Université du Québec à Rimouski
Bibliothèque

*Rouyn-Noranda*

Université du Québec en Abitibi-Témiscamingue
Bibliothèque

*Saguenay*

Université du Québec à Chicoutimi
Bibliothèque Paul-Émile-Boulet

*Sherbrooke*

Bishop's University
John Bassett Library

Université de Sherbrooke
Bibliothèque du Frère-Théode—Service des bibliothèques et archives

*Trois-Rivières*

Cégep de Trois-Rivières
Bibliothèque du Cégep de Trois-Rivières

Université du Québec à Trois-Rivières
Bibliothèque—Cartothèque

## Saskatchewan

*Regina*

University of Regina
Dr. John Archer Library—Map Library

*Saskatoon*

University of Saskatchewan
Murray Library—Data and GIS Library Services

## Newfoundland and Labrador

*St John's*

Memorial University of Newfoundland
Queen Elizabeth II Library—Map Room

## Yukon

*Whitehorse*

Yukon Archives

## U.S. LIBRARY MAP DEPOSITORIES

## Alabama

*Auburn*

Auburn University
Ralph Brown Draughon Library
Selective depository library

*Birmingham*

Birmingham-Southern College
Rush Learning Center/Miles Library
Selective depository library

City of Birmingham
Birmingham Public Library
Selective depository library

Jefferson State Community College
James B. Allen Library
Selective depository library

Samford University
University Library
Selective depository library

*Enterprise*

Enterprise State University
Learning Resources Center
Selective depository library

*Fayette*

Bevill State Community College
Fayette Learning Resource Center
Selective depository library

*Florence*

University of North Alabama
Collier Library
Selective depository library

*Gadsden*

City of Gadsden
Gadsden Public Library
Selective depository library

*Huntsville*

University of Alabama, Huntsville
M. Louis Salmon Library
Selective depository library

*Jacksonville*

Jacksonville State University
Houston Cole Library
Selective depository library

*Maxwell Air Force Base*

Air University
Muir S Fairchild Research Info Center
Selective depository library

*Mobile*

Spring Hill College
Burke Memorial Library
Selective depository library

University of South Alabama
University Libraries
Selective depository library

*Montgomery*

Alabama Supreme Court
Supreme Court and State Law Library
Selective depository library

Auburn University at Montgomery
Auburn University at Montgomery Library
Regional depository library

Faulkner University
Jones School of Law Library
Selective depository library

*Normal*

Alabama A&M University
J. F. Drake Memorial Learning Resources Center
Selective depository library

*Troy*

Troy University
Troy University Library
Selective depository library

*Tuscaloosa*

University of Alabama
Amelia Gayle Corgas Library
Regional depository library

University of Alabama
Bounds Law Library
Selective depository library

*Tuskegee*

Tuskegee University
Ford Motor Company Library/Learning Resource Center
Selective depository library

## Alaska

*Anchorage*

Alaska Court System
Alaska State Court Law Library
Selective depository library

Alaska Resources Library and Information Services
ARLIS Library
Selective depository library

University of Alaska Anchorage
UAA/APU Consortium Library
Selective depository library

*Barrow*

Iḷisaġvik College
Tuzzy Consortium Library
Selective depository library

*Fairbanks*

University of Alaska, Fairbanks
Elmer E. Rasmuson Library
Selective depository library

*Juneau*

Alaska Department of Education and Early Development
Alaska State Library—Government Publications
Selective depository library

University of Alaska Southeast
William A. Egan Library
Selective depository library

## Arizona

*Apache Junction*

City of Apache Junction
Apache Junction Public Library
Selective depository library

*Coolidge*

Central Arizona College
Central Arizona College Libraries
Selective depository library

*Flagstaff*

Northern Arizona University
Cine Library
Selective depository library

*Glendale*

Arizona State University
Fletcher Library
Selective depository library

*Phoenix*

Arizona State Library, Archives and Public Records
State of Arizona Research Library
Regional depository library

Phoenix Public Library
Burton Barr Central Library
Selective depository library

*Prescott*

Yavapai College
Yavapai College Library
Selective depository library

*Tucson*

University of Arizona
Main Library
Selective depository library

University of Arizona James E. Rogers College of Law
Cracchiolo Law Library
Selective depository library

*Winslow*

Northland Pioneer College
Little Colorado Campus Library
Selective depository library

*Yuma*

Arizona Western College
Academic Library
Selective depository library

**Arkansas**

*Arkadelphia*

Ouachita Baptist University
Riley-Hickingbotham Library
Selective depository library

*Conway*

University of Central Arkansas
Torreyson Library
Selective depository library

*Fayetteville*

University of Arkansas
University Libraries
Selective depository library

University of Arkansas School of Law
Young Law Library
Selective depository library

*Jonesboro*

Arkansas State University, Jonesboro
Dean B. Ellis Library
Selective depository library

*Little Rock*

Arkansas Baptist College
J. C. Oliver Library
Selective depository library

Arkansas State Library
Regional depository library

Arkansas Supreme Court
Arkansas Supreme Court Library
Selective depository library

University of Arkansas, Little Rock
Ottenheimer Library
Selective depository library

University of Arkansas, Little Rock
Pulaski County Law Library
Selective depository library

*Magnolia*

Southern Arkansas University
Magale Library
Selective depository library

*Monticello*

University of Arkansas, Monticello
UAM Taylor Library
Selective depository library

*Pine Bluff*

University of Arkansas at Pine Bluff
John Brown Watson Memorial Library
Selective depository library

*Russellville*

Arkansas Tech University
Ross Pendergraft Library and Technology Center
Selective depository library

**California**

*Arcadia*

Arcadia Public Library
Selective depository library

*Arcata*

Humboldt State University
University Library
Selective depository library

*Bakersfield*

California State University, Bakersfield
Walter W. Stiern Library
Selective depository library

Kern County Library
Beale Memorial Library
Selective depository library

*Berkeley*

University of California, Berkeley
Charles Franklin Doe Memorial Library
Selective depository library

University of California, Berkeley
School of Law Library
Selective depository library

*Calexico*

San Diego State University
Imperial Valley Campus Library
Selective depository library

*Carson*

California State University, Dominguez Hills
University Library
Selective depository library

*Chico*

California State University, Chico
Meriam Library
Selective depository library

*Claremont*

Claremont University Consortium
Honnold/Mudd Library
Selective depository library

*Costa Mesa*

Whittier Law School
Library
Selective depository library

*Davis*

University of California, Davis
Mabie Law Library
Selective depository library

University of California, Davis
Shields Library
Selective depository library

*Downey*

Downey City Library
Selective depository library

*Fresno*

California State University, Fresno
Henry Madden Library
Selective depository library

Fresno County Public Library
Selective depository library

*Fullerton*

California State University, Fullerton
Pollak Library
Selective depository library

*Garden Grove*

Orange County Public Libraries
Garden Grove Main Library
Selective depository library

*Hayward*

California State University East Bay
University Library
Selective depository library

*Inglewood*

City of Inglewood
Inglewood Public Library
Selective depository library

*Irvine*

University of California, Irvine
Ayala Science Library
Selective depository library

*La Jolla*

University of California, San Diego
Geisel Library
Selective depository library

*Long Beach*

California State University, Long Beach
University Library
Selective depository library

Long Beach Public Library
Selective depository library

*Los Angeles*

California State University, Los Angeles
John F. Kennedy Memorial Library
Selective depository library

Los Angeles County Law Library
LA Law Library
Selective depository library

Los Angeles Public Library
Central Library
Selective depository library

University of California, Los Angeles
Charles E. Young Research Library
Selective depository library

University of California, Los Angeles
Hugh and Hazel Darling Law Library
Selective depository library

University of Southern California
Von Kleinsmid (VKC) Library—Government Documents
Selective depository library

*Malibu*

Pepperdine University
Payson Library
Selective depository library

*Martinez*

Contra Costa County Library
Technical Services/Government Documents
Selective depository library

*Merced*

University of California, Merced
UC Merced Library
Selective depository library

*Monterey*

U.S. Naval Postgraduate School
Dudley Knox Library
Selective depository library

*Monterey Park*

City of Monterey Park
Monterey Park Bruggemeyer Library
Selective depository library

*Northridge*

California State University, Northridge
Delmar T. Oviatt Library
Selective depository library

*Norwalk*

County of Los Angeles Public Library
Norwalk Library
Selective depository library

*Oakland*

Oakland Public Library
Selective depository library

*Ontario*

University of La Verne
College of Law Library
Selective depository library

*Pasadena*

Pasadena Public Library
Selective depository library

California Institute of Technology
Caltech Library
Selective depository library

*Redding*

Shasta Public Libraries
Selective depository library

*Redlands*

University of Redlands
Armacost Library
Selective depository library

*Riverside*

Riverside Public Library
Selective depository library

University of California, Riverside
Rivera Library
Selective depository library

*Sacramento*

California State Library
Government Publications Section
Regional depository library

California State University, Sacramento
Library
Selective depository library

Sacramento County Public Law Library
Selective depository library

Sacramento Public Library
Central Library
Selective depository library

University of the Pacific McGeorge School of Law
Gordon D. Schaber Law Library
Selective depository library

*San Bernardino*

California State University, San Bernardino
John M. Pfau Library
Selective depository library

Law Library for San Bernardino County
Selective depository library

*San Diego*

San Diego County Public Law Library
Selective depository library

San Diego Public Library
Central Library
Selective depository library

San Diego State University
Library and Information Access
Selective depository library

University of San Diego
Pardee Legal Research Center
Selective depository library

*San Francisco*

California Supreme Court
California Judicial Center Library
Selective depository library

Golden Gate University
Law Library
Selective depository library

San Francisco Public Library
Selective depository library

San Francisco State University
J. Paul Leonard Library
Selective depository library

University of California
Hastings College of Law
Selective depository library

University of San Francisco
Gleeson Library/Geschke Center
Selective depository library

*San Jose*

San Jose State University
Dr. Martin Luther King Jr. Library
Selective depository library

*San Leandro*

San Leandro Public Library
Main Library
Selective depository library

*San Luis Obispo*

California Polytechnic State University
Robert E. Kennedy Library
Selective depository library

*San Marcos*

California State University San Marcos
Kellogg Library
Selective depository library

College of San Mateo
College of San Mateo Library
Selective depository library

*Santa Ana*

Orange County Public Law Library
Selective depository library

*Santa Barbara*

University of California, Santa Barbara
Davidson Library
Selective depository library

*Santa Clara*

Santa Clara University
University Library
Selective depository library

*Santa Cruz*

University of California, Santa Cruz
McHenry Library
Selective depository library

*Santa Rosa*

Sonoma County Library
Central Library
Selective depository library

*Stanford*

Stanford University
Cecil H. Green University
Selective depository library

Stanford University
Robert Crown Law Library
Selective depository library

*Stockton*

Public Library of Stockton and San Joaquin County
Selective depository library

*Turlock*

California State University, Stanislaus
University Library
Selective depository library

## Colorado

*Alamosa*

Adams State University
Nielsen Library, Suite 4010
Selective depository library

*Boulder*

University of Colorado Boulder
Norlin Library
Regional depository library

University of Colorado Boulder
School of Law Library
Selective depository library

*Broomfield*

Mamie Doud Eisenhower Public Library
Selective depository library

*Colorado Springs*

Colorado College
Charles Learning Tutt Library
Selective depository library

University of Colorado Springs
Kraemer Family Library
Selective depository library

USAF Academy
U.S. Air Force Academy
McDermott Library
Selective depository library

*Denver*

Colorado Supreme Court
Colorado Supreme Court Library
Selective depository library

Denver Public Library
Selective depository library

Regis University
Dayton Memorial Library
Selective depository library

University of Colorado Denver
Auraria Library
Selective depository library

University of Denver
Main Library
Selective depository library

University of Denver Sturm College of Law
Westminster Law Library
Selective depository library

U.S. Court of Appeals
Tenth Circuit Library
Selective depository library

*Fort Collins*

Colorado State University
William E. Morgan Library
Selective depository library

*Golden*

Colorado School of Mines
Arthur Lakes Library
Selective depository library

*Grand Junction*

Colorado Mesa University
John U. Tomlinson Library
Selective depository library

*Greeley*

University of Northern Colorado
James A. Michener Library
Selective depository library

*Gunnison*

Western State College
Leslie J. Savage Library
Selective depository library

*Pueblo*

Colorado State University–Pueblo
CSU-Pueblo Library
Selective depository library

## Connecticut

*Bridgeport*

Bridgeport Public Library
Selective depository library

*Danbury*

Western Connecticut State University
Ruth A. Haas Library
Selective depository library

*Hartford*

Connecticut State Library
Regional depository library

Trinity College
Trinity College Library
Selective depository library

University of Connecticut
School of Law Library
Selective depository library

*Middletown*

Wesleyan University
Olin Library
Selective depository library

*New Britain*

Central Connecticut State University
Elihu Burritt Library
Selective depository library

*New Haven*

Southern Connecticut State University
Hilton C. Buley Library
Selective depository library

Yale University
Center for Science and Social Science Information
Selective depository library

Yale University
Lillian Goldman Law Library
Selective depository library

*New London*

Connecticut College
C. E. Shain Library
Selective depository library

*North Haven*

Quinnipiac University
Lynne L. Pantalena Law Library
Selective depository library

*Stamford*

Ferguson Library
Selective depository library

*Storrs*

University of Connecticut
Homer Babbidge Library
Selective depository library

*Waterbury*

City of Waterbury
Silas Bronson Library
Selective depository library

Post University
Traurig Library
Selective depository library

*West Haven*

University of New Haven
Marvin K. Peterson Library
Selective depository library

*Willimantic*

Eastern Connecticut State University
J. Eugene Smith Library
Selective depository library

## Delaware

*Dover*

Delaware Division of Libraries
State Library
Selective depository library

Delaware State University
William C. Jason Library
Selective depository library

*Georgetown*

Delaware Technical and Community College
Stephen J. Betze Library
Selective depository library

*Newark*

University of Delaware
Hugh M. Morris Library
Selective depository library

*Wilmington*

Widener University, Delaware Law School
Legal Information Center
Selective depository library

## Florida

*Boca Raton*

Florida Atlantic University
S. E. Wimberly Library
Selective depository library

*Bradenton*

Manatee County Public Library
Selective depository library

*Coral Gables*

University of Miami
Otto G. Richter Library
Selective depository library

*Daytona Beach*

Volusia County Library System
Daytona Beach Regional Library
Selective depository library

*De Land*

Stetson University
duPont-Ball Library
Selective depository library

*Fort Lauderdale*

Broward County Libraries Division
Broward County Main Library
Selective depository library

Nova Southeastern University
Law Library and Technology Center
Selective depository library

*Fort Pierce*

Indian River State College
Miley Library
Selective depository library

*Gainesville*

University of Florida
George A. Smathers Libraries
Regional depository library

University of Florida
Lawton Chiles Legal Information Center
Selective depository library

*Gulfport*

Stetson University College of Law
Dolly and Homer Hand Law Library
Selective depository library

*Jacksonville*

Jacksonville Public Library
Main Library
Selective depository library

Jacksonville University
Carl S. Swisher Library
Selective depository library

University of North Florida
Thomas G. Carpenter Library
Selective depository library

*Leesburg*

Lake-Sumter State College
Library
Selective depository library

*Melbourne*

Florida Institute of Technology
Evan Library
Selective depository library

*Miami*

Florida International University
Green Library
Selective depository library

Miami-Dade Public Library System
Main Library
Selective depository library

*Miami Gardens*

St. Thomas University
University Library
Selective depository library

*Naples*

Hodges University
Terry P. McMahan Library
Selective depository library

*Orlando*

University of Central Florida
John C. Hitt Library
Selective depository library

*Pensacola*

University of West Florida
John C. Pace Library
Selective depository library

*Saint Petersburg*

Saint Petersburg Public Library
Selective depository library

*Sarasota*

New College of Florida
Jane Bancroft Cook Library
Selective depository library

Sarasota County Library System
Selby Public Library
Selective depository library

*Tallahassee*

Florida A&M University
Coleman Memorial Library
Selective depository library

Florida Division of Library and Information Services
State Library of Florida
Selective depository library

Florida State University
Research Center
Selective depository library

Florida State University
Robert Manning Strozier Library
Selective depository library

Supreme Court of Florida
Florida Supreme Court Library
Selective depository library

*Tampa*

University of South Florida
Tampa Library
Selective depository library

University of Tampa
Macdonald Kelce Library
Selective depository library

## Georgia

*Americus*

Georgia Southwestern State University
James Earl Carter Library
Selective depository library

*Athens*

University of Georgia
Map and Government Information Library
Regional depository library

University of Georgia
School of Law Library
Selective depository library

*Atlanta*

Atlanta-Fulton Public Library
Selective depository library

Atlanta University Center
Robert W. Woodruff Library
Selective depository library

Emory University
Hugh F. MacMillan Law Library
Selective depository library

Emory University
Robert W. Woodruff Library
Selective depository library

Georgia Institute of Technology
Georgia Tech Library
Selective depository library

Georgia State University
Law Library
Selective depository library

Georgia State University
University Library
Selective depository library

U.S. Court of Appeals
Eleventh Circuit Library
Selective depository library

*Augusta*

Augusta University
Reese Library
Selective depository library

*Carrollton*

University of West Georgia
Irvine Sullivan Ingram Library
Selective depository library

*Columbus*

Columbus State University
Simon Schwob Memorial Library
Selective depository library

*Dahlonega*

University of North Georgia
Library Technology Center
Selective depository library

*Dalton*

Dalton State College
Derrell C. Roberts Library
Selective depository library

*Kennesaw*

Kennesaw State University
Horace W. Sturgis Library
Selective depository library

*Macon*

Mercer University
Furman Smith Law Library
Selective depository library

Mercer University
Jack Tarver Library
Selective depository library

*Milledgeville*

Georgia College and State University
Ina Dillard Russell Library
Selective depository library

*Mount Berry*

Berry College
Memorial Library
Selective depository library

*Statesboro*

Georgia Southern University
Zach S. Henderson Library
Selective depository library

*Valdosta*

Valdosta State University
Odum library
Selective depository library

## Hawaii

*Hilo*

University of Hawaii at Hilo
Edwin H. Mookini Library
Selective depository library

*Honolulu*

Ali'iolani Hale
Supreme Court Law Library
Selective depository library

Hawaii State Public Library System
Hawaii State Library
Selective depository library

University of Hawaii at Manoa
Library
Regional depository library

University of Hawaii at Manoa
William S. Richardson School of Law Library
Selective depository library

*Kahului*

Hawaii State Public Library System
Kahului Public Library
Selective depository library

*Laie*

Brigham Young University–Hawaii
Joseph F. Smith Library
Selective depository library

*Lihue*

Hawaii State Public Library System
Lihue Public Library
Selective depository library

## Idaho

*Boise*

Boise Public Library
Selective depository library

Boise State University
Albertsons Library
Selective depository library

Idaho Supreme Court
State Law Library
Selective depository library

*Lewiston*

Lewis-Clark State College
Library
Selective depository library

*Moscow*

University of Idaho
College of Law Library
Selective depository library

University of Idaho
Library
Regional depository library

*Nampa*

Northwest Nazarene University
John E. Riley Library
Selective depository library

*Pocatello*

Idaho State University
Eli M. Oboler Library
Selective depository library

## Illinois

*Bourbonnais*

Olivet Nazarene University
Benner Library and Resource Center
Selective depository library

*Carbondale*

Southern Illinois University
Morris Library—Government Documents
Selective depository library

Southern Illinois University, Carbondale
School of Law Library
Selective depository library

*Carlinville*

Blackburn College
Lumpkin Library
Selective depository library

*Carterville*

John A. Logan College
Library
Selective depository library

*Champaign*

University of Illinois at Urbana-Champaign
Albert E. Jenner Jr. Memorial Law Library
Selective depository library

*Charleston*

Eastern Illinois University
Booth Library
Selective depository library

*Chicago*

Chicago Public Library
Harold Washington Library Center
Selective depository library

Chicago State University
Library—LIB 203
Selective depository library

DePaul University
John T. Richardson Library
Selective depository library

DePaul University
Vincent G. Rinn Law Library
Selective depository library

IIT Chicago-Kent College of Law
Chicago-Kent Law Library
Selective depository library

Illinois Institute of Technology
Paul V. Galvin Library
Selective depository library

Loyola University, Chicago
E. M. Cudahy Memorial Library
Selective depository library

Loyola University, Chicago
School of Law Library
Selective depository library

Northeastern Illinois University
Ronald Williams Library
Selective depository library

Northwestern University School of Law
Pritzker Legal Research Center
Selective depository library

The John Marshall Law School
Louis L. Biro Law Library
Selective depository library

University of Chicago
D'Angelo Law Library
Selective depository library

University of Chicago
Regenstein Library
Selective depository library

University of Illinois at Chicago
Richard J. Daley Library
Selective depository library

U.S. Court of Appeals
Seventh Circuit Library
Selective depository library

*De Kalb*

Northern Illinois University
David C. Shapiro Memorial Law Library
Selective depository library

Northern Illinois University
Founders Memorial Library
Selective depository library

*Decatur*

Decatur Public Library
Selective depository library

*Des Plaines*

Oakton Community College
Library
Selective depository library

*Edwardsville*

Southern Illinois University Edwardsville
Lovejoy Memorial Library
Selective depository library

*Elsah*

Principia College
Marshall Brooks Library
Selective depository library

*Evanston*

Northwestern University
Deering Library
Selective depository library

*Freeport*

Freeport Public Library
Selective depository library

*Jacksonville*

MacMurray College
Henry Pfeiffer Library
Selective depository library

*Lake Forest*

Lake Forest College
Donnelley and Lee Library
Selective depository library

*Lebanon*

McKendree University
Holman Library
Selective depository library

*Macomb*

Western Illinois University
University Libraries
Selective depository library

*Moline*

Black Hawk College
Library
Selective depository library

*Monmouth*

Monmouth College
Hewes Library
Selective depository library

*Mount Prospect*

Mount Prospect Public Library
Selective depository library

*Normal*

Illinois State University
Miner Library
Selective depository library

*Palos Hills*

Moraine Valley Community College
Library
Selective depository library

*Peoria*

Bradley University
Cullom-Davis Library
Selective depository library

Peoria Public Library
Selective depository library

*River Forest*

Dominican University
Rebecca Crown Library
Selective depository library

*Rockford*

Rock Valley College
Library
Selective depository library

*Romeoville*

Lewis University
Lewis University Library
Selective depository library

*South Holland*

South Suburban College
Library
Selective depository library

*Springfield*

Illinois State Library
Illinois State Library
Regional depository library

University of Illinois at Springfield
Norris L. Brookens Library
Selective depository library

*Streamwood*

Poplar Creek Public Library
Selective depository library

*University Park*

Governors State University
University Library
Selective depository library

*Urbana*

University of Illinois at Urbana-Champaign
Government Information
Selective depository library

*Wheaton*

Wheaton College
Buswell Memorial Library
Selective depository library

**Indiana**

*Anderson*

Anderson University
Robert A. Nicholson Library
Selective depository library

*Bloomington*

Indiana University, Bloomington
Herman B. Wells Library
Selective depository library

Indiana University Maurer School of Law
Jerome Hall Law Library
Selective depository library

*Evansville*

Evansville-Vanderburgh Public Library
Central Library
Selective depository library

University of Southern Indiana
David L. Rice Library
Selective depository library

*Fort Wayne*

Allen County Public Library
Selective depository library

Indiana University–Purdue University Fort Wayne
Walter E. Helmke Library
Selective depository library

*Gary*

Gary Public Library
Main Library
Selective depository library

Indiana University Northwest
John W. Anderson Library
Selective depository library

*Greencastle*

DePauw University
Roy O. West Library
Selective depository library

*Hammond*

Hammond Public Library
Selective depository library

*Hanover*

Hanover College
Duggan Library
Selective depository library

*Huntington*

Huntington University
RichLyn Library
Selective depository library

*Indianapolis*

Indiana State Library
Regional depository library

Indiana Supreme Court
Law Library
Selective depository library

Indiana University Purdue University Indianapolis
University Library
Selective depository library

Indiana University Robert H. McKinney School of Law
Ruth Lilly Law Library
Selective depository library

The Indianapolis Public Library
Central Library
Selective depository library

*Kokomo*

Indiana University, Kokomo
Library
Selective depository library

*Muncie*

Ball State University
Alexander M. Bracken Library
Selective depository library

*New Albany*

Indiana University, Southeast
Library
Selective depository library

*Notre Dame*

University of Notre Dame
Hesburgh Libraries
Selective depository library

University of Notre Dame
Kresge Law Library
Selective depository library

*Richmond*

Morrisson-Reeves Library
Selective depository library

Earlham College
Lilly Library
Selective depository library

*South Bend*

Indiana University, South Bend
Franklin D. Schurz Library
Selective depository library

*Terre Haute*

Indiana State University
Cunningham Memorial Library—Government Documents Unit
Selective depository library

*Valparaiso*

Valparaiso University
Christopher Center for Library and Information
Selective depository library

Valparaiso University
Law Library
Selective depository library

*West Lafayette*

Purdue University
HSSE Library
Selective depository library

**Iowa**

*Ames*

Iowa State University
Parks Library
Selective depository library

*Cedar Falls*

University of Northern Iowa
Rod Library
Selective depository library

*Davenport*

Davenport Public Library
Selective depository library

*Des Moines*

Drake University
Cowles Library
Selective depository library

Drake University
Law Library
Selective depository library

*Dubuque*

Loras College
Loras College Library
Selective depository library

*Fayette*

Upper Iowa University
Henderson-Wilder Library
Selective depository library

*Grinnell*

Grinnell College
Burling Library
Selective depository library

*Iowa City*

University of Iowa
Law Library
Selective depository library

University of Iowa
University Libraries
Regional depository library

*Lamoni*

Graceland University
Frederick Madison Smith Library
Selective depository library

*Sioux City*

Sioux City Public Library
Wilbur Aalfs Main Library
Selective depository library

**Kansas**

*Atchison*

Benedictine College
Library
Selective depository library

*Baldwin City*

Baker University
Collins Library
Selective depository library

*Colby*

Colby Community College
H. F. Davis Memorial Library
Selective depository library

*Dodge City*

Dodge City Community College
Library
Selective depository library

*Emporia*

Emporia State University
University Libraries and Archives
Selective depository library

*Hays*

Fort Hays State University
Forsyth Library
Selective depository library

*Hutchinson*

Hutchinson Public Library
Selective depository library

*Kansas City*

Kansas City Kansas Community College
Library
Selective depository library

*Lawrence*

University of Kansas
Anschutz Library
Regional depository library

University of Kansas
Wheat Law Library
Selective depository library

*Manhattan*

Kansas State University
Hale Library
Selective depository library

*Overland Park*

Johnson County Library
Selective depository library

*Pittsburg*

Pittsburg State University
Leonard H. Axe Library
Selective depository library

*Topeka*

Kansas State Historical Society
Library
Selective depository library

Kansas Supreme Court
Law Library
Selective depository library

State of Kansas
State Library
Selective depository library

Washburn University
School of Law Library
Selective depository library

*Wichita*

Wichita State University
Ablah Library
Selective depository library

**Kentucky**

*Ashland*

Ashland Community and Technical College
Mansbach Memorial Library
Selective depository library

*Barbourville*

Union College
Weeks-Townsend Memorial Library
Selective depository library

*Bowling Green*

Western Kentucky University
Helm-Cravens Library
Selective depository library

*Columbia*

Lindsey Wilson College
Katie Murrell Library
Selective depository library

*Crestview Hills*

Thomas More College
Library
Selective depository library

*Danville*

Centre College
Grace Doherty Library
Selective depository library

*Frankfort*

Administrative Office of the Courts
Kentucky State Law Library
Selective depository library

Kentucky State University
Paul G. Blazer Library
Selective depository library

*Hazard*

Hazard Community and Technical College
Stephens Library
Selective depository library

*Highland Heights*

Northern Kentucky University
W. Frank Steely Library
Selective depository library

*Lexington*

University of Kentucky
Law Library
Selective depository library

University of Kentucky
William T. Young Library
Regional depository library

*Louisville*

Metro Louisville
Louisville Free Public Library
Selective depository library

University of Louisville
William F. Ekstrom Library
Selective depository library

*Morehead*

Morehead State University
Camden-Carroll Library
Selective depository library

*Murray*

Murray State University
Waterfield Library
Selective depository library

*Owensboro*

Kentucky Wesleyan College
Howard Greenwell Library
Selective depository library

*Richmond*

Eastern Kentucky University
EKU Libraries
Selective depository library

*Williamsburg*

University of the Cumberlands
Hagan Memorial Library
Selective depository library

## Louisiana

*Baton Rouge*

Louisiana State University, Baton Rouge
Paul M. Hebert Law Center
Selective depository library

Louisiana State University, Baton Rouge
Troy H. Middleton Library
Selective depository library

Southern University A&M College
John B. Cade Library
Selective depository library

Southern University Law Center
Oliver B. Spellman Law Library
Selective depository library

State Library of Louisiana
Selective depository library

*Eunice*

Louisiana State University, Eunice
Arnold LeDoux Library
Selective depository library

*Hammond*

Southeastern Louisiana University
Linus A. Sims Memorial Library
Selective depository library

*Lafayette*

University of Louisiana at Lafayette
Edith Garland Dupré Library
Selective depository library

*Lake Charles*

McNeese State University
Lether Frazar Memorial Library
Selective depository library

*Leesville*

Vernon Parish Library
Selective depository library

*Monroe*

University of Louisiana at Monroe
University Library
Selective depository library

*Natchitoches*

Northwestern State University
Watson Memorial Library
Selective depository library

*New Orleans*

Louisiana Supreme Court
Law Library of Louisiana
Selective depository library

Loyola University New Orleans
College of Law Library
Selective depository library

Loyola University New Orleans
Library
Selective depository library

New Orleans Public Library
Selective depository library

Southern University at New Orleans
Leonard S. Washington Memorial Library
Selective depository library

Tulane University
Howard-Tilton Memorial Library
Selective depository library

Tulane University
School of Law Library
Selective depository library

University of Holy Cross
Blaine S. Kern Library
Selective depository library

University of New Orleans
Earl K. Long Library
Selective depository library

U.S. Court of Appeals
Fifth Circuit Library
Selective depository library

*Pineville*

Louisiana College
Richard W. Norton Memorial Library
Selective depository library

*Ruston*

Louisiana Tech University
Prescott Memorial Library
Regional depository library

*Shreveport*

Louisiana State University, Shreveport
Noel Memorial Library
Selective depository library

Shreve Memorial Library
Selective depository library

*Thibodaux*

Nicholls State University
Ellender Memorial Library
Selective depository library

## Maine

*Augusta*

Maine State Legislature
Law and Legislative Reference Library
Selective depository library

*Bangor*

Bangor Public Library
Selective depository library

*Brunswick*

Bowdoin College
Bowdoin College Library
Selective depository library

*Castine*

Maine Maritime Academy
Nutting Memorial Library
Selective depository library

*Lewiston*

Bates College
George and Helen Ladd Library
Selective depository library

*Orono*

University of Maine, Orono
Raymond H. Fogler Library
Regional depository library

*Presque Isle*

University of Maine, Presque Isle
Library
Selective depository library

*Portland*

Portland Public Library
Selective depository library

University of Maine School of Law
Garbrecht Law Library
Selective depository library

*Waterville*

Colby College
Miller Library
Selective depository library

## Maryland

*Annapolis*

Maryland State Law Library
Selective depository library

United States Naval Academy
Nimitz Library
Selective depository library

*Baltimore*

Enoch Pratt Free Library—Central Library
State Library Resource Center
Selective depository library

John Hopkins University
Eisenhower Library
Selective depository library

Morgan State University
Earl S. Richardson Library
Selective depository library

University of Baltimore
Langsdale Library
Selective depository library

University of Baltimore
Law Library
Selective depository library

University of Maryland, Baltimore County
Albin O. Kuhn Library and Gallery
Selective depository library

University of Maryland School of Law
Thurgood Marshall Law Library
Selective depository library

*Bel Air*

Harford Community College
Library
Selective depository library

*Beltsville*

U.S. Department of Agriculture
National Agricultural Library
Selective depository library

*Bethesda*

Uniformed Services University of the Health Sciences
James A. Zimble Learning Resource Center
Selective depository library

U.S. Department of Health and Human Services
National Library of Medicine
Selective depository library

*Chestertown*

Washington College
Clifton M. Miller Library
Selective depository library

*College Park*

University of Maryland, College Park
McKeldin Library
Regional depository library

*Cumberland*

Allegany College of Maryland
Library
Selective depository library

*Frostburg*

Frostburg State University
Lewis J. Ort Library
Selective depository library

*Rockville*

Montgomery County Public Libraries
Rockville Library
Selective depository library

*Salisbury*

Salisbury University
SU Libraries
Selective depository library

*Silver Spring*

U.S. Department of Commerce, NOAA
NOAA Central Library
Selective depository library

*Towson*

Towson University
Albert S. Cook Library
Selective depository library

*Westminster*

McDaniel College
Hoover Library
Selective depository library

## Massachusetts

*Amherst*

Amherst College
Robert Frost Library
Selective depository library

University of Massachusetts, Amherst
W. E. B. Du Bois Library
Selective depository library

*Boston*

Boston Public Library
Regional depository library

Northeastern University
Snell Library
Selective depository library

State Library of Massachusetts
George Fingold Library
Selective depository library

Supreme Judicial Court
Social Law Library
Selective depository library

U.S. Court of Appeals
First Circuit Library
Selective depository library

*Cambridge*

Harvard College
Lamont Library
Selective depository library

Harvard Law School
Library
Selective depository library

Massachusetts Institute of Technology
MIT Libraries
Selective depository library

*Chestnut Hill*

Boston College
Thomas P. O'Neill Jr. Library
Selective depository library

*Chicopee*

College of Our Lady of the Elms
Alumnae Library
Selective depository library

*Dartmouth*

University of Massachusetts Dartmouth
Claire T. Carney Library
Selective depository library

*Easton*

Stonehill College
MacPhaidin Library
Selective depository library

*Lowell*

University of Massachusetts, Lowell
Lydon Library
Selective depository library

*Medford*

Tufts University
Tisch Library
Selective depository library

*New Bedford*

New Bedford Free Public Library
Selective depository library

*Newton Centre*

Boston College Law School
Law Library
Selective depository library

*Springfield*

Massachusetts Trial Court
Hampden Law Library
Selective depository library

Springfield City Library
Selective depository library

Western New England University
School of Law Library
Selective depository library

*Waltham*

Brandeis University
Library and Technology Services
Selective depository library

*Wellesley*

Wellesley College
Margaret Clapp Library
Selective depository library

*Wenham*

Gordon College
Jenks Library
Selective depository library

*Williamstown*

Williams College
Sawyer Library
Selective depository library

*Worcester*

American Antiquarian Society
Library
Selective depository library

Worcester Public Library
Selective depository library

## Michigan

*Albion*

Albion College
Stockwell-Mudd Library
Selective depository library

*Allendale*

Grand Valley State University
Mary Idema Pew Library Learning and Information Commons
Selective depository library

*Ann Arbor*

University of Michigan, Ann Arbor
Law Library
Selective depository library

University of Michigan, Ann Arbor
University Library
Selective depository library

*Benton Harbor*

Benton Harbor Public Library
Selective depository library

*Big Rapids*

Ferris State University
Ferris Library for Information, Technology
Selective depository library

*Dearborn*

Henry Ford College
Eshleman Library
Selective depository library

*Detroit*

Detroit Public Library
Selective depository library

University of Detroit Mercy
Kresge Law Library
Selective depository library

University of Detroit Mercy
McNichols Campus Library
Selective depository library

Wayne State University
Arthur Neef Law Library
Selective depository library

Wayne State University
Purdy/Kresge Library
Selective depository library

*East Lansing*

Michigan State University
Main Library
Selective depository library

Michigan State University College of Law
College of Law Library
Selective depository library

*Farmington Hills*

Oakland Community College
M. L. King Library
Selective depository library

*Flint*

Flint Public Library
Selective depository library

*Grand Rapids*

Calvin College and Calvin Theological Seminary
Hekman Library
Selective depository library

Grand Rapids Public Library
Selective depository library

*Houghton*

Michigan Technological University
Van Pelt and Opie Library
Selective depository library

*Jackson*

Jackson District Library
Selective depository library

*Kalamazoo*

Kalamazoo Public Library
Central Library
Selective depository library

Western Michigan University
Dwight B. Waldo Library
Selective depository library

*Lansing*

Michigan Department of Education
Library of Michigan
Selective depository library

Western Michigan University Cooley Law School
Brennan Law Library
Selective depository library

*Livonia*

Schoolcraft College
Eric J. Bradner Library
Selective depository library

*Madison Heights*

Madison Heights Public Library
Selective depository library

*Marquette*

Northern Michigan University
Lydia M. Olson Library
Selective depository library

*Monroe*

Monroe County Library System
Ellis Library and Reference Center
Selective depository library

*Mount Pleasant*

Central Michigan University
Charles V. Park Library
Selective depository library

*Muskegon*

Hackley Public Library
Selective depository library

*Petoskey*

North Central Michigan College
Library
Selective depository library

*Pontiac*

Oakland County Library
Selective depository library

*Port Huron*

St. Clair County Library
Selective depository library

*Rochester*

Oakland University
Kresge Library
Selective depository library

*Saginaw*

Public Libraries of Saginaw
Hoyt Public Library
Selective depository library

*Sault Ste. Marie*

Lake Superior State University
Kenneth J. Shouldice Library
Selective depository library

*Traverse City*

Northwestern Michigan College
Mark and Helen Osterlin Library
Selective depository library

*University Center*

Delta College
Delta College Library
Selective depository library

*Warren*

Warren Public Library
Warren Civic Center Branch
Selective depository library

*Ypsilanti*

Eastern Michigan University
Bruce T. Halle Library
Law Library

**Minnesota**

*Bemidji*

Bemidji State University
A. C. Clark Library
Selective depository library

*Blaine*

Anoka County Library System
Northtown Central Library
Selective depository library

*Cass Lake*

Leech Lake Tribal College
Leech Lake Tribal College Library
Selective depository library

*Duluth*

Duluth Public Library
Selective depository library

University of Minnesota Duluth
Kathryn A. Martin Library
Selective depository library

*Eagan*

Dakota County Library System
Wescott Library
Selective depository library

*Mankato*

Minnesota State University, Mankato
Memorial Library
Selective depository library

*Marshall*

Southwest Minnesota State University
McFarland Library
Selective depository library

*Minneapolis*

Hennepin County Library
Minneapolis Central Library
Selective depository library

University of Minnesota
Government Publications Library
Regional depository library

University of Minnesota
Law Library
Selective depository library

*Moorhead*

Minnesota State University Moorhead
Livingston Lord Library
Selective depository library

*Morris*

University of Minnesota, Morris
Rodney A. Briggs Library
Selective depository library

*Northfield*

Carlton College
Laurence McKinley Gould Library
Selective depository library

Saint Olaf College
Rolvaag Memorial Library
Selective depository library

*Saint Cloud*

Saint Cloud State University
James W. Miller Learning Resources Center
Selective depository library

*Saint Paul*

Minnesota Supreme Court
Minnesota State Law Library
Selective depository library

Mitchell Hamline School of Law
Warren E. Burger Library
Selective depository library

Saint Paul Public Library
George Latimer Central Library
Selective depository library

University of Minnesota
Magrath Library
Selective depository library

*Saint Peter*

Gustavus Adolphus College
Folke Bernadotte Library
Selective depository library

*Winona*

Winona State University
Darrell W. Krueger Library
Selective depository library

**Mississippi**

*Alcorn*

Alcorn State University
John Dewey Boyd Library
Selective depository library

*Cleveland*

Delta State University
Roberts-LaForge Library
Selective depository library

*Columbus*

Mississippi University for Women
Fant Memorial Library
Selective depository library

*Hattiesburg*

University of Southern Mississippi
Joseph Anderson Cook Memorial Library
Selective depository library

*Jackson*

Jackson State University
Henry Thomas Sampson Library
Selective depository library

Mississippi College
Law Library
Selective depository library

Mississippi Library Commission
Selective depository library

Supreme Court of Mississippi
State Law Library
Selective depository library

**Missouri**

*Cape Girardeau*

Southeast Missouri State University
Kent Library
Selective depository library

*Columbia*

University of Missouri
School of Law Library
Selective depository library

University of Missouri, Columbia
Elmer Ellis Library
Regional depository library

*Hillsboro*

Jefferson College
Jefferson College Library
Selective depository library

*Jefferson City*

Lincoln University
Inman E. Page Library
Selective depository library

Missouri State Library
Selective depository library

Missouri Supreme Court Library
Missouri Supreme Court Building
Selective depository library

*Joplin*

Missouri Southern State University
George A. Spiva Library
Selective depository library

*Kansas City*

Kansas City Public Library
Selective depository library

Rockhurst University
Greenlease Library
Selective depository library

University of Missouri, Kansas City
Leon E. Bloch Law Library
Selective depository library

University of Missouri, Kansas City
Miller Nichols Library
Selective depository library

*Kirksville*

Truman State University
Pickler Memorial Library
Selective depository library

*Liberty*

William Jewell College
Charles F. Curry Library
Selective depository library

*Maryville*

Northwest Missouri State University
B. D. Owens Library
Selective depository library

*O'Fallon*

Saint Charles City-County Library District
Middendorf-Kredell Branch
Selective depository library

*Rolla*

Missouri University of Science and Technology
Curtis Laws Wilson Library
Selective depository library

*Saint Joseph*

St. Joseph Public Library
Selective depository library

*Saint Louis*

Maryville University of Saint Louis
University Library
Selective depository library

Saint Louis County Library
Selective depository library

Saint Louis Public Library
Selective depository library

Saint Louis University
Law Library
Selective depository library

Saint Louis University
Pius XII Memorial Library
Selective depository library

University of Missouri–Saint Louis
Thomas Jefferson Library
Selective depository library

U.S. Court of Appeals, Eighth Circuit
St. Louis Headquarters Library
Selective depository library

Washington University in St. Louis
John M. Olin Library
Selective depository library

Washington University School of Law
School of Law Library
Selective depository library

*Springfield*

Missouri State University
Duane G. Meyer Library
Selective depository library

*Warrensburg*

University of Central Missouri
James C. Kirkpatrick Library
Selective depository library

## Montana

*Billings*

Montana State University, Billings
Library
Selective depository library

*Bozeman*

Montana State University, Bozeman
Renne Library
Selective depository library

*Butte*

Montana Tech of the University of Montana
Montana Tech Library
Selective depository library

*Harlem*

Aaniih Nakoda College
Aaniih Nakoda College Library
Selective depository library

*Havre*

Montana State University–Northern
State Law Library of Montana
Selective depository library

*Helena*

Montana State Library
Selective depository library

Montana Supreme Court
State Law Library of Montana
Selective depository library

*Missoula*

University of Montana
Jameson Law Library
Selective depository library

University of Montana
Mansfield Library
Regional depository library

*Pablo*

Salish Kootenai College
D'Arcy McNickle Library
Selective depository library

*Poplar*

Fort Peck Community College
James E. Shanley Tribal Library
Selective depository library

## Nebraska

*Crete*

Doane University
Perkins Library
Selective depository library

*Kearney*

University of Nebraska at Kearney
Calvin T. Ryan Library
Selective depository library

*Lincoln*

Nebraska Library Commission
Selective depository library

Nebraska Supreme Court
Nebraska State Library
Selective depository library

University of Nebraska, Lincoln
Don L. Love Memorial Library
Regional depository library

University of Nebraska, Lincoln
Schmid Law Library
Selective depository library

*Omaha*

Creighton University
Klutznick Law Library
Selective depository library

Creighton University
Reinert/Alumni Library
Selective depository library

Omaha Public Library
W. Dale Clark Library
Selective depository library

University of Nebraska at Omaha
Criss Library
Selective depository library

*Scottsbluff*

Lied Scottsbluff Public Library
Selective depository library

*Wayne*

Wayne State College
Conn Library
Selective depository library

## Nevada

*Carson City*

Nevada State Library and Archives
Selective depository library

Nevada Supreme Court Library
Selective depository library

*Elko*

Great Basin College
Great Basin College Library
Selective depository library

*Las Vegas*

Las Vegas–Clark County Library District
Las Vegas Library
Selective depository library

University of Nevada, Las Vegas
Wiener-Rogers Law Library
Selective depository library

*Reno*

Washoe County Law Library
Selective depository library

Washoe County Library
Selective depository library

University of Nevada, Reno
Selective depository library

## New Hampshire

*Concord*

New Hampshire Department of Cultural Resources
New Hampshire State Library
Selective depository library

New Hampshire Supreme Court
New Hampshire Law Library
Selective depository library

University of New Hampshire School of Law
Library
Selective depository library

*Durham*

University of New Hampshire
Dimond Library
Selective depository library

*Hanover*

Dartmouth College
Baker-Berry Library
Selective depository library

*Manchester*

Manchester City Library
Selective depository library

Saint Anselm College
Geisel Library
Selective depository library

*Nashua*

Nashua Public Library
Selective depository library

**New Jersey**

*Camden*

Rutgers University, Camden
Law School Library
Selective depository library

Rutgers University, Camden
Paul Robeson Library
Selective depository library

*Elizabeth*

Free Public Library of Elizabeth
Selective depository library

*Galloway*

Richard Stockton College of New Jersey
Library
Selective depository library

*Glassboro*

Rowan University
Campbell Library
Selective depository library

*Jersey City*

Jersey City Free Public Library
Selective depository library

New Jersey City University
Guarini Library
Selective depository library

*Lawrenceville*

Rider University
Franklin F. Moore Library
Selective depository library

*Madison*

Drew University
Library
Selective depository library

*Montclair*

Montclair State University
Harry A. Sprague Library
Selective depository library

*New Brunswick*

Rutgers University, New Brunswick
Archibald S. Alexander Library
Selective depository library

*Newark*

Newark Public Library
Regional depository library

Rutgers University, Newark
John Cotton Dana Library
Selective depository library

Rutgers University, Newark
Law Library
Selective depository library

Seton Hall University School of Law
Peter W. Rodino Jr. Law Library
Selective depository library

Sussex County Library System
Main Library
Selective depository library

*Phillipsburg*

Phillipsburg Free Public Library
Selective depository library

*Princeton*

Princeton University
Firestone Library
Selective depository library

*Randolph*

County College of Morris
Masten Learning Resources Center
Selective depository library

*Shrewsbury*

Monmouth County
Monmouth County Library, Eastern Branch
Selective depository library

*Toms River*

Ocean County College
Library
Selective depository library

*Trenton*

New Jersey State Library
Selective depository library

*West Long Branch*

Monmouth University
Library
Selective depository library

*Woodbridge*

Free Public Library of Woodbridge
Main Library
Selective depository library

**New Mexico**

*Albuquerque*

University of New Mexico
School of Law Library
Selective depository library

University of New Mexico
University Libraries
Regional depository library

*Farmington*

City of Farmington
Farmington Public Library
Selective depository library

*Hobbs*

New Mexico Junior College
Pannell Library
Selective depository library

*Las Cruces*

New Mexico State University
Thomas C. Donnelly Library
Selective depository library

*Las Vegas*

New Mexico Highlands University
Thomas C. Donnelly Library
Selective depository library

*Portales*

Eastern New Mexico University
Golden Library
Selective depository library

*Santa Fe*

Department of Cultural Affairs
New Mexico State Library
Selective depository library

Institute of American Indian and Alaska Native Culture and
Arts Development
Library
Selective depository library

New Mexico Supreme Court
Supreme Court Law Library
Selective depository library

*Silver City*

Western New Mexico University
J. Cloyd Miller Library
Selective depository library

*Socorro*

New Mexico Institute of Mining and Technology
Joseph R. Skeen Library
Selective depository library

## New York

*Albany*

Albany Law School
Schaffer Law Library
Selective depository library

New York State Library
Cultural Education Center
Regional depository library

State University of New York, Albany
University Library
Selective depository library

*Binghamton*

Binghamton University
Glenn G. Bartle Library
Selective depository library

*Bronx*

Herbert H. Lehman College/City University of New York
Leonard Lief Library
Selective depository library

Fordham University
Walsh Family Library
Selective depository library

State University of New York Maritime College
Stephen B. Luce Library
Selective depository library

*Bronxville*

Sarah Lawrence College
Esther Raushenbush Library
Selective depository library

*Brooklyn*

Brooklyn College
Brooklyn College Library
Selective depository library

Brooklyn Law College
Library
Selective depository library

Brooklyn Public Library
Central Library
Selective depository library

*Brookville*

Long Island University
B. Davis Schwartz Memorial Library
Selective depository library

*Buffalo*

Buffalo and Erie County Public Library
Central Library
Selective depository library

University of Buffalo
Lockwood Library
Selective depository library

*Canton*

St. Lawrence University
Owen D. Young Library
Selective depository library

*Central Islip*

Touro College Jacob D. Fuchsberg Law Center
Gould Law Library
Selective depository library

U.S. Courts Library
Selective depository library

*Delhi*

State University of New York at Delhi
Resnick Library
Selective depository library

*Farmingdale*

Farmingdale State College, State University of New York
Greenley Library
Selective depository library

*Flushing*

Queens College, City University of New York
Benjamin S. Rosenthal Library
Selective depository library

*Hamilton*

Colgate University
Case Library
Selective depository library

*Hempstead*

Hofstra University
Axinn Library
Selective depository library

Maurice A. Deane School of Law at Hofstra University
Law Library
Selective depository library

*Ithaca*

Cornell University
Albert R. Mann Library
Selective depository library

Cornell University
John M. Olin Library
Selective depository library

Cornell University
Law Library
Selective depository library

*Jamaica*

Queens Public Library
Central Library
Selective depository library

Saint John's University
Rittenberg Law Library
Selective depository library

*Kings Point*

U.S. Merchant Marine Academy
Bland Memorial Library
Selective depository library

*Middletown*

Middletown Thrall Library
Selective depository library

*Mount Vernon*

Mount Vernon Public Library
Selective depository library

*New Paltz*

State University of New York, New Paltz
Sojourner Truth Library
Selective depository library

*New York*

Columbia University
Lehman Library
Selective depository library

Cooper Union for the Advancement of Science and Art
Library
Selective depository library

Fordham University School of Law
Leo T. Kissam Memorial Library
Selective depository library

New York Law Institute
Selective depository library

New York Law School
Mendik Library
Selective depository library

New York Public Library
Science, Industry and Business
Selective depository library

New York University
Elmer Holmes Bobst Library
Selective depository library

New York University
School of Law Library
Selective depository library

St. John's University—Manhattan
Davis Library
Selective depository library

The City College of New York
Cohen Library
Selective depository library

Yeshiva University
Chutick Law Library
Selective depository library

Yeshiva University
Pollack Library
Selective depository library

*Newburgh*

Newburgh Free Library
Selective depository library

*Oakdale*

Dowling College
Library
Selective depository library

*Oneonta*

State University of New York College at Oneonta
James M. Milne Library
Selective depository library

*Oswego*

State University of New York at Oswego
Penfield Library
Selective depository library

*Plattsburgh*

Plattsburgh State University
Feinberg Library
Selective depository library

*Potsdam*

Clarkson University
Burnap Memorial Library
Selective depository library

State University of New York at Potsdam
F. W. Crumb Memorial Library
Selective depository library

*Poughkeepsie*

Vassar College
Vasaar College Libraries
Selective depository library

*Purchase*

State University of New York at Purchase
College Library
Selective depository library

*Queens*

Saint John's University
St. Augustine Hall
Selective depository library

*Rochester*

Monroe County Library System
Rochester Public Library
Selective depository library

University of Rochester
Rush Rhees Library
Selective depository library

*Saint Bonaventure*

Saint Bonaventure University
Friedsam Memorial Library
Selective depository library

*Schenectady*

Union College
Schaffer Library
Selective depository library

*Stony Brook*

Stony Brook University
Frank Melville Jr. Memorial Library
Selective depository library

*Syracuse*

Syracuse University
Frank Melville Jr. Memorial Library
Selective depository library

Syracuse University College of Law
Syracuse University College of Law Library
Selective depository library

*Troy*

Troy Public Library
Selective depository library

*West Point*

U.S. Military Academy
Library
Selective depository library

*White Plains*

Elizabeth Haub School of Law at Pace University
Pace Law Library
Selective depository library

*Yonkers*

Yonkers Public Library
Riverfront Library
Selective depository library

## North Carolina

*Asheville*

University of North Carolina at Asheville
D. Hiden Ramsey Library
Selective depository library

*Boiling Springs*

Gardner-Webb University
Dover Memorial Library
Selective depository library

*Boone*

Appalachian State University
Carol Grotnes Belk Library and Information Commons
Selective depository library

*Buies Creek*

Campbell University
Wiggins Memorial Library
Selective depository library

*Chapel Hill*

University of North Carolina at Chapel Hill
Davis Library
Regional depository library

University of North Carolina at Chapel Hill
Kathrine R. Everett Law Library
Regional depository library

*Charlotte*

Charlotte Mecklenburg Library
Selective depository library

University of North Carolina at Charlotte
J. Murrey Atkins Library
Selective depository library

*Cullowhee*

Western Caroline University
Hunter Library
Selective depository library

*Davidson*

Davidson College
E. H. Little Library
Selective depository library

*Durham*

Duke University
William R. Perkins Library
Selective depository library

Duke University School of Law
J. Michael Goodson Law Library
Selective depository library

North Carolina Central University
James E. Shepard Memorial Library
Selective depository library

North Carolina Central University
School of Law Library
Selective depository library

*Elon*

Elon University
Belk Library
Selective depository library

*Fayetteville*

Fayetteville State University
Charles W. Chesnutt Library
Selective depository library

*Greensboro*

East Carolina University
J. Y. Joyner Library
Selective depository library

Elon University
Elon University School of Law Library
Selective depository library

North Carolina A&T State University
F. D. Bluford Library
Selective depository library

University of North Carolina at Greensboro
Walter Clinton Jackson Library
Selective depository library

*Laurinburg*

St. Andrews University
DeTamble Library
Selective depository library

*Lexington*

Davidson County Public Library System
Lexington Library
Selective depository library

*Pembroke*

University of North Carolina at Pembroke
Mary Livermore Library
Selective depository library

*Raleigh*

Department of Cultural Resources
State Library of North Carolina
Selective depository library

North Carolina State University
D. H. Hill Library
Selective depository library

North Carolina Supreme Court
Law Library
Selective depository library

*Rocky Mount*

North Carolina Wesleyan College
Pearsall Library
Selective depository library

*Salisbury*

Catawba College
Corriher-Linn-Black Library
Selective depository library

*Wilmington*

University of North Carolina at Wilmington
William M. Randall Library
Selective depository library

*Wilson*

Barton College
Hackney Library
Selective depository library

*Winston-Salem*

Forsyth County Public Library
Central Library
Selective depository library

Wake Forest University
Professional Center Library
Selective depository library

Wake Forest University
Z. Smith Reynolds Library
Selective depository library

## North Dakota

*Bismarck*

North Dakota State Library
Selective depository library

North Dakota Supreme Court
Law Library
Selective depository library

State Historical Society of North Dakota
State Archives
Selective depository library

*Fargo*

North Dakota State University
The Libraries
Regional depository library

*Fort Yates*

Sitting Bull College
Sitting Bull College Library
Selective depository library

*Grand Forks*

University of North Dakota
Chester Fritz Library
Regional depository library

*Minot*

Minot State University
Gordon B. Olson Library
Selective depository library

*Valley City*

Valley City State University
Allen Memorial Library
Selective depository library

## Ohio

*Ada*

Ohio Northern University
Jay P. Taggart Law Library
Selective depository library

*Akron*

Akron-Summit County Public Library
Main Library
Selective depository library

University of Akron
Bierce Library
Selective depository library

University of Akron
School of Law Library
Selective depository library

*Alliance*

University of Mount Union
University of Mount Union Library
Selective depository library

*Ashland*

Ashland University
Ashland University Library
Selective depository library

*Athens*

Ohio University
Alden Library
Selective depository library

*Bowling Green*

Bowling Green State University
University Libraries
Selective depository library

*Chardon*

Geauga County Library System
Chardon Branch Library
Selective depository library

*Cincinnati*

Public Library of Cincinnati and Hamilton County
Main Library
Selective depository library

University of Cincinnati
Langsam Library
Selective depository library

University of Cincinnati College of Law
Robert S. Marx Law Library
Selective depository library

*Cleveland*

Case Western Reserve University
Kevin Smith Library
Selective depository library

Case Western Reserve University
The Judge Ben C. Green Law Library
Selective depository library

Cleveland Public Library
Public Administration Branch
Selective depository library

Cleveland State University
Cleveland-Marshall College of Law Library
Selective depository library

Cleveland State University
Michael Schwartz Library
Selective depository library

*Columbus*

Capital University
Blackmore Library
Selective depository library

The Ohio State University
Moritz Law Library
Selective depository library

The Ohio State University Libraries
Thompson Library
Selective depository library

State Library of Ohio
Government Information Services
Regional depository library

Supreme Court of Ohio
Law Library
Selective depository library

*Dayton*

Dayton Metro Library
Selective depository library

University of Dayton
Rosech Library
Selective depository library

Wright State University
Paul Laurence Dunbar Library
Selective depository library

*Delaware*

Ohio Wesleyan University
L. A. Beeghly Library
Selective depository library

*Findlay*

The University of Findlay
Shafer Library
Selective depository library

*Gambier*

Kenyon College
Olin/Chalmers Libraries
Selective depository library

*Granville*

Denison University
William Howard Doane Library
Selective depository library

*Hiram*

Hiram College
Library
Selective depository library

*Kent*

Kent State University
University Library
Selective depository library

*Marietta*

Marietta College
Legacy Library
Selective depository library

*Marion*

Marion Public Library
Selective depository library

*New Concord*

Muskingum University
Library
Selective depository library

*New Philadelphia*

Kent State University
Tuscarawas Campus Library
Selective depository library

*Oberlin*

Oberlin College
Library
Selective depository library

*Oxford*

Miami University
King Library
Selective depository library

*Portsmouth*

Shawnee State University
Clark Memorial Library
Selective depository library

*Rio Grande*

University of Rio Grande
Jeanette Albiez Davis Library
Selective depository library

*Steubenville*

Public Library of Steubenville and Jefferson County
Selective depository library

*Tiffin*

Heidelberg University
Beeghly Library
Selective depository library

*Toledo*

Toledo-Lucas County Public Library
Selective depository library

University of Toledo
College of Law Library
Selective depository library

University of Toledo
William S. Carlson Library
Selective depository library

*Westerville*

Otterbein University
Courtright Memorial Library
Selective depository library

*Westlake*

Westlake Porter Public Library
Selective depository library

*Wooster*

The College of Wooster
Libraries
Selective depository library

*Worthington*

Worthington Libraries
Selective depository library

*Youngstown*

Public Library of Youngstown and Mahoning County
Selective depository library

Youngstown State University
William F. Maag Jr. Library
Selective depository library

**Oklahoma**

*Ada*

East Central University
Linscheid Library
Selective depository library

*Claremore*

Rogers State University
Stratton Taylor Library
Selective depository library

*Durant*

Southeastern Oklahoma State University
Henry G. Bennett Memorial Library
Selective depository library

*Edmond*

University of Central Oklahoma
Max Chambers Library
Selective depository library

*Enid*

Public Library of Enid and Garfield County
Selective depository library

*Langston*

Langston University
G. Lamar Harrison Library
Selective depository library

*Norman*

University of Oklahoma
Bizzell Memorial Library
Selective depository library

University of Oklahoma
Donald E. Pray Law Library
Selective depository library

*Oklahoma City*

Metropolitan Library System
Ronald J. Norick Downtown Library
Selective depository library

Oklahoma City University School of Law
Law Library
Selective depository library

Oklahoma Department of Libraries
U.S. Government Information Division
Selective depository library

*Shawnee*

Oklahoma Baptist University
Mabee Learning Center
Selective depository library

*Stillwater*

Oklahoma State University
Edmon Low Library
Regional depository library

*Tahlequah*

Northeastern State University
John Vaughan Library
Selective depository library

*Tulsa*

Tulsa City-County Library
Selective depository library

University of Tulsa
Mabee Legal Information Center
Selective depository library

University of Tulsa
McFarlin Library
Selective depository library

*Weatherford*

Southwestern Oklahoma State University
Al Harris Library
Selective depository library

**Oregon**

*Ashland*

Southern Oregon University
Lenn and Dixie Hannon Library
Selective depository library

*Bend*

Central Oregon Community College
Library
Selective depository library

*Corvallis*

Oregon State University
Oregon State University Libraries and Press
Selective depository library

*Eugene*

University of Oregon
Knight Library
Selective depository library

*Forest Grove*

Pacific University
Pacific University Library
Selective depository library

*Klamath Falls*

Oregon Institute of Technology
Library
Selective depository library

*La Grande*

Eastern Oregon University
Pierce Library
Selective depository library

*McMinnville*

Linfield College
Nicholson Library
Selective depository library

*Monmouth*

Western Oregon University
Hamersly Library
Selective depository library

*Pendleton*

Blue Mountain Community College
Library
Selective depository library

*Portland*

Lewis and Clark College
Aubrey R. Watzek Library
Selective depository library

Lewis and Clark Law School
Paul L. Boley Law Library
Selective depository library

Multnomah County Library
Selective depository library

Portland State University
Branford Price Millar Library
Selective depository library

Reed College
Eric V. Hauser Library
Selective depository library

U.S. Department of Energy
Bonneville Power Administration Library
Selective depository library

*Salem*

State of Oregon Law Library
Selective depository library

Oregon State Library
State Library Building
Regional depository library

Willamette University
College of Law Library
Selective depository library

Willamette University
Mark O. Hatfield Library
Selective depository library

## Pennsylvania

*Allentown*

Muhlenberg College
Trexler Library
Selective depository library

*Bethel Park*

Bethel Park Public Library
Selective depository library

*Bethlehem*

Lehigh University
Fairchild-Martindale Library
Selective depository library

*Bloomsburg*

Bloomsburg University of Pennsylvania
Harvey A. Andruss Library
Selective depository library

*Blue Bell*

Montgomery County Community College
Brendlinger Library
Selective depository library

*Carlisle*

Dickinson School of Law, Pennsylvania State University
H. Laddie Montague Jr. Law Library
Selective depository library

*Cheyney*

Cheyney University of Pennsylvania
Leslie Pinckney Hill Library
Selective depository library

*Collegeville*

Ursinus College
Myrin Library
Selective depository library

*East Stroudsburg*

East Stroudsburg University
Kemp Library
Selective depository library

*Erie*

Erie County Public Library
Raymond M. Blasco, M.D. Memorial Library
Selective depository library

*Greenville*

Thiel College
Langenheim Memorial Library
Selective depository library

*Harrisburg*

State Library of Pennsylvania
Regional depository library

Widener University Commonwealth Law School Library
Selective depository library

*Indiana*

Indiana University of Pennsylvania
Stapleton Library
Selective depository library

*Johnstown*

Cambria County Library System
Cambria County Library
Selective depository library

*Lancaster*

Franklin and Marshall College
Shadek-Fackenthal Library
Selective depository library

*Lewisburg*

Bucknell University
Ellen Clarke Bertrand Library
Selective depository library

*Meadville*

Allegheny College
Lawrence Lee Pelletier Library
Selective depository library

*Millersville*

Millersville University of Pennsylvania
Francine G. McNairy Library
Selective depository library

*Moon Township*

Robert Morris University
Library
Selective depository library

*New Castle*

New Castle Public Library
Selective depository library

*Newtown*

Bucks County Community College
Library
Selective depository library

*Norristown*

New Castile Public Library
Selective depository library

*Philadelphia*

Free Library of Philadelphia
Selective depository library

Temple University
Samuel L. Paley Library
Selective depository library

Temple University
School of Law Library
Selective depository library

University of Pennsylvania
Biddle Law Library
Selective depository library

University of Pennsylvania
Van Pelt-Dietrich Library Center
Selective depository library

*Pittsburgh*

Allegheny County
Law Library
Selective depository library

Carnegie Library of Pittsburgh
Selective depository library

Duquesne University
Center for Legal Information
Selective depository library

La Roche College
John J. Wright Library
Selective depository library

University of Pittsburgh
Barco Law Library
Selective depository library

University of Pittsburgh
Hillman Library
Selective depository library

*Pottsville*

Pottsville Free Public Library
Selective depository library

*Reading*

Reading Public Library
Selective depository library

*Scranton*

Scranton Public Library
Albright Memorial Building
Selective depository library

*Shippensburg*

Shippensburg University of Pennsylvania
Ezra Lehman Memorial Library
Selective depository library

*Slippery Rock*

Slippery Rock University
Bailey Library
Selective depository library

*Swarthmore*

Swarthmore College
McCabe Library
Selective depository library

*University Park*

Pennsylvania State University
Paterno Library
Selective depository library

*Upper Burrell*

Penn State New Kensington
Elisabeth S. Blissell Library
Selective depository library

*Villanova*

Villanova University
Law School Library
Selective depository library

*Warren*

Warren Library Association
Law School Library
Selective depository library

*West Chester*

West Chester University of Pennsylvania
Francis Harvey Green Library
Selective depository library

*Williamsport*

Lycoming College
John G. Snowden Memorial Library
Selective depository library

## Rhode Island

*Bristol*

Roger Williams University
Library
Selective depository library

*Kingston*

University of Rhode Island
Robert L. Carothers Library
Selective depository library

*Newport*

Newport Public Library
Selective depository library

U.S. Naval War College
Eccles Library
Selective depository library

*Providence*

Brown University
John D. Rockefeller Jr. Library
Selective depository library

Rhode Island College
James P. Adams Library
Selective depository library

Rhode Island Secretary of State
Rhode Island State Library
Selective depository library

Rhode Island State Law Library
Providence County Courthouse
Selective depository library

*Westerly*

Westerly Public Library
Selective depository library

## South Carolina

*Aiken*

University of South Carolina, Aiken
Gregg-Graniteville Library
Selective depository library

*Charleston*

Charleston Southern University
L. Mendel Rivers Library
Selective depository library

College of Charleston
Addlestone Library
Selective depository library

The Citadel Military College
Daniel Library
Selective depository library

*Clemson*

Clemson University
Robert Muldrow Cooper Library
Selective depository library

*Columbia*

Benedict College
Benjamin F. Payton Learning Resource Center
Selective depository library

South Carolina State Library
Selective depository library

University of South Carolina, Columbia
Law Library
Selective depository library

University of South Carolina, Columbia
Thomas Cooper Library
Regional depository library

*Conway*

Coastal Carolina University
Kimbel Library
Selective depository library

*Due West*

Erskine College and Seminary
McCain Library
Selective depository library

*Florence*

Florence County Library
Selective depository library

Francis Marion University
James A. Rogers Library
Selective depository library

*Greenville*

Furman University
James B. Duke Library
Selective depository library

Greenville County Library System
Selective depository library

*Greenwood*

Lander University
Larry A. Jackson Library
Selective depository library

*Lancaster*

University of South Carolina, Lancaster
Medford Library
Selective depository library

*Orangeburg*

South Carolina State University
Miller F. Whittaker Library
Selective depository library

*Rock Hill*

Winthrop University
Dacus Library
Selective depository library

*Spartanburg*

Spartanburg County Public Libraries
Headquarters Library
Selective depository library

**South Dakota**

*Aberdeen*

Northern State University
Williams Library
Selective depository library

*Brookings*

South Dakota State University
Hilton M. Briggs Library
Selective depository library

*Kyle*

Oglala Lakota College
Woksape Tipi Library
Selective depository library

*Pierre*

South Dakota State Library
Selective depository library

South Dakota Supreme Court
Law Library
Selective depository library

*Rapid City*

South Dakota School of Mines and Technology
Devereaux Library
Selective depository library

*Sioux Falls*

Augustana College
Mikkelsen Library
Selective depository library

*Spearfish*

Black Hills State University
E. Y. Berry Library Learning Center
Selective depository library

*Vermillion*

University of South Dakota
I. D. Weeks Library
Selective depository library

**Tennessee**

*Bristol*

King University
E. W. King Library
Selective depository library

*Clarksville*

Austin Peay State University
Felix G. Woodward Library
Selective depository library

*Cleveland*

Cleveland State Community College
Library
Selective depository library

*Cookeville*

Tennessee Technological University
Angelo and Jennette Volpe Library and Media Center
Selective depository library

*Jefferson City*

Carson-Newman University
Stephens-Burnett Library
Selective depository library

*Johnson City*

East Tennessee State University
Sherrod Library
Selective depository library

*Knoxville*

Knoxville County Public Library System
Lawson McGhee Library
Selective depository library

University of Tennessee, Knoxville
Joel A. Katz Law Library
Selective depository library

University of Tennessee, Knoxville
John C. Hodges Library Govdocs
Selective depository library

*Martin*

University of Tennessee, Martin
Paul Meek Library
Selective depository library

*Memphis*

City of Memphis
Memphis Public Library and Information Center
Selective depository library

The University of Memphis
Cecil C. Humphreys School of Law Library
Selective depository library

University of Memphis
McWherter Library
Regional depository library

*Murfreesboro*

Middle Tennessee State University
James E. Walker Library
Selective depository library

*Nashville*

Fisk University
Franklin Library
Selective depository library

Public Library of Nashville and Davidson County
Nashville Public Library
Selective depository library

State of Tennessee
Tennessee State Library and Archives
Selective depository library

Tennessee State University
Brown-Daniel Library
Selective depository library

Vanderbilt University
Jean and Alexander Heard Library
Selective depository library

Vanderbilt University Law School
Alyne Queener Massey Law Library
Selective depository library

*Sewanee*

University of the South
Jessie Ball duPont Library
Selective depository library

**Texas**

*Abilene*

Abilene Christian University
Brown Library
Selective depository library

Hardin-Simmons University
Richardson Library
Selective depository library

*Arlington*

Arlington Public Library System
Central Express
Selective depository library

University of Texas at Arlington
Library
Selective depository library

*Austin*

Texas State Library and Archives Commission
Regional depository library

Texas State Law Library
Tom C. Clark Building
Selective depository library

University of Texas, Austin
Perry-Castaneda Library
Selective depository library

University of Texas, Austin
Tarlton Law Library
Selective depository library

*Baytown*

Lee College
Lee College Library
Selective depository library

*Beaumont*

Lamar University
Mary and John Gray Library
Selective depository library

*Brownsville*

University of Texas, Rio Grande Valley
University Library
Selective depository library

*Brownwood*

Howard Payne University
Walker Memorial Library
Selective depository library

*Canyon*

West Texas A&M University
Cornette Library
Selective depository library

*College Station*

Texas A&M University
Sterling C. Evans Library
Selective depository library

*Commerce*

Texas A&M University–Commerce
James Gilliam Gee Library
Selective depository library

*Corpus Christi*

Texas A&M University–Corpus Christi
Mary and Jeff Bell Library
Selective depository library

*Corsicana*

Navarro College
Richard M. Sanchez
Selective depository library

*Dallas*

Dallas Baptist University
Vance Memorial Library
Selective depository library

Dallas Public Library System
J. Erik Jonsson Library
Selective depository library

Southern Methodist University
Central University Libraries
Selective depository library

University of North Texas
UNT Libraries
Selective depository library

*El Paso*

City of El Paso
El Paso Public Library
Selective depository library

University of Texas, El Paso
Library
Selective depository library

*Fort Stockton*

Fort Stockton Public Library
Selective depository library

*Fort Worth*

Texas Christian University
Mary Couts Burnett Library, Government Information
Selective depository library

*Houston*

Houston Public Library
Selective depository library

Lone Star College-North Harris
Library—Government Documents
Selective depository library

Rice University
Fondren Library
Selective depository library

South Texas College of Law Houston
Fred Parks Law Library
Selective depository library

Texas Southern University
Thurgood Marshall School of Law
Selective depository library

University of Houston
John O'Quinn Law Library
Selective depository library

University of Houston
M. D. Anderson Library
Selective depository library

University of Houston, Clear Lake
Alfred R. Neumann Library
Selective depository library

*Huntsville*

Sam Houston State University
Newton Gresham Library
Selective depository library

*Kingsville*

Texas A&M University–Kingsville
James C. Jernigan Library
Selective depository library

*Laredo*

Texas A&M International University
Sue and Radcliffe Killam Library
Selective depository library

*Longview*

Longview Public Library
Selective depository library

*Lubbock*

Texas Tech University
Library
Regional depository library

Texas Tech University
School of Law Library
Selective depository library

*Nacogdoches*

Stephen F. Austin State University
Steen Library
Selective depository library

*Prairie View*

Prairie View A&M University
John B. Coleman Library
Selective depository library

*Richardson*

University of Texas, Dallas
Eugene McDermott Library
Selective depository library

*San Angelo*

Angelo State University
Porter Henderson Library
Selective depository library

Palo Alto College
Ozuna Library
Selective depository library

*San Antonio*

San Antonio College
San Antonio College Library
Selective depository library

Saint Mary's University
Louis J. Blume Library
Selective depository library

Saint Mary's University
Sarita Kenedy East Law Library
Selective depository library

Trinity University
Coates Library
Selective depository library

University of Texas at San Antonio
UTSA Libraries
Selective depository library

*San Marcos*

Texas State University–San Marcos
Albert B. Alkek Library
Selective depository library

*Seguin*

Texas Lutheran University
Blumberg Memorial Library
Selective depository library

*Texarkana*

Texarkana College
Palmar Memorial Library
Selective depository library

*Victoria*

University of Houston, Victoria
VC/UHC Library
Selective depository library

*Waco*

Baylor University
Jesse H. Jones Library
Selective depository library

Baylor University Law School
Williams Legal Research Center
Selective depository library

*Wichita Falls*

Midwestern State University
Moffett Library
Selective depository library

## Utah

*Cedar City*

Southern Utah University
Gerald R. Sherratt Library
Selective depository library

*Logan*

Utah State University
Merrill Cazier Library
Regional depository library

*Ogden*

Weber State University
Stewart Library
Selective depository library

*Provo*

Brigham Young University
Harold B. Lee Library
Selective depository library

Brigham Young University
Howard W. Hunter Law Library
Selective depository library

*Salt Lake City*

University of Utah
Eccles Health Science Library
Selective depository library

University of Utah
James E. Faust Law Library
Selective depository library

University of Utah
Marriott Library
Selective depository library

Utah Appellate Courts
Utah State Law Library
Selective depository library

**Vermont**

*Burlington*

University of Vermont
Bailey/Howe Library
Selective depository library

*Johnson*

Johnson State College
Willey Library
Selective depository library

*Middlebury*

Middlebury College
Davis Family Library
Selective depository library

*Montpelier*

Vermont Department of Libraries
Information and Law Services
Selective depository library

*Northfield*

Norwich University
Kreitzberg Library
Selective depository library

*South Royalton*

Vermont Law School
Cornell Library
Selective depository library

**Virginia**

*Alexandria*

U.S. Patent and Trademark Office
Scientific and Technical Information Center
Selective depository library

*Arlington*

George Mason University
School of Law Library
Selective depository library

*Blacksburg*

Virginia Tech
Newman Library
Selective depository library

*Bridgewater*

Bridgewater College
Alexander Mack Memorial Library
Selective depository library

*Charlottesville*

University of Virginia
Alderman Library
Regional depository library

University of Virginia
Arthur J. Morris Law Library
Selective depository library

*Emory*

Emory and Henry College
Kelly Library
Selective depository library

*Fairfax*

George Mason University
Fenwick Library
Selective depository library

*Fredericksburg*

University of Mary Washington
Simpson Library
Selective depository library

*Grundy*

Appalachian School of Law
Library
Selective depository library

*Hampden-Sydney*

Hampden-Sydney College
Bortz Library
Selective depository library

*Hampton*

Hampton University
Harvey Library
Selective depository library

*Harrisonburg*

James Madison University
Carrier Library
Selective depository library

*Lexington*

Virginia Military Institute
Preston Library
Selective depository library

Washington and Lee University
James G. Leyburn Library
Selective depository library

Washington and Lee University
Wilbur C. Hall Law Library
Selective depository library

*Norfolk*

National Defense University
Ike Skelton Library—Joint Forces Staff College
Selective depository library

Norfolk Public Library System
Norfolk Main Library
Selective depository library

Old Dominion University
Perry Library
Selective depository library

*Petersburg*

Virginia State University
Johnston Memorial Library
Selective depository library

*Quantico*

Federal Bureau of Investigation
FBI Library
Selective depository library

Gen. Alfred M. Gray Marine Corps Research Center
Breckinridge Research Library
Selective depository library

*Reston*

U.S. Department of the Interior
U.S. Geological Survey Library
Selective depository library

*Richmond*

Library of Virginia
Selective depository library

Supreme Court of Virginia
State Law Library
Selective depository library

University of Richmond
Boatwright Memorial Library
Selective depository library

University of Richmond
Muse Law Library
Selective depository library

Virginia Commonwealth University
James Branch Cabell Library
Selective depository library

*Roanoke*

Hollins University
Wyndham Robertson Library
Selective depository library

*Salem*

Roanoke College
Fintel Library
Selective depository library

*Williamsburg*

College of William and Mary
Earl Gregg Swem Library
Selective depository library

College of William and Mary
Wolf Law Library
Selective depository library

*Wise*

University of Virginia's College at Wise
John Cook Wyllie Library
Selective depository library

## Washington

*Bellevue*

King County Library System
Bellevue Regional Library
Selective depository library

*Bellingham*

Northwest Indian College
Lummi Library
Selective depository library

Western Washington University
Mabel Zoe Wilson Library
Selective depository library

*Cheney*

Eastern Washington University
John F. Kennedy Library
Selective depository library

*Des Moines*

Highline College
Library
Selective depository library

*Ellensburg*

Central Washington University
James E. Brooks Library
Selective depository library

*Everett*

Everett Public Library
Selective depository library

*Olympia*

Washington State Law Library
Temple of Justice
Selective depository library

*Pullman*

Washington State University
Holland and Terrell Libraries
Selective depository library

*Seattle*

Seattle Public Library
Selective depository library

Seattle University
School of Law Library
Selective depository library

University of Washington
Gallagher Law Library
Selective depository library

University of Washington
Suzzallo Library
Selective depository library

*Spokane*

Gonzaga University School of Law
Chastek Library
Selective depository library

Spokane Public Library
Selective depository library

*Tacoma*

Tacoma Public Library
Selective depository library

*Tumwater*

Washington State Library
Regional depository library

*Vancouver*

Fort Vancouver Regional Library
Selective depository library

*Walla Walla*

Whitman College
Penrose Library
Selective depository library

**Washington, D.C.**

American University, Washington College of Law
Pence Law Library
Selective depository library

Catholic University of America
Judge Kathryn J. Dufour Law Library
Selective depository library

Department of Defense
Pentagon Library
Selective depository library

District of Columbia Court of Appeals
Library/D.C. Historic Court House
Selective depository library

District of Columbia Public Library
Martin Luther King, Jr. Memorial Library
Selective depository library

Executive Office of the President
Library
Selective depository library

Federal Election Commission
Law Library
Selective depository library

George Washington University
Gelman Library
Selective depository library

Georgetown University
Lauinger Library
Selective depository library

Howard University School of Law
Howard University Law Library
Selective depository library

Library of Congress
Congressional Research Service
Selective depository library

Library of Congress
Serial and Government Publications Division
Selective depository library

National Defense University
Library
Selective depository library

Naval Historical Center
Navy Department Library
Selective depository library

Office of the Comptroller of the Currency
OCC Library
Selective depository library

Pension Benefit Guaranty Corporation
Corporate Library
Selective depository library

Supreme Court of the United States
Supreme Court of the United States Library
Selective depository library

U.S. Court of Appeals for the D.C. Circuit
Circuit Library
Selective depository library

U.S. Court of Appeals for the Federal Circuit
Circuit Library
Selective depository library

U.S. Department of Commerce
Commerce Research library
Selective depository library

U.S. Department of Education
National Library of Education
Selective depository library

U.S. Department of Homeland Security
Main Library Depository
Selective depository library

U.S. Department of Homeland Security
U.S. Coast Guard Law Library
Selective depository library

U.S. Department of Housing and Urban Development
HUD Library
Selective depository library

U.S. Department of Labor
Wirtz Labor Library
Selective depository library

U.S. Department of State
Ralph J. Bunche Library
Selective depository library

U.S. Department of the Interior
Interior Library
Selective depository library

U.S. Department of the Treasury
Library
Selective depository library

U.S. Department of Veterans Affairs
VA Central Office Library
Selective depository library

U.S. Government Accountability Office
Library and Information Services
Selective depository library

U.S. Postal Service
Postal Service Library
Selective depository library

U.S. Senate
U.S. Senate Library
Selective depository library

## West Virginia

*Athens*

Concord University
J. Frank Marsh Library
Selective depository library

*Bluefield*

Bluefield State College
Wendell G. Hardway Library
Selective depository library

*Charleston*

Kanawha County Public Library
Selective depository library

West Virginia Library Commission
Reference Library
Selective depository library

West Virginia Supreme Court of Appeals
State Law Library
Selective depository library

*Elkins*

Davis and Elkins College
Booth Library
Selective depository library

*Fairmont*

Fairmont State University
Ruth Ann Musick Library
Selective depository library

*Huntington*

Marshall University
James E. Morrow Library
Selective depository library

*Morgantown*

West Virginia University
Downtown Campus Library
Regional depository library

*Salem*

Salem International University
Benedum library
Selective depository library

*Shepherdstown*

Shepherd University
Scarborough Library
Selective depository library

*Weirton*

Mary H. Weir Public Library
Selective depository library

## Wisconsin

*Appleton*

Lawrence University
Seeley G. Mudd Library
Selective depository library

*Beloit*

Beloit College
Col. Robert H. Morse Library
Selective depository library

*Eau Claire*

University of Wisconsin–Eau Claire
William D. McIntyre Library
Selective depository library

*Green Bay*

University of Wisconsin–Green Bay
David A. Cofrin Library
Selective depository library

*La Crosse*

La Cross Public Library
Selective depository library

University of Wisconsin–La Crosse
Murphy Library
Selective depository library

*Madison*

Wisconsin State Law Library
Selective depository library

University of Wisconsin–Madison
Law Library
Selective depository library

University of Wisconsin–Madison
Memorial Library
Regional depository library

*Milwaukee*

Marquette University
Eckstein Law Library
Selective depository library

Milwaukee Public Library
Regional depository library

University of Wisconsin–Milwaukee
UWM Libraries
Selective depository library

*Platteville*

University of Wisconsin–Platteville
Karrmann Library
Selective depository library

*Ripon*

Ripon College
Lane Library
Selective depository library

*River Falls*

University of Wisconsin–River Falls
Chalmer Davee Library
Selective depository library

*Stevens Point*

University of Wisconsin–Stevens Point
University Library
Selective depository library

*Superior*

University of Wisconsin–Superior
Jim Dan Hill Library
Selective depository library

*Waukesha*

Waukesha Public Library
Selective depository library

*Whitewater*

University of Wisconsin–Whitewater
Andersen Library
Selective depository library

## Wyoming

*Cheyenne*

Wyoming State Law Library
Selective depository library

Wyoming State Library
Selective depository library

*Gillette*

Campbell County Public Library
Selective depository library

*Laramie*

University of Wyoming
Coe Library, Department 3334
Selective depository library

University of Wyoming, College of Law
George W. Hopper Law Library
Selective depository library

*Powell*

Northwest College
John Taggart Hinckley Library
Selective depository library

*Riverton*

Central Wyoming College
Library
Selective depository library

*Rock Springs*

Western Wyoming Community College
Hay Library
Selective depository library

# Index

aboriginal, 54, 96, 98, 125, 241, 328, 368, 371, 372, 385; Bureau of Indian Affairs, 4, 8
aerial photographs, 4, 205–21, 223, 244, 248, 266, 289–94, 315, 320–22, 338, 347–51. *See also* orthoimagery
agriculture, 60, 68, 112, 242, 243, 246, 248, 250, 291, 296, 310, 313, 321, 340, 382, 386
air photos. *See* aerial photographs
Alabama, 61, 98, 126, 205, 314–16, 328; American county atlases, 66–67; cartographic collections, 16–17; GIS Data, 250–52; map depository libraries, 418–19; railroad maps, 116, 117; subject guides, 270–71, 284
Alaska, 54, 61, 126, 225, 251, 284, 314, 319, 355, 397, 419, 447; air photos, 205–6; cartographic collections, 17; nautical charts, 166–75
Alberta, 94, 95, 115, 124, 223, 268, 315, 317, 322, 325, 343, 397, 415; atlases, 56, 57–58; cartographic collections, 9–10; GIS data, 243–44
Alberta Geological Survey, 124, 243
American Library Association, 7, 337
American Map and GIS Associations, 344–46
ArcCAD, 302
ArcEditor, 304, 305
ArcGIS, 340, 341; handbooks and manuals, 295, 297, 302–8; used in libraries, 267–83. *See also* ArcMap; ESRI; GIS Software
ArcHydro, 303
ArcInfo, 303, 304, 305, 308
ArcMap. *See* ArcGIS
ArcObjects, 303, 304, 305
ArcPy, 302, 305

ArcSDE, 303
Archaeology, 294, 295, 298, 313, 321
Arizona, 18, 67, 116, 126–27, 206, 271, 314, 315, 318, 328, 419–20; cartographic collections, 17–18; GIS data, 251–252
Arizona State University, 18, 271, 421
Arkansas, 18–19, 62, 67, 98, 116, 126, 127, 206, 225, 237, 252, 271, 314, 328, 420; cartographic collections, 18–19
Association of Canadian Map Libraries and Archives (ACMLA), 7, 337, 343
Auburn University, 16, 270, 418

Ball State University, 29, 275, 433
bathymetric charts, 9, 264–65, 314, 316, 318, 320. *See also* nautical charts
bibliographies, 311–22
bird's eye-view maps, 223–40. *See also* panoramic maps
Boise State University, 26, 285, 429
bridges, 65, 263
Brigham Young University, 26, 48, 274, 282, 429, 464, 465
British Columbia, 93, 95, 116, 124, 325, 344; atlases, 54, 56, 58; bird's eye view, 223–24; cartographic collections, 9, 10–11; GIS data, 244–45; map depository libraries, 415–16; subject guides, 268–69, 283
Brock University, 14, 337, 417

cadastral, 2, 93, 95, 97, 107–8, 110, 244, 245, 255, 257, 259, 260, 292. *See also* land ownership; parcels
California, 98, 116, 126, 127, 225, 268, 328, 339, 345; air photos, 206–7; atlases, 62, 67–68; bibliographies,

312, 314, 327, 317, 318, 319, 320, 321; cartographic collections, 9, 19–22; depository libraries, 420–24; GIS data, 250, 252–53; subject guides, 271–72, 284
California Polytechnic State University, 22, 271, 423
California State University, 9, 19, 20, 271, 284, 420, 421, 422, 423, 424
CanMatrix, 94, 242
CanTopo, 94
CanVec, 94, 242
Carleton University, 14, 269, 416
cartographer, 93, 289, 290, 292, 293, 309, 318
cartographic bibliographies, 312–13
cartographic citations, 337–41
cartographic collections, 7–52
cartographic dictionaries, 309–10
cartographic handbooks and manuals, 289–94
cartographic resources, 1–5
cartographic subject guides, 268–88
cartography, 1, 9, 278, 289–94, 297, 303, 305
cataloging, 293, 312
census, 8, 45, 61, 79, 242, 247, 248, 250, 251, 254, 257, 324, 327. *See also* statistics
Charles E. Goad, 4, 203
citations, 290, 300, 311–12, 337–41
city and town maps, 93–113
city plans, 57, 112
Claremont University, 19, 268, 421
climate, 242, 243, 246, 248, 249, 250, 252, 264, 266
Colorado, 62, 69, 98–99, 116–17, 126, 190, 207, 315, 316, 328; bird's eye view, 225–26; cartographic

collections, 23–24; depository libraries, 426–27; GIS data, 252–54; subject guides, 272–73, 284
Colorado State University, 23, 272, 425
Connecticut, 62, 68, 99, 117, 126, 127, 207, 226, 318, 328; cartographic collections, 24; depository libraries, 425–26; GIS data, 254; subject guides, 273, 284
Cornell University, 38, 278, 449

Delaware, 24, 126, 315, 318, 328; air photos, 207–8; depository libraries, 426; GIS data, 254; subject guides, 273
Department of Fisheries and Oceans, 3, 131
DePauw University, 28, 285, 434
depository map program, 7, 8, 93, 94, 291, 415–71
dictionaries, 5, 309–10. *See also* gazetteers
digital elevation models (DEM), 245, 249, 264
District of Columbia, 24–25, 107–12, 121, 126, 220, 238, 254, 284, 315, 317; cartographic collections, 24–25; city and town maps, 107–12; depository libraries, 468–69
drones, 4, 291

East Carolina University, 40, 286, 451
Eastern Illinois University, 27, 274, 430
Eastern Michigan University, 34, 276, 441
Eastern Washington University, 50, 288, 467
Emory University, 26, 285, 428
Emporia State University, 30, 285, 434
ERDAS Imagine, 267, 275, 276, 277, 278, 304
ESRI, 289, 291, 294, 297, 298, 301, 302, 310; ESRI data, 264, 276; ESRI handbooks and manuals, 302–8; ESRI software, 271, 280, 295, 297. *See also* ArcGIS

Federal Geographic Data Committee (FGDC), 248, 345
fire insurance plans, 3–4, 203–4, 273, 294, 317, 352–55; in library collections, 281–86

gazetteers, 4–5, 323–36. *See also* dictionaries
geocoding, 302, 304
geodatabase, 294, 295, 296, 302, 303, 305, 306
geodesy, 290, 292, 294, 295, 309, 310, 343

geographic coordinates, 2, 5, 53, 323, 324
Geographic Information Systems (GIS), 1, 7, 241, 289; GIS associations, 343–46; GIS bibliographies, 313–20; GIS citations, 341; GIS data, 2, 3, 94, 95, 124–30, 241–66, 268–88; GIS handbooks and manuals, 294–308; GIS software, 93, 289, 302–7, 268–88 (*see also* ArcGIS; ESRI; QGIS); GIS subject guide, 267–88
Geography Network, 242, 246, 265
Geogratis, 93, 94, 242, 243
Geoinformatics, 298, 309, 310
Geological Survey of Canada, 3, 123–26
geology, 7, 9, 53, 304, 309, 311, 312, 314, 315, 319; geological data, 124, 243, 244, 245, 246, 247, 248, 251, 253, 258, 266; geological maps, 2–3, 57, 61, 70, 123–30, 242, 250, 292, 315, 316, 317, 319, 320; geology department, 13, 19, 28, 36, 37, 38, 41, 43, 47, 49, 51, 52
George Mason University, 49, 282, 465
Georgia, 63, 69, 99, 116, 117, 126, 127, 285, 329; air photos, 208–9; bird's eye view, 226–27; cartographic collections, 25–26; depository libraries, 428–29; GIS data, 254–55; subject guides, 273–74
Georgia State University, 26, 273, 428
*Geoserver, 272, 304*
geospatial data. *See* Geographic Information Systems
GIS. *See* Geographic Information Systems
GIS data. *See* Geographic Information Systems
GIS software. *See* Geographic Information Systems
Google Earth, 4, 94, 124, 205, 250, 267, 268–83, 302, 304, 307. *See also* KML
Google Maps, 4, 205, 241, 341
Grand Valley State University, 33, 286, 439
GRASS, 268, 272, 303, 305, 306

handbooks, 289–318; cartographic, 289–94; GIS, 294–302; GIS software, 302–8
Hawaii, 26, 63, 126, 209, 255, 274, 317, 329, 429; nautical charts, 200–203
Humboldt State University, 19, 272, 420
hydrological maps, 9, 246, 248, 255, 305, 316, 330. *See also* bathymetric charts; nautical charts; water resources
hydrology, 2, 9, 53, 95, 127, 330, 345; GIS layer, 244–64; manuals and handbooks, 291, 296, 303, 305, 316. *See also* water resources

Idaho, 63, 69, 117, 126, 127, 285, 317; air photos, 209; cartographic collections, 26–27; depository libraries, 429; GIS data, 255–56
Illinois, 60, 63, 99, 117, 126, 127, 227, 314, 315, 317, 329; air photos, 209; atlases, 69–72; cartographic collections, 27–28; depository libraries, 429–32; GIS data, 255–56; subject guides, 274–75, 285
Indian. *See* aboriginal
Indiana, 60, 63, 99, 117, 127, 227, 256, 275, 285, 292, 317, 329; air photos, 209–10; atlases, 72–73; cartographic collections, 28–29; depository libraries, 432–33
infrared, 291, 321, 322, 349, 350, 352
Insurers Advisory Organization of Canada, 4, 203
interactive maps, 1, 125–29, 241–65, 304, 305, 341
Iowa, 60, 63, 117, 127, 210, 275, 285, 317, 321, 329; atlases, 73–77; bird's eye view, 227–28; cartographic collections, 29–30; depository libraries, 433–34; GIS data, 256–57

Kansas, 63, 77, 99–100, 106, 117, 127, 228, 285, 317, 329; air photos, 210; cartographic collections, 30; depository libraries, 434–35; GIS data, 256–57; subject guides, 275–76
Kansas State University, 30, 275, 434
Kent State University, 42, 280, 454
Kentucky, 63, 77, 100, 117, 121, 127, 228, 257, 315, 329–30, 435; air photos, 210, 211; bibliographies, 311–12; cartographic collections, 30–31; subject guides, 276, 285
Keyhole Markup Language. *See* KML
KML, 1, 250. *See also* Google Earth

Lake Champlain, 61, 115, 164, 186, 298
Lake Ontario, 1, 144, 145, 186, 224
Lakehead University, 14, 269, 416
land ownership, 2, 3, 53, 78, 93, 95, 98, 262, 323. *See also* cadastral; parcels
land use, 2, 53, 244–65, 294, 296, 298, 307, 321, 339
Library and Archives Canada, 9, 14, 95, 223, 416
Library of Congress, 4, 9, 24, 95, 115, 203, 223, 284, 291, 311, 317, 318, 319, 337, 468
Louisiana, 63, 77, 100, 127, 257, 317, 318, 330; air photos, 211; cartographic collections, 31–32; depository libraries, 436–37; subject guides, 285–86
Louisiana State University, 31, 285, 436, 437

Maine, 32, 63, 100, 115, 117, 128, 286, 318, 330, 437; air photos, 211; bird's eye view, 228–29; GIS data, 257–258

Manitoba, 11, 56, 58, 97, 115, 124–25, 224, 269, 325, 416; city and town maps, 95–96; GIS data, 245

manuals, *See* handbooks

map associations, 343–46

map design, 289, 290, 291, 297, 299

map projections, 290, 291, 293, 297, 301, 306, 312, 315

Maryland, 32, 64, 77, 109, 110, 111, 118, 128, 229, 258, 311, 330; air photos, 211; city and town maps, 100–101; depository libraries, 437–38; subject guides, 276

Massachusetts, 64, 78, 101, 118, 128, 212, 258, 276, 286, 318, 319, 330, 331; bird's eye view, 229–30; depository libraries, 438–39; library collections, 32–33

Massachusetts Institute of Technology, 33, 268, 276

McGill University, 15, 270, 317, 417

McMaster University, 13, 269, 337, 416

Miami University, 42, 280, 454

Michigan, 60, 101, 118, 128, 212, 224, 230, 258, 286, 318, 329, 330; atlases, 63–64, 78–79; cartographic collections, 9, 33–34; depository libraries, 439–441; subject guides, 276–77

Michigan Technological University, 33, 277, 440

Minnesota, 60, 64, 118, 122, 128, 212, 277, 286, 318, 330; atlases, 79–81; bird's eye view, 230–31; cartographic collections, 34–35; city and town maps, 101–2; depository libraries, 441–42; GIS data, 258–59

Minnesota State University, 34, 277, 442

Mississippi, 35, 61, 64, 102, 128, 213, 259, 318, 331; atlases, 81–82; depository libraries, 442–43

Mississippi river, 61, 100, 190–92

Missouri, 35, 60, 64, 102, 118, 128, 213, 231, 259, 277, 286, 318, 331; atlases, 82–85; depository libraries, 443–44

Missouri State University, 35, 286, 443, 444

Montana, 64, 118, 128, 213, 231, 259, 286, 315, 317, 318, 319; depository libraries, 444–45; library collections, 35–36

Montana State University, 35, 286, 444

National Map, 94, 203, 248–49, 250, 267

National Oceanic and Atmospheric Administration. *See* NOAA

National Topographic System (NTS), 2, 93, 94, 123, 243

Natural Earth, 249, 267

Natural Resources Canada, 3, 93, 94, 123, 242, 243, 323

nautical charts, 3, 131–202, 292, 293, 316. *See also* bathymetric charts; hydrological maps

Nebraska, 64, 85, 118, 128, 231, 318, 331, 445; air photos, 213–14; library collections, 36–37; subject guides, 277–78

Nevada, 37, 64, 116, 128, 214, 259, 278, 315, 318, 331, 445

New Brunswick, 57, 94, 96, 115, 125, 224, 245, 325, 416; cartographic collections, 11; subject guides, 283–84

New England, 60, 101, 119, 331

New Hampshire, 37, 61, 64, 66, 102, 118, 128, 228, 259, 278, 318, 331; air photos, 214–15; atlases, 85–86; bird's eye view, 231–32; depository libraries, 445–46

New Jersey, 64, 86, 102, 118, 128, 215, 232, 278, 315, 319, 331–32; cartographic collections, 37–38; depository libraries, 446–47; GIS data, 259–60

New Mexico, 38, 86, 128, 215, 260, 278, 314, 319, 332; depository libraries, 447–48

New York, 64, 86, 103–4, 119, 128–29, 286, 319; bird's eye view, 232–34; cartographic collections, 38–39; depository libraries, 448–51; gazetteer, 332–33; subject guides, 278–79

New York Public Library, 9, 39, 449

Newfoundland and Labrador, 12, 55, 57, 94, 96, 125, 132, 224, 245, 318, 344, 418

NOAA, 3, 8, 50, 131, 249, 438. *See also* nautical charts

North Carolina, 65, 86, 119, 121, 129, 216, 324, 260, 279, 286, 289, 315, 319, 333; cartographic collections, 39–40; depository libraries, 451–52

North Carolina State University, 40, 337, 452

North Dakota, 40, 86, 119, 129, 216, 260, 319; depository libraries, 452–53; subject guides, 286–87

Northern Illinois University, 27, 285, 315, 430

Northern Kentucky University, 31, 285, 435

Northwest Territories (NWT), 57, 93, 94, 125, 245, 323, 325

Northwestern University, 27, 274, 430, 431

Nova Scotia, 57, 58, 96, 115, 125, 224, 324, 325, 344; cartographic collections, 12–13; depository libraries, 416; GIS data, 245–246

Nunavut, 94, 125, 246, 323

Ohio, 65, 104, 104, 119, 129, 216, 260, 268, 314, 315, 316, 319; atlases, 86–88; bird's eye view, 234–35; cartographic collections, 40–42; depository libraries, 454–55; gazetteers, 333–34; subject guides, 279–80

Oklahoma, 42, 65, 88, 129, 235, 319, 334; air photos, 216–17; depository libraries, 457–58; GIS data, 260–61

Ontario, 96–97, 115, 224, 323, 325–26; atlases, 57–59; cartographic collections, 13–15; depository libraries, 416–17; geological maps, 123–24; GIS data, 246–47; nautical charts, 144–45; subject guides, 269–70

Ontario Geological Survey, 123, 124

Open data. *See* GIS data

Oregon, 65, 88, 104, 120, 121, 129, 217, 235, 261, 280, 287, 300, 317, 319, 334; cartographic collections, 42–43; depository libraries, 456–57

Oregon State University, 42, 287, 456

orthoimagery, 94, 244, 249, 255, 260. *See also* aerial photographs

panoramic map, 4, 223. *See also* bird's eye-view maps

parcels, 95, 244, 251, 252, 253, 255, 256, 258, 259, 260, 262, 263, 302. *See also* cadastral; land ownership

Pennsylvania, 65, 104–5, 120, 126, 129, 217, 261, 287, 300, 316, 317, 319, 334; atlases, 88–89; bird's eye view, 235–37; cartographic collections, 43–44; depository libraries, 457; subject guides, 280–81

Pennsylvania State Archives, 43, 317

Pennsylvania State University, 44, 281, 457, 459

planners, 2, 245, 294, 295, 296, 300, 307

Portland State University, 43, 280, 287, 457

*Potomac River,* 107, 108, 109, 110, 180

Prince Edward Island, 15, 57, 94, 125, 247, 325, 326, 417

Princeton University, 38, 278, 447

projections. *See* map projections

public land surveys, 3

Purdue University, 29, 268, 275, 432, 433

Python, 272, 276, 302, 303, 305, 306

QGIS, 267–68, 269, 270, 272, 276, 281, 303, 305, 306, 307. *See also* geographic information systems

Quantum GIS. *See* QGIS
Quebec, 270, 284, 325, 326; depository libraries, 417
Queen's University, 13, 224, 269, 337, 416

railroads, 54, 60, 61, 88, 98, 99, 104, 105, 106, 115–22, 233, 260, 262, 263, 264, 319, 325, 326, 327, 330, 335
relief, 9, 93, 94, 290, 291
remote sensing, 265–83, 289, 290, 292–96, 299–302, 305, 309, 311, 313, 314, 320–22, 343–45
Rhode Island, 44, 120, 126, 129, 287, 312, 318, 328, 334, 459; air photos, 217–18; GIS data, 261
road maps, 2, 53, 63, 83, 347
Rutgers University, 37, 278, 446
Ryerson University, 14, 269, 417

Saint Louis University, 35, 277, 444
Saint Mary's University, 48, 287
Sanborn Map Company, 4, 203, 294, 317
San Diego State University, 21, 284, 421, 423
San Jose State University, 22, 284, 423
Saskatchewan, 16, 56, 57, 94, 95, 96, 97, 116, 125, 247, 270, 326–27, 418
satellite imagery, 2, 9, 53, 248, 264, 266, 282, 289, 339
shapefile (shp), 3, 124, 125, 128, 129, 250, 251, 255, 260, 262. *See also* geographical information systems
Simon Fraser University, 10, 269, 417
Soil, 123, 242, 243, 245, 246, 248, 249, 250, 252, 257, 261, 262, 263, 265, 266, 289, 293, 296, 301, 314, 317, 320, 321, 323, 331
South Carolina, 65, 89, 105, 120, 129, 218, 262, 287, 291, 294, 319, 334, 336; cartographic collections, 44–45; depository libraries, 461–62
South Dakota, 90, 105–6, 126, 129, 218, 237, 287, 315, 319, 334; cartographic collections, 45–46; depository libraries, 460–61
Southern Illinois University, 27, 274, 285, 429, 430
spatial analysis, 241, 242, 259, 294, 296–307, 344
spatial databases, 301, 309, 310, 313
Stanford University, 22, 272, 424
State University of New York, 38, 286, 448, 449, 450
State Library of North Carolina, 40, 260, 452
State Library of Ohio, 41, 315, 454
State Library of Pennsylvania, 43, 287, 457

statistics, 68, 87, 242, 243, 247, 248, 249, 257, 265, 266, 295, 300, 301, 309, 310, 314, 317, 321, 327, 328, 331, 333. *See also* census
Stetson University, 25, 426, 427
Stony Brook University, 39, 279, 450
subject guides. *See* cartographic subject guides. *See* geographical information systems
surveying, 289–94, 301, 309, 310, 343, 344, 345
Syracuse University, 39, 268, 279, 450

Tennessee, 46, 65, 90, 106, 116, 120, 129, 237, 262, 281, 311, 319; air photos, 218–19; depository libraries, 461–62; gazetteers, 334–35
Texas, 61, 65, 90, 94, 106, 120, 129, 219, 225, 237, 262, 290, 301, 313, 318, 320, 335, 337; atlases, 65–66; cartographic collections, 46–48; depository libraries, 462–64; subject guides, 281–82, 287
Texas A&M University, 47, 268, 282, 462
topographic maps, 93–95, 123, 131, 241–50, 255–57, 263–86, 292, 294, 314–20, 327, 338
toponyms, 4, 93, 242, 323, 324, 326
Toporama, 94, 243
Transportation, 95, 108, 115, 118, 121, 241, 244–65, 294–95, 305–7, 321, 324
Trent University, 14, 269, 319, 417

Underwriters' Survey Bureau, 4, 203
union list, 8, 203, 204, 317, 320
United States Geological Survey, 1, 3, 8, 9, 20, 23, 49, 93, 123, 126–30
Université de Sherbrooke, 16, 284, 417
Université du Québec à Montréal, 16, 284, 417
Université Laval, 16, 284, 417
University of Alabama, 17, 271, 418, 419
University of Alaska, 17, 284, 419
University of Alberta, 9, 268, 415
University of Akron, 40, 280, 453
University of Arizona, 18, 271, 420
University of Arkansas, 19, 268, 271, 420
University of British Columbia, 9, 10, 223, 268, 269, 415
University of Calgary, 9, 268, 415
University of California, 9, 19, 20, 21, 22, 203, 268, 272, 284, 321, 421, 422, 423, 424
University of Chicago, 27, 274, 430
University of Cincinnati, 41, 280, 453
University of Colorado, 23, 273, 422

University of Connecticut, 24, 284, 425, 426
University of Delaware, 24, 273, 426
University of Florida, 9, 25, 284, 426
University of Georgia, 25, 255, 273, 428
University of Hawaii, 26, 268, 274, 429
University of Idaho, 26, 285, 429
University of Illinois, 27, 28, 274, 322, 430, 431, 432
University of Iowa, 30, 275, 434
University of Kansas, 30, 268, 275, 318, 434
University of Kentucky, 31, 285, 435
University of Louisville, 31, 276, 435
University of Maine, 32, 286, 437
University of Maryland, 32, 276, 437, 438
University of Massachusetts, 32, 33, 286, 438, 439
University of Miami, 25, 273, 426
University of Michigan, 33, 277, 320, 440
University of Minnesota, 34, 277, 321, 441, 442
University of Montana, 36, 444
University of Nebraska, 36, 37, 277, 445
University of Nevada, 37, 278, 445
University of New Brunswick, 11, 283, 416
University of New Hampshire, 37, 446
University of New Mexico, 38, 278, 447
University of North Carolina, 39, 40, 279, 451, 452
University of North Texas, 47, 287, 463
University of Northern Iowa, 29, 285, 433
University of Oregon, 43, 456
University of Ottawa, 14, 270, 337, 343, 416
University of Pennsylvania, 281, 457, 458, 459
University of Pittsburgh, 44, 268, 281, 458
University of Regina, 16, 268, 270, 418
University of Rhode Island, 44, 287, 459
University of South Alabama, 17, 284, 418
University of South Carolina, 45, 287, 459, 460
University of South Dakota, 45, 287, 461
University of Tennessee, 46, 281, 461
University of Texas, 47, 48, 94, 282, 287, 337, 462, 463, 464
University of Toronto, 14, 15, 270, 416, 417
University of Utah, 48, 282, 465
University of Vermont, 49, 282, 465
University of Victoria, 11, 283, 416
University of Virginia, 49, 283, 465, 467

University of Washington, 50, 268, 283, 467
University of Waterloo, 15, 270, 347–414, 419
University of Windsor, 15, 270, 417
University of Wisconsin, 9, 51, 52, 283, 288, 320, 337, 469, 470
University of Wyoming, 52, 283, 470
U.S. Geological Survey (USGS), 1, 3, 4, 8, 93, 126–30, 324
U.S. Topo Quadrangle Series, 2, 93
USGS. *See* United States Geological Survey
Utah, 90, 129, 237, 262, 282, 287, 320, 335; cartographic collections, 48–49; depository libraries, 464–65
Utah State University, 48, 287, 464

Vanderbilt University, 46, 281, 462
Vermont, 49, 64, 106, 120–21, 129, 219, 237–38, 262, 282, 335; depository libraries, 465–67

Virginia, 49, 66, 107, 117, 121, 238; air photos, 219–20; depository libraries, 465–67; gazetteers, 335–36; GIS data, 262–63; subject guides, 282–83
Virginia Tech University, 49, 282, 465

Washington, 65, 66, 90, 107, 121, 130, 220, 238, 263, 283, 334, 336; cartographic collections, 50–51; depository libraries, 467; subject guides, 287–88
Washington D.C. *See* District of Columbia
Washington State University, 50, 283, 467
watershed, 53, 55, 56, 61, 64, 65, 66, 252, 253, 254, 263, 301, 305, 308
West Virginia, 51, 66, 112–113, 121, 130, 220, 263, 288, 316, 320, 336, 469; air photos, 238–39
West Virginia University, 51, 288, 469
Western Association of Map Libraries, 7, 52

Western Carolina University, 39, 279, 286
Western Illinois University, 27, 274, 431
Western Michigan University, 33, 286, 440
Western University, 13, 270, 416
Western Washington University, 50, 287, 337, 467
Wichita State University, 30, 276, 435
Wisconsin, 9, 51, 60, 66, 118, 121–22, 130, 239, 283, 288, 312, 314, 317, 320, 336; air photos, 220–21; atlases, 90–92; cartographic collections, 51–52; depository libraries, 469–70; GIS data, 263
wildfire, 248, 252, 259, 295
wind, 57, 63, 66, 242

Yale University, 24, 273, 425
York University, 15, 268, 270, 417
Yukon, 16, 54, 93, 94, 97, 125–26, 225, 247, 325, 327, 418

# About the Author

**Eva H. Dodsworth** is the geospatial data services librarian at the University of Waterloo library, where she specializes in teaching GIS and map-related content to the university community. Her interests include historical cartographic research, digital mapping projects, and teaching geoweb applications. Dodsworth is also a part-time online instructor for a number of library schools and continuing education organizations, where she teaches the use of GIS technology in libraries. Her list of publications is available at http://www.evadodsworth.com/.

CPSIA information can be obtained
at www.ICGtesting.com
Printed in the USA
BVHW05*2117090818
523504BV00005B/6/P